# MODERN NUCLEAR CHEMISTRY

# MODERN NUCLEAR CHEMISTRY

**WALTER D. LOVELAND**
Oregon State University

**DAVID J. MORRISSEY**
Michigan State University

**GLENN T. SEABORG**
University of California, Berkeley

A JOHN WILEY & SONS, INC., PUBLICATION

Copyright © 2006 by John Wiley & Sons, Inc. All rights reserved

Published by John Wiley & Sons, Inc., Hoboken, New Jersey
Published simultaneously in Canada

No part of this publication may be reproduced, stored in a retrieval system, or transmitted in any form or by any means, electronic, mechanical, photocopying, recording, scanning, or otherwise, except as permitted under Section 107 or 108 of the 1976 United States Copyright Act, without either the prior written permission of the Publisher, or authorization through payment of the appropriate per-copy fee to the Copyright Clearance Center, Inc., 222 Rosewood Drive, Danvers, MA 01923, (978) 750-8400, fax (978) 750-4470, or on the web at www.copyright.com. Requests to the Publisher for permission should be addressed to the Permissions Department, John Wiley & Sons, Inc., 111 River Street, Hoboken, NJ 07030, (201) 748-6011, fax (201) 748-6008, or online at http://www.wiley.com/go/permission.

Limit of Liability/Disclaimer of Warranty: While the publisher and author have used their best efforts in preparing this book, they make no representations or warranties with respect to the accuracy or completeness of the contents of this book and specifically disclaim any implied warranties of merchantability or fitness for a particular purpose. No warranty may be created or extended by sales representatives or written sales materials. The advice and strategies contained herein may not be suitable for your situation. You should consult with a professional where appropriate. Neither the publisher nor author shall be liable for any loss of profit or any other commercial damages, including but not limited to special, incidental, consequential, or other damages.

For general information on our other products and services or for technical support, please contact our Customer Care Department within the United States at (800) 762-2974, outside the United States at (317) 572-3993 or fax (317) 572-4002.

Wiley also publishes its books in a variety of electronic formats. Some content that appears in print may not be available in electronic formats. For more information about Wiley products, visit our web site at www.wiley.com.

*Library of Congress Cataloging-in-Publication Data.*

Loveland, Walter D.
  Modern nuclear chemistry / Walter D. Loveland, David J. Morrissey, Glenn T. Seaborg.
     p. cm.
  Includes bibliographical references and index.
  ISBN-13 978-0-471-11532-8 (cloth: alk. paper)
  ISBN-10 0-471-11532-0 (cloth: alk. paper)
  1. Nuclear chemistry--Textbooks. I. Morrissey, David J. II. Seaborg, Glenn Theodore, 1912– III. Title.
  QD601.3.L68 2005
  541'.38--dc22
                                                                    2005022036

Printed in the United States of America

10  9  8  7  6  5  4  3  2  1

# CONTENTS

**PREFACE** xv

**CHAPTER 1  INTRODUCTORY CONCEPTS** 1

    1.1  Introduction / 1
    1.2  The Atom / 2
    1.3  Atomic Processes / 3
        1.3.1  Ionization / 3
        1.3.2  X-ray Emission / 4
    1.4  The Nucleus Nomenclature / 6
    1.5  Survey of Nuclear Decay Types / 8
    1.6  Modern Physical Concepts Needed in Nuclear Chemistry / 11
        1.6.1  Types of Forces in Nature / 11
        1.6.2  Elementary Mechanics / 12
        1.6.3  Relativistic Mechanics / 13
        1.6.4  De Broglie Wavelength, Wave–Particle Duality / 17
        1.6.5  Heisenberg Uncertainty Principle / 19
        1.6.6  Units and Conversion Factors / 19
    1.7  Particle Physics / 20
    1.8  Exchange Particles and Force Carriers / 24
    Problems / 24
    Bibliography / 26

## CHAPTER 2  NUCLEAR PROPERTIES  29

2.1 Introduction / 30
2.2 Nuclear Masses / 30
2.3 Terminology / 32
2.4 Binding Energy Per Nucleon / 33
2.5 Separation Energy Systematics / 35
2.6 Abundance Systematics / 36
2.7 Semiempirical Mass Equation / 36
2.8 Nuclear Sizes and Shapes / 42
2.9 Quantum Mechanical Properties / 44
  2.9.1 Nuclear Angular Momenta / 44
2.10 Electric and Magnetic Moments / 47
  2.10.1 Magnetic Dipole Moment / 47
  2.10.2 Electric Quadrupole Moment / 50
Problems / 53
References / 56
Bibliography / 56

## CHAPTER 3  RADIOACTIVE DECAY KINETICS  57

3.1 Basic Decay Equations / 58
3.2 Mixture of Two Independently Decaying Radionuclides / 65
3.3 Radioactive Decay Equilibrium / 67
3.4 Branching Decay / 75
3.5 Natural Radioactivity / 77
3.6 Radionuclide Dating / 81
Problems / 87
References / 89
Bibliography / 89

## CHAPTER 4  RADIOTRACERS  91

4.1 Introduction / 91
4.2 Design of a Radiotracer Experiment / 92
  4.2.1 Basic Design Criteria / 92
  4.2.2 Practical Considerations / 95
4.3 Preparation of Radiotracers and Their Compounds / 97
  4.3.1 Chemical Synthesis / 99
  4.3.2 Biosynthesis / 100
  4.3.3 Tritium Labeling / 100
  4.3.4 Radiolysis of Labeled Compounds / 101

4.4 Tracing of Physical Process / 101
4.5 Chemical Applications of Tracers / 102
4.6 Isotope Effects / 104
4.7 Biological Applications / 107
4.8 Environmental Applications / 109
4.9 Industrial Use of Radiotracers / 113
4.10 Nuclear Medicine / 113
4.11 Isotope Dilution Analysis / 122
    4.11.1 Direct IDA / 122
    4.11.2 Inverse IDA / 123
    4.11.3 General Comments / 124
    4.11.4 Special IDA Techniques / 124
4.12 Radiometric Techniques / 125
Problems / 127
References / 128
Bibliography / 128

## CHAPTER 5 NUCLEAR FORCES      129

5.1 Introduction / 129
5.2 Characteristics of the Strong Force / 130
5.3 Charge Independence of Nuclear Forces / 132
Problems / 134
Reference / 135

## CHAPTER 6 NUCLEAR STRUCTURE      137

6.1 Nuclear Potentials / 139
6.2 Schematic Shell Model / 140
6.3 Independent Particle Model / 152
6.4 Collective Model / 154
6.5 Nilsson Model / 160
6.6 Nucleus as a Fermi Gas / 163
Problems / 171
References / 174
Bibliography / 174

## CHAPTER 7   $\alpha$ DECAY      177

7.1 Energetics of $\alpha$ Decay / 179
7.2 Theory of $\alpha$ Decay / 183
7.3 Hindrance Factors / 192

7.4  Heavy Particle Radioactivity / 193
7.5  Proton Radioactivity / 195
Problems / 197
References / 198
Bibliography / 198

## CHAPTER 8  β DECAY                                         199

8.1  Introduction / 199
8.2  Neutrino Hypothesis / 200
8.3  Derivation of Spectral Shape / 203
8.4  Kurie Plots / 207
8.5  β-Decay Rate Constant / 208
8.6  Electron Capture Decay / 213
8.7  Parity Nonconservation / 214
8.8  Neutrinos / 215
8.9  β-Delayed Radioactivities / 216
8.10 Double-β Decay / 217
Problems / 219
References / 220
Bibliography / 220

## CHAPTER 9  γ-RAY DECAY                                     221

9.1  Introduction / 221
9.2  Energetics of γ Decay / 222
9.3  Classification of Decay Types / 223
9.4  Electromagnetic Transition Rates / 226
9.5  Internal Conversion / 232
9.6  Angular Correlations / 235
9.7  Mössbauer Effect / 241
Problems / 247
References / 248
Bibliography / 248

## CHAPTER 10  NUCLEAR REACTIONS                              249

10.1  Introduction / 249
10.2  Energetics of Nuclear Reactions / 250
10.3  Reaction Types and Mechanisms / 254
10.4  Nuclear Reaction Cross Sections / 255

10.5   Reaction Observables / 264
10.6   Rutherford Scattering / 265
10.7   Elastic (Diffractive) Scattering / 268
10.8   Direct Reactions / 270
10.9   Compound Nucleus Reactions / 272
10.10  Photonuclear Reactions / 278
10.11  Heavy Ion Reactions / 279
      10.11.1  Coulomb Excitation / 280
      10.11.2  Elastic Scattering / 281
      10.11.3  Fusion Reactions / 282
      10.11.4  Deep Inelastic Scattering / 286
      10.11.5  Incomplete Fusion / 286
      10.11.6  Reactions Induced by Radioactive Projectiles / 287
10.12  High-Energy Nuclear Reactions / 288
      10.12.1  Spallation/Fragmentation / 288
      10.12.2  Multifragmentation / 291
      10.12.3  Quark–Gluon Plasma / 292
Problems / 293
References / 296
Bibliography / 297

## CHAPTER 11   FISSION   299

11.1   Introduction / 299
11.2   Probability of Fission / 302
      11.2.1  Liquid Drop Model / 302
      11.2.2  Shell Corrections / 304
      11.2.3  Spontaneous Fission / 306
      11.2.4  Spontaneously Fissioning Isomers / 308
      11.2.5  Transition Nucleus / 310
11.3   Fission Product Distributions / 316
      11.3.1  Total Kinetic Energy (TKE) Release in Fission / 316
      11.3.2  Fission Product Mass Distributions / 316
      11.3.3  Fission Product Charge Distributions / 318
11.4   Excitation Energy of the Fission Fragments / 322
11.5   Dynamical Properties of the Fission Fragments / 325
Problems / 329
References / 329

## CHAPTER 12  NUCLEAR REACTIONS IN NATURE: NUCLEAR ASTROPHYSICS    331

12.1  Introduction / 331
12.2  Elemental and Isotopic Abundances / 332
12.3  Primordial Nucleosynthesis / 336
12.4  Stellar Evolution / 338
12.5  Thermonuclear Reaction Rates / 342
12.6  Stellar Nucleosynthesis / 344
    12.6.1  Introduction / 344
    12.6.2  Hydrogen Burning / 345
    12.6.3  Helium Burning / 348
    12.6.4  Synthesis of Nuclei with $A < 60$ / 349
    12.6.5  Synthesis of Nuclei with $A > 60$ / 351
12.7  Solar Neutrino Problem / 354
    12.7.1  Introduction / 354
    12.7.2  Expected Solar Neutrino Sources, Energies, and Fluxes / 355
    12.7.3  Detection of Neutrinos / 357
    12.7.4  Solar Neutrino Problem / 359
    12.7.5  Solution of the Problem—Neutrino Oscillations / 359
12.8  Synthesis of Li, Be, and B / 361
Problems / 362
References / 363
Bibliography / 363

## CHAPTER 13  ANALYTICAL APPLICATIONS OF NUCLEAR REACTIONS    365

13.1  Activation Analysis / 366
    13.1.1  Basic Description of Method / 366
    13.1.2  Advantages and Disadvantages of Activation Analysis / 367
    13.1.3  Practical Considerations in Activation Analysis / 368
    13.1.4  Applications of Activation Analysis / 372
13.2  Particle-Induced X-ray Emission / 373
13.3  Rutherford Backscattering (RBS) / 376
Problems / 379
References / 380
Bibliography / 380

CONTENTS    xi

## CHAPTER 14   REACTORS AND ACCELERATORS                         383

    14.1  Nuclear Reactors / 384
        14.1.1  Neutron-Induced Reactions / 384
        14.1.2  Neutron-Induced Fission / 387
        14.1.3  Neutron Inventory / 388
        14.1.4  Light Water Reactors / 390
        14.1.5  The Oklo Phenomenon / 395
    14.2  Neutron Sources / 395
    14.3  Neutron Generators / 396
    14.4  Accelerators / 397
        14.4.1  Ion Sources / 397
        14.4.2  Electrostatic Machines / 399
        14.4.3  Linear Accelerators / 403
        14.4.4  Cyclotrons, Synchrotrons, and Rings / 406
    14.5  Charged Particle Beam Transport and Analysis / 412
    14.6  Radioactive Ion Beams / 417
    14.7  Nuclear Weapons / 421
    Problems / 426
    References / 427
    Bibliography / 427

## CHAPTER 15   THE TRANSURANIUM ELEMENTS                          429

    15.1  Introduction / 429
    15.2  Limits of Stability / 429
    15.3  Element Synthesis / 431
    15.4  History of Transuranium Element Discovery / 438
    15.5  Superheavy Elements / 447
    15.6  Chemistry of the Transuranium Elements / 449
    15.7  Environmental Chemistry of the Transuranium Elements / 457
    Problems / 462
    References / 463
    Bibliography / 464

## CHAPTER 16   NUCLEAR REACTOR CHEMISTRY                          465

    16.1  Introduction / 465
    16.2  Fission Product Chemistry / 466

16.3 Radiochemistry of Uranium / 470
    16.3.1 Uranium Isotopes / 470
    16.3.2 Metallic Uranium / 470
    16.3.3 Uranium Compounds / 470
    16.3.4 Uranium Solution Chemistry / 471
16.4 Nuclear Fuel Cycle—The Front End / 472
    16.4.1 Mining and Milling / 472
    16.4.2 Refining and Chemical Conversion / 475
    16.4.3 Enrichment / 475
    16.4.4 Fuel Fabrication / 478
16.5 Nuclear Fuel Cycle—The Back End / 479
    16.5.1 Properties of Spent Fuel / 479
    16.5.2 Fuel Reprocessing / 481
16.6 Radioactive Waste Disposal / 483
    16.6.1 Classification of Radioactive Waste / 483
    16.6.2 Amounts and Associated Hazards / 484
    16.6.3 Storage and Disposal of Nuclear Waste / 485
16.7 Chemistry of Operating Reactors / 492
    16.7.1 Radiation Chemistry of Coolants / 493
    16.7.2 Corrosion / 493
    16.7.3 Coolant Activities / 494
Problems / 494
References / 495
Bibliography / 496

## CHAPTER 17 INTERACTION OF RADIATION WITH MATTER     497

17.1 Introduction / 497
17.2 Heavy Charged Particles ($A \geq 1$) / 499
17.3 Electrons / 514
17.4 Electromagnetic Radiation / 518
    17.4.1 Photoelectric Effect / 520
    17.4.2 Compton Scattering / 522
    17.4.3 Pair Production / 524
17.5 Neutrons / 526
17.6 Radiation Exposure and Dosimetry / 530
Problems / 533
References / 535
Bibliography / 535

CONTENTS    xiii

## CHAPTER 18   RADIATION DETECTORS                                         537

    18.1   Detectors Based on Ionization / 540

           18.1.1  Gas Ionization Detectors / 540

           18.1.2  Semiconductor Detectors (Solid-State Ionization Chambers) / 548

    18.2   Scintillation Detectors / 558

    18.3   Nuclear Track Detectors / 564

    18.4   Nuclear Electronics and Data Collection / 565

    18.5   Nuclear Statistics / 567

           18.5.1  Rejection of Abnormal Data / 574

           18.5.2  Setting Upper Limits When No Counts are Observed / 576

Problems / 576

References / 577

Bibliography / 577


## CHAPTER 19   RADIOCHEMICAL TECHNIQUES                                    579

    19.1   Unique Aspects of Radiochemistry / 580

    19.2   Availability of Radioactive Material / 584

    19.3   Targetry / 584

    19.4   Measuring Beam Intensity and Fluxes / 589

    19.5   Recoils, Evaporation Residues (EVRs), and Heavy Residues / 591

    19.6   Radiochemical Separation Techniques / 595

           19.6.1  Precipitation / 595

           19.6.2  Solvent Extraction / 596

           19.6.3  Ion Exchange / 599

           19.6.4  Extraction Chromatography / 602

           19.6.5  Rapid Radiochemical Separations / 602

    19.7   Low-Level Measurement Techniques / 603

           19.7.1  Introduction / 603

           19.7.2  Blanks / 604

           19.7.3  Low-Level Counting—General Principles / 605

           19.7.4  Low-Level Counting—Details / 605

           19.7.5  Limits of Detection / 608

Problems / 609

References / 610

Bibliography / 611

| | | |
|---|---|---|
| APPENDIX A | FUNDAMENTAL CONSTANTS AND CONVERSION FACTORS | 613 |
| APPENDIX B | NUCLEAR WALLET CARDS | 617 |
| APPENDIX C | PERIODIC TABLE OF ELEMENTS | 639 |
| APPENDIX D | LIST OF ELEMENTS | 641 |
| APPENDIX E | ELEMENTS OF QUANTUM MECHANICS | 643 |
| INDEX | | 665 |

# PREFACE

There are many fine textbooks of nuclear physics and chemistry in print at this time. So the question can be raised as to why we would write another textbook, especially one focusing on the smaller discipline of nuclear chemistry. When we began this project over 5 years ago, we felt that we were at a unique juncture in nuclear chemistry and technology and that, immodestly, we had a unique perspective to offer to students.

Much of the mainstream of nuclear chemistry is now deeply tied to nuclear physics, in a cooperative endeavor called *nuclear science*. At the same time, there is a large, growing, and vital community of people who use the applications of nuclear chemistry to tackle a wide-ranging set of problems in the physical, biological and environmental sciences, medicine, and engineering. We thought it was important to bring together, in a single volume, a rigorous, detailed perspective on both the "pure" and "applied" aspects of nuclear chemistry. As such, one might find more detail about any particular subject than one might like. We hope this encourages instructors to summarize the textbook material and present it in a manner most suitable to a particular audience. The amount of material contained in this book is too much for a one-quarter or one-semester course and a bit too little for a year-long course. Instructors can pick and choose which material seems most suitable for their course.

We have attempted to present nuclear chemistry and the associated applications at a level suitable for an advanced undergraduate or beginning graduate student. We have assumed the student has prior or, concurrent instruction in physical chemistry or modern physics and has some skills in handling differential equations. We have attempted to sprinkle solved problems throughout the text, as we believe that one learns by working problems. The end-of-the-chapter homework problems are largely examination questions used at Oregon State University. They should be considered an integral part of the textbook as they are intended to illustrate or amplify the main points of each chapter. We have taken some pains to use quantum mechanics in a schematic way, that is, to use the conclusions of such considerations without using or demanding a rigorous, complete approach. The use of hand-waving quantum mechanics, we believe, is appropriate for our general audience. We summarize, in the appendices, some salient features of quantum mechanics that may be useful for those students with limited backgrounds.

Our aim is to convey the essence of the ideas and the blend of theory and experiment that characterizes nuclear and radiochemistry. We have included some more advanced material for those who would like a deeper immersion in the subject. Our hope is that the reader can use this book for an introductory treatment of the subject of interest and can use the end-of-chapter references as a guide to more advanced and detailed presentations. We also hope the practicing scientist might see this volume as a quick refresher course for the rudiments of relatively unfamiliar aspects of nuclear and radiochemistry and as an information booth for directions for more detailed inquiries.

It is with the deep sense of loss and sadness that the junior authors (WDL, DJM) note the passing of our dear friend, colleague, and co-author, Prof. Glenn T. Seaborg before the completion of this work. Glenn participated in planning and development of the textbook, wrote some of the text, and reviewed much of the rest. We deeply miss his guidance and his perspective as we have brought this project to conclusion. We regret not paying closer attention to his urging that we work harder and faster as he would remark to us "You know I'm not going to live forever." We hope that the thoughts and ideas that he taught us are reflected in these pages.

We gratefully acknowledge the many colleagues and students who have taught us about nuclear chemistry and other things. Special thanks are due to Darrah Thomas and the late Tom Sugihara for pointing out better ways to discuss some material. We acknowledge the efforts of Einar Hagebø who used an early version of this book in his classes and gave us important feedback. We gratefully acknowledge the helpful comments of D. Peterson, P. Mantica, A. Paulenova, and R. A. Schmitt on various portions of the book. One of us (WDL) wishes to acknowledge the hospitality of the National Superconducting Cyclotron Laboratory at Michigan State University for its hospitality in the fall of 1999 during which time a portion of this book was written.

WALTER LOVELAND
Corvallis, Oregon

DAVID MORRISSEY
East Lansing, Michigan

# CHAPTER 1

# INTRODUCTORY CONCEPTS

## 1.1 INTRODUCTION

*Nuclear chemistry* consists of a four-pronged endeavor made up of (a) studies of the chemical and physical properties of the heaviest elements where detection of radioactive decay is an essential part of the work, (b) studies of nuclear properties such as structure, reactions, and radioactive decay by people trained as chemists, (c) studies of macroscopic phenomena (such as geochronology or astrophysics) where nuclear processes are intimately involved, and (d) the application of measurement techniques based upon nuclear phenomena (such as nuclear medicine, activation analysis or radiotracers) to study scientific problems in a variety of fields. The principal activity or "mainstream" of nuclear chemistry involves those activities listed under part (b).

As a branch of chemistry, the activities of nuclear chemists frequently span several traditional areas of chemistry such as organic, analytical, inorganic, and physical chemistry. Nuclear chemistry has ties to all branches of chemistry. For example, nuclear chemists are frequently involved with the synthesis and preparation of radiolabeled molecules for use in research or medicine. Nuclear analytical techniques are an important part of the arsenal of the modern analytical chemist. The study of the actinide and transactinide elements has involved the joint efforts of nuclear and inorganic chemists in extending knowledge of the periodic table. Certainly, the physical concepts and reasoning at the heart of modern nuclear chemistry are familiar to physical chemists. In this book we will touch on many of these interdisciplinary topics and attempt to bring in familiar chemical concepts.

*Modern Nuclear Chemistry*, by W.D. Loveland, D.J. Morrissey, and G.T. Seaborg
Copyright © 2006 John Wiley & Sons, Inc.

**2** INTRODUCTORY CONCEPTS

A frequently asked question is "What are the differences between *nuclear physics* and *nuclear chemistry*?" Clearly, the two endeavors overlap to a large extent, and in recognition of this overlap, they are collectively referred to by the catchall phrase "nuclear science." But we believe that there are fundamental, important distinctions between these two fields. Besides the continuing close ties to traditional chemistry cited above, nuclear chemists tend to study nuclear problems in different ways than nuclear physicists. Much of nuclear physics is focused on detailed studies of the fundamental interactions operating between subatomic particles and the basic symmetries governing their behavior. Nuclear chemists, by contrast, have tended to focus on studies of more complex phenomena where "statistical behavior" is important. Nuclear chemists are more likely to be involved in applications of nuclear phenomena than nuclear physicists, although there is clearly a considerable overlap in their efforts. Some problems, such as the study of the nuclear fuel cycle in reactors or the migration of nuclides in the environment, are so inherently chemical that they involve chemists almost exclusively.

One term that is frequently associated with nuclear chemistry is that of *radiochemistry*. The term radiochemistry refers to the chemical manipulation of radioactivity and associated phenomena. All radiochemists are, by definition, nuclear chemists, but not all nuclear chemists are radiochemists. Many nuclear chemists use purely nonchemical, that is, physical techniques, to study nuclear phenomena, and thus their work is not radiochemistry.

## 1.2 THE ATOM

Before beginning a discussion of nuclei and their properties, we need to understand the environment in which most nuclei exist, that is, in the center of atoms. In elementary chemistry, we learn that the atom is the smallest unit a chemical element can be divided into that retains its chemical properties. As we know from our study of chemistry, the radii of atoms are approximately $1-5 \times 10^{-10}$ m, or $1-5$ Å. At the center of each atom we find the *nucleus*, a small object ($r \sim 1-10 \times 10^{-15}$ m) that contains almost all the mass of the atom (Fig. 1.1). The atomic nucleus contains $Z$ protons, where $Z$ is the *atomic number* of the element under study, $Z$ being number of protons and is thus the number of positive charges in the nucleus. The *chemistry* of the element is controlled by $Z$ in that all nuclei with the same $Z$ will have similar chemical behavior. The nucleus also contains $N$ neutrons, where $N$ is the *neutron number*. Neutrons are uncharged particles with masses approximately equal to the mass of a proton ($\approx 1$ u). Each proton has a positive charge equal to that of an electron. The overall charge of a nucleus is $+Z$ electronic charge units.

Most of the atom is empty space in which the electrons surround the nucleus. (Electrons are small, negatively charged particles with a charge of $-1$ electronic charge units and a mass of about $1/1840$ of the proton mass.) The negatively charged electrons are bound by an electrostatic (Coulombic) attraction to the positively charged nucleus. In a neutral atom, the number of electrons in the atom equals the number of protons in the nucleus.

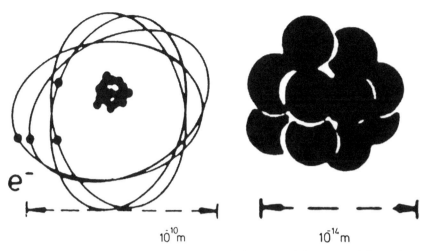

**Figure 1.1** Schematic representation of the relative sizes of the atom and the nucleus.

Quantum mechanics tells us that only certain discrete values of $E$, the total electron energy, and $J$, the angular momentum of the electrons are allowed. These discrete states have been depicted in the familiar semiclassical picture of the atom (Fig. 1.1) as a tiny nucleus with electrons rotating about it in discrete orbits. In this book, we will examine nuclear structure and will develop a similar semiclassical picture of the nucleus that will allow us to understand and predict a large range of nuclear phenomena.

## 1.3 ATOMIC PROCESSES

The sizes and energy scales of atomic and nuclear processes are very different. These differences allow us to consider them separately.

### 1.3.1 Ionization

Suppose one atom collides with another atom. If the collision is *inelastic* (the kinetic energies of the colliding nuclei are not conserved), one of two things may happen. They are (a) *excitation* of one or both atoms to an excited state involving a change in electron configuration or (b) *ionization* of one or both atoms, that is, removal of one or more of the atom's electrons to form a positively charged ion. For ionization to occur, an atomic electron must receive an energy that is at least equivalent to its binding energy, which for the innermost or K electrons is $(Z_{\text{eff}}/137)^2(255.5)$ keV, where $Z_{\text{effective}}$ is the effective nuclear charge felt by the electron (and includes the effects of screening of the nuclear charge by other electrons). This effective nuclear charge for K electrons can be approximated by the expression $(Z - 0.3)$. As one can see from these expressions, the energy necessary to cause ionization far exceeds the kinetic energies of gaseous atoms at room temperature. Thus, atoms must be moving with high speeds (as the result of nuclear decay processes or acceleration) to eject tightly bound electrons from other atoms through collisions.

### 1.3.2 X-ray Emission

The term *X-ray* refers to the electromagnetic radiation produced when an electron in an outer atomic electron shell drops down to fill a vacancy in an inner atomic electron shell (Fig. 1.2), such as going from the M shell to fill a vacancy in the L shell. The electron loses potential energy in this transition (in going to a more tightly bound shell) and radiates this energy in the form of X-rays. (X-rays are not to be confused with generally more energetic $\gamma$ rays, which result from transitions made by the neutrons and protons in the nucleus of the atom, not in the atomic electron shells.) The energy of the X-ray is given by the difference in the binding energies of the electrons in the two shells, which, in turn, depends on the atomic number of the element. Thus X-ray energies can be used to determine the atomic number of the elemental constituents of a material and are also regarded as conclusive proof of the identification of a new chemical element.

In X-ray terminology, X-rays due to transitions from the L to K shell are called $K_\alpha$ X-rays; X-rays due to transitions from the M to K shells are called $K_\beta$ X-rays. [In a further refinement, the terms $K_{\alpha_1}$, $K_{\alpha_2}$ refer to X-rays originating in different subshells ($2p_{3/2}$, $2p_{1/2}$) of the L shell.] X-rays from M to L transitions are $L_\alpha$ X-rays, and so forth. For each transition, the changes in orbital angular momentum, $\Delta l$, and total angular momentum, $\Delta j$, are required to be

$$\Delta l = \pm 1$$
$$\Delta j = 0, \pm 1 \qquad (1.1)$$

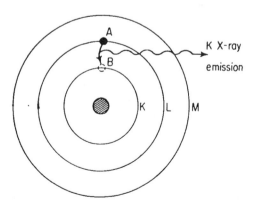

**Figure 1.2** Schematic diagram to show X-ray emission to fill vacancy caused by nuclear decay. An L-shell electron (*A*) is shown filling a K-shell vacancy (*B*). In doing so, it emits a characteristic K X-ray.

The simple Bohr model of the hydrogen-like atom (one electron only) predicts that the X-ray energy or the transition energy, $\Delta E$, is given as

$$\Delta E = E_{\text{initial}} - E_{\text{final}} = R_\infty hcZ^2 \left( \frac{1}{n_{\text{initial}}^2} - \frac{1}{n_{\text{final}}^2} \right) \tag{1.2}$$

where $R_\infty$, $h$, $c$, and $n$ denote the Rydberg constant, Planck constant, the speed of light, and the principal quantum number for the orbital electron, respectively. Since the X-ray energy, $E_x$, is actually $-\Delta E$, we can write (after substituting values for the physical constants)

$$E_x = 13.6 Z^2 \left( \frac{1}{n_{\text{final}}^2} - \frac{1}{n_{\text{initial}}^2} \right) \text{eV} \tag{1.3}$$

where $E_x$ is given in units of electron volts (eV).

For $K_\alpha$ X-rays from ions with only one electron

$$E_x^K = 13.6 \left( \frac{1}{1^2} - \frac{1}{2^2} \right) Z^2 \text{ eV} \tag{1.4}$$

while for $L_\alpha$ X-rays, we have

$$E_x^L = 13.6 \left( \frac{1}{2^2} - \frac{1}{3^2} \right) Z^2 \text{ eV} \tag{1.5}$$

In reality, many electrons will surround the nucleus, and we must replace $Z$ by $Z_{\text{effective}}$ to reflect the screening of the nuclear charge by these other electrons. This correction was done by Moseley who showed that the frequencies, $v$, of the $K_\alpha$ series X-rays could be expressed as

$$v^{1/2} = \text{const}(Z - 1) \tag{1.6}$$

while for $L_\alpha$ series X-rays

$$v^{1/2} = \text{const}(Z - 7.4) \tag{1.7}$$

Moseley thus demonstrated the X-ray energies ($=hv$) depend on the square of some altered form (due to screening) of the atomic number. Also, the relative intensities of the $K_{\alpha_1}$, $K_{\alpha_2}$, and so on, X-rays will be proportional to the number of possible ways to make the transition. Thus, we expect the $K_{\alpha_1}/K_{\alpha_2}$ intensity ratio to be $\sim 2$ as the maximum number of electrons in the $2p_{3/2}$ level is 4 while the maximum number of electrons in the $2p_{1/2}$ level is 2. The relative intensities of different X-rays depend on the chemical state of the atom, its oxidation state, bonding with ligands, and other factors that affect the local electron density. These relative intensities are, thus, useful in chemical speciation studies. We should also note, as discussed extensively in Chapters 7–9, that X-ray production can accompany radioactive decay. Radioactive decay modes, such as electron capture or internal conversion, directly result in vacancies in the atomic electron shells. The resulting X-rays are signatures that can be used to characterize the decay modes and/or the decaying species.

## 1.4 THE NUCLEUS NOMENCLATURE

A nucleus is said to be composed of *nucleons*. There are two "kinds" of nucleons, the neutrons and the protons. A nucleus with a given number of protons and neutrons is called a *nuclide*. The *atomic number* $Z$ is the number of protons in the nucleus, while $N$, the *neutron number*, is used to designate the number of neutrons in the nucleus. The total number of nucleons in the nucleus is $A$, the *mass number*. Obviously $A = N + Z$. Note that $A$, the number of nucleons in the nucleus, is an integer while the actual mass of that nucleus, $m$, is not an integer.

Nuclides with the same number of protons in the nucleus but with differing numbers of neutrons are called *isotopes*. (This word comes from the Greek *iso + topos*, meaning "same place" and referring to the position in the periodic table.) Isotopes have very similar chemical behavior because they have the same electron configurations. Nuclides with the same number of neutrons in the nucleus, $N$, but differing numbers of protons, $Z$, are referred to as *isotones*. Isotones have some nuclear properties that are similar in analogy to the similar chemical properties of isotopes. Nuclides with the same mass number, $A$, but differing numbers of neutrons and protons are referred to as *isobars*. Isobars are important in radioactive decay processes. Finally, the term *isomer* refers to a nuclide in an excited nuclear state that has a measurable lifetime ($>10^{-9}$ s). These labels are straightforward, but the term isotope is frequently misused. For example, radioactive nuclei (radionuclides) are often *incorrectly* referred to as radioisotopes, even though the nuclides being referenced do not have the same atomic numbers.

The convention for designating a given nuclide (with $Z$ protons, $N$ neutrons) is to write

$$^A_Z \text{Chemical symbol}_N$$

with the relative positions indicating a specific feature of the nuclide. Thus, the nucleus with six protons and eight neutrons is $^{14}_{6}C_8$, or completely equivalently, $^{14}C$. (The older literature used the form $_Z$Chemical symbol$^A$, so $^{14}C$ was designated as $C^{14}$. This nomenclature is generally extinct.) Note that sometimes the atomic charge of the entity containing the nuclide is denoted as an upper-right-hand superscript. Thus, a doubly ionized atom containing a Li nucleus with three protons and four neutrons and only one electron is designated sometimes as $^7Li^{2+}$.

**Sample Problem 1.1** Consider the following nuclei:

$^{60}C^m$, $^{14}C$, $^{14}N$, $^{12}C$, $^{13}N$

Which are isotopes? Isotones? Isobars? Isomers?

**Answer** $^{60}C^m$ is the isomer, $^{14}C$ and $^{12}C$ are isotopes of C, $^{13}N$ and $^{14}N$ are isotopes of N, $^{14}C$ and $^{14}N$ are isobars ($A = 14$), while $^{12}C$ and $^{13}N$ are isotones ($N = 6$).

We can now make an estimate of two important quantities, the size and the density of a typical nucleus. We can write

$$\rho \equiv \text{density} = \frac{\text{mass}}{\text{volume}} \approx \frac{A \text{ amu}}{\frac{4}{3}\pi R^3} \tag{1.8}$$

if we assume that the mass of each nucleon is about 1 u and the nucleus can be represented as a sphere. It turns out (Chapter 2) that an empirical rule to describe the radii of stable nuclei is that radius $R$ is

$$R = 1.2 \times 10^{-13} A^{1/3} \text{ cm} \tag{1.9}$$

Thus, we have

$$\rho = \frac{(A \text{ u})(1.66 \times 10^{-24} \text{ g/u})}{\frac{4}{3}\pi (1.2 \times 10^{-13} A^{1/3} \text{cm})^3}$$

where we have used the value of $1.66 \times 10^{-24}$ g for 1 u (Appendix A). Before evaluating the density $\rho$ numerically, we note that the $A$ factor cancels in the expression, leading us to conclude that all nuclei have approximately the same density. This is similar to the situation with different sized drops of a pure liquid. All of the molecules in a drop interact with each other with the same short-range forces, and the overall drop size grows with the number of molecules. Evaluating this expression and converting to convenient units, we have

$$\rho \approx 200,000 \text{ metric tons/mm}^3$$

A cube of nuclear matter that is 1 mm on a side contains a mass of 200,000 tonnes. WOW! Now we can realize what all the excitement about the nuclear phenomena is about. Think of the tremendous forces that are needed to hold matter together with this density. Relatively small changes in nuclei (via decay or reactions) can release large amounts of energy. (From the point of view of the student doing calculations with nuclear problems, a more useful expression of the nuclear density is 0.14 nucleons/fm$^3$.)

## 1.5 SURVEY OF NUCLEAR DECAY TYPES

Nuclei can emit radiation spontaneously. The general process is called *radioactive decay*. While this subject will be discussed in detail in Chapters 3, 7, 8, and 9, we need to know a few general ideas about these processes right away (which we can summarize below).

Radioactive decay usually involves one of three basic types of decay, $\alpha$ decay, $\beta$ decay, or $\gamma$ decay in which an unstable nuclide spontaneously changes into a more stable form and emits some radiation. In Table 1.1, we summarize the basic features of these decay types.

The fact that there were three basic decay processes (and their names) was discovered by Rutherford. He showed that all three processes occur in a sample of decaying natural uranium (and its daughters). The emitted radiations were designated $\alpha$, $\beta$, and $\gamma$ to denote the penetrating power of the different radiation types. Further research has shown that in $\alpha$ decay, a heavy nucleus spontaneously emits a $^4$He nucleus (an $\alpha$ particle). The emitted $\alpha$ particles are monoenergetic, and, as a result of the decay, the parent nucleus loses two protons and two neutrons and is transformed into a new nuclide. All nuclei with $Z > 83$ are unstable with respect to this decay mode.

Nuclear $\beta$ decay occurs in three ways, $\beta^-$, $\beta^+$, and electron capture (EC). In these decays, a nuclear neutron (or proton) changes into a nuclear proton (or neutron) with the ejection of a neutrino (or antineutrino) and an electron (or positron). In electron capture, an orbital electron is captured by the nucleus, changing a proton into a neutron with the emission of a neutrino. The total number of nucleons, $A$, in the nucleus does not change in these decays, only the relative number of neutrons and protons. In a sense, this process can "correct" or "adjust" an imbalance between the number of neutrons and protons in a nucleus. In $\beta^+$ and $\beta^-$ decays, the decay energy is shared between the emitted electron, the neutrino, and the recoiling daughter nucleus. Thus, the energy spectrum of the emitted electrons and neutrinos is continuous, ranging from zero to the decay energy. In EC decay, essentially all the decay energy is carried away by the emitted neutrino. Neutron-rich nuclei decay by $\beta^-$ decay, whereas proton-rich nuclei decay by $\beta^+$ or EC decay. $\beta^+$ decay is favored in the light nuclei and requires the decay energy to be greater than 1.02 MeV (for reasons to be discussed later), whereas EC decay is found mostly in the heavier nuclei.

Nuclear electromagnetic decay occurs in two ways, $\gamma$ decay and internal conversion (IC). In $\gamma$-ray decay a nucleus in an excited state decays by the emission of a photon. In internal conversion the same excited nucleus transfers its energy radiationlessly to an orbital electron that is ejected from the atom. In both types of decay, only the excitation energy of the nucleus is reduced with no change in the number of any of the nucleons.

**Sample Problem 1.2** Because of the conservation of the number of nucleons in the nucleus and conservation of charge during radioactive decay (Table 1.1),

**TABLE 1.1 Characteristics of Radioactive Decay**

| Decay Type | Emitted Particle | $\Delta Z$ | $\Delta N$ | $\Delta A$ | Typical Energy of Emitted Particle | Example | Occurrence |
|---|---|---|---|---|---|---|---|
| $\alpha$ | $^4\text{He}^{2++}$ | $-2$ | $-2$ | $-4$ | $4 \leq E_\alpha \leq 10\,\text{MeV}$ | $^{238}\text{U} \rightarrow {}^{234}\text{Th} + \alpha$ | $Z > 83$ |
| $\beta^-$ | Energetic $e^-$, $\bar{\nu}_e$ | $+1$ | $-1$ | $0$ | $0 \leq E_{\beta^-} \leq 2\,\text{MeV}$ | $^{14}\text{C} \rightarrow {}^{14}\text{N} + \beta^- + \bar{\nu}_e$ | $N/Z > (N/Z)_\text{stable}$ |
| $\beta^+$ | Energetic $e^+$, $\nu_e$ | $-1$ | $+1$ | $0$ | $0 \leq E_{\beta^+} \leq 2\,\text{MeV}$ | $^{22}\text{Na} \rightarrow {}^{22}\text{Ne} + \beta^+ + \nu_e$ | $(N/Z) < (N/Z)_\text{stable}$; light nuclei |
| EC | $\nu_e$ | $-1$ | $+1$ | $0$ | $0 \leq E_\nu \leq 2\,\text{MeV}$ | $e^- + {}^{207}\text{Bi} \rightarrow {}^{207}\text{Pb} + \nu_e$ | $(N/Z) < (N/Z)_\text{stable}$; heavy nuclei |
| $\gamma$ | Photon | $0$ | $0$ | $0$ | $0.1 \leq E_\gamma \leq 2\,\text{MeV}$ | $^{60}\text{Ni}^* \rightarrow {}^{60}\text{Ni} + \gamma$ | Any excited nucleus |
| IC | Electron | $0$ | $0$ | $0$ | $0.1 < E_e < 2\,\text{MeV}$ | $^{125}\text{Sb}^m \rightarrow {}^{125}\text{Sb} + e^-$ | Cases where $\gamma$-ray emission is inhibited |

it is relatively easy to write and balance nuclear decay equations. For example, consider the

$\beta^-$ decay of $^{90}$Sr
$\alpha$ decay of $^{232}$Th
$\beta^+$ decay of $^{62}$Cu
EC decay of $^{256}$Md

These decay equations can be written, using Table 1.1, as

$$^{90}_{38}\text{Sr} \longrightarrow {}^{90}_{39}\text{Y}^+ + \beta^- + \bar{\nu}_e$$

$$^{232}_{90}\text{Th} \longrightarrow {}^{228}_{88}\text{Ra} + {}^{4}_{2}\text{He}$$

$$^{62}_{29}\text{Cu} \longrightarrow {}^{62}_{28}\text{Ni}^- + \beta^+ + \nu_e$$

$$e^- + {}^{256}_{101}\text{Md}^+ \longrightarrow {}^{256}_{100}\text{Fm} + \nu_e$$

Besides its qualitative description, radioactive decay has an important quantitative description. Radioactive decay can be described as a first-order reaction, that is, the number of decays is proportional to the number of decaying nuclei present. It is described by the integrated rate law

$$N = N_0 e^{-\lambda t} \qquad (1.10)$$

where $N$ is the number of nuclei present at time $t$ while $N_0$ is the number of nuclei present at time $t = 0$. The decay constant $\lambda$, a characteristic of each nucleus, is related to the half-life, $t_{1/2}$, by

$$\lambda = \ln 2 / t_{1/2} \qquad (1.11)$$

The half-life is the time required for the number of nuclei present to decrease by a factor of 2. The number of decays that occur in a radioactive sample in a given amount of time is called *the activity* $A$ of the sample. The activity is equal to the number of nuclei present, $N$, multiplied by the probability of decay per nucleus, $\lambda$, that is, $A = \lambda N$. Therefore, the activity will also decrease exponentially with time,

$$A = A_0 e^{-\lambda t} \qquad (1.12)$$

where $A$ is the number of disintegrations per unit time at time $t$, and $A_0$ is the activity at time $t = 0$. The half-lives of nuclei with respect to each decay mode are often used to identify the nuclei.

**Sample Problem 1.3** $^{14}$C decays to $^{14}$N by $\beta^-$ decay with a half-life of 5730 y. If a 1-g sample of carbon contains 15.0 disintegrations per minute, what will be its activity after 10,000 y?

*Solution*

$$A = A_0 e^{-\lambda t}$$

$$\lambda = \frac{\ln 2}{5730 \text{ y}} = 1.210 \times 10^{-4} \text{ y}^{-1}$$

$$A = (15 \text{ dis/min}) \, e^{-(1.210 \times 10^{-4})(10,000)} = 4.5 \text{ dis/min}$$

All living things maintain a constant level of $^{14}$C per gram of carbon through exchange with their surroundings. When they die, this exchange stops and the amount of $^{14}$C present decreases exponentially with time. A measurement of the $^{14}$C content of a dead object can be used to determine the age of the object. This process and other geologically important decay processes are discussed in Chapter 3.

## 1.6 MODERN PHYSICAL CONCEPTS NEEDED IN NUCLEAR CHEMISTRY

While we shall strive to describe nuclear chemistry without using extensive mathematics and physics, there are several important concepts from modern physics that we need to review because we will use these concepts in our discussions.

### 1.6.1 Types of Forces in Nature

Let us review briefly some physical concepts that we shall use in our study of nuclear chemistry. First, we should discuss the types of forces found in nature. There are *four* fundamental forces in nature (Table 1.2). As far as we know, all the interactions in the universe are the result of these forces. The weakest force is *gravity*, which is most significant when the interacting objects are massive, such as planets, stars, and the like. The next strongest force is *the weak interaction*, which is important in nuclear β decay. The familiar *electromagnetic force*, which governs most behavior in our sensory world, is next in strength while the *nuclear or strong interaction* is the

**TABLE 1.2 Types of Force Encountered in Nature**

| Force | Range (m) | Relative Strength | Force Carrier |
|---|---|---|---|
| Gravitational | $\infty$ | $10^{-38}$ | Graviton |
| Weak | $10^{-18}$ | $10^{-5}$ | $W^\pm, Z^0$ |
| Electromagnetic | $\infty$ | $\alpha = 1/137$ | Photon |
| Strong | $10^{-15}$ | 1 | Gluon |

**12** INTRODUCTORY CONCEPTS

strongest force. Please note, as indicated earlier in our discussion of nuclear densities, that the strong or nuclear force is more than 100 times stronger than the electromagnetic force holding atoms together.

In the 19th century, electricity and magnetism were linked together. The 20th century has seen the demonstration that the electromagnetic and weak forces are just two different aspects of the same force, called the electroweak force. Current efforts are directed at unifying the strong and electroweak forces in a so-called *grand unified theory*, or GUT. The final step in this direction would be to include gravity in a *theory of everything*. Discussion of these unified theories is beyond the scope of this book; however, the relative strength and character of the forces will form an important part of our discussion of nuclear phenomena.

### 1.6.2 Elementary Mechanics

Let us recall a few elementary relationships from classical physics that we shall use. Force can be represented as a vector, **F**, that describes the rate of change of the momentum with time:

$$\mathbf{F} = \frac{d\mathbf{p}}{dt} \tag{1.13}$$

where the momentum $p = mv$ and where $m$ is the mass and $v$ is the velocity of the particle. Neglecting relativistic effects (Section 1.6.3) that are important for particles whose velocity approaches the speed of light, we can say that the kinetic energy of a moving body $T$ is given as

$$T = \frac{1}{2}mv^2 \tag{1.14}$$

For the situation depicted in Figure 1.3 for the motion of a particle past a fixed point, we can say that the orbital angular momentum of the particle, $l$, with mass $m$ with respect to the point $Q$ is

$$\mathbf{l} = \mathbf{r} \times \mathbf{p} \tag{1.15}$$

The quantity **l** is a vector whose magnitude is $mvr$ for circular motion. For motion past a stationary point, the magnitude is $mvb$ where $b$ is the distance of closest approach called the *impact parameter*.

Let us also recall the relationship between the magnitude of a force $F(r)$ that depends on the distance between two objects, $r$, and the potential energy, $V(r)$,

$$F = \frac{-\partial V}{\partial r} \tag{1.16}$$

Thus, if the Coulomb potential energy between two charged objects is given as

$$V = \frac{+kq_1q_2}{r_{12}} \tag{1.17}$$

## 1.6 MODERN PHYSICAL CONCEPTS NEEDED IN NUCLEAR CHEMISTRY

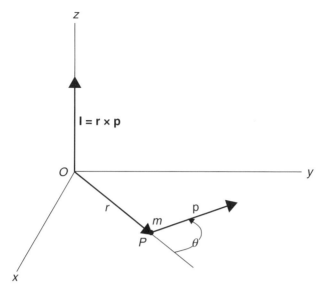

**Figure 1.3** Particle of mass $m$, moving with velocity $v$ has a linear momentum $p = mv$. Relative to the point $O$, the particle has an angular momentum of $\mathbf{l} = \mathbf{r} \times \mathbf{p}$, where $\mathbf{r}$ is a vector connecting the point $O$ and the particle. At the point of closest approach, $r$ is equal to the impact parameter, $b$.

where $r_{12}$ is the distance separating charges $q_1$ and $q_2$ (and where $k$ is a constant), we can say the magnitude of the Coulomb force, $F_C$, is

$$F_C = \frac{-\partial V}{\partial r} = \frac{kq_1q_2}{r_{12}^2} \tag{1.18}$$

Since forces are usually represented as vectors, it is more convenient when discussing nuclear interactions to refer to the scalar, potential energy. From the above discussion, we should always remember that a discussion of potential energy $V(r)$ is also a discussion of force $F(r)$.

### 1.6.3 Relativistic Mechanics

As Einstein demonstrated, when a particle moves with a velocity approaching that of light, the classical relations (Section 1.6.2) describing its motion in a stationary system are no longer valid. Nuclear processes frequently involve particles with such high velocities. Thus, we need to understand the basic elements of relativistic mechanics. According to the special theory of relativity, the mass of a moving particle changes with speed according to the equation

$$m^* = \gamma m_0 \tag{1.19}$$

where $m^*$ and $m_0$ are the mass of a particle in motion and at rest, respectively. The Lorentz factor $\gamma$ is given as

$$\gamma = (1 - \beta^2)^{-1/2} \tag{1.20}$$

where $\beta$ is the speed of the particle, $v$, relative to the speed of light, $c$, so that $\beta = v/c$. Thus, as the speed of the particle increases, the mass also increases; making further increases in speed more difficult. Since the mass $m^*$ cannot be imaginary, no particle can go faster than the speed of light. The total energy of a particle, $E_{TOT}$, is given as

$$E_{TOT} = m^* c^2 \tag{1.21}$$

Since the total energy equals the kinetic energy plus the rest mass energy, we can write

$$E_{TOT} = T + m_0 c^2 \tag{1.22}$$

where $T$ is the particle's kinetic energy. Thus,

$$T = (\gamma - 1) m_0 c^2 \tag{1.23}$$

The space–time coordinates $(x, y, z, t)$ of a point in a stationary system are, according to the special theory of relativity, related to the space–time coordinates in a system moving along the $x$ axis $(x', y', z', t')$ by the relations

$$\begin{aligned} x' &= \gamma(x - \beta c t) \\ y' &= y \\ z' &= z \\ t' &= \gamma [t - (\beta/c) x] \end{aligned} \tag{1.24}$$

These transformations from the stationary to the moving frame are called the Lorentz transformations. The inverse Lorentz transformation is obtained by reversing the sign of $v$, so that

$$\begin{aligned} x &= \gamma (x' + \beta c t') \\ y &= y' \\ z &= z' \\ t &= \gamma [t + (\beta/c) x'] \\ \Delta t &= t_1 - t_2 = \gamma [\Delta t' + (\beta/c) \Delta x] \\ \Delta x &= \Delta x'/\gamma \end{aligned} \tag{1.25}$$

Since $\gamma > 1$, time is slowed down for the stationary observer, and distance in the $x$ direction is contracted.

One application of these equations in nuclear chemistry involves the decay of rapidly moving particles. The muon, a heavy electron, has a lifetime, $\tau$, *at rest*, of 2.2 μs. When the particle has a kinetic energy of 100 GeV (as found in cosmic rays), we observe a lifetime of $\gamma\tau$ or about $10^3\tau$. (This phenomenon is called *time dilation* and explains why such muons can reach the surface of Earth.)

A series of relationships have been derived between the stationary coordinate system (the scientist in his or her laboratory) and a moving (intrinsic, invariant) coordinate system that can be compared to classical calculations of dynamic variables (Table 1.3).

Note that for a particle at rest

$$E_{\text{TOT}} = m_0 c^2 \tag{1.26}$$

where $m_0$ is the rest mass and $c$ is the speed of light. For a massless particle, such as a photon, we have

$$E_{\text{TOT}} = pc \tag{1.27}$$

where $p$ is the momentum of the photon. These equations make it clear why the units of MeV/$c^2$ for mass and MeV/$c$ for momentum can be useful in nuclear calculations.

An important question is when do we use classical expressions and when do we use relativistic expressions? A convenient, but arbitrary, criterion for making this decision is to use the relativistic expression when $\gamma \geq 1.1$. This corresponds roughly to the point at which a 13% error occurs in the classical expression. What does this criterion mean, in practice? In Table 1.4, we indicate the values of the kinetic energy at which $\gamma = 1.1$ for different particles. Thus, one should always

**TABLE 1.3 Comparison of Relativistic and Classical Expressions for a Free Particle Moving in $x$ Direction**

| Classical Expression | Relativistic Expression |
|---|---|
| $x$ | $x = \gamma(x' + \beta c t')$ |
| $y$ | $y = y'$ |
| $z$ | $z = z'$ |
| $t$ | $t = \gamma(t' + \beta/cx')$ |
| $\Delta t = t_2 - t_1$ | $\Delta t' = \gamma \Delta t$ |
| Mass $m$ | $m = \gamma m_0$ ($m_0 \equiv$ rest mass) |
| Momentum $p = mv$ | $p = \gamma m v$ |
| $T \equiv$ kinetic energy $= \frac{1}{2}mv^2$ | $T = (\gamma - 1)m_0 c^2$ |
| Total energy $E_{\text{TOT}} = E_k$ (free particle) | $E_{\text{TOT}} = \gamma m_0 c^2$ |
| Energy momentum relationship $E = p^2/2m$ | $E_{\text{TOT}}^2 = p^2 c^2 + m_0^2 c^4$ |

**TABLE 1.4 When Does One Use Relativistic Expressions?**

| Particle | $T$ (MeV) when $\gamma = 1.1$ |
|---|---|
| $\gamma, \nu$ | 0 |
| e | 0.051 |
| $\mu$ | 11 |
| $\pi$ | 14 |
| p, n | 94 |
| d | 188 |
| $\alpha$ | 373 |

use the relativistic expressions for photons, neutrinos, and electrons (when $T_e > 50$ keV) or for nucleons when the kinetic energy/nucleon exceeds 100 MeV.

**Sample Problem 1.4   Relativistic Mechanics**   Consider a $^{20}$Ne ion with a kinetic energy of 1 GeV/nucleon. Calculate its velocity, momentum, and total energy.

*Solution*   Total kinetic energy = $20 \times 1$ GeV/nucleon = 20 GeV = 20,000 MeV. But we know: $T = (\gamma - 1)m_0c^2$
The rest mass is approximately 20 u or $(20)(931.5)$ MeV/$c$ or 18,630 MeV. So we can say

$$\gamma = \frac{T}{m_0c^2} + 1 = 1 + \frac{20,000}{18,630} = 2.07$$

But we know

$$\gamma = (1 - \beta^2)^{-1/2}$$

So we can say

$$\beta = \left(1 - \frac{1}{\gamma^2}\right)^{1/2} = 0.88$$

So the velocity $v$ is $0.88c$ or $(0.88)(3.00 \times 10^8 \text{ m/s}) = 2.6 \times 10^8$ m/s. The momentum is given as

$$p = \frac{mv}{\sqrt{1-\beta^2}} = \gamma mv$$
$$= (2.07)(20)(1.67 \times 10^{-27} \text{ g})(2.6 \times 10^8)$$
$$= 1.8 \times 10^{-17} \text{ kg} \cdot \text{m/s}$$

or in other units

$$pc = \frac{mcv}{\sqrt{1-\beta^2}} = (931.5)(20)(0.88)(2.07)$$
$$= 33.9 \, \text{GeV}$$
$$p = 33.9 \, \text{GeV}/c$$

The total energy

$$E_{\text{TOT}} = E_k + m_0 c^2$$
$$= \gamma m_0 c^2 = (2.07)(20)(931.5) = 38.6 \, \text{GeV}$$

### 1.6.4 De Broglie Wavelength, Wave–Particle Duality

There is no distinction between wave and particle descriptions of matter. It is simply a matter of convenience, which we choose to use in a given situation. For example, it is quite natural to describe matter in terms of particles with values of momenta, kinetic energies, and so forth. It is also natural to use a wave description for light. However, associated with each material particle, there is a wave description in which the particle is assigned a wavelength (the de Broglie wavelength $\lambda$) whose magnitude is given as

$$\lambda = \frac{h}{p} \tag{1.28}$$

where $p$ is the momentum of the particle and $h$ is Planck's constant. (Note that Planck's constant is extremely small, $6.6 \times 10^{-34}$ J s. Thus, the wavelength of a particle is only important when the momentum is extremely small, such as with electrons whose mass is $9 \times 10^{-31}$ kg.) The expression for the de Broglie wavelength may be written in rationalized units

$$\lambdabar = \frac{\hbar}{p} \tag{1.29}$$

where $\hbar$ is $h/2\pi$. The above expressions are classical and should be replaced by their relativistic equivalents where appropriate, that is,

$$\lambdabar = \frac{\hbar c}{[E_k(E_k + 2m_0 c^2)]^{1/2}} \tag{1.30}$$

We can calculate typical magnitudes of these wavelengths of particles encountered in nuclear chemistry (Table 1.5). Given typical nuclear dimensions of $10^{-13}$ cm, the data of Table 1.5 indicate the energies at which such particles might have a

**TABLE 1.5 Typical Magnitudes of de Broglie Wavelengths**

| Energy (MeV) | Wavelength (cm) | | |
|---|---|---|---|
| | Photon | Electron | Proton |
| 0.1 | $1.2 \times 10^{-9}$ | $3.7 \times 10^{-10}$ | $9.0 \times 10^{-12}$ |
| 1 | $1.2 \times 10^{-10}$ | $8.7 \times 10^{-11}$ | $2.9 \times 10^{-12}$ |
| 10 | $1.2 \times 10^{-11}$ | $1.2 \times 10^{-11}$ | $0.9 \times 10^{-12}$ |
| 100 | $1.2 \times 10^{-12}$ | $1.2 \times 10^{-12}$ | $2.8 \times 10^{-13}$ |
| 1000 | $1.2 \times 10^{-13}$ | $1.2 \times 10^{-13}$ | $0.7 \times 10^{-13}$ |

wavelength similar or smaller than nuclear dimensions. These particles can be used as probes of nuclear sizes and shapes.

In a similar manner, it is quite natural to associate a wave description to photons (Table 1.4). Here we recall that

$$\lambda = \frac{c}{v} = \frac{hc}{E_\gamma} \tag{1.31}$$

where $v$ is the frequency associated with the wave of length $\lambda$. A convenient form of this equation is

$$\lambda \text{ (cm)} = \frac{1.2397 \times 10^{-10}}{E_\gamma \text{ (MeV)}} \tag{1.32}$$

which was used to calculate the values in Table 1.5. But it is often useful to speak of photons as particles particularly when they are emitted or absorbed by a nucleus, when we write

$$E_\gamma = hv = pc \tag{1.33}$$

**Sample Problem 1.5   de Broglie Wavelength**   Consider the case of a beam of 1 eV neutrons incident on a crystal. First-order Bragg reflections are observed at 11.8°. What is the spacing between crystal planes?

**Solution**   Low-energy neutrons are diffracted like X-rays. The Bragg condition is that $n\lambda = 2d \sin \Theta$ where the index $n = 1$ for first-order diffraction.

$$\lambda = 2d \sin \Theta$$

$$d = \frac{\lambda}{2 \sin \Theta} = \frac{\frac{h}{p}}{2 \sin \Theta} = \frac{\frac{h}{(2mE_k)^{1/2}}}{2 \sin \Theta}$$

$$d = \frac{\frac{6.63 \times 10^{34} \text{ J s}}{(2 \ 1.67 \times 10^{-27} \text{ kg } 1.60 \times 10^{-19} \text{ J})^{1/2}}}{2 \sin (11.8°)} = 7.0 \times 10^{-11} \text{ m}$$

### 1.6.5 Heisenberg Uncertainty Principle

Simply put, the Heisenberg uncertainty principle states that there are limits on knowing both where something is and how fast it is moving. Formally, we can write

$$\Delta p_x \cdot \Delta x \geq \hbar$$
$$\Delta p_y \cdot \Delta y \geq \hbar$$
$$\Delta p_z \cdot \Delta z \geq \hbar \quad (1.34)$$
$$\Delta E \cdot \Delta t \geq \hbar$$

where $\Delta p_x$, $\Delta x$ are the uncertainties in the $x$ component of the momentum and the $x$ coordinate, respectively, whereas $\Delta t$ is the lifetime of a particle and $\Delta E$ is the uncertainty in its total energy. These limits on our knowledge are not due to the limitations of our measuring instruments. They represent fundamental limits even with ideal or perfect instruments.

It is instructive to consider a practical example to see the effect of these limits. Consider an electron with a kinetic energy of 5 eV. Its speed can be calculated (nonrelativistically):

$$v = \left(\frac{2E_k}{m}\right)^{1/2} = \left(\frac{(2)(5)(1.602 \times 10^{-19} \text{ J/eV})}{9.11 \times 10^{-31} \text{ kg}}\right)^{1/2}$$
$$= 1.33 \times 10^6 \text{ m/s}$$

Its momentum is then

$$p = mv = 1.21 \times 10^{-24} \text{ kg m/s}$$

Assume the uncertainty in its measured momentum is 1%. The uncertainty principle then tells us

$$\Delta x = \frac{\hbar}{\Delta p} = \frac{1.06 \times 10^{-34} \text{ J/s}}{1.21 \times 10^{-26} \text{ kg m/s}} = 8.8 \times 10^{-9} \text{ m}$$

which is about 40 atomic diameters. In short, if you know the momentum relatively well, you do not know where the electron is in space.

### 1.6.6 Units and Conversion Factors

Every field has its own special units of measure and nuclear chemistry is no different. The unit of length is the femtometer ($10^{-15}$ m), which is called a *fermi*. The unit of mass is the *atomic mass unit* (amu or u), which has a numerical value of $\sim 1.66 \times 10^{-24}$ g or expressed in units of MeV/$c^2$, it is 931.5 MeV/$c^2$. The unit of energy is MeV ($10^6$ eV), which is $\sim 1.602 \times 10^{-13}$ J, the energy gained when a proton is accelerated through a potential of $10^6$ V. Appendix A contains a list of the exact numerical values of these and other convenient units. Special attention is called to five very useful quantities:

$$\frac{e^2}{4\pi\varepsilon_0} = 1.43998 \, \text{MeV fm}$$

$$\hbar = 6.58212 \times 10^{-22} \, \text{MeV s}$$

$$c = 2.9979 \times 10^{23} \, \text{fm s}^{-1} = 29.979 \, \text{cm/ns}$$

$$\hbar c = 197.3 \, \text{MeV fm}$$

$$1 \text{ year (sidereal)} = 3.1558 \times 10^7 \, \text{s} \approx \pi \times 10^7 \, \text{s}$$

## 1.7 PARTICLE PHYSICS

Elementary particle physicists ("high-energy physicists") study the fundamental particles of nature and the symmetries found in their interactions. The study of elementary particle physics is an important endeavor in its own right and beyond the scope of this book. But we need to use some of the concepts of this area of physics in our discussion of nuclei.

Particles can be classified as *fermions* or *bosons*. Fermions obey the Pauli principle and have antisymmetric wave functions and half-integer spins. (Neutrons, protons, and electrons are fermions.) Bosons do not obey the Pauli principle and have symmetric wave functions and integer spins. (Photons are bosons.)

Particle groups, like fermions, can also be divided into the *leptons* (such as the electron) and the *hadrons* (such as the neutron and proton). The hadrons can interact via the nuclear or strong interaction while the leptons do not. (Both particle types can, however, interact via other forces, such as the electromagnetic force.) Figure 1.4 contains artistic conceptions of the standard model, a theory that describes these fundamental particles and their interactions. Examples of bosons, leptons, hadrons, their charges, and masses are given in Table 1.6.

There are six different kinds of leptons (light particles) (Table 1.6), and they can be arranged in three pairs. The electron (e), the muon ($\mu$), and the tau lepton ($\tau$) each carry a charge of $-e$ and have associated with them the electron ($\nu_e$), muon ($\nu_\mu$), and tau neutrinos ($\nu_\tau$). These neutrinos are electrically neutral and have small or zero rest mass. The actual mass of the neutrinos is a subject of current research (see Chapter 12). The electron neutrino is seen in nuclear phenomena such as $\beta$ decay, whereas the other neutrinos are involved in higher energy processes.

One important aspect of leptons is that their number is conserved in nuclear processes. Consider, for example, the decay of the free neutron

$$n \rightarrow p^+ + e^- + \bar{\nu}_e$$

## (a) The Standard Model

### PARTICLES

All the *matter* in the Universe, including atoms, stars, rocks, plants and animals is made of...

The particles *carrying* the forces between the matter are...

**Fermions**
2 TYPES

**STRUCTURE OF THE ATOM**

**Bosons**
A family of particles called **gauge bosons** transmit the forces between the fermions. There is a different kind of particle for each force:
- **Photons** (the particles of light) carry the electromagnetic force;
- **Gluons** carry the strong force;
- **W** and **Z** bosons carry the weak force;
- **Gravitons**—not yet observed—are believed to be responsible for gravity, which is not a part of the Standard Model.

**QUARKS**
The **protons** and **neutrons** of an atom's nucleus are themselves complex structures, made up of groups of three basic particles called **quarks**. Quarks can also bind with **antiquarks** to make other particles called **mesons**.

**LEPTONS**
Leptons are not made of quarks, and include the **electrons** that orbit the atomic nucleus, and their more esoteric relatives, like **muons, taus** and **neutrinos**.

### FORCES

There appear to be four basic forces at work:
- **Strong force** is responsible for holding together protons and neutrons.
- **Weak force** causes certain forms of radioactivity.
- **Electromagnetic force** holds atoms and molecules together.
- **Gravity** is responsible for the large-scale structure of the Universe, binding stars and galaxies together.

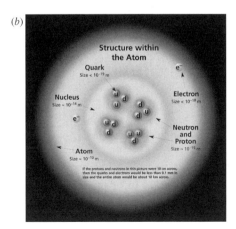

(b)

**Figure 1.4** Two artists' conceptions of the standard model. (*a*) From New York Times, 22 September, 1998. Reprinted by permission of the New York Times. (*b*) From "Nuclear Science," Contemporary Physics Education Project (CPEP), LBNL. (Figure 1.4(*b*) also appears in color figure section.)

**TABLE 1.6 Table of Leptons and Their Properties**

| Flavor | Mass (GeV/$c^2$) | Electric Charge |
|---|---|---|
| $\nu_e$ electron neutrino | $<1 \times 10^{-11}$ | 0 |
| e electron | 0.000511 | $-1$ |
| $\nu_\mu$ muon neutrino | $<0.0002$ | 0 |
| $\mu$ muon | 0.106 | $-1$ |
| $\nu_\tau$ tau neutrino | $<0.02$ | 0 |
| $\tau$ tau | 1.7771 | $-1$ |

(The symbol $\bar{\nu}_e$ indicates the antiparticle of the electron neutrino.) In this equation, the number of leptons on the left is zero, so the number of leptons on the right must also be zero. This equivalence can only be true if we assign a lepton number L of 1 to the electron (by convention) and $L = -1$ to the $\bar{\nu}_e$ (being an antiparticle). Consider the reaction

$$\bar{\nu}_e + p^+ \longrightarrow e^+ + n$$

Here $L = -1$ on both sides of the equation where we assign lepton numbers of $+1$ for every lepton and $-1$ for every antilepton ($e^+$ is an antilepton). By contrast, the reaction

$$\nu_e + p^+ \neq e^+ + n$$

is forbidden by lepton conservation. The law of lepton conservation applies separately to electrons, muons, and tau muons.

**Sample Problem 1.6** Is the reaction $\mu^- \longrightarrow e^- + \bar{\nu}_e + \nu_\mu$ possible?

**Solution**
Left-hand side  $L_\mu = 1$   $L_e = 0$
Right-hand side  $L_e = 1 + (-1) = 0$
          $L_\mu = +1$
Yes, the reaction is possible.

If we focus our attention on the neutrons and protons (the nucleons), we note they have similar masses ($\sim$1 u). We also note the neutron is slightly ($\sim$0.14%) more massive than the proton with the mass difference being $\sim$1.29 MeV/$c^2$ (Appendix A). (This energy difference causes a free neutron to decay to a proton with a half-life of approximately 10 min.) As remarked earlier, the neutron has no net electric charge, whereas the proton has a positive charge equal in magnitude to the charge on the electron. The electric charge on the proton is uniformly and symmetrically distributed about the center of the proton with a charge radius of about 0.8 fm. The neutron, although electrically neutral, also has an extended charge distribution with a positive charge near the center being canceled out by a negative charge at larger values of the radius. The values of the magnetic dipole moment of the neutron and proton are also indications of their complex structure (Chapter 2). As far as their interaction via the nuclear or strong force, the neutrons and protons behave alike (the "charge independence" of the nuclear force). They can be regarded collectively as "nucleons." The nucleon can be treated as a physical entity with a mass of 938 MeV/$c^2$. One can speak of excited states of the nucleon such as the one with a mass of 1232 MeV/$c^2$ (which is called the $\Delta$ state).

The fermionic hadrons (called *baryons*) are thought to be made up of three fundamental particles called *quarks*. There are six different kinds (or *flavors*) of quarks: u (up), d (down), s (strange), c (charm), t (top), and b (bottom). The masses and charges of the quarks are given in Table 1.7. The size of each quark is thought to be $<10^{-18}$ m. The lightest two quarks, the u and d quarks, are thought to make up the nucleons. The proton is a *uud* combination with a charge of $(\frac{2}{3}+\frac{2}{3}-\frac{1}{3})e$, whereas the neutron is a *udd* combination with a charge of $(\frac{2}{3}-\frac{1}{3}-\frac{1}{3})e$. The up and down quarks are light ($m \sim$ 5–10 MeV/$c^2$) and pointlike. The quarks account for $\sim$2% of the mass of the proton. The rest of the mass is in gluons, which "connect" the quarks. The most massive of the quarks is the top quark with a mass approximately equivalent to that of a $^{197}$Au nucleus and a short lifetime ($\sim 10^{-24}$ s).

Like the leptons, there is a number conservation law for baryons. To each baryon, such as the neutron or proton, we assign a baryon number $B = +1$ while we assign $B = -1$ to each antibaryon, such as the antiproton. Our rule is that the total baryon number must be conserved in any process. Consider the reaction

$$p^+ + p^+ \longrightarrow p^+ + n + \pi^+$$

On the left, $B = 2$ as it does on the right (the $\pi^+$ is a meson and has $B = 0$).

As well as binding three quarks (antiquarks) together to make baryons (antibaryons), the nuclear or strong interaction can bind a quark and an antiquark to form unstable particles called *mesons* (q, q̄). The $\pi^+$ and $\pi^-$ mesons (u$\bar{\text{d}}$, d$\bar{\text{u}}$) are of special importance in nuclear science. The quark/antiquark pairs in the $\pi$ mesons couple to have zero spin, and thus these mesons are bosons. In fact, all mesons have integer spins and are thus bosons.

**TABLE 1.7 Table of Quarks and Their Properties**

| Flavor | Approx. Mass (GeV/$c^2$) | Electric Charge |
|---|---|---|
| u up | 0.003 | 2/3 |
| d down | 0.006 | −1/3 |
| c charm | 1.3 | 2/3 |
| s strange | 0.1 | −1/3 |
| t top | 175 | 2/3 |
| b bottom | 4.3 | −1/3 |

## 1.8 EXCHANGE PARTICLES AND FORCE CARRIERS

The force carrier (or "exchange") particles are all bosons. These particles are responsible for carrying the four fundamental forces. This family includes the strong interaction carrier, the gluon; the weak interaction carriers, the $W^{\pm}$ and $Z^0$; the carrier of the electromagnetic force, the photon; and the postulated but unobserved carrier of the gravitational force, the graviton.

To understand how these force carriers work, let us consider the electromagnetic force acting between two positively charged particles. Quantum electrodynamics tells us that the force between these two particles is caused by photons passing between them. At first one may find that idea nonsensical because the emission of a photon should change the energy of the emitter/source (but exchange of a force-carrier does not). The trick is that the uncertainty principle allows the emission of *virtual* particles (which violate energy conservation) if such emission and absorption occur within a time $\Delta t$ that is less than that allowed by the uncertainty principle

$$\Delta t = \frac{\hbar}{\Delta E} \quad (1.35)$$

where $\Delta E$ is the extent to which energy conservation is violated. We will consider the range of forces in Chapter 5.

## PROBLEMS

1. Define or describe the following terms or phenomena: radiochemistry, isotone, internal conversion, gluon, lepton.

2. Define or describe the following phenomena: electron capture, exchange forces, time dilation.

3. Define or describe the following terms: quark, hadron, baryon, lepton, meson.

4. In an experiment one observes the characteristic $K_\alpha$ X-rays of two elements at energies of 6.930 and 7.478 eV. The higher energy line is due to Ni. What element is responsible for the lower energy line?

5. Using the Bohr theory, calculate the ratio of the energies of the $K_\alpha$ X-rays of I and Xe.

6. Given the following energies of the $K_\alpha$ X-rays for the following elements, make a Moseley plot of the data

| | |
|---|---|
| V | 4.952 eV |
| Cr | 5.415 |
| Mn | 5.899 |
| Fe | 6.404 |

7. Predict the mode of decay of the following nuclei: $^{14}C$, $^{3}H$, $^{11}C$, $^{233}U$, $^{138}La$.

8. Write complete, balanced equations for the following decays:

   a. $\alpha$ decay of $^{230}$Th
   b. $\beta^-$ decays of $^{95}$Zr
   c. $\beta^+$ decay of $^{17}$F
   d. EC decay of $^{192}$Au

9. Consider the decay of $^{238}$U to $^{206}$Pb. How many $\alpha$ particles and $\beta^-$ particles are emitted in this decay?

10. If a rock has a ratio of $^{206}$Pb to $^{238}$U of 0.6, what is the age of the rock?

11. How long will it take for a sample of $^{239}$Pu ($t_{1/2}$ = 24, 119 y) to decay to $\frac{1}{10}$ its original amount?

12. If a radioactive sample of $^{59}$Fe ($t_{1/2}$ = 44.496 d) has an activity of 1000 disintegrations per minute, what weight of $^{59}$Fe is present?

13. The environmental concentration of $^{239}$Pu ($t_{1/2}$ = 24, 119 y) in a lake is $3.7 \times 10^{-6}$ disintegrations/s/liter. What is the molarity of the solution?

14. $^{32}$P ($t_{1/2}$ = 14.262 d) is a popular tracer in biochemistry. If I need to have $0.1 \times 10^6$ disintegrations/s 60 days from now, how much $^{32}$P tracer must I purchase today?

15. Calculate the speed of a particle whose kinetic energy is three times its rest energy.

16. Calculate the speed parameter $\beta$ and the Lorenz factor $\gamma$ for the following particles: an electron with $E_K$ = 1 MeV; a proton with $E_K$ = 1 MeV; a $^{12}$C nucleus with $E_K$ = 12 MeV.

17. Consider the following free particles: a 1-eV photon, a 1-MeV electron, and a 10-MeV proton. Which is moving the fastest? Slowest? Has the most momentum? The least momentum?

18. How much energy is necessary to increase the speed of a proton from $0.2c$ to $0.3c$? From $0.98c$ to $0.99c$?

19. A nonrelativistic particle is moving five times as fast as a proton. The ratio of their de Broglie wave lengths is 10. Calculate the mass of the particle.

20. What are the wavelengths of a 500-MeV photon, a 500-MeV electron, and a 500-MeV proton?

21. What is the wavelength of a "thermal" neutron? (The kinetic energy of the neutron can be taken to be $\frac{3}{2}kT$ where $T$ is the absolute room temperature.)

22. Consider a nuclear excited state with a lifetime of 10 ps that decays by the emission of a 2-MeV $\gamma$ ray. What is the uncertainty in the $\gamma$-ray energy?

**23.** Which of the following decays are allowed by conservation laws?

a. $p \rightarrow e^+ + \gamma$
b. $p \rightarrow \pi^+ + \gamma$
c. $n \rightarrow p + \gamma$
d. $p + n \rightarrow p + p + \pi^-$
e. $p + p \rightarrow p + p + p + \bar{p}$

**24.** What is the quark composition of the antiproton and the antineutron?

## BIBLIOGRAPHY

There are many fine textbooks for nuclear and radiochemistry that cover the material covered in this book. A limited selection of some of the authors' favorites appears below.

### Simple Introductions to Nuclear Chemistry

Ehmann, W. D. and D. E. Vance. *Radiochemistry and Nuclear Methods of Analysis*, Wiley, New York, 1991. An up-to-date survey of nuclear chemistry that emphasizes its applications in analytical chemistry.

Harvey, B. G. *Nuclear Chemistry*, Prentice-Hall, Englewood Cliffs, NJ, 1965. A dated but elegant summary of the essential features of nuclear science.

Loveland, W. Nuclear Chemistry, in *Encyclopedia of Physical Science and Technology*, Vol. 11, Academic, Orlando, FL, 1992. A microversion of this text.

Wang, C. H., D. L. Willis, and W. D. Loveland. *Radiotracer Methodology in the Biological, Environmental and Physical Sciences*, Prentice-Hall, Englewood Cliffs, NJ, 1975. A somewhat out-of-date survey of radiotracer methods that includes an introduction to nuclear science for life scientists.

### History

Romer, A. *Radiochemistry and the Discovery of Isotopes*, Dover, New York, 1970. An intriguing view of the beginning of nuclear chemistry.

Romer, A. *The Discovery of Radioactivity and Transmutation*, Dover, New York, 1964. A presentation of the earliest explorations of radioactivity.

Seaborg, G. T. and W. Loveland. *Nuclear Chemistry*, Hutchinson-Ross, Stroudsberg, 1982. Reprints of the most significant papers in nuclear chemistry from the earliest work to present with annotations and English translations.

### Intermediate-Level Textbooks—Similar to This Book

Choppin, G. R., J. O. Liljenzin, and J. Rydberg. *Radiochemistry and Nuclear Chemistry*, 3rd ed., Butterworth-Heineman, Oxford, 2001. A very good, broad discussion of nuclear chemistry that is oriented toward nuclear power and nuclear power applications.

Cohen, B. L. *Concepts of Nuclear Physics*, McGraw-Hill, New York, 1971. This book is especially noted for its discussion of the shell model and direct reactions.

Evans, R. *The Atomic Nucleus*, McGraw-Hill, New York, 1955. A dated, but encyclopedic, treatment of nuclear science that has set the standard for its successors.

Friedlander, G., J. Kennedy, J. M. Miller, and E. S. Macias. *Nuclear and Radiochemistry*, Wiley, New York, 1981. The bible of nuclear chemistry.

Harvey, B. G. *Introduction to Nuclear Physics and Chemistry*, 2nd ed., Prentice-Hall, Englewood Cliffs, NJ, 1969. A wonderful, clear description of the physics of nuclei and their interaction that is somewhat dated.

Keller, C. *Radiochemistry*, Harwood, 1981. A very condensed presentation of radioactivity and its applications.

Krane, K. S. *Introductory Nuclear Physics*, Wiley, New York, 1987. A clear, relatively up-to-date discussion from the point of view of a practicing experimental nuclear physicist.

Meyerhof, W. *Elements of Nuclear Physics*, McGraw-Hill, New York, 1967. A very concise summary of the essential ideas of nuclear science.

Mukhin, K. N. *Experimental Nuclear Physics, Vol. I*, Mir, Moscow, 1987. Nuclear physics described from a more formal Russian viewpoint.

Valentin, L. *Subatomic Physics: Nuclei and Particles*, North-Holland, Amsterdam, 1981. An eclectic treatment at a slightly more advanced level.

## More Advanced Textbooks

Burcham, W. E. and M. Jobes. *Nuclear and Particle Physics*, Longman, Brnt Mill, 1995. A more comprehensive treatment with extensive discussion of particle physics.

de Shalit, A. and H. Feshbach. *Theoretical Nuclear Physics, Vol. I: Nuclear Structure, Vol. II: Nuclear Reactions*, Wiley, New York, 1974. A comprehensive treatment of the theory of nuclear structure and reactions.

Frauenfelder, H. and E. M. Henley. *Subatomic Physics*, 2nd ed., Prentice-Hall, Englewood Cliffs, NJ, 1991. A treatment of both nuclear and elementary particle physics.

Heyde, K. *Basic Ideas and Concepts in Nuclear Physics*, 2nd ed., IOP, Bristol, 1999. An excellent treatment of many newer aspects of nuclear physics.

Hodgson, P. E., E. Gadioli, and E. Gadiolo Erba. *Introductory Nuclear Physics*, Clarendon, Oxford, 1997. Emphasis on nuclear reactions.

Marmier, P. and E. Sheldon. *Physics of Nuclei and Particles*, Vol. I and Vol. II, Academic, New York, 1969. A dated, but accessible, treatment aimed at experimentalists.

Segre, E. *Nuclei and Particles*, 2nd ed., Benjamin, Reading, 1977. Remarkable for its breadth and insight in nuclear physics.

Wong, S. S. M. *Introductory Nuclear Physics*, 2nd ed., Prentice-Hall, Englewood Cliffs, NJ, 1998. A very up-to-date, readable treatment of nuclear physics.

## General Physics Textbooks

Halliday, D., R. Resnick, and K. S. Krane. *Physics*, 4th ed., Vol. I and Vol. II, *Extended*, Wiley, New York, 1992. A remarkable encyclopedic treatment of introductory physics.

## General References

Anderson, H. L. Ed. *A Physicist's Desk Reference*, AIP, New York, 1989. Helpful summaries of many types of data and directions as to their proper use.

Browne, E. and R. B. Firestone. *Table of Radioactive Isotopes*, Wiley, New York, 1986. An authoritative compilation of radioactive decay properties. Do note that the spontaneous fission half-lives are missing for several heavy nuclei.

Firestone, R. B. and V. S. Shirley. *Table of Isotopes*, 8th ed., Wiley, New York, 1996. Although available on the Web, this reference is still useful because it contains simplified energy level schemes not easily found in other places.

## Web References

*Living Textbook for Nuclear Chemistry* (http://livingtextbook.orst.edu) A compilation of supplemental materials related to nuclear and radiochemistry.

# CHAPTER 2

# NUCLEAR PROPERTIES

Chapter 7. THE STRUCTURE OF THE NUCLEUS OF THE ATOM
"What?" exclaimed Roger, as Karen rolled over on the bed and rested her warm body against his. "I know some nuclei are spherical and some are ellipsoidal, but where did you find out that some fluctuate in between?"

Karen pursed her lips. "They've been observed with a short-wavelength probe . . ." From S. Harris, Chalk Up Another one. Copyright © 1992 by AAAS Press. Reprinted by permission of AAAS Press.

*Modern Nuclear Chemistry*, by W.D. Loveland, D.J. Morrissey, and G.T. Seaborg
Copyright © 2006 John Wiley & Sons, Inc.

## 2.1 INTRODUCTION

In this chapter we will turn to a systematic look at the general properties of nuclei, including their masses and matter distributions. A very large number of nuclei have been studied over the years, and the general size, shape, mass, and relative stability of these nuclei follow patterns that can be understood and interpreted with two complimentary models of nuclear structure. The average size and stability of a nucleus can be described by the average binding of the nucleons to each other in a *macroscopic* model while the detailed energy levels and decay properties can be understood with a quantum mechanical or *microscopic* model. We will consider the average behavior in this chapter, and a detailed description of nuclear structure is given later in Chapter 6.

## 2.2 NUCLEAR MASSES

One of the most important nuclear properties that can be measured is the mass. Nuclear or atomic masses are usually given in atomic mass units (amu or u) or their energy equivalent. The mass unit u is defined so that the mass of one *atom* of $^{12}$C is equal to 12.0000 ... u. Note we said *atom*. For convenience, the masses of atoms rather than nuclei are used in all calculations. When needed, the nuclear mass $m^{\text{nucl}}$ can be calculated from the relationship

$$m^{\text{nucl}} c^2 = M^{\text{atomic}} c^2 - [Zm_0 c^2 + B_e(Z)]$$

where $m_0$ is the rest mass of the electron, and $B_e(Z)$ is the total binding energy of all the electrons in the atom. $B_e(Z)$ can be estimated on the basis of the Thomas–Fermi model of the atom as

$$B_e(Z) = 15.73 Z^{7/3} \text{ eV}$$

Because the values of $B_e(Z)$ are generally small relative to the masses of the nuclei and electrons, we shall neglect this factor in most calculations. Let us make a few calculations to illustrate the use of masses in describing nuclear phenomena. Consider the $\beta^-$ decay of $^{14}$C, that is, $^{14}$C $\rightarrow$ $^{14}$N$^+$ + $\beta^-$ + $\bar{\nu}_e$ + energy. Neglecting the mass of the electron antineutrino, thought to be a few electron volts or less, we have

$$\text{Energy} = [(m(^{14}\text{C}) + 6m_0) - (m(^{14}\text{N}) + 6m_0) - m(\beta^-)]c^2$$

where $m_0$ is the electron rest mass and $m(X)$ is the mass of the *nucleus* X. Substituting in atomic masses as appropriate, recognizing that the $\beta^-$ is an electron, we have

$$\text{Energy} = [M(^{14}\text{C}) - M(^{14}\text{N})]c^2$$

where $M(X)$ is the atomic mass of X.

## 2.2 NUCLEAR MASSES

Let us now consider the case of the $\beta^+$ decay of $^{64}$Cu, $^{64}$Cu $\rightarrow$ $^{64}$Ni$^-$ + $\beta^+$ + $\nu_e$ + energy. Writing the equation for the energy release in the decay, we have

$$\text{Energy} = [(m(^{64}\text{Cu}) + 29m_0) - (m(^{64}\text{Ni}) + 28m_0) - (m_0) - m(\beta^+)]c^2$$

Substituting in atomic masses, and noting that the mass of a position equals the mass of an electron

$$\text{Energy} = [M(^{64}\text{Cu}) - M(^{64}\text{Ni}) - 2m_0]c^2$$

Our straightforward bookkeeping has shown us that for $\beta^+$ decay, the difference between the initial and final nuclear masses, must be at least $2m_0c^2$ (1.02 MeV) for the decay to be energetically possible. This energy represents the cost of creating the positron.

To complete our survey of the energy release in $\beta$ decay, let us consider the case of electron capture, that is, the electron capture decay of $^{207}$Bi: e$^-$ + $^{207}$Bi$^+$ $\rightarrow$ $^{207}$Pb + $\nu_e$ + energy. For the energy release in the decay, we have

$$\text{Energy} = [(m(^{207}\text{Bi}) + 83m_0) - (m(^{207}\text{Pb}) + 82m_0)]c^2$$

where we have recognized that $^{207}$Bi captures one of its orbital electrons. Substituting in atomic masses, we have

$$\text{Energy} = [M(^{207}\text{Bi}) - M(^{207}\text{Pb})]c^2$$

The energy release in nuclear reactions is called the *Q value* of the reaction, and its calculation is strictly a matter of bookkeeping. If we consider the reaction

$$^{56}\text{Fe} + {}^{4}\text{He} \longrightarrow {}^{59}\text{Co} + {}^{1}\text{H} + \text{energy}$$

we can use the atomic masses to get:

$$Q = [M(^{56}\text{Fe}) + M(^{4}\text{He}) - M(^{59}\text{Co}) - M(^{1}\text{H})]c^2$$

*Note that the sign convention used in nuclear chemistry and physics that assigns a positive Q value for exoergic reactions is opposite to that used in chemistry where exoergic reactions have negative values of $\Delta H$ and $\Delta E$.*

Calculate the energy release in the $\beta^-$ and $\beta^+$ decay of $^{64}$Cu:

$$\begin{aligned}(\text{Energy release in } {}^{64}\text{Cu } \beta^- \text{ decay}) &= \left[M({}^{64}\text{Cu}) - M({}^{64}\text{Zn})\right]c^2 \\ &= [-65.421 - (-65.999)] \\ &= 0.578 \text{ MeV}\end{aligned}$$

$$\begin{aligned}(\text{Energy release in } {}^{64}\text{Cu } \beta^+ \text{ decay}) &= \left[M({}^{64}\text{Cu}) - M({}^{64}\text{Ni}) - 2m_0\right]c^2 \\ &= [-65.421 - (-67.096) - 1.022] \\ &= 0.653 \text{ MeV}\end{aligned}$$

By convention $Q_{\beta^-}$ = energy release and $Q_{\beta^+}$ = energy release + 1.022 MeV. So

$$Q_{\beta^-} = 0.578 \text{ MeV} \quad \text{and} \quad Q_{\beta^+} = 1.675 \text{ MeV}$$

## 2.3 TERMINOLOGY

The difference between the actual nuclear mass and the mass of all the individual nucleons, which must be assembled to make the nucleus, is called the *total binding energy*, $B_{\text{tot}}(A, Z)$. It represents the work necessary to dissociate the nucleus into separate nucleons or the energy that would be released if all the nucleons came together to form the nucleus. We write

$$B_{\text{tot}}(A, Z) = [ZM({}^1\text{H}) + (A - Z)M(\text{n}) - M(A, Z)]c^2$$

where $M(A, Z)$ is the atomic mass of $^A Z$, $M(\text{n})$ and $M({}^1\text{H})$ are the mass of a neutron and a hydrogen atom, respectively. The *average binding energy per nucleon*, $B_{\text{ave}}(A, Z)$ is given by

$$B_{\text{ave}}(A, Z) = B_{\text{tot}}(A, Z)/A$$

In many tabulations of nuclear properties, such as that in Appendix B, the quantity that is tabulated is the *mass excess* or *mass defect* rather than the mass. The *mass excess*, $\Delta$, is defined as $M(A, Z) - A$, usually given in units of the energy equivalent of mass. Since in most, if not all calculations, the number of nucleons will remain constant, the use of mass excesses in the calculations will introduce an arithmetic simplification. Another term that is sometimes used is the mass excess per nucleon or the *packing fraction* $[\equiv (M - A)/A]$.

The work necessary to separate a neutron, proton, or $\alpha$ particle from a nucleus is called the (neutron, proton, or $\alpha$ particle) *separation energy* $S$. For a neutron

$$S_{\text{n}} = [M(A - 1, Z) + M(\text{n}) - M(A, Z)]c^2$$

Such separation energies can be expressed in terms of the total binding energy by

$$S_n = B_{tot}(A, Z) - B_{tot}(A - 1, Z)$$

Calculate the neutron separation energy of $^{236}$U and $^{239}$U.

$$S_n = [M(A - 1, Z) + M(n) - M(A, Z)]c^2$$

For $^{236}$U

$$S_n = [M_{235_U} + M_n - M_{236_U}]c^2$$
$$= 40.914 + 8.071 - 42.441$$
$$= 6.544 \text{ MeV}$$

For $^{239}$U

$$S_n = [M_{238_U} + M_n - M_{239_U}]c^2$$
$$= 47.304 + 8.071 - 50.596$$
$$= 4.779 \text{ MeV}$$

Notice that the neutron separation energy of $^{A}Z$ is the excitation energy of the nucleus $^{A}Z$ produced when $^{A-1}Z$ is irradiated with "zero energy" neutrons. Thus, when even–odd $^{235}$U is irradiated with neutrons, the $^{236}$U is produced at excitation energy of 6.5 MeV while the same process with $^{238}$U gives an excitation energy of 4.8 MeV. If it takes 5–6 MeV to cause these nuclei to fission, $^{235}$U is "fissionable" with zero energy neutrons while $^{238}$U is not.

## 2.4 BINDING ENERGY PER NUCLEON

The binding energy per nucleon is a measure of the relative stability of a nucleus. The more tightly bound a nucleus is, the greater the binding energy per nucleon is. A plot of the average binding energy per nucleon as function of the mass number is shown in Figure 2.1. Several features of this plot are worth noting. The greatest stability is associated with medium mass nuclei, with the most stable nucleus being $^{62}$Ni. The heaviest nuclei could increase their stability by fissioning while the lightest nuclei could increase their stability by fusing to make nuclei in the Fe–Ni region.

The most striking feature of Figure 2.1 is the approximate independence of $A$ as the average binding energy per nucleon for most nuclei (ranging from 7.4 to 8.8 MeV). This is a direct consequence of the short range, saturation character of the nuclear force. Suppose that the nuclear force was long range and not saturated. Suppose further that the binding energy of one nucleon to every other nucleon was some constant $K$. In a nucleus with $A$ nucleons, there would be $A(A - 1)/2$ "bonds"

**34** NUCLEAR PROPERTIES

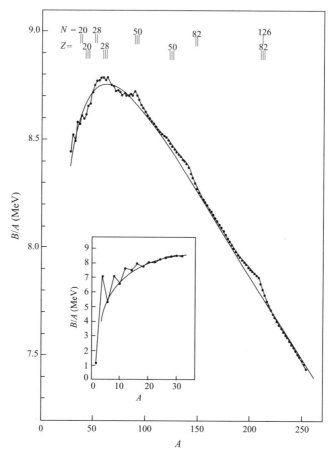

**Figure 2.1** Average binding energy per nucleon vs. mass number $A$ for the stable nuclei from Valentin, Subatomic physics: nuclei and particles. Copyright © 1981 by North-Holland Publishing Company. Reprinted by permission of North-Holland Publishing Company.

and thus the total binding energy would be $KA(A-1)/2$ with the binding energy per nucleon being $K(A-1)/2$. In other words, one would predict that the average binding energy per nucleon would increase linearly with $A$. That does not happen, as shown in Figure 2.1. Thus, one concludes that the nuclear force saturates and is short range.

In Figure 2.1, one also observes definite peaks in the average binding energy per nucleon for certain values of $A$. This is quite reminiscent of the plots of ionization potential vs. Z for atoms and suggests that there are certain special stable nucleonic configurations similar to the inert gas structures of atoms. The general decrease of $B_{\text{ave}}$ at higher values of $A$ is due to the increasing influence of the Coulomb force.

**Example of Binding Energies** Contrast the binding energy per nucleon in a $^{16}$O nucleus with the binding energy per molecule in liquid water.

$$B(16,8)/16 = [8 * M(^1H) + 8 * M(n) - M(16,8)] * 931.5/16$$
$$= 7.97 \text{ MeV}$$

For water:

$$\Delta H_{\text{vaporization}}/N_A = 40700 \text{ J/mol}/6.02 \times 10^{23}/\text{mol}/1.602 \times 10^{-19}\text{J/eV}$$
$$= 0.42 \text{ eV}$$

Note that these are both constants per particle, only different by approximately 7 orders of magnitude.

## 2.5 SEPARATION ENERGY SYSTEMATICS

Figure 2.2 shows a plot of the neutron separation energy for several isotopes of lead. For a given Z, $S_n$ is larger for even N compared to that for odd N. Similarly for a given N, $S_p$ is larger for even Z compared to that for odd Z. This effect is caused by that part of the nuclear force that likes to have neutrons paired with neutrons (with antiparallel spin) and to have protons paired with protons but have no n–p pairing. This pairing of like nucleons causes even–even nuclei (Z even, N even) to be more stable than even–odd or odd–even nuclides which, in turn, are more stable than odd–odd nuclei.

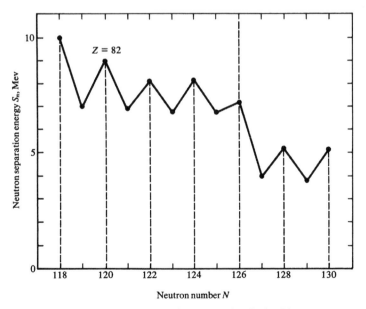

**Figure 2.2** Neutron separation energy for the lead isotopes.

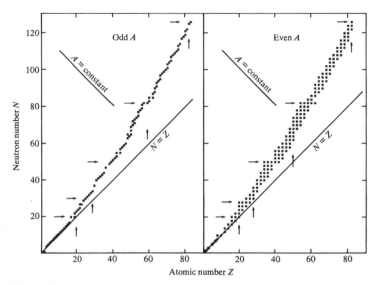

**Figure 2.3** Positions of the stable odd A and even A nuclei in a Segre chart from W. E. Meyerhof, Elements of Nuclear Physics. Copyright © 1967 by McGraw-Hill Book Company, Inc. Reprinted by permission of McGraw-Hill Book Company, Inc.

## 2.6 ABUNDANCE SYSTEMATICS

In Figure 2.3, we compare the positions of the known stable nuclides of odd $A$ with those of even $A$ in the chart of the nuclides. Note that as $Z$ increases, the line of stability moves from $N = Z$ to $N/Z \sim 1.5$ due to the influence of the Coulomb force. For odd $A$ nuclei, only one stable isobar is found while for even $A$ nuclei there are, in general, no stable odd–odd nuclei. This is further demonstrated by the data of Table 2.1 showing the distribution of stable isotopes.

## 2.7 SEMIEMPIRICAL MASS EQUATION

C. F. von Weizsäcker developed a crude theory of nuclear masses in 1935. The theory takes as its basis the idea that nuclei behave like incompressible uniformly charged liquid drops. How can we account for the variation of nuclear masses? We begin by stating that

$$M(Z,A)c^2 = [Z * M(^1H) + (A - Z) * M(n)]c^2 - B_{\text{tot}}(Z,A)$$

**TABLE 2.1 Distribution of Stable Nuclides**

| N | Even | Odd | Even | Odd |
|---|------|-----|------|-----|
| Z | Even | Even | Odd | Odd |
| Number | 160 | 53 | 49 | 4 |

## 2.7 SEMIEMPIRICAL MASS EQUATION

Weizsäcker's mass equation has evolved into what is called the semiempirical mass equation, which begins by parameterizing the total binding energy of species $Z, A$ as

$$B_{\text{tot}}(A, Z) = a_v A - a_s A^{2/3} - a_c \frac{Z^2}{A^{1/3}} - a_a \frac{(A - 2Z)^2}{A} \pm \delta$$

The justification of this representation of the total binding energy of the nucleus is as follows:

1. Since there are $A$ nucleons in the nucleus and the nuclear force saturates, we expect each nucleon to contribute to the total binding energy. This term is known as the volume term. The coefficient $a_v$ is the energy by which a nucleon in the interior of the nucleus is bound to its nearest neighbors and is a parameter to be determined experimentally.

2. However, not all nucleons are in the interior. Those nucleons on the surface are less tightly bound because they do not have a full complement of neighbors. We need a correction term to the binding energy proportional to the surface area of the nucleus. The surface area of the nucleus can be taken to be $4\pi R^2$. If, as asserted earlier, $R \propto A^{1/3}$, then $4\pi R^2 \propto A^{2/3}$. (Notice that the volume is $\frac{4}{3}\pi R^3$, which is proportional to $A$. Hence, the form of the first term.) The $A^{2/3}$ factor is multiplied by another coefficient, $a_s$, that is to be determined experimentally.

3. The third term reflects the decrease in binding due to the Coulomb repulsion between the protons. The Coulomb energy of a uniform sphere can be written as

$$E_{\text{Coul}} = \frac{3}{5} \frac{Z^2 e^2}{R} \quad (2.1)$$

If we denote $R$ as $r_0 A^{1/3}$, we can substitute for $R$ in the above equation, arriving at the point that $E_{\text{Coul}} = 0.72 \, Z^2/A^{1/3}$ MeV. (The usual fitted or adjusted value of the coefficient $a_c$ is 0.7 rather than 0.72.)

4. The fourth term along with the fifth term represent effects on the binding energy that are quantum mechanical in origin. The fourth term, the asymmetry energy, is the difference in energy of a nucleus with $N$ neutrons and $Z$ protons ($N \neq Z$) and one where $Z = N = A/2$. To evaluate this term, we remember that neutrons and protons occupy orbitals in the nucleus at well-defined energies and that the neutrons and protons obey the Pauli principle for fermions. A simple model (Fig. 2.4) should suffice to calculate this energy. Assume the neutron and proton levels of a nucleus are equidistant with spacing $\Delta$ and that we can have only one nucleon per level. To make the nucleus $^A Z$ (where $Z \neq N$) from the nucleus with $N = Z = A/2$, we must take $q$ protons and transform them into neutrons. Thus, we have $N = q + A/2$, $Z = A/2 - q$ and therefore $q = (N - Z)/2$. Each of the $q$ protons must be raised in energy an amount $q\Delta$. The work needed to transform the $N = Z$ nucleus into the nucleus $^A Z$ is $q^2 \Delta = (N - Z)^2 \Delta/4$. Note that we could have made exactly the same argument by replacing neutrons with protons. We finish the argument by noting that the

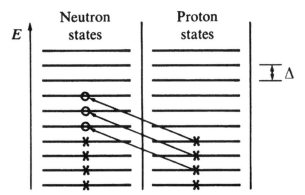

**Figure 2.4** Schematic model of how the nucleus $^A Z$ is assembled from an $N = Z$ nucleus from W. E. Meyerhof, Elements of Nuclear Physics. Copyright © 1967 by McGraw-Hill Book Company, Inc. Reprinted by permission of McGraw-Hill Book Company, Inc.

energy levels get closer together as the total number of nucleons increases and thus $\Delta \propto 1/A$. As a final matter of notation, we replace $(N - Z)$ by $(A - 2Z)$.

5. The last term represents the special stability associated with completely paired spins in a nucleus. The pairing energy term is chosen to be zero for odd $A$ nuclides; for even–even nuclides, use the positive form, for odd–odd nuclides, use the negative form.

The constants of the semiempirical binding energy equation can be determined by fitting the data on the masses of nuclei. A recent set of values of the coefficients are $a_v = 15.56$ MeV, $a_s = 17.23$ MeV, $a_c = 0.7$ MeV, $a_a = 23.285$ MeV, and $\delta = 11/A^{1/2}$ MeV. The relative contribution of each term to the binding energy per nucleon is shown in Figure 2.5. Note the large constant contribution of the volume energy to the average binding energy per nucleon. The surface energy correction is most important for the lighter nuclei where the fraction of nucleons in the surface is greatest. Similarly, the Coulomb energy correction is most important for the heaviest nuclei since it depends on $Z^2$. The asymmetry energy is a smaller effect that is most important in the heaviest nuclei where the $N/Z$ ratio is the greatest.

**Example Problem** Calculate the average binding energy per nucleon of $^{58}$Fe using the semiempirical mass equation:

**Solution**

$$B_{tot}(A, Z) = a_v A - a_s A^{2/3} - a_c \frac{Z^2}{A^{1/3}} - a_a \frac{(A - 2Z)^2}{A} \pm \delta$$

$$B_{tot}(58, 26) = 15.56(58) - 17.23(58^{2/3}) - 0.7\left(\frac{26^2}{58^{1/3}}\right) - 23.285\left(\frac{(58 - 52)^2}{58}\right) + \frac{11}{58^{1/2}}$$

$$B_{tot}(58, 26) = 902.48 - 258.17 - 122.25 - 14.45 + 1.44 = 509.05 \text{ MeV}$$

$$B_{tot}/A = 509.05/58 = 8.78 \text{ MeV}$$

Notice the relative contribution of the various terms of the binding energy.

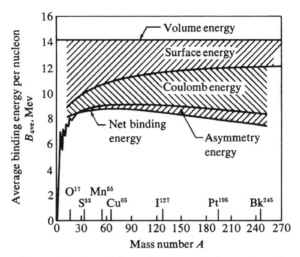

**Figure 2.5** Relative contributions of the various terms in the semiempirical mass equation to the average binding energy per nucleon from W. E. Meyerhof, Elements of Nuclear Physics. Copyright © 1967 by McGraw-Hill Book Company, Inc. Reprinted by permission of McGraw-Hill Book Company, Inc.

Myers and Swiatecki (1966) have proposed a modification of the semiempirical mass equation that gives a better description of the experimental masses. This modification can be summarized in the following equation:

$$BE_{tot}(Z,A) = c_1 A \left[1 - k\left(\frac{N-Z}{A}\right)^2\right] - c_2 A^{2/3} \left[1 - k\left(\frac{N-Z}{A}\right)^2\right] - \frac{c_3 Z^2}{A^{1/3}} + \frac{c_4 Z^2}{A + \delta} \tag{2.2}$$

where $c_1 = 15.677$ MeV, $c_2 = 18.56$ MeV, $c_3 = 0.717$ MeV, $c_4 = 1.211$ MeV, $k = 1.79$, and $\delta = 11/A^{1/2}$. What Myers and Swiatecki (1966) have done is to add an asymmetry energy correction term to the surface energy and to add a term (the $c_4$ term) that is a correction to the Coulomb energy term due to the diffuseness of the nuclear surface.

We will now look at some of the predictions of the semiempirical mass equation. The first question we pose is what happens if we hold $A$ constant and vary $Z$ (neglecting for a moment the pairing term). We can write

$$M(Z,A) = Z * M(^1H) + (A - Z)M(n) - B_{tot}(Z,A) \tag{2.3}$$

$$B_{tot}(Z,A) = a_v A - a_s A^{2/3} - a_c Z^2/A^{1/3} - a_a(A - 2Z)^2/A \tag{2.4}$$

## NUCLEAR PROPERTIES

Let us expand the asymmetry energy term as follows:

$$\frac{a_a(A-2Z)^2}{A} = a_a \frac{A^2 - 4AZ + 4Z^2}{A} = a_a\left(A - 4Z + \frac{4Z^2}{A}\right) \quad (2.5)$$

Substituting back into the equation for the mass and collecting terms, we have

$$M = A\left[M(\text{n}) - a_v + \frac{a_s}{A^{1/3}} + a_a\right] + Z[(M(^1\text{H}) - M(\text{n}) - 4Za_a)] + Z^2\left(\frac{a_c}{A^{1/3}} + \frac{4a_a}{A}\right) \quad (2.6)$$

Thus, the mass equation at constant $A$ takes on the form of a parabola ($\alpha + \beta Z + \gamma Z^2$) in $Z$. The third term, $\gamma$, is positive, and so the parabola goes through a minimum for some value of $Z$, which is termed $Z_A$. Note that $Z_A$ is not necessarily an integer. We can now ask ourselves what is the value of $Z_A$? We can evaluate this by minimizing $M$ with respect to $Z$ at constant $A$, that is, we get a simple partial differential equation:

$$\left(\frac{\partial M}{\partial Z}\right)_{Z/A} = 0 = \beta + 2\gamma Z_A \quad (2.7)$$

$$Z_A = \frac{-\beta}{2\gamma} = -\frac{M(^1\text{H}) - M(\text{n}) - 4a_a}{2\left(\frac{a_c}{A^{1/3}} + \frac{4a_a}{A}\right)} \quad (2.8)$$

Substituting numerical values for the coefficients in this expression, we can show that

$$\frac{Z_A}{A} \sim \frac{1}{2}\frac{81}{80 + 0.6A^{2/3}} \quad (2.9)$$

Thus, as $A$ goes to 0, $Z_A/A$ becomes equal to $\frac{1}{2}$, or $Z = N = A/2$. As $A$ gets large, $Z_A/A$ is less than $\frac{1}{2}$, typically, about 0.4. The underlying physics behind this trend is that, in the absence of the Coulomb repulsion between the protons, we would expect equal numbers of neutrons and protons due to the asymmetry energy term. When $Z$ gets large, the Coulomb energy becomes large. The nucleus can gain stability by converting protons into neutrons. Stability results when we have a balance between the excess Coulomb energy and the asymmetry energy.

Let us now consider the case where $A = 111$. From the above relations, we can calculate $Z_A = 47.90$. In Figure 2.6, we show the actual masses of the nuclei with $A = 111$. The expected parabolic dependence of mass upon $Z$ is observed. The most stable nucleus has $Z = 48$ (Cd). All the $A = 111$ nuclei that have more neutrons than $^{111}$Cd release energy when they decay by $\beta^-$ decay while the nuclei with fewer neutrons than $^{111}$Cd release energy when they decay by $\beta^+$ or EC decay.

Now let us consider the case of the even $A$ nuclei with $A = 112$. We calculate that $Z_A = 48.29$. Plotting the actual masses of the $A = 112$ nuclei vs. $Z$ (Fig. 2.7) gives us two parabolas, one for the even–even nuclei and one for the odd–odd nuclei,

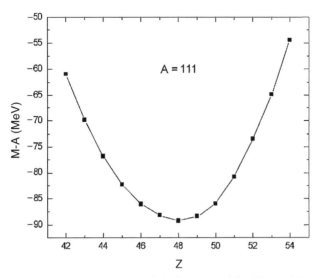

**Figure 2.6** Mass excesses of the known nuclei with $A = 111$.

displaced from one another by the energy $2\delta$. Since all nuclei on the upper parabola (the odd–odd nuclei) can decay to a nucleus on the lower parabola (the even–even nuclei), we conclude that there are no stable odd–odd nuclei. (The only known exceptions to this rule occur in the light nuclei where nuclear structure effects make $^2$H, $^6$Li, $^{10}$B, and $^{14}$N stable). Note that some odd–odd nuclei can thus decay by both $\beta^-$ or $\beta^+$ emission. Note that *double* $\beta$ decay is energetically possible

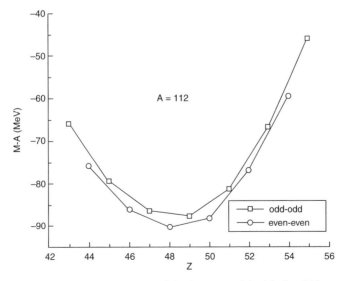

**Figure 2.7** Mass excesses of the known nuclei with $A = 112$.

**42** NUCLEAR PROPERTIES

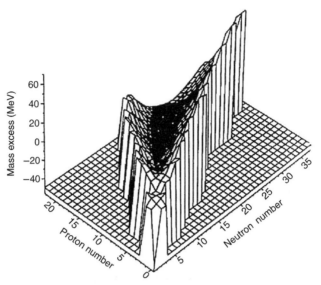

**Figure 2.8** Plot of the nuclear mass excesses vs. neutron number $N$ and atomic number $Z$ for the light nuclei showing the nuclear mass surface and the valley of β stability from Halliday, et al., 1992; reprinted by permission of John Wiley & Sons, Inc.

in some cases ($^{112}$Pd → $^{112}$Cd + 2β$^-$ + 2$\bar{\nu}_e$). This mode of decay has only been observed for $^{130}$Te and $^{82}$Se, and the half-lives for this mode of decay are very long ($t_{1/2} = 10^{20}$–$10^{21}$ y). Note that we can have more than one stable isotope for a given $A$ and that all of them will be even–even.

This parabolic dependence of the nuclear mass upon $Z$ for fixed $A$ can be used to define a nuclear mass surface (Fig. 2.8). The position of the minimum mass for each $A$ value defines what is called the valley of β stability. β decay is then depicted as falling down the walls of the valley toward the valley floor.

## 2.8 NUCLEAR SIZES AND SHAPES

We ask how big are nuclei? Our zero-order answer is that the radii of nuclei are in the range of 1–10 fm. Our first-order answer to this question begins by assuming the nucleus is spherical with a uniform density out to some sharp cutoff radius, that is, the nucleus has the shape and density distribution of a billiard ball. This density distribution is shown in Figure 2.9.

One can parameterize this distribution by saying that the nuclear radius $R$ can be written as

$$R = r_0 A^{1/3} \tag{2.10}$$

where the nuclear radius constant can be taken to be 1.2 fm for the "charge radius" and 1.4 fm for the "matter radius." What do we mean by this dichotomy? What we

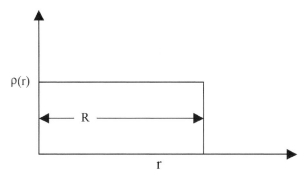

**Figure 2.9** Schematic diagram of a sharp cutoff, constant density model for nuclei.

mean is that when one *measures* the nuclear radius by scattering high-energy electrons from the nucleus or when one *measures* the radius by scattering low-energy ions from the nucleus, one gets slightly different answers for the nuclear size. The electron probes the charge distribution via the electromagnetic force, that is, the distribution of the protons while other particles probe the matter distribution or the range over which nuclear forces are acting. Which value of $r_0$ should one use in calculations? The answer depends upon the nuclear property being calculated and whether it is sensitive to the distribution of charge or matter.

A somewhat more sophisticated approach to the problem of defining the nuclear size and density is to assume the nuclear density distribution, $\rho(r)$, assumes the form of a Fermi distribution, that is,

$$\rho(r) = \frac{\rho_0}{1 + e^{(r-R)/a}} \tag{2.11}$$

where $\rho_0$ is the density in the interior of the nucleus ($\rho_0 = 0.172$ nucleons/fm$^3$), $a$ is a measure of the diffuseness of the nuclear surface, and $R$ is the half-density radius of the nucleus (Fig. 2.10). The half-density radius is given by the expression

$$R = r_0 A^{1/3}$$

where $r_0 = 1.12$ fm. The thickness of the nuclear skin, $t$, indicated in Figure 2.10, can be related to the diffuseness parameter as $t = 4a \ln 3 \sim 4.4a$. Most nuclei show a skin thickness $t$ of 2.4–2.5 fm. The meaning of this value of $t$ can be ascertained by calculating the fraction of the nucleons that lies in the skin region of the nucleus as a function of the nuclear size (Table 2.2).

Thus, the lighter nuclei are mostly "skin" and the heaviest nuclei still have substantial "skin" regions. These approximate models for the nuclear size and density distribution compare favorably to the measured distributions for typical nuclei (Fig. 2.11).

Up to this point, we have assumed that all nuclei are spherical in shape. That is not true. There are regions of large stable nuclear deformation in the chart of nuclides, that is, the rare earths ($150 < A < 180$) and the actinides ($220 < A < 260$). We shall discuss these cases in more detail later in this chapter when we discuss the electric moments of nuclei.

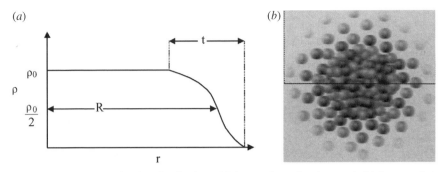

**Figure 2.10** Nuclear density distribution: (*a*) in a schematic view and (*b*) in an artist's conception from R. Mackintosh, J. Al-Khalili, B. Jonson and T. Pena, Nucleus: A Trip into the Heart of Matter. Copyright © 2001 by The Johns Hopkins University Press, 2001; reprinted by permission of Johns Hopkins. (Figure also appears in color figure section.)

Another question we might pose to ourselves is whether the neutron and proton distributions in nuclei are the same? Modern models for the nuclear potential predict the nuclear skin region to be neutron-rich. The neutron potential is predicted to extend out to larger radii than the proton potential. Extreme examples of this behavior are the *halo nuclei*. A halo nucleus is a very n-rich (or p-rich) nucleus (generally with low A) where the outermost nucleons are very weakly bound. The density distribution of these weakly bound outermost nucleons extends beyond the radius expected from the $R \propto A^{1/3}$ rule. Examples of these nuclei are $^{11}$Be, $^{11}$Li, and $^{19}$C. The most well-studied case of halo nuclei is $^{11}$Li. Here the two outermost nucleons are so weakly bound (a few hundred keV each) as to make the size of $^{11}$Li equal to the size of a $^{208}$Pb nucleus (see Fig. 2.12).

## 2.9 QUANTUM MECHANICAL PROPERTIES

### 2.9.1 Nuclear Angular Momenta

It is well known and, in fact, an essential underlying part of chemical behavior that the electron has an intrinsic angular momentum, $s = \frac{1}{2}\hbar$. That is, the electron

**TABLE 2.2 Fraction of Nucleons in Nuclear "Skin"**

| Nucleus | Fraction of Nucleons in the "Skin" |
|---|---|
| $^{12}$C | 0.90 |
| $^{24}$Mg | 0.79 |
| $^{56}$Fe | 0.65 |
| $^{107}$Ag | 0.55 |
| $^{139}$Ba | 0.51 |
| $^{208}$Pb | 0.46 |
| $^{238}$U | 0.44 |

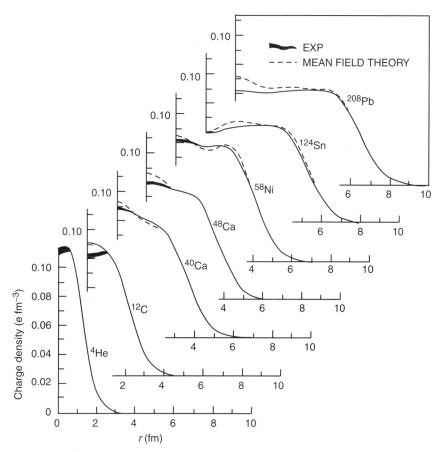

**Figure 2.11** Nuclear ground-state charge distributions as measured for a sample of nuclei throughout the periodic table from B. Frois, Proc. Int. Conf. Nucl. Phys., Florence, 1983, eds. P. Blasi and R.A. Ricci (Tipografia Compositori Bologna) Vol. 2, p. 221.

behaves as if it is rotating or spinning about an internal axis. The electron spin angular momentum provides an important criterion for assigning quantum numbers to atomic electrons through the Pauli principle and thus has far-reaching consequences. The electrons occupy quantum mechanical states or orbitals that are labeled by the principal quantum number, $N$, which is one more than the number of radial nodes in the atomic wave function and the angular momentum quantum number, $l$, the number of angular nodes in the wave function. The electrons distribute themselves among those states with degenerate energies so that their spin-angular momenta, $s$, are aligned (Hund's rules). The resulting atomic energy level can be characterized by a single total angular momentum, $J$, that is made up from the total orbital motion of all the electrons, $L$, and a total intrinsic spin, $S$. These values of $L$ and $S$ are calculated by two separate vector couplings of the two

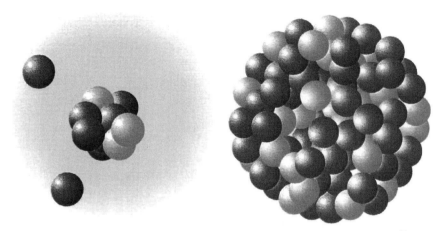

**Figure 2.12** Schematic representation of the relative sizes of the halo nucleus $^{11}$Li and $^{208}$Pb. (Figure also appears in color figure section.)

types of angular momenta of the electrons. The electrons in all but the heaviest atoms exhibit such "*LS*" coupling.

The neutron and the proton also have an intrinsic angular momentum, $s = \frac{1}{2}\hbar$, and so each appears as if it is spinning about an internal axis. Thus, we can expect that a large nucleus, which contains some number of neutrons and protons, will have a total intrinsic angular momentum, or a nuclear spin, $I$, just from the combination of the intrinsic spins of the neutrons and protons. Similarly, we can imagine that neutrons and protons will occupy discrete states in the nucleus, and some of these states will have orbital angular momenta in a manner similar to the orbital angular momenta of electronic states ($l = 1$, p-states, etc.) While there are such similarities, the fact that the potential well for nucleons is dramatically different from the central Coulomb potential for electrons introduces several important differences in the concepts used to describe nuclear states and levels. A detailed discussion of the quantum mechanical structure of nuclei is presented in Chapter 6. At this point we only need to address the gross features.

The orbital angular momenta of the nuclear (and atomic) states are all integer multiples of $\hbar$ starting with zero. Nucleons exhibit a strong coupling of the orbital and spin angular momenta of *individual* nucleons such that $j = l + s$ is the appropriate quantum number for a nucleon. We can immediately see that the combination of the intrinsic spins of the nucleons with their orbital motion will always give half-integer values for the total spin, $I$, of any odd-$A$ nucleus and integer values for any even-$A$ nucleus:

$$\text{odd-}A \text{ nuclei} \quad I = \frac{1}{2}, \frac{3}{2}, \frac{5}{2}, \ldots$$

$$\text{even-}A \text{ nuclei} \quad I = 0, 1, 2, \ldots$$

The numerical value will depend on the filling of the nuclear states with angular momenta $j$ and on the coupling of all of those angular momenta. At first glance we might expect that a large nucleus might have a very large intrinsic angular momentum. However, recall that the nuclear force has a short range and that the nucleons are more strongly bound when they are in close proximity. Two nucleons will be in the closest proximity when they are in the same orbital. If the two nucleons in the same orbital are both neutrons or both protons, then their spins must be opposed in order to satisfy the Pauli principle and each have a unique set of quantum numbers. So we find that the nuclear force tends to put pairs of nucleons into the same orbitals and their orbital angular momenta and intrinsic spins cancel, summing to zero. (This behavior is opposite from that of atomic electrons.) Thus, the angular momenta of the ground states of nuclei tend to be small, even for nuclei with hundreds of nucleons in states with high angular momenta.

*Parity*, as used in nuclear science, refers to the symmetry properties of the wave function for a particle or a system of particles. If the wave function that specifies the state of the system is $\Psi(r,s)$ where $r$ represents the position coordinates of the system $(x, y, z)$ and $s$ represents the spin orientation, then $\Psi(r,s)$ is said to have positive or even parity when

$$\Psi(r,s) = +\Psi(-r,-s) \qquad (2.12)$$

where the minus sign indicates the sign of the spatial coordinates has been reversed as well as the direction of the spin. When

$$\Psi(r,s) = -\Psi(-r,-s) \qquad (2.13)$$

the system is said to have negative or odd parity. For a central potential $[V = V(r)]$, that is, the potential energy depends only on the distance and not the spatial orientation $(\vartheta, \phi)$, the parity, denoted as $\pi$, is given as

$$\pi = (-1)^l \qquad (2.14)$$

where $l$ is the orbital angular momentum of the system (s, d orbitals have positive parity while p or f orbitals have negative parity). The spin and parity of a given nuclear state are usually used as labels for that state. Thus, a state with $I = \frac{7}{2}$ and negative parity is referred to as a $\frac{7}{2}-$ state.

## 2.10 ELECTRIC AND MAGNETIC MOMENTS

### 2.10.1 Magnetic Dipole Moment

The magnetic moments of nuclei are measures of the distribution of electric currents in the nucleus while the electric moments are measures of the distribution of electric charges. Because the magnetic moment may not be a familiar concept, we will begin

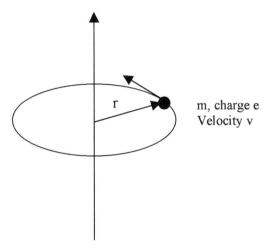

**Figure 2.13** Classical magnetic dipole moment.

by discussing a simple example of a "classical" magnetic moment. Consider an electron moving with a velocity $v$ in a circular orbit of radius $r$ about a point (Fig. 2.13). The magnetic dipole moment is defined as the product of the area of the loop made by the electron, $A$, and the current, $i$. The area of the circle is $\pi r^2$ and the current $i$ is (electron charge)/(time to make a loop) or $e/(2\pi r/v)$. Thus, we have

$$|\mu| = iA = \left(\frac{ev}{2\pi r}\right)(\pi r^2) = \frac{evr}{2} \tag{2.15}$$

Remember that the angular momentum of the electron moving in a circle, $l$, is $m_0 vr$. Thus,

$$|\mu| = \frac{evr}{2} \times \frac{m_0}{m_0} = \frac{el}{2m_0} \tag{2.16}$$

The gyromagnetic ratio $\gamma$ is defined as $|\mu|/l = e/2m_0$.

In quantum mechanics, the projection of the angular momentum $l$ is $m_l \hbar$. We, therefore, expect the magnetic dipole moment due to the orbital motion of the electron to be given by

$$\mu = (e/2m_0)m_l \hbar = m_l \mu_B \tag{2.17}$$

where $\mu_B$ is called the Bohr magneton ($= e\hbar/2m_0$) and has a magnitude of $5.78 \times 10^{-5}$ eV/tesla or $9.27 \times 10^{-21}$ erg/gauss. The electron has an intrinsic spin, $m_s = \frac{1}{2}$. It also has a component of the magnetic moment due to this spin.

Extending these ideas to nucleons, we can define the nuclear magneton, $\mu_N$ as $(e\hbar/2m_p)$, which has the numerical value of $3.15 \times 10^{-8}$ eV/tesla or

$5.50 \times 10^{-24}$ erg/gauss:

$$\mu_l^{\text{proton}} = m_l \mu_N \quad (2.18)$$

Note that the nuclear magneton is smaller than the Bohr magneton by the factor of the ratio of the proton to electron masses, $\sim$1840.

It is traditional to rewrite the definition of the nuclear magnetic moment in terms of magnetons and include a constant of proportionality called the gyromagnetic ratio or simply g factor:

$$\mu = g_l m_l \mu_N \quad (2.19)$$

By adding a constant of proportionality we are anticipating that the magnetic moment will be the net result of a complicated cancellation process. For example, we would expect $g_l = 1$ for the orbital motion of a proton due to its charge and $g_l = 0$ for a neutron because they are uncharged.

Both the neutrons and the protons have an intrinsic spin, and so by extension we can expect additional contributions to the magnetic moment of the form:

$$\mu = g_s m_s \mu_N \quad (2.20)$$

where the projection $m_s$ is $\frac{1}{2}$ for fermions like the proton and neutron. The spin g factor, $g_s = 2.0023$, as calculated with the relativistic Dirac equation for *electrons*, including known higher order correction terms, is in very good agreement with measurements. However, the measured values of $g_s$ for both the proton and the neutron are surprisingly large:

Proton $\quad g_s = 5.5856912(22)$
Neutron $\quad g_s = -3.8260837(18)$

Notice that the neutron with exactly zero *net* charge has a nonzero magnetic moment. Thus, both the proton and the neutron do not appear to be elementary particles. Rather they both seem to have internal moving constituents. It was noted some time ago that the magnetic moment of the proton is larger than the expected value of "2" and that of the neutron is smaller than its expected value of "0" by about 3.6 units. Older models of the nuclear force attributed these differences to "clouds" of mesons surrounding the nucleons. In the modern theory of quantum chromodynamics the nucleons are made up from three quarks each with their own magnetic moments and electronic charges.

The presence of a magnetic dipole moment in many nuclei that have an intrinsic spin has found enormous application in nuclear magnetic resonance (NMR) and magnetic resonance imaging (MRI). NMR is extensively used in chemical laboratories to identify the structural and chemical environments of the nuclei in molecules, whereas MRI uses a tomographic technique to locate specific molecules

**50** NUCLEAR PROPERTIES

on a microscopic scale. Both techniques rely on the splitting of the energies of the magnetic substates by a (strong) magnetic field. NMR measures tiny shifts in the relative energies of the magnetic substates due to induced magnetization of the local electron density to provide information on the chemical environment. These states have a fine structure or splitting due to the presence of neighboring magnetic nuclei that provides information on the structure of the molecule. MRI applies a spatially varying magnetic field to detect the resonance of a single type of nucleus, usually the hydrogen nuclei in water and aliphatic compounds, and to measure the concentration in a three-dimensional space. Both techniques are nondestructive and can be applied to living systems. The concentration of water molecules varies widely in tissues and other biological media and can provide detailed microscopic images for medical purposes.

### 2.10.2 Electric Quadrupole Moment

Imagine the nucleus is an extended charged object as sketched in Figure 2.14. Consider trying to calculate the potential energy at some point $P$, which is at a distance $D$ from the center of the charged object (nucleus). Suppose we evaluate the potential, $d\Phi$, at point $P$ due to a charge at a distance $r$ from the center of the charge object. Assume further the line from the center of the object to the charge makes an angle $\theta$ with the line connecting the center of the object with the point $P$ (Fig. 2.14). If the density of charge in the object is $\rho(\theta, \phi, r)$, then the total charge at the indicated point is $\rho\, d\tau$ or $\rho\,(r^2\, dr\, \sin\theta\, d\theta\, d\phi)$. We can write

$$d\Phi = \rho\, d\tau/\delta = \rho\, d\tau [D^2 + r^2 - 2Dr \cos\theta]^{-1/2} \quad (2.21)$$

Factoring out $D$ and substituting the Legendre polynomials into the equation where

$$P_1(\cos\theta) = \cos\theta \quad (2.22)$$

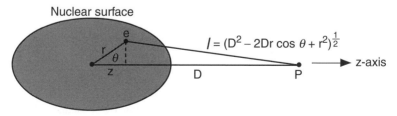

**Figure 2.14** Potential at a point due to an extended charge object from B. G. Harvey, Introduction to Nuclear Physics and Chemistry, 2nd Edition. Copyright © 1969 by Prentice-Hall, Inc. Reprinted by permission of Pearson Education.

and

$$P_2(\cos\theta) = \tfrac{3}{2}\cos^2\theta - \tfrac{1}{2} \qquad (2.23)$$

we get

$$d\Phi = \frac{\rho\, d\tau}{D}\left[1 + \frac{r}{D}P_1(\cos\theta) + \left(\frac{r}{D}\right)^2 P_2(\cos\theta) + \cdots\right] \qquad (2.24)$$

If we set up the integrals over the entire volume of the charged object (nucleus), we get

$$V = \frac{1}{D}\left[\int_{\text{volume}}\rho\, d\tau\right] + \frac{1}{D^2}\left[\int_{\text{volume}}\rho r\cos\theta\, d\tau\right] + \frac{1}{D^3}\left[\int_{\text{volume}}\rho r^2(\tfrac{3}{2}\cos^2\theta - \tfrac{1}{2})d\tau + \cdots\right] \qquad (2.25)$$

The first term in the square bracket in this equation is the electric monopole moment, which is equal to the nuclear charge, $Ze$. The second term in the square bracket is the electric dipole moment while the third term in the square bracket is the electric quadrupole moment. For a quantum mechanical system in a well-defined quantum state, the charge density $\rho$ is an even function, and because the dipole moment involves the product of an even and an odd function, the corresponding integral is identically zero. Therefore, there should be no electric dipole moment or any other odd electric moment for nuclei. For spherical nuclei, the charge density $\rho$ does not depend on $\theta$, and thus the quadrupole moment $Q$ is given by

$$Q = \iiint r^2 \rho(r)(\tfrac{3}{2}\cos^2\theta - \tfrac{1}{2})r^2 dr \sin\theta\, d\theta\, d\phi \qquad (2.26)$$

The quadrupole moment will differ from zero only if the nucleus is not spherical. Thus, the quadrupole moment is a measure of the nonsphericity or shape of the nucleus. We can further elaborate on this by making a simple model (Fig. 2.15) for nonspherical nuclei. We shall assume such nuclei are

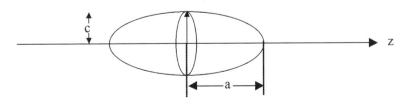

**Figure 2.15** Sketch of the $a$ spheroidal nucleus.

spheroids with a shape generated by rotating an ellipse about one of its axes. We can define a semiminor axis of the ellipse, $c$, and a semimajor axis, $a$ ($a$ is the axis about which the ellipse rotates). If $a$ is the long axis, we have a prolate spheroid (the shape of an American football). If $a$ is the short axis (pancake shape), we have an oblate spheroid. We can show that

$$Q = \tfrac{2}{5} Z e (a^2 - c^2) \tag{2.27}$$

Since we also know the square of the mean radius $R$ of the spheroid is given by

$$R^2 = \tfrac{1}{2}(a^2 + c^2) = (r_0 A^{1/3})^2 \tag{2.28}$$

we can solve for $a$ and $c$, the two axes of the spheroid. Thus, the quadrupole moment gives us a direct measure of the shape of nuclei. Note further that $Q$ has the dimensions of charge × area. It is common to tabulate $Q/e$, which has the dimension of area. The nuclear dimension of area is the *barn*, which is equal to $10^{-24}$ cm$^2$. Hence quadrupole moments are frequently given in barns. A plot of the experimental values of the electric quadrupole moments is shown in Figure 2.16. Note (Fig. 2.16) that the rare earth and actinide

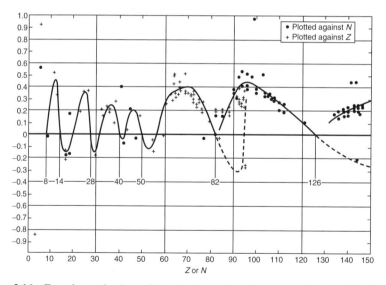

**Figure 2.16** Experimental values of the electric quadrupole moment of nuclei. The lines are drawn through the data to emphasize the trends from M. A. Preston, Physics of the Nucleus. Copyright © 1962 by Addison-Wesley Publishing Company. Reprinted by permission of Pearson Education, Inc.

nuclei have prolate shapes ($Q = +$) while there are other nuclei with oblate shapes ($Q = -$).

**Example Problem**  For $^{177}$Hf, $Q = +3.0$ e-barns. Calculate the ratio of the semimajor to semiminor axes of this prolate nucleus.

*Solution*

$$Q/e = \tfrac{2}{5} Z(a^2 - c^2)$$
$$a^2 - c^2 = \frac{Q/e}{\tfrac{2}{5} * Z} = \frac{+3.0 \times 10^{-24}}{\tfrac{2}{5}(72)} = 1.042 \times 10^{-25}$$
$$R^2 = \tfrac{1}{2}(a^2 + c^2) = (r_0 A^{1/3})^2$$
$$a^2 + c^2 = 2(r_0 A^{1/3})^2 = 2(1.2 \times 10^{-13} * 177^{1/3})^2 = 9.079 \times 10^{-25}$$
$$2a^2 = 1.012 \times 10^{-24} \qquad a = 7.11 \times 10^{-13} \text{ cm} \qquad c = 6.34 \times 10^{-13} \text{ cm}$$
$$a/c = 1.12$$

## PROBLEMS

1. Define or describe the following terms or phenomena in your own words: nuclear surface energy, parity, asymmetry energy, packing fraction, nuclear magneton, Schmidt limits, mass defect, magnetic dipole moment,

2. The total nuclear binding energies of $^{27}$Mg, $^{27}$Al, and $^{27}$Si are 244.2667, 246.8741, and 241.6741 MeV, respectively. Determine the values of the Coulomb energy and asymmetry energy coefficients of the semiempirical mass equation using these data.

3. The ground-state quadrupole moment of $^{152}$Eu is $+3.16 \times 10^2$ fm$^2$. Deduce the ratio of semimajor to semiminor axes for $^{152}$Eu.

4. Some nuclei can decay by either $\beta^-$ or $\beta^+$ emission. Show that such nuclei must have even $A$, odd $N$.

5. For $^{181}$Ta, $Q/e = 4.20$ barns. Calculate the ratio of the semimajor to semiminor axes of this nucleus.

6. Calculate the electric quadrupole moment of a charge of magnitude $Ze$ distributed over a ring of radius $R$ with an axis along the $z$ axis.

7. Use the semiempirical mass equation to compute, for given $A$, the relation between $Z$ and $N$ for a nucleus that has $S_n = 0$ (the neutron "drip line"). Compute $N/Z$ for $A = 100$.

8. Find the electric dipole moment and electric quadrupole moment of two positive point charges $+q$, one at $z = a/2$ and the other at $z = -a/2$.

9. Show that the quadrupole moment $Q$ of a uniformly charged spheroid about the axis of symmetry is $\frac{2}{5}Z(b^2 - a^2)$, where $a$, $b$ are the semiaxes, $b$ being along the axis of symmetry. Show that the quadrupole moment about an axis making an angle $\beta$ with the axis of symmetry is $[\frac{3}{2}\cos^2\beta - \frac{1}{2}]Q$.

10. The quadrupole moments of $^{176}$Lu and $^{127}$I are 7.0 and $-0.6$ barns, respectively. Assume that $^{176}$Lu and $^{127}$I are ellipsoids of revolution obtained by deforming (without volume change) a sphere of radius $R = 1.4\,A^{1/3}$ fm. Calculate the ratio $a/b$ of the semimajor to semiminor axes.

11. Explain why we expect that there are no stable odd–odd nuclei. What are the exceptions to this rule?

12. Use the semiempirical mass equation to derive an expression for the energy released in $\alpha$ decay. For fixed $Z$, how should the energy released depend on $A$?

13. Assume that $a_v = 15.835$, $a_s = 18.33$, $a_{\text{asym}} = 23.20$, and $a_c = 0.714$ in the semiempirical mass equation. Show the binding energy per nucleon reaches a maximum for $Z \sim 26$ (iron). Assume $Z = N = A/2$. Neglect pairing.

14. The red giant stars, which are cooler than the sun, produce energy from reactions such as

$$^9\text{Be} + {}^1\text{H} \longrightarrow {}^6\text{Li} + {}^4\text{He} + \text{energy}$$

From the masses tabulated in the Appendices, calculate the energy release for this reaction and the percentage of the initial mass of the reactants converted to energy.

15. Consider the nuclei $^{15}$C, $^{15}$N, and $^{15}$O. Which of these nuclei is stable? What types of radioactive decay would the other two undergo? Calculate the binding energy difference between $^{15}$N and $^{15}$O. Assuming this difference comes from the Coulomb term in the semiempirical binding energy equation, calculate the nuclear radius.

16. By computing the appropriate parabolas, predict the types of decay and their energies for the isobars of $A = 180$. Which isobar or isobars are stable?

17. Assume that all the sun's energy is produced by the reaction

$$4{}^1\text{H} \longrightarrow {}^4\text{He} + 2\beta^+$$

The sun yields 2 cal/min-cm$^2$ at the surface of Earth. The distance of Earth from the sun is $1.49 \times 10^6$ km. How much helium does the sun produce per year?

18. Calculate the electric monopole, dipole, and quadrupole moments of the following arrangements of charge:

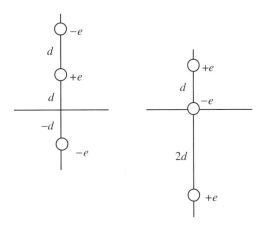

19. Suppose that the density of nucleons $\rho$ in a nucleus varies with radial distance $r$ from the center of the nucleus as shown below. What fraction of the nucleons lie in the surface region in the nuclei $^{28}$Si, $^{132}$Sn, and $^{208}$Pb if $\rho_0 = 0.17$ nucleons/fm$^3$, $c = 1.2\, A^{1/3}$ fm, and $a = 2.4$ fm?

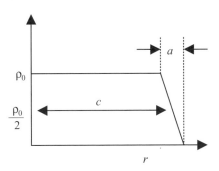

20. Use the semiempirical mass equation to calculate the percentage contribution to the average binding energy per nucleon of the volume energy, the surface energy, the Coulomb energy, and the asymmetry energy for $A = 60$ and $A = 240$.

21. Explain why in the decay of $^{238}$U to $^{206}$Pb by successive $\alpha$ and $\beta^-$ decays, one sees one or two successive $\alpha$ decays followed by $\beta^-$ decays, and so on. Why are there no $\beta^+$ or EC decays in this chain?

22. (a) What regions of the periodic table are characterized by large permanent prolate nuclear deformations?
    (b) What nuclei in the periodic table have the highest binding energy per nucleon?

## REFERENCES

Resnick, R., D. Halliday, and K. S. Krane. *Physics*, 4th ed., Wiley, New York, 1992.
Meyerhof, W. E. *Elements of Nuclear Physics*, McGraw-Hill, New York, 1967.
Myers, W. D. and W. J. Swiatecki. *Nucl. Phys.* **81**, 1 (1966).

## BIBLIOGRAPHY

Evans, R. D. *The Atomic Nucleus*, McGraw-Hill, New York, 1955.
Hasse, R. W. and W. D. Myers. *Geometrical Relationships of Macroscopic Nuclear Physics*, Springer, Berlin, 1988.
Krane, K. S. *Introductory Nuclear Physics*, Wiley, New York, 1988.
Mackintosh, R., J. Al-Khalili, B. Jonson, and T. Pena. *Nucleus: A Trip to the Heart of Matter*, Johns Hopkins University Press, Baltimore, 2001.
Wong, S. S. M. *Introductory Nuclear Physics*, 2nd ed., Wiley, New York, 1998.

# CHAPTER 3

# RADIOACTIVE DECAY KINETICS

The number of nuclei in a radioactive sample that disintegrate during a given time interval decreases exponentially with time. Because the nucleus is insulated by the surrounding cloud of electrons, this rate is essentially independent of pressure, temperature, the mass action law, or any other rate-limiting factors that commonly affect chemical and physical changes.[1] As a result, this decay rate serves as a very useful means of identifying a given nuclide. Since radioactive decay represents the transformation of an unstable radioactive nuclide into a more stable nuclide, which may also be radioactive, it is an irreversible event for each nuclide.

The unstable nuclei in a radioactive sample do not all decay simultaneously. Instead, the decay of a given nucleus is an entirely random event. Consequently, studies of radioactive decay events require the use of statistical methods. With these methods, one may observe a large number of radioactive nuclei and predict with fair assurance that, after a given length of time, a definite fraction of them will have disintegrated but not which ones or when.

---

[1]In the case of electron capture and internal conversion, the chemical environment of the electrons involved may affect the decay rate. For L-electron capture in $^7$Be ($t_{1/2} = 53.3$ d), the ratio of $t_{1/2}^{BeF_2}/t_{1/2}^{Be}$ is 1.00084. Similarly, a fully stripped radioactive ion cannot undergo either EC or IC decay, a feature of interest in astrophysics.

*Modern Nuclear Chemistry*, by W.D. Loveland, D.J. Morrissey, and G.T. Seaborg
Copyright © 2006 John Wiley & Sons, Inc.

**58** RADIOACTIVE DECAY KINETICS

## 3.1 BASIC DECAY EQUATIONS

Radioactive decay is what chemists refer to as a first-order reaction; that is, the rate of radioactive decay is proportional to the number of each type of radioactive nuclei present in a given sample. So, if we double the number of a given type of radioactive nuclei in a sample, we double the number of particles emitted by the sample per unit time.[2] This relation may be expressed as follows:

$$\begin{pmatrix} \text{Rate of} \\ \text{particle emission} \end{pmatrix} \equiv \begin{pmatrix} \text{rate of} \\ \text{disintegration of} \\ \text{radioactive nuclei} \end{pmatrix} \propto \begin{pmatrix} \text{number of} \\ \text{radioactive nuclei} \\ \text{present} \end{pmatrix}$$

Note that the foregoing statement is only a proportion. By introducing the decay constant, it is possible to convert this expression into an equation, as follows:

$$\begin{pmatrix} \text{Rate of} \\ \text{disintegration of} \\ \text{radioactive nuclei} \end{pmatrix} = \begin{pmatrix} \text{decay} \\ \text{constant} \end{pmatrix} \times \begin{pmatrix} \text{number of} \\ \text{radioactive} \\ \text{nuclei present} \end{pmatrix} \quad (3.1)$$

The decay constant $\lambda$ represents the average probability per nucleus of decay occurring per unit time. Therefore, we are taking the probability of decay per nucleus, $\lambda$, and multiplying it by the number of nuclei present so as to get the rate of particle emission. The units of rate are (disintegration of nuclei/time) making the units of the decay constant (1/time), that is, probability/time of decay.

To convert the preceding word equations to mathematical statements using symbols, let $N$ represent the number of radioactive nuclei present at time $t$. Then, using differential calculus, the preceding word equations may be written as

$$-\frac{dN}{dt} \propto N$$
$$-\frac{dN}{dt} = \lambda N \quad (3.2)$$

Note that $N$ is constantly reducing in magnitude as a function of time. Rearrangement of Equation (3.2) to separate the variables gives

$$\frac{dN}{N} = -\lambda \, dt \quad (3.3)$$

---

[2] In order to make this statement completely correct, we should say that as we double the number of nuclei present, we double the rate of particle emission. This rate is equal to the number of particles emitted per unit time, provided that the time interval is small.

If we say that at time $t = 0$ we have $N_0$ radioactive nuclei present, then integration of Equation (3.3) gives the radioactive decay law

$$N = N_0 e^{-\lambda t} \tag{3.4}$$

This equation gives us the number of radioactive nuclei present at time $t$. However, in many experiments, we want to know the counting rate that we will get in a detector as a function of time. In other words, we want to know the *activity* of our samples.

Still, it is easy to show that the counting rate in one's radiation detector, $C$, is equal to the rate of disintegration of the radioactive nuclei present in a sample, $A$, multiplied by a constant related to the efficiency of the radiation measuring system. Thus

$$C = \varepsilon A = \varepsilon \left(-\frac{dN}{dt}\right) = \varepsilon \lambda N \tag{3.5}$$

where $\varepsilon$ is the efficiency. Substituting into Equation (3.4), we get

$$C = C_0 e^{-\lambda t} \tag{3.6}$$

where $C$ is the counting rate at some time $t$ due to a radioactive sample that gave counting rate $C_0$ at time $t = 0$. Equations (3.4) and (3.6) are the basic equations governing the number of nuclei present in a radioactive sample and the number of counts observed in one's detector as a function of time. Equation (3.6) is shown graphically as Figure 3.1. As seen in Figure 3.1, this exponential curve flattens out and asymptotically approaches zero. If the same plot is made on a semilogarithmic scale (Fig. 3.2), the decay curve is a straight line, with a slope equal to the value of $-(\lambda/2.303)$.

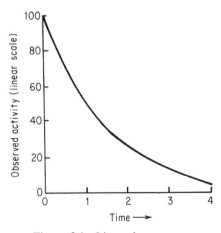

**Figure 3.1** Linear decay curve.

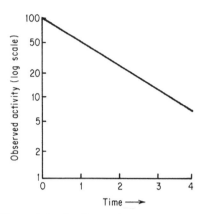

**Figure 3.2** Semilogarithmic decay curve.

The *half-life* ($t_{1/2}$) is another representation of the decay constant. The half-life of a radionuclide is the time required for its activity to decrease by exactly one-half. Thus, after one half-life, 50% of the initial activity remains. After two half-lives, only 25% of the initial activity remains. After three half-lives, only 12.5% is yet present and so forth. Figure 3.3 shows this relation graphically.

The half-life for a given nuclide can be derived from Equation (3.6) when the value of the decay constant is known. In accordance with the definition of the term half-life, when $A/A_0 = 1/2$, then $t = t_{1/2}$. Substituting these values into Equation (3.6) gives

$$\frac{A}{A_0} = \frac{1}{2} = e^{-\lambda t_{1/2}} \tag{3.7}$$

Hence

$$t_{1/2} = \frac{\ln 2}{\lambda} = \frac{0.693}{\lambda} \tag{3.8}$$

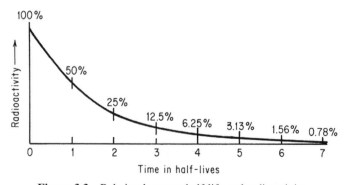

**Figure 3.3** Relation between half-life and radioactivity.

Note that the value of the expression for $t_{1/2}$ has the units of $1/\lambda$ or dimensions of (time).

The half-lives for different nuclides range from less than $10^{-6}$ s to $10^{10}$ y. The half-life has been measured for all the commonly used radionuclides. When an unknown radioactive nuclide is encountered, a determination of its half-life is normally one of the first steps in its identification. This determination can be done by preparing a semilog plot of a series of activity observations made over a period of time. A short-lived nuclide may be observed as it decays through a complete half-life and the time interval observed directly (Fig. 3.4).

**Example Problem** Given the data plotted below for the decay of a single radionuclide, determine the decay constant and the half-life of the nuclide.

*Solution*

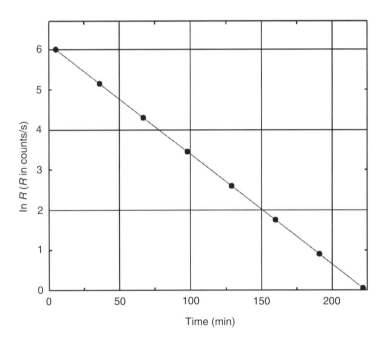

The data is plotted above. The slope $(-\lambda)$ is given as

$-\lambda = -(6.06 - 0)/(220 \text{ min} - 0)$

$\lambda = 0.0275 \text{ min}^{-1}$

$t_{1/2} = \ln 2/\lambda = 0.693/0.0275 = 25.2 \text{ min}$

What nuclide might this be?

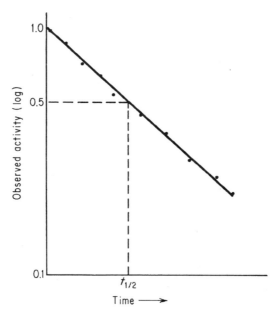

**Figure 3.4** Direct graphic determination of half-life.

It is difficult to measure the half-life of a very long-lived radionuclide. Here variation in disintegration rate may not be noticeable within a reasonable length of time. In this case, the decay constant must be calculated from the absolute decay rate according to Equation (3.2). The absolute number of atoms of the radioisotope present ($N$) in a given sample can be calculated according to

$$N = \frac{6.02 \times 10^{23} \text{ (Avogadro's number)}}{\text{atomic weight radionuclide}} \times \text{mass of the radionuclide} \qquad (3.9)$$

The total mass of the radioisotope in the given sample can be determined once the isotopic composition of the sample is ascertained by such means as mass spectrometry. When the decay constant is known, the half-life can then be readily calculated. A table of the half-lives of a number of the known nuclei can be found in Appendix B.

Although the half-life of a given radionuclide is a defined value, the actual moment of disintegration for a particular atom can be anywhere from the very beginning of the nuclide's life to infinity. The average or mean life of a population of nuclei can, however, be calculated. The mean life $\tau$ is naturally related to the decay constant and is, in fact, simply the reciprocal of the decay constant:

$$\tau = \frac{1}{\lambda} \qquad (3.10)$$

or the mean life can be expressed in terms of the half-life:

$$\tau = 1.443 t_{1/2} \tag{3.11}$$

One can understand the preceding relationship by recalling that the decay constant, $\lambda$, was defined as the average probability of decay per unit time, so the $1/\lambda$ is the average time between decays. The concept of average life allows us to calculate of the total number of particles emitted during a defined decay period. This number is essential in determining total radiation dose delivered by a radioisotope sample, as in medical research and therapy. During the time equal to one mean life, $\tau$, the activity falls to $1/e$ of its original value. For a sample of $N_0$ nuclei with lifetimes $t_i$, we can write for the mean life $\tau$

$$\tau = \frac{\sum_1^{N_0} t_i}{N_0} = -\frac{1}{N_0} \int_{t=0}^{t=\infty} t\, dN = \frac{1}{N_0} \int_0^\infty t\lambda N\, dt = \lambda \int_0^\infty t e^{-\lambda t}\, dt$$

$$= \left[ \frac{-\lambda t + 1}{\lambda} e^{-\lambda t} \right]_0^t = \frac{1}{\lambda} \tag{3.12}$$

The average or mean life is also of fundamental physical significance because it is the time to be substituted in the mathematical statement of the Heisenberg uncertainty principle, that is,

$$\Delta E \cdot \Delta t \geq \hbar$$

In this expression relating the uncertainty in energy of a system, $\Delta E$, to its lifetime $\Delta t$, $\tau \equiv \Delta t$.

$$\Delta E = \frac{\hbar}{\tau} = \frac{0.658 \times 10^{-15} \text{eV}}{\tau \text{ (s)}}$$

The quantity $\Delta E$ is called the width, $\Gamma$.

The natural unit of radioactivity is disintegrations/time, such as disintegration per second (dps) or disintegrations per minute (dpm), and so on. The SI (International System) unit of radioactivity is the Becquerel (Bq) where

$$1 \text{ Becquerel (Bq)} \equiv 1 \text{ disintegration/s}$$

Counting rates in a detection system are usually given in counts per second (cps), counts per minute (cpm), and so on, and differ from the disintegration rates by a factor representing the detector efficiency, $\varepsilon$. Thus

$$(\text{dpm})\, \varepsilon = (\text{cpm})$$

An older unit of radioactivity that still finds some use is the curie (Ci). It is defined as

$$1 \text{ curie (Ci)} = 3.7 \times 10^{10} \text{ Bq} = 3.7 \times 10^{10} \text{ dis/s}$$

The curie is a huge unit of radioactivity and is approximately equal to the activity of one gram of radium. The inventories of radioactivity in a nuclear reactor upon shutdown are typically $10^9$ Ci, whereas radiation sources used in tracer experiments have activities of μCi and the environmental levels of radioactivity are nCi or pCi.

Note also that because radionuclides, in general, have different half-lives, the number of nuclei per curie will differ from one species to another. For example, let us calculate how many nuclei are in 1 MBq (~27 μCi) of tritium ($t_{1/2} = 12.33$ y). We know that

$$N = \frac{(-dN/dt)}{\lambda} = \frac{10^6/s}{\lambda}$$

But

$$\lambda = \frac{0.693}{t_{1/2}} = \frac{(0.693)}{(12.33 \text{ y})(\pi \times 10^7 \text{ s/y})} = 1.789 \times 10^{-9} \text{ s}^{-1}$$

Thus

$$N = \frac{A}{\lambda} = \frac{(10^6/s)}{(1.789)(10^{-9}/s)} = (5.59)(10^{14}) \text{ nuclei}$$

The same calculation carried out for $^{14}$C ($t_{1/2} = 5730$ y) would give $2.60 \times 10^{17}$ nuclei/MBq. It is also interesting to calculate the mass associated with 1 MBq of tritium. We have

$$M = \frac{(N)(\text{atomic weight})}{(\text{Avogadro's number})} = \frac{(5.59)(10^{14})(3)}{(6.02)(10^{23})} = 2.78 \times 10^{-9} \text{g}$$

In other words, 1 MBq of tritium contains about 3 ng of tritium. Thus, an important feature of radionuclides becomes apparent—we routinely work with extremely small quantities of material. Pure samples of radioisotopes are called "carrier free."

Unless a radionuclide is in a carrier-free state, it is mixed homogeneously with the stable nuclides of the same element. It is, therefore, desirable to have a simple expression to show the relative abundances of the radioisotope and the stable isotopes. This specification is readily accomplished by using the concept of *specific activity*, which refers to the amount of radioactivity per given mass or other similar units of the total sample. The SI unit of specific activity is Bq/kg. Specific activity can also be expressed in terms of the disintegration rate (Bq or dpm), or counting rate (counts/min, cpm, or counts/s, cps), or curies (or mCi, μCi) of the specific radionuclide per unit mass of the sample.

## 3.2 MIXTURE OF TWO INDEPENDENTLY DECAYING RADIONUCLIDES

Where two or more radioisotopes with different half-lives are present in a sample and one does not or cannot distinguish the particles emitted by each isotope, a composite decay rate will be observed. The decay curve, in this situation, drawn on a semilogarithmic plot, will not be a straight line. The decay curves of each of the isotopes present usually can be resolved by graphical means if their half-lives are sufficiently different and if not more than three radioactive components are present. In the graphic example shown in Figure 3.5, line $C$ represents the total observed activity. Only the activity of the longer-lived component $A$ is observed after the shorter-lived component $B$ has become exhausted through decay. Extrapolation of this long-time portion of the curve back to zero time gives the decay curve for component $A$ and the activity of component $A$ at $t = 0$. The curve for component $B$ is obtained by subtracting out, point by point, the activity values of component $A$ from the total activity curve. If the half-lives of the two components in such samples are not sufficiently different to allow graphic resolution, a differential detection method may be applicable. If the radiation characteristics of the nuclides in the mixture are suitably distinct, that is, emission of different particles or γ rays, it may be possible to measure the activity of one component without interference from the radiation emitted by the other component. A case in point would be where one

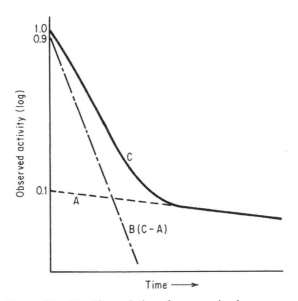

**Figure 3.5** Graphic resolution of a composite decay curve.

nuclide was a pure β emitter, while the other emitted both β and γ rays. In the case where the half-lives of the components are known but are not sufficiently different to allow graphical resolution of the decay curve, computer techniques that utilize least-squares fitting to resolve such a case are also available.

**Example Problem** Given the following decay data, determine the half-lives and initial activities of the radionuclides present:

| $t$ (h) | $A$ (cpm) |
|---|---|
| 0.1 | 270 |
| 0.5 | 210 |
| 1.0 | 170 |
| 1.5 | 130 |
| 2.0 | 110 |
| 2.5 | 90 |
| 3 | 80 |
| 4 | 65 |
| 5 | 55 |
| 7 | 44 |
| 10 | 34 |
| 15 | 22 |

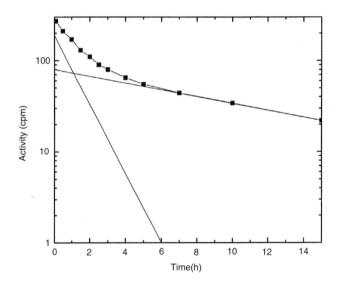

**Solution** From the graph, we see

$t_{1/2}(b) = 8.0\,\text{h}$   $A_0(b) = 80\,\text{cpm}$

$t_{1/2}(c) = 0.8\,\text{h}$   $A_0(c) = 190\,\text{cpm}$

## 3.3 RADIOACTIVE DECAY EQUILIBRIUM

When a radionuclide decays, it does not disappear but is transformed into a new nuclear species of higher binding energy and often differing $Z$, $A$, $J$, $\pi$, and so on. The equations of radioactive decay discussed so far have focused on the decrease of the parent radionuclides but have ignored the formation (and possible decay) of daughter, granddaughter, and so forth, species. It is the formation and decay of these "children" that is the focus of this section.

Let us begin by considering the case when a radionuclide 1 decays with decay constant $\lambda_1$, forming a daughter nucleus 2, which in turn decays with decay constant $\lambda_2$. Schematically, we have

$$1 \rightarrow 2 \rightarrow$$

We can write terms for the production and depletion of 2, that is,

Rate of change of 2 = rate of production − rate of decay of
nuclei present                of 2                              2
at time $t$

$$\frac{dN_2}{dt} = \lambda_1 N_1 - \lambda_2 N_2 \tag{3.13}$$

where $N_1$ and $N_2$ are the numbers of 1 and 2 present at time $t$.

Rearranging and collecting similar terms

$$dN_2 + \lambda_2 N_2\, dt = \lambda_1 N_1\, dt \tag{3.14}$$

Remembering that

$$N_1 = N_1^0 e^{-\lambda_1 t} \tag{3.15}$$

we have

$$dN_2 + \lambda_2 N_2\, dt = \lambda_1 N_1^0 e^{-\lambda_1 t}\, dt \tag{3.16}$$

This is a first-order linear differential equation and can be solved using the method of integrating factors that we show below. Multiplying both sides by $e^{\lambda_2 t}$, we have

$$e^{\lambda_2 t} dN_2 + \lambda_2 N_2 e^{\lambda_2 t} dt = \lambda_1 N_1^0 e^{(\lambda_2 - \lambda_1)t}\, dt \tag{3.17}$$

The left-hand side is now a perfect differential

$$d(N_2 e^{\lambda_2 t}) = \lambda_1 N_1^0 e^{(\lambda_2 - \lambda_1)t}\, dt \tag{3.18}$$

Integrating from $t = 0$ to $t = t$, we have

$$N_2 e^{\lambda_2 t}\Big|_0^t = \frac{\lambda_1 N_1^0 e^{(\lambda_2 - \lambda_1)t}}{\lambda_2 - \lambda_1}\Big|_0^t \quad (3.19)$$

$$N_2 e^{\lambda_2 t} - N_2^0 = \frac{\lambda_1}{\lambda_2 - \lambda_1} N_1^0 (e^{(\lambda_2 - \lambda_1)t} - 1) \quad (3.20)$$

Multiplying by $e^{-\lambda_2 t}$ and rearranging gives

$$N_2(t) = \frac{\lambda_1}{\lambda_2 - \lambda_1} N_1^0 (e^{-\lambda_1 t} - e^{-\lambda_2 t}) + N_2^0 e^{-\lambda_2 t} \quad (3.21)$$

where $N_2^0$ is the number of species 2 present at $t = 0$. The first term in Equation (3.21) represents the growth of the daughter due to the decay of the parent, whereas the second term represents the decay of any daughter nuclei that were present initially. Remembering that $A_2 = \lambda_2 N_2$, we can write an expression for the activity of 2 as

$$A_2 = \frac{\lambda_1 \lambda_2}{\lambda_2 - \lambda_1} N_1^0 (e^{-\lambda_1 t} - e^{-\lambda_2 t}) + A_2^0 e^{-\lambda_2 t} \quad (3.22)$$

These two equations, (3.21) and (3.22), are the general expressions for the number of daughter nuclei and the daughter activity as a function of time, respectively.

The general behavior of the activity of parent and daughter species, as predicted by Equation (3.22), is shown in Figure 3.6. As one expects qualitatively for the case

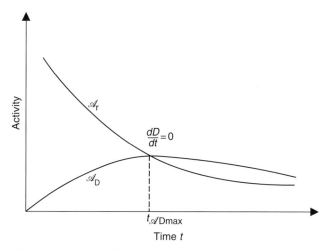

**Figure 3.6** Illustration of the conditions at ideal equilibrium, as would apply to the case of a fairly long-lived parent nuclide.

with $N_2^0 = 0$, the initial activity of the daughter begins at zero, rises to a maximum, and if one waits long enough, eventually decreases.

Thus, there must be a time when the daughter activity is the maximum. We can calculate this time by noting the condition for a maximum in the activity of 2 is

$$\frac{dN_2}{dt} = 0 \tag{3.23}$$

Taking the derivative of Equation (3.21) and simplifying,

$$\lambda_1 e^{-\lambda_1 t} = \lambda_2 e^{-\lambda_2 t} \tag{3.24}$$

Solving for $t$,

$$t_{\max} = \frac{\ln(\lambda_2/\lambda_1)}{\lambda_2 - \lambda_1} \tag{3.25}$$

All of this development may seem like something that would be best handled by a computer program or just represents a chance to practice one's skill with differential equations. But that is not true. It is important to understand the mathematical foundation of this development to gain insight into practical situations. Let us consider some cases that illustrate this point.

Consider the special case where $\lambda_1 = \lambda_2$. Plugging into Equation (3.21) or (3.22) or a computer program based upon them leads to a division by zero. Does nature therefore forbid $\lambda_1$ from equaling $\lambda_2$ in a chain of decays? Nonsense! One simply understands that one must redo the derivation [Equations (3.13) through (3.21)] of Equations (3.21) and (3.22) for this special case (see Problems).

Let us now consider a number of other special cases of Equations (3.21) and (3.22) that are of practical importance. *Suppose the daughter nucleus is stable* ($\lambda_2 = 0$). Then we have

$$\frac{dN_2}{dt} = \lambda_1 N_1 \tag{3.26}$$

$$dN_2 = \lambda_1 N_1 dt = \lambda_1 N_1^0 e^{-\lambda_1 t} dt \tag{3.27}$$

$$N_2 = \frac{\lambda_1 N_1^0}{-\lambda_1} \left(e^{-\lambda_1 t}\right)\Big|_0^t = N_1^0 \left(1 - e^{-\lambda_1 t}\right) \tag{3.28}$$

These relations are shown in Figure 3.7. They represent the typical decay of many radionuclides prepared by neutron capture reactions, the type of reaction that commonly occurs in a nuclear reactor.

In Figure 3.8, we show the activity relationships for parent and daughter [as predicted by Equation (3.22)] for various choices of the relative values of the half-lives of the parent and daughter nuclides. In the first of these cases, we have $t_{1/2}$ (parent) < $t_{1/2}$ (daughter), that is, the parent is shorter lived than the daughter.

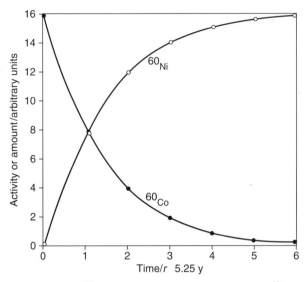

**Figure 3.7** Decay of $^{60}$Co (period 5.25 y) and the growth of $^{60}$Ni (stable).

This is called the *no equilibrium* case because the daughter buildup (due to the decay of the parent) is faster than its loss due to decay. Essentially, all of the parent nuclides are converted to daughter nuclides, and the subsequent activity is due to the decay of the daughters only. Thus, the name "no equilibrium" is used. Practical examples of this decay type are $^{131}$Te → $^{131}$I, $^{210}$Bi → $^{210}$Po, and $^{92}$Sr → $^{92}$Y. This situation typically occurs when one is very far from stability and the nuclei decay by β decay toward stability.

A second special case of Equations (3.21) and (3.22) is called *transient equilibrium* (Figs. 3.8c and 3.9). In this case, the parent is significantly (∼10×) longer lived than the daughter and thus controls the decay chain. Thus

$$\lambda_2 > \lambda_1 \tag{3.29}$$

In Equation (3.21), as $t \to \infty$,

$$e^{-\lambda_2 t} \ll e^{-\lambda_1 t}$$
$$N_2^0 e^{-\lambda_2 t} \to 0 \tag{3.30}$$

and we have

$$N_2 \approx \frac{\lambda_1}{\lambda_2 - \lambda_1} N_1^0 e^{-\lambda_1 t} \tag{3.31}$$

Substituting

$$N_1 = N_1^0 e^{-\lambda_1 t} \tag{3.32}$$

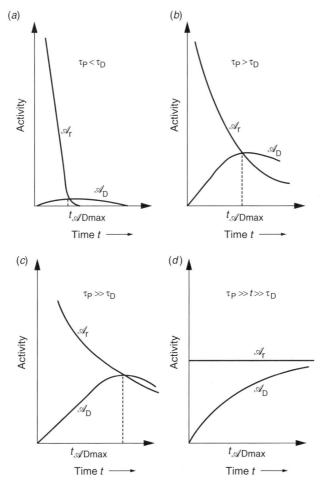

**Figure 3.8** Activity curves for various parent–daughter relationships: (*a*) short-lived parent ($\tau_P < \tau_D$); (*b*) long-lived parent ($\tau_P > \tau_D$); (*c*) very long-lived parent ($\tau_P \gg \tau_D$); (*d*) almost stable, or constantly replenished, parent ($\tau_P \gg t \gg \tau_D$).

we have

$$\frac{N_1}{N_2} = \frac{\lambda_2 - \lambda_1}{\lambda_1} \qquad (3.33)$$

At long times, the ratio of daughter to parent activity becomes constant, and both species disappear with the effective half-life of the parent. The classic examples of this decay equilibrium are the decay of $^{140}$Ba ($t_{1/2} = 12.8$ d) to $^{140}$La ($t_{1/2} = 40$ h) or the equilibrium between $^{222}$Rn ($t_{1/2} = 3.8$ d) and its short-lived decay products.

**72** RADIOACTIVE DECAY KINETICS

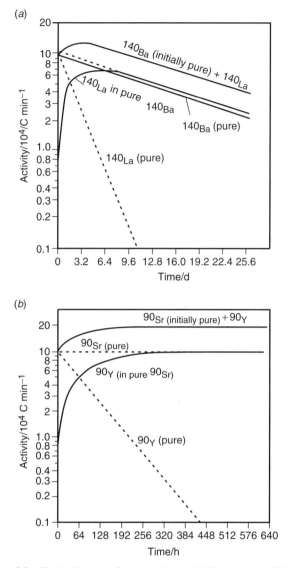

**Figure 3.9** Typical cases of (a) transient and (b) secular equilibrium.

A third special case of Equations (3.21) and (3.22) is called *secular equilibrium* (Figs. 3.8d, and 3.9). In this case, the parent is very much longer lived ($\sim 10^4 \times$) than the daughter or the parent is constantly being replenished through some other process. During the time of observation, there is no significant change in the number of parent nuclei present, although several half-lives of the daughter may occur. In the previous case of transient equilibrium, we had

$$\frac{N_1}{N_2} = \frac{\lambda_2 - \lambda_1}{\lambda_1} \tag{3.34}$$

Since we now also have

$$\lambda_1 \ll \lambda_2 \tag{3.35}$$

we can simplify even more to give

$$\frac{N_1}{N_2} = \frac{\lambda_2}{\lambda_1} \tag{3.36}$$

$$\lambda_1 N_1 = \lambda_2 N_2 \tag{3.37}$$

$$A_1 = A_2$$

In short, the activity of the parent and daughter are the same, and the total activity of the sample remains effectively constant during the period of observation.

The naturally occurring heavy element decay chains (see below) where $^{238}$U → $^{206}$Pb, $^{235}$U → $^{207}$Pb, and $^{232}$Th → $^{208}$Pb and the extinct heavy element decay series $^{237}$Np → $^{209}$Bi are examples of secular equilibrium because of the long half-lives of the parents. Perhaps the most important cases of secular equilibrium are the production of radionuclides by a nuclear reaction in an accelerator, a reactor, a star, or the upper atmosphere. In this case, we have

$$\text{Nuclear reaction} \longrightarrow (2) \longrightarrow \tag{3.38}$$

which produces the radionuclide 2 with rate $R$. If the reaction is simply the decay of a long-lived nuclide, then $R = \lambda_1 N_1^0$ and $N_2^0 = 0$. Substitution into Equation (3.21) gives the expression

$$N_2 = \frac{\lambda_1}{\lambda_2 - \lambda_1} N_1^0 \left(e^{-\lambda_1 t} - e^{-\lambda_2 t}\right) \tag{3.39}$$

If the reaction is slower than the decay or

$$\lambda_1 \ll \lambda_2 \tag{3.40}$$

It is most appropriate to say (since $\lambda_1 \approx 0$)

$$N_2 \approx \frac{\lambda_1}{\lambda_2} N_1^0 \left(1 - e^{-\lambda_2 t}\right) \tag{3.41}$$

or in terms of the activities

$$A_2 = \lambda_2 N_2 = R\left(1 - e^{-\lambda_2 t}\right) \tag{3.42}$$

Equation (3.42) is known as the activation equation and is shown in Figure 3.10.

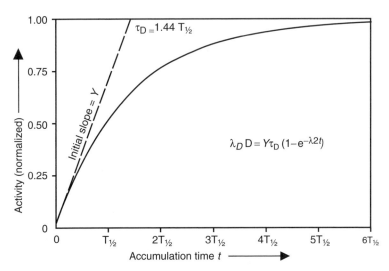

**Figure 3.10** Growth of the activity of a primary reaction product induced by a constant bombardment.

Initially, the growth of the product radionuclide activity is nearly linear (due to the behavior of $(1 - e^{-\lambda_2 t})$ for small values of $\lambda t$), but eventually the product activity becomes "saturated" or constant, decaying as fast as it is produced. At an irradiation time of one half-life, half the maximum activity is formed; after two half-lives, three fourths of the maximum activity is formed, and so on. This situation gives rise to the rough rule that irradiations that extend for periods that are greater than twice $t_{1/2}$ of the desired radionuclide are usually not worthwhile.

Equation (3.21) may be generalized to a chain of decaying nuclei of arbitrary length by using the Bateman equations (Bateman, 1910). If we assume that at $t = 0$, none of the daughter nuclei are present, $N_2^0 = N_3^0 = , \ldots, N_n^0 = 0$, we get

$$(1) \rightarrow (2) \rightarrow (3), \ldots, (n) \rightarrow$$

$$N_n = C_1 e^{-\lambda_1 t} + C_2 e^{-\lambda_2 t} + C_3 e^{-\lambda_3 t} + , \ldots, C_n e^{-\lambda_n t}$$

where

$$C_1 = \frac{\lambda_1 \lambda_2, \ldots, \lambda_{n-1}}{(\lambda_2 - \lambda_1)(\lambda_3 - \lambda_1), \ldots, (\lambda_n - \lambda_1)} N_1^0$$

$$C_2 = \frac{\lambda_1 \lambda_2, \ldots, \lambda_{n-1}}{(\lambda_1 - \lambda_2)(\lambda_3 - \lambda_2), \ldots, (\lambda_n - \lambda_2)} N_1^0 \quad (3.43)$$

$$C_n = \frac{\lambda_1 \lambda_2, \ldots, \lambda_{n-1}}{(\lambda_1 - \lambda_n)(\lambda_2 - \lambda_n), \ldots, (\lambda_{n-1} - \lambda_n)} N_1^0$$

These equations describe the activities produced in new fuel in a nuclear reactor. No fission or activation products are present when the fuel is loaded, and they grow in as the reactions take place.

**Example Problem** Consider the decay of a 1 μCi sample of pure $^{222}$Rn ($t_{1/2} =$ 3.82 d). Use the Bateman equations to estimate the activity of its daughters ($^{218}$Po, $^{214}$Pb, $^{214}$Bi, and $^{214}$Po) after a decay time of 4 h.

The decay sequence is

$$^{222}\text{Rn} \xrightarrow{\alpha} {}^{218}\text{Po} \xrightarrow{\alpha} {}^{214}\text{Pb} \xrightarrow{\beta^-} {}^{214}\text{Bi} \xrightarrow{\beta^-} {}^{214}\text{Po} \xrightarrow{\alpha}$$

| | | | | | |
|---|---|---|---|---|---|
| $t_{1/2}$ | 3.82 d | 3.1 m | 26.8 m | 19.9 m | 164 μs |
| activity | A | B | C | D | E |
| λ ($10^{-4}$s) | 0.021 | 37.3 | 4.31 | 5.81 | $4.3 \times 10^7$ |

*Solution*

$$A = A_0 e^{-\lambda_A t} = 1\mu\text{Ci}\left(\exp -\frac{\ln 2 \cdot 4}{24 \cdot 3.82}\right)$$

$$A = 0.97 \, \mu\text{Ci}$$

$$B = \lambda_B\left(C_1 e^{-\lambda_1 t} + C_2 e^{-\lambda_2 t}\right)$$

$$C_1 = \frac{\lambda_A N_A^0}{\lambda_B - \lambda_A} = \frac{A_0}{\lambda_B - \lambda_A} \quad C_2 = \frac{\lambda_A N_A^0}{\lambda_A - \lambda_B} = \frac{A_0}{\lambda_A - \lambda_B}$$

$$B = \lambda_B\left(\frac{A_0 e^{-\lambda_A t}}{\lambda_B - \lambda_A} + \frac{A_0 e^{-\lambda_B t}}{\lambda_A - \lambda_B}\right)$$

$$B = 37.3\left(\frac{0.97}{37.3 - 0.021} + \frac{\exp\frac{-\ln 2 \cdot 4}{3.1/60}}{0.021 - 37.3}\right)$$

$$B = 0.97 \, \mu\text{Ci}$$

(Actually $B/A = 1.00056$)

The reader should verify that for C, D, and E, the only significant term is the term multiplying $e^{-\lambda_A t}$ as it was for B. Thus, for $D/A$, we have

$$\frac{D}{A} = \frac{\lambda_B}{\lambda_B - \lambda_A} \cdot \frac{\lambda_C}{\lambda_C - \lambda_A} \cdot \frac{\lambda_D}{\lambda_D - \lambda_A} = 1.0091$$

The reader should, as an exercise, compute the quantities of C and E present.

## 3.4 BRANCHING DECAY

Some nuclides decay by more than one mode. Some nuclei may decay by either $\beta^+$ decay or electron capture; others by α decay or spontaneous fission; still others by γ-ray emission or internal conversion, and so on. In these cases, we can characterize each competing mode of decay by a separate decay constant $\lambda_i$ for each type of decay where the total decay constant $\lambda$ is given by the sum

$$\lambda = \lambda_1 + \lambda_2 + \cdots = \sum_{i=1}^{N} \lambda_i \tag{3.44}$$

Corresponding to each partial decay constant $\lambda_i$, there is a partial half-life $t_{1/2}^i$ where

$$t_{1/2}^i = \frac{0.693}{\lambda_i} \tag{3.45}$$

and the total half-life, $t_{1/2}$, is the sum of the reciprocals

$$\frac{1}{t_{1/2}} = \frac{1}{t_{1/2}^1} + \frac{1}{t_{1/2}^2} + \cdots = \sum_{i=1}^{N} \frac{1}{t_{1/2}^i} \tag{3.46}$$

The fraction of decays proceeding by the $i$th mode is given by the obvious expression

$$f_i = \frac{\lambda_i}{\Sigma \lambda_i} = \frac{\lambda_i}{\lambda} \tag{3.47}$$

By analogy, the energy uncertainty associated with a given state, $\Delta E$, through the Heisenberg uncertainty principle can be obtained from the lifetime contributed by each decay mode. If we use the definition $\Delta E \equiv \Gamma$, the level width, then we can express $\Gamma$ in terms of the partial widths for each decay mode $\Gamma_i$ such that

$$\Gamma = \Gamma_1 + \Gamma_2 + \Gamma_3 + \cdots = \sum_{i=1}^{N} \Gamma_i \tag{3.48}$$

where

$$\Gamma_i = \frac{1}{\tau_i} \tag{3.49}$$

where $\tau_i$ is the partial mean life associated with each decay mode. This approach is especially useful in treating the decay of states formed in nuclear reactions in which a variety of competing processes such as α emission, p emission, n emission, and so on, may occur as the nucleus de-excites. In such cases, we can express the total width as

$$\Gamma = \Gamma_\alpha + \Gamma_p + \Gamma_n \tag{3.50}$$

**Example Problem** Consider the nucleus $^{64}$Cu ($t_{1/2} = 12.700$ h). $^{64}$Cu is known to decay by electron capture (61%) and $\beta^-$ decay (39%). What are the partial half-lives for EC and $\beta^-$ decay? What is the partial width for EC decay?

*Solution*

$$\lambda = \ln 2/12.700 \text{ h} = 5.46 \times 10^{-2} \text{ h}^{-1}$$
$$\lambda = \lambda_{EC} + \lambda_\beta = \lambda_{EC} + (39/61)\lambda_{EC}$$
$$\lambda_{EC} = 3.329 \times 10^{-2} \text{ h}^{-1}$$
$$t^{EC}_{1/2} = (\ln 2)/\lambda_{EC} = 20.8 \text{ h}$$
$$\lambda_\beta = (39/61)\lambda_{EC} = 2.128 \times 10^{-2} \text{ h}^{-1}$$
$$t^\beta_{1/2} = (\ln 2)/\lambda_\beta = 32.6 \text{ h}$$
$$\tau^{EC} = t^{EC}_{1/2}/\ln 2 = 30.0 \text{ h} = 108131 \text{ s}$$
$$\Gamma_{EC} = \hbar/\tau^{EC} = 6.582 \times 10^{-22} \text{Me V} \cdot \text{s}/108131 \text{ s} = 6.1 \times 10^{-27} \text{ MeV}$$

All naturally occurring radioactive nuclei have extremely small partial widths. Did you notice that $^{64}$Cu can decay into $^{64}$Zn and $^{64}$Ni? This is unusual but can occur for certain odd–odd nuclei (see Chapter 2).

## 3.5 NATURAL RADIOACTIVITY

There are approximately 70 naturally occurring radionuclides on Earth. Most of them are heavy-element radioactivities present in the natural decay chains, but there are several important light-element activities, such as $^3$H, $^{14}$C, $^{40}$K, and so forth. These radioactive species are ubiquitous, occurring in plants, animals, the air we breathe, the water we drink, the soil, and so forth. For example, in the 70-kg "reference man," one finds $\sim$4400 Bq of $^{40}$K and $\sim$3600 Bq of $^{14}$C, that is, about 8000 dis/s due to these two radionuclides alone. In a typical U.S. diet, one ingests $\sim$1 pCi/day of $^{238}$U, $^{226}$Ra, and $^{210}$Po. The air we breathe contains $\sim$0.15 pCi/L of $^{222}$Rn, the water we drink contains $>$10 pCi/L of $^3$H while the Earth's crust contains $\sim$10 and $\sim$4 ppm of the radio elements Th and U, respectively. One should not forget that the interior heat budget of planet Earth is dominated by the contributions from the radioactive decay of uranium, thorium, and potassium.

The naturally occurring radionuclides can be classified as: (a) *primordial*, that is, nuclides that have survived since the time the elements were formed, (b) *cosmogenic*, that is, shorter-lived nuclides formed continuously by the interaction of cosmic rays with matter, and (c) *anthropogenic*, that is, a wide variety of nuclides introduced into the environment by the activities of humans, such as nuclear weapons tests, the operation (or misoperation) of nuclear power plants, and the like. The primordial radionuclides have half-lives greater than $10^9$ y or are the decay products of these nuclei. This class includes $^{40}$K ($t_{1/2} = 1.277 \times 10^9$ y), $^{87}$Rb ($t_{1/2} = 47.5 \times 10^9$ y), $^{238}$U ($t_{1/2} = 4.467 \times 10^9$ y), $^{235}$U ($t_{1/2} = 0.704 \times 10^9$ y), and $^{232}$Th ($t_{1/2} = 14.05 \times 10^9$ y) as its most important members. (Additional members of this group are $^{115}$In, $^{123}$Te, $^{138}$La, $^{144}$Nd, $^{147}$Sm, $^{148}$Sm, $^{176}$Lu, $^{174}$Hf, $^{187}$Re, and $^{190}$Pt.)

$^{40}$K is a $\beta^-$-emitting nuclide that is the predominant radioactive component of normal foods and human tissue. Due to the 1460-keV $\gamma$ ray that accompanies the $\beta^-$ decay, it is also an important source of background radiation detected by $\gamma$-ray spectrometers. The natural concentration in the body contributes about 17 mrem/y to the whole body dose. The specific activity of $^{40}$K is approximately 855 pCi/g potassium. Despite the high specific activity of $^{87}$Rb of $\sim$2400 pCi/g, the low abundance of rubidium in nature makes its contribution to the overall radioactivity of the environment small.

There are three naturally occurring decay series. They are the uranium ($A = 4n + 2$) series, in which $^{238}$U decays through 14 intermediate nuclei to form the stable nucleus $^{206}$Pb, the actinium or $^{235}$U ($A = 4n + 3$) series in which $^{235}$U decays through 11 intermediate nuclei to form stable $^{207}$Pb, and the thorium ($A = 4n$) series in which $^{232}$Th decays through a series of 10 intermediates to stable $^{208}$Pb (Fig. 3.11).

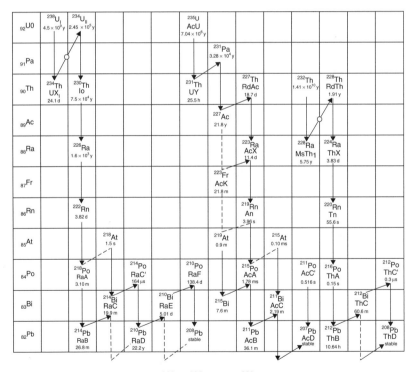

**Figure 3.11** The decay series of $U^{238}$, $U^{235}$, and $Th^{232}$. Not shown are several intermediate daughter products of little significance in geochemical applications. For the sake of completeness, old notations still referred to frequently in present-day texts, e.g., RaA for Po$^{218}$, Io for Th$^{230}$, are given in the scheme.

Because the half-lives of the parent nuclei are so long relative to the other members of each series, all members of each decay series are in secular equilibrium, that is, the activities of each member of the chain are equal at equilibrium if the sample has not been chemically fractionated. Thus, the activity associated with $^{238}$U in secular equilibrium with its daughters is 14× the activity of the $^{238}$U. The notation $4n + 2$, $4n$, $4n + 3$ refers to the fact that the mass number of each member of a given chain is such that it can be represented by $4n$, $4n + 2$, $4n + 3$ where $n$ is an integer. (There is an additional decay series, the $4n + 1$ series, that is extinct because its longest lived member, $^{237}$Np, has a half-life of only $2.1 \times 10^6$ y, a time that is very short compared to the time of element formation.)

The uranium series contains two radionuclides of special interest, $^{226}$Ra ($t_{1/2} = 1600$ y) and its daughter, 38 d $^{222}$Rn. $^{226}$Ra (and its daughters) are responsible for a major fraction of the radiation dose received from internal radioactivity. Radium is present in rocks and soils, and as a consequence in water, food, and human tissue. The high specific activity and gaseous decay products of radium also make it difficult to handle in the laboratory.

$^{226}$Ra decays by α emission to $^{222}$Rn. This latter nuclide is the principal culprit in the radiation exposures from indoor radon. Although radon is an inert gas and is not trapped in the body, the short-lived decay products are retained in the lungs when inhaled if the $^{222}$Rn decays while it is in the lungs. Indoor radon contributes about 2 mSv/y (200 mrem/y) to the average radiation exposure in the United States, that is, about two-thirds of the dose from natural sources. Under normal circumstances, radon and its daughters attach to dust particles and are in their equilibrium amounts. These dust particles can also deposit in the lungs. It has been estimated that in the United States, 5000–10,000 cases of lung cancer (6–12% of all cases) are due to radon exposure.

The second class of naturally occurring radionuclides is the *cosmogenic* nuclei, produced by the interactions of primary and secondary cosmic radiation with nuclei in the stratosphere. The most important of these nuclei are $^3$H (tritium), $^{14}$C, and $^7$Be. Less importantly, $^{10}$Be, $^{22}$Na, $^{32}$P, $^{33}$P, $^{35}$S, and $^{39}$Cl are also produced. These nuclei move into the troposphere through normal exchange processes and are brought by Earth's surface by rainwater. Equilibrium is established between the production rate in the primary cosmic ray interaction and the partition of the radionuclides among the various terrestrial compartments (atmosphere, surface waters, biosphere, etc.) leading to an approximately constant specific activity of each nuclide in a particular compartment. When an organism dies after being in equilibrium with the biosphere, the specific activity of the nuclide in that sample will decrease since it is no longer in equilibrium. This behavior allows these nuclides to act as tracers for terrestrial processes and for dating.

$^{14}$C ($t_{1/2} = 5730$ y) is formed continuously in the upper atmosphere by cosmic rays that produce neutrons giving the reaction

$$\text{n (slow)} + {}^{14}\text{N} \longrightarrow {}^{14}\text{C} + \text{p}$$

or, in a shorthand notation, $^{14}$N(n, p)$^{14}$C. $^{14}$C is a soft $\beta^-$ emitter ($E_{\max} \sim 158$ keV).

This radiocarbon ($^{14}$C) reacts with oxygen and eventually exchanges with the stable carbon (mostly $^{12}$C) in living things. If the cosmic ray flux is constant, and the terrestrial processes affecting $^{14}$C incorporation into living things are constant, and there are no significant changes in the stable carbon content of the atmosphere, then a constant level of $^{14}$C in all living things is found (corresponding to ~1 atom of $^{14}$C for every $10^{12}$ atoms of $^{12}$C or about 227 Bq/kg C). When an organism dies, it ceases to exchange its carbon atoms with the pool of radiocarbon, and its radiocarbon content decreases in accord with Equation (3.6). Measurement of the specific activity of an old object allows one to calculate the age of the object (see below).

$^{14}$C reaches the Earth's surface at the rate of ~2.3 atoms/cm$^2$/s after production by cosmic ray interaction in the atmosphere, corresponding to a total production of ~$1.4 \times 10^{15}$ Bq/y. $^{14}$C is also formed by the $^{14}$N(n, p) reaction by atmospheric tests of nuclear weapons. About $2.2 \times 10^{17}$ Bq were made in the atmospheric test "spike" of the 1950s and 1960s that has been primarily transferred to the oceans and the biosphere. This means that $^{14}$C is the most significant fallout nuclide from the point of view of population dose. Nuclear power plants also release $^{14}$C as part of their normal operation contributing ~$0.1 \times 10^{15}$ Bq/y.

Tritium ($^3$H) is produced naturally through atmospheric cosmic ray interactions via the reaction

$$\text{n (fast)} + {}^{14}\text{N} \rightarrow {}^{12}\text{C} + {}^{3}\text{H}$$

Tritium is also produced in ternary fission and by neutron-induced reactions with $^6$Li and $^{10}$B. Tritium is a very low energy $\beta^-$ emitter with a half-life of 12.33 y. The global inventory of naturally produced tritium is $9.6 \times 10^{17}$ Bq. Tritium is readily incorporated in water and is removed from the atmosphere by rain or snow. Its residence time in the stratosphere is 2–3 y; after reaching the troposphere it is removed in 1–2 months. The "natural" concentration of $^3$H in streams and freshwater is ~10 pCi/L.

The nuclear weapons tests of the late 1950s and early 1960s also injected a huge spike of tritium into the atmosphere along with $^{14}$C. The tritium levels in the troposphere increased by a factor of 100 at this time. Estimates of $2.4 \times 10^{20}$ Bq for this spike have been made. Assuming that there will not be more atmosphere testing of nuclear weapons, the tritium from fallout should decrease with a half-life of 12.3 y. At present the fallout tritium in surface waters is approximately equal to that generated from nuclear power plant operation (as a ternary fission product or from n reactions with $^{10}$B). (Nuclear plant operation generates ~$10^{16}$ Bq/y.) As a result of all of these developments, the current tritium content of surface waters is ~10× the "natural" level.

The third principal component of environmental radioactivity is that due to the activities of humans, the anthropogenic radionuclides. This group of nuclides includes the previously discussed cases of $^3$H and $^{14}$C along with the fission products and the transuranium elements. The primary sources of these nuclides are nuclear weapons tests and nuclear power plant accidents. These events and the gross nuclide releases associated with them are shown in Table 3.1. Except for $^{14}$C and

**TABLE 3.1 Events Leading to Large Injections of Radionuclides into Atmosphere (From Choppin, Rydberg and Liljenzin)**

| Source | Country | Time | Radioactivity (Bq) | Important Nuclides |
|---|---|---|---|---|
| Hiroshima & Nagasaki | Japan | 1945 | $4 \times 10^{16}$ | Fission Products Actinides |
| Atmospheric weapons tests | USA USSR | 1963 | $2 \times 10^{20}$ | Fission Products Actinides |
| Windscale | UK | 1957 | $1 \times 10^{15}$ | $^{131}$I |
| Chelyabinsk (Kysthym) | USSR | 1957 | $8 \times 10^{16}$ | Fission Products $^{90}$Sr, $^{137}$Cs |
| Harrisburg | USA | 1979 | $1 \times 10^{12}$ | Noble gases, $^{131}$I |
| Chernobyl | USSR | 1986 | $2 \times 10^{18}$ | $^{137}$Cs |

*Source*: From Choppin et al. (1995).

$^3$H (T), the anthropogenic contributions from nuclear weapons testing or use (which is the most significant source of man-made environmental exposure) are negligible compared to other sources of natural radioactivity. (The principal component of these large releases of radioactivity was shorter-lived fission products such as $^{131}$I, which have decayed, leaving $^{137}$Cs, $^{90}$Sr, and the Pu isotopes as the nuclides of most concern. For further descriptions of these events and their environmental consequences, the reader is referred to the material in the Bibliography.)

## 3.6 RADIONUCLIDE DATING

An important application of the basic radioactive decay law is that of radionuclide dating. From Equation (3.6), we have

$$N = N_0 e^{-\lambda t} \quad (3.51)$$

We can solve this equation for $t$:

$$t = \frac{\ln (N_0/N)}{\lambda} \quad (3.52)$$

where $N_0$ and $N$ are the number of radionuclides present at times $t = 0$ and $t = t$ and $\lambda$ is the decay constant. The quantity $t$ is the age of the object, and it can be determined from a knowledge of the nuclear decay constant ($t_{1/2}$) and the number of radioactive nuclei present in the object now, $N$, and initially, $N_0$. Clearly, $N$ can be determined by counting the sample ($A = \lambda N$), but the trick is to determine $N_0$. One obvious approach is to recognize that for a decay of parent $P$ to daughter $D$, the total number of nuclei is constant:

$$D(t) + P(t) = P(t_0) \equiv P^0 \quad (3.53)$$

and

$$P(t) = P^0 e^{-\lambda t} \tag{3.54}$$

so that

$$t = \frac{1}{\lambda} \ln\left(1 + \frac{D(t)}{P(t)}\right) \tag{3.55}$$

Thus, by measuring the current ratio of daughter to parent atoms ($D(t)/P(t)$) one can deduce the age of the sample. (This assumes, of course, that there are no daughter atoms present at $t = 0$, that they are all due to the parent decay, and that none have been lost.)

**Example Problem**  In a rock, one finds a nuclidic ratio of $^{206}$Pb to $^{238}$U of 0.60. What is the age of the rock?

**Solution**

$$t = \frac{1}{\lambda} \ln\left(1 + \frac{D(t)}{P(t)}\right) = \frac{1}{\ln 2/(4.5 \times 10^9 \text{ y})} \ln(1 + 0.60)$$

$$t = 3.1 \times 10^9 \text{ y}$$

If we want to relax this latter condition that no daughter atoms were present at $t = 0$ [$D(t=0) = 0$], then we need an additional term:

$$D(t) + P(t) = D^0 + P^0 \tag{3.56}$$

and we need to make an estimate of $D^0$. Suppose there is another isotope of the daughter element that is stable and is not formed in the decay of anything else. We can assume that

$$D_s(t) = D_s^0 \equiv D_s \tag{3.57}$$

where $D_s$ is the number of such stable atoms. Then, dividing by $D_s$

$$\frac{D(t)}{D_s} + \frac{P(t)}{D_s} = \frac{D^0}{D_s} + \frac{P^0}{D_s} \tag{3.58}$$

Substituting $P^0 = P e^{\lambda t}$ and rearranging,

$$\frac{D(t)}{D_s} = \frac{D^0}{D_s} + \frac{P(t)}{D_s}\left(e^{\lambda t} + 1\right) \tag{3.59}$$

Thus, if we plot a set of measurements of $D(t)/D_s$ vs. $P(t)/D_s$, we will get a straight line with the intercept $D^0/D_s$ and a slope of $(e^{\lambda t} - 1)$. Figure 3.12 shows such a plot

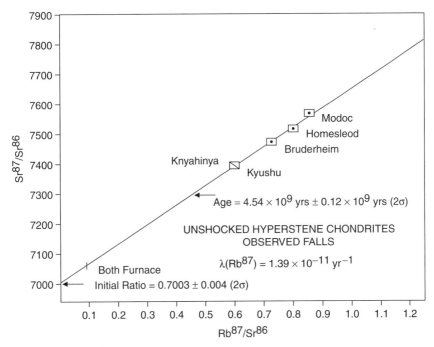

**Figure 3.12** $^{87}$Rb–$^{87}$Sr evolution diagram for six hypersthene chondrite meteorites. The data can be interpreted as showing that 4.54 billion years ago all of these rocks had the same $^{87}$Sr/$^{86}$Sr ratio of 0.7003 (from Wetherill, 1975).

of a set of meteorite samples using the $^{87}$Rb → $^{87}$Sr decay as a chronometer ($t_{1/2} = 4.75 \times 10^{10}$ y).

Other geochronometers that can be used in a similar manner involve the decay of $1.277 \times 10^9$ y $^{40}$K to $^{40}$Ar (K/Ar dating) or the decay of $^{235}$U or $^{238}$U to their $^{207}$Pb and $^{206}$Pb daughters. Each chronometer poses special problems with regard to the loss of daughter species over geologic time by diffusion, melting, or chemical processes. The "normalizing" stable nuclide in the case of the uranium decay series is $^{204}$Pb, and in the case of K/Ar dating it is $^{36}$Ar. (See Problems at end of chapter for a further discussion of these methods.)

The dating methods discussed up to now have been based on the use of long-lived radionuclides that are present in nature. Dating is also possible using "extinct radionuclides," that is, nuclei whose half-lives are so short that if they existed at the time of formation of our solar system, they would have decayed away essentially completely by now. The nuclides $^{129}$I ($t_{1/2} = 1.57 \times 10^7$ y) and $^{244}$Pu ($t_{1/2} = 8.08 \times 10^7$ y) are noteworthy examples of this type of nuclide.

The decay of extinct radionuclides is measured by measuring anomalies in the isotopic abundance of their stable daughters. For example, $^{129}$I decays to $^{129}$Xe and its decay will lead to an anomalously high concentration of $^{129}$Xe in the mass spectrum of Xe isotopes found in a rock system. What is dated is the "formation age" of the rock, that is, the time interval between the isolation of the solar system material from galactic nucleosynthesis and the time at which the rock cooled enough to retain its Xe. Formally, this formation age, $\Delta$, may be calculated as from the isotopic ratios in a fashion similar to that of Equation (3.59):

$$\Delta = \frac{1}{\lambda} \ln \frac{(^{129}\text{I}/^{127}\text{I})_0}{(^{129}\text{Xe}^*/^{127}\text{I})} \tag{3.60}$$

where $^{129}\text{Xe}^*$ is the excess Xe attributed to the decay of $^{129}\text{I}$, $^{127}\text{I}$ is the concentration of stable, nonradiogenic $^{127}\text{I}$, $\lambda$ is the decay constant for $^{129}\text{I}$, and $(^{129}\text{I}/^{127}\text{I})_0$ is the ratio of the abundance of the iodine isotopes at the time of isolation from galactic nucleosynthesis. This latter ratio is derived from theories of nucleosynthesis and is $\sim 10^{-4}$.

The decay of extinct $^{244}\text{Pu}$ is deduced from excess abundances of the nuclides $^{136}\text{Xe}$, $^{134}\text{Xe}$, and $^{132}\text{Xe}$, produced by the spontaneous fission of $^{244}\text{Pu}$. Uncertainties arise because there is no stable isotope of Pu that can be used in the way that $^{127}\text{I}$ is used in Equation (3.60) and the use of other heavy nuclides $^{238}\text{U}$ or $^{252}\text{Th}$ as "substitutes" leads to difficulties due to differences in primordial production and chemistry.

By far the most important dating method involves the decay of $^{14}\text{C}$ ($t_{1/2} = 5730$ y). As indicated previously, $^{14}\text{C}$ is formed continuously by the cosmic ray induced $^{14}\text{N}(n, p)$ $^{14}\text{C}$ reaction in the upper atmosphere. This radiocarbon ($^{14}\text{C}$) exchanges with stable carbon ($^{12}\text{C}$) in living things leading to the existence of a constant level of $^{14}\text{C}$ in living systems as indicated schematically in Figure 3.13. When an organism dies, it will cease to exchange its carbon atoms with the pool of radiocarbon and its radiocarbon will decay. Measurement of the specific activity $\left(\text{dpm} - {}^{14}\text{C}/\text{g}\,{}^{12}\text{C}\right)$ of an old object allows the determination of the age. When organic matter has decayed for 10 or more half-lives of $^{14}\text{C}$, it is no longer possible to directly measure the $^{14}\text{C}$ radioactivity of an object. In these cases, one can use *accelerator mass spectrometry* (AMS) to count the atoms of $^{14}\text{C}$ directly. An accelerator, such as a cyclotron or tandem Van de Graaff, is used as a mass spectrometer to separate the $^{14}\text{C}$ atoms from the more prevalent $^{12}\text{C}$ or $^{13}\text{C}$. Another difficulty is the separation of $^{14}\text{C}$ from the ubiquitous $^{14}\text{N}$ isobar and various molecular ions; thus, accelerators are used to provide energetic ions that can be identified with standard nuclear techniques. Using this technique, it has been possible to determine ages as long as 100,000 y.

**Example Problem** Consider a sample of organic material that contains 1 mg of C. Suppose it has a $^{14}\text{C}/^{12}\text{C}$ atom ratio of $1.2 \times 10^{-14}$.

a. How many $^{14}\text{C}$ atoms are present?
b. What would be the expected $^{14}\text{C}$ disintegration rate for this sample?
c. What is the age of this sample?

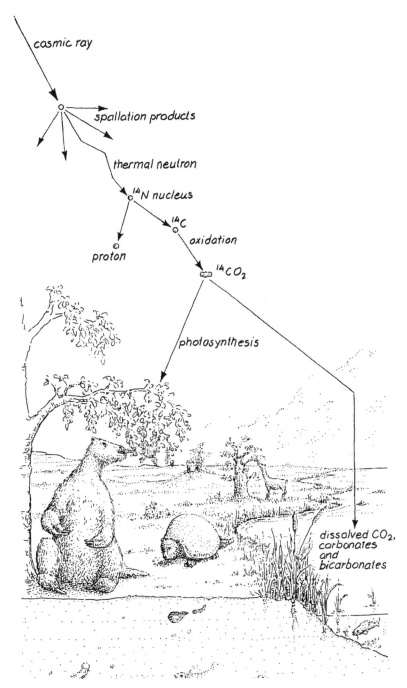

**Figure 3.13** Artist's conception of how $^{14}C$ is generated and incorporated into living things (reprinted by permission from Taylor, 2000).

*Solution*

a.
$$N = \frac{10^{-3}\,\text{g}}{12\,\text{g/g-atom}} \times 6.02 \times 10^{23}\,\text{atoms/g-atom} = 5.02 \times 10^{19}\,\text{atoms}$$

\# $^{14}$C atoms $\approx (1.2 \times 10^{-14})(5.02 \times 10^{19}) = 6.02 \times 10^{5}$

b.
$$A = \lambda N = \frac{\ln 2}{(5730\,\text{y})(3.15 \times 10^{7}\,\text{s/y})}(6.02 \times 10^{5})$$
$$= 2.3 \times 10^{-6}\,\text{Bq} = 0.2\,\text{dis/d}$$

c. Note that a typical AMS facility would collect several thousand of these $^{14}$C atoms in one hour. The assumed constant specific activity of $^{14}$C in nature in the prenuclear era is 227 Bq/kgC or $227 \times 10^{-6}$ Bq/mg. From Equation (3.52), the age would be

$$\text{Age} = \frac{\ln\left(227 \times 10^{-6}/2.3 \times 10^{-6}\right)}{\ln 2/5730\,\text{y}} = 38{,}000\,\text{y}$$

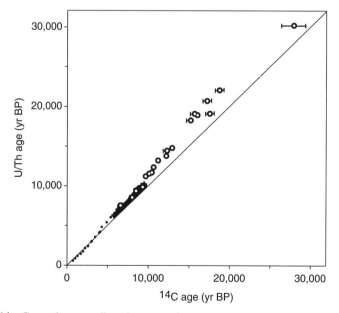

**Figure 3.14** Correction to radiocarbon ages for coral samples based upon age estimates made for the same samples using $^{234}$U/$^{230}$Th ages.

As noted earlier, the fundamental assumption in radiocarbon dating is that the specific activity of $^{14}C$ in nature $(\text{dpm}\ ^{14}C/\text{g}\ ^{12}C)$ is and has remained constant. This assumes the cosmic ray flux that generates the $^{14}C$ has been constant, and there are no sources of $^{14}C$ or $^{12}C$ that would change its equilibrium specific activity. Neither of these assumptions is strictly true, and corrections must be used to obtain correct ages from radiocarbon dating. In Figure 3.14, we show the typical magnitude of these corrections. The primary cosmic ray flux is moderated by fluctuations in solar activity or the Earth's magnetic field over time. Since the Industrial Revolution, the global carbon cycle is out of balance due to fossil fuel burning (of "old" nonactive fossil carbon). This has caused a 1–3% dilution of the pre-historic $^{14}C/^{12}C$ ratio. As noted earlier, atmospheric testing of nuclear weapons contributed a spike to the global $^{14}C$ inventory that perturbed the $^{14}C/^{12}C$ ratio by a factor of 2 in the opposite direction. Continued operation of nuclear power plants also contributes an amount that is $\sim 10\%$ of the "natural" $^{14}C$ production rate. Similarly, a dating scheme for water containing objects, such as wines, based upon the equilibrium production of tritium ($^3H$) and its decay has been similarly perturbed by an injection of thousands of times the natural levels due to atmospheric testing.

## PROBLEMS

1. Calculate the expected activity in Bq and in Ci for the following radionuclides (see Appendix for nuclear data):

    a. 1.0 g $^{239}Pu$
    b. 1.0 g $^{14}C$
    c. 1.0 g $^{137}Cs$
    d. spontaneous fission activity for 1.0 g $^{252}Cf$
    e. 1 g $^{226}Ra$

2. Consider the decay sequence $^{239}U \rightarrow\ ^{239}Np \rightarrow\ ^{239}Pu \rightarrow$ . If you start with 1 mCi of initially pure $^{239}U$, what is the activity of $^{239}Pu$ after (a) 1 day, (b) 1 month, and (c) 1 year?

3. Calculate the time necessary to reduce the activities of the following nuclei to 1% of their initial values:

    a. $^{131}I$   b. $^3H$   c. $^{137}Cs$   d. $^{14}C$   e. $^{239}Pu$

4. What is the mass (g) of the following activities:

    a. 1 µCi $^{241}Am$
    b. 1 pCi $^{239}Pu$
    c. 5000 Bq $^{252}Cf$

5. What is the partial half-life for decay by spontaneous fission for $^{252}Cf$?

6. If $^{222}$Rn is initially purified from its daughters, how long does it take for them to grow back to 50% of their values at secular equilibrium?

7. What are the partial half-lives of $^{22}$Na for decay by (a) EC and (b) $\beta^+$ emission?

8. Calculate the relative mass ratios of $^{238}$U, $^{226}$Ra, and $^{222}$Rn in an old uranium ore.

9. Consider the decay of $^{140}$Ba to $^{140}$La. At what time does the $^{140}$La activity reach a maximum?

10. Consider a reactor in which the production rate of $^{239}$U via the $^{238}$U (n, $\gamma$) $^{239}$U reaction is $10^5$ atoms/s. Calculate the activity of $^{239}$Pu after an irradiation of (a) 1 day, (b) 1 month, and (c) 1 year.

11. What is the probability of a $^{222}$Rn atom decaying in our lungs? The atmospheric concentration of $^{222}$Rn may be assumed to be 1 pCi/L. In an average breath we inhale 0.5 L of air and exhale it 3.5 s later.

12. Consider a radionuclide (decay constant $\lambda$) with activity $A$ Bq at time $t_1$. Calculate the number of nuclei that decay between times $t_1$ and $t_2$.

13. Consider the following decay scheme (Evans, 1955):

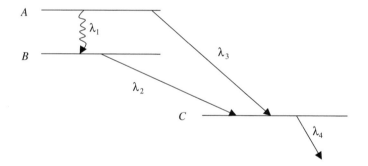

   a. Derive expressions for the activity of $B$ and $C$ as a function of time if at $t = 0$, $A = A$, $B = C = D = 0$.
   b. What happens when the cross over transition $\lambda_3 = 0$?

14. If one "milks" a sample of $^{99}$Mo to remove the daughter nuclide $^{99}$Tc, how long does it take before the $^{99}$Mo "cow" has an equilibrium amount of $^{99}$Tc present?

15. Consider the case where $A \rightarrow B \rightarrow C$ and where $\lambda_A = \lambda_B$. Derive an expression for the activity of $B$ as a function of time. Calculate the time $t_m$ when the activity of $B$ reaches a maximum. Show that $t_m \approx (\tau_A \tau_B)^{1/2}$ where $\tau_A$, $\tau_B$ are the mean lives for $A$ and $B$.

16. A uranium mineral was found to contain the Pb isotopes $^{204}$Pb, $^{206}$Pb, and $^{207}$Pb in the ratio of 1 : 1000 : 400. Estimate the age of this mineral (Choppin et al., 1995).

17. What was the rate of production of $^{24}$Na in a 30-m reactor irradiation of $^{23}$Na if the activity of $^{24}$Na was found to be 1.0 μCi 3 h after the end of irradiation?

18. Calculate the heat generated per kilogram of natural uranium by the $^{238}$U and the $^{235}$U in secular equilibrium with their decay products. Assume all emitted radiation is absorbed.

19. Given the following data:

| Sample | Rb/Sr Weight Ratio | $^{87}$Sr/$^{86}$Sr Atom Ratio |
|---|---|---|
| 1 | 1.06 | 0.7597 |
| 2 | 3.51 | 0.8248 |
| 3 | 6.61 | 0.9085 |
| 4 | 9.33 | 0.9796 |
| 5 | 10.67 | 1.0200 |

Determine the age of the rock and the initial $^{87}$Sr/$^{86}$Sr ratio.

## REFERENCES

Choppin, G., J. Rydberg, and J. O. Liljenzin. *Radiochemistry and Nuclear Chemistry*, Butterworth, Oxford, 1995.

Evans, R. D. *The Atomic Nucleus*, McGraw-Hill, New York, 1955.

Taylor, R. E. Fifty Years of Radiocarbon Dating, *Am. Scient.* **88**, 60 (2000).

Wetherill, G. W. Radiometric Chronolgy of the Early Solar System, *Ann. Rev. Nucl. Sci.* **25**, 283 (1975).

## BIBLIOGRAPHY

Arnikar, H. J. *Essentials of Nuclear Chemistry*, 2nd ed., Wiley, New York, 1982.

Ehmann, W. D. and D. E. Vance. *Radiochemistry and Nuclear Methods of Analysis*, Wiley, New York, 1991.

Friedlander, G., J. W. Kennedy, E. S. Macias, and J. M. Miller. *Nuclear and Radiochemistry*, Wiley, New York, 1981.

Heyde, K. *Basic Ideas and Concepts in Nuclear Physics*, IOP, Bristol, 1994.

Krane, K. S. *Introductory Nuclear Physics*, Wiley, New York, 1988.

### Monographs and Specialized Articles

Eisenbud, M. *Environmental Radioactivity*, 3rd ed., Academic, Orlando, 1987.

Lal, D. and H. Suess. The Radioactivity of the Atmosphere and Hydrosphere, *Ann. Rev. Nucl. Sci.* **18**, 407 (1968).

UNSCEAR 93, *Sources and Effects of Ionizing Radiation*, UN, New York, 1993.

# CHAPTER 4

# RADIOTRACERS

## 4.1 INTRODUCTION

The basic idea behind the use of radiotracers is that all the isotopes of a given element will behave the same chemically. Thus, the atoms of radioactive $^{24}$Na will behave in the same way as those of stable $^{23}$Na in a chemical system except for the effects due to the small difference in mass. So what? The point is that it is easier to follow the radioactive $^{24}$Na atoms than it is to detect the nonradioactive $^{23}$Na atoms. Using conventional chemical analysis techniques, one can typically detect nanogram to microgram amounts of a substance. For $^{23}$Na, these amounts would correspond to needing $10^{13}$–$10^{16}$ atoms to get an analytical response. For a radioactive atom, such as $^{24}$Na, one detected disintegration corresponds to the decay of a single atom. Routine radioanalytical techniques will allow detection of quantities that are $10^5$ times smaller than those needed for chemical analysis. This high sensitivity is of great importance in any number of studies. For example, the specific activity of pure tritium is $\sim 30$ Ci/mmol. Thus, one can tolerate a dilution factor of $10^{12}$ and still detect tritium-labeled compounds. It is thus possible to detect the occurrence of metabolic substances that are normally present at such low concentrations to defy the most sensitive chemical methods of identification.

One can measure the activity of enzymes by following the rate of disappearance of a labeled substrate or the rate of appearance of a labeled product. Biological compounds, such as vitamins or hormones, which are normally in such low

---

*Modern Nuclear Chemistry*, by W.D. Loveland, D.J. Morrissey, and G.T. Seaborg
Copyright © 2006 John Wiley & Sons, Inc.

concentrations as to make detection difficult, can be measured using radiotracer techniques such as radioimmunoassay (RIA).

The unique advantages of radiotracer experiments include their high sensitivity, their simplicity, and small expense (compared to competing technologies such as mass spectrometry). In a well-designed experiment, the presence of radiotracers does not affect the system under study and any analysis is nondestructive. Interference from other species that may be present is not important (as compared to conventional methods of analysis where interferences may thwart the analysis).

Perhaps the most outstanding advantage of the use of radioisotopes is the opportunity offered to *trace dynamic mechanisms*. Such biological phenomena as ion transport across cell membranes, turnover, intermediary metabolism, or translocation in plants could, before the advent of radiotracer methods, be approached only indirectly.

The use of the *isotope effect* to study *rate-determining steps* in a sequence of chemical reactions represents an additional advantage of radiotracer methodology. The term *isotope effect* (to be discussed more fully later) refers to the influence on a reaction rate of the difference in the masses of isotopes. This effect may create significant problems in the use of radioisotopes as tracers but can, nevertheless, be used to advantage in a limited number of cases in order to understand the kinetics of certain chemical reactions.

## 4.2 DESIGN OF A RADIOTRACER EXPERIMENT

### 4.2.1 Basic Design Criteria

The use of radiotracers is dependent on certain basic assumptions being fulfilled. *The first assumption, mentioned above, is that the radioactive isotopes of a given element behave identically as the stable isotopes of the same element.* Actually, this assumption is not exactly true. The difference in masses between radiotracer nuclei and stable nuclei can cause a shift in the reaction rate or equilibria (the isotope effect). It is true, however, that in most cases the isotope effect does not significantly affect the utility of the radioisotope method. Since the degree of chemical bond stability due to vibrational motion is directly related to the square root of the masses of the isotopes involved, it is apparent that an isotope effect will be of significance only for elements of low atomic weight (at wt $\leq 25$).

The isotopes of hydrogen present the extreme case. Thus, $^1$H, $^2$H(D), and $^3$H(T) could scarcely be expected to act as the same substances chemically since the relative mass differences are so great. Hence, *tritium* cannot be employed uncritically as a tracer for hydrogen in regard to reaction rates, although, of course, its use in determining hydrogen location in an organism is not precluded.

In the case of radioactive $^{14}$C and stable $^{12}$C, there is a $\sim 15\%$ difference in mass that may affect the outcome of some studies. In general, one should note that the so-called isotopic effect, in reality, should be considered from the standpoint of the two basic types: the *intramolecular* and the *intermolecular* isotopic effects.

In regard to the first type, the decarboxylation (loss of $CO_2$ from the carbonyl group) of malonic acid (HOO$^{14}$C—$^{12}$CH$_2$—$^{12}$COOH) is a good example. Here the pyrolytic decarboxylation of malonic acid can occur at either end of the molecule and gives rise to $CO_2$, and acetic acid. Consequently, the reaction is concerned with the relative bond strength of $^{12}$C—$^{14}$C and $^{12}$C—$^{12}$C. Since the former bond is relatively more stable (owing to the greater mass of the $^{14}$C), we would expect that the $CO_2$ resulting from the reaction would be comparatively enriched in $^{12}$C and that the acetic acid would be comparatively enriched in $^{14}$C from the —COOH. Such is the case. Inasmuch as the competitive reactions occur within the same molecule, the isotopic effect will be observed even if the reaction goes to completion.

In the case of the *intermolecular* isotope effect, the decarboxylation of benzoic acid–7–$^{14}$C can be cited as an example. Here, one is usually dealing with the decarboxylation of two compounds: $C_6H_5$—$^{14}$COOH and $C_6H_5$—$^{12}$COOH. Again, since the $^{12}$C—$^{14}$C bond is relatively more stable, we would expect that during the initial phase of the decarboxylation reaction there would be an enrichment of $^{12}CO_2$ compared to the specific activity of the labeled carbon in the benzoic acid. As the reaction approaches completion, however, the reactant (benzoic acid) will be relatively enriched with the $^{14}$C-labeled variety, and, consequently, the decarboxylation product ($CO_2$) will have a specific activity higher than the labeled carbon of the original starting material. When the reaction is driven to completion, the overall specific activity of $CO_2$ will naturally be the same as the labeled carbon atom in the benzoic acid; that is, no isotope effect can be observed. This is an important concept for it indicates that the significance of the isotope effect can be minimized or ignored in radiotracer studies if the intramolecular type of reaction is not involved.

*The second basic assumption is that the radioactivity does not change the chemical and physical properties of the experimental system.* It is essential that the radiation dose from the tracer does not elicit a response from the experimental system that would distort the experimental results. The amount of activity employed should be restricted to the minimum necessary to permit reasonable counting rates in the samples to be assayed. The possibility of excessive concentration of the tracer compound in certain biological tissues and the degree of radiosensitivity of these tissues must also be carefully considered, especially when $\alpha$- or $\beta^-$-emitting tracers are used. Fortunately, the excellent sensitivity of most radioactivity assay methods minimizes the need to employ tracer doses of such a magnitude that any detectable radiation damage occurs. The possibilities of interference due to physiological response to radiation are further minimized because most studies are short term and thus completed before any latent radiation effects appear. In physical tracer studies, the radiation damage produced by the decaying nuclei must not so destroy the crystal structure and similar features as to modify the experimental results.

In addition, one must remember that as the result of radioactive decay, the daughter atom is usually a different chemical element than the mother atom. One must be sure that the presence of these "foreign species" and any related equilibria does not affect the observations. (For example, if both the mother and daughter atoms are $\beta^-$ emitters, then one might see a increase in the count rate due to the inability to distinguish the $\beta^-$ particles from the tracer and its daughter.)

*A third basic assumption, for biological studies, is that there is no deviation from the normal physiological state.* If, to produce the required tracer activity, the chemical level of the compound given to an organism greatly exceeds the normal physiological or chemical level, the experimental results are open to question. The specific activity of the tracer compound must be high enough for the total chemical level to be administered to be within the normal range. As an example, $^{36}$Cl might be quite useful for biological investigations, but the maximum specific activity obtainable in the inorganic form is about 100 μCi/g Cl. This situation is in contrast to the specific activities of $^{14}$C (as Ba$^{14}$CO$_3$) of up to 2.2 Ci/g of carbon.

*A fourth basic assumption is that the chemical and physical form of the radionuclide-labeled compound is the same as the unlabeled variety.* Herein we deal with certain subtle effects related to the low concentration of the tracer species. What are these effects? One involves the question of adsorption on surfaces, such as the walls of the container for the system. For example, it is easy to show that to cover the surface of a 1-L flask may require $\sim 10^{-7}$ mole. For a 1-L solution containing 1 μCi of a radioactive species with a half-life of 30 days, one quickly calculates the number of moles present is $\sim 10^{-13}$, and, thus, one could think that all the radioactive atoms might be adsorbed on the container walls (if no other atoms were present). This adsorption can be prevented by adding (or already having present) enough molecules of the nonradioactive compound to fill all the adsorption sites, that is, a *carrier*. Similarly, the presence of a significantly larger concentration of nonradioactive species is necessary to prevent the tracer from co-precipitating or being adsorbed on the surface of any precipitates that form. (Because the number of tracer atoms is frequently too small to satisfy the $K_{sp}$ for any precipitate, the tracer can only be precipitated in the presence of a large number of nonradioactive species.)

Most importantly, the radionuclide and the stable nuclide must undergo isotopic exchange. In practice, this means that the tracer and the stable atom must be in the same redox state. By heating or using redox cycles, the experimenter must assure this to be true. Anomalous experimental results have frequently been traceable to the chemical form of the administered radiotracer. Since reactor production of radionuclides often results in side reactions (see Chapter 10), various oxidation states may be present when the sample is produced. In one case involving phosphate–$^{32}$P uptake in plants, the unexpected experimental results were explained by the fact that a large percentage of the tracer dose was actually in the form of phosphite–$^{32}$P.

The *radiochemical purity* of a compound cannot be assumed. The presence of other radioactive species in low chemical concentration but high specific activity is frequently encountered. This situation is particularly true in the labeling of compounds with $^3$H by means of the Wilzbach direct-exposure method (discussed below). Thus, for example, direct hydrogenation of a double bond with $^3$H during the Wilzbach operation may give rise to a small amount of impurity (saturated form of the compound in question) having a specific activity many times higher than the $^3$H-labeled compound derived from the recoil-labeling operation.

The problem of radiochemical purity with respect to the chemical state of aged tritium or $^{14}$C-labeled compounds is still more acute. Because of the short range of the low-energy β particles associated with these two isotopes, the sizable radiation dose delivered to such compounds by their own radiation leads to self-decomposition *(radiolysis)* and hence a variable concentration and a number of labeled products.

*The fifth basic assumption is that only the labeled atoms are traced.* Never assume that the appearance of the radioactive label in a given sample indicates the presence of the administered compound. It is the labeled atoms that are being followed, not the intact compound. Not only may metabolic reactions involve the cleavage of the labeled atom from the original compound, but exchange reactions may also occur, thus removing labile atoms from the labeled compound. Such chemical exchanges particularly plague many experiments with tritium-labeled compounds. The extent of chemical exchange is strongly dependent on the molecular species involved, the position of the label in the molecule, and the environmental factors (such as the pH of a biological fluid).

### 4.2.2 Practical Considerations

The feasibility of radiotracer experiments is usually dependent on certain practical matters. These factors include:

***Availability of the Radiotracer***  A primary factor is whether a radioisotope of the element to be traced is available with the proper characteristics (half-life, particle energy, etc.). For example, although radioisotopes of oxygen and nitrogen would be highly desirable in many investigations, the longest-lived radionuclides available of these elements have half-lives of 2 and 10 min, respectively. Clearly, such short half-lives severely limit the use of such isotopes for many tracer experiments. On the other hand, for some elements, a choice of usable radioisotopes may be available such as $^{22}$Na or $^{24}$Na and $^{57}$Co or $^{60}$Co. Of equal importance is the available specific activity of a given radionuclide. There are radionuclides, such as $^{36}$Cl, that cannot readily be made with desirable specific activity. Ideally, the radiotracer should have a half-life that is several times the duration of the experiment to lessen or avoid corrections for decay, but short enough to not cause long-term contamination or disposal issues. If possible, the emitted radiation should be relatively easy to detect. A second factor is whether the tagged compound desired is commercially available or can be easily synthesized. The number of labeled compounds available is large indeed, and most radiochemical suppliers will attempt custom syntheses of unstocked compounds. In some cases, however, it is not economically feasible or even possible to introduce a given radioactive atom into the molecular structure under consideration. Furthermore, the available specific activity may be too low for the proposed experimental use of the tagged compound.

***Calculation of the Amount of Tracer Needed for the Experiment***  Radiotracer experiments should, in general, involve easily detected quantities of radioactivity. There is little excuse for doing a tracer experiment where the result is uncertain due to a lack of observed counts or one that requires the ultimate in low-level counting systems. As an example of how such calculations are done, let us consider the laboratory exercise to determine the solubility product constant of silver iodide.

To begin with, we must have some idea of the basic chemistry/science that we are attempting to measure. Let us assume we do not know the $K_{sp}$ of AgI but do know the $K_{sp}$ for the analogous compounds, AgCl and AgBr. Thus, we have

$$AgCl \longleftrightarrow Ag^+ + Cl^-$$
$$AgBr \longleftrightarrow Ag^+ + Br^-$$

with $K_{sp} = 1.8 \times 10^{-10}$ and $5.4 \times 10^{-13}$, respectively. From this we will guesstimate the $K_{sp}$ of AgI as $10^{-15}$. (The actual value is $8.5 \times 10^{-17}$, but that sort of error in experimental design must be tolerated.) From this, we can calculate the solubility of silver iodide, $S$, as

$$S = (K_{sp})^{1/2} = (10^{-15})^{1/2} = 3 \times 10^{-8} \, M$$

What we will do in the experiment is to take a solution containing labeled $I^-$ and precipitate the AgI from the solution. We will measure the activity of the AgI precipitate, suspend it in a known volume of water, and measure the activity of the water. From this measurement, we can calculate the solubility of AgI and thus $K_{sp}$.

What iodine radiotracer should we use? Considering that we will have to assay the activity of a liquid under a suitable counting geometry, we need a tracer that emits energetic photons (to minimize absorption corrections). This consideration as well as the others outlined above, causes us to choose $^{131}$I ($t_{1/2} = 8$ d) as the tracer. (Solutions of $^{131}$I-labeled sodium iodide are available commercially.)

Let us assume we precipitate a convenient amount of AgI for counting and handling, say 20 mg. The fraction iodine in this precipitate is 54% ≈ 10.8 mg. Upon equilibrating this precipitate with water (2 mL), one will have $(0.002)(3 \times 10^{-8}) = 6 \times 10^{-11}$ mole of iodine in solution or $7.6 \times 10^{-9}$ g.

Assume we count this water solution using a NaI well detector (see Chapter 18) that has an efficiency of ~20% for the 0.365-MeV photons from $^{131}$I. Assuming one wants to collect $10^3$ counts in a 10-min count (and 81% of the $^{131}$I decays result in a 0.365-MeV photon), one estimates a radioactivity in the solution as

$$D = (\text{counts/time})(\text{branching ratio})(\text{detection efficiency})$$
$$= (10^3/10)(1/0.81)(1/0.2) \approx 600 \, \text{dpm} \sim 10 \, \text{Bq}$$

The fraction of the iodine in the precipitate that dissolved is $(7.6 \times 10^{-9}/10.8 \times 10^{-3}) = 7 \times 10^{-7}$. Therefore the activity of the precipitate must be ~400 µCi ~ 15 MBq. Since the initial precipitation of AgI is quantitative, this is the nominal amount of tracer needed in the solution. However, it is common to build in a "safety factor" of 5–10 in the amount of tracer used (to account for the misestimates such as we made in the $K_{sp}$ of AgI).

The exact amount of $^{131}$I in the original precipitate must be known, but the activity of this precipitate is so high as to preclude its direct measurement in the well detector. Therefore, what one does is to prepare a standard dilution of the original I$^-$ solution that can be measured in the well detector.

**Evaluation of Hazard** The first item to be considered is the possibility of harm to the experimenter or to co-workers. In the great majority of radiotracer experiments, the hazard from direct external radiation does not pose a serious problem. However, there are situations where such is not the case, for example, where high levels (millicuries) of γ-ray emitters are utilized. For instance, 10 mCi of $^{24}$Na will deliver a dose of about 204 milliroentgens per hour (mR/h) (at a 1-ft distance). One should also be quite cautious about the radiation dose delivered to the hands and fingers while handling radioactive materials. Another item of concern in the use of α- or β$^-$-emitting tracers is the possibility of ingestion of the labeled compounds, particularly those known to have a long turnover time in the human body. This problem is made acute where the sample is in the form of an aerosol or a dry powder at some stage of the experiment.

Radiation damage to a biological system under study may occur at two levels: the physiological and the histological. In general, higher radiation doses are required to elicit the latter type of damage. Whenever it is suspected that radiation damage is influencing the physiological response of the organism, it is advisable to repeat the experiment with lower levels of radioactivity, while maintaining the same total chemical level of the administered compound. Biological effects of radiation from radiotracer doses have been reported to occur at the following dose levels: 0.045 μCi $^{131}$I/g of body weight in mice, 0.8 μCi $^{32}$P/g of body weight in mice, 47 μCi $^{24}$Na/g body weight in mice and rats, 0.5 μCi $^{89}$Sr/g body weight in mice and rats, 0.05 μCi $^{32}$P/mL of rearing solution for mosquito larvae, and 2 μCi $^{32}$P/L of nutrient solution for barley plants.

Attention must also be given to the disposal of *radioactive wastes* resulting from the experiment, such as excreta, carcasses, or large volumes of solutions. The possible method of disposal will depend on the specific radioisotope present, its concentration and activity, and the nature of the waste.

## 4.3 PREPARATION OF RADIOTRACERS AND THEIR COMPOUNDS

There are several hundred radionuclides that have been used as radiotracers. A partial list of the properties of these nuclides and their production methods are shown in Table 4.1. The three common production mechanisms for the primary radionuclides are (n,γ) or (n,p) or (n,α) reactions in a nuclear reactor (R), charged-particle-induced reactions usually involving the use of a cyclotron (C), and fission product nuclei (F), typically obtained by chemical separation from irradiated uranium. The neutron-rich nuclei are generally made using reactors or

**TABLE 4.1 Commonly Used Tracers**

| Nuclide | Method of Production[a] | Half-Life | Tracer Radiations/Energy (MeV) |
|---|---|---|---|
| $^3$H(T) | R | 12.33 y | $\beta^-$ 0.018 |
| $^{14}$C | R | 5730 y | $\beta^-$ 0.156 |
| $^{22}$Na | C | 2.60 y | $\beta^+$, $\gamma$ 1.274 |
| $^{24}$Na | R | 15.0 h | $\gamma$ 1.369 |
| $^{32}$P | R | 14.3 d | $\beta^-$ 1.71 |
| $^{33}$P | R | 25.3 d | $\beta^-$ 0.249 |
| $^{35}$S | R | 87.4 d | $\beta^-$ 0.167 |
| $^{36}$Cl | R | $3.0 \times 10^5$ y | $\beta^-$ 0.71 |
| $^{45}$Ca | R | 162.6 d | $\beta^-$ 0.257 |
| $^{47}$Ca | R | 4.54 d | $\beta^-$ 1.99; $\gamma$ 1.297 |
| $^{51}$Cr | R | 27.7 d | $\gamma$ 0.320 |
| $^{54}$Mn | R | 312 d | $\gamma$ 0.835 |
| $^{55}$Fe | R | 2.73 y | EC |
| $^{59}$Fe | R | 44.5 d | $\gamma$ 1.292, 1.099 |
| $^{57}$Co | C | 271.7 d | $\gamma$ 0.122 |
| $^{60}$Co | R | 5.27 y | $\gamma$ 1.173, 1.332 |
| $^{63}$Ni | R | 100.1 y | $\beta^-$ 0.067 |
| $^{65}$Zn | C, R | 244.3 d | $\gamma$ 1.116 |
| $^{75}$Se | R | 119.8 d | $\gamma$ 0.265, 0.136 |
| $^{86}$Rb | R | 18.6 d | $\beta^-$ 1.77 |
| $^{85}$Sr | R, C | 64.8 d | $\gamma$ 0.514 |
| $^{99}$Mo/$^{99}$Tc$^m$ | F | 65.9 h/6.01 h | $\gamma$ 0.143 |
| $^{106}$Ru | F | 373.6 d | $\beta^-$ 0.039 |
| $^{110}$Ag$^m$ | R | 249.8 d | $\beta^-$ 3.0 |
| $^{109}$Cd | C | 461 d | $\gamma$ 0.088 |
| $^{111}$In | C | 2.80 d | $\gamma$ 0.171 |
| $^{125}$I | R | 59.4 d | $\gamma$ 0.035 |
| $^{131}$I | R | 8.02 d | $\beta^-$ 0.606, $\gamma$ 0.365 |
| $^{137}$Cs | F | 30.1 y | $\gamma$ 0.662 |
| $^{153}$Gd | R | 240.4 d | $\gamma$ 0.103 |
| $^{201}$Tl | C | 72.9 h | $\gamma$ 0.167 |
| $^{210}$Pb | R | 22.3 y | $\beta^-$ 0.017, 0.064 |

[a] R, reactor; C, cyclotron; F, fission product.

as fission products, while the proton-rich nuclei are produced in cyclotrons. [Not shown in Table 4.1 are the short-lived positron emitters, $^{11}$C, $^{13}$N, $^{15}$O, and $^{19}$F commonly used in positron emission tomography (PET), that are produced in cyclotrons.]

In certain experiments, the primary radionuclides may be used directly, but usually the investigator wants to secure a specific labeled compound for use in radiotracer experiments. Before considering the details of the production of these labeled compounds, let us discuss the nomenclature and rules used in referring to them.

1. The position of a single labeled atom in a molecule is shown following the chemical name of the compound. Thus, acetic–1–$^{14}$C acid is CH$_3^{14}$COOH, whereas acetic–2–$^{14}$C acid is $^{14}$CH$_3$COOH.
2. Certain terms are used to indicate the distribution of material with more than one labeled atom. These terms and their meanings are as follows:
   a. *Specifically labeled.* Chemicals are designated as specifically labeled when all labeled positions are included in the name of the compound and 95% or more of the radioactivity of the compound is at these positions. Thus, specifically labeled aldosterone-1, 2-$^3$H implies that $\geq 95\%$ of the tritium label is in the 1 and 2 positions.
   b. *Uniformly labeled* (*U*). Uniformly labeled compounds are labeled in all positions in a uniform or nearly uniform pattern. Thus, L-valine-$^{14}$C (U) implies that all of the carbon atoms in L-valine are labeled with approximately equal amounts of $^{14}$C.
   c. *Nominally labeled* (*N*). This designation means that some part of the label is at a specific position in the material, but no further information is available as to the extent of labeling at other positions. Thus, cholestrol-7-$^3$H (N) implies that some tritium is at position 7, but it may also be at other positions in the molecule.
   d. *Generally labeled* (*G*). This designation is for compounds (usually tritium labeled) in which there is a random distribution of labeled atoms in the molecule. Not all positions in a molecule are necessarily labeled.

Since the greatest number of labeled compounds are $^{14}$C labeled, our discussion on the preparation of labeled compounds deals mainly with them. However, many of the general principles can be applied to other nuclides and molecules as well.

## 4.3.1 Chemical Synthesis

A $^{14}$C label may be introduced into a wide variety of compounds by the standard synthetic procedures of organic chemistry. In addition, some new methods have been devised to conserve the radionuclides being used. When chemical synthesis is at all possible, *it is usually the method of choice*. Synthetic methods give the greatest control over yield, position of the label, and purity of the product. For all syntheses involving $^{14}$C, Ba$^{14}$CO$_3$ is usually taken as the starting material. Quite often this is converted to $^{14}$CO$_2$ for the synthesis.

Chemical synthesis of labeled compounds suffers from some limitations and problems, though. One limitation concerns the amount and cost of the radioactive starting material. This factor necessitates devising synthetic routes to the desired compounds in which the radiolabel can be introduced near the end of the sequence of reactions, so as to secure as high an overall yield of labeled material as possible. At present, numerous labeled compounds are available commercially as starting materials for syntheses. Still, in planning a new synthetic route, it is necessary to consider its compatibility with the specific starting material available.

Another disadvantage of chemical synthesis is that when it is used to produce certain biologically important compounds, such as amino acids, a racemic mixture of D and L isomers results. Since organisms, by and large, metabolize the L-form selectively, as in the case of amino acids, the use of such racemates in biological investigations is somewhat unphysical and may lead to undesirable confusion. Methods for the resolution of racemic mixtures are available. Most of these are tedious and not suited for small-scale operation.

### 4.3.2  Biosynthesis

Living organisms, or active enzyme preparations, offer a biochemical means of synthesizing certain labeled compounds that are not available by chemical synthesis. These include both the macromolecules (proteins, polysaccharides, nucleic acids, etc.) and many simpler molecules (vitamins, hormones, amino acids, and sugars). The successful use of biosynthesis for the production of a given labeled compound depends on several factors. First, an organism must be selected that will synthesize and accumulate practical quantities of the desired compound. Culture conditions must be established so as to provide optimal yields of high specific activity. Last and most important, you must plan to isolate and purify the labeled compound, as well as determining the distribution pattern of the label, if a specific labeling is desired.

*Photosynthetic methods* offer the advantage of using the relatively cheap $^{14}CO_2$ (from $Ba^{14}CO_3$) as the starting material. Carbon-14-labeled starch, glucose, fructose, and sucrose can be isolated in good yields from green leaves or algal suspensions that have been exposed to $^{14}CO_2$ and illuminated for a prolonged period.

*Microorganisms* or enzyme systems prepared from them have been used to produce organic acids labeled with $^{14}C$, either by direct synthesis or transformation of labeled substrates. Several species of microorganisms have been used to produce higher fatty acids by condensation.

In general, biosynthetic procedures are likely to be laborious and limited to small-scale operations. One often encounters purification problems when attempting to isolate specific biological compounds in a typical system.

### 4.3.3  Tritium Labeling

Compounds may be labeled with tritium by several methods. The classic synthetic methods utilizing labeled intermediates have the advantage of yielding products that have predictable specific activities, are specifically labeled, and have a minimum of aged by-products. Among the methods of tritium labeling are:

1. *By Reduction of Unsaturated Precursors*  The method of choice for labeling with tritium is the reduction of a suitable unsaturated precursor (containing a double bond, carbonyl group, etc.) with carrier-free tritium gas or tritiated metal hydrides. The major limitation of this method is the availability of a suitable unsaturated precursor of the desired compound. It is essential to carry out the synthesis in a nonhydroxylic solvent (dioxane, ethyl acetate, etc.). Reductions carried out in alcohol or water will lead to almost complete exchange of the tritium gas with the solvent.

2. *By Exchange Reactions* Random tritium labeling may be secured by simple exchange methods, with or without catalytic action. Although high specific activities may be obtained by this method, some of the introduced tritium may be labile. Removal of this labile tritium and purification of the product are necessary.

3. *By Gas Exposure* In the mid-1950s, Wolfgang and Rowland described tritium recoil labeling of organic compounds. Wilzbach, in 1957, first described the simplified approach to random labeling with tritium that has come to be called the *Wilzbach gas exposure method.* In this method, the compound to be labeled is exposed to Curie amounts of carrier-free tritium gas in a sealed reaction vessel for a period of a few days to several weeks. The energy released in the disintegration of tritium and absorbed by the system provides the activation energy necessary to effect labeling. Compounds labeled by the Wilzbach method are "generally labeled (G)." Specific activities of $1-125$ mCi/g of purified compound have been reported.

Unfortunately, Wilzbach labeling is often accompanied by the formation of tritiated by-products of high specific activity. As in the case of exchange labeling, a considerable portion of the tritium in the labeled compound is often labile. This formation of labeled by-products is the major problem of the gas exposure method since the specific activity of the by-products may be several orders of magnitude greater than the desired compound. *In general, because of the magnitude of the purification procedures required and the random nature of the labeling, we suggest that all other synthetic routes be explored before the gas exposure method is chosen.*

### 4.3.4 Radiolysis of Labeled Compounds

In many situations, the experimenter will prefer to buy labeled compounds from commercial suppliers rather than attempt to synthesize them. The radiochemical purity of such purchased compounds cannot be assumed. Radiation-induced self-decomposition (radiolysis) can result in the formation of a variety of labeled degradation products, which must be removed before experimental use of the compounds. The extent of radiolysis depends on the nature of the labeled compound, how long it has been stored, and the manner of storage. Radiolysis is most significant with low-energy $\beta^-$ emitters (especially tritium) since the decay energy is dissipated almost entirely with the compound itself. Furthermore, impurities involving other radionuclides may be present.

## 4.4 TRACING OF PHYSICAL PROCESS

In many cases in which radiotracers are used, the chemical identity of the tracer is not important. These applications can be referred to as *tracing physical processes.* For example, consider those experiments that seek to *locate an object in some system* by labeling it with radioactivity and then measuring the position of the radioactivity in the system. Quite often a tracer that decays by $\gamma$-ray emission is selected so that the radiation from the source will penetrate large masses of tissue, pipe, earth, and the like. *Mixing studies* are frequently carried out using radiotracers. Here the objective is usually to see if proper mixing has taken place between two components of a system. Generally, one of the components is labeled with a radiotracer and its distribution in the system is monitored as a function of time. A short-lived tracer is often used so that it can quickly decay away at the conclusion of the experiment and leave an essentially "nonradioactive" mixed system for further use.

A form of isotope dilution (see below) is frequently used to measure the *volume of an inaccessible container*. A small volume $V_1$ of tracer solution is assayed to give its activity $A$. The tracer is added to the liquid in the container and mixed, and a sample of size $V_1$ is removed and assayed to show its activity $A_2$. The volume of the original container is given as

$$V = V_1 \left( \frac{A_1}{A_2} - 1 \right) \tag{4.1}$$

*Leak testing* can also be done using radiotracers. Here the basic idea is simple—namely, to inject radiotracer into a pipe, flask, or whatever is suspected to be leaking and look for activity that appears outside the container. Suitable caution must be exercised, of course, to be sure that the "leaked radioactivity" is not a hazard.

For studies involving water, it may only be important to assure that the tracer remains fluid-bound. One can use tritium or almost any metal atom that can be complexed with EDTA (ethylenediaminetetraacetic acid) or DTPA (diethylenetriaminepeutacetic acid) (with a large stability constant).

## 4.5 CHEMICAL APPLICATIONS OF TRACERS

One of the most important uses of radiotracers in chemistry has been to *test separation procedures* in analytical chemistry. Tracers furnish a specific, easy-to-apply, quick method of following the path of a given material in a chemical separation. Physical chemical data or separation parameters can also be determined. An example of this type of application is the work of Sunderman and Meinke (1957) in studying separations by precipitation. The scavenging efficiency of $Fe(OH)_3$ was evaluated by seeing how many other radiolabeled ions would coprecipitate with $Fe(OH)_3$ and the quantitative extent to which they were incorporated in the $Fe(OH)_3$ precipitate. A famous example of the use of radiotracers in evaluating separation procedures is the work of Kraus and Nelson (1957) in which they studied the pH dependence, eluant volume and similar factors in separation of metal ions by ion exchange. Tracers have also been used to locate the position of a particular fraction in column, thin-layer, and paper chromatography. When used in thin-layer chromatography (TLC), paper chromatography, and electrophoresis, autoradiography is frequently utilized to locate the position of the activity.

The use of radiotracers is an excellent technique for *measuring the solubility product constant* of sparingly soluble salts or for making other studies of substances present in low concentrations. Another very important and classic example of the use of radiotracers is that of studying the occurrence and properties of isotopic *exchange reactions*—reactions of the type

$$AX^* + BX \Longleftrightarrow BX^* + AX$$

where X, X* represent stable and radioactive atoms, respectively, of the same element.

Perhaps the most significant of the numerous applications of radiotracers in chemistry has been the study of *chemical reaction mechanisms*. In fact, most of the proposed reaction mechanisms have been "verified" by means of a radiotracer study. One of the simplest mechanistic experiments using radiotracers is to *test the equivalence of various atoms in molecules* in chemical reactions. An example of this type of study is the work of Volpin et al. (1959) on the equivalence of the seven carbon atoms in the tropylium ring. Volpin et al. reacted labeled diazomethane with benzene and brominated the cyclohepatriene product to form a labeled tropylium bromide, as shown below:

$$C_6H_6 \xrightarrow{^{14}CH_2N_2} C_7H_8 \xrightarrow{Br_2} {}^{14}C_7H_7Br \qquad (4.2)$$

Then the tropylium bromide was subjected to a Grignard reaction and the product oxidized to give a labeled benzoic acid. Thus, one had

$$^{14}C_7H_7Br \xrightarrow{C_6H_5MgBr} {}^{14}C_7H_7C_6H_5 \xrightarrow{HNO_3} C_6H_5{}^{14}COOH \qquad (4.3)$$

The specific activity of the labeled benzoic acid was found to be one-seventh that of the initial labeled diazomethane, thus showing the equivalence of the seven carbon atoms of the tropylium ring.

Another popular use of radiotracers in studying chemical reaction mechanisms is the study of *molecular rearrangments*. An example of this class of reactions that illustrates the use of radiotracers is the cyclization of ω-phenoxyacetophenone (**I**) to 2-phenylbenzofuran (**II**). Two possible mechanisms are shown below:

$$\underset{(\mathbf{I})}{\text{Ph-O-CH}_2\text{-}^{14}COC_6H_5} \longrightarrow {}^{14}-C_6H_5^- \longrightarrow \underset{(\mathbf{II})}{{}^{14}\text{-benzofuran-C}_6H_5} \qquad (4.4)$$

**104** RADIOTRACERS

Or

$$\text{(I)} \quad \underset{\text{O}}{\text{C}_6\text{H}_5}\text{—O—CH}_2\text{—}^{14}\text{COC}_6\text{H}_5 \longrightarrow \left[ \text{C}_6\text{H}_5\text{—OH} + \overset{+}{\text{CH}_2}\text{—}^{14}\text{C(=O)—C}_6\text{H}_5 \right] \xrightarrow{-\text{H}^+} \quad \longrightarrow \text{(II)} \qquad (4.5)$$

Note that the two possible mechanisms can be distinguished by the position of the labeled atom in the 2-phenylfuran.

Two problems that are common in tracer studies of the reaction mechanism occurred in this study. The first was the synthesis of the labeled starting material (**I**) for the reaction. A chemical synthesis, shown below, was used to produce (**I**):

$$^{14}\text{CH}_3\text{COBr} \xrightarrow[\text{Craft Reaction}]{\text{Freidel}} \text{C}_6\text{H}_5{}^{14}\text{COCH}_3 \xrightarrow[\text{acetone}]{\text{Br}_2 \text{ in}} \text{C}_6\text{H}_5{}^{14}\text{COCH}_2\text{Br} \xrightarrow[\text{Na}_2\text{CO}_3]{\substack{\text{Phenol in} \\ \text{acetone} \\ \text{with}}} \text{C}_6\text{H}_5\text{—}^{14}\overset{\text{O}}{\underset{\|}{\text{C}}}\text{—CH}_2\text{—O—C}_6\text{H}_5 \quad \text{(I)} \qquad (4.6)$$

After obtaining the labeled starting material and carrying out the cyclization reaction, the second problem was to degrade the product (**II**) to reveal the position of the label. This was done by using the following steps:

$$\text{(benzofuran-C}_6\text{H}_5) \xrightarrow[+\text{CH}_3\text{I in ethanol}]{\text{Metallic Na}} \text{(ArCH}_2\text{—CH}_2\text{—C}_6\text{H}_5, \text{OCH}_3) \xrightarrow[\text{and triethylamine}]{N\text{-bromosuccinimide}} \text{(ArCH=CH—C}_6\text{H}_5, \text{OCH}_3) \xrightarrow[\text{permanganate}]{\text{Oxidation with}} \text{(ArCOOH + C}_6\text{H}_5\text{COOH, OCH}_3) \qquad (4.7)$$

Note that the reaction above the two carbon atoms in the five-membered ring of **II** end up in different reaction products. Thus, one can check to see which product contains the label and decide on the correct mechanism.

## 4.6 ISOTOPE EFFECTS

Up to now in our discussion of radiotracers, we have assumed that all isotopes of a given element, stable or radioactive, would behave alike chemically and physically. We will now examine this point more critically to see how different isotopes behave and how this difference in behavior (the isotope effect) can be detected and used to our advantage. The difference in behavior between different isotopes of the same element is due to the different masses. This mass difference will affect the kinetic energy of the molecules (giving rise to *physical isotope effects*) or will change the vibrational and rotational properties of molecules (giving rise to *chemical isotope effects*).

Some examples of physical isotope effects follow:

1. *Gaseous Effusion*   Graham's law states that

$$R = \sqrt{\frac{m_1}{m_2}} \quad (4.8)$$

    where $R$ is the relative rate of effusion of two isotopes of mass $m_1$ and $m_2$, respectively, through a hole. Thus, a labeled molecule will effuse at a different rate than an unlabeled molecule.

2. *Distillation*   For a given temperature, the velocity of a light isotope will be greater than that of a heavy isotope, so that the lighter isotope will have a greater vapor pressure.

Chemical isotope effects are divided into two classes—those affecting the position of the equilibrium in a chemical reaction and those affecting the rate of a chemical reaction. *Equilibrium* isotope effects have their origin in the fact that the extent to which any chemical reaction "goes" is governed by the number of possible ways it can proceed (the phase space available). The more equally probable reaction paths available, the more likely the reaction will go. To illustrate this point, consider the exchange reaction

$$AX + BX^* \Longleftrightarrow AX^* + BX$$

The equilibrium constant $K$ for the reaction is given by

$$K = \frac{f_{AX^*} f_{BX}}{f_{AX} f_{BX^*}} \quad (4.9)$$

where $f_i$ is the partition function for the $i$th species. ($f_i$ reflects the probability of occurrence of the $i$th species and is formally the probability of occurrence of a set of vibrational, rotational, and translational energy levels for that species.) The partition functions can be calculated by using statistical mechanics, and they may depend in a complicated way on the relative masses of AX, AX*, and so on. Table 4.2 shows some typical values of $K$ for various exchange reactions and

**TABLE 4.2   Typical Equilibrium Isotope Effects**

| Reacting System | K |
|---|---|
| $H^2H(g) + H_2O(l) \Leftrightarrow H_2(g) + {}^2HOH(l)$ | 3.2 |
| ${}^{14}CO_2(g) + {}^{12}COCl_2(g) \Leftrightarrow {}^{14}COCl_2(g) + {}^{12}CO_2(g)$ | 1.0884 |
| $[Co(NH_3)_4{}^{12}CO_3]^+ + {}^{14}CO_3^{2-} \Leftrightarrow {}^{12}CO_3^{2-} + [Co(NH_3)_4{}^{14}CO_3]^+$ | 0.8933 |

illustrates the point that chemical equilibrium is shifted when a radioisotope is substituted for another isotope. (Note that when $X = X^*, K = 1$). Table 4.2 also shows the important fact that the greater the mass difference between isotopes, the larger the equilibrium isotope effects are. In general, it has been found that such effects can be neglected when the atomic number is greater than 10.

*Kinetic isotope effects* are very important in the study of chemical reaction mechanisms. The substitution of a labeled atom for an unlabeled one in a molecule will cause a change in reaction rate for $Z < 10$, and this change can be used to deduce the reaction mechanism. The change in reaction rate due to changes in the masses of the reacting species is due to differences in vibrational frequency along the reaction coordinate in the transition state or activated complex.

Experimentally, it is relatively straightforward to measure the existence and magnitude of kinetic isotope effects. Consider the reaction shown below, proceeding through a reaction intermediate, AB, as

$$A + B \longrightarrow AB \longrightarrow \text{products}$$

Let $S_0$ and $S_\gamma$ be the specific activities of B at time $t = 0$ and after a fraction $\gamma$ of the reaction has been completed, respectively. If B is the only labeled reactant, we have

$$\frac{d[B]}{dt} = k[A]^a[B]^b \tag{4.10}$$

And assuming B* is present as a tracer

$$\frac{d[B^*]}{dt} = k^*[A]^a[B^*][B]^{b-1} \tag{4.11}$$

where $a$ and $b$ are the reaction order with respect to A and B. Dividing the previous two equations, we have

$$\frac{d[B]}{d[B^*]} = \frac{k[B]}{k^*[B^*]} \tag{4.12}$$

Separation of variables and integration gives

$$\log_{10}\left(\frac{S_\gamma}{S_0}\right) = \left(\frac{k^*}{k} - 1\right)\log_{10}(1 - \gamma) \tag{4.13}$$

Thus, plotting $\log_{10}(S_\gamma/S_0)$ versus $\log(1 - \gamma)$ gives a straight line of slope $(k^*/k - 1)$. If there was no isotope effect, $k = k^*$, then the slope will be zero. Any finite slope in the preceding plot will give $k^*/k$. This effect is shown in Figure 4.1 for the study of the Cannizzaro reaction by Downes and Harris (1952).

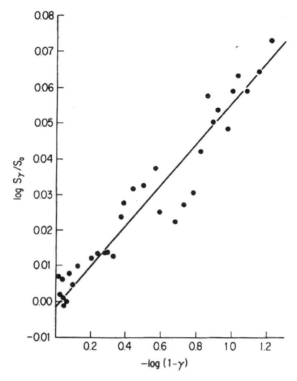

**Figure 4.1** Plot of $\log(S_\gamma/S_0)$ as a function of $\log(1-\gamma)$ for the Cannizzaro reaction, showing the isotope effect. [From Downes and Harris (1952).]

## 4.7 BIOLOGICAL APPLICATIONS

Today the largest number of applications of radiotracers is in biology and medicine. Because of the large number of applications, it is beyond the scope of this work to review them in any detail. Instead, we shall focus on three specific applications, one that is very old, one that is "middle-aged," and one that is "adolescent' in its scope, if not in its development. We refer to autoradiography, radioimmunoassay, and DNA (deoxyribonucleic acid) analysis, respectively.

1. *Autoradiography* Autoradiography is the oldest method of detecting radioactivity. In autoradiography, a radioactive sample is placed on a photographic emulsion. After a period of time, the film is developed and the precise location of the radioactive matter in the sample is determined from the pattern of darkening on the film. Thus, autoradiography is used to *locate* radionuclides in a tissue section, gross sample, or chromatogram. Special techniques (Tolgyessy, 1972) are used to get optimum spatial resolution in the image and to prevent *artifacts* (non-radiation-induced darkening of the film). Optical or electron microscopes can be used to examine the images since the grain size of film is of the order of microns.

2. *Radioimmunoassay (RIA)*   Radioimmunoassay is a highly sensitive method of determining the amounts of hormones, drugs, vitamins, enzymes, viruses, tumor antigens, and serum proteins in biological samples. It is based on the immunological reaction of antibodies and antigens. One starts with either an antigen or antibody labeled with a radiotracer such as $^{3}$H, $^{14}$C, or $^{125}$I. In a radioimmunoassay for antigen, one uses the idea that if a limited amount of antibody is available, the antigen molecules will compete for binding sites on the antibody. If one starts with a certain amount of radiolabeled antigen, any additional antigen added will displace some the radiolabeled antigen and will not allow it to bind to the antibodies.

Thus, the procedure might be to mix antibody and radiolabeled antigen together and let them bind to each other. Precipitate the complex from the solution. Measure the activity of the supernatant. This is the amount of unbound antigen. Now mix the same amounts of antibody and radiolabeled antigen together along with the unknown stable antigen sample. The stable antigen will compete with the radiolabeled antigen for binding sites on the antibody molecules. Some of the radiolabeled antigen will not be able to bind as it did before. Precipitate the complexes from solution. Measure the activity of the supernatant. This activity is a measure of the amount of stable, unlabeled antigen present in the sample. By constructing a calibration curve that shows the amount of radioactivity present in the supernatant after adding known amounts of unlabeled antigen, one can determine the amount of antigen present in a sample. Thus, RIA is a special form of isotope dilution analysis (IDA) discussed below.

More than $10^{7}$ immunoassays are performed in medicine and biochemistry in the United States per year. The important advantages of this technique are high sensitivity and high specificity. In some cases, picogram quantities can be measured.

3. *DNA Analysis*   DNA analysis is widely discussed in the news media because of its use in identifying criminals, establishing paternity, detecting genetic diseases, and so on. The DNA in each living cell can be used as a fingerprint to identify individuals in a large population. To obtain a DNA fingerprint of an individual, one extracts the DNA from a sample of blood, skin, hair, semen, and the like (Fig. 4.2). (The cell walls are destroyed by osmosis or other techniques and the double-stranded DNA is decomposed into single-stranded pieces, which are collected.) This DNA is then cut into pieces using enzymes that cut either side of a repeated sequence. The result is a DNA mixture of segments of differing size. Electrophoresis is used to sort the fragments by size spatially. The spatially separated fragments are allowed to react with radiolabeled "gene probes." These gene probes contain radiolabeled specific fragments of DNA that bind only to those DNA segments containing a nucleotide sequence that is complementary to its own (a sequence that would be its matching strand in the DNA double helix). The original DNA fragments are then identified by the radiolabeled DNA that has reacted with them, usually by autoradiography. The physical pattern on the autoradiograph is a pattern of the DNA sequences and sizes.

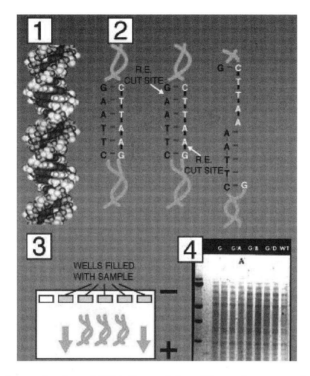

**Figure 4.2** Schematic view of DNA fingerprinting. (Figure also appears in color figure section.)

## 4.8 ENVIRONMENTAL APPLICATIONS

In recent years a great deal of applied research has centered on the study of problems related to the environment and environmental processes. In some of these studies, radiotracers have been used as primary tools to measure the dynamics of many physical and biological processes. In the best studies, the use of radiotracers to measure flow patterns, dispersion, and similar features is closely coupled to tests of theoretical models of the processes involved. This modeling is important because in environmental studies the experimental conditions are difficult to control and, in general, only a few of the many possible conditions in a given experiment will be sampled. Therefore, it is important to have some way (i.e., a model) to correlate experimental results measured under special conditions to general statements regarding an environmental process.

Radiotracers have been employed in studies of physical and biological processes in the atmosphere and the hydrosphere. Among the quantities that have been measured in atmospheric studies are the natural airflow patterns in large- and small-scale investigations, the dispersion of atmospheric pollutants from various sources, and the identification of the sources of various pollutants. In studies of the hydrosphere, radiotracers have been utilized to measure general water circulation patterns and various features of the hydrologic cycle, including precipitation, runoff, and stream flow; total water inventories; infiltration; groundwater problems, such as the origin and age of water, its flow velocity and direction, evaporative transport, and aerosol production. Many studies of a biological nature, such as pollutant dispersal, uptake, and concentration in the ecosystem, have involved the use of radiotracers.

The controlled use of radioactivity to study processes occurring in our environment is a well-established and respected technique. Radioactive tracers have several significant advantages over conventional tracers for environmental studies:

1. The detectability of the tracer is not influenced by the physico-chemical nature of the environment (i.e., factors such as watercolor and pH).
2. Because nuclear radiation, particularly $\gamma$-radiation, is highly penetrating, the tracer can be detected while part of a living organism or when deeply buried in the ground.
3. Because only a small number of atoms are necessary to give a significant disintegration rate, there is a much better detection sensitivity in radiotracer experiments compared to conventional tracer experiments. This point is particularly important for environmental studies where high dilution factors are commonly encountered.
4. Because tracers of short half-life can be used that will rapidly disappear after the completion of an experiment, experiments can be repeated several times without damaging the environment or getting erroneous results due to persistence of tracers from previous experiments in the environment.
5. In many instances, radiotracers are the cheapest method of tracing pollutant flow.

The principal disadvantage of using radiotracers in environmental studies is the actual (or imagined) problem of nuclear safety. Public concern over possible harmful effects of ionizing radiation has increased in recent years. This fact, together with the lack of control over experimental conditions found in many environmental studies, has caused many regulatory agencies to establish extremely strict rules concerning the use of radiotracers in environmental research. Many research workers now find it difficult to demonstrate that the radiotracer concentrations will not exceed the regulatory agency's maximum permissible radionuclide concentration limits at all times and places. One must be especially aware, in this connection, of "concentration effects" present in food chains. At a minimum, the experimenter can usually look forward to a significant amount of red tape before performing environmental studies with radiotracers.

The criteria for choosing a particular radionuclide as a tracer for environmental studies are similar to those encountered in radiotracer laboratory experiments. Such items as the nature of the radiation emitted, the half-life of the radionuclide, the ease of obtaining it, the ease of detecting its radiations, and its cost play important roles. Tables 4.3a, 4.3b, and 4.3c show some typical radiotracers that have been used in environmental studies.

**TABLE 4.3a  Radioactive Tracers for Gaseous Material**

| Nuclide | Half-Life | Radiation of Interest (MeV) | Chemical Form |
|---|---|---|---|
| $^{35}$S | 87 d | β: 0.167 (100%) | $H_2S$ |
| $^{41}$Ar | 110 min | γ: 1.37 | Gas |
| $^{76}$As | 26.5 h | γ 0.55–2.02 | $AsH_3$ |
| $^{82}$Br | 36 h | γ 0.55–1.32 | $CH_3Br$ |
| $^{85}$Kr | 10 y | β 0.7<br>γ 0.54 | Gas |
| $^{133}$Xe | 5.27 d | β: 0.34<br>γ: 0.03, 0.08 | Gas |

*Source*: *Radioisotope Tracers in Industry and Geophysics.* Vienna: International Atomic Energy Agency, 1967.

Many examples of the application of radiotracers to environmental problems exist, and so we have selected only a few of the more interesting ones to discuss. Barry (1971) used $^{41}$Ar, a short-lived gas found in reactor stack effluents, to trace the dispersion of stack effluent from the Chalk River nuclear reactor and relate it to conventional dispersion models. The $^{41}$Ar concentrations in the air at various distances from the emitting stack were measured by circulating the air through a counter consisting of layers of plastic scintillator. The β particles emitted in the decay of $^{41}$Ar were detected by the plastic scintillators. The detectors were

**TABLE 4.3b  Radioactive Tracers for Solid Material**

| Nuclide | Half-Life | Radiation of Interest | Chemical Form |
|---|---|---|---|
| $^{24}$Na | 15 h | γ 1.37 (100%)<br>γ 2.75 (100%) | $Na_2CO_3$ in polypropylene balls |
| $^{46}$Sc | 84 d | γ 0.89 (100%)<br>γ 1.48 (100%) | $Sc_2O_3$ |
| $^{51}$Cr | 27.8 d | γ 0.325 (9%) | Absorbed on quartz |
| $^{64}$Cu | 12.8 | β 0.57 (38%)<br>γ: 0.51 (19%) | CuO |
| $^{65}$Zn | 274 d | γ: 1.11 (49%) | ZnO |
| $^{82}$Br | 36 h | γ 0.55 (70%)<br>γ 1.32 (27%) | CaBr in polyethylene containers |
| $^{110m}$Ag | 253 d | γ 0.66 (100%)<br>γ 1.50 (13%) | Absorbed on solid grains |
| $^{140}$La | 40 h | γ: 0.33–2.54 | $La_2O_3$ polypropylene balls |
| $^{144}$Ce | 285 d | β: 3.1 (98%) | $Ce_2O_3$ |
| $^{182}$Ta | 115 d | γ: 1.19<br>γ: 1.12 | $Ta_2O_3$ |
| $^{198}$Au | 2.7 d | γ: 0.41 (91%) | $AuCl_3$ absorbed on powder |

*Source*: *Radioisotope Tracers in Industry and Geophysics.* Vienna: International Atomic Energy Agency, 1967.

**TABLE 4.3c  Radioactive Tracers for Organic Materials**

| Nuclide | Half-Life | Radiation of Interest | Chemical Form |
|---|---|---|---|
| $^3$H | 12.26 y | β: 0.018 (100%) | Various organic compounds |
| $^{14}$C | 5568 y | β: 0.155 (100 %) | Various organic compounds |
| $^{24}$Na | 15 h | γ: 1.37 (100%) | Naphthenate |
|  |  | γ: 2.75 (100%) | Salicylate |
| $^{35}$S | 97 d | β: 0.167 (100%) | Various organic compounds |
| $^{38}$Cl | 0.3 min | γ: 1.60 (31%) | Chlorobenzene |
|  |  | γ: 2.15 (47%) |  |
| $^{59}$Fe | 44.5 d | γ: 1.1 (57%) | Ferrocene |
|  |  | γ 1.29 (43%) | Dicyclopentadienyl-iron |
| $^{60}$Co | 5.3 y | γ 1.17 (100%) | Naphthenate |
|  |  | γ 1.33 (100%) |  |
| $^{64}$Cu | 12.8 h | γ 0.51 (19%) | Naphthenate |
| $^{65}$Ni | 2.56 h | γ: 0.37, 1.11, | Stearate |
|  |  | 1.49 | Oxalate |
| $^{77}$Ge | 11 h | γ: 0.21–2.02 | Various organic compounds |
| $^{82}$Br | 36 h | γ: 0.55–1.48 | Bromobenzene |
|  |  |  | Paradibromo-benzene |
| $^{124}$Sb | 60 d | γ 0.61 (99%) | Triphenylstibine |
|  |  | γ: 0.72 (14%) |  |
| $^{131}$I | 8.04 d | γ 0.36 (80%) | I-Kerosene |
|  |  | γ 0.64 (9%) | Iodobenzene |
| $^{140}$La | 40 h | γ: 0.33–2.54 | Naphthenate |
| $^{198}$Au | 2.7 d | γ: 0.41 (99%) | Sodium cyanide solution |

*Source*: *Radioisotope Tracers in Industry and Geophysics.* Vienna: International Atomic Energy Agency, 1967.

connected to automatic recording equipment that could monitor the stack effluent dispersion continuously.

A typical example of a radiotracer study in an aqueous system is the study of the concentration dynamics of soluble material in the Eshkol Reservoir of Israel by Gilath and Stuhl (1971). The Eshkol Reservoir is a shallow lake (depth, −7 in.; volume, ∼3.5 × $10^6$ m$^3$) with $10^6$ m$^3$ of water flowing through the reservoir per day. As is typically done in studies of this type, a "spike" of radiotracer was injected into the reservoir and its dispersal was measured. $^{82}$Br was chosen as the tracer because of its high maximum permissible concentration limit (100 μCi/m$^3$), its low detection limit (∼2 × $10^{-6}$ μCi/mL), and its general solubility in water. A spike of ∼15 μCi $^{82}$Br was injected into the water over a 1-h period, allowing the tracer concentration to be measured for ∼3 to 4 times the typical residence time of the reservoir (∼70 h). The $^{82}$Br activity was measured by using NaI detectors stationed at the reservoir outlet and in probes lowered into the water from boats.

An interesting set of radiotracer studies of the deposition of pesticides was made by Atkins and Eggleton (1971). The pesticides were labeled with $^{14}$C, which was assayed using liquid scintillation counting. Results of the study showed that direct absorption on soil or vegetation was a more effective way of removing the pesticides from the atmosphere than deposition in rain. Furthermore, the mean residence times of the pesticides in the atmosphere were shown to be long enough to allow global distribution of a pesticide from any given location.

## 4.9 INDUSTRIAL USE OF RADIOTRACERS

There are a large number of industrial uses for radiotracers. Table 4.4 shows some typical industrial uses of radiotracers.

Several industrial uses involve the determination of flow rates and liquid volumes. In the flow rate applications, one introduces the tracer into the liquid stream by either pulse or continuous injection and measures the activity as a function of distance. (One must be careful to begin sampling beyond the mixing region, where tracer and fluid mix. This is done by beginning sampling at a distance of $\sim 100 \times$ the stream diameter from the point of injection.) When a pulse with activity $A$ Bq is injected into a stream with flow rate $F$, then $F$ is given simply as

$$F = \frac{A}{\int_0^\infty r(t)\, dt} \tag{4.14}$$

where $r$ is the efficiency-corrected counting rate in a downstream detector.

Another interesting and important use of radiotracers in industry is in wear and corrosion studies. In studies of wear (tribology) one labels the part under study with a radionuclide. The radioactivity is concentrated on the surface undergoing wear by plating, diffusion, or ion implantation with a low-energy accelerator. The labeled part is put in service and, typically, one measures the radioactivity released in the lubricant as a measure of wear. Calibrations of the technique can be done to get absolute measures of wear.

## 4.10 NUCLEAR MEDICINE

The most rapidly expanding area of tracer use is in nuclear medicine. Nuclear medicine deals with the use of radiation and radioactivity to diagnose and treat disease. The two principal areas of endeavor, diagnosis and therapy, involve different methods and considerations for radiotracer use. (As an aside, we note that radiolabeled drugs that are given to patients are called *radiopharmaceuticals*.) A list of radionuclides commonly used in diagnosis is shown in Table 4.5. Most nuclear medicine procedures (>90%) use either $^{99}$Tc$^m$ or one of the iodine isotopes. Most diagnostic use of radiotracers is for imaging of specific organs, bones, or tissue. Typical administered quantities of tracer are 1–30 mCi for adults. Nuclides used for imaging should emit photons with an energy between 100 and 200 keV, have small decay branches for particle emission (to minimize radiation damage), have

**TABLE 4.4  Industrial Uses of Radionuclides**

| Nuclide | Application |
|---|---|
| $^{3}$H | Self-luminous aircraft and exit signs |
| | Luminous dials, gauges, wrist watches |
| | Luminous paint |
| $^{24}$Na | Location of pipeline leaks |
| | Oil well studies |
| $^{46}$Sc | Oil exploration tracer |
| $^{55}$Fe | Analysis of electroplating solutions |
| | Defense power source |
| $^{60}$Co | Surgical instrument and medicine sterilization |
| | Safety and reliability of oil burners |
| $^{63}$Ni | Detection of explosives |
| | Voltage regulators and current surge protectors |
| | Heat power source |
| $^{85}$Kr | Home appliance indicator lights |
| | Gauge thickness of various thin materials |
| | Measurement of dust and pollutant levels |
| $^{90}$Sr | Survey meters |
| $^{109}$Cd | XRF of metal alloys |
| $^{124}$Sb | Oil exploration tracer |
| $^{126}$Sb | Oil exploration tracer |
| $^{131}$I | Petroleum exploration |
| $^{136}$Cs | Oil exploration tracer |
| $^{137}$Cs | Measure and control liquid flow in pipes |
| | Measure of oil well plugging by sand |
| | Measure fill level of consumer products |
| $^{140}$Ba | Oil exploration tracer |
| $^{147}$Pm | Used in electric blanket thermostats |
| | Gauge thickness of thin materials |
| | Oil exploration tracer |
| $^{151}$Sm | Heat source |
| $^{156}$Eu | Oil exploration tracer |
| $^{192}$Ir | Pipeline, boiler and aircraft weld radiography |
| | Oil exploration tracer |
| $^{198}$Au | Oil exploration tracer |
| $^{204}$Tl | Thickness gauge |
| $^{210}$Po | Reduction of static charge |
| $^{229}$Th | Extend life of fluorescent lights |
| $^{230}$Th | Coloring and fluorescence in glazes and glass |
| $^{232}$Th | With W, electric arc welding rods |
| $^{234}$U | Natural color, brightness in dentures |

*(continued)*

**TABLE 4.4** *Continued*

| Nuclide | Application |
|---|---|
| $^{235}$U | Nuclear reactor fuel |
| | Fluorescent glassware, glazes, and wall tiles |
| $^{238}$Pu | Radioisotope thermal generator |
| $^{241}$Am | Smoke detectors |
| $^{244}$Cm | Analysis of pit mining and drilling slurries |
| $^{252}$Cf | Luggage inspection for explosives |
| | Soil moisture content |

a half-life that is $\sim 1.5\times$ the duration of the test procedure, and be inexpensive and readily available. $^{99}$Tc$^m$ is used in more than 80% of nuclear medicine imaging because its 143-keV $\gamma$ rays produce excellent images with today's $\gamma$ cameras, it has a convenient 6-h half-life and it can be obtained "generators" involving a transient equilibrium with $^{99}$Mo.

Most people are familiar with the medical or dental use of X-rays for providing images of tissue or bones where an external radiation source is used to do the imaging. To improve contrast, agents such as barium sulfate, which attenuate the X-rays, are frequently administered to the patient. In the 1970s, a significant improvement in medical imaging occurred with the advent of computerized tomography (CT). In this technique, photographic plates are replaced by one or more radiation detectors, and an apparatus is used to move the source of imaging radiation relative to the patient with a digital computer system with appropriate software to provide on-line images from observed changes in counting rates as the source–patient geometry changes. [Tomography is from the Greek words, "to cut or section" (*tomos*) and "to write" (*graphein*). Tomography shows slices of the body with typical resolution of <1 mm.] A simple diagram of such apparatus is shown in Figure 4.3.

Tomography can involve images generated by the transmission of radiation through the body (Fig. 4.3) or by incorporating radionuclides into the body and detecting the emitted radiation (emission tomography). For emission tomography, the imaging techniques can involve PLANAR (see below) images where a two-dimensional view of an organ is obtained (Fig. 4.4a), SPECT (single-photon emission computerized tomography) (Fig. 4.4b) where a three-dimensional computer reconstructed image is obtained, or PET. In PET (Fig. 4.5), positron-emitting nuclides, such as $^{18}$F, $^{11}$C, $^{15}$O, or $^{13}$N, are introduced into a region to be studied. The two 0.511-MeV photons, produced when the $\beta^+$ annihilates, emerge in opposite directions and define a line passing through the point where the decay occurred. The two photons are detected in coincidence by an array of scintillation detectors. After the observation of many decays, computer techniques are used to reconstruct a three-dimensional image of the area where the decays occurred.

**TABLE 4.5 Commonly Used Diagnostic Radionuclides**

| Nuclide | Application |
|---|---|
| $^{11}C$ | PET brain scans |
| $^{14}C$ | Radiolabeling |
| $^{13}N$ | PET scans |
| $^{15}O$ | PET scans of cerebral blood flow |
| $^{18}F$ | PET brain scans |
| $^{32}P$ | Bone disease diagnosis |
| $^{33}P$ | Radiolabeling |
| $^{35}S$ | Heart disease diagnosis |
| | Nucleic acid labeling |
| $^{47}Ca$ | Cell function and bone formation |
| $^{46}Sc$ | Blood flow studies |
| $^{47}Sc$ | Cancer diagnosis |
| $^{51}Cr$ | Red blood cell survival studies |
| | Intestinal blood loss |
| $^{51}Mn$ | Myocardial localizing agent |
| $^{52}Mn$ | PET scans |
| $^{59}Fe$ | Bone marrow scanning |
| | Iron metabolism studies |
| $^{57}Co$ | Scanning of various organs |
| $^{58}Co$ | Tracer for pernicious anemia |
| $^{64}Cu$ | PET scans |
| $^{67}Cu$ | Cancer diagnosis |
| $^{67}Ga$ | Tumor and inflammatory lesion imaging |
| $^{68}Ga$ | Thrombosis and atherosclerous studies |
| $^{72}Se$ | Brain imaging |
| $^{75}Se$ | Protein studies |
| | Liver and pancreas imaging |
| $^{81}Kr^m$ | Lung imaging |
| $^{82}Rb$ | Myocardial localizing agent |
| $^{85}Sr$ | Measurement of bone metabolism |
| $^{99}Tc^m$ | Brain, heart, lung, thyroid, gall bladder, skin, lymph node, bone, liver, spleen, and kidney imaging |
| | Blood flow studies |
| $^{109}Cd$ | Cancer detection |
| | Pediatric imaging |
| | Heart disease diagnostics |
| $^{111}In$ | Detection of heart transplant rejections |
| | Imaging of abdominal infections |
| | Imaging of metastatic melanoma |
| $^{123}I$ | Thyroid disorders |
| $^{125}I$ | Osteoporosis detection |
| | Tracer for drugs |
| $^{131}I$ | Thyroid disorders |
| | Brain biochemistry in disease |

(*continued*)

**TABLE 4.5** *Continued*

| Nuclide | Application |
|---|---|
| $^{127}Xe$ | Lung imaging |
| | Neuroimaging for brain disorders |
| $^{133}Xe$ | Lung ventilation studies |
| $^{169}Yb$ | Gastrointestinal tract diagnosis |
| $^{191}Ir^m$ | Cardiovascular angiography |
| $^{195}Pt^m$ | Pharmacokinetic studies of antitumor agents |

Most imaging is of the PLANAR type in which a stationary γ-ray detector is used. Typically, a single picture is taken of a patient's liver, heart, and the like to determine the presence and distribution of the radionuclide. Sometimes, multiple images are taken over a short time to study the dynamic behavior of an organ through its radionuclide uptake. SPECT is used mostly for brain and cardiac imaging with typical resolutions of 3–5 mm. The radionuclides used are $^{99}Tc^m$, $^{201}Tl$, $^{67}Ga$, $^{111}In$, and $^{123}I$, all single-photon emitters. PET imaging is used primarily for dynamic studies of the brain, heart, and lungs but its use is still expanding. $^{11}C$-labeled glucose has been used extensively for the study of brain metabolism. Other nuclides, such as $^{15}O$, can be used to study blood flow and volume. Figure 4.6 shows a set of PET pictures using both $^{11}C$ and $^{15}O$ to study the use of TPA (tissue plasminogen activator), a clot-busting drug, in a patient with an acute myocardial infarction.

$^{99}Tm^m$ is the most widely used radionuclide for diagnostic purposes. It is used in 10 million procedures per year in the United States and 20 million/per year worldwide. As discussed previously, its single 142.7-keV photon is ideal for imaging, and its 6-h half-life will accommodate most procedures with excessive radiation dose to

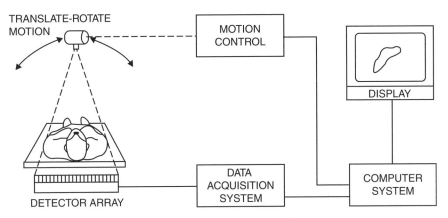

**Figure 4.3** Schematic diagram of a CT system.

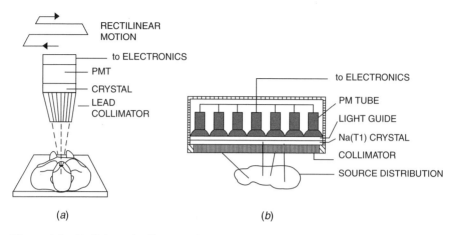

**Figure 4.4** (*a*) Schematic diagram of PLANAR imaging and (*b*) schematic diagram of SPECT.

the patients. It is easily obtained from a Mo/Tc generator ("cow") and thus is available for continuous use at a reasonable cost.

How does a $^{99}$Mo/$^{99}$Tc$^m$ generator work? $^{99}$Mo [which can be produced as a fission product or from the $^{98}$Mo(n, γ) reaction] decays to $^{99}$Tc$^m$ as follows:

$$^{99}\text{Mo} \xrightarrow{\beta^-} {}^{99}\text{Tc}^m \xrightarrow{\text{IT}(142.7\,\text{keV}\gamma)} {}^{99}\text{Tc}$$

The decay of $^{99}$Mo goes about 91% of the time to the isomeric state of $^{99}$Tc. This state decays to the ground state of $^{99}$Tc ($t_{1/2} = 2.1 \times 10^5$ g) by the emission of a single 142.7-keV photon. $^{99}$Mo, as a reaction product, is purified and dissolved in acid media to form the anionic species molybdate (MoO$_4^{2-}$) and paramolybdate (Mo$_7$O$_{24}^{6-}$). The molybdate anions are adsorbed on an aluminum oxide column. This column can be "milked" at will to extract the [$^{99}$Tc$^m$O$_4^-$] ion formed by the

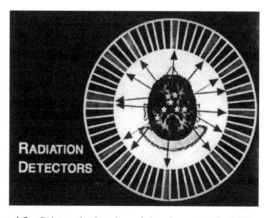

**Figure 4.5** Schematic drawing of the elements of a PET system.

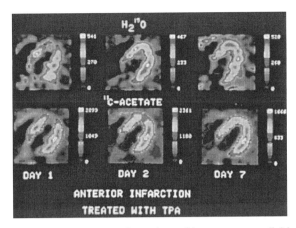

**Figure 4.6** PET pictures of the heart of a patient with acute myocardial infarction treated with a thrombolytic agent. Top row shows scans after administration of water containing $^{15}$O to trace blood flow. Bottom row shows tomograms obtained after administration of acetate containing $^{11}$C to trace the heart's metabolism, that is, its rate of oxygen use. The defects are clearly visible on day 1, both in the impaired blood flow (top left) and the impaired metabolic use of oxygen (bottom left). Recovery of blood circulation has taken place on day 2 and is maintained. (Figure also appears in color figure section.)

decay of $^{99}$Mo. The daughter $^{99}$Tc$^m$ is eluted from the column with 0.9% NaCl. The $^{99}$Mo remains bound to the column as it is insoluble in 0.9% NaCl (see Fig. 4.7). After separation from the molybdenum, the technetium is converted to a suitable complex prior to use in a patient.

The $^{99}$Mo/$^{99}$Tc$^m$ system represents a case of transient equilibrium. We can use Equations (3.6) and (3.22) to trace out the activity of the 69.5 h $^{99}$Mo and 6.0 h $^{99}$Tc$^m$ as a function of time (Fig. 4.8). The $^{99}$Tc$^m$ activity grows in after each milking of the cow, with a maximum amount being present approximately 22 h after separation. If all the $^{99}$Mo decayed to $^{99}$Tc$^m$, then the activity of $^{99}$Tc$^m$ would exceed that of the $^{99}$Mo after equilibrium. Since only 91% of the $^{99}$Mo decays to $^{99}$Tc$^m$, then there is slightly less $^{99}$Tc$^m$ than $^{99}$Mo.

As stated earlier, the nuclides used most for PET are $^{11}$C, $^{13}$N, $^{15}$N, and $^{18}$F. $^{15}$O is used for studies of blood volume and flow in the form of $^{15}$O-labeled carbon monoxide and carbon dioxide. These gases are administered by inhalation with CO binding to the hemoglobin in the blood. It is also possible to use $^{15}$O-labeled H$_2$O. The most widely used $^{18}$F-labeled radiotracer is [$^{18}$F] 2-deoxy-2-fluoro-D-glucose (FDG). When taken into tissue, the fluoroglucose is converted in fluoroglucose-6-phosphate, which cannot be metabolized further and is trapped in the tissue. The trapped tracer is used then for imaging those organs that metabolize glucose most rapidly.

The therapeutic uses of radiation and radioactivity are no less important than the diagnostic uses. One is most familiar with external sources of radiation being used to destroy diseased tissue. A problem with these radiation therapies is similar to that

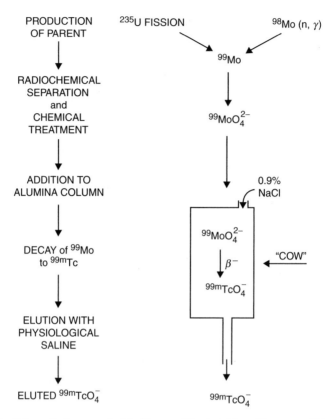

**Figure 4.7** Schematic diagram of a Mo/Tc cow. [Reprinted from Ehmann and Vance (1991).]

encountered in chemotherapy, how to kill the diseased cells without killing so many normal cells that the organism does not survive.

One approach to this problem is to use an internal source of radiation in the form of a physically or chemically implanted radionuclide. $^{131}$I, for example, has been used to treat disorders of the thyroid because of the ability of this gland to concentrate iodine. A promising avenue is the development of monoclonal antibodies that seek out particular cancer cells and bind to them. If one can radiolabel these antibodies with nuclides such as $^{211}$At, $^{131}$I, $^{186,188}$Re, $^{125}$I, or $^{90}$Y, then one can deliver a large dose to the cancer cells with reduced damage to the normal tissue.

An alternative approach, using external radiation, is to deposit, by various means, a large amount of energy into the tumor cell with as little loss of energy elsewhere as possible. One straightforward way to do this is to deposit a radiation-absorbing compound preferentially in the cancer cell and to irradiate the organism, thus localizing the dose. One such material is boron, which can undergo the $^{10}$B(n, α) reaction, splitting into two large fragments with short ranges in tissue. This therapy is being developed at present. Another approach, that of using protons to irradiate the diseased tissue, takes advantage of the fact that the energy deposition

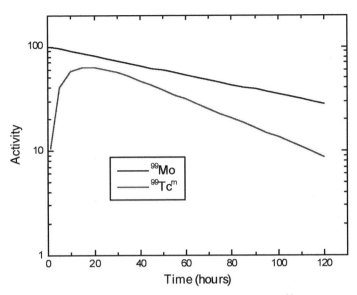

**Figure 4.8** Decay of an initially pure sample of $^{99}$Mo.

of protons in matter is concentrated near the end of its range in matter (the Bragg peak). In theory, carefully choosing the energy of the incident proton, one could localize the damage to the tumor tissue. It turns out (Fig. 4.9) that the sizes and positions of real tumors are much larger than the typic beam dimensions so that the proton beams have to be smeared to match the tumor size.

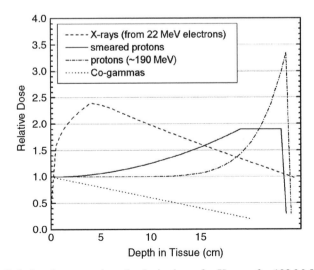

**Figure 4.9** Relative dose at various depths in tissue for X-rays, for 190-MeV protons, for protons with a smeared energy distribution, and from $^{60}$Co γ rays.

## 4.11 ISOTOPE DILUTION ANALYSIS

There are certain analytical methods that can be applied to all fields of tracer use. Foremost among these is *isotope dilution analysis (IDA)*. In this section we summarize the variants of this analytical method.

### 4.11.1 Direct IDA

The basic idea of isotope dilution analysis is to measure the changes in specific activity of a substance upon incorporation into a system containing an unknown amount of that substance. There are several types of IDA. We begin by considering *direct IDA*.

In direct IDA, we are faced with the problem of determining the amount of some inactive material A in a system. Let us define this unknown amount as $x$ grams. To the system containing $x$ grams of inactive A, we add $y$ grams of active material A* of known activity $D$. Thus, we know the specific activity of the added active material, $S_1$. That is,

$$S_1 = \frac{D}{y} \tag{4.15}$$

After thoroughly mixing the active material A* with the inactive A in the system, one isolates, not necessarily quantitatively, and purifies a sample of the mixture of A and A* and measures its specific activity, $S_2$. Clearly, conservation of material says that

$$S_2 = \frac{D}{x+y} \tag{4.16}$$

Since

$$y = \frac{D}{S_1} \tag{4.17}$$

we can substitute for $y$, obtaining

$$S_2 = \frac{D}{x + (D/S_1)} \tag{4.18}$$

Rearranging, we get

$$x = \frac{D}{S_2} - \frac{D}{S_1} = \frac{D}{S_1}\left(\frac{S_1}{S_2} - 1\right) = y\left(\frac{S_1}{S_2} - 1\right) \tag{4.19}$$

This is the basic equation of direct isotope dilution analysis. The unknown amount $x$ of material A is given in terms of the amount $y$ of added labeled material A* and the two measured specific activities $S_1$ and $S_2$.

**Example Problem** Let us consider a practical problem to illustrate the use of this technique. A protein hydrolysate is to be assayed for aspartic acid. Exactly 5.0 mg of aspartic acid, having a specific activity of 0.46 µCi/mg, is added to the hydrolysate. From the hydrolysate, 0.21 mg of highly purified aspartic acid, having a specific activity of 0.01 μCi/mg, can be isolated. How much aspartic acid was present in the original hydrolysate?

**Solution** We say that

$x$ = number of mg aspartic acid in original hydrolysate
$y = 5.0$ mg
$S_1 = 0.46$ μCi/mg
$S_2 = 0.01$ μCi/mg

$$x = y\left(\frac{S_1}{S_2} - 1\right) = (5.0)\left(\frac{0.46}{0.01} - 1\right)$$

$x = 225$ mg aspartic acid

Thus, by isolating a small fraction of the added aspartic acid and measuring the diminution in its specific activity, the aspartic acid content of the original sample can be determined. Note this example involved a large change in specific activity upon dilution. Poor experimental design or other circumstances may lead to a small change in specific activity upon dilution. In such cases, the results obtained from IDA involve a small difference between two large numbers and are quite uncertain.

### 4.11.2 Inverse IDA

Inverse IDA is a simple variant on the basic direct IDA. In inverse IDA, we measure the change in specific activity of an unknown radioactive material A* after diluting it with inactive A. Specifically, let us assume that we have $q$ milligrams (where $q$ is unknown) of a radioactive substance A* whose specific activity is known (i.e., $S_q = D/q$). ($S_q$ can be measured by isolating a small portion of A*, weighing it, and measuring its activity.) Let us add $r$ milligrams of inactive A to A* and thoroughly mix the A and A*. Suppose that we then isolate and purify some of the mixture and measure its specific activity $S_r$. Note that $S_r = D/(q+r)$. And so, we have

$$S_r = \frac{D}{q+r} = \frac{qS_q}{q+r} \qquad (4.20)$$

by substitution. Rearranging, we have

$$S_r(q+r) = qS_q \qquad (4.21)$$

$$\frac{r}{q} = \frac{S_q}{S_r} - 1 \qquad (4.22)$$

$$q = \frac{r}{(S_q/S_r) - 1} \qquad (4.23)$$

The above equation is the basic equation of inverse isotope dilution analysis and indicates that the unknown amount $q$ of active material A* can be deduced by adding $r$

grams of inactive material A to A* and measuring the specific activities before and after the addition $S_q$ and $S_r$, respectively.

### 4.11.3 General Comments

Certain general comments can be made about the experimental techniques used in isotope dilution analysis. First, reagents and tracers of high purity are necessary. They must not contain any spurious activity or any unknown compounds, for the presence of either could affect the specific activity of substances being analyzed. Although pure reagents and tracers are generally available commercially, it would be wise to check for contaminants before use.

One of the key steps in any isotope dilution analysis concerns the isolation and purification of the diluted activity, plus the measurement of its specific activity. Two techniques are usually preferred for the separation: precipitation and solvent extraction. As a purification step, precipitation has the advantage that the precipitate can easily be weighed at the time of separation, thereby allowing a quick determination of the specific activity. The main problem with the use of precipitation techniques involves the occurrence of co-precipitation phenomena, in which unwanted materials are precipitated along with the desired substance, thus altering the sample specific activity. Precipitation techniques are used for the isolation of inorganic components.

Solvent extraction is a frequently employed technique in isotope dilution analysis. It gives very clean separations, resulting in high-purity samples. It has the disadvantage of requiring further chemical processing to determine the mass of material isolated and the specific activity.

One must be aware of the possible occurrence of certain problems in isotope dilution analysis. One of these is *incomplete isotopic exchange*, in which the active and inactive atoms do not mix. This lack of exchange can be due to differing physical and chemical states of tracer and inactive materials. Steps must be taken to ensure complete exchange. One must also be sure that the labeled position in any compound is relatively inert. If the atom in question is very labile, one can get a reduction in specific activity without any dilution having taken place. To compare specific activities, all samples must be counted under identical conditions with proper corrections for self-absorption in samples of varying mass.

In summary, we can say that isotope dilution analysis is a highly sensitive, selective analytical method capable of high precision. It offers the opportunity to determine the amount of material present in a system without the need for a quantitative separation of the material from the system. The applications of isotope dilution analysis cited in the literature are myriad. Perhaps the best summary of these applications is the book by Tolgyessy, Braun, and Krys (1972).

### 4.11.4 Special IDA Techniques

*Substoichiometric isotope dilution analysis* was first developed by Ruzicka and Stary (1968) as another variation on the basic IDA technique. The basic idea of substoichiometric IDA is to isolate equal but substoichiometric amounts of both the diluted and the undiluted substance being analyzed and count these samples. Since the mass of the samples is the same, the specific activities in Equations (4.3) and (4.7) can be replaced by the activities. In this way, the sometimes tricky task of measuring the specific activities is avoided. The key to the technique is obviously whether the analyst can isolate exactly equal quantities of both diluted and undiluted samples.

Another variant on the basic isotope dilution technique is that of *double isotope dilution*, or variants thereof, as first proposed by Block and Anker (1948). It is used in reverse IDA where the specific activity of the original unknown radioactive material A* cannot be measured for some reason. Hence a second dilution is made to determine the specific activity of the original sample.

Consider a system containing an unknown amount $q$ of some active substance A* whose specific activity A* cannot be measured. Take two equal aliquots of this unknown substance A*. Add $r$ milligrams of inactive A to one aliquot and $p$ milligrams of inactive A to the other aliquot. Measure the specific activities of the two aliquots, $S_r$ and $S_p$, respectively. For the first sample of specific activity $S_r$, we have

$$q = \frac{r}{(S_q/S_r) - 1} \qquad S_q = S_r + \frac{r}{q}S_r \tag{4.24}$$

For the second sample we have

$$q = \frac{p}{(S_q/S_p) - 1} \qquad S_q = S_p + \frac{p}{q}S_p \tag{4.25}$$

Setting the two expressions for $S_q$ equal and rearranging yields

$$q = \frac{rS_r - pS_p}{S_p - S_r} \tag{4.26}$$

A major difficulty with double isotope dilution analysis is that, because of the double dilution, the specific activities involved become low and therefore more uncertain.

## 4.12 RADIOMETRIC TECHNIQUES

The central idea in all radiometric techniques of analysis is to have a radioactive reagent R* of known activity combine quantitatively with some unknown amount of material U to form a radioactive addition product R*U. By measuring the activity of the product R*U, the original amount of unknown material U is deduced. The advantages of such techniques are the high sensitivity due to the use of radioactivity and the requirement that the product R*U need not be chemically pure. All that is required is the R*U not contain any spurious radioactivity. The disadvantages of these techniques are that the reaction between R* and U must be quantitative, and

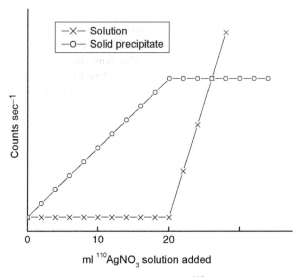

**Figure 4.10** Radiometric titration of $^{110}$AgNO$_3$ with NaCl.

there must not be another species in the system besides U that will react with R*. Several discussions of radiometric techniques are in the literature, including the monograph on radiometric titrations by Braun and Tolgyessy (1948) and the survey of radiometric techniques by Bowen (1969).

One of the radiometric techniques is precipitation with a radioactive reagent. For example, the halides can be precipitated with $^{110}$Ag, silver precipitated with $^{131}$I, the sulfates and chromates precipitated with $^{212}$Pb or $^{131}$Ba, while Al, Be, Bi, Ga, In, Th, U, Zr, and the rare-earth elements can be precipitated with $^{32}$PO$_4^{3-}$. This method suffers from the lack of selectivity and the need to make careful corrections for self-absorption in counting the samples.

One of the most popular radiometric techniques is that of radiometric titrations. In a radiometric titration, the unknown is titrated with a radioactive reagent, and the radioactivity of the product or supernate, formed by the chemical reaction of the reagent and the unknown material, is monitored as a function of titrant volume to determine the endpoint. This means that the radioactive component being followed must be isolated during the titration and its activity measured—that is, a discontinuous titration.

The classic example of a radiometric titration concerns the titration of some unknown material with a radioactive reagent to give a radioactive precipitate. In this case, the activity of the supernatant or the precipitate can be followed as a function of titrant volume, as shown in Figure 4.10. In this type of titrations, the tracer must have a long half-life and must emit high-energy $\beta^-$ or $\gamma$ rays so as to minimize self-absorption corrections (assuming, as is common practice, that the supernatant or precipitate is removed from the system and counted in an external sample counter after the addition of each volume of titrant).

## PROBLEMS

1. Compute the amount of a radionuclide necessary to perform an experiment with a sample count rate of 1000 cpm, a detector efficiency of 33%, a sample aliquot for counting consisting of 10% of the total isolated sample and where the percent incorporation of the nuclide into the total isolated sample was 0.5%.

2. Isotope X, with a half-life of 5 d, is to be used in an experiment that includes the following factors: (a) sample count rate of 100 cpm, (b) detector efficiency of 10%, (c) assume the sample with the lowest count rate will represent a 0.5% incorporation, and (d) assume all samples will represent only 5% of the total isotope administered. What amount of X must be used?

3. Three tracers, 90-y $^{151}$Sm (0.076 MeV $\beta^-$, 100% of the disintegrations and 0.022 MeV X-ray, 4%), 244.3-d $^{65}$Zn (0.33 MeV $\beta^-$, 1.7% of disintegrations and 0.511 MeV $\gamma$ rays, 3.45), and 14.3-d $^{32}$P (1.71 MeV $\beta^-$, 100% of the disintegrations) will be used simultaneously in a multitracer experiment. Suppose you wish to measure the uptake of these three elements in the blood of a rat and the loss of these elements to the rest of the rat's organs and tissue from the blood. What levels of the tracer will you inject into the rat? Why? (Assume that you will withdraw 0.1-mL blood volumes every hour for 24 h. The total blood volume of an adult rat is about 15 mL.)

4. A 10-mL sample of blood is withdrawn from a patient, and the red cells are labeled with $^{51}$Cr, a 27-d $\gamma$ emitter. One milliliter of the labeled blood diluted to 15 mL with water gave a net counting rate of 33,000 cpm (background corrected). The remaining labeled blood is injected back into the patient, and after several hours 10 mL of blood is withdrawn and counted as before. The net counting rate (background corrected) was 500 cm. What is the total volume of the patient's blood?

5. Isotope dilution analysis is applied to the following analysis. Calculate the amount of the compound Y present in the sample and express your answer as percent by weight. A 1-g sample is analyzed for compound Y, molecular weight of 150. A derivative is formed of compound Y and the added radioactive Y (1.5 mCi at a specific activity of 3 mCi/mmol). The derivative, molecular weight of 150 (1 mol of compound Y per mole of derivative), is recrystallized until pure. It has a specific activity of 4.44 × 103 dpm/mg.

6. A 5-kg batch of crude penicillin was assayed by isotope dilution analysis: To a 1-g sample of the batch was added 10 mg of pure penicillin having an activity of 10,500 cpm; only 1.40 mg of pure penicillin having an activity of 290 cpm was recovered. What is the penicillin content of the batch?

7. Isotope dilution analysis permits one to determine the purity of a radiochemical. Compound X, molecular weight of 150 (specific activity 1.0 mCi/mmol), was checked for purity by carefully weighing 1.5 mg of the radiochemical and mixing with 1000 mg of unlabeled compound X and recrystallizing until a

constant specific activity. Radioassay gave a value of 2500 dpm/mg. What was the purity of the radiochemical in percent?

## REFERENCES

Atkins, D. H. F. and A. E. J. Eggleton. In *Nuclear Techniques in Environmental Pollution*, IAEA, Vienna, 1971, pp. 521–534.

Barry, P. J. *Nuclear Techniques in Environmental Pollution*, IAEA, Vienna, 1971, pp. 241–255.

Block, K. and H. S. Ancker. *Science* **107**, 228 (1948).

Bowen, H. J. M. *Chemical Applications of Radioisotopes*, Methuen, London, 1969.

Downs, A. M. and G. M. Harris. *J. Chem. Phys.* **20**, 196 (1952).

Ehmann, W. D. and D. E. Vance. *Radiochemistry and Nuclear Methods of Analysis*, Wiley, New York, 1991.

Gilath, C. and Z. Stuhl. In *Nuclear Techniques in Environmental Pollution*, IAEA, Vienna, pp. 483–496.

Kraus, K. A. and F. Nelson. *Ann. Rev. Nucl. Sci.* **7**, 31 (1957).

Ruzicka, J. and J. Stary. *Substoichiometry in Radiochemical Analysis*, Pergamon, Oxford, 1968.

Sunderman, D. N. and W. W. Meinke. *Anal. Chem.* **29**, 1578 (1957).

Tolgyessy, J., T. Braun, and T. Kyrs. *Isotope Detection Analysis*, Pergamon, Oxford, 1972.

Volpin, M. E. et al. *Zh. Obshch. Khim.* **29**, 3711 (1959).

## BIBLIOGRAPHY

Braun, T. and J. Tolgyessy. *Radiometric Titrations*, Pergamon, Oxford, 1967.

Choppin, G. R., J. Rydberg, and J. O. Liljenzin. *Nuclear Chemistry*, 2nd ed., An excellent treatment that emphasizes the practical aspects of nuclear chemistry.

Collins, C. J. and O. K. Nevill. *J. Am. Chem. Soc.* **73**, 2471 (1951).

Downs, J. and K. E. Johnson. *J. Am. Chem. Soc.* **77**, 2098 (1955).

Groh, J. and G. Hevesy. *Ann. Phys.* **65**, 216 (1921).

Lieser, K. H. *Nuclear and Radiochemistry*, VCH, New York, 1997. An excellent account of the modern aspects of radiochemistry.

Petti, P. L. and A. J. Lenox. *Ann. Rev. Nucl. Sci.* **44**, 155 (1994). Excellent survey of radiotherapy.

Wang, C. H., D. L. Willis, and W. Loveland. *Radiotracer Methodology in the Biological, Environmental and Physical Sciences*, Prentice-Hall, Englewood Cliffs, NJ, 1975. Much of this chapter has been summarized from this book.

Welch, M. J. *J. Chem. Ed.* **71**, 830 (1994). Up-to-date survey of nuclear medicine.

# CHAPTER 5

# NUCLEAR FORCES

## 5.1 INTRODUCTION

In Chapter 1, we discussed the four forces of nature, the electromagnetic, the strong (nuclear), the weak, and the gravitational force. In dealing with the structure, reactions, and decay of nuclei, we shall be dealing with the electromagnetic, strong, and weak interactions. The principal force we shall concern ourselves with is the strong or nuclear force. In this chapter, we shall summarize some important features of the nuclear force.

One basic characteristic of all the fundamental forces is their *exchange* character. They are thought to operate through the *virtual* exchange of particles that act as force carriers. What do we mean by the term *virtual?* We mean that the exchange particles only exist for a short time consistent with the Heisenberg uncertainty principle and cannot be detected experimentally.

How is this possible? Consider the familiar electromagnetic interaction. Two charged particles can be imagined to interact electromagnetically by the emission of virtual photons that are continuously emitted and absorbed by the particles (i.e., exchanged). The Heisenberg uncertainty principle tells us that

$$\Delta E \cdot \Delta t \geq \hbar$$

or that we can "violate" the law of conservation of energy by an amount of energy $\Delta E$ for a time $\Delta t$ such that

$$\Delta t \approx \hbar / \Delta E$$

---

*Modern Nuclear Chemistry*, by W.D. Loveland, D.J. Morrissey, and G.T. Seaborg
Copyright © 2006 John Wiley & Sons, Inc.

(The emission of a virtual photon by a charged particle violates the law of conservation of energy by the photon energy $\Delta E$.) If this photon is traveling at the speed of light, it can travel a distance $R$ such that

$$R = c\,\Delta t \approx \hbar c/\Delta E = \hbar c/E_v$$

where $E_v$ is the photon energy.

If the exchanged particle is not a photon but has mass $m$, its minimum energy is its rest mass $mc^2$, so

$$\Delta t \leq \hbar/mc^2$$

and the range $R$ of the interaction is

$$R \leq \hbar/mc$$

The exchange particles are the graviton for the gravitational force, the pion for the strong interaction between nucleons, the photon for the electromagnetic force, and the $W^\pm$ and $Z$ bosons for the weak interaction. For an exchange particle of zero mass (the photon), the range of the force is essentially infinite. In the case of the strong interaction between nucleons, the range of the force is less than 1.4 fm, so $m_{exchange} \geq 140$ MeV/$c^2$. In the case of the weak interaction, the exchange particles, the $W^\pm$ and $Z$ bosons have masses $m \approx 90$ GeV/$c^2$, so $R \approx 10^{-3}$ fm.

When dealing with atoms and molecules and their interactions, one is dealing primarily with the electromagnetic interaction, which is well known. In principle, the problems of atomic and molecular structure are thus soluble, albeit sometimes with a great deal of mathematical complexity. For the nuclear or strong interaction that is not the case. Although we know much about nucleons and their interactions, there are some features of the nuclear force that are poorly understood even today. Since each nucleon is a composite particle, it is not surprising that the interaction between nucleons is complicated. Nonetheless, an exploration of some of the features of the nuclear force will greatly aid us in understanding nuclear phenomena.

## 5.2 CHARACTERISTICS OF THE STRONG FORCE

As discussed earlier, the range of the nuclear force $R$ is thought to be short with $R \leq 1.4$ fm. What evidence do we have for this? The fact that the strong force plays no role in atomic or molecular structure restricts its range to less than the nuclear radius. In our discussion of the semiempirical binding energy equation, we showed that nuclear forces "saturate" and that nucleons only interact with their nearest neighbor. Thus, the range of the nucleon–nucleon interaction must be of the order of the size of a nucleon, that is, a few femtometers ($10^{-15}$ m).

We know that the nuclear force is strongly attractive, binding nucleons together to form a densely packed nucleus. Experiments involving the scattering of high-energy particles from nuclei have shown the nuclear force also has a repulsive core. What we mean by this statement is that below some value of the separation between nucleons (~0.5 fm), the nuclear force becomes repulsive instead of attractive. (This feature, due to the quark substructure of the nucleon, prevents the nucleus from collapsing on itself.)

The simplest bound nuclear system, the deuteron, consists of a neutron and a proton. The deuteron is known to have a quadrupole moment, 0.00286 barns, which tells us that the deuteron is not perfectly spherical and that the force between two nucleons is not spherically symmetric. Formally, we say the force between two nucleons has two components, a spherically symmetric central force and an asymmetric tensor force that depends on the angles between the spin axis of each nucleon and the line connecting them.

The deuteron has only one bound state, a triplet angular momentum state, in which the spins of the neutron and proton are parallel, adding to make a $I = 1$ state. The singlet $^1S$ state in which the nucleon spins are antiparallel is unbound. Thus, the nuclear force is spin dependent. Also we shall see that the nuclear force depends on the coupling of the nucleon spin and nucleon orbital angular momentum. The deuteron magnetic moment, 0.857 $\mu_n$, is close to the sum of the neutron ($-1.913$) and proton magnetic moments (2.793). Detailed studies show a small portion (~4%) of the time, the neutron and proton are in a $^3D$ state ($L = 2$, $S = 1$, $I = 1$) rather than ground state $^3S$ configuration ($L = 0$, $S = 1$, $I = 1$).

Using the relationship between force and potential energy discussed earlier, we can represent the nuclear force in terms of a simple plot of the nuclear potential energy as a function of distance to the center (Fig. 5.1). Since low-energy particles

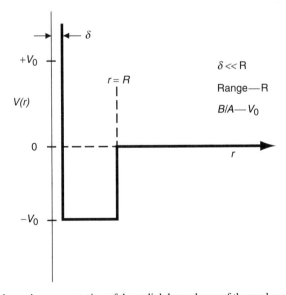

**Figure 5.1** Schematic representation of the radial dependence of the nuclear potential energy.

cannot probe the interior of nucleons or the nucleus, we can usually ignore the repulsive core in most problems involving low-energy nuclear structure and just use a square well potential ($V = -V_0$ for $r < R$, $V = 0$ for $r > R$). Occasionally, the Yukawa form of the potential is used where $V = -V_0 \exp(-r/R)/(r/R)$ or the Woods–Saxon form where $V = V_0/\{1 + \exp[(r-R)/a]\}$. The typical values of $R$ for these potentials are 1.5–2 fm with $V_0 = 30$–60 MeV. Important additional components of the nuclear force are discussed as they become important in our discussions of nuclear structure in Chapter 6.

## 5.3 CHARGE INDEPENDENCE OF NUCLEAR FORCES

The nuclear force between two nucleons is *charge independent*. By this we mean that the strong interaction between two protons or two neutrons or a neutron and a proton is the same. (Of course, there will be differing electromagnetic forces in these cases.) Evidence for the charge independence of nuclear forces can be found in nucleon–nucleon scattering and in the binding energies of light mirror nuclei (Table 5.1). (Mirror nuclei are isobars where the number of protons in one nucleus is equal to the number of neutrons in the other nucleus and vice versa.) In Table 5.1, we tabulate the total nuclear binding energy of some light mirror nuclei, the difference in Coulomb energy between the nuclei, and the resulting net "nuclear" binding energy. The latter quantity is remarkably similar for these mirror nuclei, supporting the idea of charge independence of nuclear forces.

**Example Problem**  Consider the mirror nuclei $^{25}$Mg and $^{25}$Al. What is the energy difference between their ground states?

***Solution***  Note the "conversion" of $^{25}$Mg into $^{25}$Al will involve the change of one neutron into one proton. The neutron and proton have slightly different masses, of course. The extra proton will interact electromagnetically with the other 12 protons giving a second term in the energy difference:

$$\Delta E = E(A, Z+1) - E(A, Z) = \Delta E_{coul} - (m_n - m_H)c^2$$
$$\Delta E_{coul} = \tfrac{6}{5} * e^2/R * Z = \tfrac{6}{5}(1.44/1.2 \times 25^{1/3})12$$
$$(m_n - m_H)c^2 = 0.782 \text{ MeV}$$
$$\Delta E = 5.910 - 0.782 = 5.128 \text{ MeV}$$

So we would expect the ground state of $^{25}$Al to be 5.128 MeV above the ground state of $^{25}$Mg.

The observation of the masses of mirror nuclei suggests the strong or nuclear force between a neutron and a proton is the same. This equivalence leads naturally to considering the neutron and the proton as corresponding to two states of the same particle, the nucleon. (A similar situation holds for the $\pi$ meson, where the $\pi^0$, $\pi^+$,

## 5.3 CHARGE INDEPENDENCE OF NUCLEAR FORCES

**TABLE 5.1  Properties of Light Nuclei**

| A | Nucleus | Total Binding Energy (MeV) | Coulomb Energy (MeV) | Net Nuclear Binding Energy (MeV) |
|---|---|---|---|---|
| 3 | $^3$H | −8.486 | 0 | −8.486 |
|   | $^3$He | −7.723 | 0.829 | −8.552 |
| 13 | $^{13}$C | −97.10 | 7.631 | −104.734 |
|    | $^{13}$N | −94.10 | 10.683 | −104.770 |
| 23 | $^{23}$Na | −186.54 | 23.13 | −209.67 |
|    | $^{23}$Na | −181.67 | 27.75 | −209.42 |
| 41 | $^{41}$Ca | −350.53 | 65.91 | −416.44 |
|    | $^{41}$Sc | −343.79 | 72.84 | −416.63 |

and $\pi^-$ mesons show the same strong force behavior.) To express this idea, we say there is a quantum number $T$ for the nucleon (or the $\pi$ meson) called the *isospin*. In analogy to spin angular momentum, we say that for the nucleon $T = \frac{1}{2}$, and in this hypothetical isospin space there are two projections of $T$, $T_3 = +\frac{1}{2}$ (the proton) and $T_3 = -\frac{1}{2}$ (the neutron). (An alternate notation system refers to the isospin projection as $T_z$.) For a system with isospin $T$, there are $2T + 1$ members of the isospin multiplet. In a nucleus of $N$ neutrons and $Z$ protons,

$$T_3 = (Z - N)/2$$

For even nuclei, $0 \leq T \leq A/2$, while for odd nuclei, $\frac{1}{2} \leq T \leq A/2$.

Isospin is a useful concept in that it is conserved in processes involving the strong interaction between hadrons. The use of isospin can help us to understand the structure of nuclei and forms the basis for selection rules for nuclear reactions and nuclear decay processes. While a detailed discussion of the effects of isospin upon nuclear structure, decay, and reactions is reserved for later chapters, a few simple examples will suffice to demonstrate the utility of this concept.

Consider the $A = 14$ isobars, $^{14}$C, $^{14}$N, and $^{14}$O. $^{14}$C and $^{14}$O are mirror nuclei and have ground states with $T_3 = \pm 1$. As such they must be part of an isospin triplet with $T = 1$ ($T_3 = 0, \pm 1$). Thus, in the $T_3 = 0$ nucleus, $^{14}$N, there must be a state with $T = 1$, $T_3 = 0$ that is the analog of the $T_3 = 0$ ground states of $^{14}$C and $^{14}$O. (See Problems section for further details.) We expect the three members of this multiplet to have approximately the same energy levels after correction for the Coulomb effect and the neutron–proton mass difference.

In heavy nuclei, the Coulomb energy shift between members of an isospin multiplet can be large due to the large number of protons in the nucleus. Thus, the *isobaric analog* of the ground state of one member of an isospin multiplet may lie at several MeV excitation. When Fox et al. (1964) were doing routine excitation function measurements for the $^{89}$Y(p,n) $^{89}$Zr reaction, which essentially converts a neutron in the traget nucleus into a proton, they observed two sharp peaks in the neutron yields near $E_p = 5$ MeV, as shown in Figure 5.2. This observation was unexpected, as the reaction was populating levels in the $^{90}$Zr

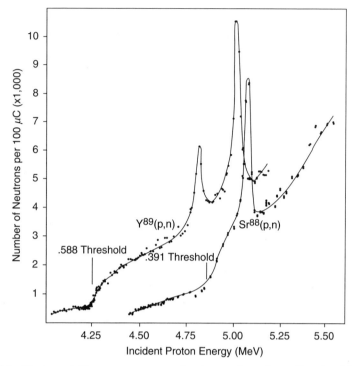

**Figure 5.2** Neutron yields vs. proton energy for the reactions shown. Reprinted from J. D. Fox, C. F. Moore, and D. Robson, *Phys. Rev. Lett.* **12**, 198 (1964). Copyright 1964 by the American Physical Society.

compound nucleus at an excitation energy of ~10 MeV where the spacing between levels was small and no states were known that produced such large resonances. Angular distributions quickly showed the $(J, \pi)$ of these states were $2^-$ and $3^-$. It was pointed out that the ground and first excited state of $^{90}$Y had $2^-$ and $3^-$ and were separated by ~200 keV. Calculations of the Coulomb energies showed these resonances to correspond to the isobaric analogs of the ground state and first excited state of $^{90}$Y. Their yields were enhanced because they represent particularly simple nuclear configurations in contrast to the normal states found at excitation energies of 10 MeV.

## PROBLEMS

1. Make a table such as Table 5.1 showing the total binding energy, the Coulomb energy, and the net nuclear binding energy for $^{14}$C, $^{14}$O, and the 2.31-MeV level of $^{14}$N.

2. If the difference in energy of the ground state of $^{14}$C and the $T = 1$ analog of the $^{14}$C ground state in $^{14}$N is 2.15 MeV is due to the Coulomb energy difference between the nuclei, calculate an average radius $R$ for these $A = 14$ nuclei.

3. For a Yukawa nuclear potential with $V_0 = 40$ MeV and $r_0 = 1.5$ fm, calculate the ratio between the nuclear and Coulomb potential for $r = 1, 2, 4, 8,$ and 16 fm.

## REFERENCE

1. Fox, J. D., C. F. Moore, and D. Robson. *Phys. Rev. Lett.* **12**, 198 (1964).

# CHAPTER 6

# NUCLEAR STRUCTURE

Nuclei have a very regular structure with many general and simple properties that are predicted by quantum mechanical treatment of particles moving in a potential well. This situation is very similar to descriptions of the electrons in atoms. The potential energy well due to the Coulomb force (atoms) is substantially different from that due to the nuclear force (nuclei). The previous chapter contained a discussion of the basic properties of the nuclear force and how we have been able to determine its features. In summary, the exact form of the nuclear or strong force is unknown, but the force is known to be short ranged ($\approx 1$ fm) with a repulsive core and is known to saturate. That is, the force acts primarily between nearest neighbors to hold them together without letting them interpenetrate one another. We use these features to form the basis of our study of nuclear structure.

We can also learn a great deal about the basic features of nuclear structure and the nature of the force that holds the nucleus together if we simply look carefully at the lightest stable and unstable nuclei. The building blocks of nuclei are the nucleons, protons, and neutrons, of course. The proton is stable and is usually found as a hydrogen atom bound to a single electron. The mass of the electron is small compared to that of the proton (511 keV/939,000 keV or $\sim 1/1800$) and the binding energy is even smaller (13 eV/939,000,000 eV or $\sim 10^{-8}$). The electrons are almost always carried along by the nuclei, so it is most convenient to imagine building nuclides up from hydrogen atoms, $^1$H, rather than bare protons. On the other hand, the free neutron is unstable and decays with a half-life of $\sim 10$ min into a proton, an electron, and an antineutrino. Thus, imagining that we will construct

*Modern Nuclear Chemistry*, by W.D. Loveland, D.J. Morrissey, and G.T. Seaborg
Copyright © 2006 John Wiley & Sons, Inc.

nuclei from these constituents, we should not expect to be able to make arbitrary heavy isotopes of any given chemical element because eventually the neutrons will decay as if they were independent.

If we now bring together two nucleons, we find a rather important and interesting fact, only one combination produces a stable (bound) nucleus. One proton and one neutron will combine to form a deuteron, or one hydrogen atom plus one neutron will form a deuterium atom with its atomic electron. Both of the other combinations, two protons that can be labeled $^2$He and two neutrons, are unbound and come apart almost as rapidly as the constituents come together. It is easy to see that the diproton, or $^2$He, is more unstable than the dineutron due to the Coulomb repulsion between the two positively charged protons. Thus, we find a preference for equal numbers of neutrons and protons even in the smallest nucleus.

If we look more carefully at the deuteron, we expect that there should be two possible combinations of the spins of the two nucleons. Both the proton and neutron have $S = \frac{1}{2}$, and we can have the parallel combination $S_p + S_n = 1$ and the antiparallel combination $S_p + S_n = 0$. Both of these states exist in a deuterium nucleus, and the $S = 1$ state is the ground state (lowest energy), and the $S = 0$ state is an excited state and is, in fact, unbound. Therefore, the alignment of the spins of the two unlike nucleons has an important effect on the total binding energy. This provides part of the explanation as to why the dineutron is unbound. Notice that the intrinsic spins of two neutrons in an $l = 0$, or s, state must be paired (according to the Pauli principle). However, the nuclear force prefers the parallel alignment. In order to align the spins in the same direction, the neutrons have to be in a ($l = 1$, or p, state, which requires more relative energy. In addition, the fact that the deuteron is not spherical having an intrinsic electric quadrupole moment tells us that there is a noncentral component of the nuclear force.

We can continue our survey of the lightest nuclei with $A = 3$. Only the combinations of two protons and one neutron, $^3$He, and one proton with two neutrons, $^3$H, are bound, while the combinations of three protons, $^3$Li, and three neutrons are unbound. Again we see a balance between the numbers of neutrons and protons with the extreme cases being unbound. The nuclear spins of both bound $A = 3$ nuclei are $\frac{1}{2}$ indicative of a pair of nucleons plus one unpaired nucleon; three unpaired nucleons would have had a total spin of $\frac{3}{2}$. In the $A = 3$ system the more neutron-rich nucleus, tritium, $^3$H, is very slightly less stable than $^3$He and, it decays by $\beta^-$ emission with a 12.3-y half-life.

Only one combination of four nucleons is bound, $^4$He, with two protons and two neutrons. All other combinations of four nucleons are unbound. Moreover, $^4$He, or the $\alpha$ particle, is especially stable (very strongly bound), and the nucleons are paired to give a total spin $S = 0$. Interestingly, if we add a nucleon of either type to the $\alpha$ particle, we produce an unbound nucleus! Thus, there are no stable nuclei with $A = 5$ as both $^5$He and $^5$Li break apart very rapidly after formation. This creates a gap in the stable masses and poses a problem for the building up of the elements in stars, which is discussed in Chapter 12. There are two bound nuclei with $A = 6$, $^6$He and $^6$Li, with the helium isotope decaying into the lithium isotope, the others are unbound. Continuing on, between mass 6 and 209, all mass numbers have at least one bound nucleus although there are no stable nuclei at $A = 8$. There is generally one stable nucleus for each odd mass number, and in heavy nuclei there are often two stable nuclei for each even mass number. There are at most three stable isobars for a given mass number.

We can summarize our observations about light nuclei and the nuclear force as follows: The nuclear force acts between nucleons in a uniform way, protons have an additional Coulombic repulsion that can destabilize proton-rich nuclei, but very neutron-rich nuclei are also unstable. The symmetric nuclei with equal numbers of neutrons and protons are favored (at least in light nuclei); and finally the nuclear force depends on the spin alignment of the nucleons. Because the underlying nature of the nuclear force is unknown at present, several parameterizations of an effective force have been developed. A detailed discussion of these effective forces or equivalently the nucleon potentials is beyond the scope of this book. Now imagine the complexity of describing a nucleus in which each nucleon is interacting with its nearest neighbors through the nuclear force, and at the same time all the protons are pushing on each other with the Coulomb force! This problem and the closely related problem of molecular motion in a liquid drop have not been solved in detail yet, and so we will present models of the average behavior of the nucleons in effective energy potentials.

## 6.1 NUCLEAR POTENTIALS

The combined interactions of the neutrons and protons can be described in terms of a "nuclear potential well." Because the protons are charged particles, we generally treat the neutrons and protons as if they move inside separate potential wells (superimposed on one another). It is useful to imagine in a very schematic and simple way the forces that would act on a neutron as it is brought up to a nucleus. At large distances ($>$few femtometers) then there will be no force (no change in the potential energy). When the neutron reaches the "surface" of the nucleus (or comes within the range of the nuclear force $\sim 1$ fm from the "edge"), there will be an attraction from the nearest-neighbor nucleons, and the neutron will be pulled into the nucleus. This attraction will increase rapidly in the surface region as the nucleon comes in contact with other nucleons until it is surrounded by nucleons and is in the interior of the nucleus. The potential energy will stay approximately constant if the neutron is moved inside the nucleus and is not near the edge. This behavior is summarized in the potential energy function shown as a function of distance from the center of the nucleus shown on the left side of Figure 6.1.

On the other hand, if we bring a proton up to the same nucleus, we will have a slightly different behavior. At first, the nucleus will repel the proton due to the long-range Coulomb force. Then, as we bring the proton very near to the surface, the same nuclear attraction will begin to overcome the repulsion. The nuclear attraction will increase until the proton is surrounded by nucleons as in the neutron case, but there will always be a net repulsion from the other protons. The repulsion decreases the overall attraction, and the proton potential energy well will not be

# 140  NUCLEAR STRUCTURE

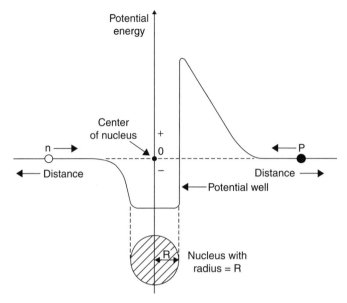

**Figure 6.1** Schematic representations of a general neutron–nucleus potential and a proton–nucleus potential as a function of radius.

as deep as the neutron well. The models that we will describe later in this chapter will rely on the ideas behind these simple schematic potentials.

Before going on to describe the models of nuclear structure in detail, it is useful to make a short comparison of the characteristics of the atomic and the nuclear potential energies. The atomic potential is in some sense, easier to describe because it is created by a central nucleus that can be ignored in almost all atomic calculations. The nucleus supplies the overall attraction for the electrons, but it does not interact with the electrons. In the nuclear case the potential is created by the nucleons themselves, and, if we disturb the nucleons (add or subtract one), then the overall potential will have to be readjusted. Fortunately, the changes in the potential energy for a large nucleus are often relatively small and the general behavior of the whole nucleus remains the same. There can be, of course, major disruptions that cannot be described with a simple potential energy.

## 6.2 SCHEMATIC SHELL MODEL

With a general understanding of the form of nuclear potentials, we can begin to solve the problem of the calculation of the properties of the quantum mechanical states that will "fill" the energy well. One might imagine that the nucleons will have certain finite energy levels and exist in stationary states or orbitals in the nuclear well similar to the electrons in the atomic potential well. This interpretation is quite valid and forms the basis of the "shell model" of the nucleus. The potential well for nucleons has a very different shape from that for atomic electrons and so we should expect that the energy levels and their filling patterns would be different.

As a very first approximation we could model the nucleus as a spherical rigid container (also called a square-well potential). The potential energy is assumed to be exactly zero when the particle is inside the walls of the container and the walls are so strong and high that the particle can never get out. An analogy would be a gaseous atom inside a very small spherical balloon. The energy levels for a particle in such a potential well are shown in Figure 6.2. We could compare these energy levels to the known nuclei, but the potential is so unrealistic that we would not expect to have much success. For example, notice that this potential goes to infinity at the edge of the nucleus, but the nuclear potential felt by a neutron goes to zero at the edge.

A much more useful potential is the harmonic oscillator potential, which has a parabolic shape. As indicated in Figure 6.2, this potential also has steep sides that continue upward and will be useful only for the low-lying energy levels. The harmonic oscillator potential has the feature of equally spaced energy levels. This potential does not "saturate," rather it has a rounded bottom and so will not be very good for large nuclei with large central volumes. Nevertheless, the harmonic oscillator potential is used extensively for light nuclei, and harmonic oscillator wave functions are often used in reaction calculations. The harmonic oscillator states are labeled by their total angular momentum starting at 0. Each principal quantum number level is said to form a shell of orbitals. The energy gap between each shell will be exactly the same, and all the sublevels with a given principal quantum number will be degenerate. The number of orbitals is given by the expression $2N + 1$ where $N = 0, 1, 2, \ldots$. The Pauli principle states that the number of nucleons (fermions) needed to fill each orbital is 2, as for electrons in atomic orbitals, so the number of nucleons needed to fill the shells are $2[2N + 1] = 2, 6, 10, \ldots$. This filling agrees with the enhanced stability of the lightest nuclei ($^4$He, $^{16}$O), taking the neutrons and protons in separate orbits, but does not agree with that of heavier nuclei.

A dramatic improvement was made to the simple harmonic oscillator potential by the addition of a spin–orbit correlation. It is known that relativistic particles have a tendency to align their orbital and intrinsic angular momenta (spins). This alignment is the basis of the familiar change in the chemistry of the bottom-row elements in the periodic table. For example, thallium favors the 1+ oxidation state even though thallium is the heaviest member of Group 13 or IIIA. This comes about because the three atomic p states separate (or split apart) into two groups in energy according to the alignment of the orbital ($l = 1$) and intrinsic spin ($s = \frac{1}{2}$) angular momentum. The $p_{1/2}$ state with the spin and angular momentum coupled in opposite directions comes lower in energy and holds two electrons, while the third electron lies in the $p_{3/2}$ state and is easily ionized.

The addition of the spin–orbit term to the nuclear harmonic oscillator potential causes a separation or removal of the degeneracy of the energy levels according to their total angular momentum ($j = l + s$). In the nuclear case, the states with

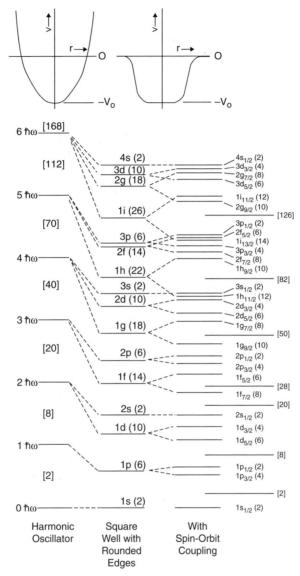

**Figure 6.2** Energies of single-particle orbitals in harmonic-oscillator and "rounded square-well" potentials, the latter with and without spin–orbit coupling. Numbers in parentheses indicate orbital capacities and those in square brackets give cumulative capacity up to the given point. [Reproduced by permission from Gordon and Coryell (1967).]

the parallel coupling and larger total angular momentum values are favored and move lower in energy than those with smaller total spin values for a given combination. The ordering of the energy levels from a spin–orbit/harmonic oscillator shell model is shown in Figure 6.3 with their spectroscopic notation. Each total

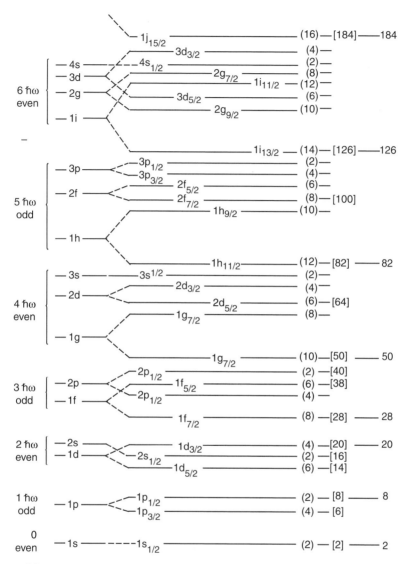

**Figure 6.3** Energy level pattern and spectroscopic labeling of states from the schematic shell model. The angular momentum coupling is indicated at the left side and the numbers of nucleons needed to fill each orbital and each shell are shown on the right side. From M. G. Mayer and J. H. D. Jenson, Elementery Theory of Nuclear Shell Structure, Wiley, New York, 1955.

angular momentum state has $2j + 1$ suborbitals or projections of the angular momentum ($m_j = -j, \ldots, 0, \ldots, +j$) just like the $l$ values of atomic electrons. Recall that we always have separate neutron states and proton states, and the Pauli principle will put a maximum of two neutrons or protons into each orbital.

Let us consider placing nucleons into these shell model states. The lowest level is called the $1s_{1/2}$, s for $l = 0$, and $j = l + s = \frac{1}{2}$. This level has only $2l + 1 = 1$ $m$ value and can hold only two protons in the proton well and two neutrons in the neutron well. The next levels are the $1p_{3/2}$ and $1p_{1/2}$ pair in the next highest shell ($N = 1\hbar\omega$). Thus, $^4$He represents the smallest nucleus with exact filling of both $N = 0$ harmonic oscillator shells for neutrons and protons and might be expected to have an enhanced stability. The next shell filling occurs when the $N = 0\hbar\omega$ and $N = 1\hbar\omega$ shells are filled. This requires eight protons and eight neutrons, so $^{16}$O should be an especially stable nucleus. The other shell closures occur at 20, 28, 50, 82, and 126 nucleons. These values correspond to places in the nuclidic table with unusually large numbers of isotopes and isotones due to their enhanced stability. A few stable nuclei have both closed neutron and proton shells and are very strongly bound (relative to their neighbors), such as $^4$He, $^{16}$O, $^{40}$Ca, $^{48}$Ca, and $^{208}$Pb. A few doubly closed shell nuclei have been produced outside the range of stable nuclei such as $^{56}$Ni, $^{100}$Sn, and $^{132}$Sn, and others have been sought such as $^{10}$He and $^{28}$O but have been shown to be unbound.

### Example of Shell Model Filling: $^7$Li

1. Place the three protons into the lowest available orbital. The protons in the $1s_{1/2}$ state must be paired according to the Pauli principle, so we have a configuration $(1s_{1/2})^2(1p_{3/2})^1$.
2. Place the four neutrons into their lowest available orbitals. The neutrons should be paired in the partially filled orbital (i.e., in contrast to the case for atomic electrons), giving a configuration of $(1s_{1/2})^2(1p_{3/2})^2$.

*Prediction* All nucleons are paired except for the $1p_{3/2}$ proton. Therefore, the spins and angular momenta will cancel except for this proton. The nuclear spin should be $\frac{3}{2}\hbar$ and the nuclear parity should be negative corresponding to the parity of a p state (odd $l$ value).

*Question* What would this model predict for an excited state of $^7$Li? Two possibilities should be apparent. We could promote the $p_{3/2}$ proton to the $p_{1/2}$ state or we could uncouple the $p_{3/2}$ neutrons giving three unpaired neutrons in the $p_{3/2}$ level. $^7$Li has only one bound excited state, and it corresponds to promotion $p_{3/2} \to p_{1/2}$ of the proton. The breaking of pairs has a significant energy cost and causes the nucleus to become unbound.

Notice that the light nuclei are extremely fragile due to the large level spacing and relatively small number of levels. The small numbers of nucleons are very sensitive to small changes in the configurations and have relatively few excited states. Heavy nuclei are much more "resilient" due to the large number of nearby energy levels with slightly different configurations, and these nuclei almost always have very large numbers of bound excited states.

The reality of this scheme of assigning nucleons to various simple shell model states can be checked very directly by nuclear reactions that give or take a nucleon from the nucleus. The (p, 2p) reaction is such a reaction that removes a proton from the nucleus. The energy required to remove a given proton is thus a measure of the energy of the corresponding nuclear state. In Figure 6.4, we show the results of such a study of the $^{16}$O(p, 2p) reaction. Three peaks in the cross section are seen corresponding to the removal of protons from the $1p_{1/2}$, $1p_{3/2}$, and $1s_{1/2}$ orbitals.

The energy level diagram for the schematic shell model, shown in Figure 6.3, allows us to make a large number of predictions about the ground states of broad ranges of nuclei. *First, the strong pairing of nucleons in the individual orbitals tells us immediately that the (net) spin of all nuclei with both even numbers of protons and even numbers of neutrons will be zero. Also the parities of the wave functions of all these nuclei will be positive. Thus, the ground-state spin and parity of all even–even nuclei is $0^+$.* These predictions are exactly correct, and the fact that all even–even nuclei have no net nuclear spin is the reason why relatively few nuclei can be used in NMR studies. *Second, we expect that the ground states of odd-A nuclei, those with an even number of one kind of nucleon and an odd number of the other kind, will be described by the spin and parity of that single odd nucleon.* These predictions are often correct, particularly if we recognize that *single vacancies* or holes in subshells will give the same angular momentum and parity as a *single particle* in the same subshell. This equivalence of "particles" and "holes" can be shown by detailed angular momentum coupling calculations that we will not go into here. However, recall that a completely filled subshell will couple to a spin of 0, so by symmetry if we add one particle to get a given $j$ value, we should expect to get the same spin value when we take one particle from the completely full subshell.

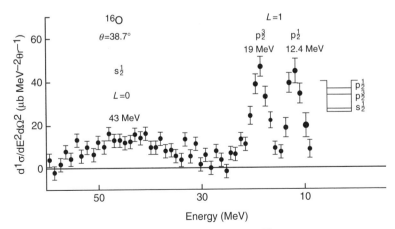

**Figure 6.4** Energy spectrum of emitted protons from the $^{16}$O(p, 2p) reaction, showing the single-particle states. [From Tyren et al. (1958).]

The shell model can also be used to predict the ground-state spins and parities of odd-proton/odd-neutron nuclei by combining the individual $j\pi$ values of the two unpaired particles. Notice that two combinations will always be possible, and we will need a way to decide which of the two alignments of the total-nucleon angular momenta will be lower in energy (i.e., be the ground state). The ground state of the deuteron with its single proton and single neutron provides the key to this selection. The spin angular momenta of the neutron and proton are aligned in the deuteron ground state, thus for the ground state of an odd–odd nucleus we should couple the total $j$ values so that the intrinsic spins of the odd particles are aligned. We can do this by inspection of the angular momenta or by applying a set of rules based on the systematics of the shell model orbitals. Brennan and Bernstein (1960) have summarized these data in the form of three rules. When the odd nucleons are both particles or holes in their respective subshells, rule 1 states that when $j_1 = l_1 \pm 1/2$ and $j_2 = l_2 \pm 1/2$, then $J = |j_1 - j_2|$. Rule 2 states that when $j_1 = l_1 \pm 1/2$ and $j_2 = l_2 \mp 1/2$, then $J = |j_1 \pm j_2|$. Rule 3 states that for configurations in which the odd nucleons are a combination of particles and holes, such as $^{36}$Cl, $J = j_1 + j_2 + 1$.

**Example Problem** Consider the odd–odd nuclei, $^{38}$Cl, $^{26}$Al, and $^{56}$Co. Predict the ground-state spin and parity for these nuclei.

*Solution*

a. $^{38}$Cl has 17 protons and 21 neutrons. The last proton is in a $d_{3/2}$ level while the last neutron is in an $f_{7/2}$ level. (Fig. 6.3).

$$j_p = 2 - \tfrac{1}{2}, \quad j_n = 3 + \tfrac{1}{2}$$
$$J = |\tfrac{7}{2} - \tfrac{3}{2}| = 2$$
$$\pi = -$$

b. $^{26}$Al has 13 protons and 13 neutrons. The last proton and the last neutron are in $d_{5/2}$ hole states, that is, $j_p = j_n = 2 + \tfrac{1}{2}$.

$$J = |\tfrac{5}{2} + \tfrac{5}{2}| = 5$$
$$\pi = +$$

c. $^{56}$Co has 27 protons and 29 neutrons. The last proton is in $f_{7/2}$ hole state and the last neutron is in a $p_{3/2}$ state $(1 + \tfrac{1}{2})$.

$$J = \tfrac{7}{2} + \tfrac{3}{2} - 1 = 4$$
$$\pi = +$$

The simple shell model is very robust and is even successful in describing nuclei at the limits of stability. For example, $^{11}$Li is the heaviest bound lithium isotope. The shell model diagram for this nucleus is indicated in Figure 6.5. Notice the prediction

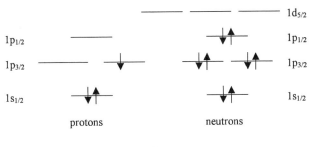

**Figure 6.5** Energy level pattern and filling for the exotic nucleus $^{11}$Li in the schematic shell model.

of two filled neutron shells. The binding energy is only ∼300 keV for the whole nucleus so it is very fragile. It is also known that $^{10}$Li, which does not have a filled p shell, is unbound. This again emphasizes the importance of pairing in nuclei. The two neutrons in the highest energy level that is very close to zero are alone in a $p_{1/2}$ state and the empty $s_{1/2}$ state is very close in energy. This nucleus has an unusually large interaction radius (or size) and a high probability to dissociate into $^9$Li + 2n that have been attributed to a large physical extent of the very weakly bound neutrons in the highest energy level. In fact, there is some debate in the literature as to the relative ordering of the s and p states.

Another nuclear parameter that can be determined experimentally that depends on nuclear structure is the magnetic moment. The magnetic moment of a nucleus is a measure of the response of that nucleus to an external magnetic field and is made up from the net effect of the motion of the protons plus the intrinsic spins of the protons and neutrons. The magnetic moment, $\mu_i$, of one particle can be written as:

$$\mu_i = g_l L_i + g_s S_i \tag{6.1}$$

where $L_i$ is the angular momentum and $S_i$ is the intrinsic spin of particle $i$. The gyromagnetic ratios, $g_l$ and $g_s$, are

$$g_l = l\mu_0 \quad g_s = 5.5845\mu_0 \quad \text{for protons}$$

and

$$g_l = 0 \quad g_s = -3.8263\mu_0 \quad \text{for neutrons}$$

where $\mu_0$ is the nuclear magneton:

$$\mu_0 = e\hbar/2m_p c$$

Due to the large amount of cancellation of the spins and angular momenta due to the strong coupling of nucleons in matching orbitals and pairing of spins, we should expect that the magnetic moments would be small and strongly dependent on the number and orbits of any unpaired particles. A relatively simple formula for the magnetic moments of nuclei with single unpaired nucleons, called the Schmidt limit, contains two forms depending on the relative orientation of the angular momentum and the spin:

For $j = l + s$,

$$\mu = lg_l + \tfrac{1}{2}g_s$$

For $j = l - s$,

$$\mu = (j/j+1)\bigl[(l+1)g_l - \tfrac{1}{2}g_s\bigr] \tag{6.2}$$

or equivalently

$$\mu = j\bigl[g_l \pm (\tfrac{1}{2}l+1)(g_s - gl)\bigr] \tag{6.3}$$

where the + sign is for $j = l + \tfrac{1}{2}$ and the − sign for $j = l - \tfrac{1}{2}$.

The measured magnetic moments of the odd-mass nuclei are similar in magnitude to the Schmidt limits as shown in Figure 6.6. Notice that the measured values fall into two groups at approximately 60% of the predicted values. The fact that the magnetic moments are less than those expected for single particles indicate that the nuclear wave function is not completely dominated by one particle. (If we were to show only the magnetic moments of nuclei that have one particle more than a closed-shell configuration, we would see better agreement with the Schmidt limits.) Also there is a large amount of variation in the magnetic moments that indicates the complexity of the underlying structure and that the cancellation effect of paired particles is not as complete as we might hope.

Up to this point we have concentrated on the properties of the ground states of nuclei predicted by the schematic shell model. However, we can use these energy levels to construct excited states by the promotion of particles and the appropriate coupling of odd (unpaired) particles. First of all, this model has already shown that odd–odd nuclei always have two possible couplings of the angular momenta of the odd particles. One coupling leads to a high spin $J = j_1 + j_2$ and a low total spin $J = j_1 - j_2$. We have already described how to decide which state will lie lower in energy, but notice the other state will always be present. This state will be an isomer that will decay to the ground state by γ-ray emission (usually with a relatively long half-life due to the large change in angular momentum between the states). The relative energy splitting of the two levels decreases as the mass increases due to the dilution effect of more and more nucleon–nucleon interactions. Examples of isomeric pairs of levels and excited states in the simple shell model are given in the accompanying examples.

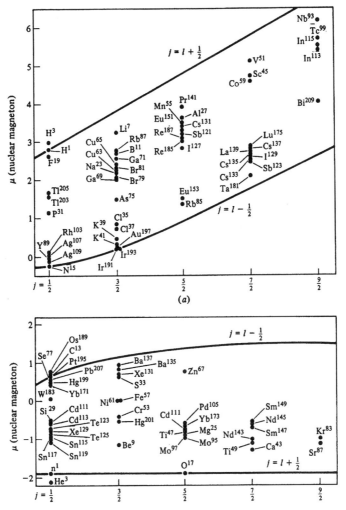

**Figure 6.6** Magnetic moments of the odd-proton (A) and of the odd-neutron nuclei plotted as a function of the nuclear spin, $j$. The Schmidt limits are shown by the solid lines. The data generally fall inside the limits and are better reproduced as 60% of the limits.

## Examples of Isomers

1. $^{26}$Al is a nucleus with 13 protons and 13 neutrons. If we fill in the shell model energy level diagram from the bottom, we find the following configurations:

$$\text{Protons} \quad (1s_{1/2})^2 (1p_{3/2})^4 (1p_{1/2})^2 (1d_{5/2})^5$$
$$\text{Neutrons} \quad (1s_{1/2})^2 (1p_{3/2})^4 (1p_{1/2})^2 (1d_{5/2})^5$$

and recall that a $1d_{5/2}$ level is filled by six particles. Therefore, the net configuration contains a proton hole coupled to a neutron hole in $1d_{5/2}$ states. This is written simply as $\pi(1d_{5/2})^{-1} \otimes \nu(1d_{5/2})^{-1}$. Coupling the proton and neutron angular momenta, we expect $j_p \pm j_n = 0$ and $5\hbar$ for the nuclear spins. The Brennan–Bernstein rules predict that the high spin isomer has the lower energy for identical orbitals, in agreement with observation. The parities of both orbitals are positive ($N = 2\hbar\omega$ shell) so the parities of both coupled states are positive.

2. $^{198}$Au is a nucleus with 79 protons and 119 neutrons. Filling in the shell model energy level diagram we should find that the highest partially filled orbitals are

$$\pi(1h_{11/2})^9 \qquad \nu(1i_{13/2})^7$$

both of which are partially filled subshells near major shell closures. If we make the simplest assumption that all the neutrons and protons are paired except the last odd particles, than we would expect a configuration: $\pi(1h_{11/2}) \otimes \nu(1i_{13/2})$ with $j_p \pm j_n = 1$ and $12\hbar$ for the nuclear spins. The parities of these orbitals are negative ($N = 5\hbar\omega$) and positive ($N = 6\hbar\omega$), respectively, making the product negative. Notice that we could add or remove a pair of neutrons from this configuration, making $^{200}$Au and $^{196}$Au, and we would leave the odd neutron in the same orbital. Therefore, we would make the same predictions for their ground and isomeric states.

An interesting subset of nuclei is those nuclear pairs in which the numbers of protons and neutrons are interchanged, for example, $^3$He and $^3$H. These sets of nuclei are called mirror pairs, and the schematic shell model predicts that they will have identical ground and excited states, after correcting for the (small) upward shift of the proton levels by the Coulomb force and the difference in mass of a neutron and a proton. This shift caused by increasing the nuclear charge by one unit while keeping the mass constant can be readily calculated from the Coulomb energy inside a uniformly charged sphere:

$$E_c = \frac{3}{5}\frac{Ze^2}{R} \qquad (6.4)$$

where $Z$ is the atomic number and $R$ is the radius. The Coulomb energy difference between a mirror pair, where $Z$ refers to the higher atomic number, is then

$$\Delta E_c = \frac{3}{5}\frac{e^2}{R}[Z^2 - (Z-1)^2] = \frac{3}{5}\frac{e^2}{R}(2Z - 1) \approx \frac{Ze^2}{R} \qquad (6.5)$$

This shift is an overestimate as it assumes the nuclei are rigid spheres but, nonetheless, is straightforward to calculate. A large number of mirror pairs have been studied, and the agreement between the energy levels in the mirrors is dramatic.

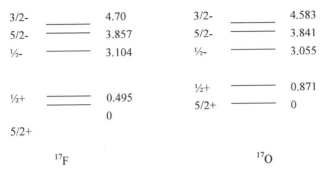

**Figure 6.7** Energy levels of the ground state and first few excited states of the mirror pair $^{17}$F, $^{17}$O are shown. The states are labeled by their intrinsic spin and parity. The matching of these mirror states is remarkable and strongly supports the idea of the neutrons and protons moving in identical orbitals.

An example of the energy level matching in the mirror pair $^{17}$F, $^{17}$O is shown in Figure 6.7. The agreement of the levels is quite remarkable and can be taken as strong evidence for the charge independence of the nuclear force, that is, the protons and neutrons move in essentially identical but separate orbitals in the nucleus.

After all these successes of the very simple shell model, we should be careful to note that there are a number of other well-established and simple properties of nuclei that it cannot describe. For example, the energy levels of essentially all nuclei, and particularly the even–even nuclei with all paired particles, have series of states that are arranged in groups (or bands) with energy spacings and state-to-state transitions that are characteristic of a collective vibration and/or rotation of the entire nucleus. Specifically, even–even nuclei have low-lying 2+ and 4+ excited states that are very strongly related to the 0+ ground state that, once excited, cascade rapidly back to the ground state by γ-ray emission. Examples of such collective states are shown in Figure 6.8. These states correspond to macroscopic vibration of the entire nucleus around the spherical ground-state shape.

Another example of collective motion that is outside the shell model is found in the rare-earth and actinide elements. These nuclei lie between the major shell closures in the shell model and the filling of the midshell high-spin orbitals causes the nuclei to be deformed (stretched like a U.S. football) in the ground state. The orbitals that are being filled in these regions have relatively large $l$ values, for example, g and h states. The angular part of these orbitals is relatively concentrated in space (due to the large number of angular nodes in the wave function) and each suborbital is relatively planar. Recall that s orbitals are spherically symmetric, and orbitals with larger $l$ values are divided by more and more planar nodes. Thus, the midshell nucleons fill relatively nonspherical suborbitals. As we have already discussed, the simple shell model was developed with a spherically symmetric potential. We should expect that the energy levels would shift if the shape of the potential were changed. We will consider the effects of just such a change later in this chapter.

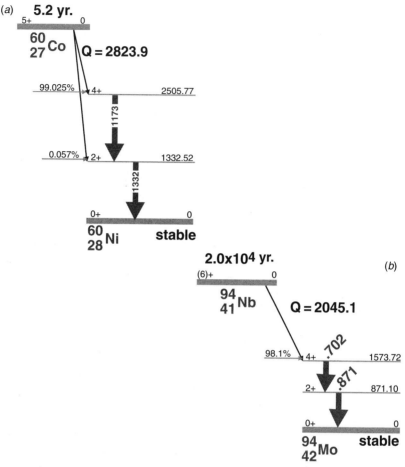

**Figure 6.8** (a) Energy level diagram showing the first (lowest energy) $2^+$ and $4^+$ states in $^{60}$Ni. The high-spin ground state, $5^+$, of $^{60}$Co β decays primarily to the $4^+$ state and initiates a well-known γ-ray cascade to the $2^+$ state and then the $0^+$ ground state. (b) For comparison, the energy level diagram showing the first (lowest energy) $2^+$ and $4^+$ states in $^{94}$Mo. The high-spin ground state, $6^+$, of $^{94}$Nb also primarily feeds the $4^+$ state initiating a γ-ray cascade. (Figure also appears in color figure section.)

## 6.3 INDEPENDENT PARTICLE MODEL

A more detailed model can be constructed for the nucleons in terms of a central potential that holds all the nucleons together plus a "residual potential" or "residual interaction" that lumps together all of the other nucleon–nucleon interactions. Other such important one-on-one interactions align the spins of unlike nucleons (p-n) and cause the pairing of like nucleons (p-p, n-n). The nucleons are then allowed to move independently in these potentials, that is, the Schrödinger equation is solved for the combined interaction to provide the energy levels and wave functions for the individual particles. Once again there will be a large amount of cancellation of the effects of the independent nucleons, and the overall properties of the nucleus can be determined by the last (few) unpaired nucleons or holes.

The central potential can be a simple harmonic oscillator potential $f(r) \sim kr^2$ or more complicated such as a Yukawa function $f(r) \sim (e^{-ar}/r)^{-1}$ or the Woods–Saxon function that has a flat bottom and goes smoothly to zero at the nuclear surface. The Woods–Saxon potential has the form

$$U(r) = \frac{U_0}{1 + \exp[(r - R_0)/a]} + \frac{U_{ls}}{r_0^2} \frac{1}{r} \frac{d}{dr} \left( \frac{1}{1 + \exp[(r - R_0)/a]} \right) \mathbf{l} \cdot \mathbf{s} \qquad (6.6)$$

where $R_0 = r_0 A^{1/3}$, with $r_0 = 1.27$ fm, $a = 0.67$ fm, and the potentials are given by

$$U_0 = [-51 + 33(N - Z)/A] \text{ MeV}$$
$$U_{ls} = -0.44 U_0$$

The spin–orbit strength (second term) is peaked on the nuclear surface as shown in Figure 6.9.

A residual interaction that is also quite simple has been developed and applied with good results. Recall that the nucleon–nucleon force is attractive and very short ranged, so one might image that the nucleons must be in contact to interact. Thus, the simplest residual interaction is an attractive force that only acts when the nucleons touch or a $\delta$ interaction (in the sense of a Kronecker $\delta$ from quantum mechanics). This can be written as $V(r_1, r_2) = a \delta_{12}$, where $a$ is the strength of the interaction, and the $\delta$ function only allows the force to be positive when the nucleons are at exactly the same point in space. In practice, the strength of the potential must be determined by comparison to experimental data. Notice,

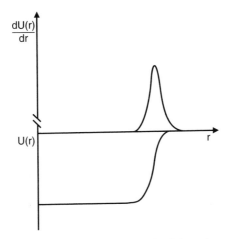

**Figure 6.9** Radial dependence of the strength of the spin–orbit potential.

however, such models have a very small number of parameters to be adjusted to give an overall or average agreement with the data. The "best-fit" values are then used to calculate the properties of other nuclei and their excited states.

## 6.4 COLLECTIVE MODEL

As we have seen, the nucleons reside in well-defined orbitals in the nucleus that can be understood in a relatively simple quantum mechanical model, the shell model. In this model, the properties of the nucleus are dominated by the wave functions of the one or two unpaired nucleons. Notice that the bulk of the nucleons, which may even number in the hundreds, only contribute to the overall central potential. These core nucleons cannot be ignored in reality and they give rise to large-scale, macroscopic behavior of the nucleus that is very different from the behavior of single particles. There are two important collective motions of the nucleus that we have already mentioned that we should address: collective or overall rotation of deformed nuclei and vibrations of the nuclear shape about a spherical ground-state shape.

Rotational motion is characteristic of nonspherical nuclei, and the deformation can be permanent (i.e., the ground state remains deformed) or it can be induced by centrifugal stretching of a nucleus under rapid rotation. The nuclei with masses in the region $150 < A < 190$ and $A > 220$ lie between the major shells and generally have permanent deformations. On the other hand, the rapid rotation of a nucleus can be dynamically induced by nuclear reactions. It is common to create rapidly rotating nuclei in compound nuclear reactions that decay by $\gamma$-ray emission, eventually slowing down to form spherical ground states.

The deformation can be very complicated to describe in a single-particle framework, but a good understanding of the basic behavior can be obtained with an overall parameterization of the shape of the whole nucleus in terms of quadrupole distortions with cylindrical symmetries. If we start from a (solid) spherical nucleus, then there are two cylindrically symmetric quadrupole deformations to consider. The deformations are indicated schematically in Figure 6.10 and give the nuclei ellipsoidal shapes (an ellipsoid is a three-dimensional object formed by the rotation of an ellipse around one of its two major axes). The prolate deformation in which one axis is longer relative to the other two produces a shape that is similar to that of a U.S. football but more rounded on the ends. The oblate shape with one axis shorter than the other two becomes a pancake shape in the limit of very large deformations.

The surface of the ellipsoid can be written in terms of the expansion:

$$R(\theta, \vartheta) = R_{\text{avg}}[1 + \beta Y_{20}(\theta, \vartheta)] \tag{6.7}$$

where $R_{\text{avg}}$, is the average radius of the three major axes, $\beta$ is the dimensionless measure of the deformation, and $Y_{20}$ is the spherical harmonic function. Formally

$$\beta = \frac{4}{3}\sqrt{\frac{\pi}{5}} \frac{b-a}{R_{\text{average}}}$$

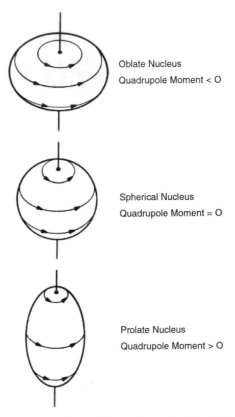

**Figure 6.10** Schematic representations of the prolate and oblate deformations of a uniform sphere. A prolate deformation corresponds to the stretching of the distribution along only one axis while the distribution shrinks equally along the other two axes. An oblate deformation corresponds to the compression of the distribution along one axis with increases along the other two axes.

where $b$ and $a$ are the semimajor and semiminor axes of the ellipsoid and $R_{\text{average}}$ is the average radius, $R_{\text{av}}^2 = \frac{1}{2}(a^2 + b^2)$. The deformation parameter can be positive (prolate shapes) or negative (oblate shapes) and is generally a small number. For example, the superdeformed prolate shape with an axis ratio of 2:1 has $\beta \sim 0.6$.

The energy levels from the quantum mechanical solution of the rotation of a rigid body have the characteristic feature of increasing separation with angular momentum. The energy levels are given by the expression:

$$E_{\text{rot}} = \frac{J(J+1)\hbar^2}{2\Im} \quad (6.8)$$

where $J$ is the rotational quantum number describing the amount of rotation and $\Im$ is the moment of inertia of the rigid body. The moment of inertia of a solid sphere with mass $m$ is $\Im_{rigid} = 2mR^2/5$. Substituting in constants and using $R = 1.2A^{1/3}$ fm, we find that the rotational energy levels of a sphere at $E_{rot} = 36.29\, J(J+1)/A^{5/3}$ MeV for $J$ in $\hbar$ units. Note the large power of $A$ in the denominator, which causes the expression for the rotational constant, $\hbar^2/2\Im$, to be on the order of keV-s. This expression is called the rigid-body limit, and spherical rigid-body values for rotational energies are generally smaller than those observed, meaning that the real nuclear moment of inertia is smaller. These spherical moment of inertia expressions can be readily extended to nuclei with static or rigid deformations by substituting the appropriate moment of inertia, thus, $\Im = 2mR_{avg}^2/5(1 + 0.31\beta)$. The result is similar in that the deformed rigid-body estimate of the moment of inertia is too large and the rotational energy is too small.

We have already seen that nuclei have some properties that are similar to those of a liquid drop; in fact, the overall binding energy is well represented in these terms. The moment of inertia for the rotation of the liquid in a rigid deformed container, for example, a large water balloon with a negligible mass wall, is $\Im_{irro} = \left(\frac{9}{8}\pi\right)mR^2\beta^2$. This moment of inertial is smaller than that of a rigid body because the liquid can "flow" inside the container to follow the motion of the walls and the moment of inertia goes to zero at $\beta = 0$ as expected by symmetry. This rotational behavior is called irrotational flow. The irrotational flow moment of inertia gives a value that is usually smaller than the experimental value, leading to rotational energies that are larger than the experimental data. Thus, we have the situation that:

$$\Im_{irro} < \Im_{exp} < \Im_{rigid}$$

which allows us to bracket the experimental value with numerical estimates.

**Example of a Rotational Constant** The ground-state rotational band of $^{152}$Gd is shown in Figure 6.11. Use the energy separation between the 2+ and 0+ levels to estimate the rotational constant in keV, the moment of inertia in amu-fm$^2$, and then compare your result to that obtained to the rigid-body result with a deformation parameter of $\beta = 0.2$. Finally, evaluate the irrotational flow moment of inertia for this nucleus.

$$E_{rot} = \frac{J(J+1)\hbar^2}{2\Im}$$

$$\Delta E_{rot}(2 \to 0) = (6-0)\frac{\hbar^2}{2\Im} = 344.3 \text{ keV}$$

$$\frac{\hbar^2}{2\Im} = 57.3 \text{ keV} \qquad \Im_{exp} = 364.7 \text{ amu-fm}^2$$

$$\Im_{rigid} = \frac{2}{5}mR_{avg}^2(1 + 0.31\beta)$$

$$\Im_{rigid} = 2494 \text{ amu-fm}^2[1 + 0.31(0.2)] = 2648 \text{ amu-fm}^2$$

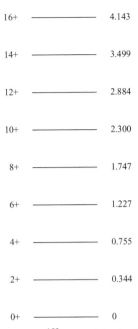

**Figure 6.11** Low-lying energy levels of $^{152}$Gd, which clearly fit the pattern of a rotational band. The rotational constant $\hbar^2/2\Im = 68.9$ keV can be extracted.

$$\frac{\Im_{exp}}{\Im_{rigid}} = \frac{1}{7.3}$$

$$\Im_{irro} = \frac{9}{8\pi} mR_{avg}^2 \beta^2$$

$$\Im_{irro} = 2232 \text{ amu-fm}^2 (0.2)^2 = 89.3 \text{ amu-fm}^2$$

Different rotational bands in a given nucleus can have different effective moments of inertia. This could reflect a larger deformation or a change in the number of paired nucleons or a different alignment of a pair of nucleons of high spin. The result is that each band can have a different pattern of energy vs. spin (Fig. 6.12a). If one plots $2\Im/\hbar^2$ vs. the rotational frequency $\hbar^2\omega^2$ for a given nucleus, then one observes a kink or "backbend" in the plot corresponding to the region where the two bands cross (Fig. 6.12b).

A special class of quantum rotors are the *superdeformed nuclei*. The moments of inertia, after scaling by $A^{5/3}$, are all similar due to the fact that the shape of these nuclei is largely independent of mass with an axis ratio of 2 : 1 due to shell stabilization effects discussed below.

Another interesting case of nuclear rotation occurs in the spherical nuclei. The observation of equally spaced γ-ray transitions implies collective rotation, but such bands have been observed in near spherical $^{199}$Pb. It has been suggested that these bands arise by a new type of nuclear rotation, called the "shears mechanism." A few

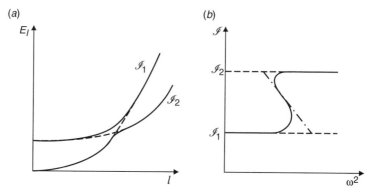

**Figure 6.12** Schematic picture of two intersecting bands with different moments of inertia, $\Im_1$ and $\Im_2$, and the corresponding backbending plot.

valence neutron and proton holes couple to form "long" angular momenta, $j_n$ and $j_p$, which couple to give the total spin $j$. By varying the angle between these "blades of the shears," states of differing spin are created. This gives rise to a magnetic moment, and the radiation associated with the γ-ray transition between the states is $M1$.

The other important macroscopic motions of nuclei are the vibrations of the nuclear volume around the spherical ground state. Recall that the great majority of nuclei have spherical ground states, but they also can behave like liquid drops; so we might imagine that the surface of the nucleus could be caused to vibrate harmonically, back and forth, around the spherical ground state. In this picture we could parameterize the shape vibrations, also called surface oscillations, in terms of the spherical harmonic functions with their characteristic multipolarities. We should also be careful to differentiate between the characteristic motion labeled by the *multipolarity*, or "shape symmetry," of the mode and the *number of quanta*, or phonons, in each vibrational mode. One might imagine multiple excitation of a single mode, single excitation of several symmetries simultaneously, or any other combination.

The lowest order macroscopic vibration is a swelling/compression of the whole nucleus with $\lambda = 0$. This is sometimes called the "breathing" mode. The next macroscopic vibration, labeled $\lambda = 1$, is a dipole motion. However, such a motion of the entire surface, first in one direction and then back in the other, simply corresponds to translation of the nucleus and not internal vibration. This motion would have to be caused by a "restoring force" that was outside the nucleus and so there is no intrinsic dipole motion of a (whole) nucleus. The next order vibration, labeled $\lambda = 2$, is the quadrupole motion in which the nucleus symmetrically stretches in and then out without moving its center of mass. This is clearly a vibrational motion with a "restoring force" generated by the nuclear potential. The third-order vibration, labeled $\lambda = 3$, is the octupole motion in which the nucleus asymmetrically expands on one end while pinching on the other. This vibration creates pear-shaped figures and requires significantly more energy to excite compared to the more symmetric quadrupole shapes.

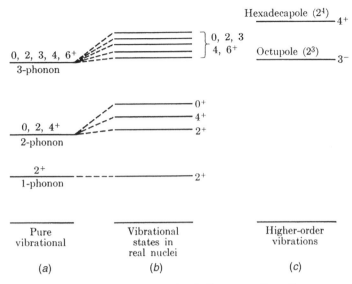

**Figure 6.13** Schematic vibrational energy-level diagrams of medium-mass even–even nuclei. [From Eichler (1964).]

Recall that the energy levels of the quantum mechanical harmonic oscillator are all equally spaced, having energies $E_N = (N + 1/2)\hbar\omega_0$, $N = 0, 1, 2$ (see Fig. 6.13). The fundamental frequency of the oscillation, $\omega_0$, is equal to the square root of the force constant divided by the effective mass. Considering even–even nuclei with 0+ ground states, single excitation of quadrupole motion with $\lambda = 2$ will require an $N = 2$ state as $N = 1$ is not allowed because it would break the symmetry of the nuclear wave function. This $N = 2$ excitation gives rise to a $2^+$ state with two $\hbar\omega_0$ of energy. We would expect the subsequent multiple excitation of this mode would create a $4^+$ state with twice the excitation energy and so on. Because there are three ways to couple two quadrupole phonons together ($J^\pi = 4^+, 2^+,$ and $0^+$), the two-phonon state is triply degenerate. The three-phonon state at an energy of $3\hbar\omega_0$ will include $6^+, 4^+, 3^+, 2^+,$ and $0^+$ states. From a global perspective, we observe that the ratios of the energy of the $4^+$ to $2^+$ states in even–even nuclei are approximately 2:1 with two strong deviations. First, the deformed rotational nuclei have $4^+/2^+$ ratios of 10:3 as discussed above. And, second, when the number of neutrons or protons are close to the magic numbers for closed spherical shells, the nucleus becomes more resistant to oscillation, and the energies of the $2^+$ and $4^+$ states increase dramatically as well as their ratio.

It is interesting to note that the vibrational model of the nucleus predicts that each nucleus will be continuously undergoing zero-point motion in all of its modes. This zero-point motion of a quantum mechanical harmonic oscillator is a formal consequence of the Heisenberg uncertainty principle and can also be seen in the fact that the lowest energy state, $N = 0$, has the finite energy of $\hbar\omega/2$.

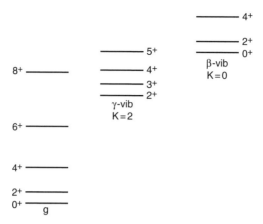

**Figure 6.14** Positive parity levels of a typical deformed nucleus.

From another standpoint, the superposition of all of these shape oscillations can be viewed as a natural basis for the diffuseness of the nuclear surface.

The energy of rotational states built on vibrations is given by

$$E = \frac{\hbar^2}{2\Im}[J(J+1) - K^2] \tag{6.9}$$

where $K$ is the projection of $J$ on the nuclear symmetry axis. For $\beta$ ($\lambda = 2$, $K = 0$) vibrations, $J^\pi = 0^+, 2^+, 4^+, \ldots$ for $\gamma$ ($\lambda = 2$, $K = 2$) vibrations, $J^\pi = 2^+, 3^+, 4^+, \ldots$ A typical sequence of states is shown in Figure 6.14.

## 6.5 NILSSON MODEL

Up to now, we have discussed two extremes of nuclear structure, those aspects that can be explained by the properties of single or individual particles moving in a spherically symmetric central potential and those aspects corresponding to large-scale collective motions of groups of nucleons away from spherical symmetry. Additional insight into the structure of nuclei can be obtained by considering the states of single particles moving in a deformed nuclear potential. S. G. Nilsson extensively studied this problem, and the resulting model of nuclear structure is referred to as the Nilsson model.

Using a deformed harmonic oscillator potential, one can make several useful observations about the nuclear structure of deformed nuclei. In Figure 6.15, we show the energies of single-particle states of such a potential as a function of the deformation of the potential. At spherical symmetry, one observes the gaps in the level spacings corresponding to the major harmonic oscillator shells that we have already discussed. But as the deformation changes, the levels move in energy and new magic numbers (shell gaps) occur when the ratio of the semimajor

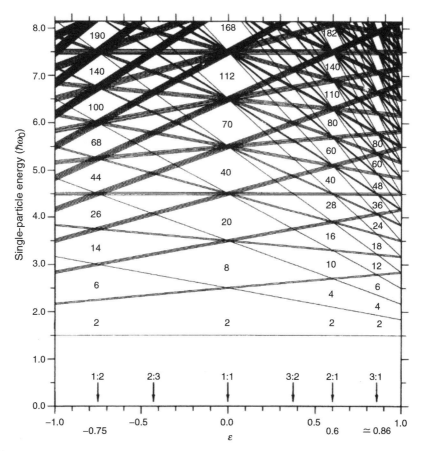

**Figure 6.15** Single particle levels of a deformed harmonic oscillator as a function of the deformation of the potential. The ratios of the semimajor to semiminor axes of the oscillator are shown also.

to the semiminor axes of the nucleus is a simple whole number. Thus, nuclei with axes ratios of 2:1 have special stability (*the superdeformed nuclei*). In addition, each spherical shell model state, for example, an $f_{7/2}$ state, is split into $(2j + 1)/2$ levels that can be labeled with a new quantum number $\Omega$ defined as the projection of the single-particle angular momentum on the nuclear symmetry axis (Fig. 6.16). For prolate deformation, states of highest $\Omega$ lie the highest in energy.

The angular momentum of an odd $A$ deformed nucleus, $J$, is the vector sum of the angular momentum of the last unpaired nucleon and the rotational angular momentum, $R$, of the core of remaining nucleons, as shown schematically in Figure 6.17. The projection of the total nuclear angular momentum $J$ upon the nuclear symmetry axis is given the symbol $K$. For axially symmetric nuclei, the direction of $R$ is perpendicular to the symmetry axis and $J = \Omega = K$. Each Nilsson single-particle level may be the ground state of a rotational band. For the ground state of such

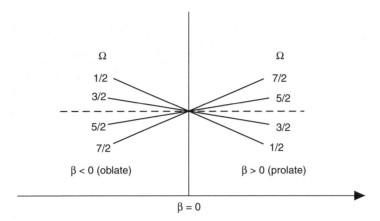

**Figure 6.16** Schematic diagram of the splitting of the $f_{7/2}$ spherical shell model level as the potential deforms. Positive deformations correspond to prolate shapes while negative deformations correspond to oblate shapes.

bands, $J = \Omega = K$. When $J = \frac{3}{2}$ or greater, the allowed nuclear spins of the members of the band are $J_0$, $J_0 + 1$, $J_0 + 2$, etc. The energies of the members of the band are given as

$$E = \frac{\hbar^2}{2\Im}[J(J+1) - J_0(J_0+1)] \qquad (6.10)$$

The Nilsson model is able to predict the ground state and low-lying states of deformed odd $A$ nuclei. Figure 6.18 is a more detailed picture of how the energies of the Nilsson levels vary as a function of the deformation parameter $\beta_2$ for the

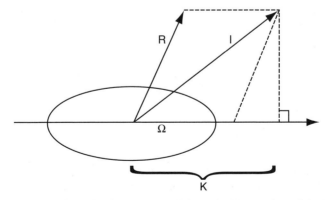

**Figure 6.17** Addition of angular momenta in a deformed odd $A$ nucleus. $\Omega$ is the projection of the total angular momentum of the odd nucleon. It is added vectorially to the rotational angular momentum of the core, $R$, to give the total angular momentum $J$ whose projection on the symmetry axis is $K$.

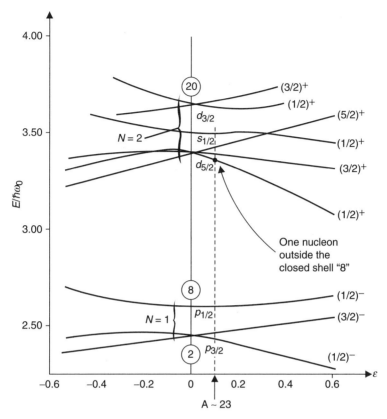

**Figure 6.18** Energy of a single nucleon in a deformed potential as a function of deformation $\varepsilon$. The is diagram pertains to either $Z < 20$ or $N < 20$. Each state can accept two nucleons.

first 20 nucleons in the nucleus. Let us consider the nuclei $^{19}$F, $^{19}$Ne, $^{21}$Ne, and $^{23}$Na. According to the simple shell model (Fig. 6.3), the last odd nucleon in these nuclei should be in a $d_{5/2}$ state giving rise to a ground-state spin and parity $J^\pi$ of $\frac{5}{2}^+$. One observes $J^\pi$ for these nuclei to be $\frac{1}{2}^+, \frac{1}{2}^+, \frac{3}{2}^+$, and $\frac{3}{2}^+$, respectively. The Nilsson model allows us to understand these observations. From the quadrupole moment of these nuclei, 0.1 barns, we can deduce $\beta = 0.1$. Thus, from the levels shown in Figure 6.18, we expect the 9th particle to have $J^\pi = \frac{1}{2}^+$, while the 11th particle will have $J^\pi = \frac{3}{2}^+$ in agreement with the observations. The low-lying excited states of many nuclei can also be explained by considering the positions of the nearby levels with small deformations in the Nilsson diagrams.

## 6.6 NUCLEUS AS A FERMI GAS

The preceding discussion of nuclear structure and models was mostly aimed at explaining the detailed properties of the ground states and small excitations of nuclei. These nuclei are produced and take part in nuclear reactions that are usually very complicated dynamical situations compared to the (simple) situation of nucleons confined in a static central potential. Detailed calculations with wave functions in dynamical potentials associated with scattering and absorption have only been carried out in a few cases. On the other hand, a very wide variety of nuclear reactions have been studied experimentally and often exhibit amazing simplicity. Models have been developed that treat the average behavior of the large number of nucleons in a nucleus on a statistical basis. An important statistical model developed to describe the average behavior of medium and large nuclei, particularly useful in reactions, treats the nucleus as a fluid of fermions. This approximation, called the Fermi gas model, uses the now familiar concept of confining the nucleons to a fixed spherical shape with a central potential, but in this case the nucleons are assumed to be all equivalent and independent. This situation loosely corresponds to an ideal gas confined to a fixed volume with the addition of Fermi–Dirac statistics and Pauli blocking to prevent spatial overlap of the particles.

An important feature of the Fermi gas model is that it allows us to describe the average behavior of a nucleus with thermodynamical functions. The fact that the energy levels in a (large) nucleus with a finite excitation energy are so closely spaced allows us to use entropy to predict the evolution of the system. (This idea is somewhat opposite to the ground-state situation that is dominated by the wave functions of individual particles.) The concept of thermodynamic entropy is closely linked to a thermodynamic nuclear temperature. Nuclear reactions are often described in terms of the imagined temperature of the internal particles, and excited nuclei emit light particles and $\gamma$ rays as they lose their excitation energy and "cool" as they approach the ground state.

The first step in developing the Fermi gas model is to determine the highest level that is occupied by nucleons. Next the average energies and momenta are calculated because we will assume that all the lower levels are exactly filled. The nucleons are confined to a fixed total volume and are assumed to have a uniform density. When quantum mechanical particles are confined in a rigid container, then they occupy fixed states that can be labeled with appropriate quantum numbers, that is, $n_x$, $n_y$, $n_z$ for a rectangular box with three dimensions, $L_x$, $L_y$, $L_z$. The particle will have a specific momentum in each state so, alternatively, we could label the states by their momenta, $p_x$, $p_y$, $p_z$, or by their wavenumbers, $k_x$, $k_y$, $k_z$ where $k_i = (n_i \pi / L_i)$, which explicitly incorporates the dimensions of the box. We would like to know what is the highest quantum number, the largest momentum, $p_f$, or the wavenumber, $k_f$, of the highest filled level, called the Fermi level. The Fermi level wavenumber can be written as:

$$k_f^2 = k_x^2 + k_y^2 + k_z^2 \tag{6.11}$$

or in terms of the quantum numbers:

$$\frac{k_f^2 L^2}{\pi^2} = n_x^2 + n_y^2 + n_z^2 \tag{6.12}$$

## 6.6 NUCLEUS AS A FERMI GAS

The number of different combinations of the positive integer quantum numbers that fulfill this equality is given by the volume of one-octant of a sphere:

$$N_{\text{states}} = \left(\frac{1}{8}\right)\frac{4\pi}{3}\left(\frac{k_f L}{\pi}\right)^3 \tag{6.13}$$

Remember that the Pauli principle allows us to put particles with two spins (up/own) into each level, and if the nucleons are all in their lowest possible states, the number of filled states can be assumed to be equal to the number of each type of nucleon. Thus, the Fermi wavenumber for protons is

$$k_f = \frac{\pi}{L}\left(\frac{2N_{\text{states}}}{3\pi}\right)^{1/3} = \frac{\pi}{L}\left(\frac{2Z}{3\pi}\right)^{1/3} = \frac{\pi}{r_0}\left(\frac{2Z}{3\pi A}\right)^{1/3} \tag{6.14}$$

and similarly for neutrons where we have taken $L$ to the nuclear radius, $r_0 A^{1/3}$. Notice that we have obtained an expression that depends only on $Z/A$ (or $N/A$) and the radius constant, $r_0$, so the value of the Fermi energy will be similar for most nuclei because the variation of $Z/A$ is small for stable nuclei and enters via the cube root. The Fermi energy for nucleons in those nuclei with $Z/A = \frac{1}{2}$ taking $r_0 = 1.2$ fm is

$$E_f = \frac{k_f^2 \hbar^2}{2m} \approx 32 \text{ MeV} \tag{6.15}$$

If the number of neutrons is greater than the number of protons, as in heavy nuclei, then the Fermi energies will be slightly different for the two kinds of particles. An approximate representation of the Fermi energy for protons and neutrons is

$$E_f^{\text{protons}} \cong 53\left(\frac{Z}{A}\right)^{2/3} \text{ MeV}$$

$$E_f^{\text{neutrons}} \cong 53\left(\frac{A-Z}{A}\right)^{2/3} \text{ MeV} \tag{6.16}$$

The average kinetic energy of the nucleons in the well can be shown to be $\frac{3}{5}E_f$, or approximately 20 MeV. Notice that the nucleons are moving rapidly inside the potential well but not extremely fast.

**Example Problem** What is the deBroglie wavelength of a neutron moving with the average Fermi energy in a $^{208}$Pb nucleus? You can assume that the neutron is nonrelativistic and use $r_0 = 1.2$ fm.

## Solution

$$k_f = \frac{\pi}{r_0}\left(\frac{2Z}{3\pi A}\right)^{1/3} = 1.145 \text{ fm}^{-1}$$

$$E_f = \frac{k_f^2 \hbar^2}{2m} = 27 \text{ MeV}$$

$$E_{avg} = \tfrac{3}{5} E_f = 16 \text{ MeV}$$

$$\lambda = \frac{h}{p} = \frac{2\pi}{k_f} = 5.487 \text{ fm}$$

Note that this wavelength is similar to the lead radius $R \sim 1.2 A^{1/3} = 7.1$ fm.

A schematic version of the Fermi gas potential energy well for a large nucleus is shown in Figure 6.19. Recall that nucleons are bound by approximately 8 MeV, on average, so the uppermost filled energy level (Fermi level) should be approximately at $-8$ MeV. The lowest level is then approximately 32 MeV below this, which makes the Fermi gas potential energy well relatively shallow. The levels between the Fermi level and zero potential energy are assumed to be completely empty in the ground state and become occupied when the nucleus absorbs excitation energy.

As a nucleus absorbs energy nucleons are promoted from the filled levels into the unfilled region between the Fermi level and zero potential energy. Each promotion leads to a specific excitation energy, and combinations of multiple excitations can lead to the same or similar energies. At high excitations the number of combinations of different possible promotions for a specific excitation energy grows dramatically. The tremendous growth of the number of energy levels with excitation energy is

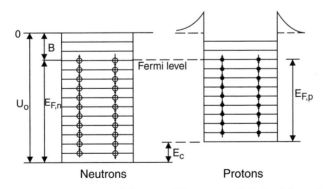

**Figure 6.19** A schematic version of the potential energy well derived from the Fermi gas model. The highest filled energy levels reach up to the Fermi level of approximately 32 MeV. The nucleons are bound by approximately 8 MeV, so the potential energy minimum is relatively shallow.

one of the interesting features of nuclei. The number of levels is so large that we can describe the system by an average level density, $\rho(E^*, N)$, which is simply the number of levels per unit excitation energy, $E^*$, for a fixed number of nucleons, $N$. The fact that excited nuclei, even with a finite number of particles, have very dense and nearly continuous distributions of levels is the feature that allows us to describe their deexcitation with statistical techniques.

The connection between the microscopic description of any system in terms of individual states and its macroscopic thermodynamical behavior was provided by Boltzmann through statistical mechanics. The key connection is that the entropy of a system is proportional to the natural logarithm of the number of levels available to the system, thus:

$$S(E, N) = k_B \ln \Gamma(E, N) = k_B \ln [\rho(E, N)\Delta E] \tag{6.17}$$

where $\Gamma$ is the total number of levels. The entropy of an excited nucleus is thus proportional to the level density in some energy interval, $\Delta E$, and goes to zero as the excitation energy goes to zero. (Recall that there is only one nuclear ground state.) The thermodynamic temperature can be calculated from the entropy as:

$$\frac{1}{T} = \frac{\partial S(E^*, N)}{\partial E} = k_B \frac{\partial \ln \rho(E^*, N)}{\partial E} \tag{6.18}$$

In statistical mechanics the Boltzmann constant, $k_B$, with dimensions of energy per degree is included in expressions so that the temperatures can be given in degrees Kelvin. The numerical values of nuclear temperatures in Kelvin are very large, for example $10^9$ K, so the product of $k_B$ and $T$ is usually quoted in energy units (MeV) and the Boltzmann factor is often not written explicitly.

At this point we have not distinguished between nuclear systems and macroscopic systems on the basis of their size. There is, however, at least one important difference between the two. The difference is the way the entropy $S(E^*, N)$ should be evaluated. In statistical mechanics one has different physical situations (ensembles) for evaluating thermodynamic quantities: fixed energy and particle number (microcanonical ensemble), fixed temperature and particle number (canonical ensemble), and fixed temperature and chemical potential (grand canonical ensemble). In the evaluation of thermodynamic quantities for macroscopic systems, each of these approaches provides essentially the same result. Thus, the entropy may be evaluated by calculating any of the following: $S_{\text{microcanonical}}$, $S_{\text{canonical}}$, or $S_{\text{grandcanonical}}$. This is not the case for nuclear systems because the only appropriate ensemble is the microcanonical ensemble of isolated systems. The fundamental definition of nuclear temperature should be written

$$\frac{1}{T} = \frac{\partial S_{\text{microcanonical}}(E^*, N)}{\partial E} \tag{6.19}$$

and it is not correct to substitute an entropy obtained with a different ensemble into this expression.

Standard procedures permit the evaluation of the entropy of a Fermi gas under the conditions of a *grand canonical ensemble*, which we will have to adjust to obtain the *microcanonical* entropy. For low excitation energies, $E^*$, the entropy is

$$S_{\text{grandcanonical}}(E^*, N) = 2(aE^*)^{1/2} \tag{6.20}$$

where $a$ is a constant proportional both to the number of particles and to the density of the single-particle levels of the Fermi gas at the Fermi energy, $E_f$. If $S_{\text{grandcanonical}}$ is used to replace $S_{\text{microcanonical}}$, one obtains $T = (E/a)^{1/2}$ as the link between temperature and excitation energy. This result would be appropriate for macroscopic systems, but as we said it must be modified for isolated nuclear systems. For small systems

$$S_{\text{microcanonical}} = S_{\text{grandcanonical}} + \Delta S \tag{6.21}$$

where $\Delta S$ becomes vanishingly small compared to $S_{\text{grandcanonical}}$ as the number of particles or the excitation energy becomes large. An approximate expression for $\Delta S$ for a Fermi gas at relatively low energy is

$$\Delta S \approx -\gamma \ln(E^*) \tag{6.22}$$

with $\gamma$ being a number of the order of unity, ranging from 1 to 2 depending on whether isospin and angular momentum are explicitly considered in the labeling of the states. When the appropriate $S_{\text{microcanonical}}$ is used to evaluate the nuclear temperature, one finds

$$\frac{1}{T} = \frac{\partial S_{\text{grandcanonical}}}{\partial E} + \frac{\partial \Delta S}{\partial E} \tag{6.23}$$

For the moderately low energies this provides

$$\frac{1}{T} \approx \left(\frac{a}{E^*}\right)^{1/2} - \left(\frac{\gamma}{E^*}\right) \tag{6.24}$$

as the link between excitation energy and nuclear temperature. For large excitation energies, $E^*$, and large particle number, the correction term proportional to $\gamma$ vanishes and $E^* \simeq aT^2$.

The density of nuclear states can then be written as:

$$\rho(E^*) \propto \frac{a}{(aE^*)^\gamma} \exp[2(aE^*)^{1/2}] \tag{6.25}$$

The factor $a$ here is called the level density parameter and is adjusted to correspond to level densities measured at low excitation energies. The analyses of data over a broad mass range suggest that $a$ is proportional to the mass of the nuclear system $A$ being $a \approx A/8.5 \text{ MeV}^{-1}$. The level densities can be corrected for angular momentum by including preexponential statistical factors and subtracting the collective energy that is involved in rotation. The rotational energy is often included with an effective moment of inertia, a parameter adjusted to match experimental spectra and yields.

We can extend the Fermi gas level density analysis to predict the relative probability of various decay modes of excited nuclei if we make the assumption that the nuclei are in full thermal equilibrium. That is, we assume that all of the energy levels corresponding to a given excitation energy are fully populated. It is not possible for a single nucleus to be in many states simultaneously, it can only be in one. So the thermal equilibrium that we require must apply to a set of nuclei created in many (identical) reactions. This is, of course, how chemical reactions take place when Avogadro's number of atoms or molecules with various kinetic energies but one temperature follow a path from reactant to products based on a specific reaction mechanism. Nuclear reactions are usually detected by producing large numbers of nuclei, $\gg 10^3$, and then observing various reaction products and determining the probabilities of each process.

Excited nuclei that have attained statistical equilibrium will decay into different products in proportion to the number of states available to the whole system after the decay. The different decays are often called channels, and we speak of the probability to decay into a given channel. A very schematic representation of the energy levels and the energies involved in the decay of an excited nucleus into various channels is shown in Figure 6.20. The total sum of the probabilities for decay into all channels is, of course, one. We can simply count the number of states available for a decay channel and obtain a general expression for the relative probability, $P(\varepsilon, n)$, for an excited nucleus to emit a portion with size $n$, requiring an energy $\varepsilon$. The expression is

$$P(\varepsilon, n) \propto \Gamma(\varepsilon, n) \cdot \Gamma(E - \varepsilon, N - n) \tag{6.26}$$

where $\Gamma(E, N)$ is the number of states in the vicinity of energy $E$ for a system of mass number $N$. The first factor on the right-hand side is contributed by the states in the

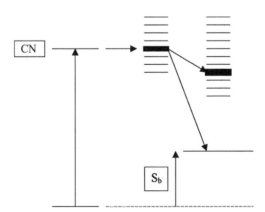

**Figure 6.20** A representation of the branching decays from a highly excited compound nucleus. In the statistical model, the relative probability for the excited nucleus to decay into a specific channel is proportional to the number of possibilities or *statistical weight* of that channel divided by the sum of all of the statistical weights of all of the channels.

emitted piece, and the second is contributed by the states in the (large) daughter nucleus. The number of internal states can be taken to be the log of the level densities used to define the entropy, above, and we will need to include a term for the kinetic energy of the emitted piece. However, we need to integrate the emission rates over the whole course of the nuclear reaction to obtain the total yields that can be measured in the laboratory.

Focusing on comparisons to measurable quantities, the relative probability of a reaction (exit) channel can be written as the ratio of the cross section for that channel, $\sigma_i$, to the total reaction cross section, $\sigma_r$. The ratios are labeled as the relative decay widths, $\Gamma_i$, in a notation that is, unfortunately, easy to confuse with the number of states discussed above. The sum of the decay widths is the total width of the state and can be used to calculate the lifetime of the excited state. Thus,

$$\sigma_i/\sigma_T = \Gamma_i/\Gamma_T \qquad (6.27)$$

and notice that the double ratio of the relative probability of two decay channels does not depend on the total reaction cross section:

$$\sigma_i/\sigma_j = \Gamma_i/\Gamma_j \qquad (6.28)$$

The width for the emission of a particle with a binding energy of $B_n$ and no internal states has been shown to have the form:

$$\Gamma_i \propto \Gamma(E, N) \int_0^{E-B_i} \varepsilon \rho(E - B_i - \varepsilon)\, d\varepsilon \qquad (6.29)$$

with $E$ the excitation of the excited parent nucleus. Therefore, the relative intensities of the channels change because the binding energies of the emitted particles change the density of states through the exponential dependence of the level density. This can be extended to the case of fission decay leading, in the simplest approximation, to a slightly different integral:

$$\Gamma_f \propto \Gamma(E, N) \int_0^{E-E_f} \rho(E - E_f - \varepsilon)\, d\varepsilon \qquad (6.30)$$

in terms of the fission barrier, $E_f$. The ratio $\Gamma_n/\Gamma_f$ is very important in determining the survival of the very heaviest elements when they are synthesized in nuclear reactions. Notice that if the nucleus emits a neutron to remove excitation, it retains its large atomic number; however, if it fissions, then it is converted into two nuclei with much smaller atomic numbers. The integrals in the above expressions can be evaluated in the Fermi gas approximation with the following approximate result (Vandenbosch and Huizenga, 1973):

$$\frac{\Gamma_n}{\Gamma_f} \approx \frac{2TA^{2/3}}{K_0} \exp\left(\frac{E_f - B_n}{T}\right) \qquad (6.31)$$

where $K_0 = \hbar^2/2mr_0^2 \sim 15$ MeV and $T$ is the nuclear temperature created by the initial reaction. The exponential function contains the difference between the fission barrier and the neutron separation energy. Therefore, this ratio is only near one when these two values are nearly equal. If there is a large difference between the fission barrier and the separation energy, then the ratio will be very large or very small.

**Example Problem** In a certain nuclear reaction, a beam of $^{18}$O was combined with $^{233}$U nuclei to form a compound nucleus of $^{256}$Fm. The nuclei were produced with an excitation energy of 95 MeV. Calculate the nuclear temperature assuming that $\gamma = 1$, and then the relative probability of neutron to fission decay of the excited system.

*Solution*

$$\frac{1}{T} \approx \left(\frac{a}{E^*}\right)^{1/2} - \left(\frac{1}{E^*}\right)$$

with $a = A/8.5 = 256/8.5 = 30.1$ MeV$^{-1}$

$$\frac{1}{T} \approx \left(\frac{30.1}{95}\right)^{1/2} - \left(\frac{1}{95}\right) \qquad T \approx 1.8 \text{ MeV}$$

We need to find the neutron separation energy and the fission barrier for this nucleus in order to evaluate the ratio:

$$\frac{\Gamma_n}{\Gamma_f} \approx \frac{2TA^{2/3}}{K_0} \exp\left(\frac{E_f - B_n}{T}\right)$$

The neutron separation energy is 6.38 MeV, and the fission barrier is 5.90 MeV.

$$\frac{\Gamma_n}{\Gamma_f} \approx \frac{2 \times 1.8 \times 256^{2/3}}{15} \exp\left(\frac{5.90 - 6.38}{1.8}\right)$$

$$\frac{\Gamma_n}{\Gamma_f} \approx 9.68 \exp(-0.266) = 7.4$$

## PROBLEMS

1. Predict the ground-state spins and parities for $^{41}$K, $^{44}$Ca, and $^{60}$Co.

2. An odd $A$ nucleus has a $J^\pi = \frac{7}{2}^+$ ground state. (a) What is $J$, $K$ for the first two excited states? (b) If the energy of the first excited state is 100 keV, what is the energy of the second excited state?

3. Define or describe the following terms or phenomena: island of isomerism, spin–orbit coupling, β vibration, Schmidt limits, and Nilsson states.

172    NUCLEAR STRUCTURE

4. What is the energy of the $2^+ \to 0^+$ γ-ray transition in $^{172}$Hf assuming $^{172}$Hf is a perfect rotor with its moment of inertia $I$ given as $\frac{2}{5}mR^2$?

5. The α decay of $^{241}$Am ($t_{1/2} = 420$ y, $J^\pi = \frac{5}{2}^+$) populates members of at least two rotational bands (A and B) in $^{237}$Np shown in the table below. (a) Using the collective model, predict the energies of the $\frac{9}{2}^+$, $\frac{9}{2}^-$, and $\frac{11}{2}^-$ levels. (b) Calculate the effective moment of inertia of $^{237}$Np.

| Band | $J, \pi$ | $E$ (keV) |
|---|---|---|
| B | $\frac{11}{2}^-$ | ? |
| B | $\frac{9}{2}^-$ | ? |
| B | $\frac{7}{2}^-$ | 103 |
| A | $\frac{9}{2}^+$ | ? |
| B | $\frac{5}{2}^-$ | 59.6 |
| A | $\frac{7}{2}^+$ | 33.2 |
| A | $\frac{5}{2}^+$ | 0 |

6. Calculate the energy of the $4^+$, $6^+$, $8^+$, and $10^+$ members of the ground-state rotational band of an even–even nucleus if the energy of the $2^+$ member of the band is 0.044 MeV above the ground state.

7. $^{237}$Np has levels at 0.033, 0.060, 0.076, 0.103, and 0.159 MeV. Which of these, if any, would you expect to be members of a rotational band whose first member is the $\frac{5}{2}^+$ state of $^{237}$Np?

8. Analyze the following level schemes in terms of the collective and Nilsson models:

| | | | | | |
|---|---|---|---|---|---|
| 0.425 | _____ | $\frac{7}{2}^-$ 6.000 | _____ | $4^+$ | |
| | | 5.220 | _____ | $3^+$ | |
| 0.129 | _____ | $\frac{7}{2}^+$ 4.230 | _____ | $2^+$ | |
| 0.117 | _____ | $\frac{5}{2}^+$ 4.113 | _____ | $4^+$ | |
| 0.005 | _____ | $\frac{3}{2}^+$ 1.369 | _____ | $2^+$ | |
| 0 | _____ | $\frac{1}{2}^+$ 0 | _____ | $0^+$ | |
| $^{171}$Tm, $\beta = 0.2–0.3$ | | $^{24}$Mg | | | |

9. A deformed even–even nuclide has energy levels characterized by the following values of spin, parity, and $K$ value. You will note that not all of the information is given for each level. Fill in the blanks with the required values. In the appropriate space, assign each of the levels to a particular mode of excitation, for example, vibrational. Assume all bands are characterized by the same value of the moment of inertia.

| Energy | J | π | K | Assignment |
|---|---|---|---|---|
| — | — | — | — | — |
| 0.400 | 1 | − | 0 | |
| 0.376 | 3 | + | — | |
| — | 4 | + | 0 | |
| 0.349 | 6 | + | 0 | |
| 0.310 | 2 | + | 2 | |
| — | 2 | + | 0 | |
| 0.200 | 0 | + | 0 | |
| 0.166 | 4 | + | — | |
| — | — | + | 0 | |
| 0 | — | — | — | |

10. Using the shell model, calculate the ground-state spins, parities, and magnetic moments for $^{32}$S, $^{33}$S, and $^{41}$K. Predict the following characteristics of the ground states of $^{25}$Mg and $^{63}$Cu: the state of the odd nucleon, the total nuclear angular momentum, nuclear magnetic dipole moment, the sign of the nuclear quadrupole moment, and the parity. Explain the probable cause of any important discrepancies between your predictions and the following measured values:

$$^{25}\text{Mg} \quad I = \frac{5}{2} \quad \mu = -0.96 \quad Q = +0.2$$

$$^{63}\text{Cu} \quad I = \frac{3}{2} \quad \mu = +2.22 \quad Q = -0.1$$

11. The energies (MeV) and spins of the lowest excited states of $^{182}$W are for $J = 2, 4, 6$, we have $E = 0.100, 0.329$, and $0.680$. Do these values agree with a rotational model?

12. For the nucleus $^{235}$U at an excitation energy of 30 MeV, what is the ratio of the density of levels of spin $J$ to the total density of levels at that excitation energy?

13. $^{249}$Bk is known to have the following level scheme. Fill in the missing energies and $J, \pi$ values.

| | | |
|---|---|---|
| 93.7 | _____ | ? |
| 82.6 | _____ | $\frac{7}{2}^-$ |
| 41.8 | _____ | $\frac{9}{2}^+$ |
| 39.6 | _____ | ? |
| 8.8 | _____ | $\frac{3}{2}^-$ |
| 0 | _____ | $\frac{7}{2}^+$ |

14. Given the following shell model state, $k_{17/2}$, show qualitatively how it might split as a function of increasing prolate deformation. Label each

state as to its $\Omega$ value and indicate the maximum number of particles in each $\Omega$ state.

15. Show that the Brennan–Bernstein rules forbid the existence of odd nuclei with ground states $0^+$ or $1^-$. Find some exceptions.

16. Given the following level scheme for $^{110}$Cd, predict the character of each state.

    | | |
    |---|---|
    | 1.783 _____ | $2^+$ |
    | 1.542 _____ | ? |
    | 1.473 _____ | $0^+$ |
    | 0.656 _____ | $2^+$ |
    | 0 _____ | $0^+$ |

17. $^{121}$Sb has a spin of $\frac{5}{2}$ and a magnetic moment of 3.36 nm. What is the state of the 51st proton? What would the shell model predict?

## REFERENCES

Brennan, M. H. and A. M. Bernstein. *Phys. Rev.* **120**, 927 (1960).
Eichler, E. *Rev. Med. Phys.* **36**, 809 (1964).
Gordan, G. E. and C. D. Coryell. *J. Chem.–Ed.* **44**, 636 (1967).
Heyde, K. *Basic Ideas and Concepts in Nuclear Physiscs*, 2nd ed., IOP, Bristol, 1999.
Tyren, H., P. Hillman, and Th. Maris. *Nucl. Phys.* **7**, 10 (1958).
Vandenbosch, R. and J. R. Huizenga. *Nuclear Fission,* Academic, New York, 1973.

## BIBLIOGRAPHY

### General

Most, if not all, textbooks in nuclear science have a chapter discussing nuclear structure. Among the favorites of the authors are:

Das, A. and T. Ferbel. *Introduction to Nuclear and Particle Physics*, Wiley, New York, 1994.
Enge, H. A. *Introduction to Nuclear Physics*, Addison-Wesley, Reading, MA, 1966.
Friedlander, G., J. W. Kennedy, E. S. Macias, and J. M. Miller. *Nuclear and Radiochemistry*, Wiley, New York, 1981.
Hodgson, P. E., E. Gadioli, and E. Gadioli-Erba. *Introductory Nuclear Physics*, Clarendon, Oxford, 1997.
Krane, K. S. *Introductory Nuclear Physics*, Wiley, New York, 1987.

### More Specialized, Advanced Treatments

Bohr, A. and B. R. Mottelson. *Nuclear Structure, Vols. I and II*, Benjamin, New York, 1969.
Casten, R. F. *Nuclear Structure from a Simple Perspective*, Oxford, New York, 1990.

de Shalit, A. and H. Feshbach. *Theoretical Nuclear Physics, Vol. I. Nuclear Structure*, Wiley, New York, 1974.

Nilsson, S. G. and I. Ragnarsson. *Shapes and Shells in Nuclear Structure*, Cambridge, Cambridge, 1995.

# CHAPTER 7

# α DECAY

In a series of seminal experiments Ernest Rutherford and his collaborators established the important features of α decay. The behavior of the radiations from natural sources of uranium and thorium and their daughters was studied in magnetic and electric fields. The least penetrating particles, labeled "α rays" because they were the first to be absorbed, were found to be positively charged and quite massive in comparison to the more penetrating negatively charged "β rays" and the most penetrating neutral "γ rays." In a subsequent experiment the α rays from a needlelike source were collected in a very small concentric discharge tube, and the emission spectrum of helium was observed in the trapped volume. Thus, α rays were proven to be energetic helium nuclei. The α particles are the most ionizing radiation emitted by natural sources (with the extremely rare exception of the spontaneous fission of uranium) and are stopped by as little as a sheet of paper or a few centimeters of air. The particles are quite energetic ($E_\alpha = 4-9$ MeV) but interact very strongly with electrons as they penetrate into material and stop within 100 μm in most condensed materials.

Understanding these features of α decay allowed early researchers to use the emitted α particles to probe the structure of nuclei in scattering experiments and later, by reaction with beryllium, to produce neutrons. In an interesting dichotomy, the α particles from the decay of natural isotopes of uranium, radium, and their daughters have sufficient kinetic energies to overcome the Coulomb barriers of light elements and induce nuclear reactions but are not energetic enough to induce reactions in the heaviest elements.

*Modern Nuclear Chemistry*, by W.D. Loveland, D.J. Morrissey, and G.T. Seaborg
Copyright © 2006 John Wiley & Sons, Inc.

α Particles played an important role in nuclear physics before the invention of charged particle accelerators and were extensively used in research. Therefore, the basic features of α decay have been known for some time. The process of α decay is a nuclear reaction that can be written as:

$$^{A}_{Z}(Z)_N \longrightarrow {}^{A-4}_{Z-2}(X)^{2-}_{N-2} + {}^{4}_{2}\mathrm{He}^{2+}_2 + Q_\alpha \tag{7.1}$$

where we have chosen to write out all of the superscripts and subscripts. Thus the α decay of $^{238}$U can be written

$$^{238}\mathrm{U} \longrightarrow {}^{234}\mathrm{Th}^{2-} + {}^{4}\mathrm{He}^{2+} + Q \tag{7.2}$$

The $Q_\alpha$ value is positive (exothermic) for spontaneous α decay. The helium nucleus emerges with a substantial velocity and is fully ionized, and the atomic electrons on the daughter are disrupted by the sudden change, but the whole process conserves electrical charge. We can rewrite the equation in terms of the masses of the neutral atoms:

$$^{A}_{Z}(Z)_N \longrightarrow {}^{A-4}_{Z-2}(X)_{N-2} + {}^{4}_{2}\mathrm{He}_2 + Q_\alpha \tag{7.3}$$

and then calculate the $Q_\alpha$ value because the *net* change in the atomic binding energies ($\sim 65.3 Z^{7/5} - 80 Z^{2/5}$ eV) is very small compared to the nuclear decay energy.

What causes α decay? (or, what causes $Q_\alpha$ to be positive?) In the language of the semiempirical mass equation, the emission of an α particle lowers the Coulomb energy of the nucleus, which increases the stability of heavy nuclei while not affecting the overall binding energy per nucleon because the tightly bound α particle has approximately the same binding energy/nucleon as the original nucleus.

Two important features of α decay are that the energies of the α particles are known to generally increase with the atomic number of the parent, but yet the kinetic energy of the emitted particle is less than that of the Coulomb barrier in the reverse reaction between the α particle and the daughter nucleus. In addition, all nuclei with mass numbers greater than $A \approx 150$ are thermodynamically unstable against α emission ($Q_\alpha$ is positive), but α emission is the dominant decay process only for the heaviest nuclei, $A \geq 210$. The energies of the emitted α particles can range from 1.8 MeV ($^{144}$Nd) to 11.6 MeV ($^{212}$Po$^m$) with the half-life of $^{144}$Nd being $5 \times 10^{29}$ times as long as that of $^{212}$Po$^m$. Typical heavy-element α decay energies are in the range from 4 to 9 MeV, as noted earlier.

In general, α decay leads to the ground state of the daughter nucleus so that the emitted particle carries away as much energy as possible and as little angular momentum as possible. The ground-state spins of even–even parents and daughters (including the α particle, of course) are zero, which makes $l = 0$ α-particle emission the most likely process for these nuclei. Small branches are seen to higher excited states, but such processes are strongly suppressed. Some decays of odd-$A$ heavy nuclei populate low-lying excited states that match the spin of the parent so that the orbital angular momentum of the α particle can be zero. For example, the strongest branch (83%) of the α decay of $^{249}$Cf goes to the 9th excited state of $^{245}$Cm because this is the lowest lying state with the same spin and parity as that of the parent. α Decay to several different excited states of a daughter nucleus is called *fine structure*; α decay from an excited state of a parent nucleus to the ground state of the daughter nucleus is said to be *long-range α emission* because these α particles are more energetic and thus have longer ranges in matter. The most famous case of long-range α emission is that of $^{212}$Po$^m$ where a 45-s isomeric level at 2.922 MeV decays to the ground state of $^{208}$Pb by emitting a 11.65-MeV α particle.

We will consider the general features of α emission, and then we will describe them in terms of a simple quantum mechanical model. It turns out that α emission is a beautiful example of the quantum mechanical process of tunneling through a barrier that is forbidden in classical mechanics.

## 7.1 ENERGETICS OF α DECAY

As we have seen in the overview of the nuclear mass surface in Chapter 2, the α particle, or $^4$He nucleus, is an especially strongly bound particle. This combined with the fact that the binding energy per nucleon has a maximum value near $A \approx 56$ and systematically decreases for heavier nuclei creates the situation that nuclei with $A \geq 150$ have positive $Q_\alpha$ values for the emission of α particles. This behavior can be seen in Figure 7.1.

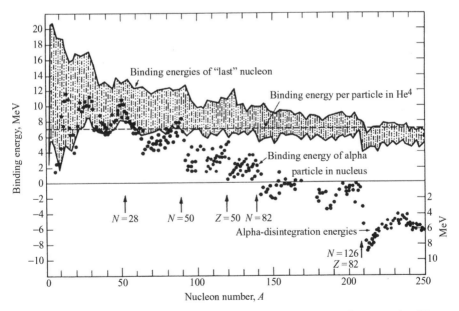

**Figure 7.1** Variation of the α-particle separation energy as a function of mass number. [From Valentin, Subatomic physics: nuclei and particles. Copyright © 1981 by North-Holland Publishing Company. Reprinted by permission of North-Holland Publishing Company.]

For example, one of the heaviest naturally occurring isotopes, $^{238}$U (with a mass defect, $\Delta$ of +47.3070 MeV) decays by $\alpha$ emission to $^{234}$Th ($\Delta = +40.612$ MeV) giving a $Q_\alpha$ value of:

$$Q_\alpha = 47.3070 - (40.612 + 2.4249) = 4.270 \text{ MeV}$$

Note that the decay energy will be divided between the $\alpha$ particle and the heavy recoiling daughter so that the kinetic energy of the $\alpha$ particle will be slightly less. (The kinetic energy of the recoiling $^{234}$Th nucleus produced in the decay of $^{238}$U is $\sim 0.070$ MeV.) Conservation of momentum and energy in this reaction requires that the kinetic energy of the $\alpha$ particle, $T_\alpha$, is

$$T_\alpha = \frac{234}{238} Q_\alpha = 4.198 \text{ MeV}$$

The kinetic energies of the emitted $\alpha$ particles can be measured very precisely so we should be careful to distinguish between the $Q_\alpha$ value and the kinetic energy, $T_\alpha$. The very small recoil energy of the heavy daughter is very difficult to measure, but it is still large compared to chemical bond energies and can lead to interesting chemistry. For example, the daughter nuclei may recoil out of the original $\alpha$ source. This can cause serious contamination problems if the daughters are themselves radioactive.

The $Q_\alpha$ values generally increase with increasing atomic number, but the variation in the mass surface due to shell effects can overwhelm the systematic increase (Fig. 7.2).

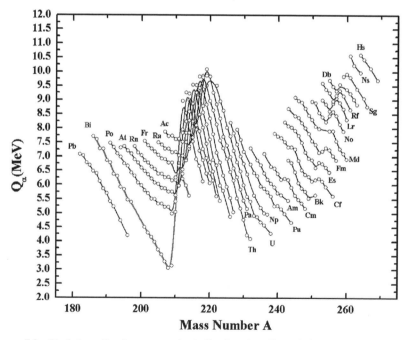

**Figure 7.2** Variation of $\alpha$-decay energies indicating the effect of the $N = 126$ and $Z = 82$ shell closures along with the $N = 152$ subshell.

The sharp peaks near A = 214 are due to the effects of the N = 126 shell. When $^{212}$Po decays by α emission, the daughter nucleus is doubly magic—$^{208}$Pb (very stable) with a large energy release. The α decay of $^{211}$Pb and $^{213}$Po will not lead to such a large $Q_\alpha$ because the products are not doubly magic. Similarly, the presence of the 82 neutron closed shell in the rare-earth region causes an increase in $Q_\alpha$, allowing observable α-decay half-lives for several of these nuclei (with N = 84). Also one has observed short-lived α emitters near doubly magic $^{100}$Sn, including $^{107}$Te, $^{108}$Te, and $^{111}$Xe. And, in addition, α emitters have been identified along the proton dripline above A = 100. For a set of isotopes (nuclei with a constant atomic number) the decay energy generally decreases with increasing mass. These effects can be seen in Figure 7.2. For example, the kinetic energy of α particles from the decay of uranium isotopes is typically 4 to 5 MeV, those for californium isotopes are ≈6 MeV, and those for rutherfordium isotopes are ≈8 MeV. However, the kinetic energy from the decay of $^{212}_{84}$Po to the doubly magic $^{208}_{82}$Pb daughter is 8.78 MeV.

The generally smooth variation of $Q_\alpha$ with Z, A of the emitting nucleus and the two-body nature of α decay can be used to deduce masses of unknown nuclei. One tool in this effort is the concept of closed decay cycles (Fig. 7.3). Consider the α and β decays connecting $^{237}_{93}$Np, $^{241}_{95}$Am, $^{241}_{94}$Pu, and $^{237}_{92}$U. By conservation of energy, one can state that the sum of the decay energies around the cycle connecting these nuclei must be zero (within experimental uncertainty). In those cases where experimental data or reliable estimates are available for three branches of the cycle, the fourth can be calculated by difference.

Even though the energies released by the decay of a heavy nucleus into an α particle and a lighter daughter nucleus are quite substantial, the energies are paradoxically small compared to the energy necessary to bring the α particle back into nuclear contact with the daughter. The electrostatic potential energy between the

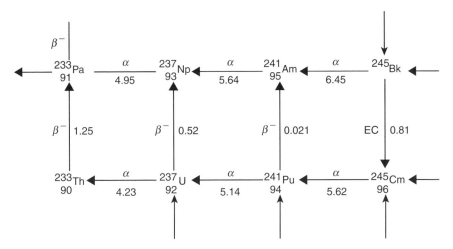

**Figure 7.3** Decay cycles for part of the 4n + 1 family. Modes of decay are indicated over the arrows; the numbers indicate total decay energies in MeV.

two positively charged nuclei, called the Coulomb potential, can be written as:

$$V_C = \frac{2Z}{R} \frac{e^2}{4\pi\varepsilon_0} \tag{7.4}$$

where $Z$ is the atomic number of the daughter and $R$ is the separation between the centers of the two nuclei. (As pointed out in Chapter 1, $e^2/4\pi\varepsilon_0$ is 1.440 MeV · fm.) To obtain a rough estimate of the Coulomb energy, we can take $R$ to be $1.2(A^{1/3} + 4^{1/3})$ fm, where $A$ is the mass number of the daughter. For the decay of $^{238}$U we get

$$V_C = \frac{(2)(90)(1.440 \text{ MeV} \cdot \text{fm})}{1.2(234^{1/3} + 4^{1/3}) \text{ fm}} \approx \frac{259 \text{ MeV} \cdot \text{fm}}{9.3 \text{ fm}} = 28 \text{ MeV} \tag{7.5}$$

which is 6 to 7 times the decay energy. This factor is typical of the ratio of the Coulomb barrier to the $Q$ value. If we accept for the moment the large difference between the Coulomb barrier and the observed decay energy, then we can attribute the two general features of increasing decay energy with increasing atomic number $Z$ and decreasing kinetic energy with increasing mass among a set of isotopes to the Coulomb potential. The higher nuclear charge accelerates the products apart, and the larger mass allows the daughter and $\alpha$ particle to start further apart.

**Example Problem** Calculate the $Q_\alpha$ value, kinetic energy $T_\alpha$, and the Coulomb barrier $V_C$, for the primary branch of the $\alpha$ decay of $^{212}$Po to the ground state of $^{208}$Pb.

**Solution** Using tabulated mass defects we have

$$Q_\alpha = -10.381 - (-21.759 + 2.4249) = 8.953 \text{ MeV}$$

$$T_\alpha = \frac{208}{212} Q = 8.784 \text{ MeV}$$

and

$$V_C = \frac{(2)(84)(1.440 \text{ MeV} \cdot \text{fm})}{1.2(208^{1/3} + 4^{1/3}) \text{ fm}} \approx 27 \text{ MeV}$$

The $^{212}$Po parent also decays with a 1% branch to the first excited state of $^{208}$Pb at an excitation energy of 2.6146 MeV. What is the kinetic energy of this $\alpha$ particle?

$$Q_\alpha = 8.953 - 2.6146 = 6.339 \text{ MeV}$$

$$T_\alpha = \frac{208}{212} \cdot 6.1339 = 6.22 \text{ MeV}$$

As discussed previously, many heavy nuclei ($A \geq 150$) are unstable with respect to $\alpha$ decay. Some of them also undergo $\beta^-$ decay. In Chapter 3, we discussed the natural decay series in which heavy nuclei undergo a sequence of $\beta^-$ and $\alpha$ decays until they form one of the stable isotopes of lead or bismuth, $^{206,207,208}$Pb

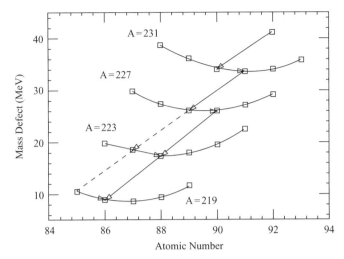

**Figure 7.4** Mass parabolas for some members of the $4n + 3$ natural decay series. The main decay path is shown by a solid line while a weak branch is indicated by a dashed line.

or $^{209}$Bi. We are now in a position to understand why a particular sequence occurs. Figure 7.4 shows a series of mass parabolas (calculated using the semiempirical mass equation) for some members of the $4n + 3$ series, beginning with $^{235}$U. Each of the mass parabolas can be thought of as a cut through the nuclear mass surface at constant $A$. $^{235}$U decays to $^{231}$Th. $^{231}$Th then decays to $^{231}$Pa by $\beta^-$ decay. This nucleus, being near the bottom of the mass parabola, cannot undergo further $\beta^-$ decay, but decays by $\alpha$ emission to $^{227}$Ac. This nucleus decays by $\beta^-$ emission to $^{227}$Th, which must $\alpha$ decay to $^{223}$Ra, and so forth.

## 7.2 THEORY OF α DECAY

The allowed emission of α particles could not be understood in classical pictures of the nucleus. This fact can be appreciated by considering the schematic potential energy diagram for $^{238}$U shown in Figure 7.5. Using simple estimates we have drawn a one-dimensional potential energy curve for this system as a function of radius. At the smallest distances, inside the parent nucleus, we have drawn a flat-bottomed potential with a depth of $\sim -30$ MeV (as discussed in Chapter 6). The potential rapidly rises at the nuclear radius and comes to the Coulomb barrier height of $V_C \approx +28$ MeV at 9.3 fm. At larger distances the potential falls as $1/r$ according to Coulomb's law.

Starting from a separated α particle and the daughter nucleus, we can determine that the distance of closest approach during the scattering of a 4.2-MeV α particle will be ∼62 fm. This is the distance at which the α particle stops moving toward the daughter and turns around because its kinetic energy has been converted into potential energy of repulsion. Now the paradox should be clear: The α particle

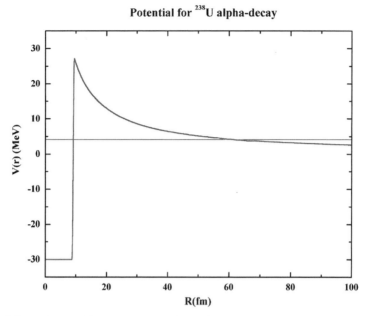

**Figure 7.5** A (reasonably accurate) one-dimensional potential energy diagram for $^{238}$U indicating the energy and calculated distances for α decay into $^{234}$Th. Fermi energy ≈30 MeV, Coulomb barrier ≈28 MeV at 9.3 fm, $Q_\alpha$ 4.2 MeV, distance of closest approach 62 fm. (Figure also appears in color figure section.)

should not get even remotely near to the nucleus; or from the decay standpoint, the α particle should be trapped behind a potential energy barrier that it cannot get over.

The solution to this paradox was found in quantum mechanics. A general property of quantum mechanical wave functions is that they are only completely confined by potential energy barriers that are infinitely high. Whenever the barrier has a finite size, the wave function solution will have its main component inside the barrier plus a small but finite part *inside* the barrier (generally exponentially decreasing with distance) and another finite piece outside the barrier. This phenomenon is called tunneling because the classically trapped particle has a component of its wave function outside the potential barrier and has some probability to go through the barrier to the outside without going over the top. The details of these calculations are discussed in Appendix E and in many quantum mechanics textbooks. Some features of tunneling should be obvious: The closer the energy of the particle to the top of the barrier, the more likely that the particle will get out. Also, the more energetic the particle is relative to a given barrier height, the more frequently the particle will "assault" the barrier and the more likely that the particle will escape.

It has been known for some time that half-life for α decay, $t_{1/2}$, can be written in terms of the square root of the α-particle decay energy, $Q_\alpha$, as follows:

$$\log(t_{1/2}) = A + \frac{B}{\sqrt{Q_\alpha}} \tag{7.6}$$

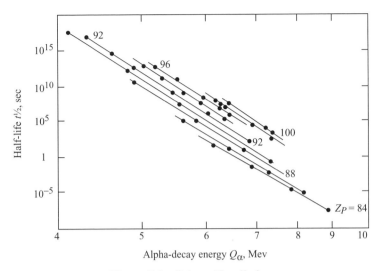

**Figure 7.6** Geiger–Nuttall plot.

where the constants A and B have a Z dependence. This relationship, shown in Figure 7.6, is known as the Geiger–Nuttall law of α decay (Geiger and Nuttall, 1911, 1912) due to the fact that Geiger and Nuttall found a linear relationship between the logarithm of the decay constant and the logarithm of the range of α particles from a given natural radioactive decay series. This simple relationship describes the data on α decay, which span over 20 orders of magnitude in decay constant or half-life. Note that a 1-MeV change in α-decay energy results in a change of $10^5$ in the half-life. A modern representation of this relationship due to Hatsukawa, Nakahara, and Hoffman has the form

$$\log_{10}(t_{1/2}) = A(Z) \times \left[\frac{A_d}{A_p Q_\alpha}\right]^{1/2} \times \left[\arccos\sqrt{X} - \sqrt{X(1-X)}\right] - 20.446 + C(Z,N)$$
(7.7)

where $C(Z,N) = 0$ for ordinary regions outside closed shells and

$$C(Z,N) = [1.94 = -0.020(82 - Z) - 0.070(126 - N)]$$

for $78 \leq Z \leq 82$, $100 \leq N \leq 126$, and

$$C(Z,N) = [1.42 - 0.105(Z - 82) - 0.067(126 - N)]$$

for $82 \leq Z \leq 90$, $110 \leq N \leq 126$. In these equations, $A_p$, $Z$ refer to the parent nuclide, $A_d$, and $Z_d$ refer to the daughter nuclide, and $X$ is defined as

$$X = 1.2249(A_d^{1/3} + 4^{1/3}) \times \frac{Q_\alpha}{2Z_d e^2}$$

This relationship is useful for predicting the expected α-decay half-lives for unknown nuclei.

The theoretical description of α emission relies on calculating the rate in terms of two factors. The overall rate of emission consists of the product of the rate at which an α particle appears at the inside wall of the nucleus times the (independent) probability that the α particle tunnels through the barrier. Thus, the rate of emission, or the partial decay constant $\lambda_\alpha$, is written as the product of a frequency factor, $f$, and a transmission coefficient, $T$, through the barrier:

$$\lambda_\alpha = fT$$

Some investigators have suggested that this expression should be multiplied by an additional factor to describe the probability of preformation of an α particle inside the parent nucleus. Unfortunately, there is no clear way to calculate such a factor, but empirical estimates have been made. As we will see below, the theoretical estimates of the emission rates are higher than the observed rates, and the preformation factor can be estimated for each measured case. However, there are other uncertainties in the theoretical estimates that contribute to the differences.

The frequency with which an α particle reaches the edge of a nucleus can be estimated as the velocity divided by the distance across the nucleus. We can take the distance to be twice the radius (something of a maximum value), but the velocity is a little more subtle to estimate. A lower limit for the velocity could be obtained from the kinetic energy of emitted α particle, but the particle is moving inside a potential energy well, and its velocity should be larger and correspond to the well depth plus the external energy. Therefore, the frequency can be written as:

$$f = \frac{v}{2R} = \frac{\sqrt{2(V_0 + Q)/\mu}}{2R} \qquad (7.8)$$

where we have assumed that the α particle is nonrelativistic, $V_0$ is the well depth indicated in Figure 7.5 of approximately 30 MeV, $\mu$ is the reduced mass, and $R$ is the radius of the daughter nucleus (because the α particle needs only to reach this distance before it is emitted). We use the reduced mass because the α particle is moving inside the nucleus, and the total momentum of the nucleus must be zero. The frequency of assaults on the barrier is quite large, usually on the order of $10^{21}$/s.

The quantum mechanical transmission coefficient for an α particle to pass through a barrier is derived in Appendix E. Generalizing the results summarized in Equation (E.48) to a three-dimensional barrier shown in Figure 7.5, we have

$$T = e^{-2G} \qquad (7.9)$$

where the Gamow factor ($2G$) can be written as:

$$2G = \frac{2}{\hbar} \int_R^b \left[ 2\mu \left( \frac{Z_\alpha Z_D e^2}{r} - Q_\alpha \right) \right]^{1/2} dr \qquad (7.10)$$

where

$$Q_\alpha = \frac{Z_\alpha Z_D e^2}{b} \qquad (7.11)$$

and the classical distance of closest approach, $b$, is given as

$$b = \frac{Z_\alpha Z_D e^2}{Q_\alpha} \qquad (7.12)$$

In these equations, $e^2 = 1.440$ MeV-fm, $Q_\alpha$ is given in MeV, $Z_\alpha$, $Z_D$ are the atomic numbers of the $\alpha$ particle and daughter nucleus, respectively.

Rearranging we have

$$2G = \frac{2}{\hbar}\sqrt{2\mu Q_\alpha} \int_R^b \left(\frac{b}{r}\right)^{1/2} dr \qquad (7.13)$$

This can be integrated to give

$$2G = \frac{2b}{\hbar}\sqrt{2\mu Q_\alpha}\left(\cos^{-1}\sqrt{\frac{R}{b}} - \sqrt{\frac{R}{b}}\sqrt{1 - \frac{R}{b}}\right) \qquad (7.14)$$

Substituting back for $b$ and simplifying, we have

$$2G = 2\sqrt{\frac{2\mu}{\hbar^2 Q_\alpha}}(Z_\alpha Z_D e^2)\left(\cos^{-1}\sqrt{\frac{R}{b}} - \sqrt{\frac{R}{b}}\sqrt{1 - \frac{R}{b}}\right) \qquad (7.15)$$

For thick barriers,

$$\frac{R}{b} \ll 1 \left(\frac{Q_\alpha}{V_C} \ll 1\right)$$

we can approximate

$$\cos^{-1}\sqrt{\frac{R}{b}} \approx \frac{\pi}{2} - \sqrt{\frac{R}{b}} \qquad (7.16)$$

We get

$$2G = 2\sqrt{\frac{2\mu}{\hbar^2 Q_\alpha}}(Z_\alpha Z_D e^2)\left(\frac{\pi}{2}\right) \qquad (7.17)$$

where $B$ is the "effective" Coulomb barrier, that is,

$$B = \frac{Z_\alpha Z_D e^2}{r_\alpha + R_D} \qquad (7.18)$$

Typically, the Gamow factor is large ($2G \sim 60\text{--}120$), which makes the transmission coefficient $T$ extremely small ($\sim 10^{-55}\text{--}10^{-27}$). Combining the various equations, we have

$$t_{1/2} = \frac{\ln 2}{\lambda} = \frac{\ln 2}{fT} = \frac{\ln 2}{\dfrac{[2(V_0 + Q_\alpha)/\mu]^{1/2}}{2R} e^{-2G}} \tag{7.19}$$

or

$$\log t_{1/2} = a + \frac{b}{\sqrt{Q_\alpha}} \tag{7.20}$$

that is, we get the Geiger–Nuttall law of $\alpha$ decay, where $a + b$ are constants that depend on Z, and so forth.

This simple estimate tracks the general behavior of the observed emission rates over the very large range in nature. The calculated emission rate is typically one order of magnitude larger than that observed, meaning that *the observed half-lives are longer than predicted.* This has led some researchers to suggest that the probability to find a "preformed" $\alpha$ particle inside a heavy nucleus is on the order of $10^{-1}$ or less. One way to obtain an estimate of the "preformation factor" is to plot, for even–even nuclei undergoing $l = 0$ decay, the ratio of the calculated half-life to the measured half-life. This is done in Figure 7.7. The average preformation factor is $\sim 10^{-2}$.

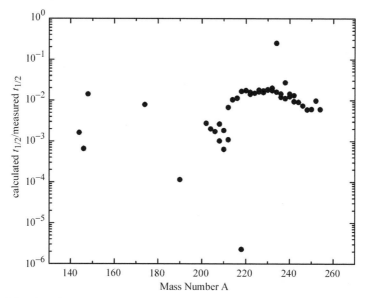

**Figure 7.7** Plot of the ratio of the calculated partial $\alpha$-decay half-life for ground-state $l = 0$ transitions of even–even nuclei to the measured half-lives. The calculations were made using the simple theory of $\alpha$ decay.

## 7.2 THEORY OF α DECAY

**Example Problem** Calculate the emission rate and half-life for $^{238}$U decay from the simple theory of α decay. Compare this to the observed half-life.

$$\lambda = fT$$

where

$$f = \frac{\sqrt{2(V_0 + Q)/\mu}}{2R}$$

$$R = r_0\left(A_1^{1/3} + A_1^{1/3}\right) = 1.2\left(4^{1/3} + 234^{1/3}\right) = 9.3\,\text{fm}$$

**Solution** Note that since we previously calculated $b \approx 62$ fm, $R/b = 8.63/62 \ll 1$:

$$\mu = 4 * \frac{234}{238} = 3.933\,\text{amu}$$

$$f = \frac{c * \sqrt{2(30 + 4.2)/(3.933 * 931.5)}}{2 * 9.3} = (2.26 \times 10^{21})/\text{s}$$

We know that

$$T = e^{-2G}$$

where

$$2G \approx 2\left(\frac{2\mu}{\hbar^2 Q_\alpha}\right)^{1/2} (Z_\alpha Z_D e^2)\left(\frac{\pi}{2} - 2\sqrt{\frac{Q}{B}}\right)$$

$$2\left(\frac{2\mu}{\hbar^2 Q_\alpha}\right) = 2\left(\frac{2\mu\,(\text{amu})\,931.5\left(\frac{\text{MeV}}{\text{amu}}\right)}{(\hbar c)^2\,(\text{MeV-fm})^2 Q_\alpha\,(\text{MeV})}\right)^{1/2}$$

$$= 2\left(\frac{(2)(3.933)(931.5)}{(197.3)^2(4.27)}\right)^{1/2} = 0.420\,(\text{MeV-fm})^{-1}$$

$$(Z_\alpha Z_D e^2) = (2)(90)(1.440) = 259.2\,\text{MeV-fm}$$

$$\left(\frac{\pi}{2} - 2\sqrt{\frac{Q}{B}}\right) = \frac{\pi}{2} - 2\left(\frac{4.27}{27.9}\right)^{1/2} = 0.788$$

$$T = e^{-85.8} = 5.43 \times 10^{-38}$$

$$\lambda = fT = (2.26 \times 10^{21})(5.43 \times 10^{-38}) = 1.23 \times 10^{-16}\,\text{s}^{-1}$$

$$t_{1/2} = \frac{\ln 2}{\lambda} = 5.65 \times 10^{15}\,\text{s} = 1.8 \times 10^8\,\text{y}$$

The observed half-life of $^{238}$U is $4.47 \times 10^9$ y, which is a factor of $\sim 25$ times longer than the calculated value. Note the qualitative aspects of this calculation. The α particle must hit the border of the parent nucleus $\sim 10^{38}$ times before it can escape. Also note the extreme sensitivity of this calculation to details of the nuclear radius. A 2% change in $R$ changes $\lambda$ by a factor of 2. In our example, we approximated $R$ as $R_{\text{Th}} + R_\alpha$. In reality, the α particle has not fully separated from the daughter nucleus when they exit the barrier. One can correct for this by approximating $R \approx 1.4 A^{1/3}$.

The theory presented above neglects the effects of angular momentum in that it assumes the α particle carries off no orbital angular momentum ($l = 0$). If α decay takes place to or from an excited state, some angular momentum may be carried off by the α particle with a resulting change in the decay constant. In quantum mechanics, we say that the α particle has to tunnel through a barrier that is larger by an amount called the centrifugal potential:

$$V_l = \frac{l(l+1)\hbar^2}{2\mu R^2} \qquad (7.21)$$

where $l$ is the orbital angular momentum of the α particle, $\mu$ is the reduced mass, and $R$ is the appropriate radius. This centrifugal potential is added to the potential energy $V_{(r)}$ resulting in a thicker and higher barrier, increasing the half-life (Fig. 7.8).

One can evaluate the effect of this centrifugal potential upon α-decay half-lives by simply adding this energy to the Coulomb barrier height. If we define

$$\sigma = \frac{\text{centrifugal barrier height}}{\text{Coulomb barrier height}} \qquad (7.22)$$

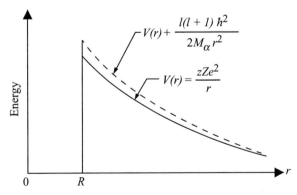

**Figure 7.8** Modification of the potential energy in α decay due to the centrifugal potential. [From W. E. Meyerhof, Elements of Nuclear Physics, Copyright 1967 by McGraw-Hill Book Company, Inc. Reprinted by permission of McGraw-Hill Book Company, Inc.]

we can say

$$\sigma = \frac{l(l+1)\hbar^2}{2\mu R^2}\frac{R}{Z_\alpha Z_D e^2} = \frac{l(l+1)\hbar^2}{2\mu R Z_\alpha Z_D e^2} \quad (7.23)$$

Then all we need to do is to replace all occurrences of $B$ by $B(1+\sigma)$. A simple pocket formula that does this is

$$\lambda_{l\neq 0} \approx \lambda_{l=0}\exp[-2.027l(l+1)Z^{-1/2}A^{-1/6}] \quad (7.24)$$

This centrifugal barrier correction is a very small effect compared to the effect of $Q_\alpha$ or $R$ upon the decay rate.

We should also note that conservation of angular momentum and parity during the α-decay process places some constraints on the daughter states that can be populated. Since the α particle has no intrinsic spin, the total angular momentum of the α particle must equal its orbital angular momentum $l$, and the α particle parity must be $(-1)^l$. If parity is conserved in α decay, the final states are restricted. If the parent nucleus has $J^\pi = 0^+$, then the allowed values of $J^\pi$ of the daughter nucleus are $0^+$ ($l=0$), $1^-$ ($l=1$), $2^+$ ($l=2$), and so on. These rules only specify the required spin and parity of the state in the daughter, while the energy of the state is a separate quantity. Recall from Chapter 6 that the heaviest elements are strongly deformed and are good rotors. The low-lying excited states of even–even nuclei form a low-lying rotational band with spins of 2, 4, 6, and so forth, while odd angular momenta states tend to lie higher in energy. Because of the decrease in the energy of the emitted α particle when populating these states, decay to these states will be inhibited. Thus, the lower available energy suppresses these decays more strongly than the centrifugal barrier.

**Example Problem** $^{241}$Am is a long-lived α emitter that is used extensively as an ionization source in smoke detectors. The parent state has a spin and parity of $\frac{5}{2}^-$ and cannot decay to the $\frac{5}{2}^+$ ground state of $^{237}$Np because that would violate parity conservation. Rather, it decays primarily to a $\frac{5}{2}^-$ excited state (85.2%, $E^* = 59.5$ keV) and to a $\frac{7}{2}^-$ higher lying excited state (12.8%, $E^* = 102.9$ keV). Estimate these branching ratios and compare them to the observed values.

$$Q_\alpha(\tfrac{5}{2}^-) = 5.578\,\text{MeV} \qquad Q_\alpha(\tfrac{7}{2}^-) = 5.535\,\text{MeV}$$
$$f(\tfrac{5}{2}^-) = 2.29\times 10^{21}/\text{s} \qquad f(\tfrac{7}{2}^-) = 2.29\times 10^{21}/\text{s}$$
$$G(\tfrac{5}{2}^-) = 33.01 \qquad G(\tfrac{7}{2}^-) = 33.84$$
$$\lambda(\tfrac{5}{2}^-) = 4.89\times 10^{-8}/\text{s} \qquad \lambda(\tfrac{7}{2}^-) = 9.2\times 10^{-9}/\text{s}$$

***Solution*** Assuming that the branches to other states are small and do not contribute to the sum of partial half-lives, we can write

$$\text{Branching ratio } \frac{5}{2^-} \approx \frac{\lambda(5/2^-)}{\lambda(5/2^-) + \lambda(7/2^-)} = 0.84$$

Note that the observed half-life of 433 y is again significantly longer than the predicted half-life of ~3 y. This difference is attributed to the combined effects of the preformation factor and the hindrance effect of the odd proton in the americium parent ($Z = 95$); see below.

## 7.3 HINDRANCE FACTORS

The one-body theory of $\alpha$ decay applies strictly to even–even $\alpha$ emitters only. The odd-nucleon $\alpha$ emitters, especially in ground-state transitions, decay at a slower rate than that suggested by the simple one-body formulation as applied to even–even nuclei. Consider the data in Figure 7.9 that shows the $\alpha$-decay half-lives of the even–even and odd $A$ uranium isotopes. The odd $A$ nuclei have substantially longer half-lives than their even–even neighbors do.

The decays of the odd $A$ nuclei are referred to as *hindered decays*, and a *hindrance factor* may be defined as the ratio of the measured partial half-life for a given $\alpha$ transition to the half-life that would be calculated from the simple one-body theory applied to even–even nuclides.

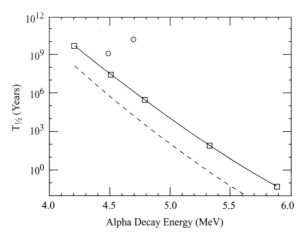

**Figure 7.9** The $\alpha$-decay half-lives of the even–even (squares) and odd $A$ (diamonds) isotopes of uranium. The measured values are connected by the solid line; the medians are shown by the dashed line.

In general, these hindrances for odd A nuclei may be divided into five classes:

1. If the hindrance factor is between 1 and 4, the transition is called a "favored" transition. In such decays, the emitted α particle is assembled from two low-lying pairs of nucleons in the parent nucleus, leaving the odd nucleon in its initial orbital.

   To form an α particle within a nucleus, two protons and two neutrons must come together with their spins coupled to zero and with zero orbital angular momentum relative to the center of mass of the α particle. These four nucleons are likely to come from the highest occupied levels of the nucleus. In odd A nuclei, because of the odd particle and the difficulty of getting a "partner" for it, one pair of nucleons is drawn from a lower lying level, causing the daughter nucleus to be formed in an excited state.

2. A hindrance factor of 4–10 indicates a mixing or favorable overlap between the initial and final nuclear states involved in the transition.

3. Factors of 10–100 indicate that spin projections of the initial and final states are parallel, but the wave function overlap is not favorable.

4. Factors of 100–1000 indicate transitions with a change in parity but with projections of initial and final states being parallel.

5. Hindrance factors of >1000 indicate that the transition involves a parity change and a spin flip, that is, the spin projections of the initial and final states are antiparallel, which requires substantial reorganization of the nucleon in the parent when the α is emitted.

## 7.4 HEAVY PARTICLE RADIOACTIVITY

As an academic exercise, one can calculate the $Q$ values for the emission of heavier nuclei than α particles and show that it is energetically possible for a large range of heavy nuclei to emit other light nuclei. For example, contours of the $Q$ values for carbon ion emission by a large range of nuclei are shown in Figure 7.10 calculated with the smooth liquid drop mass equation without shell corrections. Recall that the binding energy steadily decreases with increasing mass (above $A \sim 60$), and several light nuclei have large binding energies relative to their neighbors similar to the α particle. As can be seen in Figure 7.10, there are many nuclei with positive $Q$ values for carbon ion emission. Such emission processes or heavy particle radioactivity have been called *heavy cluster emission*.

We should also note that the double-shell closures at $Z = 82$ and $N = 126$ lead to especially large positive $Q$ values, as already shown in Figure 7.2. Thus, the emission of other heavy nuclei, particularly $^{12}$C, has been predicted or at least anticipated for a long time. Notice also that $^{12}$C is an even–even nucleus and $s$-wave emission without a centrifugal barrier is possible. However, the Coulomb barrier will be significantly larger for higher $Z$ nuclei than that for α particles.

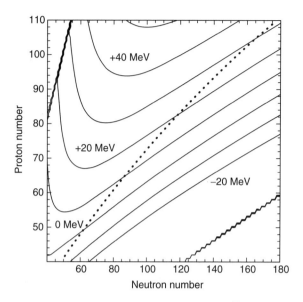

**Figure 7.10** Contours of the $Q$ value for the emission of a $^{12}$C nucleus as a function of neutron and proton numbers calculated with the liquid drop model mass formula. The contour lines are separated by 10 MeV. The dotted curve indicates the line of β stability [Eq. (2.9)].

We can use the simple theory of α decay to make an estimate of the relative branching ratios for α emission and $^{12}$C emission from $^{220}$Ra, a very favorable parent that leads to the doubly magic $^{208}$Pb daughter. In this case we find $Q_\alpha = 7.59$ MeV and $Q_C = 32.02$ MeV. Using the simple theory and ignoring differences in the preformation factor, the predicted half-life for $^{12}$C emission is only longer by a factor of 2!

$$^{220}\text{Ra} \implies {}^{216}\text{Rn} + {}^{4}\text{He} \qquad Q = 7.59 \qquad \lambda_{\text{calc}} = 5.1 \times 10^4 \text{ s}$$

$$^{220}\text{Ra} \implies {}^{208}\text{Pb} + {}^{12}\text{C} \qquad Q = 32.02 \qquad \lambda_{\text{calc}} = 3.34 \times 10^4 \text{ s}$$

The encouraging results from simple calculations such as these have spurred many searches for this form of radioactivity.

It was relatively recently that heavy cluster emission was observed at a level enormously lower than these estimates. Even so, an additional twist in the process was discovered when the radiation from a $^{223}$Ra source was measured directly in a silicon surface barrier telescope. The emission of $^{14}$C was observed at the rate of $\sim 10^{-9}$ times the α-emission rate, and $^{12}$C was not observed. Thus, the very large neutron excess of the heavy elements favors the emission of neutron-rich light products. The fact that the emission probability is so much smaller than the simple barrier penetration estimate can be attributed to the very small probability to "preform" a $^{14}$C residue inside the heavy nucleus. This first observation has been confirmed in subsequent measurements with magnetic spectrographs. The more rare emission of other larger neutron-rich light nuclei have been reported in very sensitive studies with nuclear track detectors.

## 7.5 PROTON RADIOACTIVITY

For very neutron-deficient (i.e., proton-rich) nuclei, the $Q$ value for proton emission, $Q_p$, becomes positive. One estimate, based on the semiempirical mass equation, of the line that describes the locus of the nuclei where $Q_p$ becomes positive for ground-state decay is shown in Figure 7.11. This line is known as the *proton-drip line.* Our ability to know the position of this line is a measure of our ability to describe the forces holding nuclei together. Nuclei to the left of the proton dripline in Figure 7.11 can decay by proton emission.

Proton decay should be a simple extension of α decay with the same ideas of barrier penetration being involved. A simplification with proton decay relative to α decay is that there should be no preformation factor for the proton. The situation is shown in Figure 7.12 for the case of the known proton emitter $^{151}$Lu. One notes certain important features/complications from this case. The proton energies, even for the heavier nuclei, are low ($E_p \sim 1$–$2$ MeV). As a consequence, the barriers to be penetrated are quite thick ($R_{\text{out}} = 80$ fm), and one is more sensitive to the proton energy, angular momentum changes, and so forth.

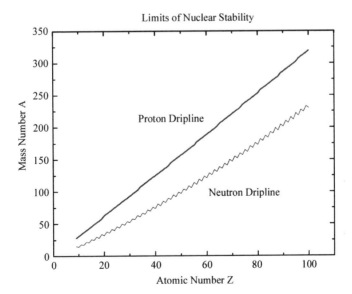

**Figure 7.11** Locus of neutron and proton driplines as predicted by the liquid drop model.

**196**   α DECAY

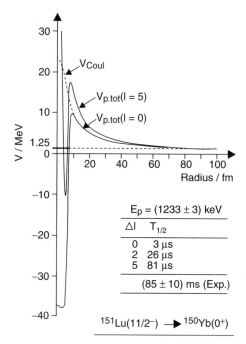

**Figure 7.12**   Proton–nucleus potential for the semiclassical calculation of the $^{151}$Lu partial proton half-life. [From S. Hofmann, In D. N. Poenaru (Ed.), *Nuclear Decay Modes*, Copyright 1996 by IOP Publishing. Reprinted by permission of IOP Publishing.]

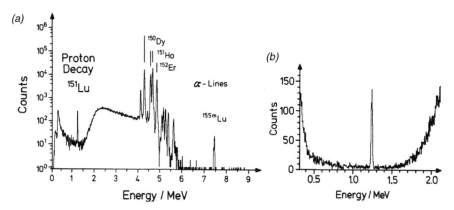

**Figure 7.13**   (*a*) Energy spectrum obtained during the irradiation of a $^{96}$Ru target with 261 MeV $^{58}$Ni projectiles. (*b*) Expanded part of the spectrum showing the proton line from $^{151}$Lu decay. [From S. Hofmann, In D. N. Poenaru (Ed.), *Nuclear Decay Modes*, Copyright 1996 by IOP Publishing. Reprinted by permission of IOP Publishing.]

The measurements of proton decay are challenging due to the low energies and short half-lives involved. Frequently, there are interfering $\alpha$ decays (Fig. 7.13). To produce nuclei near the proton dripline from nuclei near the valley of $\beta$ stability requires forming nuclei with high excitation energies that emit neutrons relative to protons and $\alpha$ particles to move toward this proton dripline. This, along with difficulties in studying low-energy proton emitters, means that the known proton emitters are mostly in the medium mass—heavy nuclei. A review article by Hofmann (1996) summarizes the details of proton decay.

## PROBLEMS

1. Using the conservation of momentum and energy, derive a relationship between $Q_\alpha$ and $T_\alpha$.

2. All nuclei with $A > 210$ are $\alpha$ emitters, yet very few emit protons spontaneously. Yet both decays lower the Coulomb energy of the nucleus. Why is proton decay not more common?

3. Use the Geiger–Nuttall rule to estimate the expected $\alpha$ decay half-lives of the following nuclei: $^{148}$Gd, $^{226}$Ra, $^{238}$U, $^{252}$Cf, and $^{262}$Sg.

4. Use the one-body theory of $\alpha$ decay to estimate the half-life of $^{224}$Ra for decay by emission of a $^{14}$C ion or a $^{4}$He ion. The measured half-life for the $^{14}$C decay mode is $10^{-9}$ relative to the $^{4}$He decay mode. Estimate the relative preformation factors for the $\alpha$ particle and $^{14}$C nucleus in the parent nuclide.

5. $^{212}$Po$^m$ and $^{269}$110 both decay by the emission of high-energy $\alpha$ particles ($E_\alpha = 11.6$ and $11.1$ MeV, respectively). Calculate the expected lifetime of these nuclei using the one-body theory of $\alpha$ decay. The observed half-lives are 45.1 s and 170 μs, respectively. Comment on any difference between the observed and calculated half-lives.

6. Consider the decay of $^{278}$112 to $^{274}$110. The ground-state $Q_\alpha$ value is 11.65 MeV. Calculate the expected ratio of emission to the $2^+$, $4^+$, and $6^+$ states of 110.

7. What is the wavelength of an $\alpha$ particle confined to a $^{238}$U nucleus?

8. $^{8}$Be decays into two $\alpha$ particles with $Q_\alpha = 0.094$ MeV. Calculate the expected half-life of $^{8}$Be using one-body theory and compare this estimate to the measured half-life of $2.6 \times 10^{-7}$ s.

9. Calculate the kinetic energy and velocity of the recoiling daughter atom in the $\alpha$ decay of $^{252}$Cf.

10. Calculate the hindrance factor for the $\alpha$ decay of $^{243}$Bk to the ground state of $^{239}$Am. The half-life of $^{245}$Bk is 4.35 h, the decay is 99.994% EC and 0.006% $\alpha$ decay. Also, 0.0231% of the $\alpha$ decays lead to the ground state of $^{241}$Am. $Q_\alpha$ for the ground state decay is 6.874 MeV.

11. Calculate $Q_\alpha$ for gold. Why do we not see $\alpha$ decay in gold?

12. The natural decay series starting with $^{232}$Th has the sequence $\alpha\beta\beta\alpha$. Show why this is the case by plotting the mass parabolas (or portions thereof) for $A = 232$, 228, and 224.

13. Using the semiempirical mass equation, verify that $Q_\alpha$ becomes positive for $A \geq 150$.

14. Calculate the heights of the centrifugal barrier for the emission of $\alpha$ particles carrying away two units of angular momentum in the decay of $^{244}$Cm. Assume $R_0 = 1 \times 10^{-13}$ cm. What fraction of the Coulomb barrier height does this represent?

15. Use one-body theory to calculate the expected half-life for the proton decay of $^{185}$Bi.

## REFERENCES

Hofmann, S. Proton Radioactivity, In D. N. Poenaru (Ed.), *Nuclear Decay Modes*, IOP, Bristol, 1996.

Meyerhof, W. *Elements of Nuclear Physics*, McGraw-Hill, New York, 1967, pp. 135–145.

## BIBLIOGRAPHY

Textbook discussions of $\alpha$ decay that are especially good.

Evans, R. *The Atomic Nucleus*, McGraw-Hill, New York, 1953.

Heyde, K. *Basic Ideas and Concepts in Nuclear Physics*, IOP, Bristol, 1994, pp. 82–103.

Krane, K. S. *Introductory Nuclear Physics*, Wiley, New York, 1988, pp. 246–271.

Rasmussen, J. O. Alpha Decay, In K. Siegbahn (Ed.), *Alpha-, Beta-, and Gamma-Ray Spectroscopy*, North-Holland, Amsterdam, 1965, Chapter XI.

Wong, S. S. M. *Introductory Nuclear Physics*, 2nd ed., Wiley, New York, 1998.

# CHAPTER 8

# β DECAY

## 8.1 INTRODUCTION

We have seen that many thousands of nuclei can be produced and studied in the lab. However, only less than 300 of these nuclei are stable; the rest are radioactive. We have also seen that the degree of instability grows with the "distance" a given nuclide is from the stable nuclide with the same mass number. In the previous chapter we considered the process of α decay in which heavy nuclei emit α particles to reduce their mass and move toward stability. The Coulomb barrier limits this process to those regions where the $Q$ value provides sufficient energy to tunnel through the barrier. The vast majority of unstable nuclei lie in regions in which α decay is not important and the nuclei undergo one or another form of β decay in order to become more stable. In a certain sense, the stable nuclei have a balance between the numbers of neutrons and protons. Nuclei are said to be unstable with respect to β decay when these numbers are "out of balance." In a very qualitative way β decay "converts" a neutron into a proton (or vice versa) inside a nucleus, which becomes more stable while maintaining a constant mass number. The β decay process is more complicated than α emission, and we will provide an overview and a discussion of its basic features in this chapter.

β Decay is named for the second most ionizing rays that were found to emanate from uranium samples. The naturally occurring β rays were identified as fast moving (negative) electrons relatively easily, but it took many years to obtain a full understanding of the emission process. The difficulty lies in the fact that two particles are

*Modern Nuclear Chemistry*, by W.D. Loveland, D.J. Morrissey, and G.T. Seaborg
Copyright © 2006 John Wiley & Sons, Inc.

"created" during the decay as compared to the "disruption" of a heavy nucleus in α decay. In contrast to α decay, angular momentum plays a crucial role in understanding the process. Let us consider the simplest form of β decay to illustrate the difficulties. The proton and the neutron are the two possible isobars for $A = 1$. We know that the neutron has a larger mass than the proton and is thus unstable with respect to the combination of a proton and an electron. A free neutron will undergo β decay with a half-life of approximately 10 min. We might expect to write the decay equation as:

$$ {}_0^1 n \rightarrow {}_1^1 p + {}_{-1}^0 e + Q \text{ (incomplete)} $$

However, all three particles in this equation are fermions with intrinsic spins $S = \frac{1}{2}\hbar$. Therefore, we *cannot* balance the angular momentum in the reaction as written. The spins of the proton and the electron can be coupled to 0 or $1\hbar$ and can also have relative angular momenta with any integral value from the emission process. This simple spin algebra will never yield the half-integral value on the left-hand side of the equation. Another fermion must be present among the products.

Another feature of β decay that was puzzling at first but really pointed to the incompleteness of the previous equation is that the β rays have a continuous energy distribution. That is, electrons are emitted from a source with a distribution of energies that extends from a maximum at the $Q$ value down to zero. Recall that if there are only two products from a reaction then they will precisely share the decay energy according to conservation of momentum. We have clearly seen such sharp energy spectra in α decay. (The continuous energy distribution is not an instrumental artifact, nor does it come from electron scattering.) Quite dramatic pictures of the tracks of charged particles from beta decay show events in which the particles move in one direction in clear violation of conservation of linear momentum. The way out of this mounting paradox with violations of very strongly held conservation laws is to introduce another conservation law and recognize that another unseen particle must be created and emitted. The conservation law is conservation of the number of "particles" in a reaction, and the unseen particle is a form of neutrino, literally "the little neutral one" in Italian.

## 8.2 NEUTRINO HYPOTHESIS

Enrico Fermi on his voyage to the new world postulated that a third particle was needed to balance the emission of the electron in β decay. However, the existing conservation laws also had to be satisfied, so there were a number of constraints on the properties of this new particle. Focusing on the decay of a neutron as a specific example, the reaction is already balanced with respect to electric charge, so any additional particle must be neutral. The electrons were observed with energies up to the maximum allowed by the decay $Q$ value so the mass of the particle must be smaller that the instrumental uncertainties. Initially, this instrumental limit was <1 keV, but this has been reduced to <10 eV in recent work. Recent experiments have shown that the neutrinos have a very small mass (Chapter 12). The third constraint on the neutrino from the decay is that it must be an "antiparticle" in order to cancel or compensate for the creation of the electron, a "particle." The fourth constraint is that the neutrino must have half-integral spin and be a fermion in order to couple the total final angular momentum to the initial spin of $\frac{1}{2}\hbar$

Combining all of these constraints we can now rewrite the previous equation properly as:

$$ {}_0^1 n \longrightarrow {}_1^1 p + {}_{-1}^0 e + {}_0^0 \bar{\nu}_e + Q $$

where we have used the notation of placing a bar over the Greek character nu to indicate that the neutrino is an antiparticle and a subscript indicating the neutrino is an electron neutrino (Chapter 1). As indicated in Chapter 1, the existence of antiparticles and antimatter extends quite generally, and we produce and observe the decays of antielectrons (positrons), antiprotons, antineutrons, and so forth, and even combine positrons and antiprotons to make antihydrogen!

The spins of all of the final products can be combined in two ways and still couple to the initial spin of the neutron. Focusing on the spins of the created particles, they can vector couple to $S_\beta = 1$ in a parallel alignment or to $S_\beta = 0$ in an antiparallel alignment. Both of these can combine with $S = \frac{1}{2}$ of the neutron for a resultant vector of $\frac{1}{2}$. The two possible relative alignments of the "created" spins are labeled as Fermi (F) ($S_\beta = 0$) and Gamow–Teller (GT) ($S_\beta = 1$) decay modes after the people that initially described the mode. Both modes are very often possible and a source will produce a mixture of relative spins. In some cases, particularly the decay of even–even nuclei with $N = Z$ (the so-called self mirror nuclei), the neutron and protons are in the same orbitals so that $0^+$ to $0^+$ decay can only take place by a Fermi transition. In heavy nuclei with protons and neutrons in very different orbitals (shells), the GT mode dominates. In complex nuclei, the rate of decay will depend on the overlap of the wave functions of the ground state of the parent and the state of the daughter. The final state in the daughter depends on the decay mode. Notice that in the example of neutron decay, the difference between the two modes is solely the orientation of the spin of the bare proton relative to the spins of the other products. The decay constant can be calculated if these wave functions are known. Alternatively, the observed rate gives some indication of the quantum mechanical overlap of the initial and final state wave functions.

The general form of $\beta^-$ decay of a heavy parent nucleus, $^A Z$, can be written as:

$$ {}^A Z_N \longrightarrow {}^A (Z+1)^+_{N-1} + e^- + \bar{\nu}_e + Q_{\beta^-} $$

where we have written out the charges on the products explicitly. Notice that the electron can be combined with the positive ion to create a neutral atom (with the release of very small binding energy). This allows us to use the masses of the neutral atoms to calculate the $Q$ value, again assuming that the mass of the antineutrino is very small. Thus,

$$Q_{\beta^-} = M[^A Z] - M[^A(Z+1)]$$

Up to this point we have concentrated on the β-decay process in which a neutron is converted into a proton. There are a large number of unstable nuclei that have more protons in the nucleus than the stable isobar and so will decay by converting a proton into a neutron. We can write an equation for β$^+$ *decay* that is exactly analogous to the previous equation:

$$^A Z_N \rightarrow {}^A(Z-1)^-_{N+1} + e^+ + \nu_e + Q_{\beta^+}$$

where we have replaced both the electron and the electron antineutrino with their respective antiparticles, the positron and the electron neutrino. Note in this case, in contrast to β$^-$ decay, the charge on the daughter ion is negative. This means that there is an extra electron present in the reaction compared to that with a neutral daughter atom. Thus, the $Q$ value must reflect this difference:

$$Q_{\beta^+} = M[^A Z] - (M[^A(Z-1)] + 2m_e c^2)$$

where $m_e$ is the electron mass. Recall that particles and antiparticles have identical masses. This equation shows that spontaneous β$^+$ decay requires that the mass difference between the parent and daughter atoms be greater than $2m_e c^2 = 1.022$ MeV. Nature takes this to be an undue restriction and has found an alternative process for the conversion of a proton into a neutron (in an atomic nucleus). The process is the capture of an orbital electron by a proton in the nucleus. This process, called *electron capture* (EC), is particularly important for heavy nuclei. The reaction is written:

$$^A Z_N \rightarrow {}^A(Z-1)_{N+1} + \nu_e + Q_{EC}$$

where all of the electrons are implicitly understood to be present on the atoms. This process also has the property that the final state has only two products so conservation of momentum will cause the neutrino to be emitted with precise energies depending on the binding energy of the captured electron and the final state of the daughter nucleus.

To summarize, there are three types of decay, all known as β decay. They are

$$^A_Z P \rightarrow {}^A_{Z+1} D + \beta^- + \bar{\nu}_e$$
$$^A_Z P \rightarrow {}^A_{Z-1} D + \beta^+ + \nu_e$$
$$e^- + {}^A_Z P \rightarrow {}^A_{Z-1} D + \nu_e$$

indicating $\beta^-$ decay of neutron-rich nuclei, $\beta^+$ decay of proton-rich nuclei, and electron capture decay of proton-rich nuclei. Neglecting the electron binding energies in computing the decay energetics, we have

$$Q_{\beta^-} = (M_P - M_D)c^2$$
$$Q_{\beta^+} = (M_P - M_D)c^2 - 2m_e c^2$$
$$Q_{EC} = (M_P - M_D)c^2$$

where $M$ is the atomic mass of the nuclide involved and $m_e$ is the electron mass. Typical values of $Q_{\beta^-}$ near stability are $\sim$0.5–2 MeV, $Q_{\beta^+} \sim$ 2–4 MeV, and $Q_{EC} \sim$ 0.2–2 MeV.

As a final point in the introduction, it is interesting to note that the analogous process of positron capture by neutron excessive nuclei should be possible in principle. However, such captures are hindered by two important facts: First, the number of positrons available for capture is vanishingly small in nature, and second, both the nucleus and the positron are positively charged and will repel one another. Compare this to the situation for electron capture in which the nucleus is surrounded by (negative) electrons that are attracted to the nucleus, of course, and the most probable position to find any s electrons is at the nucleus ($r = 0$).

**Example Problem** Write the balanced equation for positron capture on the $\beta$-unstable nucleus, $^{24}$Na. Calculate the $Q$ value for this process.

**Solution** On the left-hand side of the equation we assume that we have a $^{24}$Na nuclide (with 11 electrons) and a single positron, which is an antilepton. The conservation rules imply that the mass number of the product will be 24, the atomic number will be $Z = 11 + 1$, the 11 electrons will carry over, and an antilepton has to be created to conserve lepton number. Thus,

$$^{24}\text{Na} + e^+ \longrightarrow {}^{24}\text{Mg}^+ + \bar{\nu}_e + Q_{pc}$$

We must be careful about the number of electrons on both sides of the equation when we calculate the $Q$ value. If we use mass defects rather than the masses and assume a zero-mass neutrino, then

$$Q_{pc} = (\Delta(^{24}\text{Na}) + m_e c^2) - (\Delta(^{24}\text{Mg}) - m_e c^2)$$

or

$$Q_{pc} = (\Delta(^{24}\text{Na}) + 2m_e c^2) - \Delta(^{24}\text{Mg})$$
$$Q_{pc} = (-8.418 + 1.022) + 13.933 = 6.537 \text{ MeV}$$

## 8.3 DERIVATION OF SPECTRAL SHAPE

$\beta$ Decay is clearly a process that follows first-order kinetics, and the rate of decay should be described by a single decay constant. Experimentally, $\beta$ decay has been observed with a huge range of half-lives, from a few milliseconds (and no shorter) to $\sim 10^{16}$ y. This large range is reminiscent of the range for $\alpha$ decay, and we should expect that the nuclear structure of the parent, ground state, and the available daughter states will play important roles in determining the half-life. We should also recognize that the calculation of the rate will require a full quantum mechanical approach because the decay process involves the creation of two particles, and the kinetic energy spectrum is relativistic for the electron because $Q_\beta \sim m_e c^2$.

Fermi developed a quantum mechanical theory of $\beta$ decay building on the foundation of the theory for the spontaneous emission of photons by systems in excited states. At first blush these may seem unrelated, but in both cases a system in a very well-defined single state that has excess energy releases the energy spontaneously by the creation of a particle (or particles). The decay constant for the emission of a photon was shown in the appendix E to be given by the general expression:

$$\lambda = \frac{2\pi}{\hbar} \left| \int \Psi^*_{\text{final}} V_p \Psi_{\text{initial}} \, d\tau \right|^2 \rho(E_f)$$

which is also called Fermi's golden rule. The wave functions, $\Psi$, represent the complete initial and final states of the *entire* system, and $V_p$ is a (very) small perturbative interaction that stimulates the transition. The form and the strength of the perturbation will have to be determined. Fermi assumed that the interaction responsible for $\beta$ decay is different from the gravitational, Coulomb, and nuclear forces. This interaction between the nucleons, electron, and neutrino is called the *weak interaction*, and a new constant expressing its strength, like e and $G$, can be defined. This constant, $g$, has the numerical value of $0.88 \times 10^{-4}$ MeV/fm$^3$, which is approximately $10^{-3}$ of the electromagnetic force constant. The last factor, $\rho(E_f)$, is the density of states that are available to the system after the transition and is often written as $dn/dE$ where $n$ is the number of states per unit energy interval. In this case the final energy is the decay $Q$ value. The initial wave function contains only the parent nucleus, whereas the final wave function will have parts for all the resultant particles. Specifically, for $\beta$ decay $\Psi_{\text{initial}} = \phi_{\text{gs}}(^A Z)$, the complete wave function for the parent in its ground state. The final wave function will have three parts, $\Psi^*_{\text{final}} = \phi^*_j(^A Z)\phi^*(e)\phi^*(\nu)$, a part for the daughter nucleus in the appropriate state $j$, a part for the traveling wave of the electron, and a part for the corresponding traveling wave of the neutrino, all of which must be coupled so that energy is conserved.

The quantum mechanical problem can be separated into two parts, the determination of $\rho(E_f)$ and the matrix element $\left| \int \varphi^*_{\text{final}} V_p \varphi_{\text{initial}} d\tau \right|^2$, to make the calculation tractable. The determination of the density of final states, $dn/dE$, is done using quantum statistical mechanics. It is basically the problem of counting the number of ways the decay energy can be divided among the electron and the neutrino, neglecting for the moment, the recoiling daughter nucleus. Classically, the number of states of a free electron with momentum between $p_e$ and $p_e + dp_e$ in a volume $V$ is $(V 4\pi p_e^2 \, dp_e)/h^3$. (This is the volume of a spherical shell in phase

## 8.3 DERIVATION OF SPECTRAL SHAPE

space where the volume of a unit cell is $h^3$.) Similarly for the neutrino, the number of states of the free neutrino with momentum between $p_v$ and $p_v + dp_v$ in a volume $V$ is $(V 4\pi p_v^2 dp_v)/h^3$. The total number of states is the product of these two factors:

$$dn = \frac{16\pi^2 V^2 p_e^2 p_v^2 \, dp_e \, dp_v}{h^6}$$

If we assume the neutrino has zero rest mass

$$p_v = \frac{T_v}{c} = \frac{Q - T_e}{c}$$

$$dp_v = \frac{dQ}{c}$$

Then, substituting, we get

$$dn = \frac{16\pi^2 V^2}{h^6 c^3} (Q - T_e)^2 p_e^2 \, dp_e \, dQ$$

$$\frac{dn}{dQ} = \frac{16\pi^2 V^2}{h^6 c^3} (Q - T_e)^2 p_e^2 \, dp_e$$

(One must understand that this equation expresses the variation of the number of final states with changes in the $Q$ value of the decay and does not represent differentiation with respect to a constant $Q$.)

The electron and neutrino wave functions can be written as plane waves as:

$$\phi_e(r) = A e^{ik_e r} = \frac{1}{\sqrt{V}} e^{ik_e r}$$

$$\phi_v(r) = B e^{ik_v r} = \frac{1}{\sqrt{V}} e^{ik_v r}$$

where we have applied a normalization condition to determine the constants $A$ and $B$. We can expand the exponentials for $r \sim 0$ (the nuclear volume) as:

$$e^{ikr} = 1 + ikr + \cdots \cong 1$$

Thus

$$\phi_e(r \sim 0) \cong \frac{1}{\sqrt{V}}$$

$$\phi_v(r \sim 0) \cong \frac{1}{\sqrt{V}}$$

The probability of emitting an electron with a momentum $p_e$ between $p_e$ and $dp_e$ becomes

$$\lambda(p_e) \, dp_e = \frac{1}{2\pi^3 \hbar^7 c^3} |M_{if}|^2 g^2 (Q - T_e)^2 p_e^2 \, dp_e$$

where $|M_{if}|^2$ is a *nuclear* matrix element representing the overlap between the initial and final *nuclear* states. This matrix element must be evaluated with the detailed nuclear wave functions, for example, those available from the shell model.

Collecting all constants for a given decay, the probability of a decay as a function of the electron momentum is

$$\lambda(p_e)\, dp_e = (\text{constants})(Q - T_e)^2 p_e^2\, dp_e$$

This form (even though it is mixed with a momentum part and an energy part for the electron) clearly goes to zero at $p_e = 0$ and also at $T_e = Q$ and thus has a maximum in between. The shape of this function is shown in Figure 8.1. This function is often called the statistical or phase space factor for the decay.

We should be sure to note that we have made a big approximation in ignoring the charge on the emitted electron. Positively charged β particles (positrons) will be repelled by the nucleus and shifted to higher energies, whereas negatively charged β particles (electrons) will be attracted by the nucleus and slowed down. These effects were incorporated by Fermi by using Coulomb-distorted wave functions and are contained in a spectrum distortion expression called the Fermi

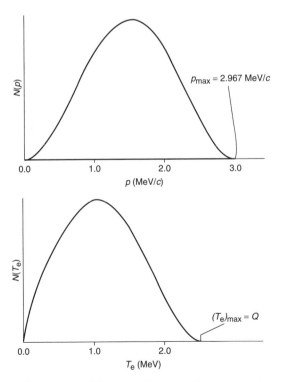

**Figure 8.1** Shape of the statistical factor for β decay, which represents the expected shape of the electron momentum distribution before distortion by the Coulomb potential.

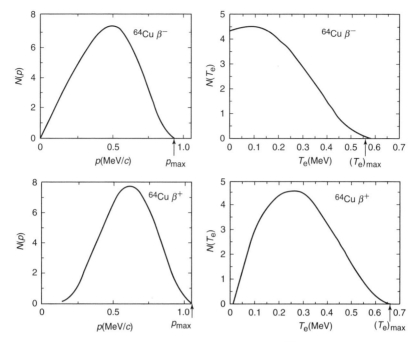

**Figure 8.2** Momentum and energy spectra from the decay of $^{64}$Cu for $\beta^-$ and $\beta^+$ decay. The $Q$ values for these decays are 0.5782 and 0.6529 MeV, respectively.

function, $F(Z_D, p_e)$, where $Z_D$ is the atomic number of the daughter nucleus. The $\beta$ spectrum thus has the form:

$$\lambda(p_e)\, dp_e = (\text{constants}) F(Z_D, p_e) p_e^2 (Q - T_e)^2\, dp_e$$

The effects of the Coulomb distortion can be seen in the measured spectra from the decay of $^{64}$Cu shown in Figure 8.2. This odd–odd nucleus undergoes both $\beta^-$ and $\beta^+$ decay to its even–even neighbors with very similar $Q$ values.

Relaxing the restriction that the neutrino rest mass is zero, we get (Heyde, 1999)

$$\lambda(p_e)\, dp_e = \frac{|M_{\text{if}}|^2}{2\pi^3 \hbar^7 c^3} g^2 F(Z_D, p_e) p_e^2 (Q - T_e)^2 \left(1 - \frac{m_\nu^2 c^4}{(Q - T_e)^2}\right)^{1/2} dp_e$$

## 8.4 KURIE PLOTS

We have seen that the $\beta$ spectrum has an endpoint at the $Q$ value, but the form of equation for the spectrum does not allow us to easily identify the endpoint.

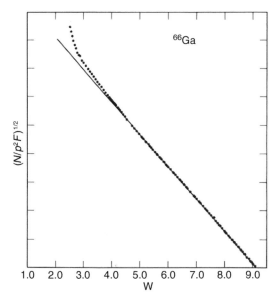

**Figure 8.3** Example of a Kurie plot. (From Camp and Langer, 1963.)

Notice that with a little rearrangement this spectrum can be represented as:

$$\left(\frac{\lambda(p_e)}{p_e^2 F(Z_D, p_e)}\right)^{1/2} \propto (Q - T_e)|M_{if}|^2$$

If the nuclear matrix element does not depend on the electron kinetic energy, as we have assumed so far, then a plot of the reduced spectral intensity, the left-hand side, versus the electron kinetic energy will be a straight line that intercepts the abscissa at the $Q$ value. Such a graph is called a Kurie plot, and an example is shown in Figure 8.3. This procedure applies to allowed transitions (see below). There are correction terms that need to be taken into account for forbidden transitions.

## 8.5 β-DECAY RATE CONSTANT

The differential form of the spectrum can be integrated over all electron momenta to obtain the total decay constant. The expression, for a constant nuclear matrix element, to be integrated is

$$\lambda = \frac{g^2 |M_{if}|^2}{2\pi^3 \hbar^7 c^3} \int_0^{p_{max}} F(Z_D, p_e) p_e^2 (Q - T_e)^2 \, dp$$

Note that an appropriate relativistic substitution for $T$ in terms of the momentum is still needed. This integral has been shown to only depend on the atomic number of the daughter and the maximum electron momentum. The integral, called the Fermi integral, $f(Z_D, Q)$, is complicated but numerical expressions or tables of the solutions are available. Note that the differential *Fermi function*, $F(Z_D, p_e)$, contains the momentum and the *Fermi integral*, $f(Z_D, Q)$, contains the $Q$ value. The Fermi integral is a constant for a given β decay and has been presented in many forms. For example, curves of the Fermi integral are shown in Figure 8.4.

The decay constant is now reduced to an expression with the nuclear matrix element, $M(\equiv |M_{if}|)$, and the strength parameter, $g$, written:

$$\lambda = \frac{g^2 |M|^2 m_e^5 c^4}{2\pi^3 \hbar^7} f(Z_D, Q)$$

or in terms of the half-life of the parent, $t_{1/2}$

$$ft_{1/2} = \ln 2 \frac{2\pi^3 \hbar^7}{g^2 |M|^2 m_e^5 c^4} \propto \frac{1}{g^2 |M|^2}$$

The left-hand side of this equation is called the comparative half-life, or "*ft* value" because this value can be readily measured in experiments and should only depend on the nuclear matrix element and the β-decay strength constant. Recall that β decay half-lives span many orders of magnitude so the *ft* values will span a similarly large range. It is therefore convenient to use the common logarithm of the *ft* value (with $t_{1/2}$ in seconds) to characterize observed β decays.

Values of log *ft* may be calculated from the nomograph and curves in Figure 8.4, which are due to Moszkowski (1951). Log *ft* values can be calculated for β⁻, β⁺, and EC decay. These *ft* values fall into groups that can be correlated with the spin and parity change in the decay (see below) and can, then, be used to assign spins and parities in nuclei whose structure is not known (see Figure 8.5).

**Example Problem** Using the graph of the Fermi integral in Figure 8.4, estimate the log *ft* value for the decay of $^{32}$P ($t_{1/2} = 14.28$ d).

*Solution*
1. This is a neutron-rich nucleus and undergoes β⁻ decay, thus:

$$Q_{\beta^-} = M(^{32}\text{P}) - M(^{32}\text{S}) = \Delta(^{32}\text{P}) - \Delta(^{32}\text{S})$$
$$Q_{\beta^-} = (-24.305) - (-26.015) \text{ MeV} = +1.71 \text{ MeV}$$

2. From the figure, $Z = 15$, $Q = 1.71$ MeV, $\log(ft) = \log(f_0 t) + \log(C) = 7.8 + 0.2 = 8.0$

The creation of relative angular momentum in β decay is even more difficult than that in α decay and causes more severe "hindrance" for each unit of relative angular momentum. The difficulty is easy to see with a simple calculation. We can write the relative angular momentum for two bodies as the *cross product* $L = r \times p$ where $r$ is the radius of emission and $p$ is the momentum. Taking a typical nuclear radius of 5 fm and a typical β-decay energy of 1 MeV, we find the maximum of the cross product to

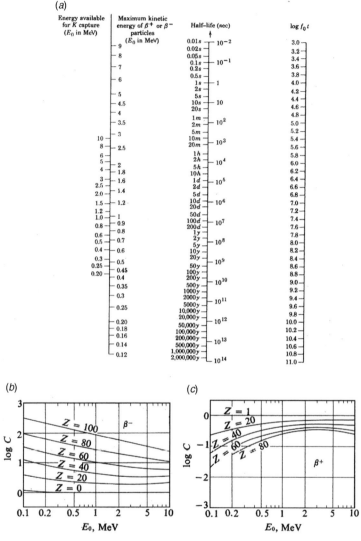

**Figure 8.4** Rapid method for determining $\log_{10}(ft)$ values. (From Moszowski, 1951.) The above figures permit the rapid calculation of $\log(ft)$ for a given type of decay, given energy, branching ratio, etc. Notation: $E_0$ for $\beta^{\pm}$ emission is the maximum kinetic energy of the particles in MeV; $E_0$ for K electron capture is the $Q$ value in MeV. When a $\beta^+$ emission and K electron capture go from and to the same level, $E_0$ for the K capture = $E_0$ for $\beta^+$ emission + 1.02 MeV. Z is the atomic number of the parent, $t$ is the total half life, and $p$ is the percentage of decay occurring in the mode under consideration. When no branching occurs, $p = 100$. To obtain $\log(ft)$, obtain $\log(f_0 t)$ using part (a). Read off $\log(c)$ from parts (b), (c), and (d) for $\beta^-$, $\beta^+$, and K EC, respectively. Get $\Delta \log(ft)$ from part (e) if $p < 100$. For $p = 100$, $\Delta \log(ft) = 0$. $\log(ft) = \log(f_0 t) + \log(C) + \Delta \log(ft)$.

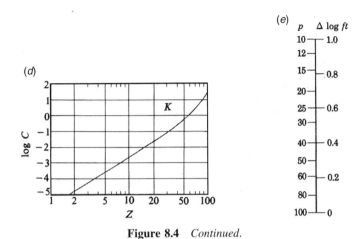

**Figure 8.4** *Continued.*

be $L = 5$ fm $(1.4 \text{ MeV}/c) = 7.90$ MeV fm/$c$ or $0.035\ \hbar$ units. Log $ft$ values increase by an average of 3.5 units for each unit, of orbital angular momentum or degree of forbiddenness. Such an increase in the lifetime indicates a hindrance of $\sim 3 \times 10^{-4}$ for each unit of angular momentum. There is a large spread in the values, however, due to the strong effect of the nuclear matrix element for each decay.

The quantum mechanical selection rules for β decay with no relative angular momentum in the exit channel ($l = 0$) are $\Delta I = 0, 1$ and $\Delta \pi = 0$. The two values

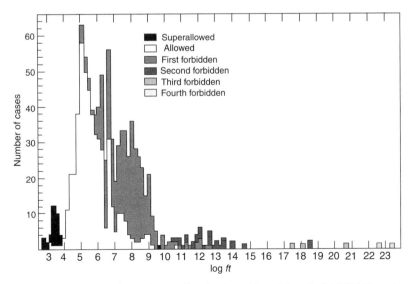

**Figure 8.5** Systematics of experimental log $ft$ values. (From Meyerhof, 1967.) Copyright © McGraw-Hill Book Company, Inc. Reprinted by permission of McGraw-Hill Book Company.

**TABLE 8.1 Representative Allowed β Decays**

| Parent | Daughter | Half-Life (s) | $Q_\beta$ (MeV) | log $ft$ | Character |
|---|---|---|---|---|---|
| $^6$He (0$^+$) | $^6$Li (1$^+$, gs) | 0.808 | 3.5097 | 2.42 | Gamow–Teller |
| $^{14}$O (0$^+$) | $^{14}$N (0$^+$, 2.313) | 71.1 | 1.180 | 2.81 | Fermi |
| n ($\frac{1}{2}^+$) | p ($\frac{1}{2}^+$) | 612 | 0.7824 | −0.27 | Mixed |
| $^{14}$O (0$^+$) | $^{14}$N (1$^+$, gs) | $1.16 \times 10^4$ | 4.123 | 7.36 | Gamow–Teller |

for the spin change come directly from the two possible couplings of the spins of the electron and neutrino. Some representative "allowed" β decays are described in Table 8.1 along with their $ft$ values and the character of the decay.

The decay of $^{14}$O to the 0$^+$ excited state of $^{14}$N can only take place by a Fermi decay where the created spins couple to zero. This parent nucleus also has a weak branch to the 1$^+$ ground state that takes place by a Gamow–Teller transition. In contrast, the decay of $^6$He to the ground state of $^6$Li must take place by a Gamow–Teller transition in order to couple the total resultant angular momentum to zero. As mentioned earlier, the decay of the neutron into a proton can take place with no change in angular momentum between the spin $\frac{1}{2}$ particles, and the angular momentum coupling rules allow both decay modes.

The decay of the neutron into the proton is an important example of decay between mirror nuclei. In the β decay of mirror nuclei, the transformed nucleons (neutron → proton or proton → neutron) must be in the same shell and have very similar wave functions. This gives rise to a large matrix element $|M_{if}|^2$ and a very small log $ft$ value. For the β decay of mirror nuclei to their partners, log $ft$ values are about 3, which is unusually small. Such transitions are called *superallowed* transitions.

When the initial and final states in β decay have opposite parities, decay by an *allowed* transition cannot occur. However, such decays can occur, albeit with reduced probability compared to the allowed transition. Such transitions are called *forbidden* transitions even though they do occur. The forbidden transitions can be classified by the spin and parity changes (and the corresponding observed values of log $ft$) as in Table 8.2.

Remember that in β decay,

$$\mathbf{J}_P = \mathbf{J}_D + \mathbf{L}_\beta + \mathbf{S}_\beta$$

$$\pi_P = \pi_D(-1)^{L_\beta}$$

**TABLE 8.2 Classifications of β-Decay Transitions**

| Transition Type | log $ft$ | $L_\beta$ | $\Delta\pi$ | Fermi $\Delta I$ | Gamow–Teller $\Delta I$ |
|---|---|---|---|---|---|
| Superallowed | 2.9–3.7 | 0 | No | 0 | 0 |
| Allowed | 4.4–6.0 | 0 | No | 0 | 0, 1 |
| First forbidden | 6–10 | 1 | Yes | 0, 1 | 0, 1, 2 |
| Second forbidden | 10–13 | 2 | No | 1, 2 | 1, 2, 3 |
| Third forbidden | >15 | 3 | Yes | 2, 3 | 2, 3, 4 |

## 8.6 ELECTRON CAPTURE DECAY

When the decay energy is less than 1.02 MeV ($2m_ec^2$), the β decay of a proton-rich nucleus to its daughter must take place by electron capture (EC). For decay energies greater than 1.02 MeV, EC and $\beta^+$ decay compete. In EC decay, only one particle, the neutrino, is emitted with an energy $M_Pc^2 - M_Dc^2 - B_e$ where $B_e$, is the binding energy of the captured electron. The decay constant for electron capture can be written, assuming a zero neutrino rest mass, as:

$$\lambda_{EC} = \frac{g^2 |M_{if}|^2 T_\nu^2}{2\pi^2 c^3 \hbar^3} |\varphi_K(0)|^2$$

where we have assumed that the capture of a 1's ($K$) electron will occur because the electron density at the nucleus is the greatest for the $K$ electrons. The $K$ electron wave function can be written as:

$$\varphi_K(0) = \frac{1}{\sqrt{\pi}} \left( \frac{Z m_e e^2}{4\pi\varepsilon_0 \hbar^2} \right)^{3/2}$$

Thus

$$\lambda_{K-EC} = \frac{g^2 Z^3 |M_{if}|^2 T_\nu^2}{\text{constants}}$$

Comparison of the decay constants for EC and $\beta^+$ decay shows

$$\frac{\lambda_K}{\lambda_{\beta^+}} = (\text{constants}) \frac{Z^3 T_\nu^2}{f(Z_D, Q)}$$

Thus EC decay is favored for high Z nuclei. Of course, the decay energy must be greater than 1.02 MeV for $\beta^+$ decay, a situation found mostly in low Z nuclei where the slope of the walls of the valley of β stability is large (see Fig. 2.8) and decay energies of >1.02 MeV occur.

Electron capture decay produces a vacancy in the atomic electron shells and secondary processes that lead to filling that vacancy by the emission of X-rays and Auger electrons occur. These X-rays permit the detection of EC decays.

## 8.7 PARITY NONCONSERVATION

In Chapter 1, we introduced the concept of parity, the response of the wave function to an operation in which the signs of the spatial coordinates were reversed. As we indicated in our discussion of α decay, parity conservation forms an important selection rule for α decay. Emission of an α particle of orbital angular momentum $l$ carries a parity change $(-1)^l$ so that $1^+ \to 0^+$ or $2^- \to 0^+$ α decays are forbidden. In general, we find that parity is conserved in strong and electromagnetic interactions.

In the late 1950s, it was found (Wu et al., 1957) that parity was not conserved in weak interaction processes such as nuclear β decay. Wu et al. (1957) measured the spatial distribution of the $\beta^-$ particles emitted in the decay of a set of polarized $^{60}$Co nuclei (Fig. 8.6). When the nuclei decay, the intensity of electrons emitted in two directions, $I_1$ and $I_2$, was measured. As shown in Figure 8.6, application of the parity operator will not change the direction of the nuclear spins but will reverse the electron momenta and intensities, $I_1$ and $I_2$. If parity is conserved, we should not be able to tell the difference between the "normal" and "parity reversed" situations, that is, $I_1 = I_2$. Wu et al. (1957) found that $I_1 \neq I_2$, that is, that the β particles were preferentially emitted along the direction opposite to the $^{60}$Co spin. (God is "left-handed.") The effect was approximately a 10–20% enhancement.

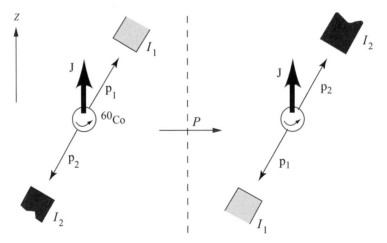

**Figure 8.6** Schematic diagram of the Wu et al. apparatus. (From H. Frauenfelder and E. M. Henley, Subatomic Physics, 2nd Edition. Copyright 1991 by Prentice-Hall, Inc. Reprinted by permission of Pearson Prentice-Hall.) A polarized nucleus emits electrons with momenta $p_1$ and $p_2$ that are detected with intensities $I_1$ and $I_2$. The left figure shows the "normal" situation while the right figure shows what would be expected after applying the parity operator. Parity conservation implies the two situations cannot be distinguished experimentally (which was not the case).

## 8.8 NEUTRINOS

A number of studies have been undertaken of the interaction of neutrinos with nuclei, to determine the neutrino mass, and to show that neutrinos and antineutrinos are produced in $\beta^+$ and $\beta^-$ decay, respectively. Neutrinos also provide important information about stellar nuclear reactions because they have a very low probability for interacting with matter and come directly out from the stellar interior.

Starting with the simple equation for the $\beta^-$ decay of the neutron and the $\beta^+$ decay of the proton, we can write two closely related reactions that are induced by neutrinos:

$$\bar{\nu}_e + p^+ \rightarrow n + e^+$$
$$\nu_e + n \rightarrow p^+ + e^-$$

These reactions, called inverse $\beta$ decay, were obtained by adding the antiparticle of the electron in the normal $\beta$ decay equation to both sides of the reaction. When we did this we also canceled (or annihilated) the antiparticle/particle pair. Notice that other neutrino-induced reactions such as $\bar{\nu}_e + n \rightarrow p^+ + e^-$ do not conserve lepton number because an antilepton, $\bar{\nu}_e$, is converted into a *lepton*, $e^-$. Proving that this reaction does not take place, for example, would show that there is a difference between neutrinos and antineutrinos. One difficulty with studying these reactions is that the cross sections are extremely small, of order $10^{-19}$ barns, compared to typical nuclear reaction cross sections, of order 1 barn ($10^{-24}$ cm$^2$).

The combination of two studies of inverse $\beta$ decay clearly showed that the neutrinos emitted in $\beta^-$ and $\beta^+$ decay were different. Both used nuclear reactors to provide strong sources of antineutrinos. Recall that nuclear fission produces very neutron-rich products that undergo a series of rapid $\beta$ decays emitting antineutrinos. In the first experiment, performed by Reines and Cowen (1953), a large volume of liquid scintillator was irradiated, and protons in the organic solution were changed into a neutron and a positron. The positron was rapidly annihilated with an electron providing the first signal of an interaction. The neutron was captured within a few microseconds by Cd nuclei that were added to the scintillator and provided a second correlated signal. The flux of neutrinos from the reactor was sufficient to produce a few events per hour in a 1-m$^3$ volume of scintillator.

In the second study, Ray Davis and co-workers, irradiated a large volume of liquid CCl$_4$ with antineutrinos from a reactor. The putative reaction, $\bar{\nu}_e + {}^{37}\text{Cl} \rightarrow {}^{37}\text{Ar} + e^-$, could be detected by periodic purging of the liquid, collection of the noble gas, and then detection of the induced activity ($^{37}$Ar is unstable, of course). The reaction was not observed to occur. Thus, they concluded that the reactor emits antineutrinos and that lepton number is conserved in the reactions.

**Example Problem** Estimate the flux of antineutrinos from an operating nuclear power reactor. For this estimate assume the power plant produces 1 GW of thermal power, that fission produces 200 MeV per event, and that there are approximately 6 rapid $\beta$ decays per fission.

**Solution** There is one antineutrino per $\beta^-$ decay, of course, so this is really a problem in dimensional analysis:

$$\text{Rate} = 1 \text{ GW}(10^6 \text{ J/s})/\text{GW})(1 \text{ fission}/200 \text{ MeV})$$
$$\times (1 \text{ MeV}/1.602 \times 10^{-13})(6\bar{\nu}_e/\text{ fission})$$
$$\text{Rate} = 2 \times 10^{17} \text{ antineutrinos/s}$$

## 8.9 β-DELAYED RADIOACTIVITIES

The central feature of β decay is that, for example, in the $\beta^-$ direction, the decay converts a neutron into a proton at a constant mass number. This conversion will clearly change the number of pairs of like nucleons in the nucleus, and we have already seen that unpaired nucleons influence the overall stability. β Decay in even mass chains will convert odd–odd nuclei into the even–even isobar with potentially large $Q$ values due to a gain of twice the pairing energy. The large $Q$ values lead to high-energy β particles and rapid decays, but the relative stability of the daughter may be less than that of the parent. The large $Q$ values also allow the population of higher lying states in the daughter. If the nuclei are far from the (most) stable isobar, the decay may have sufficient energy to populate states in the daughter that are above the separation energy.

$^{90}$Sr provides an example of a change in relative stability following β decay. This even–even parent is an important fission product that has a 29-y half-life. It decays to the odd–odd $^{90}$Y, which then decays to the stable isobar $^{90}$Zr with a half-life of only 64 h. Thus, a pure preparation of $^{90}$Sr will come into equilibrium with its daughter after about a week, and the observed activity will be the sum of the two decays. A chemical separation can be used to strip out the daughter activity. The daughter will decay away in the separated sample and will grow back into the parent sample. There are several examples of these parent–daughter pairs that provide convenient sources of short-lived activities. For example, the 66-h $^{99}$Mo decays predominantly to a 6-h excited state in $^{99}$Tc because the decay to ground state would require a very large spin change. The daughter, $^{99}$Tc$^m$, is used extensively in nuclear medicine.

The natural decay chains have several examples of short-lived α activities that are "delayed" by a longer-lived parent. In fact, the existence of these activities on Earth is possible by the fact that the "head" of the chain has a half-life on the order of the age of the Earth. Another more practical example near the end of the 4n chain is $^{212}$Pb with a half-life of 10.6 h that decays to $^{212}$Bi. The daughter rapidly decays by α or β emission. The lead nucleus is also preceded by a short-lived Rn parent, which can produce very thin sources of α particles by emanation.

The β decay of nuclei far from the bottom of the valley of β stability can feed unbound states and lead to direct nucleon emission. This process was first recognized during the discovery of fission by the fact that virtually all the neutrons are emitted promptly, but on the order of 1% are delayed in time with respect to the fission event. These delayed neutrons play a very important role in the control of nuclear reactors. The fission products are very neutron rich and have large β-decay energies. For example, $^{87}$Br is produced in nuclear fission and decays with a half-life of 55 s to $^{87}$Kr with a $Q$ value of 6.5 MeV. The decay populates some high-lying states in the krypton daughter. Notice that $^{87}$Kr has 51 neutrons, one more than the magic number 50, and the neutron separation energy of 5.1 MeV is less than the $Q$ value. Thus, any states that lie above the neutron separation energy will be able to rapidly emit a neutron and form $^{86}$Kr.

**Example Problem**   An important delayed neutron emitter in nuclear fission is $^{137}$I. This nuclide decays with a half-life of 25 s and emits neutrons with an average energy of 0.56 MeV and a total probability of approximately 6%. Estimate the energy of an excited state in $^{137}$Xe that would emit a 0.56-MeV neutron.

**Solution**   First obtain the $Q$ value for the neutron emission reaction. This is the minimum amount of energy necessary to "unbind" the 83rd neutron and should be negative, of course.

$$^{137}\text{Xe} \rightarrow \text{n} + ^{136}\text{Xe} + Q_n$$

$$Q_n = \Delta(^{137}\text{Xe}) - [\Delta(^{136}\text{Xe}) + \Delta(\text{n})]$$

$$Q_n = -82.218 - [-86.425 + 8.0714] = -3.86 \text{ MeV}$$

The average energy of the excited state will be $Q_n$ plus the kinetic energies of the particles, that is, the neutron plus the energy of the recoil. In this case the recoil energy is very small and could have been ignored. The recoil energy is obtained by conservation of momentum in the two-body decay.

$$E^* = -Q_n + T_n + T_n\left(\tfrac{1}{137}\right) = 3.86 + 0.57 = 4.43 \text{ MeV}$$

Now as a check, obtain the $Q$ value for the β decay and verify that it is more than the excitation energy:

$$^{137}\text{I} \rightarrow ^{137}\text{Xe} + \bar{\nu} + Q_\beta$$

$$Q_\beta = \Delta(^{137}\text{I}) - \Delta(^{137}\text{Xe}) = -76.72 - 82.21 = 5.49 \text{ MeV}$$

The population of high-lying unbound states by β decay is an important feature of nuclei near the driplines. β-Delayed proton emission and β-delayed neutron emission have been studied extensively and provide important insight into the structure of exotic nuclei.

## 8.10  DOUBLE-β DECAY

The periodic variation of the mass surface caused by the pairing energy also causes a large number of even–even nuclei to be unstable with respect to two successive β decays. This process is called double-β decay, and extensive searches have been carried out for it. The difficulty is that the probability of a double transition is extremely low. A gross estimate can be made by squaring the rate constant obtained above, and the number of decays from even large samples is at best one per day and at worst a few per year for the systems that have been considered for study.

Two reactions have been studied as possible candidates for double-β decay. The first reaction is simply two times the normal β decay process:

$$^{A}Z \rightarrow {}^{A}(Z-2) + 2e^{-} + 2\bar{\nu}$$

and thus follows the conservation laws. A second, more exotic reaction has been proposed as a test of weak interaction theory and proceeds without creation of neutrinos:

$$^{A}Z \rightarrow {}^{A}(Z-2) + 2e^{-}$$

Instrumental searches for this latter neutrinoless process have been made, but there is no strong evidence for its existence. The former two-neutrino decay has been observed with a variety of techniques that were carefully tuned to detect the rare products.

As an example of the process, the $^{86}$Kr nucleus just mentioned above as the daughter in delayed neutron emission is stable with respect to single $\beta^{-}$ decay to $^{86}$Rb having a $Q$ value of $-0.526$ MeV. However, $^{86}$Kr is unstable with respect to the double-β decay to $^{86}$Sr as it has a $Q$ value of 1.249 MeV. In this case decay to the intermediate state is energetically forbidden, and only the simultaneous emission of two β particles can take place. To obtain the gross estimate, we can rewrite the expression for the decay constant:

$$\lambda = \left(\frac{m_e c^2}{\hbar}\right)\left(\frac{|M|^2 m_e^4 c^2}{2\pi^3 \hbar^6} g^2 f(Z_D, Q)\right)$$

The first term is the constant $8 \times 10^{20}$/s and the second term reflects the details of the decay. Using $|M| = \sqrt{2}$ for the decay from the $0^+$ ground state, to the $0^+$ ground state of the daughter, the second term is $1.5 \times 10^{-25} f$. For this case, $\log(f) \sim 1.5$, then taking the first term times the square of the second for double-β decay, we get $\lambda \sim 10^{-26}$/s, or $\sim 10^{-19}$ per year! Given that a mole of this gas has $\sim 10^{24}$ atoms, we expect about one decay per day in the entire sample.

The techniques used to observe double-β decay fall into three general categories: geochemical, radiochemical, and instrumental. The geochemical studies rely on assumptions that are similar to those used in geochemical dating (see Chapter 3). A sample of an ore containing the parent nuclide is processed; the daughter atoms are chemically extracted and then assayed, for example, with a mass spectrometer. The number of daughter atoms is then compared to the number of parent atoms and with an estimate of the lifetime of the ore, the double-β decay half-life can be calculated. Difficulties with this technique are discussed in the Chapter 3. The radiochemical searches for double-β decay relied on chemically separating and identifying a radioactive daughter of the process. Such cases are relatively rare but the decay $^{238}$U $\rightarrow$ $^{238}$Pu was observed by chemically separating a uranium ore and observing the characteristic α decay of the plutonium isotope. The successful instrumental searches for double-β decay have used time projection chambers in which samples of the parent were introduced into the active volume of the detector. The tracks of the two coincident β particles can be observed providing a clear signal for the exotic process.

## PROBLEMS

1. The $\beta^-$ decay of $^{144}$Ce is shown below.

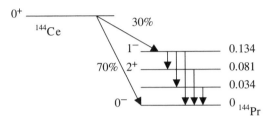

   a. What log $ft$ value should we expect for the β decay to the $1^-$ state of $^{141}$Pr?
   b. Why is there no β decay to the $2^+$ level?

2. Sketch quantitatively the shape of the neutrino energy spectrum for the following types of decay. Label all axes carefully and indicate the types of neutrinos involved.

   a. The electron capture decay of $^{207}$Bi, $Q_{EC} = 2.40$ MeV.
   b. The $\beta^+$ decay of $^{22}$Na, $Q_\beta = 3.86$ MeV.
   c. The $\beta^-$ decay of $^{14}$C, $Q_\beta = 0.156$ MeV.

3. Suppose a state in a Bi isotope decays by EC to the $2^+$ state of an even–even Pb nucleus in which the three lowest states are the $0^+$, $2^+$, and $4^+$, with $E_{EC} = 1.0$ MeV. Assume $Q_{EC} = 4$ MeV, $t_{1/2} = 4$ s. Calculate $J\pi$ for the initial state of the Bi nucleus.

4. Given the β decay scheme shown below for the decay of a pair of isomers to three excited states A, B, and C of the daughter nucleus. List the spins and parities of the three levels A, B, and C.

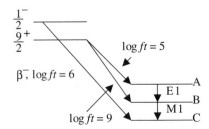

5. The results of some measurements with a β-ray spectrometer of the radiation coming from a given radionuclide are shown below.

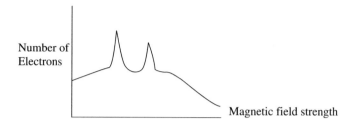

The two sharp peaks were labeled $K$ and $L$ by the experimenter. Explain what the labels $K$ and $L$ mean. Which peak is the $K$ peak? Why?

6. A $1^-$ excited state of a Lu isotope decays to a $0^+$ state of a Yb isotope with a maximum $\beta^+$ energy of 4.6 MeV. Estimate $t_{1/2}$ for the transition. Do not neglect electron capture.


## REFERENCES

Camp, D. C. and L. M. Langer. *Phys. Rev.* **129**, 1782 (1963).
Meyerhof, W. E. *Elements of Nuclear Physics*, McGraw-Hill, New York, 1967.
Moszkowski, S. A. *Phys. Rev.* **82**, 35 (1951).
Wu C. S. et al. *Phys. Rev.* **105**, 1413 (1957).


## BIBLIOGRAPHY

Evans, R. D. *The Atomic Nucleus*, McGraw-Hill, New York, 1956.
Krane, K. S. *Introductory Nuclear Physics*, Wiley, New York, 1988.
Lamarsh, J. R. *Introduction to Nuclear Reactor Theory*, Addison-Wesley, Reading, MA, 1967.
Moe, M. and P. Vogel. *Ann. Rev. Nucl. Sci.* **44**, 247 (1994).
Remies, T. and C. L. Cowen, Jr. *Phys. Rev.* **92**, 830 (1953).
Siegbahn, K. *Alpha, Beta and Gamma Ray Spectroscopy*, North-Holland, Amsterdam, 1966.
Wu, C. S. and S. A. Moszkowski. *Beta Decay*, Wiley, New York, 1966.

# CHAPTER 9

# γ-RAY DECAY

## 9.1 INTRODUCTION

γ-Ray decay occurs when a nucleus in an excited state releases its excess energy by emission of electromagnetic radiation, that is, a photon. Thus, we have

$$^{A}X^{*} \rightarrow {}^{A}X + \gamma$$

where the symbol * indicates an excited state of the nucleus. Note that there is no change in Z or A during this type of decay, only the release of energy. One can also get γ-ray emission from a high-lying excited state to a lower-lying state of the same nucleus. Thus, γ-ray transitions do not have to go to the ground state of the nucleus. Figure 9.1 depicts a typical situation in which a series of γ rays deexcite the levels of a nucleus with so-called crossover transitions also occurring (4 → 1, 4 → 2, etc.). Also note that the γ-ray energy spectrum shows discrete lines corresponding to each transition. Note that the energies of the γ rays can vary from a few keV to many MeV. Any nucleus with bound excited states can decay by γ-ray emission.

In some unusual cases a nucleus can have two configurations of nucleons that have very similar low-lying energies but very different total angular momenta. One of these states will be lower in energy, of course, but the transition between the two states will be strongly hindered, due to the fact that the photon will have to balance the large change in angular momentum. This hindered decay is similar to the hindrance of the decay of triplet states in atomic and molecular systems. These long-lived nuclear states are called *isomeric states*, and their γ-ray decay is

*Modern Nuclear Chemistry*, by W.D. Loveland, D.J. Morrissey, and G.T. Seaborg
Copyright © 2006 John Wiley & Sons, Inc.

**222** γ-RAY DECAY

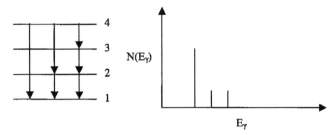

**Figure 9.1** Schematic diagram of the γ-ray transitions for a nucleus and the resulting γ-ray energy spectrum.

called an *isomeric transition*, or simply *IT decay*. An example of an isomeric state is shown in Figure 9.2 for $^{69}$Zn. The ground state of the zinc nucleus is unstable with respect to β decay with a half-life of 56 min. The lowest excited state of this nucleus has an energy of only 439 keV, but it has a much larger spin and the opposite parity compared to the ground state. The transition from the excited state to the ground state is hindered by the large change in angular momentum, $4\hbar$, combined with a change in parity (discussed below), which leads to an IT half-life of 14 h.

## 9.2 ENERGETICS OF γ DECAY

Imagine a γ transition between two nuclear states. Applying the law of conservation of energy, we have

$$M_0^* c^2 = M_0 c^2 + E_\gamma + T_r$$

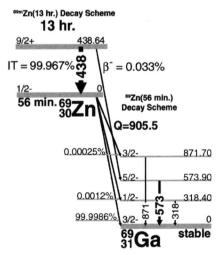

**Figure 9.2** Decay scheme for $^{69}$Zn$^m$.

where $E_\gamma$ is the photon energy, $T_r$ the kinetic energy of the recoiling nucleus after γ-ray emission, and $M_0^*$ and $M_0$ are the masses of the higher and lower nuclear states, respectively. Applying the law of conservation of momentum, we have

$$p_\gamma = p_r$$

where $p_\gamma$ and $p_r$ are the momenta of the photon and recoiling nucleus, respectively. The kinetic energy of the recoil is so small that nonrelativistic mechanics can be used. Hence we have

$$T_r = p_r^2/2M_0$$

Substituting

$$T_r = p_\gamma^2/2M_0$$

$$T_r = E_\gamma^2/2M_0^2$$

If $E_\gamma = 2$ MeV, and $A = 50$, the recoil energy is about 40 eV, which is negligible except for Mossbauer studies (see later).

**Example Problem** (a) Calculate the recoil energy for the IT decay of $^{69}$Zn$^m$ to the ground state of $^{69}$Zn. Using the energy of the excited state from Figure 9.2, we have

$$(M_0^* - M_0)c^2 = E_\gamma = 0.439 \text{ MeV}$$

**Solution** Recall that $M_0c^2 =$ amu*931.5 MeV/amu. Using the mass defect of 68.417 MeV found in the Wallet Cards (see Appendix B), the mass of $^{69}$Zn is 68.297 amu.

$$T_r = \frac{E_\gamma^2}{2M_0c^2} = \frac{(0.439 \text{ MeV})^2}{2(68.297 \times 931.5)}$$

$$T_r = 1.50 \times 10^{-6} \text{ MeV} = 1.5 \text{ eV}$$

(b) Calculate the recoil energy from the emission of a 15.1-MeV photon by an excited $^{12}$C nucleus. Recall that the mass defect of $^{12}$C is 0 so that the mass of $^{12}$C is 12.000 amu.

$$T_r = \frac{E_\gamma^2}{2M_0c} = \frac{(15.1 \text{ MeV})^2}{2(12 \times 931.5 \text{ MeV})}$$

$$T_r = 1.02 \times 10^{-2} \text{ MeV} = 10.2 \text{ keV}$$

## 9.3 CLASSIFICATION OF DECAY TYPES

The conservation of angular momentum has provided an enormous amount of information on the structure of nuclei and plays a controlling role in the γ-ray decay process. From a schematic viewpoint, a stationary nucleus in a definite quantum mechanical state makes a transition to a lower energy state during $\gamma$ decay and emits a single photon. Both the initial and final states of the nucleus will have a definite angular momentum and parity, and so the photon must connect the two states and conserve both parity and angular momentum. Photons each carry an exact integer number of angular momentum units ($\hbar$) and each has a definite parity. The conservation of angular momentum and parity are different, of course, and conservation of each has a different effect on the possible properties of the emitted photon. The angular momenta of the initial and final states of final nucleus can be labeled as $I_i \hbar$ and $I_f \hbar$ and change in intrinsic angular momentum, $\Delta I(\hbar)$ is, of course, $l = \Delta I = |(I_i - I_f)|\hbar$. A photon must carry at least one unit of angular momentum so that $\Delta I = 0$ is forbidden for single-photon emission. The emitted photon should have a minimum intrinsic spin of $l\hbar$ units to connect the two nuclear states. However, the standard coupling rules for angular momenta allow the photon to carry away up to a maximum of $l = |(I_i + I_f)|\hbar$ units. Therefore, given known values of the spins of initial and final states of the nucleus, the angular momentum carried by the photon can take any value in the range:

$$|(I_i - I_f)| \leq l \leq (I_i + I_f)\hbar$$

The *multipolarity* of the photon is a quantification of the amount of angular momentum carried by the photon. The nomenclature is that a photon with $l$ units of angular momentum is called a $2^l$-pole photon. (The nomenclature comes from the classical radiation patterns of electromagnetic radiation and the design of the antennas used to create those patterns.) For example, a photon with $l = 1$ is called a dipole photon, $l = 2$ a quadrupole photon, and so on as indicated in Table 9.1. The transition rate depends strongly on the angular momentum change so that the smallest value of $l = |(I_i - I_f)|\hbar$ is usually observed although conservation of parity plays a role. Transitions with the maximum change in the angular momentum of the nuclear states are called *stretched* transitions.

To understand the parity of electromagnetic transitions, we need to recall that each of the initial and final states of the nucleus undergoing the transition can be

TABLE 9.1  $\gamma$-Ray Selection Rules and Multipolarities

| Radiation Type | Name | $l = \Delta I$ | $\Delta \pi$ |
|---|---|---|---|
| E1 | Electric dipole | 1 | Yes |
| M1 | Magnetic dipole | 1 | No |
| E2 | Electric quadrupole | 2 | No |
| M2 | Magnetic quadrupole | 2 | Yes |
| E3 | Electric octupole | 3 | Yes |
| M3 | Magnetic octupole | 3 | No |
| E4 | Electric hexadecapole | 4 | No |
| M4 | Magnetic hexadecapole | 4 | Yes |

viewed as having a definite distribution of matter and charge. When the excited nucleus makes a transition from the excited state to a lower energy state, the distribution of matter and charge will change in some way. For example, a nucleus that is spinning with a certain value of angular momentum will slow down as it excites and reaches the ground state. Thus, the emission of the photon can be associated with the change in the overall distribution of neutrons and protons, but we can identify two different changes that are analogous to classical antennas. A shift in the distribution of charge (e.g., the transition of a proton from one orbital to another) will give rise to an electric field, but a shift in the distribution of current in the nucleus (e.g., the shift of the direction of a proton orbital) will give rise to a magnetic field. The parity of the photon depends on both the angular momentum and the type (electric or magnetic) of transition indicated in Table 9.1. Notice that electric and magnetic radiation with a given multipole character have opposite parities.

With the list of properties of photons we can generalize the procedure to identify the probable type of photon for a given transition between nuclear states. First, the parity of the photon will be given by the difference in parities of the two nuclear states. Then the angular momentum of the photon will be limited to be in the range of $|I_i - I_f|$ to $I_i + I_f$. The combination of allowed angular momenta and parity will determine the character of the electromagnetic radiation. For example, the first excited state in $^7$Li has spin and parity $\frac{1}{2}^-$ and the ground state is $\frac{3}{2}^-$. Possible electromagnetic transitions between the two states must have $\Delta \pi =$ No and $1 \leq l \leq 2$. Consulting Table 9.1, we find that the only candidates are Ml ($l = 1$, $\Delta \pi =$ No) and E2 ($l = 2, \Delta \pi =$ No) while the other combinations E1 and M2 are ruled out by parity. As we will see in the next section, all of the allowed radiation types will be emitted but at substantially different rates so that the overall radiation usually has one predominant character.

**Example Problem** Use the electromagnetic selection rules to identity the character of the crossover transitions that could link the second excited state at 2.080 MeV ($\frac{7}{2}^+$) in $^{23}$Na with the ground state ($\frac{3}{2}^+$).

**Solution** First, we should note that $\Delta \pi =$ No. Then,

$$|I_i - I_f| \leq l \leq I_i + I_f \implies |2| \leq l \leq 5$$

| $l$ | $\Delta \pi$ | Type |
|---|---|---|
| 2 | No | E2 |
| 3 | No | M3 |
| 4 | No | E4 |
| 5 | No | M5 |

As a final point on the topic of selection rules, we noted that $\Delta l = 0$ is forbidden for the emission of a single photon. The electric monopole distribution (E0) corresponds to the static electric charge of the nucleus and is constant. Similarly, the M0 distribution corresponds to the nonexistent magnetic monopole moment. Nonetheless, there are a few examples of even–even nuclei that have first excited and ground states that are both $0^+$. Once populated, these states decay by internal conversion processes in which the atomic electrons, particularly s electrons with significant penetration into the nucleus, or an electron–positron pair (if there is sufficient energy) are directly emitted from the atom.

**Example Problem** Calculate the ratio of the wavelength of the 439-keV photon emitted when the isomeric state of $^{61}Zn^m$ makes an IT to the diameter of this nucleus.

**Solution** For a photon,

$E_\gamma = h\nu$

$\lambda \nu = c$

$\lambda = \dfrac{hc}{E_\gamma} = \dfrac{(6.626 \times 10^{-34} \text{ js})(3.0 \times 10^8 \text{ m/s})}{(439 \times 10^3 \text{ eV})(1.602 \times 10^{-19} \text{ J/eV})}$

$\lambda = 2.82 \times 10^{-12} \text{m}$

Recall that $R = 1.2 \times A^{1/3}$ fm, so that

$\dfrac{\lambda}{2R} = \dfrac{2.82 \times 10^{-12} \text{ m}}{2 \times 4.92 \times 10^{-15} \text{ m}} = 287$

Thus, the nucleus is not an effective antenna due to its small size compared to the wavelength of the radiation. γ Rays are in the long wavelength limit and are not very sensitive to the detailed internal structure of the emitting nucleus.

## 9.4 ELECTROMAGNETIC TRANSITION RATES

Determining the rate at which an excited state will decay by the emission of a photon is a very general quantum mechanical problem that is not limited to the world of nuclei. The detailed derivation of the transition rate is beyond the scope of this text, and we will only sketch out the results. The decay constant for the emission of a photon by a very well defined single state that has excess energy is shown in Appendix E to be given by the general expression:

$$\lambda = \dfrac{2\pi}{\hbar} \left| \int \varphi_{\text{final}}^* V_p \, \varphi_{\text{initial}} \, dv \right|^2 \rho(E_f)$$

which is also called Fermi's golden rule. The wave functions, φ, represent the complete initial and final states of the entire system and $V_p$ is a (very) small perturbative interaction between the nuclear and electric fields that stimulates the transition. The form and the strength of the perturbation will depend on the multipolarity of the transition. The last factor, $\rho(E_f)$ is the product of the density of nuclear and electromagnetic states that are available to the system after the transition. The initial wave function contains only the nuclear excited state, whereas the final wave function will have parts for the electromagnetic wave and the daughter nuclear state. After some extensive calculus and input from the theory of electromagnetism, we come to an expression for the electromagnetic decay rate:

$$\lambda(l, I_i, \pi \to I_f, \pi) = \frac{8\pi(l+1)}{l[(2l+1)!!]^2} \frac{k^{2l+1}}{\hbar} B(l, I_i, \pi \to I_f, \pi)$$

where $k$ is the photon wave number ($k = E_\gamma/\hbar c$). The symbol !! calls for the double factorial of its argument, which for the case of $l = 2$ and $2l + 1 = 5$ would be the product of the odd integers: $5!! = 5 \times 3 \times 1 = 15$. The reduced transition probability $B(l, I_i, \pi \to I_f, \pi)$ is the matrix element for the reduced nuclear wave functions (i.e., summed over magnetic orientations) and the multipole operator (either electric or magnetic in character),

$$B(l, I_i, \pi \to I_f, \pi) = \frac{1}{2I_i + 1} |\langle I_f \xi \| O_l \| I_i \xi \rangle|^2$$

in which the symbols $\xi$ in the nuclear wave functions are meant to represent all the other relevant quantum numbers. (As an aside, we should note that the two "types" of electric and magnetic radiation are only different in terms of their parity and in the orientation of their plane of polarization.) This expression is still somewhat complicated and is difficult to evaluate. Victor Weisskopf derived a general expression for the reduced transition probability with the assumption that the transition results from the change of a single particle inside a nucleus with a uniform density with the familiar radius function, $R = r_0 A^{1/3}$. His expression for electric multipole radiation, called the Weisskopf single-particle limit, is

$$B_{sp}(E, l) = \frac{1}{4\pi} \left[ \frac{3}{(l+3)} \right]^2 (r_0)^{2l} A^{2l/3} e^2 \text{ fm}^{2l}$$

The single-particle limit for magnetic multipole radiation obtained by assuming that the change in current is due to a single nucleon is

$$B_{sp}(M, l) = \frac{10}{\pi} \left[ \frac{3}{l+3} \right]^2 (r_0)^{(2l-2)/2} \mu_n^2 \text{ fm}^{2l-2}$$

One of the nagging features of these expressions is that the radial integral from the multipole expansion introduces a factor of $r^{2l}$, and thus the dimensions of $B(E, l)$ and $B_{sp}(E, l)$ depend on $l$.

Either of the single-particle limits for the reduced electric or magnetic transition probability can be substituted into the expression for the transition rate to obtain numerical estimates of the deexcitation rates under the assumption that one particle was responsible for the change in electric charge distribution or electric current associated with the change in nuclear states. The transition rates vary over an enormous range, as shown in Figure 9.3 depending most strongly on the value of $l$. Electric transitions are faster than magnetic transitions by about two orders of magnitude. Looking back to the discussion of the fact that several different types of photons can be associated with a given nuclear transition, we now see that we expect the rates of emission to favor the lowest multipolarity. This fact can be simply demonstrated by evaluating the expressions for the transition rate for electric dipole, $l = 1$, and electric quadrupole, $l = 2$, radiation with a typical nuclear radius parameter of $r_0 = 1.2$ fm. Combining the expressions for the transition rate and the reduced transition probability for an E1 transition we get:

$$\lambda_{sp}(E,l) = \frac{8\pi(l+1)}{l[(2l+1)!!]^2} \frac{k^{2l+1}}{\hbar} \frac{1}{4\pi} \left[\frac{3}{l+3}\right]^2 e^2 (r_0 \text{ fm})^{2l} A^{2l/3}$$

Substituting in $l = 1$,

$$\lambda_{sp}(E1) = \frac{8\pi(2)}{1[(2l+1)!!]^2} \left(\frac{E_\gamma}{\hbar c}\right) \frac{e^2}{4\pi\hbar} \left[\frac{3}{1+3}\right]^2 (1.2 \text{ fm})^2 A^{2/3} \text{s}^{-1}$$

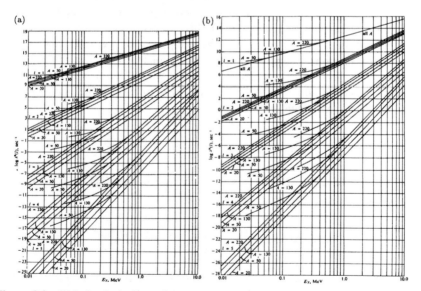

**Figure 9.3** Weisskopf single-particle estimates of the transition rates for (a) electric multipoles and (b) magnetic multipoles. From Condon and Odishaw, Handbook on Physics, 2nd Edition. Copyright © 1967 by McGraw-Hill Company, Inc. Reprinted by permission of McGraw-Hill Book Company, Inc.

Recall that $\hbar c = 197.3$ MeV-fm and $e^2 = 1.440$ MeV-fm, so that

$$\lambda_{sp}(E1) = \frac{16\pi}{9}\left(\frac{E(\text{MeV})}{197.3 \text{ MeV-fm}}\right)^3 \frac{1.440 \text{ fm}}{4\pi\hbar}\left[\frac{3}{4}\right]^2 (1.2 \text{ fm})^2 A^{2/3} \text{ s}^{-1}$$

$$\lambda_{sp}(E1) = 1.03 \times 10^{14} E_\gamma^3 A^{2/3} \text{ s}^{-1}$$

Similar substitution into the expression for $\lambda_{sp}(El)$ with $l = 2$ for electric quadrupole radiation will eventually yield

$$\lambda_{sp}(E2) = 7.28 \times 10^7 E_\gamma^5 A^{4/3}$$

So we see that the rates depend very strongly on the energy of the photon and on the size (mass number) of the emitting nucleus. If we consider the specific hypothetical case of a 1-MeV transition in a medium mass nucleus, $A = 100$, the ratio of transition rates is

$$\frac{\lambda_{sp}(E1)}{\lambda_{sp}(E2)} = \frac{1.03 \times 10^{14} E_\gamma^3 A^{2/3}}{7.28 \times 10^7 E_\gamma^5 A^{4/3}} = 1.41 \times 10^6 E_\gamma^{-2} A^{-2/3}$$

$$\frac{\lambda_{sp}(E1)}{\lambda_{sp}(E2)} = 6.54 \times 10^4$$

The formulas for the transition rates are summarized in Table 9.2 for the lowest five multipoles of each character. The transition rates always increase with a high power of the $\gamma$-ray energy so that low-energy transitions, say below 100 keV, are much slower than high-energy transitions, say above 1 MeV. Table 9.2 also shows that in some cases, particularly in heavy nuclei, an $l + 1$ electric transition can compete favorably with an $l$ magnetic transition.

The Weisskopf estimates are usually good to within a factor of 10, which is remarkable given the large number of orders of magnitude that they span, and provide important references for comparison to the observed transition rates. Notice that if a transition occurs more rapidly than the single-particle rate then

**TABLE 9.2 Weisskopf Single-Particle Transition Rates ($E_\gamma$ is in MeV)**

| Multipole | E | M |
|---|---|---|
| $l$ | $\lambda$ (s$^{-1}$) | $\lambda$ (s$^{-1}$) |
| 1 | $1.03 \times 10^{14} A^{2/3} E_\gamma^3$ | $3.15 \times 10^{13} E_\gamma^3$ |
| 2 | $7.28 \times 10^7 A^{4/3} E_\gamma^5$ | $2.24 \times 10^7 A^{4/3} E_\gamma^5$ |
| 3 | $3.39 \times 10^1 A^2 E_\gamma^7$ | $1.04 \times 10^1 A^{4/3} E_\gamma^7$ |
| 4 | $1.07 \times 10^{-5} A^{8/3} E_\gamma^9$ | $3.27 \times 10^{-6} A^2 E_\gamma^9$ |
| 5 | $2.40 \times 10^{-12} A^{10/3} E_\gamma^{11}$ | $7.36 \times 10^{-13} A^{8/3} E_\gamma^{11}$ |

the transition is more collective, that is, more particles participate in the change. If the transition is significantly slower than the Weisskopf estimate, then the nuclear matrix element must be smaller than the single-particle limit, that is, the overlap of the initial and final states is smaller. The ratio of the observed decay rate to the Weisskopf estimated rate is often quoted in the literature as the transition rate in Weisskopf units (W.u.).

**Example Problem** Use the electromagnetic selection rules to identity the character of the isomeric transition from the first excited state at 0.439 MeV($\frac{9}{2}^+$) in $^{69}Zn^m$ with the ground state ($\frac{1}{2}^-$). Then calculate the Weisskopf single-particle rates for the allowed transitions.

**Solution** First, we should note that $\Delta\pi$ = Yes. Then,

$$|I_i - I_f| \leq l \leq I_i + I_f \implies |4| \leq l \leq 5$$

Thus, only M4 and E5 transitions are allowed. Using the expressions in Table 9.2:

$\lambda(M4) = 3.27 \times 10^{-6} \, A^2 E_\gamma^9 \, (s^{-1})$

$\lambda(M4) = 7.66 \times 10^{-6} \, (s^{-1})$

and

$\lambda(E5) = 2.40 \times 10^{-12} \, A^{10/3} E_\gamma^{11} \, (s^{-1})$

$\lambda(E5) = 3.77 \times 10^{-10} \, (s^{-1})$

so we expect that the transition will be predominantly M4 in character. This example also shows that the observed transition occurs about twice as fast as the single-particle estimate:

$$\lambda = \ln 2/(14 \, h \times 3600 \, s/h) = 1.37 \times 10^{-5} (s^{-1})$$

which indicates that the "current" from more than one particle contributes to the magnetic transition.

It should be noted that E2 transitions are often enhanced by an order of magnitude compared to the single-particle estimates. This enhancement of these specific transitions stems from collective nuclear motion, and the enhancement is particularly strong for nuclei that lie in between major shell closures. An example of a set of E2 transitions to the ground state of $^{160}$Dy and the first three excited (collective) states is shown in Figure 9.4. The excited nucleus cascades down from the $6^+$ level in a series of three E2 transitions with no crossover transitions. The lifetimes of the states, indicated in the figure, were used to calculate the transition rates in Weisskopf units, also indicated in the figure. Notice that the rate of emission in this case ranges from 200 to 1100 times the single-particle rate. If we take a

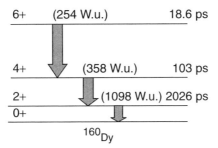

**Figure 9.4** Schematic diagram of the ground-state rotational band transitions for $^{160}$Dy. (Figure also appears in color figure section.)

closer look at the transition rate for electric quadrupole transitions, then we would find that the reduced transition probability could be written in terms of the quadrupole moment, $Q_0$, as:

$$B(E2, J_i \to J_f) = \frac{5}{16\pi} e^2 Q_0^2 \langle J_i, K, 2, 0 | J_f, K \rangle^2$$

in which the last term is a Clebsch–Gordon coefficient, which is a very general normalization coefficient for the values of the spins and the spin projections that can be found in standard reference tables. In the present case of transitions between the states of a rotational band, $K = 0$, $J_i = J$, and $J_f = J - 2$, the normalization coefficient only contains values of $J$:

$$\langle J_i, K, 2, 0 | J_f, K \rangle^2 = \langle J, 0, 2, 0 | (J-2), 0 \rangle^2 = \frac{3J(J-1)}{2(2J+1)(2J-1)}$$

so that the reduced transition probability is

$$B(E2, J \to J-2) = \frac{15}{32\pi} e^2 Q_0^2 \frac{J(J-1)}{(2J+1)(2J-1)}$$

Thus, the experimental transition rate provides a measurement of the quadrupole moment of the nucleus, and we should not be surprised that a strongly deformed nucleus with a large quadrupole moment will have a larger E2 transition rate because the whole nucleus can participate in the transition compared to a single particle.

In the single-particle estimates of $\gamma$-ray decay, one presumes a single nucleon interacts with a photon. This means there is an isospin selection rule

$$\Delta T = 0 \quad \text{or} \quad 1$$

for $\gamma$-ray decay between two pure isospin states. Also we note that E1 $\gamma$ transitions cannot occur when $\Delta T = 0$ in a self-conjugate nucleus ($N = Z$).

## 9.5 INTERNAL CONVERSION

*Internal conversion* (IC) is a competing process to γ-ray decay and occurs when an excited nucleus interacts electromagnetically with an orbital electron and ejects it. This transfer of the nuclear excitation energy to the electron occurs radiationlessly (without the emission of a photon). The energy of the internal conversion electron, $E_{IC}$, is given by

$$E_{IC} = E_{transition} - E_{electron\ binding\ energy}$$

Thus, if a transition has $E_{transition} = 0.412$ MeV, you would expect to see the spectrum of emitted internal conversion electrons shown in Figure 9.5. Note the different lines corresponding to the ejection of electrons from the K, L, and M shells. The nucleus will interact more readily with the K electrons than with the L electrons,

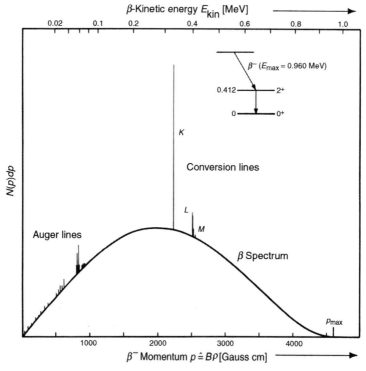

**Figure 9.5** Kinetic energy spectrum of internal conversion electrons for a 412-keV nuclear transition in $^{198}$Hg. Superimposed on this spectrum is the accompanying spectrum of β$^-$ particles from the β decay that feeds the excited state. The peaks labeled K, L, and M represent conversion of electrons with principal quantum numbers of 1, 2, or 3, respectively. (From Marmier and Sheldon, 1969, p. 332.) Copyright © Academic Press. Reprinted by permission of Elsevier.

than with the M electrons, and so forth because the K electrons spend more time in the nucleus than the L electrons than the M electrons, and so forth.

To characterize this decay process and its competition with γ-ray emission, we define the *internal conversion coefficient*, α, by the relationship

$$\alpha = \frac{\text{number of internal conversion decays}}{\text{number of }\gamma\text{-ray decays}} = \frac{\lambda_{IC}}{\lambda_\gamma}$$

where α can take on values from zero to infinity. Note further that

$$\lambda = \lambda_{IC} + \lambda_\gamma = \lambda_\gamma(1 + \alpha)$$

One can define this ratio, the internal conversion coefficient, for electrons from the K shell only for electrons from the M shell only, and so on, giving rise to $\alpha_K$, $\alpha_M$, and so on. Since the total probability of decay must equal the sum of the probabilities of decay via various paths, we have

$$\alpha_{total} = \alpha_K + \alpha_L + \alpha_M + \cdots$$

The internal conversion coefficient depends primarily on the density of the atomic electrons at the center of the nucleus, and thus it can be calculated using principles from atomic physics. Large tables and nomographs of internal conversion coefficients exist, such as those shown in Figure 9.6.

Rough approximate formulas for the internal conversion coefficients are

$$\alpha(EL) = \frac{Z^3}{n^3}\left(\frac{L}{L+1}\right)\left(\frac{e^2}{4\pi\varepsilon_0 \hbar c}\right)^4 \left(\frac{2m_e c^2}{E}\right)^{L+5/2}$$

$$\alpha(ML) = \frac{Z^3}{n^3}\left(\frac{e^2}{4\pi\varepsilon_0 \hbar c}\right)^4 \left(\frac{2m_e c^2}{E}\right)^{L+3/2}$$

where Z is the atomic number of the atom in which the conversion is taking place, n is the principal quantum number of the bound electron being ejected, and $e^2/4\pi\varepsilon_0 \hbar c$, the fine structure constant is 1/137. Note that the internal conversion coefficient, α, increases approximately as $Z^3$, making internal conversion most important for heavy nuclei. The last factor in the equations gives the energy and multipolarity dependence with more internal conversion for low energies and higher transition multipolarities. The $\alpha_K/\alpha_L$ ratio is approximately 8 due to the $n^3$ factor.

**Example Problem**  Use a standard reference such as the *Table of Isotopes*, 8th ed., 1996, to determine the internal conversion coefficients for each shell for the transition from the first excited state at 0.08679 keV ($2^+$) in $^{160}$Dy to the ground state ($0^+$). Then calculate the decay rates for internal conversion and for γ-ray emission.

*Solution*  We have already identified this transition as E2, using Appendix F, pages 3, 7, in the *Table of Isotopes*, 8th ed., 1996. We have to interpolate in a graph to find:

$\alpha_K$ (E2, $Z \approx 65$, 0.090 MeV) = 1.5
$\alpha_{L1}$ (E2, $Z \approx 65$, 0.090 MeV) = 0.1
$\alpha_{L2}$ (E2, $Z \approx 65$, 0.090 MeV) = 5.
$\alpha_{L3}$ (E2, $Z \approx 65$, 0.090 MeV) = 2.5
$\alpha = \alpha_K + \alpha_{L1} + \alpha_{L2} + \alpha_{L3} = 9.1$
$\lambda = (\ln 2)/2.02 \times 10^{-9} = 3.34 \times 10^8 \text{ s}^{-1} = \lambda_\gamma(1 + \alpha)$
$\lambda_\gamma = \dfrac{3.34 \times 10^8 \text{ s}^{-1}}{1 + \alpha} = 3.4 \times 10^7 \text{ s}^{-1}$
$\lambda_{IC} = \lambda - \lambda_\gamma = 3 \times 10^8 \text{ s}^{-1}$

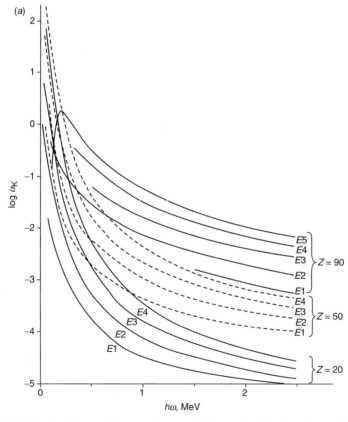

**Figure 9.6** Calculated internal conversion coefficients for (*a*) electric transitions and (*b*) magnetic transitions. (From M. A. Preston, 1962, p. 307.) Copyright © 1962 by Addison-Wesley Publishing Company. Reprinted by permission of Pearson Education.

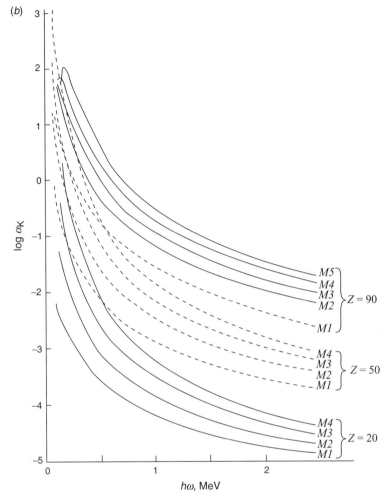

**Figure 9.6** *Continued.*

Note that internal conversion occurs approximately 10 times *faster* than γ-ray emission for this transition in this nucleus.

## 9.6 ANGULAR CORRELATIONS

One of the features of the derivation of the emission rate for γ rays that we glossed over is that the angular distribution of the emitted radiation from a single state must be isotropic. The isotropy comes from the fact that the nuclei are oriented at random, and the process sums over all the internal magnetic substates and thus includes all possible angular distributions. We used this fact in the derivation by using the "reduced (or double-barred) matrix elements." Anisotropic angular distributions can only be observed when a preferred direction or nuclear orientation is established prior to the emission of the photon. There are several techniques to establish such preferred orientations that rely on observing an angular correlation with either an external magnetic field or another particle or photon emitted in "cascade" from the same nucleus. All of these techniques rely on unequal populations of the magnetic substates of the emitting nuclear state. Two of these techniques are shown schematically in Figure 9.7. Another important application of angular correlations is to determine the multipolarity of the electromagnetic transition. We have seen that the selection rules often provide a range of possibilities for the spin change, and the lifetimes of the states depend on the nuclear matrix elements as well as the multipolarity. In order to reliably identify the multipolarity, we have to measure the angular distribution of the radiation; however, we need a reference axis.

The conceptually simplest technique to observe an angular correlation is to measure the angular distribution of radiation from an excited nucleus relative to an external, applied, magnetic field. The magnetic substates of nuclear excited states that have angular momenta, $I$, greater than 0 will split in proportion to the strength of the external magnetic field, $B_{ext}$, and the magnetic field provides the reference axis. This substate splitting provides the basis for NMR and MRI techniques, of course. The difficulty with this correlation technique is that the (Zeeman) splitting of the nuclear spin substates, $\Delta E_m$, given by the simple expression

$$\Delta E_m = gIB_{ext}\mu_0$$

is a very small energy. In this expression $g$ is the gyromagnetic ratio or $g$ factor for the state, and $\mu_0 = e\hbar/2m_p c$ is the nuclear magneton. We should note that this energy splitting is much too small compared to the energy of a nuclear transition so that

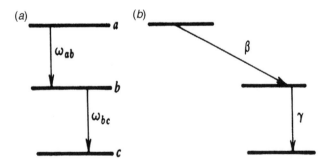

**Figure 9.7** Schematic examples of two techniques to prepare a nuclear state with unequal populations of the internal magnetic substates: (*a*) correlating the sequential emission of two γ rays and (*b*) correlating the emission of a β particle with a subsequent γ ray. (From de Shalit and Feshbach, 1974, p. 693.) Copyright © 1974 John Wiley & Sons, Inc. Reprinted by permission of John Wiley & Sons, Inc.

we could not expect to directly observe different energy transitions with different angular distributions. Rather, we can obtain unequal populations of the substates through the Boltzmann distribution of thermal energy when the sample is cooled to a temperature where $k_BT$ is small compared to the energy splitting. The typical temperature to maintain nuclear orientation in an external magnetic field is on the order of 10 mK. Cooling small samples to such low temperatures is possible, but it requires special techniques such as $^3$He dilution refrigeration.

**Example Problem** Determine the temperature at which $k_BT$ is equal to the energy level splitting for the metastable state of $^{123}$Te at 247.6 keV in an external magnetic field of 4.0 tesla (T).

**Solution** This state $I = \frac{11}{2}$ decays by (M4) IT to the ground state $\frac{1}{2}$ with a half-life of 119.7 d. The gyromagnetic ratio, or $g$ factor, for this state is 0.1685:

$$\Delta E_m = gIB_{ext}\mu_N = k_BT$$

$$T = \frac{gIB_{ext}\mu_N}{k_B}$$

$$T = \frac{0.1685\left(\frac{11}{2}\right)(4.0T)\left(5.05084 \times 10^{-27} \text{ J/T}\right)}{1.38066 \times 10^{-23} \text{ J/K}}$$

$$T = 1.4 \times 10^{-3} \text{ K}$$

A much more common technique for observing angular correlations relies on detecting the direction of radiation from a process that feeds the excited state and then observing the angular distribution relative to that direction. As indicated in Figure 9.7 this process could be a γ-ray transition from a higher lying excited state, or it could be a β or α particle emitted by a parent nucleus. The first particle provides the reference axis, but it must also introduce an unequal population of the magnetic substates of the intermediate state in order for the second transition to have an anisotropic angular distribution.

The angular distribution of the intensity of electromagnetic radiation is given by specific analytic functions written in terms of an angle, $W(\theta, m_I)$, relative to the quantization axis, Z, and the magnetic quantum number, $m_I$. The patterns depend on the order of the multipole, dipole, quadrupole, and so forth, but they are the same for electric and magnetic transitions with the same order. For example, the angular distributions for dipole radiation are

$$W_{dipole}(\theta, m_I = 0) = \frac{3}{8\pi}\sin^2\theta$$

$$W_{dipole}(\theta, m_I = +1) = \frac{3}{16\pi}\left(1 + \cos^2\theta\right)$$

$$W_{dipole}(\theta, m_I = -1) = \frac{3}{16\pi}\left(1 + \cos^2\theta\right)$$

**238** γ-RAY DECAY

A schematic representation of these angular distributions is shown in Figure 9.8. First, we should notice that these functions depend on only one angle, and thus they are cylindrically symmetric. Therefore, we will not find any asymmetry in radiation from systems with only two substates, that is, $I = \frac{1}{2}, m_I = \pm \frac{1}{2}$. Notice also that

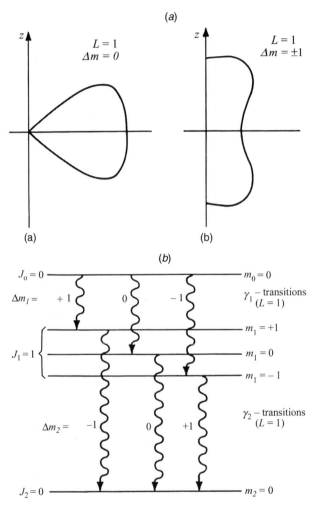

**Figure 9.8** Schematic diagram of how angular correlations occur. Panel (*a*) shows the angular distribution of dipole radiation for $\Delta m = 0$ and $\Delta m = \pm 1$. Panel (*b*) shows the magnetic substates populated in a $\gamma_1\gamma_2$ cascade from $J = 0$ to $J = 1$ to $J = 0$. When $\gamma_1$ defines the Z axis, then the $m_1 = 0$ state cannot be fed and one has only $\Delta m_1 = \pm 1$ and $\Delta m_2 = \mp 1$, causing $\gamma_2$ to have an anisotropic distribution relative to $\gamma_1$ shown in panel (*c*). [From Marmier and Sheldon, 1969.] Copyright © 1969 Academic Press. Reprinted by permission of Elsevier.

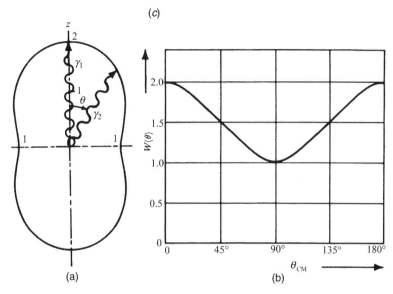

**Figure 9.8** *Continued.*

the intensity of $m = 0$ for dipole radiation is exactly zero along the Z axis because the sine function becomes zero, whereas the $m = \pm 1$ distributions have nonzero minima perpendicular to the Z axis.

Let us consider the specific case of the cascade of two electromagnetic dipoles from nuclear states with $I_a = 0$ to $I_b = 1$ to $I_{gs} = 0$. The energy level diagram is shown in Figure 9.8. The radiation pattern of the second photon will have the $(1 + \cos^2 \theta)$ form relative to the direction of the first photon in this case. The fact that the radiation will not be isotropic can be seen from a classical as well as from a quantum mechanical standpoint. The simple definition that the Z axis is the direction of this first photon forces the observed photon to have $m_{\mathrm{I}} = \pm 1$ because $W_{\mathrm{dipole}}(\theta, m_{\mathrm{I}} = 0) = 0$ at $\theta = 0$, and such a photon will not be observed along the Z axis. The first dipole transition from $I_a = 0$ to $I_b = 1$ will carry away one unit of angular momentum, and the angular momentum of the residual nucleus must be oriented in such a way as to be equal and opposite to that of the photon. Then, to conserve angular momentum in the second transition, the next photon must also have $m_{\mathrm{I}} = \mp 1$ and will follow the $(1 + \cos^2 \theta)$ distribution relative to the first photon and the Z axis. From a quantum mechanical standpoint the argument depends simply on spin algebra. The only allowed magnetic substates for the initial and final states are $m_a = m_{gs} = 0$ (because $I_a = I_{gs} = 0$). The only allowed value for the multipolarity of both photons, $l_1 = l_2$, is one by the relation $|0 - 1| \le l \le (0 + 1)$. The coupling of the angular momentum of the photon with that of the initial state to create the intermediate state requires that only allowed magnetic substate of $I_b = 1$ be $M_b = \pm 1$. Thus, both photons must have $m = \pm 1$ and follow the $(1 + \cos^2 \theta)$ distribution.

The angular distributions for γ-ray cascades have been worked out, but each case requires substantial and sophisticated algebra that will not be presented here. The general result is that the angular distributions can be written in terms of a sum of Legendre polynomials that depends on the multipolarities of the photons, $l_1, l_2$ and the spin of the intervening state. It is common to analyze the observed angular correlations in terms of a power series of cos θ that is normalized with $W(\theta = 90°) = 1$ so

$$W(\theta) = (1 + a_2 \cos^2 \theta + a_4 \cos^4 \theta + a_6 \cos^6 \theta + \cdots + a_{2L} \cos^{2L} \theta)$$

where the coefficients, $a_2$, $a_4$, and so on are fitted to the data and can be compared to predicted values for assumed values of $I_a$, $I_b$, $I_c$, $l_1$, and $l_2$. The number of radiation patterns or angular distributions may seem extensive with these five variables, but there are certain rules that simplify the situation. The highest even power of the cosine function, $2L$, is determined by the smallest value of $2I_b$, $2l_1$, and $2l_2$ and is one unit less than the smallest if the smallest of these is an odd number. For example, when $I_b = 0$ or $\frac{1}{2}$, then $2L = 0$ and $W(\theta) = 1$; when $I_b = 1$, then $2L = 2$ and $W(\theta) = (1 + a_2 \cos^2 \theta)$. The theoretical coefficients for a few types of pure dipole and pure quadrupole transitions are given in Table 9.3.

The third technique for establishing a reference axis for angular correlations can be applied to nuclear reactions when the direction of a particle involved in the reaction is detected. This direction provides a reference axis that can be related to the angular momentum axis, but each nuclear reaction has its own peculiarities and constraints on the angular momentum vector. For example, the direction of an α particle from a decay process that feeds an excited state can be detected as indicated in Figure 9.7, but, as is discussed in Chapter 7, the energetics of α decay

**TABLE 9.3 Angular Correlation Coefficients for Some γ–γ Cascades with Pure Multipolarities**

| $I_a(l_1); I_b(l_2); I_c$ | $a_2$ | $a_4$ |
|---|---|---|
| 0(1); 1(1); 0 | 1 | 0 |
| 1(1); 1(1); 0 | $-\frac{1}{3}$ | 0 |
| 1(2); 1(1); 0 | $-\frac{1}{3}$ | 0 |
| 2(1); 1(1); 0 | $\frac{1}{13}$ | 0 |
| 3(2); 1(1); 0 | $-\frac{3}{29}$ | 0 |
| 0(2); 2(2); 0 | $-3$ | 4 |
| 1(2); 2(2); 0 | $-\frac{1}{3}$ | 0 |
| 2(2); 2(2); 0 | $\frac{3}{7}$ | 0 |
| 2(2); 2(2); 0 | $-\frac{15}{13}$ | $-\frac{16}{33}$ |
| 3(2); 2(2); 0 | $-\frac{3}{29}$ | 0 |
| 4(2); 2(2); 0 | $\frac{1}{8}$ | $\frac{1}{24}$ |

*Source*: From Evans (1955).

are such that decay to excited states or decays with large orbital angular momenta are hindered. Nuclear reactions can produce nuclei with large amounts of angular momenta with characteristic distributions. The motion of the center of mass provides a good reference that coincides with the direction of the initial beam for the usual case of a target at rest in the lab system. The angular momentum vector must lie in the plane perpendicular to the beam direction in compound nuclear reactions. The angular momentum vector is further confined in two-body scattering reactions to be normal to the plane containing the beam (or center of mass vector) and the two particles. The effects of angular momentum on nuclear reactions are discussed further in Chapter 10.

At this point we have established techniques to identify the multipolarity of a transition through its angular distribution. We still have the ambiguity of the parity of the electromagnetic wave, that is, whether it was produced by an electric or a magnetic transition in the nucleus. The parity of the radiation corresponds to the plane of polarization of the electromagnetic radiation. The polarization of the wave can be determined from knowledge of the direction of the plane of the electric vector of the photons relative to the plane containing two coincident photons. The direction of the electron emitted in the Compton scattering process is sensitive to the direction of the electric vector of the incoming photon and has been used to determine the parity of electromagnetic transitions. Alternatively, the number and type of conversion electrons emitted in the decay is also sensitive to the electric or magnetic nature of the radiation. Measurements of the conversion coefficients are used to establish the character of the radiation.

## 9.7 MÖSSBAUER EFFECT

We could imagine that the inverse of $\gamma$-ray emission from an excited nuclear state to the ground state might be possible if a nucleus in its ground state was bathed in sufficient energy. A large difficulty with causing this absorption to take place is the very large amount of energy associated with nuclear transitions (MeV), compared to the amount of available (terrestrial) thermal energies ($10^{-6}$ MeV). This inverse process can occur in two situations: (a) in nuclear reactions called Coulomb excitation that take place when heavy ions pass very near to large target nuclei (Chapter 10), and (b) when there is a resonant absorption of a $\gamma$ ray emitted by nuclear deexcitation in another identical nucleus. The latter process is called the Mössbauer effect, and the process requires some special conditions in order to take place. The energies of the nuclear states are very precise so that the resonant absorption or energy matching is very sensitive to the chemical environment of the nucleus. As we will see, there are relatively few nuclei that are suitable for Mössbauer studies due to the requirements of a half-life that allows a high specific activity with a reasonable useful period, a single $\gamma$-ray transition, and the absorbing nucleus must be a stable isotope of an important/practical chemical element. The important examples are $^{57}$Fe, $^{191}$Ir, and $^{198}$Hg.

The first nucleus in which the resonant absorption of photons was observed was $^{191}$Ir. The excited states of this nucleus are fed by the electron capture decay of $^{191}$Pt, one of which decays by a 129.43-keV M1 transition to the ground state. Now we can ask what will happen if we shine γ rays from a radioactive source of $^{191}$Pt onto a set of stable $^{191}$Ir nuclei? We could use an iridium foil because iridium only has two stable isotopes 191 (37.3%) and 193 (72.7%). Without careful preparation, the answer is that very few photons will be absorbed by the $^{191}$Ir nuclei! The difficulty comes from the fact that in order to be absorbed the γ ray will have to exactly match the energy of the transition. Remember that quantum mechanics dictates that the absorption of the γ ray will move the nucleus from its ground state to a single and specific excited state that has an exact energy. A single nucleus cannot absorb a random amount of energy. Several important effects shift the energy of the emitted photon, but first we could ask how accurately do we have to match the energy of the state in order to be absorbed? This corresponds to the natural width of the state.

The measured half-life of the state is 89.4 ps, which corresponds to a energy width, Γ, or ΔE, due to the Heisenberg uncertainty principle of:

$$\Gamma = \frac{\hbar}{\tau} = \hbar\lambda = \hbar \times \frac{\ln 2}{t_{1/2}} = \frac{4.6 \times 10^{-16} \text{ eV-s}}{t_{1/2}(\text{s})}$$

where τ is the mean life or the reciprocal of the decay constant $\lambda = \ln 2/t_{1/2}$. In this case the energy width of the excited state is only the tiny amount of $5.1 \times 10^{-6}$ eV, a factor of $2 \times 10^{-10}$ less than the energy of the state. Such narrow widths are a general property of nuclear excited states that decay by γ-ray emission. Thus, the energy matching of the nuclear state and photon energy has to be incredibly exact for significant absorption to take place.

The linewidth of an observed transition is broadened by the random thermal motion of the nuclei that emit the photon. That is, the energies of photons emitted along the direction of thermal motion of the atom will be slightly higher than the average and vice versa for those emitted opposite. The value of the energy of a photon emitted by a moving source is shifted according to the expression:

$$E'_\gamma = E_{\gamma 0}(1 \pm \beta_x)$$

where $E_{\gamma 0}$ is the energy of the transition and $\beta_x = v_x/c$ is the familiar ratio of the velocity along the photon direction to the speed of light. As an upper limit, we could use the kinetic theory of gases and the Maxwell–Boltzmann velocity distribution to estimate of the width of the velocity distribution for gaseous iron nuclei. (The motion of atoms in liquids and the vibrations of atoms in solids are smaller but not zero.) The Boltzmann (thermal) probability distribution for the kinetic energy of an atom, P(KE), is always a decreasing exponential function, $P(\text{KE}) \approx e^{-mv^2/2k_BT}$, and it applies to the total kinetic energy and to the kinetic energy along one coordinate of a normal gas in a closed container. Solving the

Doppler expression for $v_x$ in terms of $E'_\gamma$

$$v_x = c\left[1 \mp (E'_\gamma/E_{\gamma 0})\right]$$

and substituting that expression into the Boltzmann probability we find

$$P(E'_\gamma) \propto e^{-mc^2(1 \mp E'_\gamma/E_{\gamma 0})^2/2k_B T}$$

Selecting one sign for the direction, multiplying through, and collecting constants, we find

$$P(E'_\gamma) \propto e^{-mc^2 {E'_\gamma}^2/(2E_{\gamma 0}^2 k_B T)}$$

This expression shows that the distribution of emitted γ-ray energies follows a Gaussian distribution with a variance something like

$$\sigma^2 \approx E_{\gamma 0}^2 k_B T/mc^2$$

In the present example of $^{191}$Ir decay at room temperature, $k_B T = 0.025$ eV, $E_{\gamma 0} = 0.1294$ MeV, and $mc^2 = 191 \times 931.5$ MeV, which combine to give $\sigma \sim 7 \times 10^{-2}$ eV, which, although small and an upper limit for gaseous atoms, is still six orders of magnitude larger than the natural linewidth of the state. Therefore, it is not very often that we will be able to actually observe the natural linewidth of a γ-ray emitting state. This broadening works in favor of the absorption of a photon because it allows the thermal motion to help match the energy of the whole system, nucleus in the atom, to the photon energy.

In addition, as we have already discussed, the emission of a photon induces a recoil by the nucleus in order to conserve momentum. The energy of the photon is less than the energy of the nuclear transition by the amount $T_r = E_\gamma^2/(2mc^2)$. Notice that to conserve energy and momentum in the reverse process of γ-ray absorption, a nucleus initially at rest will recoil with the same value of the recoil energy after absorbing a photon. In the present example of $^{191}$Ir, the recoil energy is $T_r = 4.7 \times 10^{-2}$ eV and is a similar magnitude to the thermal Doppler shift for a gas. We probably can expect the radioactive platinum atoms to be in a metal lattice so their motion would correspond to lattice vibrational motion and be somewhat less than that in a gas. The relative energy distributions expected for the emitted and absorbed photons are shown in Figure 9.9 using the estimate of the thermal widths. Notice that the recoil energy moves the peaks apart, and the thermal width provides only a partial overlap. It is these photons in the overlap region that have the proper energy to be absorbed; they must encounter a nucleus, of course, in order to actually be absorbed.

We might imagine that we could prepare a system that physically moves the source of the radiation toward the absorbing nuclei with sufficient speed that the Doppler shift compensates for the energy difference. Restricting the motion

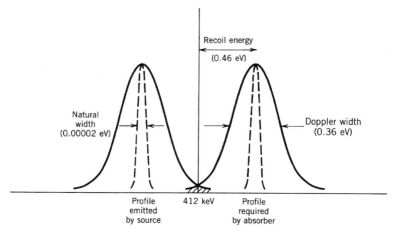

**Figure 9.9** Schematic indication of the position and widths of the emitted and absorbed radiation corresponding to the 412-keV transition in $^{198}$Hg. (From Krane, 1988, p. 364.) Copyright © John Wiley & Sons, Inc. Reprinted by permission of John Wiley & Sons, Inc.

to the approaching direction, we can rearrange the expression above to obtain the velocity in terms of the Doppler shift:

$$\Delta E = (E'_\gamma - E_{\gamma 0}) = E_{\gamma 0}\left(\frac{v_x}{c}\right)$$

The necessary velocity that would create a Doppler shift corresponding to twice the recoil energy is

$$\Delta E = 2T_r = \frac{2E_{\gamma 0}^2}{2mc^2} = E_{\gamma 0}\left(\frac{v_x}{c}\right)$$

$$\frac{v_x}{c} = \frac{E_{\gamma 0}}{mc^2}$$

For the example of $^{198}$Hg, $\beta_x = 2.2 \times 10^{-6}$ or $v_x = 670$ m/s and corresponds to a kinetic energy of 0.92 eV. The magnitude of this difference is visible in Figure 9.9 because the separation between the two peaks is about twice the thermal width. Such a high velocity is difficult to attain with any macroscopic, that is, physical source.

The Mössbauer effect relies on a very different technique for overcoming the energy mismatch of twice the recoil energy between nuclear emission and nuclear absorption. Notice that the recoil energies that we have calculated are small fractions of an electron volt per atom. You might recall that chemical bonds have energies on the order of a few electron volts per bond and are stronger in some sense than the recoil effect. Mössbauer showed that the resonant emission/absorption of photons could be strongly enhanced by binding the emitting atoms and the absorbing atoms into crystal lattices. In practice the emitter is produced by a β decay of a parent nuclide that is a different chemical element from the absorber; thus, two separate crystals are used. Due to the chemical bonds or the lattice energy of the crystal, the atom that absorbs the photon is held in place, and the entire, macroscopic, lattice "recoils" to conserve momentum. The mass of the entire lattice should be used to calculate the recoil velocity, but this mass is on the order of Avogadro's number larger than that of an atom, so that there is effectively no recoil. One analogy is to compare the difference that you would feel if you hit a single stone with a bat compared to that you would feel if you hit the same stone if it was part of a cement wall in a concrete building. The actual difference in the atomic case is orders of magnitude larger. Thus, with the atoms bound into the crystal lattice, the Doppler motion is limited to the vibrational motion of the atoms, and the linewidth shrinks to essentially the natural width of the state. In this case the energy of the emitted photon and the energy absorbed in the nuclear excitation overlap. The Mössbauer experiment is then to remove the overlap between the photon energies by moving one crystal lattice with respect to the other. The relative velocity is on the order of cm/s, which is, of course, much smaller than that necessary to compensate for the nuclear recoil. The resonance is then seen as a preferential absorption as a function of relative velocity between the emitter and absorber.

Notice that the Mössbauer effect is very sensitive to the energy of the nuclear state; changes on the order of $10^{-6}$ eV are readily detected. This is the level at which atomic orbitals can shift nuclear states through the penetration of electron density into the nucleus. As a first approximation, we could imagine that the interaction of the electron wave function with the nucleus will depend on the size, that is, radius, of the nuclear wave function. The nuclear wave function for the excited state will be (slightly) different, and thus the penetration of the electrons into the excited nucleus will be slightly different. Thus, the transition energy will be different, albeit by a very small amount, from the pure nuclear transition that would occur in a bare nucleus (no electrons). When the chemical state or environment of both the absorber and the emitter are the same, the transition will occur at a definite but different energy, but one could not perform the measurement of the pure nuclear transition (without electrons). Finally, when the chemical environment of the emitter and absorber are different, then the transition will occur at a new energy. The shift of the energy of the resonance between the identical environments and different environments is called the chemical shift in analogy to NMR work. In practice, the chemical shift in the Mössbauer resonance lines provides a probe for the overall chemical environment of the absorbing nuclei.

The most extensively used nuclide for Mössbauer studies is $^{57}$Fe due to the very low energy of the nuclear transition. Let us consider the low-lying excited states of $^{57}$Fe shown in Figure 9.10. The first excited state in $^{57}$Fe lies at only 14.4125 keV, and it decays to the ground state with a half-life of 98 ns. As shown in Figure 9.10, the β decay of the parent nucleus, $^{57}$Co, feeds this excited state of the daughter nucleus so that we can image having a strong source of the low-energy γ rays. As shown in the example calculation, the energy of this transition is so low that the recoil energy is also quite low and comparable to the thermal energy.

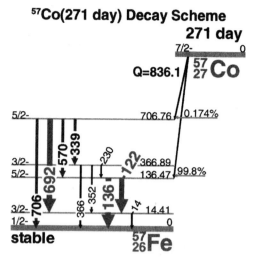

**Figure 9.10** Energy level diagram of two members of the $A = 57$ mass chain. $^{57}$Co decays to excited states of $^{57}$Fe, which result in the M1 transition from the $\frac{3}{2}^-$ state at 14.41 keV to the $\frac{1}{2}^-$ ground state. (Figure also appears in color figure section.)

Thus, studies can be performed with the source bound in a crystal lattice, but the absorber can be in solution.

**Example Problem** Calculate the natural linewidth of the state at 14.4 keV in $^{57}$Fe given that $t_{1/2} = 98$ ns. Then calculate the velocity of the source lattice that would correspond to twice the natural width and would lie outside the Mössbauer resonance effect:

*Solution*

$$\Delta E = \Gamma = \hbar/\tau = \hbar[\ln(2)/t_{1/2}]$$

$$\Delta E = \Gamma = \frac{4.135 \times 10^{-15} \text{ eV-s}}{2\pi} \frac{\ln(2)}{98 \times 10^{-9} \text{ s}}$$

$$\Delta E = \Gamma = 4.65 \times 10^{-9} \text{ eV}$$

The velocity that would correspond to twice this energy can be found from the nonrelativistic expression for the kinetic energy:

$$\text{KE} = \frac{1}{2}mv_x^2 = 2\Gamma$$

$$\frac{v_x}{c} = \sqrt{\frac{4\Gamma}{mc^2}} = \sqrt{\frac{4 \times 4.65 \times 10^{-9} \text{ eV}}{57 \times 931.5 \times 10^6 \text{ eV}}}$$

$$\frac{v_x}{c} = 5.92 \times 10^{-10} \implies v_x = 0.178 \text{ m/s}$$

## PROBLEMS

1. $^{195}$Pt has a ground-state spin and parity of $\frac{1}{2}^-$, with excited states at 0.029 MeV ($\frac{3}{2}^-$) and 0.130 MeV ($\frac{5}{2}^-$). Does the $\frac{5}{2}^-$ level decay primarily to the $\frac{3}{2}^-$ level or to the $\frac{1}{2}^-$ level? Why? What is the transition multipolarity?

2. The $\frac{1}{2}^-$ isomeric state of $^{95}$Nb decays to the $\frac{9}{2}^+$ ground state by means of an M4 transition. The half-life of the isomeric state is 90 h while the half-life of the ground state is 35 d ($\alpha_{\text{total}} = 4.5$). Calculate the partial half-life for the γ-ray decay of the isomeric state.

3. Consider the following decay schemes for $^{60}$Co and $^{60}$Co$^m$:

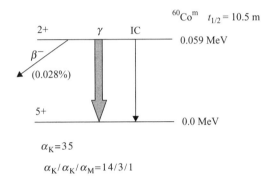

a. Classify the most likely multipolarity for the γ-ray decay of $^{60}$Co$^m$.
b. Calculate the partial decay constants for $\beta^-$, internal conversion, and γ-ray decay.
c. What is the width of $^{60}$Co$^m$ in eV?

4. $^{52}$Mn has an excited state at 0.377 MeV above the ground state. This excited state decays to the ground state with $t_{1/2} = 21.1$ min. The $J\pi$ values of initial excited state and the ground state are $2^+$ and $6^+$, respectively. (a) What is the lowest multipole order that can contribute to the transition? Calculate the decay constant and compare it to the experimental value. (b) Suppose we wanted to check whether, in the initial state, there was any mixture of other angular momenta than 2. Set a rough upper limit to the amplitude of a $J = 1$ component of the initial state, using as data only the measured half-life and transition energy. Assume parity conservation.

5. Consider $^{10}$B. The ground state has $J\pi = 3^+$ and the excited states in order of increasing excitation energy are $1^+$, $0^+$, $1^+$, $2^+$, $3^+$, $2^-$, $2^+$.... (a) Explain why $^{10}$B is stable even though it is odd–odd. (b) The first excited state is at 0.72 MeV, and the second excited state is at 1.74 MeV. What are the energies,

multipolarities, and relative intensities of the γ rays that are emitted in the deexcitation of the second excited state?

6. A 64-d isomer of an even Z, and an odd nucleus with $A \sim 90$ occurs at 105 keV above the ground state. The isomeric state decays 10% by EC and 90% by IT. If the internal conversion coefficient $\alpha = 50$, what is the γ-ray lifetime and the most likely multipolarity of the isomeric transition? If this is a magnetic transition and the isomeric state has $J\pi = \frac{1}{2}^-$, what is the $J\pi$ of the ground state?

7. $^{51}$V has a ground-state spin and parity of $\frac{7}{2}^-$ with excited states at 0.3198 MeV ($\frac{5}{2}^-$) and at 0.930 MeV ($\frac{3}{2}^-$). What is the energy and multipolarity of the principal γ ray that deexcites each excited state?

8. The ground state of $^{61}$Ni has $J\pi = \frac{3}{2}^-$. $^{61}$Co ($t_{1/2} = 1.65$ h) decays by $\beta^-$ emission with $E_{max} = 1.24$ MeV to a 0.067-MeV excited state of $^{61}$Ni. The 0.067-MeV transition has $\alpha_K = 0.10$, $\alpha_K/\alpha_L = 8$. The branching ratio for the transition from $^{61}$Co to the $^{61}$Ni ground state is $10^{-6}$. What is $J\pi$ for the ground state of $^{61}$Co and the first excited state (0.067 MeV) of $^{61}$Ni?

## REFERENCES

Condon, E. U. and H. Odishaw. *Handbook of Physics*, 2nd ed., McGraw-Hill, New York, 1967.

de Shalit, A. and H. Feshbach. *Theoretical Nuclear Physics*, Vol. 1, Wiley, New York, 1974.

Evans, R. D. *The Atomic Nucleus*, McGraw-Hill, New York, 1955.

Krane, K. S. *Introductory Nuclear Physics*, Wiley, New York, 1988.

Marmier, P. and E. Sheldon. *Physics of Nuclei and Particles*, Vol. 1, Academic, New York, 1969.

Preston, M. A. *Physics of the Nucleus*, Addison-Wesley, Reading, MA, 1962.

## BIBLIOGRAPHY

Bohr, A. and B. Mottelson. *Nuclear Structure, Vol. 1*, Benjamin, New York, 1969.

Wong, S. S. M. *Introductory Nuclear Physics*, 2nd ed., Wiley, New York, 1998.

# CHAPTER 10

# NUCLEAR REACTIONS

## 10.1 INTRODUCTION

The study of nuclear reactions is important for a number of reasons. Progress in the understanding of nuclear reactions has occurred at a faster pace, and generally a higher level of sophistication has been achieved compared to similar studies of chemical reactions. The approaches used to understand nuclear reactions are of value to any chemist who wishes a deeper insight into chemical reactions. There are certain nuclear reactions that play a preeminent role in the affairs of humans and our understanding of the natural world in which we live. For example, life on Earth would not be possible without the energy provided to us by the sun. That energy is the energy released in the nuclear reactions that power the sun and other stars. For better or worse, the nuclear reactions, fission and fusion, are the basis for nuclear weapons, which have shaped much of the geopolitical dialog for the last 50 years. Apart from the intrinsically interesting nature of these dynamic processes, their practical importance would be enough to justify their study.

To discuss nuclear reactions effectively we must understand the notation or jargon that is widely used to describe them. Let us begin by considering the nuclear reaction

$$^{4}\text{He} + {}^{14}\text{N} \rightarrow {}^{17}\text{O} + {}^{1}\text{H}$$

Most nuclear reactions are studied by inducing a collision between two nuclei where one of the reacting nuclei is at rest (the *target* nucleus) while the other nucleus (the

---

*Modern Nuclear Chemistry*, by W.D. Loveland, D.J. Morrissey, and G.T. Seaborg
Copyright © 2006 John Wiley & Sons, Inc.

*projectile* nucleus) is in motion. (Exceptions to this occur both in nature and in the laboratory in studies where both the colliding nuclei are in motion relative to one another.) But let us stick to the scenario of a moving projectile and a stationary target nucleus. Such nuclear reactions can be described generically as:

Projectile P + target T $\rightarrow$ emitted particle x and residual nucleus R

For example, the first reaction discussed above might occur by bombarding $^{14}$N with $\alpha$ particles to generate an emitted particle, the proton and a residual nucleus $^{17}$O. A shorthand way to denote such reactions is, for the general case,

$$T(P, x)R$$

or for the specific example

$$^{14}N(\alpha, p)^{17}O$$

In a nuclear reaction, there is conservation of the number of protons and neutrons (and thus the number of nucleons). Thus, the total number of neutrons (protons) on the left and right sides of the equations must be equal.

**Example Problem** Consider the reaction $^{59}$Co(p, n). What is the product of the reaction?

**Solution**

$$_1^1H + _{27}^{59}Co \rightarrow _0^1n + _X^Y Z$$

On the left side of the equation we have 27 + 1 protons. On the right side we have 0 + X protons where X is atomic number of the product. Obviously X = 28 (Ni). On the left hand side we have 59 + 1 nucleons, and on the right side we must have 1 + Y nucleons where Y = 59. So the product is $^{59}$Ni.

There is also conservation of energy, linear momentum, angular momentum, and parity, which will be discussed below.

## 10.2 ENERGETICS OF NUCLEAR REACTIONS

Consider the T (P, x) R reaction. Neglecting electron binding energies, we have, for the energy balance in the reaction,

$$m_P c^2 + T_P + m_T c^2 = m_R c^2 + T_R + m_x c^2 + T_x$$

where $T_i$ is the kinetic energy of the *i*th particle and $m_i$ represents the mass-energy of the *i*th species. (Note that since R and x may be formed in an excited state, the values of m may be different than the ground-state masses.)

## 10.2 ENERGETICS OF NUCLEAR REACTIONS

The *Q value* of the reaction is defined as the difference in mass energies of the product and reactants, that is,

$$Q = [m_P + m_T - (m_x + m_R)]c^2 = T_x + T_R - T_P$$

Note that if $Q$ is positive, the reaction is *exoergic* while if $Q$ is negative, the reaction is *endoergic*. Thus, the sign convention for $Q$ is exactly the opposite of the familiar $\Delta H$ in chemical reactions. A necessary *but not sufficient* condition for the occurrence of a nuclear reaction is that

$$Q + T_P > 0$$

Note that $Q$ is an important quantity for nuclear reactions. If the masses of both the products and reactants are known (Appendix B), the $Q$ value can be calculated using the mass defects, $\Delta$, as:

$$Q = \Delta(\text{projectile}) + \Delta(\text{target}) - \Sigma\Delta(\text{products})$$

It can be obtained by measuring the masses or kinetic energies of the reactants and products in a nuclear reaction. However, we can show, using conservation of momentum, that only $T_x$ and the angle $\theta$ of x with respect to the direction of motion of P suffice to determine $Q$ in these two-body reactions.

In the laboratory system, a typical nuclear collision can be depicted as shown in Figure 10.1. Conserving momentum in the $x$ direction, we can write

$$m_P v_P = m_x v_x \cos\theta + m_R v_R \cos\phi$$

Applying conservation momentum in the $y$ direction, we have

$$0 = -m_x v_x \sin\theta + m_R v_R \sin\phi$$

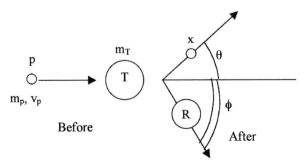

**Figure 10.1** Schematic diagram of a nuclear reaction.

## NUCLEAR REACTIONS

where $m_i$ and $v_i$ are the mass and velocity of the $i$th species. If we remember that the momentum $p = mv = (2mT)^{1/2}$, we can substitute in the above equation as

$$(m_P T_P)^{1/2} - (m_x T_x)^{1/2} \cos\theta = (m_R T_R)^{1/2} \cos\phi$$

$$(m_x T_x)^{1/2} \sin\theta = (m_R T_R)^{1/2} \sin\phi$$

Squaring and adding the equations, we have

$$m_P T_P - 2(m_P T_P m_x T_x)^{1/2} \cos\theta + m_x T_x = m_R T_R$$

Previously, we had said that

$$Q = T_x - T_P - T_R$$

Plugging in this definition of $Q$, the value of $T_R$, which we have just calculated, we get

$$Q = T_x\left(1 + \frac{m_x}{m_R}\right) - T_P\left(1 - \frac{m_P}{m_R}\right) - \frac{2}{m_R}(m_P T_P m_x T_x)^{1/2}\cos\theta$$

This is the all-important $Q$ equation. What does it say? It says that if we measure the kinetic energy of the emitted particle x and the angle at which it is emitted in a reaction, and we know the identities of the reactants and products of the reactions, we can determine the $Q$ value of the reaction. In short, we can measure the energy release for any two-body reaction by measuring the properties of one of the products. If we calculate the $Q$ value of a reaction using a mass table, then we can turn this equation around to calculate the energy of the emitted particle using the equation

$$T_x^{1/2} = \frac{(m_P m_x T_P)^{1/2}\cos\theta \pm \{m_P m_x T_P \cos^2\theta + (m_R + m_x)[m_R Q + (m_R - m_P)T_P]\}^{1/2}}{m_R + m_x}$$

For additional insight, let us now consider the same reaction as described in the *center-of-mass* (cm) coordinate system. In the cm system the total momentum of the particles is zero, before and after the collisions. The reaction as viewed in the laboratory, and cm system is shown in Figure 10.2.

The kinetic energy of the center of mass is

$$T_{cm} = \frac{(m_P + m_T) v_{cm}^2}{2}$$

where $v_{cm}[= v_P m_P/(m_P + m_T)]$ is the speed of the center of mass. Substituting, in the above equation, we have

$$T_{cm} = \frac{1}{2}(m_P + m_T)\left[\frac{m_P v_P}{m_P + m_T}\right]^2 = \frac{1}{2}m_P v_P^2\left(\frac{m_P}{m_P + m_T}\right) = T_{lab}\left(\frac{m_P}{m_P + m_T}\right)$$

## 10.2 ENERGETICS OF NUCLEAR REACTIONS

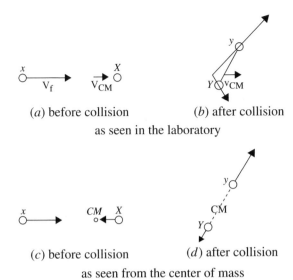

**Figure 10.2** Schematic view of a nuclear reaction in the laboratory and center-of-mass systems. [From Weidner and Sells (1973).]

where $T_{\text{lab}}$ is the kinetic energy in the lab system before the reaction, that is,

$$T_{\text{lab}} = \tfrac{1}{2} m_P v_P^2$$

The kinetic energy carried in by the projectile, $T_{\text{lab}}$, is not fully available to be dissipated in the reaction. Instead, an amount $T_{\text{cm}}$ must be carried away by the center of mass. Thus, the available energy to be dissipated is $T_{\text{lab}} - T_{\text{cm}} \equiv T_0$. The energy available for the nuclear reaction is $Q + T_0$. To make the reaction go, the sum $Q + T_0$ must be greater than or equal to zero. Thus, rearranging a few terms, the condition for having the reaction occur is that

$$T_P \geq -Q(m_P + m_T)/m_T$$

This minimum kinetic energy that the projectile must have to make the reaction go is called the *threshold energy* for the reaction.

**Example Problem**  Consider the $^{14}\text{N}(\alpha, p)^{17}\text{O}$ reaction. What is the threshold energy for this reaction?

**Solution**

$Q = [m_\alpha + m_N - (m_P + m_O)]c^2$
$\quad = 2.425 + 2.863 - 7.289 - (-0.809) = -1.19 \text{ MeV}$
$T_\alpha = -(-1.19)(4 + 14)/14 = 1.53 \text{ MeV}$

## 10.3 REACTION TYPES AND MECHANISMS

Nuclear reactions, like chemical reactions, can occur via different reaction mechanisms. Weisskopf has presented a simple conceptual model (Fig. 10.3) for illustrating the relationships between the various nuclear reaction mechanisms.

Consider a general nuclear reaction of the type A (a, b) B, bearing in mind that for some cases, the nuclei b and B may be identical to a and A. As the projectile a moves near the target nucleus A, it will have a certain probability of interacting with the nuclear force field of A, causing it to change direction but not to lose any energy ($Q = 0$) (Fig. 10.3). This reaction mechanism is called *shape elastic scattering*. If shape elastic scattering does not occur, then the projectile may interact with A via a two-body collision between the projectile and some nucleon of A, raising the nucleon of A to an unfilled level (Fig. 10.3). If the struck nucleon leaves the nucleus, a *direct reaction* is said to have occurred. If the struck nucleon does not leave the nucleus, further two-body collisions may occur, and eventually the entire kinetic energy of the projectile nucleus may be distributed between the nucleons of the a + A combination leading to the formation of a *compound nucleus* C (see Fig. 10.3). Because of the complicated set of interactions leading to the formation of the compound nucleus, loosely speaking, it "forgets" its mode of formation, and its subsequent breakup only depends on the excitation energy, angular momentum of C, and so forth and not the nature of the projectile and target nuclei. Sometimes the compound nucleus may emit a particle of the same kind as the projectile (or even the projectile itself) with the same energy as the projectile had. If this happens, we say *compound elastic scattering* has occurred. Also C may decay into reaction products that are unlike the projectile or target nuclei. We shall spend much of this chapter discussing these reaction mechanisms and some others not yet mentioned. But before doing so, let us see what general properties of nuclear reactions we can deduce from relatively simple arguments.

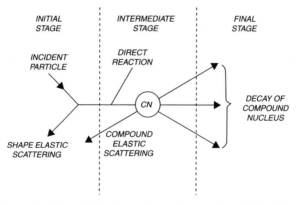

**Figure 10.3** Conceptual view of the stages of a nuclear reaction. [After Weisskopf (1959).]

## 10.4 NUCLEAR REACTION CROSS SECTIONS

Consider the situation (Fig. 10.4) where a beam of projectile nuclei of intensity $\phi_0$ particles/second is incident upon a thin foil of target nuclei with the result that the beam is attenuated by reactions in the foil such that the transmitted intensity is $\phi$ particles/second. We can ask what fraction of the incident particles disappear from the beam, that is, react, in passing through the foil. Let us assume the beam intersects an area $A$ on the foil. We can then assert that the fraction of beam particles that is blocked (reacts) is the fraction of the area $A$ that is covered by target nuclei. If the foil contains $\mathbb{N}$ atoms/cm$^2$, then the area $a$ that is covered by nuclei is $\mathbb{N}$ (atoms/cm$^2$) $\times$ $a$ (cm$^2$) $\times$ (the effective area subtended by one atom) (cm$^2$/atom). This latter term, the effective area subtended by one atom, is called the *cross section*, $\sigma$, for the reaction under study. Then the fraction of the area $A$ that is blocked is $a/A$ or $\mathbb{N}$ (atoms/cm$^2$) $\sigma$ (cm$^2$/atom). If we say the number of projectile nuclei absorbed per unit time is $\Delta\phi$, then we have

$$\Delta\phi = -\phi \mathbb{N} \sigma$$

As an aside, we note the units of $\mathbb{N}$ are atoms/cm$^2$ or thickness, $\Delta x$ (cm) $\times$ density $n$ (atoms/cm$^3$).

Expressing the above equation as a differential equation, we have

$$-d\phi = \phi \mathbb{N} \sigma$$

Thus, upon rearranging, we have

$$\frac{d\phi}{\phi} = -\mathbb{N}\sigma = -n\sigma\,dx$$

$$\int_{\phi_{\text{initial}}}^{\phi_{\text{trans}}} \frac{d\phi}{\phi} = -n\sigma \int_0^x dx$$

$$\ln \frac{\phi_{\text{trans}}}{\phi_{\text{initial}}} = -n\sigma x$$

$$\phi_{\text{trans}} = \phi_{\text{initial}} e^{-n\sigma x}$$

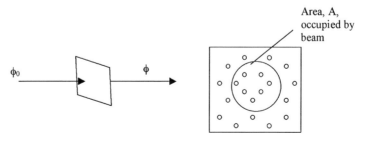

**Figure 10.4** Schematic diagram showing the attenuation of an incident projectile beam.

Thus, we see exponential absorption of the incident projectile beam (and have thus derived a form of the Lambert–Beers law). The number of reactions that are occurring is the difference between the initial and transmitted flux, that is,

$$\phi_{initial} - \phi_{trans} = \phi_{initial}\left[1 - \exp(-n\sigma x)\right]$$

The foregoing discussion focused on the attenuation of the incident beam and thus refers to *all* reactions. In many cases, we are interested in only one of several reactions that may be taking place. We can refer to the cross section for that particular reaction. In addition, we may be interested not only in a specific product but a particular product moving in a particular direction relative to the direction of the projectile beam (see Fig. 10.5 for a sketch of a typical experimental measurement).

In this case, we can speak of a *differential cross section*, or the cross section per unit solid angle, $d\sigma/d\Omega$. For a thin target in which the attenuation of the beam is not significant, we have

$$\frac{dN}{d\Omega} = \phi n \left(\frac{d\sigma}{d\Omega}\right) dx$$

where $dN/d\Omega$ is the number of particles detected moving in a particular direction per unit solid angle. The total cross section, $\sigma$, is given as

$$\sigma = \int_0^{2\pi} \int_0^{\pi} \frac{d\sigma}{d\Omega}(\theta) \sin\theta \, d\theta \, d\phi$$

The description given above is appropriate for work at accelerators, where one has a beam of particles that is smaller than the target. In this case, the beam intensity

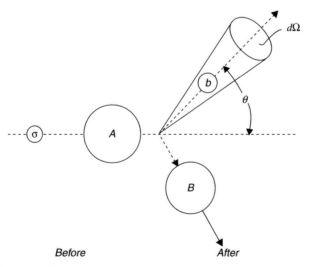

**Figure 10.5** Schematic diagram of a typical experimental setup.

ϕ is given in particles/seconds and the target density $\mathbb{N}$ is given in atoms/cm². In a nuclear reactor, we immerse a small target in a sea of neutrons. In this case, the neutron flux ϕ represents the number of neutrons passing through the target per cm² per sec, and $\mathbb{N}$ is the total number of atoms in the target. Otherwise the arithmetic is the same. For charged particles from an accelerator, the beam intensity is usually measured as a current. Thus, for a beam of protons with a current of 1 μA, we have

$$\phi = (1\,\mu A)(10^{-6}\,\text{C/s}/\mu A)\left(\frac{1}{1.602 \times 10^{-19}\,\text{C/proton}}\right)$$
$$= 6.24 \times 10^{12}\,\text{protons/second}$$

For a beam of some other ion with charge $q$, one simply divides by the charge on the ion to get the projectile beam intensity. Thus, for a beam of 4 μA of $Ar^{+17}$ ions, we have

$$\phi = (4 \times 10^{-6}\,\text{C/s})\left(\frac{1}{17 \times 1.602 \times 10^{-19}\,\text{C/Ar}}\right) = 1.47 \times 10^{12}\,\text{Ar/s}$$

To put the intensities of beams of differing charges on a common footing, it is common to divide the electric current by the charge state and quote charge particle beam intensities in units of *particle microamperes* or *particle nanoamperes* where 1 particle microampere = $6.24 \times 10^{12}$ ions/s.

It is easy to calculate the number of product nuclei produced during an irradiation, $N$. If we assume the product nuclei are stable, then the number of nuclei produced is the (rate of production) × (length of the irradiation, $t$). For a thick target irradiation, we have

$$N = \phi[1 - \exp(-n\sigma\,\Delta x)]t$$

For a thin target if we expand the function $[1 - \exp(-n\sigma\,\Delta x)]$, we have

$$N = \phi n\sigma\,\Delta x\, t$$

But, what if the products are radioactive? Then some of them will decay during the irradiation. In this case, we can set up the familiar differential equations:

$$\frac{dN}{dt} = (\text{rate of production}) - (\text{rate of decay})$$

$$\frac{dN}{dt} = n\sigma\,\Delta x\,\phi - \lambda N$$

$$\frac{dN}{n\sigma\,\Delta x\,\phi - \lambda N} = dt$$

Multiplying by $\lambda$ and rearranging

$$\frac{d(\lambda N)}{\lambda N - n\sigma\,\Delta x\,\phi} = -\lambda\,dt$$

Integrating, we have

$$\ln(\lambda N - n\sigma\,\Delta x\,\phi)\big|_0^N = -\lambda t\big|_0^t$$

$$\frac{N\lambda - n\sigma\,\Delta x\,\phi}{-n\sigma\,\Delta x\,\phi} = e^{-\lambda t}$$

$$A = \lambda N = n\sigma\,\Delta x\,\phi(1 - e^{-\lambda t})$$

where $A$ is the disintegration rate of product nuclei at the end of the irradiation. The number of product nuclei, $N$, present at the end of the irradiation is $A/\lambda$ or

$$N = \frac{n\sigma\,\Delta x\,\phi}{\lambda}(1 - e^{-\lambda t})$$

This relationship is shown in Figure 10.6.

Note that in the limit of infinitely long irradiation, $e^{-\lambda t} \to 0$, and thus the activity present is $n\sigma\,\Delta x\,\phi$, which is termed the *saturation activity*. Note also that for very short times (compared to the half-life of the product nuclei), $e^{-\lambda t} \to 1 - \lambda t + \cdots$. Thus, the activity increases linearly with time. In general, we note that we achieve one-half the saturation activity after an irradiation of one half-life, three-fourths of the saturation activity after irradiating two half-lives, seven-eighths of the saturation activity after irradiating three half-lives, and so forth. Thus, it does not pay to make the irradiation longer than one to two half-lives. (This effect can be used to tune the length of the irradiation to maximize the yield of the product of interest relative to the other reaction products.)

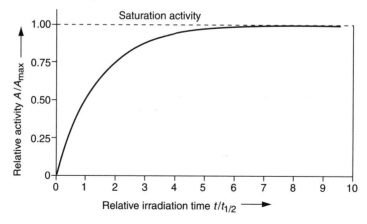

**Figure 10.6** Variation of product activity during an irradiation. [From Lieser (1997).]

**Example Problem** Calculate the activity of $^{254}$No ($t_{1/2} = 55$ s) produced in a 1-minute irradiation of $^{208}$Pb by $^{48}$Ca. Assume the $^{208}$Pb target thickness is 0.5 mg/cm$^2$, the $^{48}$Ca beam current is 0.5 particle microamperes, and the $^{208}$Pb($^{48}$Ca, 2n) reaction cross section is 3.0 µb.

*Solution*

$A = N\sigma\phi(1 - e^{-\lambda t_i})$

$N = (0.5 \times 10^{-3}$ g/cm$^2)(6.02 \times 10^{23}$ atoms/g-at.-wt.)/208 g/g-at.-wt.

$\phantom{N} = 1.44 \times 10^{18}$ atoms/cm$^2$

$\sigma = 3 \times 10^{-30}$ cm$^2$

$\phi = (0.5 \times 10^{-6}$ C/s)/1.602 $\times 10^{-19}$ C/ion $= 3.12 \times 10^{12}$ ions/s

$t_i = 60$ s

$\lambda = (\ln 2)/55$ s $= 1.26 \times 10^{-2}$ s$^{-1}$

$A = 7.2$ dis/s

Let us consider what we can learn about cross sections from some general considerations. Consider the reaction of an uncharged particle (a neutron) with a nucleus as shown in Figure 10.7. The neutron makes a grazing collision with the nucleus. The impact parameter $b$ is taken to be the sum of the radii of the projectile and target nuclei. Thus, the cross section can be written as

$$\sigma \approx \pi(R + r')^2 = \pi r_0^2 (A_P + A_T)^2$$

where $r'$ is the radius of the projectile. Applying classical mechanics to this problem, we can write for the orbital angular momentum, $l$,

$$l = \mathbf{r} \times \mathbf{p} = pb$$

In quantum mechanics, $l \rightarrow l\hbar$, and the momentum $p$ is given by

$$p = \frac{\hbar}{\lambdabar}$$

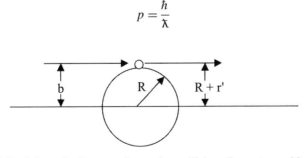

**Figure 10.7** Schematic diagram of a grazing collision of a neutron with a nucleus.

Thus, we have

$$l\hbar = \frac{\hbar b}{\lambda}$$

$$b = l\lambda$$

This is not quite right because $l$ is quantized but $b$ is not. We get around this by associating $b$ with certain rings or zones on the target (Fig. 10.8). Figure 10.8 suggests that for head-on collisions ($l = 0$), the range of $b$ is from 0 to $\lambda$, whereas for $l = 1$ collisions, the range of $b$ is from $\lambda$ to $2\lambda$. Thus, the cross section is larger for larger impact parameters, and these larger impact parameters are associated with larger angular momenta. We can write the cross section for a specific value of $l$ as:

$$\sigma_l = \pi(l+1)^2\lambda^2 - \pi l^2\lambda^2$$

$$\sigma_l = \pi\lambda^2(l^2 + 2l + 1 - l^2)$$

$$\sigma_l = \pi\lambda^2(2l + 1)$$

The total reaction cross section is obtained by summing over all $l$ values as:

$$\sigma_{\text{total}} = \sum_l \sigma_l = \sum_{l=0}^{l_{\max}} \pi\lambda^2(2l+1) = \pi\lambda^2 \sum_{l=0}^{l_{\max}} (2l+1) = \pi\lambda^2(l_{\max}+1)^2$$

We can write for the maximum angular momentum, $l_{\max}$,

$$l_{\max} = \frac{R}{\lambda}$$

$$l_{\max} + 1 = \frac{R + \lambda}{\lambda}$$

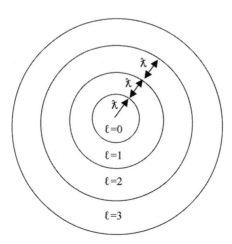

**Figure 10.8** Schematic bulls-eye view of the target nucleus.

## 10.4 NUCLEAR REACTION CROSS SECTIONS

Thus, we have for the total cross section

$$\sigma_{\text{total}} = \pi(R + \lambdabar)^2$$

The total cross section is proportional to the size of the target nucleus and the "size" of the projectile nucleus. Since the wavelength of the projectile, $\lambdabar$, goes to infinity as the projectile energy goes to zero, the cross sections for neutrons at low energies can be very large. The above discussion is based upon semiclassical mechanics. We need to indicate how the problem would look if we used quantum mechanics to treat it. In quantum mechanics, we can write a similar expression for the total reaction cross section:

$$\sigma_{\text{total}} = \pi \lambdabar^2 \sum_{l=0}^{\infty} (2l+1) T_l$$

where the transmission coefficient $T_l$ varies between 0 and 1. The transmission coefficient expresses the probability that a given angular momentum transfer $l$ will occur. At high projectile energies, $T_l = 1$ for $l \leq l_{\max}$ and $T_l = 0$ for $l \geq l_{\max}$. (This is called the "sharp cutoff limit".) At very low projectile energies, $T_l = \varepsilon^{1/2}$ for $l = 0$ and $T_l = 0$ for $l > 0$, where $\varepsilon$ is the projectile energy. Thus, at very low energies, we have

$$\sigma_{\text{total}} \propto \pi \lambdabar^2 \sqrt{\varepsilon} \propto \pi \frac{\hbar^2}{2m\varepsilon} \sqrt{\varepsilon} \propto \frac{1}{\sqrt{\varepsilon}}$$

Such behavior of the cross sections for neutron-induced reactions is referred to as "$1/v$" behavior.

Now let us consider the interaction of a charged particle with a nucleus as shown in Figure 10.9. As the projectile approaches the target nucleus, it feels the long-range Coulomb force and is deflected. As a consequence, the range of collisions corresponds to a smaller range of impact parameters. If the incident projectile has an energy $\varepsilon$ at an infinite separation from the target nucleus, at the distance of closest approach $R$, it has a kinetic energy of $\varepsilon - B$ where $B$, the Coulomb barrier, is given by

$$B = Z_1 Z_2 e^2 / R$$

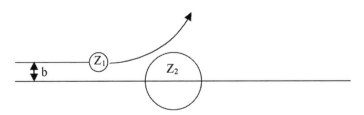

**Figure 10.9** Schematic diagram of a charged-particle-induced reaction.

At the point of closest approach, the momentum $p$ of the projectile is $(2mT)^{1/2}$. Thus, we can write

$$p = (2mT)^{1/2} = (2\mu)^{1/2}(\varepsilon - B)^{1/2} = (2\mu\varepsilon)^{1/2}(1 - B/\varepsilon)^{1/2}$$

where $\mu$ is the reduced mass of the system $[=A_1 A_2/(A_1 + A_2)]$. Classically, we have, for the orbital angular momentum,

$$l = \mathbf{r} \times \mathbf{p}$$

$$l_{\max} = R(2\mu\varepsilon)^{1/2}\left(1 - \frac{B}{\varepsilon}\right)^{1/2}$$

Quantum mechanically, we have $l \rightarrow l\hbar$. So we can write

$$\sigma_{\text{total}} = \pi\lambdabar^2(l_{\max} + 1)^2 \approx \pi\lambdabar^2 l_{\max}^2 = \pi\lambdabar^2 R^2 \frac{2\mu\varepsilon}{\hbar^2}\left(1 - \frac{B}{\varepsilon}\right)$$

$$= \pi\lambdabar^2 R^2 \frac{1}{\lambdabar^2}\left(1 - \frac{B}{\varepsilon}\right)$$

$$\sigma_{\text{total}} = \pi R^2\left(1 - \frac{B}{\varepsilon}\right)$$

Note this last classical expression is valid only when $\varepsilon > B$. The combined general properties of cross sections for charged and uncharged particles are shown in Figure 10.10.

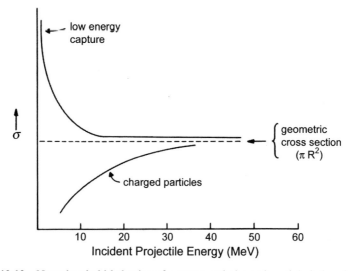

**Figure 10.10** Near threshold behavior of neutron and charged-particle-induced reactions. [From Ehmann and Vance (1991).]

## 10.4 NUCLEAR REACTION CROSS SECTIONS

**Example Problem** Calculate the energy dependence of the total reaction cross section for the $^{48}$Ca + $^{208}$Pb reaction.

***Solution***

$$\sigma_{\text{total}} = \pi R^2 \left(1 - \frac{B}{\varepsilon}\right)$$

$R = R_{\text{Pb}} + R_{\text{Ca}} = 1.2(208^{1/3} + 48^{1/3}) = 11.47 \text{ fm}$
$B = Z_1 Z_2 e^2/R = (82)(20)(1.44 \text{ MeV-fm})/11.47 \text{ fm} = 205.9 \text{ MeV}$
$\varepsilon = $ energy of the projectile in the cm system

| $\varepsilon$ (MeV) | $E_{\text{lab}}$ (McV) | $\sigma$ (mb) |
|---|---|---|
| 208 | 256   | 41.7  |
| 210 | 258.5 | 80.7  |
| 220 | 270.8 | 264.9 |
| 230 | 283.1 | 433.1 |
| 240 | 295.4 | 587.2 |
| 250 | 307.7 | 729.1 |

*Aside on Barriers* In our semiclassical treatment of the properties of charged-particle-induced reaction cross sections, we have equated the reaction barrier $B$ to the Coulomb barrier. This is, in reality, a simplification that is applicable to many but not all charged-particle-induced reactions.

The actual force (potential energy) felt by an incoming projectile is the sum of the nuclear, Coulomb, and centrifugal forces (Fig. 10.11). The Coulomb potential, $V_C(r)$, is approximated as the potential between a point charge $Z_1 e$ and a homogeneous charged sphere with charge $Z_2 e$ and radius $R_C$ as

$V_C(r) = Z_1 Z_2/r \quad \text{for } r > R_C$
$V_C(r) = (Z_1 Z_2/R_C)\left[\frac{3}{2} - \frac{1}{2}(r^2/R_C^2)\right] \quad \text{for } r < R_C$

The nuclear potential is frequently represented by a Woods–Saxon form (Chapter 5) as:

$$V_{\text{nucl}}(r) = V_0/\{(1 + \exp[(r - R/a)]\}$$

while the centrifugal potential is taken as:

$$V_{\text{cent}}(r) = \frac{\hbar^2}{2\mu} \frac{l(l+1)}{r^2}$$

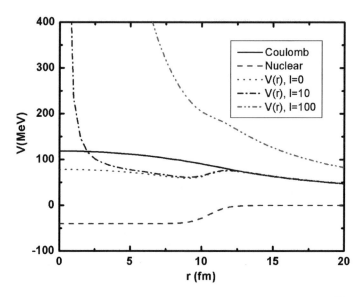

**Figure 10.11** Nuclear, Coulomb, and total potentials for the interaction of $^{16}$O with $^{208}$Pb for three values of the orbital angular momentum.

where $l\hbar$ is the orbital angular momentum of the incident projectile. The total potential, $V_{tot}(r)$, is the sum $V_C(r) + V_{nucl}(r) + V_{cent}(r)$. These different potentials are shown in Figure 10.11 using the $^{16}$O + $^{208}$Pb reaction as an example and input angular momenta of $l = 0, 10$, and $100\hbar$. Note that for the highest angular momentum, $l = 100\hbar$, the total potential is repulsive at all distances, that is, the ions do not fuse.

The actual *interaction barrier* is the value of $V_{total}(r)$ at the point when the colliding nuclei touch. That is slightly different from $V_C(r)$ at $r = R_C$, the *Coulomb barrier*.

## 10.5 REACTION OBSERVABLES

What do we typically measure when we study a nuclear reaction? We might measure $\sigma_R$, the total reaction cross section. This might be measured by a beam attenuation method ($\Phi_{transmitted}$ vs. $\Phi_{incident}$) or by measuring all possible exit channels for a reaction where

$$\sigma_R = \sum_i^{b+B} \sigma_i(b, B)$$

We might measure the cross section for producing a particular product at the end of the reaction, $\sigma(Z, A)$. We might do this by measuring the radioactivity of the reaction products. We might, as discussed previously, measure the products emerging in a particular angular range, $d\sigma(\theta, \phi)/d\Omega$. This measurement is especially relevant for experiments with charge-particle-induced reactions where the incident beam provides a reference axis for θ and ϕ. The energy spectra of the emitted particles can be measured as $d\sigma/dE$, or we might observe the products emerging at a particular angle and with a particular energy, $d^2\sigma/dE\,d\Omega$.

## 10.6 RUTHERFORD SCATTERING

One of the first possible outcomes of the collision of a charged particle with a nucleus is Rutherford or Coulomb scattering. The incident charged particle feels the long-range Coulomb force of the positively charged nucleus and is deflected from its path (Fig. 10.12).

The Coulomb force acting between a projectile of mass $m$, charge $Z_1e$, and a target nucleus with charge $Z_2e$ is given as:

$$F_{\text{Coul}} = \frac{Z_1 Z e^2}{r^2}$$

where $r$ is the distance between the projectile and target nuclei. The potential energy (PE) in this interaction is given as:

$$\text{PE} = \frac{Z_1 Z_2 e^2}{r}$$

Consider a target nucleus that is much heavier than the projectile nucleus so that we can neglect the recoil of the target nucleus in the interaction. The projectile will

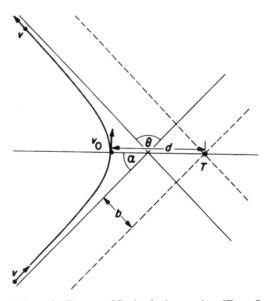

**Figure 10.12** Schematic diagram of Rutherford scattering. [From Satchler (1990).]

follow a hyperbolic orbit, as shown in Figure 10.12 where $b$ is the impact parameter, $T_P$ is the kinetic energy of the projectile and $d$ is the distance of closest approach. At infinity, the projectile velocity is $v$. At $r = d$, the projectile velocity is $v_0$. Conservation of energy gives

$$\frac{1}{2}mv^2 = \frac{1}{2}mv_0^2 + \frac{Z_1 Z_2 e^2}{d}$$

Rearranging, we have

$$\left(\frac{v_0}{v}\right)^2 = 1 - \frac{d_0}{d}$$

where $d_0$ is given as

$$d_0 = \frac{2Z_1 Z_2 e^2}{mv^2} = \frac{Z_1 Z_2 e^2}{T_P}$$

If we now invoke the conservation of angular momentum, we can write

$$mvb = mv_0 d$$

$$b^2 = \left(\frac{v_0}{v}\right)^2 d^2 = d(d - d_0)$$

It is a property of a hyperbola that

$$d = b \cot(\alpha/2)$$

Substituting from above, we have

$$\tan \alpha = 2b/d_0$$

Since $\theta = \pi - 2\alpha$, we can write

$$\cot\left(\frac{\theta}{2}\right) = \frac{2b}{d_0}$$

In Figure 10.13, we show the expected orbits of the projectile nuclei after undergoing Rutherford scattering for a typical case. Note that the most probable grazing trajectories result in projectiles being scattered to forward angles but that some nearly head-on collisions result in large angle scattering. It was these latter events that led Rutherford to conclude that there was a massive object at the center of the atom.

We can make these observations more quantitative by considering the situation where a flux of $I_0$ particles/unit area is incident on a plane normal to the beam

## 10.6 RUTHERFORD SCATTERING

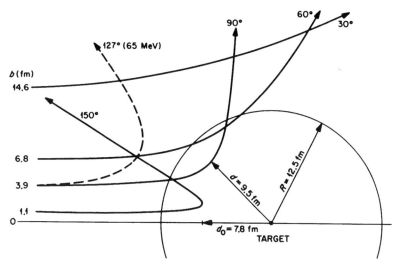

**Figure 10.13** Diagram showing some representative projectile orbits for the interaction of 130 MeV $^{16}$O with $^{208}$Pb. [From Satchler (1990).]

direction. The flux of particles passing through a ring of width $db$ and with impact parameters between $b$ and $b + db$ is given as:

$$dI = \left(\frac{\text{flux}}{\text{unit area}}\right)(\text{area of ring})$$

$$dI = I_0(2\pi b \, db)$$

Substituting from above, we have

$$dI = \frac{1}{4}\pi I_0 d_0^2 \frac{\cos(\theta/2)}{\sin^3(\theta/2)} d\theta$$

If we want to calculate the number of projectile nuclei that undergo Rutherford scattering into a solid angle $d\Omega$ at a plane angle $\theta$, we can write

$$\frac{d\sigma}{d\Omega} = \frac{dI}{I_0}\frac{1}{d\Omega} = \left(\frac{d_0}{4}\right)^2 \frac{1}{\sin^4(\theta/2)} = \left(\frac{Z_1 Z_2 e^2}{4T_P^{\text{cm}}}\right)^2 \frac{1}{\sin^4(\theta/2)}$$

if we remember that

$$d\Omega = 2\pi \sin\theta \, d\theta \text{ sr}.$$

Note the strong dependence of the Rutherford scattering cross section upon scattering angle. Remember that Rutherford scattering is not a nuclear reaction, as it does not involve the nuclear force, only the Coulomb force between the charged nuclei. Remember also that Rutherford scattering will occur to some extent in all studies of charged-particle-induced reactions and will furnish a "background" of scattered particles at forward angles.

**Example Calculation** Calculate the differential cross section for the Rutherford scattering of 215 MeV (lab energy) $^{48}$Ca from $^{208}$Pb at an angle of 20°.

**Solution**

$$\frac{d\sigma}{d\Omega} = \left(\frac{Z_1 Z_2 e^2}{4T_P^{cm}}\right)^2 \frac{1}{\sin^4(\theta/2)}$$

$$T_P^{cm} = 215 \times \frac{208}{256} = 174.7 \text{ MeV}$$

$$\frac{d\sigma}{d\Omega} = \left(\frac{20 \times 82 \times 1.44}{4 \times 174.7}\right)^2 \frac{1}{\sin^4(20/2)} = 12562 \text{ fm}^2/\text{sr} = 125.6 \, b/\text{sr}$$

## 10.7 ELASTIC (DIFFRACTIVE) SCATTERING

Suppose we picture the interaction of the incident projectile nucleus with the target nucleus as it undergoes shape elastic scattering. It is convenient to think of this interaction as that of a plane wave interacting with the nucleus as depicted in Figure 10.14.

Imagine further that all interactions take place on the nuclear surface. Assume that only points $A$ and $B$ on the nucleus scatter particles and that all other points on the surface absorb them. To get constructive interference between the incoming and outgoing wave, we must fulfill the condition that

$$CB + BD = n\lambda$$

where $\lambda$ is the wavelength of the incident particle and $n$ is an integer. Hence peaks should occur in the scattering cross section when

$$n\lambda = 2 \cdot 2R \cdot \sin\left(\frac{\theta}{2}\right)$$

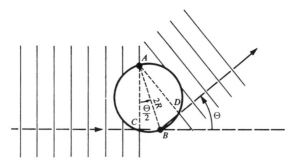

**Figure 10.14** Schematic diagram of the interaction of a plane wave with the nucleus. [From Meyerhof (1967).]

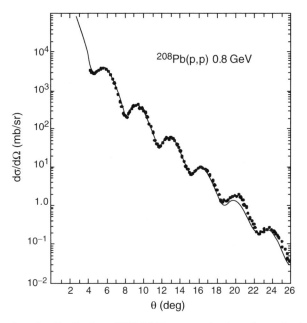

**Figure 10.15** Angular distribution of 800 MeV protons that have been elastically scattered from $^{208}$Pb. [From Blanpied et al. (1978).]

In Figure 10.15, we show the angular distribution for the elastic scattering of 800 MeV protons from $^{208}$Pb. The de Broglie wavelength of the projectile is 0.85 fm, while the nuclear radius $R$ is about 7.6 fm [$1.28(208)^{1/3}$]. We expect peaks ($n = 2, 3, 4 \ldots$) with a spacing between them, $\Delta\theta$, of $3.2°$ while one observes a spacing of $3.5°$. [This discussion of 800 MeV proton scattering is taken from Bertsch and Kashy (1993).]

*Aside on the Optical Model* The optical model is a tool to understand and parameterize studies of elastic scattering. It likens the interaction of projectile and target nucleus with that of a beam of light interacting with a glass ball. To simulate the occurrence of both elastic scattering and absorption (reactions) in the interaction, the glass ball is imagined to be somewhat cloudy.

In formal terms, the nucleus is represented by a nuclear potential that has a real and an imaginary part:

$$U_{\text{nucl}}(r) = V(r) + iW(r)$$

where the imaginary potential $W(r)$ describes absorption (reactions) as the depletion of flux into nonelastic channels and the real potential $V(r)$ describes the elastic scattering. Frequently, the nuclear potential is taken to have the Woods–Saxon form:

$$U_{\text{nucl}}(r) = -V_0 f_R(r) - i W_0 f_I(r)$$

where

$$f_{R,I}(r) = \left[1 + \exp\left(\frac{r - R_{R,I}}{a_{R,I}}\right)\right]^{-1}$$

The potential is thus described in terms of six parameters, the potential depths, $V_0$, $W_0$; the radii $R_R$, $R_I$; and the surface diffuseness $a_R$, $a_I$. By solving the Schrödinger equation with this nuclear potential (along with the Coulomb and centrifugal potentials), one can predict the cross section for elastic scattering, the angular distribution for elastic scattering, and the total reaction cross section. The meaning of the imaginary potential depth $W$ can be understood by noting that the mean free path of a nucleon in the nucleus, $\Lambda$, can be given as

$$\Lambda = \frac{v\hbar}{2W_0}$$

where $v$ is the relative velocity. By fitting measurements of elastic scattering cross sections and angular distributions over a wide range of projectiles, targets, and beam energies, one might hope to gain a universal set of parameters to describe elastic scattering (and the nuclear potential). That hope is only partially realized because only the tail of the nuclear potential affects elastic scattering, and there are families of parameters that fit the data equally well, as long as the potential energy functions agree in the exterior regions of the nucleus.

## 10.8 DIRECT REACTIONS

As we recall from our general description of nuclear reactions, a *direct reaction* is said to occur if one of the participants in the initial two-body interaction involving the incident projectile leaves the nucleus. Generally speaking, these direct reactions are divided into two classes, the *stripping reactions* in which part of the incident projectile is "stripped away" and enters the target nucleus and the *pickup reactions* in which the outgoing emitted particle is a combination of the incident projectile and one or a few target nucleons.

Let us consider stripping reactions first and, in particular, the most commonly encountered stripping reaction, the $(d, p)$ reaction. Formally, the result of a $(d, p)$ reaction is to introduce a neutron into the target nucleus, and thus this reaction should bear some resemblance to the simple neutron capture reaction. But because of the generally higher angular momenta associated with the $(d, p)$ reaction, there can be differences between the two reactions. Consider the $A(d, p)B^*$ reaction where the recoil nucleus $B$ is produced in an excited state $B^*$. We sketch out a simple picture of this reaction and the momentum relations in Figure 10.16.

The momentum diagram for the reaction shown in Figure 10.16 assumes the momentum of the incident deuteron is $k_d\hbar$, the momentum of the emitted proton is $k_p\hbar$, while $k_n\hbar$ is the momentum of the stripped neutron. From conservation of

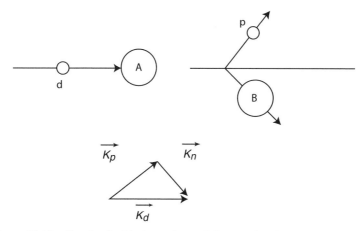

**Figure 10.16** Sketch of a $(d, p)$ reaction and the associated momentum triangle.

momentum, we have

$$k_n^2 = k_d^2 + k_p^2 - 2k_d k_p \cos\theta$$

If the neutron is captured at impact parameter $R$, the orbital angular momentum transferred to the nucleus, $l_n\hbar$, is given by

$$l_n\hbar = \mathbf{r} \times \mathbf{p} = Rk_n\hbar$$
$$l_n = Rk_n$$

Since we have previously shown that $k_n$ is a function of the angle $\theta$, we can now associate each orbital angular momentum transfer in the reaction with a given angle $\theta$ corresponding to the direction of motion of the outgoing proton. Thus, the $(d,p)$ reaction becomes a very powerful spectroscopic tool. By measuring the energy of the outgoing proton, we can deduce the $Q$ value of the reaction and thus the energy of any excited state of the residual nucleus that is formed. From the direction of motion of the proton, we can deduce the orbital angular momentum transfer in the reaction, $l_n$. If we know the ground-state spin and parity of the residual nucleus, we can deduce information about the spin and parity of the excited states of the residual nucleus using the rules

$$\left|\left(|J_A - l_n| - \tfrac{1}{2}\right)\right| \leq J_{B^*} \leq J_A + l_n + \tfrac{1}{2}$$
$$\pi_A \pi_{B^*} = (-1)^l$$

Other stripping reactions are reactions such as $(\alpha, t)$, $(\alpha, d)$, and so forth. Typical pickup reactions are $(p, d)$, $(p, t)$, $(\alpha, {}^6\text{Li})$, and so forth.

**Example Problem** Calculate the angle at which the $(d,p)$ cross section has a maximum for $l = 0, 1, 2, 3,$ and 4. Assume a deuteron energy of 7 MeV and a proton energy of 13 MeV. Use $R = 6$ fm.

**Solution**

$k_d = 0.82 \text{ fm}^{-1}$

$k_p = 0.79 \text{ fm}^{-1}$

$k_n = \dfrac{l}{R}$

Thus, for $l = 0, 1, 2, 3,$ and 4, $k_n = 0, 0.17 \text{ fm}^{-1}, 0.33 \text{ fm}^{-1}, 0.50 \text{ fm}^{-1},$ and $0.67 \text{ fm}^{-1}$. Solving the momentum triangle,

$$\cos \vartheta = \dfrac{-k_n^2 + k_d^2 + k_p}{2 k_d k_p}$$

$\theta = 0°, 12°, 24°, 36°, 49°$ for $l = 0, 1, 2, 3, 4$

(A somewhat more correct expression would say $k_n R = [l(l+1)]^{1/2}$.)

## 10.9 COMPOUND NUCLEUS REACTIONS

The compound nucleus is a relatively long-lived reaction intermediate that is the result of a complicated set of two-body interactions in which the energy of the projectile is distributed among all the nucleons of the composite system. How long does the compound nucleus live? From our definition above, we can say the compound nucleus must live for at least several times the time it would take a nucleon to traverse the nucleus ($10^{-22}$ s). Thus, the time scale of compound nuclear reactions is of the order of $10^{-18}$–$10^{-16}$ s. Lifetimes as long as $10^{-14}$ s have been observed. These relatively long times should be compared to the typical time scale of a direct reaction that takes place in one transit of the nucleus of $10^{-22}$ s.

Another important feature of compound nucleus reactions is that the mode of decay of the compound nucleus is independent of its mode of formation (the *Bohr independence hypothesis* or the *amnesia assumption*). While this statement is not true in general, it remains a useful tool for understanding certain features of compound nuclear reactions. For example, let us consider the classical work of Ghoshal (1950). Ghoshal formed the compound nucleus $^{64}$Zn in two ways, that is, by bombarding $^{63}$Cu with protons and by bombarding $^{60}$Ni with α particles. He examined the relative amounts of $^{62}$Cu, $^{62}$Zn, and $^{63}$Zn found in the two bombardments and within his experimental uncertainty of 10%, he found the amounts of the products were the same in both bombardments. (Later experiments have shown smaller scale deviations from the independence hypothesis.) Because of the long time scale of the reaction and the "amnesia" of the compound nucleus about its mode of formation, one can show that the angular distribution of the products is symmetric about 90° (in the frame of the moving compound nucleus).

## 10.9 COMPOUND NUCLEUS REACTIONS

The cross section for a compound nuclear reaction can be written as the product of two factors, the probability of forming the compound nucleus and the probability that the compound nucleus decays in a given way. As described above, the probability of forming the compound nucleus can be written as:

$$\sigma = \pi \lambdabar^2 \sum_{l=0}^{\infty} (2l+1) T_l$$

The probability of decay of the compound nucleus (CN) into a given set of products β can be written as:

$$\text{Probability} = \left[ \frac{T_l^\beta (E_\beta)}{\sum_{l_\gamma, E_\gamma} T_l^\gamma (E_\gamma)} \right]$$

where $T_l$ is the transmission coefficient for CN decay into products $i$. Figure 10.17 shows a schematic view of the levels of the compound nucleus. Note the increasing number of levels as the CN excitation energy increases. Quantitatively, the number of levels per MeV of excitation energy $E$ increases approximately exponentially as $E^{1/2}$.

The interesting categories of CN reactions can be defined by the ratio of the width of a compound nucleus level, $\Gamma$, to the average spacing between compound nuclear levels, $D$. (Recall from the Heisenberg uncertainty principle that $\Gamma \cdot \tau \geq \hbar$, where $\tau$ is the mean life of a compound nucleus level.) The categories are (a) $\Gamma/D \ll 1$, that is, the case of isolated nonoverlapping levels of the compound nucleus and (b) $\Gamma/D \gg 1$, the case of many overlapping levels in the compound nucleus (Fig. 10.17). Intuitively category (a) reactions are those in which the excitation energy of the compound nucleus is low, while category (b) reactions are those in which the excitation energy is high.

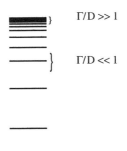

**Figure 10.17** Schematic view of the levels of a compound nucleus.

Let us first consider the case of $\Gamma/D \ll 1$. This means that at certain values of the compound nucleus excitation energy, individual levels of the compound nucleus can be excited (i.e., when the excitation energy exactly equals the energy of a given CN level). When this happens, there will be a sharp rise, or *resonance*, in the reaction cross section akin to the absorption of infrared radiation by sodium chloride when the radiation frequency equals the natural crystal oscillation frequency. In this case, the formula for the cross section (the *Breit–Wigner single-level formula*) for the reaction $a + A \to C \to b + B$ is

$$\sigma = \pi D^2 \frac{(2J_C + 1)}{(2J_A + 1)(2J_a + 1)} \frac{\Gamma_{aA}\Gamma_{bB}}{(\varepsilon - \varepsilon_0)^2 + (\Gamma/2)^2}$$

where $J_i$ is the spin of $i$th nucleus, $\Gamma_{aA}$, $\Gamma_{bB}$, and $\Gamma$ are the partial widths for the formation of $C$, the decay of $C$ into $b + B$, and the total width for the decay of $C$, respectively. The symbols $\varepsilon$ and $\varepsilon_0$ refer to the energy of the projectile nucleus and the projectile energy corresponding to the excitation of a single isolated level. Applying this formula to the case of $(n, \gamma)$ reactions gives

$$\sigma_{n,\gamma} = \pi D^2 \frac{(2J_C + 1)}{(2J_A + 1)(2)} \frac{\Gamma_n \Gamma_\gamma}{(\varepsilon - \varepsilon_0)^2 + (\Gamma/2)^2}$$

An example of this behavior is shown in Figure 10.18.

Resonances are seen in low-energy neutron-induced reactions where one is populating levels in the compound nucleus at excitation energies of the order of the neutron binding energy where the spacing between levels is of the order of electron volts. For neutron energies well below $\varepsilon_0$, so that $(\varepsilon - \varepsilon_0)^2 \approx \varepsilon_0^2$, then the cross section for the $(n, \gamma)$ reaction goes as $1/v$ where $v$ is the neutron velocity, a general behavior described earlier.

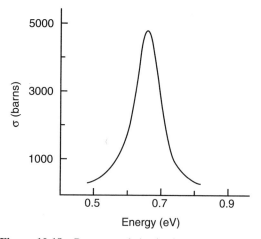

**Figure 10.18** Resonance behavior in $(n, \gamma)$ reactions.

Let us now consider the case where $\Gamma/D \gg 1$, that is, many overlapping levels of the compound nucleus are populated. (We are also tacitly assuming a large range of compound nuclear excitation energies.) The cross section for the reaction $a + A \to C \to b + B$ can be written as

$$\sigma_{ab} = \sigma_C P_C(b)$$

where $\sigma_C(a)$ is the cross section for the formation of the compound nucleus $C$ and $P_C$ is the probability that $C$ will decay to form $b + B$. Clearly $\Sigma P_C(b) = 1$. Now let us consider, in detail, the probability that emitted particle $b$ has an energy $\varepsilon_b$. First of all, we can write down that the maximum energy that $b$ can have is $E_C^* - S_b$ where $E_C^*$ is the excitation energy of the compound nucleus and $S_b$ is the separation energy of $b$ in the residual nucleus $B$. But $b$ can be emitted with a variety of energies less than this with the result that the nucleus $B$ will be left in an excited state. By using the arguments of detailed balance from statistical mechanics (see Lefort, 1968 or Friedlander, Kennedy, Miller, and Macias (FKMM)) we can write for the probability of emitting a particle $b$ with an energy $\varepsilon_b$ ($<\varepsilon_{\max}$, leaving the nucleus $B$ at an excitation energy $E_B^*$)

$$W_b(\varepsilon_b) d\varepsilon_b = \frac{(2J_b + 1)\mu}{\pi^2 \hbar^3} \varepsilon_b \sigma_{\text{inv}} \frac{\rho(E_B^*)}{\rho(E_C^*)} d\varepsilon_b$$

In this equation, $\mu$ is the reduced mass of the system, and $\sigma_{\text{inv}}$ is the cross section for the inverse process in which the particle $b$ is captured by the nucleus $B$ where $b$ has an energy, $\varepsilon_b$. The symbols $\rho(E_B^*)$ and $\rho(E_C^*)$ refer to the level density in the nucleus $B$ excited to an excitation energy $E_B^*$ and the level density in the compound nucleus $C$ excited to an excitation energy, $E_C^*$. The inverse cross section can be calculated using the same formulas used to calculate the compound nucleus formation cross section. Using the Fermi gas model, we can calculate the level densities of the excited nucleus as

$$\rho(E^*) = C \exp[2(aE^*)^{1/2}]$$

where the *level density parameter*, $a$, is $A/12 - A/8$. The nuclear temperature $T$ is given by the relation

$$E^* = aT^2 - T$$

The ratio of emission widths for emitted particles $x$ and $y$ is given as:

$$\frac{\Gamma_x}{\Gamma_y} = \frac{g_x \mu_x}{g_y \mu_y} \frac{R_x}{R_y} \frac{a_x}{a_y} \exp\left[2(a_x R_x)^{1/2} - 2(a_y R_y)^{1/2}\right]$$

where $g_i$ is the spin of the *i*th particle, $a_i$ and $R_i$ are the level density parameter and maximum exciation energy for the residual nucleus that results from the emission of the $i$th particle; $R$ is formally $E^* - S - \varepsilon_s$ where $\varepsilon_s$ is the threshold for charged particle emission ($\varepsilon_s$ for neutrons is 0).

If the emitted particles are neutrons, the emitted neutron energy spectrum has the form

$$N(\varepsilon)\,d\varepsilon = \frac{\varepsilon}{T^2}\exp\left(\frac{-\varepsilon}{T}\right)d\varepsilon$$

as shown in Figure 10.19. In other words, the particles are emitted with a Maxwellian energy distribution. The most probable energy is $T$ while the average energy is $2T$. The compound nucleus appears to "evaporate" particles like molecules leaving the surface of a hot liquid. By measuring the energy spectrum of the particles emitted in a compound nuclear reaction, we are using a "nuclear thermometer" in that

$$\frac{d[\ln N(\varepsilon)d\varepsilon]/\varepsilon\sigma_{\text{inv}}}{d\varepsilon} = \frac{1}{T}$$

Charged particles may also be evaporated, except the minimum kinetic energy is not zero as it is for neutrons. Instead, the threshold for charged particle emission $\varepsilon_s$ (which is approximately the Coulomb barrier) determines the minimum energy of an evaporated particle (see Fig. 10.10). The energy spectrum of evaporated charged particles is

$$N(\varepsilon)\,d\varepsilon = \frac{\varepsilon - \varepsilon_s}{T^2}\exp\left(\frac{-\varepsilon - \varepsilon_s}{T}\right)d\varepsilon$$

What will be the distribution in space of the reaction products? Let us assume that because the compound nucleus has "forgotten" its mode of formation, there should be no preferential direction for the emission of the decay products. Thus, we might

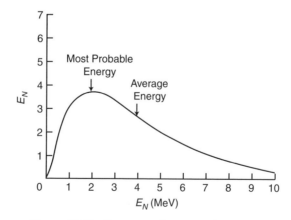

**Figure 10.19** Spectrum of evaporated neutrons.

expect that all angles of emission of the particles, θ, to be equally probable. Thus, we would expect that $P(\theta)$, the probability of emitting a particle at an angle θ, might be a constant. Then, we would expect that $d\sigma/d\Omega(\theta)$ would be given as:

$$\frac{d\sigma}{d\Omega}(\theta) = \int P(\vartheta) \frac{d\theta}{d\Omega}$$

This assumes that we are making the measurement of the emitted particles angular distribution in the frame of the moving compound nucleus. In the laboratory frame, there will appear to be more particles emitted in the forward direction (with higher energies) than are emitted in the backward direction due to the motion of the center of mass system.

The energy variation of the cross section (the *excitation function*) for processes involving evaporation is fairly distinctive, as shown in Figure 10.20, where the excitation function for the $^{209}$Bi($\alpha$, $xn$) reaction is shown. Starting from the threshold $\varepsilon_s$, the cross section rises with increasing energy because the formation cross section for the compound nucleus is increasing. Eventually, the excitation energy of the compound nucleus becomes large enough that emission of two neutrons is energetically possible. This "2n out" process will dominate over the "1n out" process, and the cross section for the "1n out" process will decrease. Eventually, the "3n out" process will dominate over the "2n out" process. We expect the peaks for the individual "xn out" processes to be at $S_{n1} + 2T, S_{n1} + S_{n2} + 4T, S_{n1} + S_{n2} + S_{n3} + 6T$, and so on (where we neglect any changes in $T$ during the emission process).

Let us recapitulate what we have said about compound nuclear reactions. We have said that they are basically nuclear reactions with a long-lived reaction intermediate, which is formed by a complicated set of two-body interactions. We can write down a set of equations that describes the overall compound nuclear cross section. We have shown how this general formula simplifies for specific

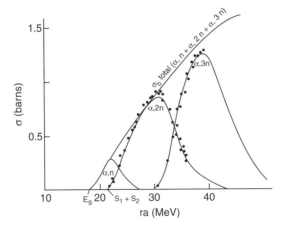

**Figure 10.20** Excitation function for the $^{209}$Bi($\alpha$, $xn$) reaction.

cases, the case of exciting a single level of the compound nucleus where we see spikes or resonances in the cross section and the case of higher excitation energies where the compound nucleus behaves like a hot liquid, evaporating particles. At all excitation energies, the angular distribution of the reaction products is symmetric with respect to a plane perpendicular to the incident particle direction.

## 10.10 PHOTONUCLEAR REACTIONS

Photonuclear reactions are a subset of nuclear reactions in which the incident projectile is a photon and the emitted particles are either charged particles or neutrons. Examples of such reactions are reactions like $(\gamma, p)$, $(\gamma, n)$, $(\gamma, \alpha)$, and so forth. The high-energy photons needed to induce these reactions can be furnished from the annihilation of positrons in flight (producing monoenergetic photons) or the energetic bremsstrahlung from slowing down high-energy electrons (producing a continuous distribution of photon energies). A special feature of the excitation function for photonuclear reactions is the appearance of a large bump in the cross section at $\sim 25$ MeV for reaction with a $^{16}$O target that slowly decreases with increasing $A$ until it is at $\sim 15$ MeV for $^{208}$Pb (Fig. 10.21).

This bump is called the *giant dipole resonance* (GDR). Goldhaber and Teller (1948) provided a model for this reaction in which the giant dipole resonance is

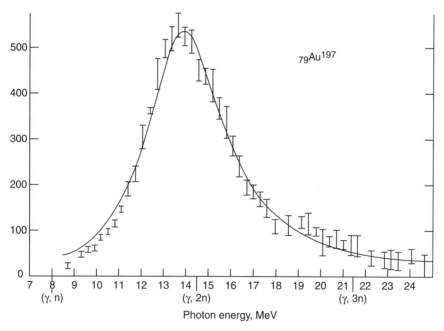

**Figure 10.21** Photonuclear cross section of $^{197}$Au. [From Fultz et al. (1962).]

due to a huge collective vibration of all the neutrons versus all the protons. This model suggests the energy of the GDR should vary as $A^{-1/6}$, in fair agreement with observations. In deformed nuclei, the GDR is split into two components, representing oscillations along the major and minor nuclear axes. One further feature of photonuclear reactions should be noted. The sum of the absorption cross section for dipole photons (over all energies) equals some constant, that is,

$$\int_0^\infty \sigma_{abs}(E_\gamma)\, dE_\gamma \propto \frac{NZ}{A} \approx 0.058 \frac{NZ}{A} \text{MeV-barns}$$

This is called the *dipole-sum rule*.

## 10.11 HEAVY ION REACTIONS

Heavy-ion-induced reactions are usually taken as reactions induced by projectiles heavier than an α particle. The span of projectiles studied is large, ranging from the light ions, C, O, and Ne, to the medium mass ions, such as S, Ar, Ca, and Kr, to the heavy projectiles, Xe, Au, and even U. Reactions induced by heavy ions have certain unique characteristics that distinguish them from other reactions. The wavelength of a heavy ion at an energy of 5 MeV/nucleon or more is small compared to the dimensions of the ion. As a result, the interactions of these ions can be described classically. The value of the angular momentum in these collisions is relatively large. For example, we can write

$$l_{max} = \frac{R}{\lambda}\left(1 - \frac{B}{E}\right)^{1/2}$$

For the reaction of 226 MeV $^{40}$Ar + $^{165}$Ho, we find that $l_{max} = 163\hbar$. This is relatively large compared to the angular momenta involved in nucleon-induced reactions. Lastly, quite often the product of the atomic numbers of the projectile and target is quite large ($>1000$), indicating the presence of large Coulomb forces acting in these collisions.

The study of heavy-ion-induced reactions is a forefront area of nuclear research. By using heavy-ion-induced reactions to make unusual nuclear species, one can explore various aspects of nuclear structure and dynamics "at its limits." Another major thrust is to study the dynamics and thermodynamics of the colliding nuclei.

In Figure 10.22, we show a cartoon of the various impact parameters and trajectories one might see in a heavy ion reaction. The most distant collisions lead to elastic scattering and Coulomb excitation. Coulomb excitation is the transfer of energy to the target nucleus via the long-range Coulomb interaction that excites the low-lying levels of the target nucleus. Grazing collisions lead to inelastic scattering and the onset of nucleon exchange. Head-on or near-head-on collisions lead to fusion of the reacting nuclei, which can lead to the formation of a compound

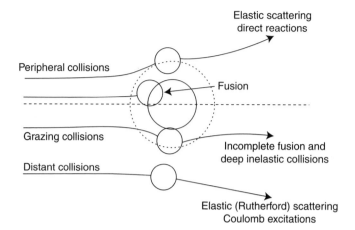

**Figure 10.22** Classification scheme of collisions based upon impact parameter. [From Hodgson et al. (1997).]

nucleus or a "quasi-fusion" reaction in which there is substantial mass and energy exchange between the projectile and target nuclei without the "true amnesia" characteristic of compound nucleus formation. For impact parameters between the grazing and head-on collisions, one observes a new type of nuclear reaction mechanism, *deep inelastic scattering*. In deep inelastic scattering, the colliding nuclei touch, partially amalgamate, exchange substantial amounts of energy and mass, rotate as a partially fused complex, and then reseparate under the influence of their mutual Coulomb repulsion before forming a compound nucleus.

The same range of reaction mechanisms can be depicted in terms of the angular momentum transfer associated with each of the mechanisms (Fig. 10.23). The most peripheral collisions lead to elastic scattering and thus have the highest values of the angular momentum transfer, $l$. The grazing collisions lead to inelastic scattering and nucleon exchange reactions, which are lumped together as "quasi-elastic" reactions. Solid-contact collisions lead to deep inelastic collisions, corresponding to intermediate values of $l$. The most head-on collisions correspond to compound nucleus formation and thus the lowest values of the angular momentum transfer, $l$. Slightly more peripheral collisions lead to the fusionlike or quasi-fusion reactions.

### 10.11.1 Coulomb Excitation

The potential energy due to the Coulomb interaction between a heavy ion and a nucleus can be written as:

$$E_C = (Z_1 Z_2 e^2 / R) \sim 1.2(Z_1 Z_2 / A^{1/3}) \text{ MeV}$$

Because of the strong, long-range electric field between projectile and target nuclei, it is possible for the incident heavy ion to excite the target nucleus

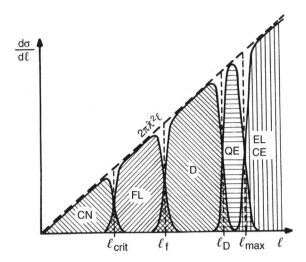

**Figure 10.23** Schematic illustration of the $l$ dependence of the partial cross section for compound nucleus (CN), fusionlike (FL), deep inelastic (D), quasi-elastic (QE), Coulomb excitation (CE), and elastic (EL) processes. [From Schroeder and Huizenga (1984, p. 242).]

electromagnetically. This is called *Coulomb excitation*, or *Coulex*. In particular, rotational bands in deformed target nuclei may be easily excited by the absorption of dipole photons. This technique is useful for studying the structure of such nuclei. Since the cross sections for these reactions are very large (involving long-range interactions with the nucleus), they are especially suitable for use when studying the structure of exotic nuclei with radioactive beams where the intensities are low. At relativistic bombarding energies, the strong electric field of the incident ion may be used to disintegrate the target nucleus (*electromagnetic dissociation*).

### 10.11.2 Elastic Scattering

In Figure 10.24, we compare elastic scattering for the collision of light nuclei with that observed in collisions involving much heavier nuclei. Collisions between the light nuclei show the characteristic Fraunhofer diffraction pattern discussed earlier in connection with the scattering of nucleons. The large Coulomb force associated with the heavier nucleus acts as a diverging lens, causing the diffraction pattern to be that of Fresnel diffraction. For the case of Fresnel diffraction, special emphasis is given to the point in the angular distribution of the scattered particle where the cross section is one-fourth that of the Rutherford scattering cross section. This "quarter-point angle" is taken to be the classical grazing angle. Note that the elastic scattering cross section equals the Rutherford scattering cross section at angles significantly less than the quarter-point angle. Since the Rutherford scattering cross section is readily calculable, experimentalists measure the number of elastically scattered particles at angles less than the quarter-point angle to deduce/monitor the beam intensity in heavy-ion-induced reaction studies.

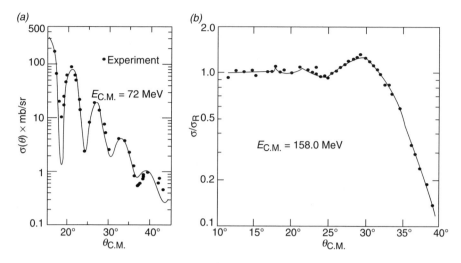

**Figure 10.24** Angular distribution for $^{12}C + ^{16}O$ elastic scattering, showing Fraunhofer diffraction and the elastic scattering of $^{16}O$ with $^{208}Pb$, which shows Fresnel diffraction. [From Valentin (1981).]

### 10.11.3 Fusion Reactions

In Figure 10.25, we show another representation of the difference between the various reaction mechanisms in terms of the energy needed to induce the reactions. The energy needed to bring the ions in contact and thus interact, is the *interaction barrier*, $V(R_{int})$. Formally, Bass (1980) has shown the reaction cross section can be expressed in terms of a one-dimensional interaction barrier as:

$$\sigma_R = \pi R_{int}^2 \left[ 1 - \frac{V(R_{int})}{E_{cm}} \right]$$

where the interaction radius is given as:

$$R_{int} = R_1 + R_2 + 3.2 \, \text{fm}$$

where the radius of the *i*th nucleus is

$$R_I = 1.12 A_i^{1/3} - 0.94 A_I^{-1/3} \, \text{fm}$$

and the interaction barrier is given as:

$$V(R_{int}) = 1.44 \frac{Z_1 Z_2}{R_{int}} - b \frac{R_1 R_2}{R_1 + R_2}$$

where $b \sim 1 \, \text{MeV/fm}$. (This expression is similar to that developed in Section 10.4 but represents a better description of the cross section.) The energy necessary to cause the ions to interpenetrate further to cause quasi-fusion is called the *extra*

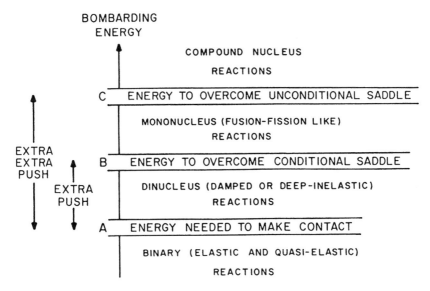

**Figure 10.25** Schematic illustration of the three critical energies and the four types of heavy ion nuclear reactions. [From S. Bjornholm and W. J. Swiatecki, *Nucl. Phys.* **A391**, 471 (1982).]

*push energy*. The energy necessary to cause the ions to truly fuse and forget their mode of formation is referred to as the *extra–extra push energy*.

The probability of fusion is a sensitive function of the product of the atomic numbers of the colliding ions. The abrupt decline of the fusion cross section as the Coulomb force between the ions increases is due to the emergence of the deep inelastic reaction mechanism. This decline and other features of the fusion cross section can be explained in terms of the potential between the colliding ions. This potential consists of three contributions, the Coulomb potential, the nuclear potential, and the centrifugal potential. The variation of this potential as a function of the angular momentum $l$ and radial separation is shown as Figure 10.26.

Note that at small values of the angular momentum, there is a pocket in the potential. Fusion occurs when the ions get trapped in this pocket. If they do not get trapped, they do not fuse. With high values of the Coulomb potential, there are few or no pockets in the potential for any value of $l$, thus no fusion occurs. For a given projectile energy and Coulomb potential, there is a value of the angular momentum above which there are no pockets in the potential (the critical value of the angular momentum) and thus no fusion occurs.

As shown in Figure 10.25, there is an $l$-dependent barrier to fusion that is the sum of the nuclear, Coulomb, and centrifugal potentials. This barrier is also a sensitive function of the relative deformation and orientation of the colliding ions. In Figure 10.27, we show the excitation function for fusion of $^{16}$O with various isotopes of Sm that span a wide range of deformations.

**284** NUCLEAR REACTIONS

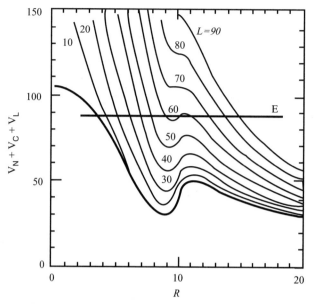

**Figure 10.26** Sum of the nuclear, Coulomb, and centrifugal potential for $^{18}$O + $^{120}$Sn as a function of radial distance for various values of the orbital angular momentum $l$.

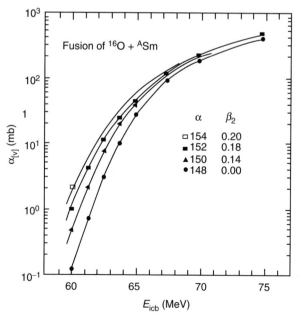

**Figure 10.27** Fusion cross sections for $^{16}$O + $^{A}$Sm.

One observes a significantly lower threshold and enhanced cross section for the case where the $^{16}$O ion interacts with deformed $^{154}$Sm compared to near-spherical $^{148}$Sm. This enhancement is the result of the lowering of the fusion barrier for the collision with the deformed nucleus due to the fact that the ions can contact at a larger value of $R$ resulting in a lower Coulomb component of the potential. Let us now consider what happens after the formation of a compound nucleus in a heavy-ion fusion reaction. In Figure 10.28, we show the predictions for the decay of the compound nuclei formed in the reaction of 147 MeV $^{40}$Ar with $^{124}$Sn to

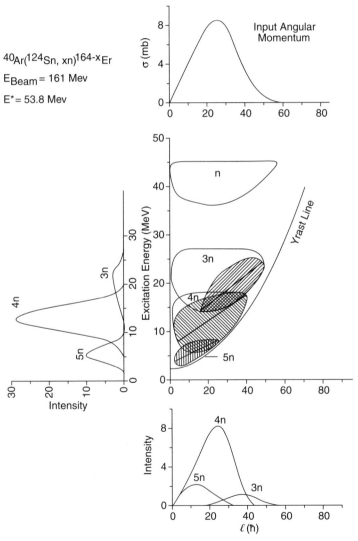

**Figure 10.28** Predicted decay of the $^{164}$Er compound nuclei formed in the reaction of $^{40}$Ar with $^{124}$Sn. [From Stokstad (1985).]

form $^{164}$Er at an excitation energy of 53.8 MeV. The angular momentum distribution in the compound nucleus shows population of states with $l = 0 - 60\hbar$. The excitation energy is such that energetically the preferred reaction channel involves the evaporation of four neutrons from the compound nucleus. As the compound nucleus evaporates neutrons, the angular momentum does not change dramatically since each neutron removes a relatively small amount of angular momentum. Eventually, the yrast line restricts the population of states in the E-J plane. The yrast line is the locus of the lowest lying state of a given angular momentum for a given J value. Below the yrast line for a given J, there are not states of the nucleus. (The word *yrast* is from the Old Norse for the "dizziest.") When the system reaches the yrast line, it decays by a cascade of γ rays. Heavy-ion reactions are thus a tool to excite levels of the highest spin in nuclei, allowing the study of nuclear structure at high angular momentum.

### 10.11.4 Deep Inelastic Scattering

Now let us turn our attention to the case of deep inelastic scattering. In the early 1970s, as part of a quest to form superheavy elements by the fusion of Ar and Kr ions with heavy target nuclei, a new nuclear reaction mechanism, deep inelastic scattering was discovered. For example, in the reaction of $^{84}$Kr with $^{209}$Bi (Fig. 10.29) instead of observing the fission of the completely fused nuclei (to form nuclei in the region denoted by the triangle), one observed projectile and targetlike nuclei and a new and unexpected group of fragments with masses near that of the target and projectile but with kinetic energies that were much lower than those expected from elastic or quasi-elastic scattering.

These nuclei appeared to be nuclei that had undergone an inelastic scattering that had resulted in the loss of a large amount of the incident projectile kinetic energy. Further measurement revealed this to be a general phenomenon in reactions where the product of the atomic numbers of the colliding ions was large (>2000). As described earlier, the ions come together, interpenetrate partially, exchange some mass, and charge with dissipation of a large amount of kinetic energy in a diffusion process, and then reseparate under the influence of their mutual Coulomb repulsion. The initial projectile energy is damped into the excitation energy of the projectile and targetlike fragments. As a consequence, the larger the kinetic energy loss, the broader the distribution of the final products becomes partly due to the evaporation of nucleons after the nuclei reseparate.

### 10.11.5 Incomplete Fusion

In the course of the fusion of the projectile and target nuclei, it is possible that one of the partners will emit a single nucleon or a nucleonic cluster prior to the formation of a completely fused system. Such processes are referred to as *preequilibrium emission* (in the case of nucleon emission) or *incomplete fusion* (in the case of cluster emission). As the projectile energy increases, these emission processes become more important and they generally dominate over fusion at projectile energies

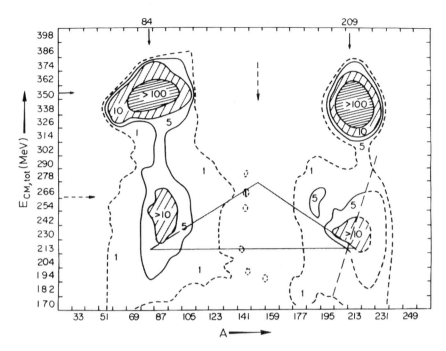

**Figure 10.29** Measurement of the product energy and mass distributions in the reaction of $^{84}$Kr with $^{209}$Bi. [From M. Lefort et al., Nucl. Phys. *A216*, 166 (1973).]

above 20 MeV/nucleon. As a consequence of these processes, the resulting product nucleus has a momentum that is reduced relative to complete fusion events. Measurement of the momentum transfer in the collision serves as a measure of the occurrence of these phenomena. In the spectra of emitted particles, a high-energy tail on the normal evaporation spectrum is another signature of preequilibrium emission.

### 10.11.6 Reactions Induced by Radioactive Projectiles

There are a few hundred stable nuclei but several thousand nuclei that are radioactive and have experimentally useful lifetimes. Since 1990, one of the fastest growing areas of research in nuclear science has been the study of nuclear reactions induced by radioactive projectiles. Using either ISOL (*I*sotope *S*eparator *O*n-*L*ine) or PF (*P*rojectile *F*ragmentation) techniques, several hundred new radioactive nuclear beams have become available (see Chapter 14).

The principal attraction in these studies is the ability to form reaction products or reaction intermediates with unusual $N/Z$ ratios. By starting with nuclei that are either very proton rich or very neutron rich, new regions of nuclei can be reached and their properties studied. At higher energies, the unusual isospin of the intermediate species allows one to determine the effect of isospin on the properties of highly excited nuclear matter. Occasionally, the radioactive beams themselves have unusual structure, for example, $^{11}$Li, and their properties and reactions are of interest.

## 10.12 HIGH-ENERGY NUCLEAR REACTIONS

A nuclear reaction is said to be a *low-energy reaction* if the projectile energy is $\leq 10$ MeV/nucleon that is, near the Coulomb barrier. A nuclear reaction is termed a *high-energy reaction* if the projectile energy is much greater than the Coulomb barrier and approaching the rest mass of the nucleon, for example, 400 MeV/nucleon. (Not surprisingly the reactions induced by 20–250 MeV/nucleon projectiles are called *intermediate-energy reactions*.)

What distinguishes low- and high-energy reactions? In low-energy nuclear collisions, the nucleons of the projectile interact with the average or mean nuclear force field associated with the entire target nucleus. In a high-energy reaction, the nucleons of the projectile interact with the nucleons of the target nucleus individually, as nucleon–nucleon collisions. To see why this might occur, let us compute the de Broglie wavelength of a 10-MeV proton and a 1000-MeV proton. We get $\lambda_{10 \text{ MeV}} = 9.0$ fm and $\lambda_{1000 \text{ MeV}} = 0.73$ fm. The average spacing between nucleons in a nucleus is $\sim 1.2$ fm. Thus, we conclude that at low energies the projectile nucleons can interact with several nucleons at once, whereas at high energies collisions occur between pairs of nucleons.

### 10.12.1 Spallation/Fragmentation

What type of reactions do we observe at high energies? Because we are dealing with processes dominated by nucleon–nucleon collisions, we do not expect any significant amount of compound nucleus formation. Instead most reactions should be direct reactions taking place on a short time scale. In Figure 10.30, we show a typical distribution of the masses of the residual nuclei from the interaction of GeV protons with a heavy nucleus, such as $^{209}$Bi.

At the highest energy, one observes a continuous distribution of product masses ranging from the target mass to very low values of $A$. Three regions can be identified in the yield distributions. One region is centered around $A_{\text{target}}/2$ ($A = 50-140$) and consists of the products of the fission of a targetlike nucleus. For larger $A$ values $[A_{\text{fragment}} \geq \frac{2}{3} A_{\text{target}}]$ the products are thought to arise from a direct reaction process termed *spallation*. The incident proton knocks out several nucleons in a series of two-body collisions, leaving behind a highly excited heavy nucleus. This nucleus decays by the evaporation of charged particles and neutrons, forming a continuous distribution of products ranging downward in $A$ from the target mass number. The term *spallation* was given to this phenomenon by one of us (GTS) after consultation with a professor of English who assured him that the verb "to spall" was a very appropriate term for this phenomenon. For the lowest mass numbers $[A_{\text{fragment}} \leq \frac{1}{3} A_{\text{target}}]$ one observes another group of fragments that are termed to be *intermediate mass fragments* (IMFs). These fragments are thought to

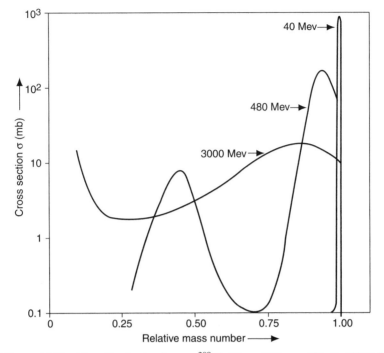

**Figure 10.30** Mass distribution for p + $^{209}$Bi. [From Miller and Hudis (1959).]

arise from the very highly excited remnants of the most head-on collisions by either sequential particle emission or a nuclear disintegration process called multifragmentation.

The course of a reaction at high energies is different than one occurring at lower energies. As mentioned earlier, collisions occur between pairs of nucleons rather than having one nucleon collide with several nucleons simultaneously. The cross section for nucleon–nucleon scattering varies inversely with projectile energy. At the highest energies, this cross section may become so small that some nucleons will pass through the nucleus without undergoing any collisions, that is, the nucleus appears to be *transparent*.

In this regard, a useful quantitative measure of the number of collisions a nucleon undergoes in traversing the nucleus is the *mean free path*, $\Lambda$. Formally, we have

$$\Lambda = 1/\rho\sigma$$

where $\sigma$ is the average nucleon–nucleon scattering cross section ($\sim$30 millibarns) and $\rho$ is the nuclear density ($\sim 10^{38}$ nucleons/cm$^3$). Thus, the mean free path is $\sim 3 \times 10^{-13}$ cm. In each collision, the kinetic energy imparted to the struck nucleon is $\sim$25 MeV, and thus the struck nucleon may collide with other nucleons, generating a *cascade* of struck particles (see Fig. 10.31). If the energy of the incident

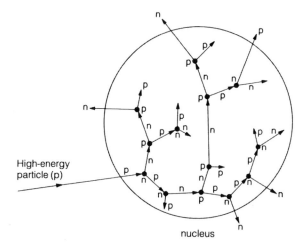

**Figure 10.31** Schematic view of nuclear cascade. [From Lieser (1997).]

nucleon exceeds ~300 MeV, then it is possible to generate $\pi$ mesons in these collisions, which, in turn, can interact with other nucleons. A typical time scale for the cascade is $10^{-22}$ s. The result of this *intranuclear cascade* is an excited nucleus, which may decay by preequilibrium emission of particles, evaporation of nucleons, sequential emission of IMFs, or disintegration into multiple fragments.

In the mid-1970s, at the Bevalac in Berkeley, heavy ion reactions at very high energies (0.250–2.1 GeV/nucleon) were studied. At these high projectile energies, a number of observations were interpreted in terms of a simple geometric model referred to as the abrasion-ablation or fireball model. (Fig. 10.32). In this model the incoming projectile sheared off a sector of the target (corresponding to the overlap region of the projectile and target nucleus—the "abrasion" step). The nonoverlapping regions of the target and projectile nuclei were assumed to be left essentially undisturbed and unheated, the so-called "spectators" to the collision. The hot overlap region (the "participants") formed a "fireball" that decayed with the release of nucleons and fragments. The wounded target nucleus was expected to have a region of extra surface area exposed by the projectile cut through it. Associated with this extra surface area is an excitation energy corresponding to the surface area term of the semiempirical mass equation of about 1 MeV per excess $fm^2$ of surface area. As the nucleus relaxes, this excess surface energy becomes available as excitation energy and results in the normal emission of nucleons and fragments (the "ablation" step).

The use of this simple model for high-energy nucleus–nucleus collisions has resulted in a general categorization of energetic nucleus–nucleus collisions as either "peripheral" or "central." In peripheral reactions, one has large impact parameters and small momentum transfer. Such reactions, which produce surviving large spectators, are referred to as *fragmentation* reactions. Such reactions are of interest in the production of new radioactive nuclei and radioactive beams.

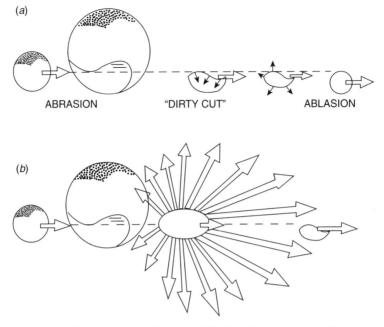

**Figure 10.32** Abrasion–ablation model of relativistic nuclear collisions.

## 10.12.2 Multifragmentation

In central collisions one has smaller impact parameters and larger energy and momentum transfer. In central nucleus–nucleus collisions at intermediate energies (20–200 MeV/nucleon), large values of the nuclear excitation energy (>1000 MeV) and temperature (>10 MeV) may be achieved for short periods of time ($10^{-22}$ s). Nuclei at these high excitation energies can decay by the emission of complex or intermediate mass fragments. (An IMF is defined as a reaction product whose mass is greater than 4 and less than that of a fission fragment.) *Multifragmentation* occurs when several IMFs are produced in a reaction. This can be the result of sequential binary processes, "statistical" decay into many fragments (described by passage through a transition state or the establishment of statistical equilibrium among fragments in a critical volume), or a dynamical process in which the system evolves into regions of volume and surface instabilities leading to multifragment production.

To investigate these phenomena, it is necessary to measure as many of the emitted fragments and particles from a reaction. As a result, various multidetector arrays have been constructed and used. Quite often these arrays consist of several hundred individual detectors to detect the emitted IMFs, light charged particles, neutrons, target fragments, and the like. As a consequence of the high granularity of these detectors, the analysis of the experimental data is time consuming and difficult. Nonetheless, several interesting developments have occurred in recent years.

One theory to describe multifragmentation postulates the formation of a hot nuclear vapor during the reaction, which subsequently condenses into droplets of liquid (IMFs) somewhere near the critical temperature. First postulated to occur in the interaction of GeV protons with Xe, recent experiments with heavy ions have resulted in deduced temperatures and excitation energies (Fig. 10.33) that resemble calculations for a liquid–gas phase transition. This somewhat speculative "caloric curve" shows an initial rise in temperature with excitation energy typical of heating a liquid, followed by a flat region (the phase transition), followed by a region corresponding to heating a vapor.

Finally, there has been an extended debate and discussion of the relative role of statistical and dynamical factors in multifragmentation. The debate has focused on the observation that the data from several reactions could be plotted such that the probability of emitting multiple fragments, $p$, could be expressed in a form, $p \exp(-B/T)$, suggestive of the dependence of the fragment emission probabilities upon a single fragment emission barrier, $B$, a feature suggesting the importance of statistical factors. Others have criticized this observation. The criticisms have focused on the details of the correlation and evidence for dynamic effects.

### 10.12.3 Quark–Gluon Plasma

The primary thrust of studies of central collisions of massive nuclei at ultrarelativistic energies ($>5$ GeV/nucleon) is to create and observe a new form of matter, the quark–gluon plasma (QGP). The modern theory of the strong interaction, quantum chromodynamics, predicts that while quarks and gluons will be confined within a nucleonic "bag" under normal conditions, deconfinement will occur at sufficiently high energies and densities. This phase transition (from normal nuclear matter to the QGP) is predicted to occur at energy densities of

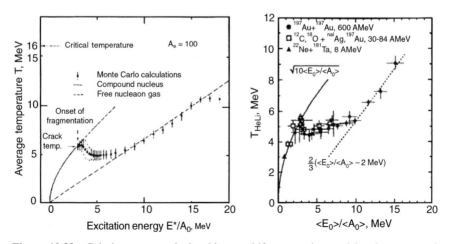

**Figure 10.33** Caloric curve as calculated by a multifragmentation model and as measured.

1–3 GeV/fm³, which can be achieved in collisions at cm energies of 17 GeV/nucleon.

The experimental signatures of a phase transition include: (a) suppression of production of the heavy vector mesons $J/\Psi$ and $\Psi'$ and the upsilon states, (b) the creation of a large number of ss quark–antiquark pairs, and (c) the momentum spectra, abundance, and direction of emission of di-lepton pairs. The first phase experiments in this field have been carried out. Energy densities of $\sim 2$ GeV/fm³ were created. Strong $J/\Psi$ suppression has been observed relative to p-A collisions along with an increase in strangeness production.

## PROBLEMS

1. Consider the reaction of $^{16}$O with $^{64}$Ni at a cm energy of 48 MeV. What is the lab kinetic energy of the $^{16}$O? What is the Coulomb barrier for the reaction? What is the total reaction cross section? What is the maximum angular momentum brought in by the $^{16}$O projectile?

2. One reaction proposed for the synthesis of element 110 is the reaction of $^{59}$Co with $^{209}$Bi at a laboratory energy of 295 MeV. Calculate the expected total reaction cross section.

3. Define or describe the following terms or phenomena: direct reaction, compound nucleus, and stripping reaction.

4. A piece of Au that is 1 mm thick is bombarded for 15 h by a slow neutron beam of intensity $10^6$/s. How many disintegrations per second of $^{198}$Au are present in the sample 24 h after the end of the bombardment? $\sigma(n,\gamma) = 98.8$ barns, $t_{1/2}$ ($^{198}$Au) = 2.7 d.

5. What was the rate of production, in atoms per second, of $^{128}$I during a constant 1-h cyclotron (neutron) irradiation of an iodine sample if the sample was found to contain 2.00 mCi of $^{128}$I activity at 15 min after end of the irradiation?

6. What is the excitation energy of the $^{116}$Sb compound nuclei formed by the bombardment of $^{103}$Rh with 50 MeV $^{13}$C ions?

7. Neutrons evaporated from a compound nucleus have an average kinetic energy of $\sim 2T$, where $T$ is the nuclear temperature of the residual nucleus. What is the optimum bombarding energy for the production of $^{66}$Ga via the $^{65}$Cu($\alpha$, 3n) reaction if the average nuclear temperature is 1.6 MeV?

8. A 100-mg/cm²-thick natural Zr target is bombarded with a beam of 11-MeV protons for one hour (beam current = 25 μA). The $^{95}$Nb$^m$ from the reaction $^{96}$Zr(p, 2n) was isolated chemically (100% yield), and the $k$ x-rays resulting from the internal conversion decay of $^{95}$Nb$^m$ were counted. In a 2-h count beginning 20 h after the end of bombardment, 1000 counts were observed in the Nb $K_\alpha$ x-ray peak. Given the $^{95}$Nb decay scheme shown below and the data given

below, calculate the cross section for the $^{95}$Zr(p, 2n) $^{95}$Nb$^m$ reaction. Fluorescence yield = 0.7, efficiency of detection of the $K$ x-ray is $10^{-3}$, $\alpha_k = 2.21$.

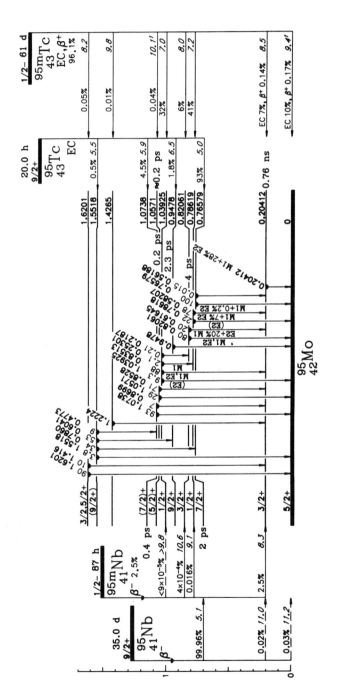

PROBLEMS 295

9. What is the number of $^{60}$Co atoms produced in a 10-mg sample of cobalt metal exposed for 2 min to a thermal neutron flux of $2 \times 10^{13}$ n/cm$^2$-s in a reactor? The cross section for producing 10.5-min $^{60}$Co$^m$ is 16 barns, while the cross section for producing 5.3 y $^{60}$Co is 20 barns. What is the disintegration rate of the cobalt sample 4 h after the end of the irradiation?

10. Consider the $^{48}$Ca + $^{248}$Cm reaction where the lab energy of the $^{48}$Ca is 300 MeV. What is the excitation energy of the putative compound nucleus $^{296}$116? What is the expected total reaction cross section?

11. Consider the reaction of 30 MeV/nucleon $^{129}$Xe with $^{238}$U. What is the energy of the elastically scattered $^{129}$Xe detected at 10° in the lab system?

12. Consider the $^{40}$Ca(d, p) reaction. What would be the most probable angle to detect the protons leading to the first excited state $\frac{3}{2}^-$ of $^{41}$Ca? What would be their energy? Assume the energy of the incident deuteron beam was 21.0 MeV.

13. Consider you want to make $^{18}$F for use in PET studies. What would be the maximum specific activity (dpm/g F) of the $^{18}$F made by irradiating 1.0 g of KF in a flux of $10^{10}$ fast neutrons/cm$^2$-s. You may assume the $^{18}$F(n, 2n) cross section is 300 millibarns. Consider you want to produce the $^{18}$F carrier free (i.e., with no stable fluorine present). Devise a synthetic scheme for producing the carrier-free $^{18}$F. Defend your choice of reactions.

14. Consider the nuclide $^{99}$Tc$^m$ that is the daughter of $^{99}$Mo. In the United States most diagnostic procedures involving radioactivity involve $^{99}$Tc$^m$. Explain how you would produce $^{99}$Mo (the 66.0-h parent of 6.0-h $^{99}$Tc$^m$). Consider two choices, production of $^{99}$Mo as a fission product or via the $^{98}$Mo(n, $\gamma$) reaction.

15. Calculate the activity of $^{254}$No ($t_{1/2} = 55$ s) present 5 min after a 10-min irradiation of a 0.001-in.-thick $^{208}$Pb foil by $^{48}$Ca projectiles ($\phi = 6.28 \times 10^{12}$ particles/second). Assume $\sigma(^{48}$Ca, 2n) is $3 \times 10^{-30}$ cm$^2$.

16. Consider the reaction $^{12}$C($\alpha$, n) where the laboratory energy of the incident projectile is 14.6 MeV. What is the excitation energy of the compound nucleus? The reaction cross section is 25 millibars. Assuming a carbon target thickness of 0.10 mg/cm$^2$ and a beam current of 25 nA, compute the $^{15}$O activity after a 4-min irradiation.

17. The cross section for the $^{60}$Ni($\alpha$, pn) reaction is 0.9 barn for 32-MeV $\alpha$ particles. Calculate the number of disintegrations per minute of $^{62}$Cu at 15 min after a 15-min bombardment of a 50-mg/cm$^2$ foil of $^{60}$Ni with 10 $\mu$A of 32 MeV $\alpha$ particles.

18. Consider the reaction $^{29}$Si($^{18}$O, p2n) which populates the metastable and ground states of $^{44}$Sc. Using the decay scheme shown below, and the fact that at end of bombardment (EOB) one observed 1000 photons/second of energy 271.2 keV and 1000 photons/second of energy 1157.0 keV, calculate the ratio of the cross

section for the production of $^{44}Sc^m$, $\sigma_m$, to the cross section for the production of $^{44}Sc$, $\sigma_g$. Neglect any decay of $^{44}Sc^m$ to $^{44}Sc$ during the irradiation and assume the length of the irradiation was 6 h.

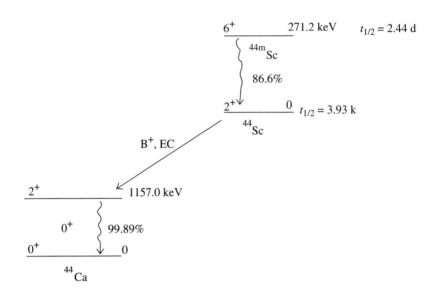

## REFERENCES

Bass, R. *Nuclear Reactions with Heavy Ions*, Springer, Berlin, 1980.

Bertsch, G. F. and E. Kashy. *Am. J. Phys.* **61**, 858 (1993).

Blanpied, G. S. et al. *Phys. Rev. C* **18**, 1436 (1978).

Ehmann, W. D. and D. E. Vance. *Radiochemistry and Nuclear Methods of Analysis*, Wiley, New York, 1991.

Fultz, C. et al. *Phys. Rev.* **127**, 1273 (1962).

Ghoshal, X. *Phys. Rev.* **80**, 939 (1980).

Goldhaber, M. and E. Teller. *Phys. Rev.* **74**, 1046 (1948).

Lefort, M. *Nuclear Chemistry*, Van Nostrand, Princeton, 1968.

Lieser, K. H. *Nuclear and Radiochemistry*, VCH, Wertheim, 1997.

Meyerhof, W. E. *Elements of Nuclear Physics*, McGraw-Hill, New York, 1967.

Miller, J. M. and J. Hudis. *Ann. Rev. Nucl. Sci.* **9**, 159 (1959).

Satchler, G. R. *Introduction of Nuclear Reactions*, 2nd ed., Oxford, New York, 1990.

Schroeder, W. U. and J. R. Huizenga. In D. A. Bromley, Ed., *Treatise on Heavy Ion Science*, Vol. 2, Plenum, New York, 1984.

Stokstad, R. In D. A. Bromley, Ed., *Treatise on Heavy Ion Science*, Vol. 3, Plenum, New York, 1985.

Valentin, L. *Subatomic Physics: Nuclei and Particles*, Vol. II, North-Holland, Amsterdam, 1981.

Weidner, R. T. and R. L. Sells. Elementary Modern Physics, Boston, Allyn and Bacon, Boston, 1973.

Weisskopf, V. F. *Rev. Mod. Phys.* **29**, 174 (1959).

## BIBLIOGRAPHY

General Electric, *Nuclides and Isotopes*, 6th ed., KAPL, 2002.

Hodgson, P. E., E. Gladioli, and E. Gladioli-Erba. *Introductory Nuclear Physics*, Clarendon, Oxford, 1997.

Jelley, N. A. *Fundamentals of Nuclear Physics*, Cambridge, Cambridge, 1990.

Krane, K. S. *Introductory Nuclear Physics*, Wiley, New York, 1988.

Lilley, J. S. *Nuclear Physics*, Wiley, Chichester, 2001.

Wong, S. S. M. *Nuclear Physics*, 2nd ed., Wiley, New York, 1998.

# CHAPTER 11

# FISSION

## 11.1 INTRODUCTION

Fission has a unique importance among nuclear reactions. Apart from the nuclear reactions that drive the sun, no other nuclear reaction has had such a profound impact on the affairs of humans. The discovery of fission, and the developments that proceeded from it, have altered the world forever and have impinged on the consciousness of every literate human being. The exploitation of nuclear energy, which followed the discovery of fission, particularly in weapons of mass destruction, has been of profound importance to humankind.

Chemists have played an important role in the study of fission. Fission was discovered by the chemists Otto Hahn and Fritz Strassman in 1938 (Hahn and Strassman, 1938). By painstakingly difficult chemical separations, they were able to show that the neutron irradiation of uranium led, not to many new elements as had been thought, but to products such as barium, lanthanum, and the like. The uranium nucleus had been split! That conclusion caused Hahn and Strassman (1938) much concern as they wrote, "As 'nuclear chemists' working very close to the field of physics, we cannot bring ourselves yet to take such a drastic step (to conclude that uranium had fissioned), which goes against all previous experience in nuclear physics." (Am. J. Phys. *32*, 15 (1964)). Nuclear chemists have continued their role in studying fission, first using chemical techniques and, more recently, using physical techniques.

*Modern Nuclear Chemistry*, by W.D. Loveland, D.J. Morrissey, and G.T. Seaborg
Copyright © 2006 John Wiley & Sons, Inc.

Knowledge of fission and its consequences is important for the nuclear power industry and the related fields of nuclear waste management and environmental cleanup. From the point of view of basic research, fission is interesting in its own right as a large-scale collective motion of the nucleus, as an important exit channel for many nuclear reactions, and as a source of neutron-rich nuclei for nuclear structure studies and use as radioactive beams.

The reader should be cautioned that understanding the fission process represents a very difficult problem. Some of the best minds in chemistry and physics have worked on the problem since the discovery of fission. Yet, while we understand many aspects of the fission process, there is no overall theoretical framework that gives a satisfactory account of the basic observations.

In Figure 11.1, we show a schematic view of the fission process. A nucleus with some equilibrium deformation absorbs energy, becoming excited, and deforms to a configuration known as the *transition state* or *saddle point* configuration. As it deforms, the nuclear Coulomb energy decreases (as the average distance between the nuclear protons increases) while the nuclear surface energy increases (as the nuclear surface area increases). At the saddle point, the rate of change of the Coulomb energy is equal to the rate of change of the nuclear surface energy. The formation and decay of this transition state nucleus is the rate-determining step in the fission process and corresponds to the passage over an activation energy barrier to the reaction. If the nucleus deforms beyond this point, it is irretrievably committed to fission. When this happens, then in a very short time the neck between the nascent fragments disappears and the nucleus divides into two fragments at the *scission point*. At the scission point, one has two highly charged, deformed fragments in contact with each other. The large Coulomb repulsion between the two fragments accelerates them to 90% of their final kinetic energy within $10^{-20}$ s. As these accelerated primary fragments move away from one another, they contract to more spherical shapes, converting their potential energy of deformation into internal excitation energy, that is, they get "hot." This excitation energy is removed by the emission of the "prompt" neutrons from the fully accelerated fragments and then, in competition with the last neutrons to be emitted, the nucleus emits γ rays. Finally, on a longer time scale the neutron-rich fragments emit β⁻ particles. Occasionally, one of these β decays populates a high-lying excited state of a daughter that is unstable with respect to neutron emission, and the resulting emitted neutrons are called the "delayed" neutrons. Note that this schematic view conflicts with some presentations of fission in elementary textbooks. For example, if the neutrons are emitted primarily from the fully accelerated fragments, their spatial distribution is along the direction of motion of the fragments. They do not emerge randomly from the nucleus as many artists' conceptions of fission depict. Also note that the energy release in fission is primarily in the form of the kinetic energies of the fragments not in the neutrons, photons, or other emitted particles. This energy is the "mass-energy" released in fission due to the increased stability of the fission fragments.

Because of the large amount of information available about fission, it is beyond the scope of this chapter to present a complete treatment of fission research. We shall

11.1 INTRODUCTION   301

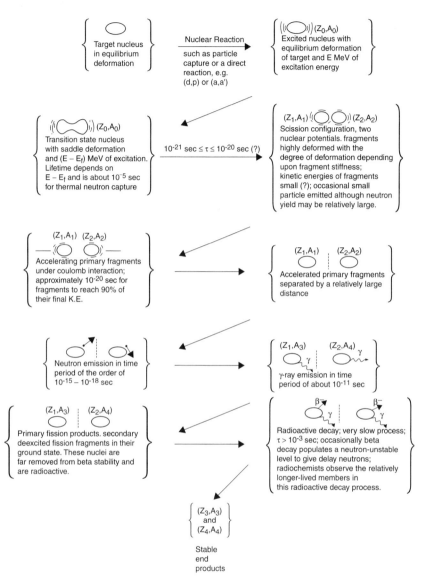

**Figure 11.1** Schematic view of the fission process. (From J. E. Gindler and J. R. Huizenga, "Nuclear Fission" in Nuclear Chemistry, Vol. II, L. Yaffe, Ed. Copyright © 1968 Academic Press, Reprinted by permission of Elsevier.)

attempt to emphasize the fundamental aspects of the subject. The reader is referred to one of the excellent monographs or reviews of fission (Vandenbosch and Huizenga, 1973; Wagemans, 1991; Oganessian and Lazarev, 1985; Hoffman et al., 1996) for further information.

## 11.2 PROBABILITY OF FISSION

### 11.2.1 Liquid Drop Model

Figure 11.1 suggests fission proceeds in two steps, the ascent to the saddle point and the passage through the scission point. We shall present our discussion of fission from this point of view. We shall assert that like chemical reactions, the reaction probability is determined by the passage through the transition state. We shall also assert, more controversially, that the distribution of fission product energies, masses, and so forth is determined at or near the scission point.

Let us begin with a discussion of the probability of fission. For the first approximation to the estimation of the fission barrier, we shall use the liquid drop model (Chapter 2). We can parameterize the small nonequilibrium deformations, that is, elongations, of the nuclear surface as

$$R(\vartheta) = R_0[1 + \alpha_2 P_2(\cos \vartheta)] \tag{11.1}$$

where $\alpha_2$ is the quadrupole distortion parameter $[=(\frac{5}{4}\pi)^{1/2}\beta_2]$ and $P_2$ is the second-order Legendre polynomial. For small distortions, the surface, $E_s$, and Coulomb, $E_C$, energies are given by

$$E_s = E_s^0\left(1 + \frac{2}{5}\alpha_2^2\right)$$

$$E_C = E_C^0\left(1 - \frac{1}{5}\alpha_2^2\right) \tag{11.2}$$

where $E_s^0$ and $E_C^0$ are the surface and Coulomb energies of the undistorted spherical drops. When the changes in the Coulomb and surface energies ($\Delta E_C = E_C^0 - E_C$, $\Delta E_s = E_s - E_s^0$) are equal, the nucleus becomes spontaneously unstable with respect to fission. At that point we find that

$$\frac{E_c^0}{2E_s^0} = 1 \tag{11.3}$$

Thus, it is natural to express the fissionability of nuclei in terms of a parameter $x$, that is, this energy ratio, and is called the *fissionability parameter*. Thus

$$x = \frac{E_C^0}{2E_s^0} = \frac{1}{2}\left(\frac{\text{Coulomb energy of a charged sphere}}{\text{surface energy of the sphere}}\right) \tag{11.4}$$

We can approximate the Coulomb and surface energies of a uniformly charged sphere by the following expressions:

$$E_C^0 = \frac{3}{5}\frac{Z^2 e^2}{R_0 A^{1/3}} = \left(a_C \frac{Z^2}{A^{1/3}}\right) \tag{11.5}$$

where $a_C = 3e^2/5R_0$ and

$$E_S^0 = 4\pi R_0^2 S A^{2/3} = a_s A^{2/3} \tag{11.6}$$

where $S$ is the surface tension per unit area and $a_s = 4\pi R_0^2 S$. Then the equation for $x$ becomes

$$x = \left(\frac{a_C}{2a_s}\right)\left(\frac{Z^2}{A}\right) = \frac{(Z^2/A)}{(Z^2/A)_{\text{critical}}} \tag{11.7}$$

where the ratio of the constants $(a_C/2a_s)^{-1}$ is referred to as $(Z^2/A)_{\text{critical}}$. The fissility of a given nucleus can be categorized relative to $(Z^2/A)_{\text{critical}}$. More sophisticated treatments of the fissionability of nuclei show that $(Z^2/A)_{\text{critical}}$ varies slightly from nucleus to nucleus (due to the isospin asymmetry) and is given by

$$\left(\frac{Z^2}{A}\right)_{\text{critical}} = 50.883\left[1 - 1.7826\left(\frac{N-Z}{A}\right)^2\right] \tag{11.8}$$

The parameters $Z^2/A$ and $x$ provide measures of the relative fissionability of nuclei. The greater the value of these parameters, the more "fissionable" the nuclei are. Very fissionable nuclei, such as $^{239}$Pu, have $Z^2/A$ values of 36.97 while less fissionable nuclei such as $^{209}$Bi have $Z^2/A$ values of 32.96. Recall that the $Z^2/A$ factor is simply proportional to the ratio of the disruptive Coulomb energy ($\propto Z^2/A^{1/3}$) to the cohesive surface (nuclear) energy ($\propto A^{2/3}$).

Note that the parameter $(Z^2/A)_{\text{critical}}$ is the ratio of two empirical constants related to the strength of the Coulomb and surface (nuclear) forces. If we take the view that the limit to the size of the periodic table is given by the point at which the heaviest nuclei spontaneously fission

$$\frac{E_C^0}{2E_S^0} = 1 \tag{11.9}$$

We can rearrange these equations to find the value of the atomic number $Z$ at which this occurs. Thus, $Z_{\text{limit}}$ is given by

$$Z_{\text{limit}}^2 = 2(a_s/a_C)A_{\text{limit}} \tag{11.10}$$

If we remember that the neutron/proton ratio in heavy nuclei is about 1.5, then $Z_{\text{limit}}$ will be about $5(a_s/a_C)$. Thus, the upper bound to the periodic table is given as a ratio of two constants relating to the strength of the nuclear and Coulomb forces. The ratio $a_s/a_C$ is about 20–25, and thus we expect about 100–125 chemical elements.

For all stable nuclei, $x$ must be less than 1. In that case, the total deformation energy of nuclei undergoing fission will increase by an amount, $(\frac{1}{5})\alpha_2^2(2E_S^0 - E_C^0)$, as the nucleus deforms toward fission. This increase in potential energy can be thought of as an activation energy barrier for the reaction. Eventually if the deformation proceeds far enough, the decrease in Coulomb energy will overwhelm the increase in surface energy and the deformation energy will decrease. (In this case, the simple deformation energy formulas used so far in our discussion become inaccurate, and more complicated formulas must be used.) One can appreciate the difficulty of these calculations by a simple example. The liquid drop fission barrier for $^{238}$U is 4.8 MeV. Equating this to $(\frac{1}{5})\alpha_2^2(2E_s^0 - E_C^0)$ and calculating values of 983 MeV for $E_C^0$ and 695 MeV for $E_s^0$ for $^{238}$U, one can calculate a value of the deformaltion parameter $\alpha_2$ of 0.243, which gives us changes in the surface and Coulomb energies ($\Delta E_s^0$ and $\Delta E_C^0$) of 16.4 and 11.6 MeV, respectively. Thus, one sees that the resulting fission barrier heights are calculated as the small differences between two large numbers that are difficult to determine. Modern calculations of the potential energy of deformation for the liquid drop model involve many deformation coordinates (not just the $\alpha_2$ used above) and represent major computational tasks.

### 11.2.2 Shell Corrections

Figure 11.2 shows some of the basic features of fission barriers. In Figure 11.2, the fission barriers as estimated from the liquid drop model for a range of actinide nuclei are shown. The fission barrier height decreases, and the maximum (saddle point)

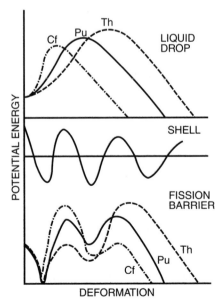

**Figure 11.2** Qualitative features of the fission barriers for actinide nuclei. (From H. C. Britt, "Fission Properties of the Actinides" in Actinides in Perspective, N. Edelstein, Ed. Copyright © 1982 Pergamon Press, Ltd. Reprinted by permission of H. C. Britt.)

moves to smaller deformations as $Z^2/A$ increases. In the lighter nuclei the saddle point and scission point configurations are more similar, that is, have a similar deformation, than in the heavier nuclei.

As we learned in Chapter 2, it is necessary to include shell effects in the liquid drop model if we want to get reasonable values for nuclear masses. Similarly, we must devise a way to include these same shell effects into the liquid drop model description of the effect of deforming nuclei. Strutinsky (1967) proposed such a method to calculate these "shell corrections" (and also corrections for nuclear pairing) to the liquid drop model. In this method, the total energy of the nucleus is taken as the sum of a liquid drop model (LDM) energy, $E_{\rm LDM}$, and the shell ($\delta S$) and pairing ($\delta P$) corrections to this energy,

$$E = E_{\rm LDM} + \sum_{\rm p,n}(\delta S + \delta P) \qquad (11.11)$$

The shell corrections, just like the liquid drop (LD) energy, are functions of the nuclear deformation. The shell corrections tend to lower the ground-state masses of spherical nuclei with magic or near-magic numbers of neutrons and protons. They also tend to lower the ground-state mass of midshell nuclei at some finite deformation ($\beta_2 \approx 0.3$), thus accounting for the deformed nature of the actinides. A large minimum in the shell correction energies occurs when the ratios of the major/minor nuclear axes are in the ratio of small whole numbers, as 3 : 2 or 2 : 1 (corresponding to level bunchings in the single-particle levels). The qualitative result of combining these deformation-dependent shell corrections with the liquid drop barriers is shown in Figure 11.2. The stable ground-state shape is predicted to have some finite deformation ($\beta_2 \sim 0.2$) rather than zero deformation (a sphere), and a secondary minimum in the barrier appears at $\beta_2 \sim 0.6$ (axes ratio of 2 : 1). In the heaviest nuclei ($Z \geq 106$), where the liquid drop fission barriers are very small or nonexistent, the fission barrier heights are enhanced relative to the liquid drop model due primarily to a lowering of the ground-state mass by shell corrections. Without these shell effects, the heaviest nuclei could not be observed as they would decay by spontaneous fission on a time scale much shorter than we can observe ($t_{1/2} < \mu$s).

Notice that this combination of macroscopic (LD) and microscopic (shell) effects predicts that for nuclei in the uranium–plutonium region, a double-humped fission barrier with equal barrier heights and a deep secondary minimum will occur. For heavier nuclei, like californium, the first barrier is predicted to be much larger than the second barrier, and passage over this first barrier is the rate determining step. In effect, these heavy nuclei ($Z \geq 100$) behave as though they have a high thin single barrier to fission. For the lighter nuclei (radium, thorium), the predicted barrier shape is triple-humped in many cases. The reader should be aware that the situation is even more complicated than this cursory description would indicate as considerations of the nuclear shapes make it clear that these fission barriers are multidimensional in character with a complicated dependence on asymmetric and symmetric deformations. In general, there is ample experimental and theoretical evidence that the lowest energy path in the fission process corresponds to having the nucleus, initially in an axially symmetric and mass (reflection) symmetric shape, pass over the first maximum in the fission barrier with an axially asymmetric but mass symmetric shape and then to pass over the second maximum in the barrier with an axially symmetric but mass (reflection) asymmetric shape. Because of the complicated multidimensional character of the fission process, there are no simple formulas for the fission barrier heights. However, the reader can find (Vandenbosch and Huizenga, 1973; Wagemans, 1991) extensive tabulations of experimental characterizations of the fission barrier heights for various nuclei.

Nuclei can be trapped in the secondary minimum of the fission barrier. Such trapped nuclei will experience a significant hindrance of their $\gamma$-ray decay back to the ground state (because of the large shape change involved) and an enhancement of their decay by spontaneous fission (due to the "thinner" barrier they would have to penetrate.) Such nuclei are called *spontaneously fissioning isomers*, and they were first observed in 1962 and are discussed below. They are members of a general class of nuclei, called *superdeformed nuclei*, that have shapes with axes ratios of 2:1. These nuclei are all trapped in a pocket in the potential energy surface due to a shell effect at this deformation.

### 11.2.3 Spontaneous Fission

In 1940 Petrzhak and Flerov discovered that $^{238}$U could decay by spontaneously fissioning into two large fragments (with a probability that was $5 \times 10^{-7}$ of that of undergoing $\alpha$ decay). Over 100 examples of this decay mode have been found since then. Spontaneous fission is a rare decay mode in the light actinides and increases in importance with increasing atomic number until it is a stability-limiting mode for nuclei with $Z \geq 98$. The spontaneous fission half-lives change by a factor of $10^{29}$ in going from the longest lived uranium nuclei to the short-lived isotopes of fermium.

It is clear from these basic facts and our picture of fission that spontaneous fission is a barrier penetration phenomenon similar to $\alpha$ or proton decay. The nucleus "tunnels" from its ground state through the fission barrier to the scission point. Therefore, we would expect the spontaneous fission (SF) half-life to have the form

$$t_{1/2}^{SF} = \frac{\ln 2}{fP} \tag{11.12}$$

where $f$ is the frequency of assaults on the fission barrier in the first minimum ($\sim 10^{20}$/s) and $P$ is the barrier penetrability. As in $\alpha$ or proton decay, the penetrability factor is the most important term. The calculation of the barrier penetrability is complicated by the double or triple humped shape of the multidimensional barrier. A simple model for the barrier (near its top) is that of an inverted harmonic oscillator potential (a parabola) (Fig. 11.3). The Hill–Wheeler formula describes the

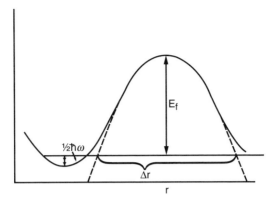

**Figure 11.3** Simple parabolic fission barrier. (From R. Vandenbosch and J. R. Huizenga, Nuclear Fission. Copyright © 1973 Academic Press. Reprinted by permission of Elsevier.)

transmission coefficient for penetration of such a barrier as:

$$P = (1 + \exp[2\pi(B_f)/\hbar\omega])^{-1} \tag{11.13}$$

where $B_f$ is the fission barrier height and $\hbar\omega$ is the barrier curvature (spacing between the levels in the corresponding "normal" harmonic oscillator potential). Large values of $\hbar\omega$ imply tall, thin barriers with high penetrabilities; low values of $\hbar\omega$ imply short, thick barriers with low penetrabilities. Combining equations gives

$$t_{1/2}^{SF} = 2.77 \times 10^{-21} \exp[2\pi B_f/\hbar\omega] \tag{11.14}$$

As an exercise, let us compare the spontaneous fission half-lives of two nuclei with barrier heights of 5 and 6 MeV, respectively, and barrier curvatures of 0.5 MeV. One quickly calculates that the spontaneous fission half-lives of these two nuclei differ by a factor of $3 \times 10^5$. The barrier heights and curvatures in this example are relevant for the actinides and illustrate the difficulty that a 1-MeV uncertainty in the fission barrier height corresponds to a factor of $10^5$ in the spontaneous fission half-life.

In our previous discussion, we showed that the fission barrier heights depend on $Z^2/A$ and thus so should the spontaneous fission half-lives. In Figure 11.4, we show the dependence of the known spontaneous fission half-lives on $x$, the fissionability parameter. There is an overall decrease in spontaneous fission half-life with increasing $x$, but clearly the spontaneous fission half-life does not depend only on $Z^2/A$. One also observes that the odd $A$ nuclei have abnormally long half-lives relative to the even–even nuclei. Also the spontaneous fission half-lives of the heaviest nuclei ($Z \geq 104$) are roughly similar with values of milliseconds.

Let us see if we can explain these observations. Swiatecki has shown that there is a correlation between the deviations of the spontaneous fission half-lives from the

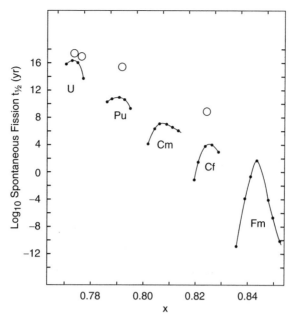

**Figure 11.4** Spontaneous fission half-lives of even–even (●) and even–odd (○) nuclides as a function of fissionability parameter, $x$. (From R. Vandenbosch and J. R. Huizenga, Nuclear Fission. Copyright © 1973 Academic Press. Reprinted by permission of Elsevier.)

smooth trend with $Z^2/A$ and the deviations of the ground-state masses from those expected from the liquid drop model. (These deviations are exactly the shell and pairing corrections discussed above.) Following the prescription developed by Swiatecki, we can plot $\log t_{1/2}(\exp) + 5\delta m$ (where $\delta m$ represents the deviation of the ground-state mass from the liquid drop model) vs. $x$ (Fig. 11.5). The correlation becomes much better, indicating we have captured the essence of the phenomenon.

We should note (Fig. 11.5) that the half-lives of the odd $A$ nuclei are still significantly longer than those of the neighboring even–even nuclei even though we have corrected for the effect of the ground-state masses. We can parameterize this difference by calculating a hindrance factor similar to that used in α-decay systematics. In the present case, the hindrance factor is defined as the log of the ratio of the observed half-life for the odd $A$ nucleus to that of the neighboring even–even nuclei. For the odd $A$ nuclei, typical hindrance factors of 5 are observed, that is, the odd $A$ half-lives are $\sim 10^5$ times longer than those of their even–even neighbors (Hoffman et al., 1996).

### 11.2.4 Spontaneously Fissioning Isomers

Since the discovery of the first spontaneously fissioning isomer, a number of other examples have been found. The positions of these nuclei in the chart of nuclides are

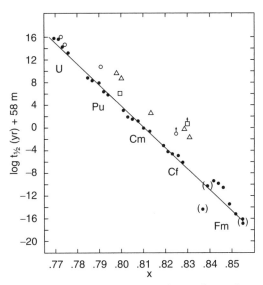

**Figure 11.5** Spontaneous fission half-lives, corrected according to the method of Swiatecki, vs. fissionability parameter $x$. (From R. Vandenbosch and J. R. Huizenga, Nuclear Fission. Copyright © 1973 Academic Press. Reprinted by permission of Elsevier.)

shown in Figure 11.6. These isomers range from thorium to berkelium, forming an island with a point of maximum stability around $^{242}$Am. γ-Ray decay back to the ground state limits the number of isomers with lower $Z$ and $N$ than those in this island, whereas spontaneous fission decay limits the number of cases with high $Z$ and $N$. The half-lives range from $10^{-9}$ to $10^{-3}$ s, whereas the ground-state half-lives are $\sim 10^{25} - 10^{30}$ times longer. The typical excitation energy of these isomers

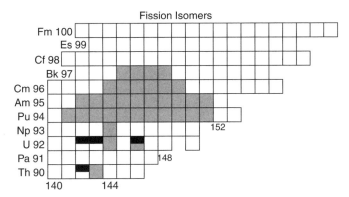

**Figure 11.6** Position of the known spontaneously fissioning isomers in the nuclide chart. (Figure also appears in color figure section.)

is 2–3 MeV. Spectroscopic studies of the transitions between the states in the second minimum has shown that the moments of inertia associated with the rotational bands are those expected for an object with an axes ratio of 2 : 1—a result confirmed in quadrupole moment studies.

### 11.2.5 Transition Nucleus

In analogy to chemical reactions, we might expect the probability of fission as expressed in the fission width, $\Gamma_f(=\hbar/\tau)$ to be given as

$$\Gamma_f = A \exp(-B_f/T) \qquad (11.15)$$

where $B_f$ is the fission barrier height. It turns out that this is an oversimplification, but it has certain pedagogical uses. For example, in an early study describing fission, Bohr and Wheeler were able to use this idea to show that a rare odd $A$ isotope of uranium, $^{235}$U, was responsible for the fission of uranium by thermal neutrons, not the more abundant even–even isotope $^{238}$U.

**Example Problem** Thermal neutrons whose kinetic energy is 0.025 eV cause $^{235}$U to fission but not $^{238}$U. Why?

*Solution* Let us calculate the energy released when a neutron is captured by $^{235}$U and $^{238}$U. (This will be equivalent to the binding energy of the last neutron in $^{236}$U and $^{239}$U.) For $^{235}$U + n, we have $E_{\text{release}} = (M_{235} + M_n - M_{236})c^2 = 40.913 + 8.071 - 42.440 = 6.5$ MeV. For $^{238}$U + n, we have $E_{\text{release}} = (M_{238} + M_n - M_{239})c^2 = 47.305 + 8.071 - 50.570 = 4.8$ MeV. The fission barrier in $^{235,238}$U is $\sim$5.7 MeV. Thus, in $^{235}$U + n, we exceed it with "zero-energy" neutrons while with $^{238}$U + n, we will need $\sim$1-MeV neutrons to cause fission. In fact, this example suggests $^{235}$U would fission even if we bombarded it with "negative kinetic energy" neutrons. Where would we find such neutrons? Consider the reaction $^{235}$U(d, p). This reaction, it turns out, is equivalent to adding a neutron to the nucleus and can even correspond, in some cases, to the addition of "negative kinetic energy" neutrons to the nucleus, allowing studies of near-barrier phenomena in the odd $A$ actinides.

The ability to cause odd $A$ actinide nuclei to undergo fission when bombarded with thermal neutrons is of great practical importance. Because of the large cross sections associated with thermal neutrons due to their long wavelength, the fission cross sections for these odd $A$ nuclei are very large. For the "big three" nuclei, $^{233}$U, $^{235}$U, and $^{239}$Pu, these cross sections have the values of 530, 586, and 752 barns, respectively. These actinides are the fuel for nuclear reactors and nuclear weapons utilizing fission by thermal neutrons.

We should note that once again, the probability of fission is more complicated than the simple relation given above would indicate. In a paper written shortly after the discovery of fission, Bohr and Wheeler showed that fission has to compete with other modes of nuclear deexcitation. They showed that $\Gamma_f$ should be written as

$$\Gamma_f = \frac{N_f}{N_f + N_n + N\gamma + N_{\text{ch.p}}} \quad (11.16)$$

where $N_i$ is a measure of the number of ways (open channels) to accomplish each possible deexcitation process ($=2\pi\Gamma/D$). When evaluating $N_f$, one must evaluate $\rho_f$, the density of levels in the transition state nucleus. $N_n$ is the principal term in this equation for heavy nuclei (why?) and is taken as the number of final states of the nucleus (after emitting a neutron) times the neutron kinetic energy. Bohr and Wheeler's predictions of the probability of fission in $^{238}$U as a function of excitation energy are shown in Figure 11.7.

In nuclear reactors one has neutrons with energies ranging from thermal (0.025 eV) to several MeV. There are a series of sharp peaks in the total cross section for neutrons with energies between 0.2 and 3000 eV that are called "resonances." These resonances correspond to exciting a specific isolated level in the compound nucleus that can decay by fission. The situation is particularly interesting for the neutron irradiation of even–even nuclei, such as $^{240}$Pu at subthreshold energies

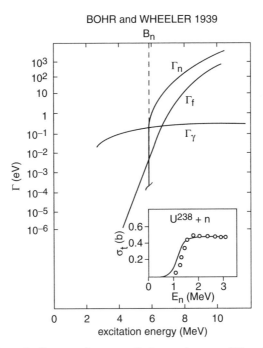

**Figure 11.7** Schematic diagram of neutron, fission, and γ-ray widths of a typical nucleus with a neutron binding energy slightly less than 6 MeV. The inset shows the predicted fission excitation function for a nucleus with $B_f - B_n = 0.75$ MeV together with more recent data.

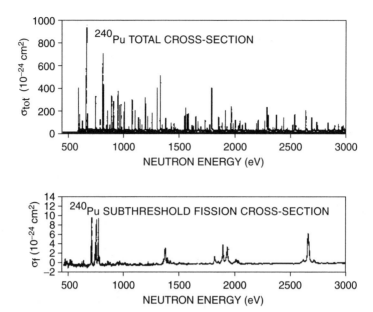

**Figure 11.8** Neutron total and subbarrier fission cross sections of $^{240}$Pu as a function of neutron energy between 0.5 and 3 keV. (From H. Weigmann, "Neutron-Induced Fission Cross Sections" in C. Wagemans, The Nuclear Fission Process. Copyright © 1991 CRC Press. Reprinted by permission of CRC Press.)

(Fig. 11.8). The resonances associated with fission appear to cluster in bunches. Not all resonances in the compound nucleus lead to fission. We can understand this situation with the help of Figure 11.9. The normal resonances correspond to excitation of levels in the compound nucleus, which are levels in the first minimum in Figure 11.9. When one of these metastable levels exactly corresponds to a level in the second minimum, then there will be an enhanced tunneling through the fission barrier and an enhanced fission cross section.

When higher energy ($E > 1$ MeV) neutrons interact with nuclei such as $^{238}$U where the fission barrier height is greater than the neutron separation energy, a stairstep pattern in the excitation function occurs (Fig. 11.10). The first rise and plateau is due to the occurrence of the $(n, f)$ reaction, the second rise and plateau is due to the $(n, nf)$ reaction ("second-chance fission"), and the third rise and plateau is due to the $(n, 2nf)$ reaction ("third-chance fission"), and so forth. For nuclei where $B_f < B_n$, the same pattern occurs but lies on top of rapidly decreasing cross sections (with increasing energy) at low energies due to $1/v$ absorption of neutrons.

How do we estimate the factors determining the fission probability when the excitation energy of the fissioning system is 10 MeV or more? (How do we calculate the various widths?) At these excitation energies, we have reached the point where the statistical model of nuclear reactions can be used. The relevant terms are only $\Gamma_f$ and $\Gamma_n$. The available experimental data on $\Gamma_n/\Gamma_f$ at excitation energies of 5–25 MeV is shown in Figure 11.11. One notes the general trend in $\Gamma_n/\Gamma_f$ with increasing $Z$ and $A$ (consistent with the qualitative dependence on $Z^2/A$ for fission). For this limited

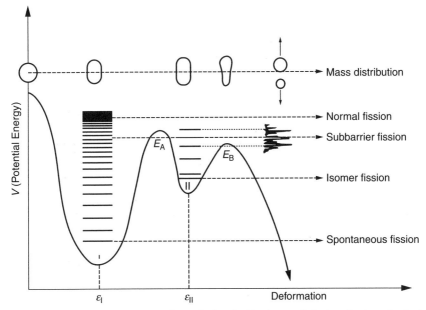

**Figure 11.9** Schematic representation of the double humped fission barrier. Intrinsic excitations in the first and second minimum are shown along with the path of fission from isomeric states and ground-state spontaneous fission.

range of energies, the ratio $\Gamma_n/\Gamma_f$ can be parameterized as:

$$\frac{\Gamma_n}{\Gamma_f} = \frac{2TA^{2/3}}{10} \exp\left(\frac{B_f - B_n}{T}\right) \tag{11.17}$$

where $B_f$, $B_n$, the nuclear temperature $T\ [=(8E^*/A)^{1/2}]$ refer to the fissioning system. A more rigorous expression that can be used over a wider range of

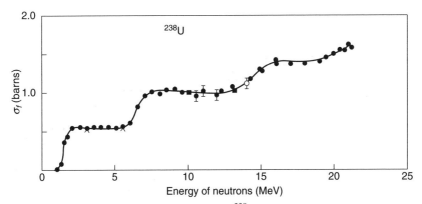

**Figure 11.10** Fission excitation function for n + $^{238}$U for $E_{\text{neutron}}$ between 1 and 22 MeV.

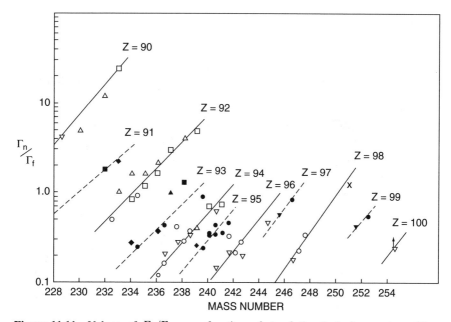

**Figure 11.11** Values of $\Gamma_n/\Gamma_f$ as a function of $A$ of the fissioning system. (From R. Vandenbosch and J. R. Huizenga, Nuclear Fission. Copyright © 1973 Academic Press. Reprinted by permission of Elsevier.)

excitation energies is

$$\frac{\Gamma_n}{\Gamma_f} = \frac{g\mu r_0^2}{\hbar^2} \frac{4A^{2/3} a_f(E^* - B_n)}{a_n[2a_f^{1/2}(E^* - B_f)^{1/2} - 1]} \exp[2a_n^{1/2}(E^* - B_f)^{1/2} - 2a_f^{1/2}(E^* - B_f)^{1/2}]$$

(11.18)

where $a_n$ is the level density parameter of the residual nucleus after emission of a neutron and $a_f$ is that of the deformed transition state nucleus. Note that $\Gamma_n/\Gamma_f$ is related to the difference $(B_f - B_n)$ as shown in Figure 11.12.

**Example Problem**  Consider the bombardment of $^{238}$U with 42-MeV α particles. What fraction of the initial nuclei undergoes first-chance fission?

**Solution**  The excitation energy of the compound nucleus $E^*$ will be

$$E^* = 42(238/242) + Q_{CN}$$
$$Q_{CN} = (M_{238} + M_\alpha - M_{242})c^2$$
$$Q_{CN} = 47.305 + 2.425 - 54.712 = -5.0$$
$$E^* = 36.3 \text{ MeV}$$

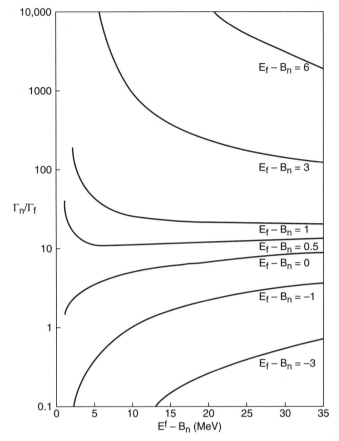

**Figure 11.12** Excitation energy dependence of $\Gamma_n/\Gamma_f$ for different values of $(B_f - B_n)$. (From R. Vandenbosch and J. R. Huizenga, Nuclear Fission. Copyright © 1973 Academic Press. Reprinted by permission of Elsevier.)

In $^{242}$Pu, $B_n = 6.3$ MeV, $B_f = 5.3$ MeV. From Figure 11.12, $\Gamma_n/\Gamma_f = 3$, that is, 25% of the initial nuclei fission at the first chance.

For reactions induced by heavy ions or high-energy charged particles, these expressions should be corrected for the effect of angular momentum. For example, there will be excitation energy tied up in rotation, which is unavailable for fission (Vandenbosch and Huizenga, 1973) and the fission barriers are lower for rotating nuclei. For reactions involving less fissionable nuclei ($x < 0.7$), especially at higher energies, one frequently sees that the primary reaction products first decay by sequential emission of neutrons or charged particles, and, then, as $Z^2/A$ increases, fission occurs at the last stages of the evaporation chains.

## 11.3 FISSION PRODUCT DISTRIBUTIONS

Up to this point, we have focused on describing the factors that control the probability of fission to occur. Now we will focus our attention on the distributions of the products in mass, energy, charge, and so forth. In doing so, we will mostly be discussing "scission point" or "postfission" phenomena. Our treatment of these phenomena is, of necessity, somewhat superficial, and the reader is referred to the excellent monograph of Vandenbosch and Huizenga (1973) for a more authoritative account.

### 11.3.1 Total Kinetic Energy (TKE) Release in Fission

To a first approximation, one can assume that the kinetic energies of the fission fragments are the result of the Coulomb repulsion of the fragments after scission. A handy pocket formula that gives the total kinetic energy is

$$\text{TKE} = \frac{Z_1 Z_2 e^2}{1.8(A_1^{1/3} + A_2^{1/3})} \text{ MeV} \tag{11.19}$$

where $Z_1, A_1, Z_2, A_2$, refer to the atomic and mass numbers of the two fragments. The factor of 1.8 (instead of the usual value of $r_0$ of 1.2) results from the fact that the fragments at scission are unusually deformed. More detailed empirical prescriptions for the TKE are available (Viola et al., 1985), but the above formula seems to work quite well over a range of excitation energies and fissioning nuclei. The most significant deviations from these formulas appear in the very heavy actinides, $^{258,259}$Fm and $^{260}$Md, where the observed kinetic energies are evidence (Hoffman et al., 1996) for an unusually compact scission configuration.

### 11.3.2 Fission Product Mass Distributions

One of the first big surprises in early studies of fission was the fission product mass distribution. Investigations of the thermal neutron-induced fission of uranium and plutonium nuclides (and later the spontaneous fission of $^{252}$Cf) showed the most probable division of mass was asymmetric ($M_H/M_L = 1.3-1.5$). The liquid drop model would predict that the greatest energy release and, therefore, the most probable mass split, would be a symmetric one, that is, $M_H/M_L = 1.0$. This situation is shown in Figure 11.13 where the mass distributions for the thermal neutron-induced fission of the "big three nuclides," $^{233}$U, $^{235}$U, and $^{239}$Pu, are shown. Symmetric fission is suppressed by at least two orders of magnitude relative to asymmetric fission. Note the peak-to-valley ratio of the distributions.

The key to understanding this situation can be seen in Figures 11.13 and 11.14. In these figures, we show that as the mass of the fissioning system increases, the position of the heavy peak in the fission mass distribution remains constant while the position of the light peak increases with increasing fissioning system mass.

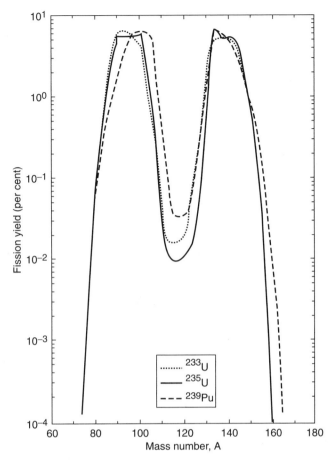

**Figure 11.13** Smoothed fragment mass distributions for the thermal neutron-induced fission of $^{233}$U, $^{235}$U, and $^{239}$Pu. [From Seaborg and Loveland (1990).]

This observation, along with the observation that the lower edge of the heavy fragment peak is anchored at $A = 132$ has suggested that the preference for asymmetric fission is due to the special stability of having one fragment with $Z = 50$, $N = 82$, a doubly magic spherical nucleus.

Further evidence for this influence of "magic" (shell model) configurations on the fission mass distributions is found in the fragment mass distributions for spontaneous fission (Fig. 11.15) and low-energy-induced fission of the "preactinides" (Fig. 11.16). One observes, in the case of spontaneous fission, a sharp transition between asymmetric fission and symmetric fission as one goes from $^{257}$Fm to $^{258}$Fm. The addition of a single neutron to the nucleus causes a large change in the fission product mass distribution. Similarly, a shift of two protons in going from $^{225}$Ac to $^{227}$Pa causes the mass distribution to shift from purely symmetric to dominantly asymmetric. These changes occur at neutron and proton numbers

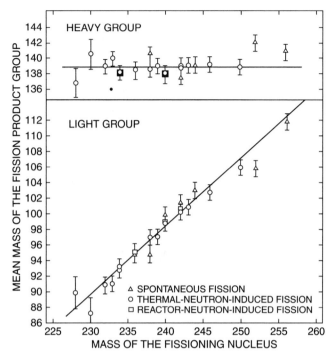

**Figure 11.14** Average masses of the light and heavy fragments as a function of the mass of the fissioning system. [From K. Flynn, et al., Phys. Rev. *C5*, 1728 (1972). Copyright (1972) by the American Physical Society.]

that are not the so-called magic numbers for spherical nuclei. The key is that the fissioning system and the fragments themselves are initially quite deformed. Thus, the relevant magic numbers, that is, configurations of special stability are those expected for deformed nuclei and as shown in Chapter 6, the actual configurations change with deformation. A detailed theory of fission scission point properties based on these ideas developed by Wilkins, Steinberg, and Chasman (1976) has been quite successful in describing the observed trends.

Qualitatively, if these explanations of the fission mass distributions for low-energy-induced fission are correct, one might expect as the excitation energy of the fissioning system is raised, the influence of the ground-state shell structure of the nascent fragments decreases, and the fission mass distributions shows a greater amount of symmetric fission. That is exactly what happens and at high energy all nuclei fission symmetrically (Fig. 11.17).

### 11.3.3 Fission Product Charge Distributions

If one were to plot the yield of fission fragments as a function of their atomic numbers (as in Fig. 11.16), the result would look very much like the fission mass

11.3 FISSION PRODUCT DISTRIBUTIONS   **319**

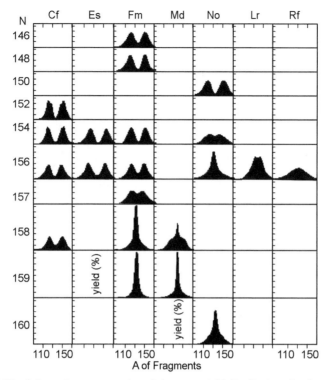

**Figure 11.15** Schematic representation of the mass yield distributions for the spontaneous fission of the trans-berkelium nuclides. (From D. C. Hoffman, et al., "Spontaneous Fission" in Nuclear Decay Modes, D. N. Poenaru, Ed. Copyright © 1996 IOP Press. Reprinted by permission of IOP Press.)

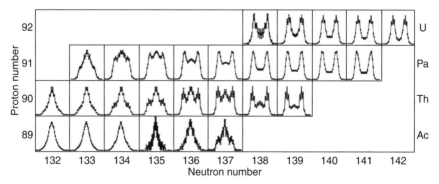

**Figure 11.16** Atomic number ($Z$) distributions for the low-energy fission of several actinide nuclei are shown in each panel. (From K.-H. Schmidt et al., in Heavy Ion Physics, Y. T. Oganessiam and R. Kalpakchieva, Eds. Copyright © 1998 World Scientific. Reprinted by permission of World Scientific.)

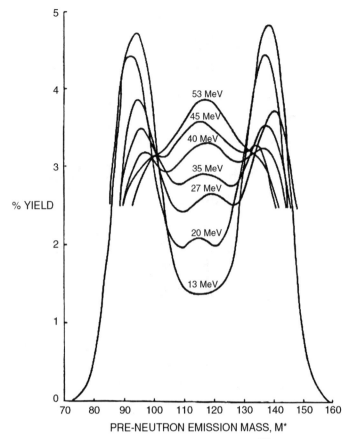

**Figure 11.17** Fission mass distributions for $^{232}$Th$(p, f)$.

distribution. Nuclear matter is not very polarizable and, to first order, the protons will divide like the neutrons. The primary fission fragments thus have neutron/proton ratios like that of the fissioning system and thus lie on the neutron-rich side of β stability. Enhanced yields for even Z nuclides relative to odd Z nuclides are observed (Fig. 11.18) due to the stabilization from proton pairing.

The yield of any given nuclide in fission is called its *independent yield*. It can be shown that the independent yield of isobars in fission has a Gaussian form:

$$P(Z) = \frac{1}{\sqrt{c\pi}} \exp\left[-\frac{(Z - Z_p)^2}{c}\right] \quad (11.20)$$

where $c$ has an average value of $0.80 \pm 0.14$ for low-energy fission and $Z_p$ is the most probable primary fragment atomic number (noninteger) for that isobar. [Large tables of $Z_p$ exist for common fissioning systems (Wahl, 1988).] One

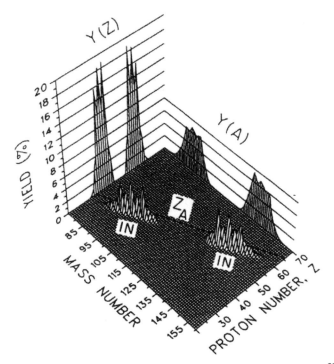

**Figure 11.18** Yields of products from the thermal neutron-induced fission of $^{235}$U. (From A. C. Wahl, "Nuclear Charge Distribution in Fission," in New Directions in Physics, N. Metropolis, D. M. Kerr, and G. C. Rota, Eds. Copyright © 1987 by Academic Press, Inc. Reprinted by permission of Elsevier.)

consequence of this small value of $c$ is that, for a given $A$, only a few isobars will have significant yields. Two effects tend to favor narrow charge distributions: (a) the high energetic cost of unfavorable charge splits and (b) the existence of ground-state correlations between neutrons and protons in the fragments.

In discussions of fission, one frequently hears the terms *cumulative yield* and *independent yield*. The *independent yield* of a nuclide is just what it appears, the yield of that nucleus as a primary fission product. Because the fission products are all $\beta^-$ emitters, they decay toward the bottom of the valley of $\beta$ stability, populating several different members of an isobaric series, as, for example, with $A = 140$ fragments:

$$^{140}\text{Xe} \xrightarrow{\beta^-} {}^{140}\text{Cs} \xrightarrow{\beta^-} {}^{140}\text{Ba} \xrightarrow{\beta^-} {}^{140}\text{La} \xrightarrow{\beta^-} {}^{140}\text{Ce}$$

The yield of each member of the isobaric series integrates, by virtue of the intervening $\beta$ decay, the yields of its precursors. Such yields are referred to as *cumulative yields*. For example, the cumulative yield of the mass 140 chain in the thermal neutron-induced fission of $^{235}$U is 6.25%.

**Example Problem** In the above example, what is the independent yield of $^{140}$Ba for the thermal neutron-induced fission of $^{235}$U and what is its cumulative yield?

**Solution** The fractional independent yield is given as:

$$P(Z) = \frac{1}{\sqrt{c\pi}} \exp\left[-\frac{(Z-Z_p)^2}{c}\right]$$

For the mass 140 chain, $Z_p = 54.55$ (Wahl, 1988). Note that this tabulated value of $Z_p/A$ (=54.55/140) is very close to that of the fissioning system, 92/236, that is, the $N/Z$ ratio of the fragments is approximately that of the fissioning system. This idea is called the UCD (unchanged charge distribution) prescription. Therefore

$$P(56) = \frac{1}{\sqrt{0.8\pi}} \exp\left[-\frac{(56-54.55)^2}{0.8}\right] = 4.56 \times 10^{-2}$$

Therefore

$$\text{Yield of } ^{140}\text{Ba} = 6.25 \times 0.0456 = 0.28\%$$

The fractional cumulative yield (FCY) of $^{140}$Ba is defined as:

$$\text{FCY} = \frac{1}{\sigma\sqrt{2\pi}} \int_{-\infty}^{Z+1/2} \exp\left[-\frac{(n-Z_p)^2}{2\sigma^2}\right] dn$$

where the familiar Gaussian parameter $\sigma$ is related to $c$ by Sheppard's relation $c = 2(\sigma^2 + \frac{1}{12})$. Evaluating this integral gives FCY = 0.9978.

## 11.4 EXCITATION ENERGY OF THE FISSION FRAGMENTS

The excitation energy of the fission fragments is equal to the difference between the total energy release, $Q$, and the total kinetic energy of the fragments. The excitation energy should be calculated for each mass split. Here we will do an average accounting to see where the energy goes. For the thermal neutron-induced fission of $^{235}$U, this corresponds to $\sim$200 MeV $-$ $\sim$172 MeV or about $\sim$28 MeV ($\sim$14% of the total energy release), averaged over all mass splits. The average number of emitted prompt neutrons is $\sim$2.4, and each neutron has a kinetic energy of $\sim$2 MeV while the emitting fragments have average neutron binding energies of $\sim$5.5 MeV. Thus about 18 MeV [=2.4 × (2 + 5.5)] of the fragment excitation energy is carried away by the prompt neutrons. Prompt photon emission carries away $\sim$7.5 MeV, which leaves about 2.5 MeV, in this crude accounting, to be emitted in the form of β particles, neutrinos, delayed neutrons, and so forth.

As noted earlier, the prompt neutrons are emitted from the fully accelerated fragments after scission. The number of these neutrons, $v_T$, (= 2.4 in the case of $^{235}$U) as

## 11.4 EXCITATION ENERGY OF THE FISSION FRAGMENTS

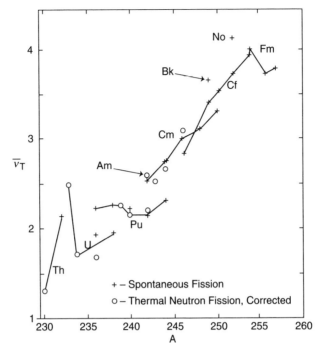

**Figure 11.19** Average total number of neutrons, $v_T$, as a function of the mass number of the fissioning system for spontaneous and thermal neutron-induced fission. The values for thermal neutron-induced fission have been corrected to zero excitation energy (spontaneous fission) assuming $dv/dt = 0.12$ MeV$^{-1}$.

a function of the mass of the fissioning system is shown in Figure 11.19. The general increase in $v_T$ with mass of the fissioning system is due to the increase in fragment excitation energy. For very heavy systems (Z ~ 114), $v_T$ is predicted to be ~7, allowing the critical mass for a self-sustaining fission reaction to be quite small.

**Example Problem** How did we get the estimate for $v_T$ for the fission of $^{298}$114? Assume this nucleus fissions symmetrically.

**Solution** The total energy released can be estimated using a modern mass formula (Liran and Zeldes, 1976) as 311 MeV. The total kinetic energy of the fragments can be calculated as

$$\text{TKE} = \frac{(57)(57)(1.440)}{(1.8)(149^{1/3})(2)} = 245 \text{ MeV}$$

This leaves a total fragment excitation energy of $311 - 245 = 66$ MeV. Since the $\gamma$ rays and $\beta^-$, are emitted only to take away the energy not emitted as neutrons, we shall assume this energy is the same as in $^{235}$U($n_{\text{th}}$, $f$) or ~10 MeV. This leaves a "neutron excitation energy" of 66 − 10 = 56 MeV. The neutron binding energy in a typical fragment is ~6 MeV (Liran and Zeldes, 1976). Thus, we would calculate $v_T = 56/(6+2) = 7$.

The average neutron kinetic energy is ~2 MeV. In the frame of the moving fragment, the distribution of fragment energies is Maxwellian, $P(E) = E_n \exp(-E_n/T)$. Transforming this spectrum into the laboratory frame gives a spectrum of the Watt form, that is,

$$P(E_n) = e^{-E_n/T} \sinh(4E_n E_f/T^2)^{1/2} \quad (11.21)$$

where $E_n$ and $E_f$ are the laboratory system energies of the neutron and fission fragment (in MeV/nucleon) and $T$ is the nuclear temperature.

Another important aspect of neutron emission is the variation of the number of emitted neutrons as a function of the fragment mass, $v(A)$ (Fig. 11.20). The striking features of these data are the nearly universal dependence of $v(A)$ upon $A$, independent of fissioning system for these actinide nuclei (which again suggests the role of *fragment* shell structure upon this property) and the sawtooth dependence of $v(A)$. Note the correlation of low values of $v(A)$ for those fragments whose structure is that of a magic nucleus, that is, a nucleus of special stability. These fragments are expected to have low excitation energies due to shell effects (Wilkins et al., 1976).

Prompt γ-ray emission competes with or follows the last stages of prompt neutron emission. These photons are emitted in times from $10^{-15}-10^{-7}$ s. Typical γ-ray multiplicities of 7–10 photons/fission are observed. These photons, as indicated earlier, carry away ~7.5 MeV. This γ-ray yield is considerably larger than one would predict if γ-ray emission followed neutron emission instead of competing with it. Because of the significant angular momentum of the fission fragments (~7–10 ℏ) even in spontaneous fission, photon emission can compete with neutron emission. The emitted γ rays are mostly dipole radiation with some significant admixture of quadrupole radiation, due to "stretched" E2 transitions ($J_f = J_i - 2$).

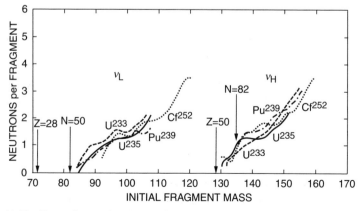

**Figure 11.20** Dependence of $v(A)$ upon fragment mass number $A$ for some actinide nuclei.

Because of the large number of possible neutron-rich fragments produced in fission, the study of the γ rays emitted by the fragments can lead to useful information about the nuclear structure of these exotic, short-lived nuclei far from stability.

As indicated in our discussion of prompt γ-ray emission, the fission fragments have a significant angular momentum. There are two origins for this angular momentum: (a) the existence of off-axis torques given the fragments during the scission process and (b) the excitation of bending and wriggling modes in the nascent fragments at the saddle point, which persist to scission and are amplified by (a).

## 11.5 DYNAMICAL PROPERTIES OF THE FISSION FRAGMENTS

One of the properties of fission fragments that is often exploited is the angular distribution. Fission is generally considered to be a "slow" process in which the fissioning nucleus is in statistical equilibrium. The angular distribution of the fission fragments will, therefore, be symmetric with respect to a plane perpendicular to the direction of motion of the fissioning system, that is, the fragment angular distributions will be symmetric about 90° in the frame of the fissioning system.

A typical fission fragment angular distribution for a heavy-ion-induced fission reaction is shown in Figure 11.21. As one can see, the fragments are emitted

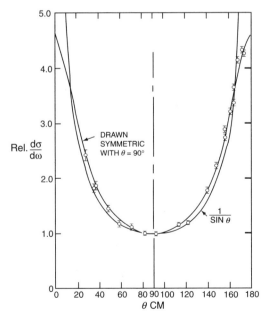

**Figure 11.21** Fission fragment angular distribution for the carbon-ion-induced fission of gold. [From G. E. Gordon et al., Phys. Rev. *120*, 1341 (1960). Copyright (1960) by the American Physical Society.]

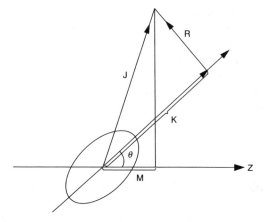

**Figure 11.22** Angular momentum coordinate system for a deformed nucleus.

preferentially forwards and backwards with respect to the direction of motion of the fissioning system. In this case involving a reaction in which the fissioning system has a significant angular momentum ($\sim 36\hbar$), the distribution closely resembles the function $1/\sin(\theta)$.

To understand these distributions, one needs to consider the fission transition nucleus. Figure 11.22 shows a coordinate system for describing this nucleus in terms of its quantum numbers $J$, the total angular momentum $M$, the projection of $J$ upon a space-fixed axis, usually taken to be the direction of motion of the fissioning system, and $K$, the projection of $J$ upon the nuclear symmetry axis.

In most collisions, the angular momentum vectors, $J$, will be concentrated in a plane perpendicular to the beam direction ($M = 0$). In this case, we can easily see a relation between $J$, $K$, and the fission fragment angular distribution. Consider the case where $J = K$, the nuclear symmetry axis is perpendicular to the beam, and the fragments emerge sidewise to the beam. Similarly, if $K = 0$, the symmetry axis of the nucleus is oriented perpendicular to $J$, that is, along the beam direction, and the fragments are emitted preferentially forward and backward. It is this extreme that leads to the $1/\sin(\theta)$ form. [If $J$ is perpendicular to the beam, and the vectors describing the possible directions of the nuclear symmetry axis are uniformly distributed over the surface of a sphere, then the probability of having a symmetry axis at angle $\theta$ with respect to the beam will go as $1/\sin(\theta)$.]

For the general case, Wheeler (1963) showed that the probability of emitting a fragment at an angle $\theta$ from a transition nucleus characterized by $J$, $K$, and $M$ is

$$P^J_{M,K}(\theta) = (2J+1)\left[\frac{2\pi R^2 \sin\theta \, d\theta}{4\pi R^2}\right]|d^J_{M,k}(\theta)|^2 \qquad (11.22)$$

The first term represents a statistical weighting factor, the second term a solid angle factor for the probability of getting the angle $\theta$, and the third term is a symmetric top wave function.

## 11.5 DYNAMICAL PROPERTIES OF THE FISSION FRAGMENTS

In low-energy fission or photofission, one can populate individual states of the fission transition nucleus and one can observe fragment angular distributions that change from forward to sidewise peaked and back again as a function of the $J$, $K$, $M$ of the transition nucleus (Vandenbosch and Huizenga, 1973). At higher energies ($E^* > 10$ MeV), one describes the states of the transition nucleus using a statistical model. One assumes there is a Gaussian distribution of $K$ values for the transition nucleus

$$\rho(K) \propto \exp(-K^2/K_0^2) \quad K < J$$
$$= 0 \quad K > J \quad (11.23)$$

Then $K_0^2$, the root-mean-square projection of $J$ upon the nuclear symmetry axis, can be calculated as:

$$K_0^2 = \frac{\Im_{\text{eff}} T}{\hbar^2} \quad (11.24)$$

where $\Im_{\text{eff}}$ is the effective moment of inertia of the transition nucleus with temperature $T$. The fission fragment angular distributions $W(\theta)$ can be written as:

$$W(\theta) \propto \sum_{J=0}^{\infty} (2J+1) T_J \sum_{K=-J}^{J} \frac{(2J+1)|d^J_{M=0,K}(\theta)|^2 \exp(-K/2K_0^2)}{\sum_{K=-J}^{J} \exp(-K^2/2K_0^2)} \quad (11.25)$$

where $T_J$ is the transmission coefficient for forming the fissioning nucleus with total angular momentum $J$. Under the assumption that $M = 0$, we get the handy "pocket" formula

$$W(\theta) \propto \sum_{J=0}^{\infty} \frac{(2J+1)^2 T_J \exp[-(J+0.5)^2 \sin^2\theta/4K_0^2]}{\text{erf}[(J+0.5)/(2K_0^2)^{1/2}]} \times J_0[i(J+0.5)^2 \sin^2\theta/4K_0^2] \quad (11.26)$$

where $J_0$ is the zero-order Bessel function with imaginary argument and $\text{erf}[(J+0.5)/(2K_0^2)^{1/2}]$ is the error function defined by

$$\text{erf}(x) = \frac{2}{\pi^{1/2}} \int_0^x \exp(-t^2)\, dt \quad (11.27)$$

If one can estimate $K_0^2$, then the fission angular distributions can be used to measure $J$ or vice versa.

One other aspect of the spatial distribution of the fission fragments that has proven to be a useful tool in studying nuclear reactions is the angular correlation between the two fission fragments. When a fission event occurs, the two fragments emerge with an angle of 180° between them (to conserve angular momentum). If the fissioning nucleus is in motion, then the initial momentum of the fissioning system must be shared between the two fragments to give the final (laboratory system) fragment momenta. Complete fusion events can thus be differentiated from incomplete fusion events by observing the mean angle between coincident fission fragments.

**Example Problem** Consider the case of 240-MeV $^{32}$S interacting with $^{181}$Ta, which fissions. What would be the laboratory correlation angle between the fragments if the full linear momentum of the projectile was transferred to the fissioning system?

**Solution**

$$^{32}S + {}^{181}Ta \rightarrow {}^{213}Ac$$

The momentum of the beam nucleus (CN) is given by

$$P = (2mT)^{1/2} = (2 \times 32 \times 240)^{1/2} = 123.9 \,(\text{MeV-amu})^{1/2}$$

For the symmetric fission of $^{213}$Ac, we expect

$$\text{TKE} = \frac{(89/2)^2 e^2}{1.8(213/2)^{1/3}(2)} = 167 \,\text{MeV}$$

The momentum of each fragment is then

$$P = (2mT)^{1/2} = \left[2\left(\frac{213}{2}\right)\left(\frac{167}{2}\right)\right]^{1/2} = 133.4 \,(\text{MeV-amu})^{1/2}$$

The half correlation angle is

$$\theta = \tan^{-1}(133.4/123.9/2)) = 65°$$

The correlation angle would be $2\theta = 130°$.

A dynamical scission point phenomenon is the violent snapping of the neck between the nascent fragments, which can result in the emission of particles into the region between the fragments. The phenomenon is rare, occurring in about 1 in 300 to 1 in 1000 of the fission events for α particles and in lesser frequency for other charged particles. (Neutrons can be emitted by this same mechanism in a few percent of all fission events.) The charged particles, being born in the region between the fragments, are strongly focused by the Coulomb field of the fragments and emerge at 90° with respect to the direction of motion of the separating fragments, with energies (~15 MeV for α particles) characteristic of the Coulomb fields of the separating fragments.

## PROBLEMS

1. Why is $^{240}$Pu not fissionable by thermal neutrons, but $^{239}$Pu is?

2. What is the expected total kinetic energy release in the fission of $^{272}$110 assuming fission occurs symmetrically?

3. What is the meaning of the terms *prompt* and *delayed* with respect to the fission neutrons?

4. Sketch the fission excitation function for the reaction of $^{232}$Th with neutrons. The fission barrier is $\sim$6.5 MeV, and the binding energies of the last neutron in $^{232}$Th and $^{233}$Th are 6.90 and 4.93 MeV, respectively.

5. What are the values of the fissionability parameter $x$ for $^{209}$Bi, $^{226}$Ra, $^{232}$Th, $^{242}$Pu, and $^{252}$Cf?

6. What is the fraction of fission neutrons with energies greater than 2 MeV (in the laboratory frame)?

7. What is the independent yield of $^{99}$Mo in the thermal neutron-induced fission of $^{239}$Pu? $Z_p$ is 39.921 and the yield of the $A = 99$ chain is 6.15%.

8. What is the value of $\Gamma_n/\Gamma_f$ for a $^{210}$Po nucleus produced in the bombardment of $^{209}$Bi with 10.5-MeV protons. $B_f = 20.4$ MeV.

## REFERENCES

Britt, H. C. In N. M. Edelstein, Ed., *Actinides in Perspective*, Pergamon, Oxford, 1982, p. 245.

Flynn, K. F., E. P. Horwitz, C. A. A. Bloomquist, R. F. Barnes, R. K. Sjoblom, P. R. Fields, and L. E. Glendenin. *Phys. Rev.* **C5**, 1725 (1972).

Gindler, J. and J. R. Huizenga. Nuclear Fission, In Vol. II, L. Yaffe, Ed., *Nuclear Chemistry*, Academic Press, New York, 1964, p. 1.

Hahn, O. and F. Strassman. *Naturwiss.* **26**, 756 (1938); see translation by H. Graetzer, *Am. J. Phys.* **32**, 10 (1964).

Hoffman, D. C., T. M. Hamilton, and M. R. Lane. Spontaneous Fission, In D. N. Poenaru, Ed., *Nuclear Decay Modes*, IOP, Bristol, 1996, pp. 393–432.

Liran, S. and N. Zeldes. *At. Data Nucl. Data Tables* **17**, 431 (1976).

Oganessian, Y. T. and Y. A. Lazarev. Heavy Ions and Nuclear Fission, In D. A. Bromley, Ed., *Treatise on Heavy-Ion Nuclear Science*, Vol. 4, Plenum, New York, 1985, pp. 1–254.

Schmidt, K.-H., et al. In Y. T. Oganessian and R. Kalpakchieva, Eds., *Heavy Ion Physics*, World, Singapore, 1998, p. 667.

Seaborg, G. T. and W. Loveland. *The Elements Beyond Uranium*, Wiley, New York, 1990.

Strutinsky, V. M. *Nucl. Phys. A* **95**, 420 (1967).

Vandenbosch, R. and J. R. Huizenga. *Nuclear Fission*, Academic, New York, 1973.

Viola, V. E., Jr., K. Kwiatkowski, and M. Walker. *Phys. Rev.* **C31**, 1550 (1985).

Wagemans, C. *The Nuclear Fission Process*, CRC Press, Boca Raton, FL, 1991.

Wahl, A. C. *At. Data Nuclear Data Tables* **39**, 1 (1988).

Wheeler, J. A. In J. B. Marion and J. L. Fowler, Eds., *Fast Neutron Physics*, Wiley, New York, 1963.

Wilkins, B. D., E. P. Steinberg, and R. R. Chasman. *Phys. Rev.* **C14**, 1832 (1976).

# CHAPTER 12

# NUCLEAR REACTIONS IN NATURE: NUCLEAR ASTROPHYSICS

## 12.1 INTRODUCTION

An important mystery that is still unfolding today is how did the chemical elements that we have here on Earth come into existence? We know that the readily available chemical elements are restricted in number to less than 90 and that they are essentially immutable by chemical reactions. The large-scale nuclear reactions that are taking place are those induced by (external) cosmic rays and radioactive decay; induced nuclear reactions such as fission take place on a tiny scale. Thus, the vast bulk of chemical elements that we have today are those that were present when the Earth was formed. The elements have undergone an enormous range of geochemical, geological, and biochemical processes, but all such processes retain the integrity of the atom. Thus, the origin of the elements is certainly extraterrestrial, but questions remain as to where and how they were formed?

These questions lie in the field of nuclear astrophysics, an area concerned with the connection of fundamental information on the properties of nuclei and their reactions to the perceived properties of astrological objects and processes that occur in space. The universe is composed of a large variety of massive objects distributed in an enormous volume. Most of the volume is very empty ($<1 \times 10^{-18}$ kg/m$^3$) and very cold ($\sim$3 K). On the other hand, the massive objects, stars, and such are very dense (sun's core $\sim 2 \times 10^5$ kg/m$^3$) and very hot (sun's core $\sim 16 \times 10^6$ K). These temperatures and densities are such that the light elements are ionized and have

---

*Modern Nuclear Chemistry*, by W.D. Loveland, D.J. Morrissey, and G.T. Seaborg
Copyright © 2006 John Wiley & Sons, Inc.

**332** NUCLEAR REACTIONS IN NATURE: NUCLEAR ASTROPHYSICS

high enough thermal velocities to occasionally induce a nuclear reaction. Thus, the general understanding of the synthesis of the heavier elements is that they were created by a variety of nuclear processes in massive stellar systems. These massive objects exert large gravitational forces, and so one might expect the new materials to remain in the stars. The stellar processing systems must explode at some point in order to disperse the heavy elements. When we look at the details of the distribution of isotopes here on Earth, we will find that some number of explosive cycles must have taken place before the Earth was formed.

In this chapter we will first consider the underlying information on the elemental abundances and some of the implications of the isotopic abundances. Then, we will consider the nuclear processes that took place to produce the primordial elements and those that processed the primordial light elements into those that we have here on Earth.

## 12.2 ELEMENTAL AND ISOTOPIC ABUNDANCES

Many students of chemistry have given little thought to the relative abundances of the chemical elements. Everyone realizes that some elements and their compounds are more common than others. The oxygen in water, for example, must be plentiful compared to mercury or gold. But what if we compare elements that are closer in the periodic table, for example, what is the amount of lead ($Z = 82$) compared to mercury ($Z = 80$) or what is the amount iron ($Z = 26$) compared to copper ($Z = 27$)? Oddly enough, the answer one gets depends on what is sampled. The relative abundances of the first 40 elements are shown in Figure 12.1 as a percentage by mass of Earth's crust and as a percentage by mass of our solar system. Notice that the scale is logarithmic and the data spans almost 11 orders of magnitude. Earth is predominantly oxygen, silicon, aluminum, iron, and calcium, which comprise more than 90% of Earth's crust. On the other hand, the mass of the solar system is dominated by the mass of the sun so that the solar system is mostly hydrogen, with some helium, and everything else is present at the trace level. The differences between the solar system abundances and those on Earth are due to geophysical and geochemical processing of the solar material. Therefore, in this text we will concentrate on understanding the solar system abundances that reflect nuclear processes. The abundances of the isotopes and the elements are the basic factual information that we have to test theories of *nucleosynthesis*. We have data from Earth, the moon, and meteorites, from spectroscopic measurements of the sun, and recently from spectroscopic measurements of distant stars. Many studies have characterized and then attempted to explain the similarities and differences from what we observe in the solar system.

The solar abundances of all of the chemical elements are shown in Figure 12.2. These abundances are derived primarily from knowledge of the elemental abundances in CI carbonaceous chrondritic meteorites and stellar spectra. Note that ~99% of the mass is in the form of hydrogen and helium. Notice that there is a general logarithmic decline in the elemental abundance with atomic number with

## 12.2 ELEMENTAL AND ISOTOPIC ABUNDANCES

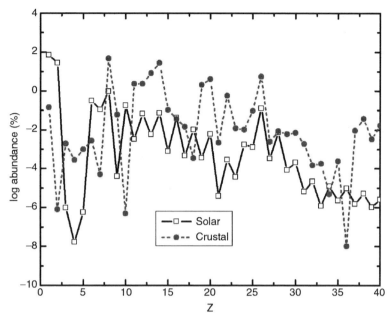

**Figure 12.1** Abundances of the first 40 elements as a percentage by mass of Earth's crust (filled squares) and in the solar system (open circles). Data from the *CRC Handbook of Chemistry and Physics*, 75th ed., 1994. (Figure also appears in color figure section.)

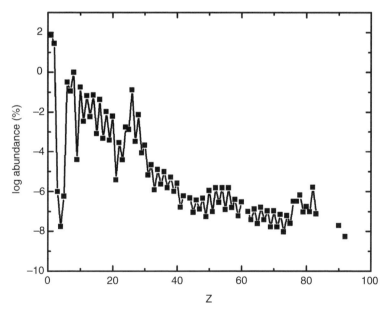

**Figure 12.2** Abundances of the elements as a percentage by mass of the solar system. Data from the *CRC Handbook of Chemistry and Physics*, 75th ed., 1994.

the exceptions of a large dip at beryllium ($Z = 4$) and of peaks at carbon and oxygen ($Z = 6-8$), iron ($Z \sim 26$) and the platinum ($Z = 78$) to lead ($Z = 82$) region. Also notice that there is a strong odd–even staggering and that all the even $Z$ elements with $Z > 6$ are more abundant than their odd atomic number neighbors. We have already encountered an explanation for this effect, that is, recall from earlier discussions on nuclear stability that there are many more stable nuclei for elements with an even number of protons than there are for elements with an odd number of protons simply because there are very few stable odd–odd nuclei. Thus, the simple number of stable product nuclei, whatever the production mechanism, will have an effect on the observed populations because nearly all radioactive decay will have taken place since the astrophysical production leaving (only) the stable remains. There are exceptions, of course, and contemporary research emphasizes observing recently produced radioactive nuclei in the cosmos.

Given what we know about nuclear structure, it is reasonable to consider the isotopic distribution rather than the elemental distribution. An example of the isotopic abundances of the top-row elements is shown in Figure 12.3. Once again the very strong staggering is seen and the depression of masses between 5 and 10 is more apparent. This mass region has gaps (no stable nuclei with $A = 5$ or 8), and the remaining nuclei are all relatively fragile and have small binding energies. For the lightest nuclei, the nuclei whose mass numbers are a multiple of 4 have the highest abundances. Again, simple nuclear stability considerations affect

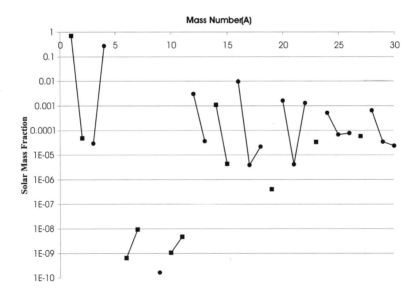

**Figure 12.3** Mass fractions of the top-row elements in the sun from Anders and Grevesse (1989). Elements with an odd atomic number are shown by square symbols, even by circles. The isotopes of a given chemical element are connected by line segments.

the amount of beryllium we find relative to the amount of carbon or oxygen, but the many orders of magnitude difference in the abundance of elements such as beryllium and carbon must be due to some effects from the production mechanisms.

The sun is a typical star (see below), and one uses the solar abundances to represent the elemental abundances in the universe (the "cosmic" abundances). It will turn out that several nucleosynthetic processes are necessary to explain the details of the observed abundances. In Figure 12.4, we jump ahead of our discussion to show a rough association between the elemental abundances and the nucleosynthetic processes that created them. Figure 12.4 is based upon a pioneering study by Burbidge, Burbidge, Fowler and Hoyle ($B^2FH$) (1957) and an independent analysis of Cameron (1957). These works have served as a framework for the discussion of nucleosynthesis since their publication in the 1950s, and we will follow a similar route in our discussion.

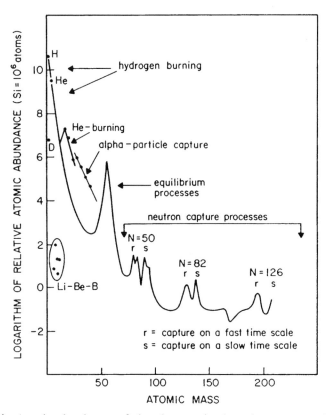

**Figure 12.4** Atomic abundances of the elements in the solar system and the major nucleosynthetic processes responsible for the observed abundances. [From E. M. Burbidge, et al., *Rev. Mod. Phys.* **29**, 547 (1957). Copyright (1957) by the American Physical Society.]

## 12.3 PRIMORDIAL NUCLEOSYNTHESIS

The universe is between 10 and 20 billion years old, with the best estimate of its age being $14 \pm 1 \times 10^9$ y old. The universe is thought to have begun with a cataclysmic explosion called the Big Bang. Since the Big Bang, the universe has been expanding with a decrease of temperature and density.

One important piece of evidence to support the idea of the Big Bang is the 2.7-K microwave radiation background in the universe. This blackbody radiation was discovered by Penzias and Wilson in 1965 and represents the thermal remnants of the electromagnetic radiation that existed shortly after the Big Bang. Weinberg (1977) tells how Penzias and Wilson found a microwave noise at 7.35 cm that was independent of direction in their radio antenna at the Bell Telephone Laboratories in New Jersey. After ruling out a number of sources for this noise, they noted a pair of pigeons had been roosting in the antenna. The pigeons were caught, shipped to a new site, reappeared, were caught again, and were "discouraged by more decisive means." The pigeons, it was noted, had coated the antenna with a "white dielectric material." After removal of this material, the microwave background was still there. It was soon realized that this 7.35-cm radiation corresponded to an equivalent temperature of the noise of about 3.5 K, which was eventually recognized as the remnants of the Big Bang. (Subsequent measurements have characterized this radiation as having a temperature of 2.7 K with a photon density of $\sim 400$ photons/cm$^3$ in the universe.)

A pictorial representation of some of the important events in the "thermal" history of the universe is shown in Figure 12.5. The description of the evolution of the universe begins at $10^{-43}$ s after the Big Bang, the so-called *Planck time*. The universe at that time had a temperature of $10^{32}$ K ($k_B T \sim 10^{19}$ GeV) and a volume that was $\sim 10^{-31}$ of its current volume. [To convert temperature in K to

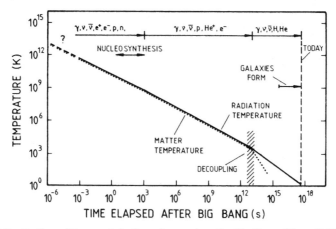

**Figure 12.5** Outline of the events in the universe since the Big Bang. (From C. E. Rolfs and W. S. Rodney, *Cauldrons in the Cosmos*, Chicago University Press, Chicago, 1988.)

energy ($k_BT$) in electron volts, note that $k_BT$ (eV) $= 8.6 \times 10^{-5}$ $T$ (K)] Matter existed in a state unknown to us, a plasma of quarks and gluons. All particles were present and in statistical equilibrium, where each particle had a production rate equal to the rate at which it was destroyed. As the universe expanded, it cooled and some species fell out of statistical equilibrium. At a time of $10^{-6}$ s ($T \sim 10^{13}$ K), the photons from the blackbody radiation could not sustain the production of the massive particles and the hadronic matter condensed into a gas of nucleons and mesons. At this point, the universe consisted of nucleons, mesons, neutrinos (and antineutrinos), photons, electrons (and positrons). The ratio of baryons to photons was $\sim 10^{-9}$.

At a time of $10^{-2}$ s ($T \sim 10^{11}$ K), the density of the universe dropped to $\sim 4 \times 10^{6}$ kg/m³. [In this photon-dominated era, the temperature $T$ (K) was given by the relation

$$T \text{ (K)} = \frac{1.5 \times 10^{10}}{\sqrt{t \text{ (s)}}}$$

where $t$ is the age in seconds.] At this time, the neutrons and protons interconvert by the weak interactions

$$\bar{\nu}_e + p \iff e^+ + n$$
$$\nu_e + n \iff p + e^-$$

[One neglects the free decay of the neutron to the proton because the half-life for that decay (10.6 m) is too long to be relevant.] The neutron–proton ratio, n : p, was determined by a Boltzmann factor, that is,

$$n : p = \exp(-\Delta mc^2/kT)$$

where $\Delta mc^2$ is the n-p mass difference (1.29 MeV). At $T = 10^{12}$ K, n/p $\sim$ 1, at $T = 10^{11}$ K, n/p $\sim$ 0.86, and so forth. At $T = 10^{11}$ K, no complex nuclei were formed because the temperature was too high to allow deuterons to form. When the temperature fell to $T = 10^{10}$ K ($t \sim 1$ s), the creation of $e^+/e^-$ pairs (pair production) ceased because $kT < 1.02$ MeV and the neutron/proton ratio was $\sim 17/83$. At a time of 225 s, this ratio was 13/87, the temperature was $T \sim 10^9$ K, then density was $\sim 2 \times 10^4$ kg/m³, and the first nucleosynthetic reactions occurred.

These primordial nucleosynthesis reactions began with the production of deuterium by the simple radiative capture process:

$$n + p \rightarrow d + \gamma$$

Notice that the deuteron can be destroyed by the absorption of a high-energy photon in the reverse process. At this time, the deuteron survived long enough to allow the subsequent reactions

$$p + d \rightarrow {}^3\text{He} + \gamma$$

and

$$n + d \longrightarrow {}^3H + \gamma$$

$^3$H and $^3$He are more strongly bound allowing further reactions that produce the very strongly bound $\alpha$ particles:

$$^3H + p \longrightarrow {}^4He + \gamma$$
$$^3He + n \longrightarrow {}^4He + \gamma$$
$$^3H + d \longrightarrow {}^4He + n$$
$$d + d \longrightarrow {}^4He + \gamma$$

Further reactions to produce the $A = 5$ nuclei do not occur because there are no stable nuclei with $A = 5$ (or $A = 8$). A small amount of $^7$Li is produced in the reactions:

$$^4He + {}^3H \longrightarrow {}^7Li + \gamma$$
$$^3He + {}^3He \longrightarrow {}^7Be + \gamma \xrightarrow{e^-} {}^7Li + \gamma$$

but the $^7$Li is also very weakly bound and is rapidly destroyed. Thus, the synthesis of larger nuclei was blocked. After about 30 m, nucleosynthesis ceased. The temperature was $\sim 3 \times 10^8$ K and the density was $\sim 30$ kg/m$^3$. (For reference, please note that water vapor at 1 atm has a density of $\sim 1$ kg/m$^3$ and liquid water has a density of $\sim 10^3$ kg/m$^3$). Nuclear matter was 76% by mass protons, 24% $\alpha$ particles with traces of deuterium, $^3$He and $^7$Li. The $\gamma$ : n : p ratio was $10^9$ : 13 : 87. The relative ratio of p : $^4$He : d : $^3$He : $^7$Li is a sensitive function of the baryon density of the universe (Fig. 12.6), a fact that is used to constrain models of the Big Bang. (The cross sections for the reactions that convert one product to another are generally known, and detailed network calculations of the reaction rates can be performed as a function of temperature and density. The resulting abundances can be compared to estimates from observations of stellar matter.) Chemistry began about $10^6$ y later, when the temperature has fallen to 2000 K and the electrons and protons could combine to form atoms. Further nucleosynthesis continues to occur in the interiors of stars.

## 12.4 STELLAR EVOLUTION

As discussed above, nucleosynthesis occurred in two steps, the primordial nucleosynthesis that occurred in the Big Bang forming only the lightest nuclei and later processes, beginning $\sim 10^6$ y after the Big Bang and continuing to the present, of nucleosynthesis in the stars. Big Bang nucleosynthesis produced hydrogen,

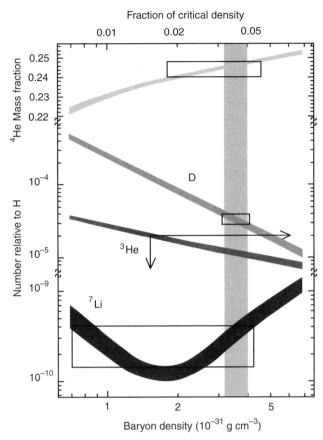

**Figure 12.6** Variation of the relative abundances of the Big Bang nuclei and the $^4$He mass fraction versus the baryon density. The boxes indicate the measured data and an estimate of the uncertainty. The curves indicate the dependence of the yield on the baryon density in the Big Bang; the vertical bar indicates the region of concordance. (Figure also appears in color figure section.)

helium, and traces of $^7$Li, whereas the rest of the elements are the result of stellar nucleosynthesis. For example, recent observations of stellar spectral lines showing the presence of $2 \times 10^5$ y $^{99}$Tc indicate ongoing stellar nucleosynthesis. To understand the nuclear reactions that make the stars shine and generate the bulk of the elements, one needs to understand how stars work. That is the focus of this section.

After the Big Bang explosion, the material of the universe was dispersed. Inhomogeneities developed, which under the influence of gravity, condensed to form the galaxies. Within these galaxies, clouds of hydrogen and helium gas can collapse under the influence of gravity. At first, the internal heat of this collapse is radiated away. As the gas becomes denser, however, the opacity increases, and the gravitational energy associated with the collapse is stored in the interior rather than being radiated into space. Eventually a radiative equilibrium is established with the development of a *protostar*. The protostar continues to shrink under the influence of gravity with continued heating of the stellar interior. When the interior temperature reaches $\sim 10^7$ K, thermonuclear reactions between the hydrogen nuclei can begin because some of the protons have sufficient kinetic energies to overcome the Coulomb repulsion between the nuclei.

The first generation of stars that formed in this way is called (for historical reasons) *population III stars*. They consisted of hydrogen and helium, were massive, had relatively short lifetimes, and are now extinct. The debris from these stars has been dispersed and was incorporated into later generation of stars.

The second generation of stars, called *population II stars*, was comprised of hydrogen, helium, and about 1% of the heavier elements such as carbon and oxygen. Finally, there was a third generation of stars, like our sun, called *population I stars*. These stars consist of hydrogen, helium, and 2–5% of the heavier elements.

Our sun is a typical population I star. It has a mass of $2.0 \times 10^{30}$ kg, a radius of $7.0 \times 10^6$ m (an average density of $1.41 \times 10^3$ kg/m$^3$), a surface temperature of $\sim 6000$ K, and a luminosity of $3.83 \times 10^{26}$ W. Our sun is $4.5 \times 10^9$ y old.

The Danish astronomer Ejnar Hertzsprung and the American astronomer Henry Norris Russell independently observed a very well defined correlation between the luminosity and surface temperature of stars. That correlation is shown in Figure 12.7 and is called a Hertzsprung–Russell, or H–R, diagram. Most stars, such as our sun, fall in a narrow band on this diagram called the *main sequence*. Stars in this main sequence have luminosities $L$ that are approximately proportional to $T_{\text{surface}}^{5.5}$, or in terms of their mass, $M$, $L \propto M^{3.5}$. How long a star stays on the main sequence will depend on its mass, which, in turn, is related to the reaction rates in its interior.

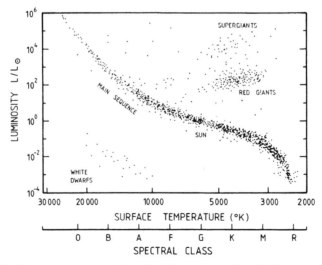

**Figure 12.7** Schematic representation of a Hertzsprung–Russell diagram. (From C. E. Rolfs and W. S. Rodney, *Cauldrons in the Cosmos*, Chicago University Press, Chicago, 1988.)

## 12.4 STELLAR EVOLUTION

In the upper-right portion of the H–R diagram, one sees a group of stars, the *red giants* or *super giants*, with large radii that are relatively cool (3000–4000 K). Stars on the main sequence move to this region when the nuclear energy liberated in the nuclear reactions occurring in the star is not enough to sustain main sequence luminosity values.

Our sun is expected to spend $\sim 7 \times 10^9$ more years on the main sequence before becoming a red giant. In a shorter time of $1.1-1.5 \times 10^9$ y, the sun will increase slowly in luminosity by $\sim 10\%$, probably leading to a cessation of life on Earth. (In short, terrestrial life has used up $\sim \frac{3}{4}$ of its allotted time since it began $\sim 3.5 \times 10^9$ y ago.)

In the lower left of the H–R diagram, one sees a group of small dense, bright stars ($T > 10^4$ K) called *white dwarfs*. The white dwarfs represent one evolutionary outcome for the red giants with masses between 0.1 and 1.4 solar masses. Red giants are helium-burning stars and as the helium is burnt, the stars become unstable, ejecting their envelopes, creating a planetary nebula, and moving across the main sequence on the H–R diagram to be white dwarfs. (See Fig. 12.7 for a schematic view of this evolution.)

For massive red giants ($M > 8$ solar masses), one finds they undergo a more spectacular death cycle, with contractions, increases in temperature leading to helium burning, carbon–oxygen burning, silicon burning, and the like with the production of the elements near iron, followed by an explosive end (Fig. 12.8).

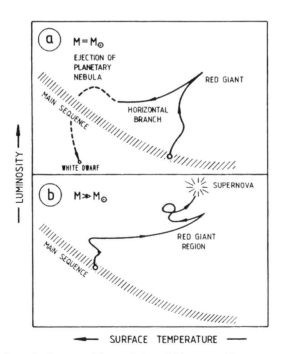

**Figure 12.8** Schematic diagram of the evolution of (*a*) a star with a mass near that of the sun and (*b*) a more massive star. (From C. E. Rolfs and W. S. Rodney, *Cauldrons in the Cosmos*, Chicago University Press, Chicago, 1988.)

The explosive end for these stars can lead to the formation of *novas and supernovas*. The name *nova* means *new* and connotes a star that undergoes a sudden increase in brightness, followed by a fading characteristic of an explosion. In this process, the outer part of the star containing only $\sim 10^{-3}$ of the stellar mass is ejected with the release of $\sim 10^{45}$ ergs. Supernovas are spectacular stellar explosions in which the stellar brightness increases by a factor of $10^6$–$10^9$, releasing $\sim 10^{51}$ ergs on a time scale of seconds. We have observed about 10 nova/year but only 2–3 supernova per century. Supernovas are classified as type I (low hydrogen, high "heavy" elements, such as oxygen to iron), and type II (primarily hydrogen, with lesser amounts of the heavy elements). Some supernovas lead to the formation of neutron stars, which are giant nuclei of essentially pure neutronic matter.

## 12.5 THERMONUCLEAR REACTION RATES

Before discussing the nuclear reactions involved in stellar nucleosynthesis, we need to discuss the rates of these reactions, which take place in a "thermal soup" as opposed to reactions studied in the laboratory. The rates of these reactions will tell us what reactions are most important. When we speak of *thermonuclear* reactions, we mean nuclear reactions in which the energy of the colliding nuclei is the thermal energy of the particles in a hot gas. Both reacting nuclei are moving, and thus it is their relative velocity (center-of-mass energy) that is important. In ordinary nuclear reactions in the laboratory, we write for the rate of the reaction $R$:

$$R = N\sigma\phi$$

where the reaction rate $R$ is in reactions/second, $\sigma$ is the reaction cross section (in cm$^2$), $\phi$ the incident particle flux in particles/second and $N$ is the number of target atoms/cm$^2$. For astrophysical reactions, we write

$$R = N_x N_y \int_0^\infty \sigma(v) v \, dv = N_x N_y \langle \sigma v \rangle$$

where $v$ is the relative velocity between nuclei $x$ and $y$, each present in a concentration of $N_i$ particles/cm$^3$, the quantity $\langle \sigma v \rangle$ is the temperature-averaged reaction rate per particle pair. To be sure that double counting of collisions between identical particles does not occur, it is conventional to express the above equation as:

$$R = \frac{N_x N_y \langle \sigma v \rangle}{1 + \delta_{xy}}$$

where $\delta_{xy}$ is the Kronecker delta (which is 0 when $x \neq y$ and 1 when $x = y$). Note the mean lifetime of nuclei $x$ is then $1/(N_x \langle \sigma v \rangle)$.

## 12.5 THERMONUCLEAR REACTION RATES

In a hot gas the velocity distribution of each component will be given by a Maxwell–Boltzmann distribution:

$$P(v) = \left(\frac{m}{2\pi kT}\right)^{3/2} \exp\left(-\frac{mv^2}{2kT}\right)$$

where $m$ is the particle mass, $k$ is Boltzmann's constant, and $T$ is the gas temperature. Integrating over all velocities for the reacting particles, $x$ and $y$, gives

$$\langle \sigma v \rangle = \left(\frac{8}{\pi \mu}\right)^{1/2} \frac{1}{(kT)^{3/2}} \int_0^\infty \sigma(E) E \exp\left(-\frac{E}{kT}\right) dE$$

where $\mu$ is the reduced mass $[m_x m_y/(m_x + m_y)]$. The rates, $R$, of stellar nuclear reactions are directly proportional to $\langle \sigma v \rangle$, which depends on the temperature $T$.

For slow neutron-induced reactions that do not involve resonances, we know (Chapter 10) that $\sigma_n(E) \propto 1/v_n$ so that $\langle \sigma v \rangle$ is a constant. For charged particle reactions, one must overcome the repulsive Coulomb force between the positively charged nuclei. For the simplest reaction, p + p, the Coulomb barrier is 550 keV. But, in a typical star such as the sun, $kT$ is 1.3 keV, that is, the nuclear reactions that occur are subbarrier, and the resulting reactions are the result of barrier penetration. (At a proton–proton center-of-mass energy of 1 keV, the barrier penetration probability is $\sim 2 \times 10^{-10}$). At these extreme subbarrier energies, the barrier penetration factor can be approximated as:

$$P = \exp\left(-\frac{2\pi Z_1 Z_2 e^2}{\hbar v}\right) = \exp\left[-31.29 Z_1 Z_2 \left(\frac{\mu}{E}\right)^{1/2}\right]$$

where $E$ is in keV and $\mu$ in amu. This tunneling probability is referred to as the *Gamow factor*. The cross section (Chapter 10) is also proportional to $\pi \lambda^2 \propto 1/E$. Thus, the cross section for nonresonant charged-particle-induced reactions can be written as:

$$\sigma(E) = \frac{1}{E} \exp\left[-31.29 Z_1 Z_2 \left(\frac{\mu}{E}\right)^{1/2}\right] S(E)$$

where the function $S(E)$, the so-called *astrophysical S factor*, contains all the constants and terms related to the nuclei involved. Substituting this expression into the equation for $\langle \sigma v \rangle$, we have

$$\langle \sigma v \rangle = \left(\frac{8}{\pi \mu}\right)^{1/2} \frac{1}{(kT)^{3/2}} \int_0^\infty S(E) \exp\left(-\frac{E}{kT} - \frac{b}{E^{1/2}}\right) dE$$

where $b$ is $0.989 Z_1 Z_2 \mu^{1/2} (\text{MeV})^{1/2}$. This equation represents the overlap between the Maxwell–Boltzmann distribution, which is peaked at low energies, and the

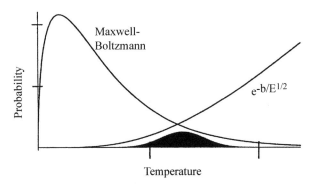

**Figure 12.9** Rate of nonresonant stellar nuclear reactions as a function of temperature. (From S. S. M. Wong, *Introductory Nuclear Physics*. Copyright © 1988 by John Wiley & Sons, Inc. Reprinted by permission of John Wiley & Sons, Inc.)

Gamow factor, which increases with increasing energy. The product of these two terms produces a peak in the overlap region of these two functions called the Gamow peak (Fig. 12.9). This peak occurs at an energy $E_0 = (bkT/2)^{2/3}$.

For reactions involving isolated single resonances or broad resonances, it is possible to derive additional formulas for $\sigma(E)$ [$R + R$] in the Breit–Wigner form, that is,

$$\sigma(E) = \pi \lambda^2 \left[ \frac{2J_r + 1}{(2J_x + 1)(2J_y + 1)} \right] \frac{\Gamma_{in}\Gamma_{out}}{(E - E_r)^2 + \frac{\Gamma_{tot}^2}{4}}$$

where $J_x$, $J_y$, $J_r$ are the spins of the interacting particles and the resonance and $\Gamma_{in}$, $\Gamma_{out}$, $\Gamma_{tot}$ are the partial widths of the entrance and exit channels and the total width, respectively.

## 12.6 STELLAR NUCLEOSYNTHESIS

### 12.6.1 Introduction

After Big Bang nucleosynthesis, we have a universe that is ~75% hydrogen and ~25% helium with a trace of $^7$Li. Stellar nucleosynthesis continues the synthesis of the chemical elements. Beginning ~$10^6$ y after the Big Bang, as described in Section 12.4, the sequence of gravitational collapse and an increasing temperature cause the onset of nuclear fusion reactions, releasing energy that stops the collapse. Starting from hydrogen and helium, fusion reactions produce the nuclei up to the maximum in the nuclear binding energy curve at $A \sim 60$. The limiting temperature of these reactions is about $5 \times 10^9$ K, where $kT \sim 0.4$ MeV. A rough outline of the nuclear reactions involved is given in Table 12.1.

The products from these reactions are distributed into the galaxies by slow emission from the red giants and by the catastrophic explosions of novas and supernovas. This dispersed material condenses in the population II and later the population I stars

**TABLE 12.1  Nuclear Reactions Involved in Stellar Nucleosynthesis**

| Fuel | T (K) | kT (MeV) | Products |
|---|---|---|---|
| $^1$H | $5 \times 10^7$ | 0.002 | $^4$He |
| $^4$He | $2 \times 10^8$ | 0.02 | $^{12}$C, $^{16}$O, $^{20}$Ne |
| $^{12}$C | $8 \times 10^8$ | 0.07 | $^{16}$O, $^{20}$Ne, $^{24}$Mg |
| $^{16}$O | $2 \times 10^9$ | 0.2 | $^{20}$Ne, $^{28}$Si, $^{32}$S |
| $^{20}$Ne | $1.5 \times 10^9$ | 0.13 | $^{16}$O, $^{24}$Mg |
| $^{28}$Si | $3.5 \times 10^9$ | 0.3 | $A < 60$ |

where additional nuclear reactions (see below) create the odd $A$ nuclei and sources of free neutrons. These neutrons allow us to get slow neutron capture reactions (s process) synthesizing many of the nuclei with $A > 60$. High-temperature photonuclear reactions and rapid neutron capture reactions in supernovas complete the bulk of the nucleosynthesis reactions.

### 12.6.2  Hydrogen Burning

The first stage of stellar nucleosynthesis, which is still occurring in stars such as our sun, is *hydrogen burning*. In hydrogen burning, protons are converted to $^4$He nuclei. Since there are no free neutrons present, the reactions differ from those of Big Bang nucleosynthesis. The first reaction that occurs is

$$p + p \longrightarrow d + e^+ + \nu_e \qquad Q = 0.42 \, \text{MeV}$$

where most of the released energy is shared between the two leptons. In our sun, $T \sim 15 \times 10^6$ K ($kT \sim 1$ keV). The proton–proton (pp) reaction is a weak interaction process and has, therefore, a very small cross section, $\sim 10^{-47}$ cm$^2$, at these proton energies. The resulting reaction rate is $5 \times 10^{-18}$ reactions/second/proton.

There is an improbable (0.4%) variant of this reaction, called the *pep reaction* that also leads to deuteron production. It is

$$p + e^- + p \longrightarrow d + \nu_e \qquad Q = 1.42 \, \text{MeV}$$

This rare reaction is a source of energetic neutrinos from the sun.

The next reaction in the sequence is

$$d + p \longrightarrow {}^3\text{He} + \gamma \qquad Q = 5.49 \, \text{MeV}$$

leading to the synthesis of $^3$He. The rate of this strong interaction is $\sim 10^{16}$ times greater than the weak p + p reaction. The product $^3$He can undergo two possible reactions. In $\sim$86% of the cases, the reaction is

$$^3\text{He} + {}^3\text{He} \longrightarrow {}^4\text{He} + 2p \qquad Q = 12.96 \, \text{MeV}$$

This reaction, when combined with the two previous reactions (p + p and d + p) corresponds to an overall reaction of

$$4p \longrightarrow {}^4\text{He} + 2e^+ + 2\nu_e \qquad A = 26.7\,\text{MeV}$$

This sequence of reactions is called the ppI chain and is responsible for 91% of the sun's energy. A schematic view of this reaction is shown in Figure 12.10.

In ~14% of the cases, the $^3$He product undergoes the side reaction with an α particle:

$$^3\text{He} + {}^4\text{He} \longrightarrow {}^7\text{Be} + \nu_e$$

The $^7$Be undergoes subsequent electron capture decay:

$$e^- + {}^7\text{Be} \longrightarrow {}^7\text{Li} + \nu_e \qquad Q = 0.86\,\text{MeV}$$

Note that this EC decay process does *not* involve capture of the orbital electron of the $^7$Be since it is fully ionized in a star but rather involves capture of a free continuum electron. As a consequence, the half-life of this decay is ~120 d rather than the terrestrial half-life of 77 d. The resulting $^7$Li undergoes proton capture as:

$$p + {}^7\text{Li} \longrightarrow 2\,{}^4\text{He}$$

This sequence of reactions [p + p, d + p, $^3$He + $^4$He, $^7$Be EC, $^7$Li(p,α)] constitutes the ppII process, which accounts for ~7% of the sun's energy.

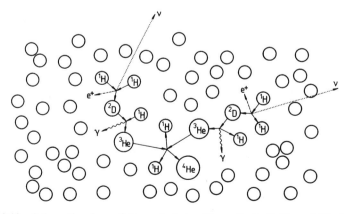

**Figure 12.10** Schematic view of the ppI chain. (From C. E. Rolfs and W. S. Rodney, *Cauldrons in the Cosmos*, Chicago University Press, Chicago, 1988.)

A small fraction of the $^7$Be from the $^3$He + $^4$He reaction can undergo proton capture, leading to

$$^7\text{Be} + p \rightarrow {}^8\text{B} + \gamma$$

$$^8\text{B} \rightarrow {}^8\text{Be}^* + e^+ + \nu_e$$

$$^8\text{Be}^* \rightarrow 2\,^4\text{He}$$

This sequence [p + p, p + d, $^3$He + $^4$He, $^7$Be(p,$\gamma$), $^8$B → $^8$Be* → 2 $^4$He] constitutes the ppIII chain (which provides about 0.015% of the sun's energy). In each of the pp processes, some energy is carried away by the emitted neutrinos. Quantitatively, in the ppI process, the loss is 2%, in the ppII process 4%, and 28.3% in the ppIII process. These pp chains are shown together in Figure 12.11.

In population II and population I stars, heavy elements such as carbon, nitrogen, and oxygen (CNO) are present, leading to the occurrence of another set of nuclear reactions whose net effect in the conversion $4p \rightarrow {}^4\text{He} + 2e^+ + 2\nu_e$. The "heavy" nuclei act as catalysts for this reaction. A typical cycle is

$$^{12}\text{C} + p \rightarrow {}^{13}\text{N} + \gamma$$

$$^{13}\text{N} \rightarrow {}^{13}\text{C} + e^+ + \nu_e$$

$$^{13}\text{C} + p \rightarrow {}^{14}\text{N} + \gamma$$

$$^{14}\text{N} + p \rightarrow {}^{15}\text{O} + \gamma$$

$$^{15}\text{O} \rightarrow {}^{15}\text{N} + e^+ + \nu_e$$

$$^{15}\text{N} + p \rightarrow {}^{12}\text{C} + {}^4\text{He}$$

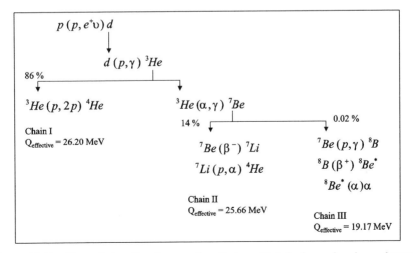

**Figure 12.11** Three chains of nuclear reactions that constitute hydrogen burning and convert protons into $^4$He. The rate-limiting step in all reactions is the first reaction to create the deuterium.

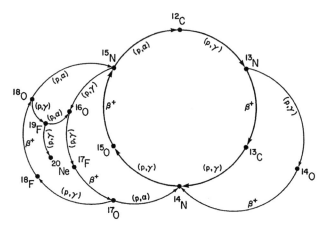

**Figure 12.12** CNO cycle with side chains. (From S. S. M. Wong, *Introductory Nuclear Physics*. Copyright © 1988 by John Wiley & Sons, Inc. Reprinted by permission of John Wiley & Sons, Inc.)

This group of reactions is referred to as *the CNO cycle* and is favored at higher temperatures where the Coulomb barrier for these reactions can be more easily overcome. In our sun, 98% of the energy comes from the pp chain and only 2% from the CNO cycle. Several side chains of this reaction cycle are possible, as illustrated in Figure 12.12.

### 12.6.3 Helium Burning

Eventually, the hydrogen fuel of the star will be exhausted and further gravitational collapse will occur. This will give rise to a temperature increase up to $\sim 1\text{--}2 \times 10^8$ K (with a density of $\sim 10^8$ kg/m$^3$). In this red giant, *helium burning* will commence.

One might think the first reaction might be

$$^4\text{He} + {}^4\text{He} \longrightarrow {}^8\text{Be} \qquad Q = -0.0191 \, \text{MeV}$$

but $^8$Be is unstable ($t_{1/2} = 6.7 \times 10^{-17}$ s), and thus that process is hindered by the short lifetime and low transient population of the Be nuclei. Instead one gets the so-called 3α process:

$$3\,{}^4\text{He} \longrightarrow {}^{12}\text{C} \qquad Q = 7.37 \, \text{MeV}$$

Three-body reactions are rare, but the reaction proceeds through a resonance in $^{12}$C at 7.65 MeV corresponding to the second excited state of $^{12}$C ($J\pi = 0^+$). This excited state has a more favorable configuration than the $^{12}$C ground state for allowing the collision to occur. (In a triumph for nuclear astrophysics, the existence of this state was postulated by astrophysicists to explain nucleosynthetic rates before it was found in the laboratory.)

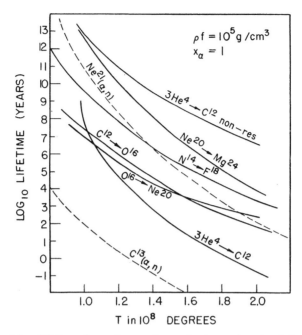

**Figure 12.13** Mean lifetimes for various nucleosynthesis reactions involving α particles as a function of temperature. The mean lifetime is inversely related to the reaction rate. [From E. M. Burbidge et al., *Rev. Mod. Phys.*, **29**, 547 (1957). Copyright (1957) by the American Physical Society.]

After a significant amount of $^{12}$C is formed, one gets the α-capture reactions:

$$^4\text{He} + ^{12}\text{C} \rightarrow ^{16}\text{O} + \gamma \quad Q = 7.16\,\text{MeV}$$

$$^4\text{He} + ^{16}\text{O} \rightarrow ^{20}\text{Ne} + \gamma \quad Q = 4.73\,\text{MeV}$$

A brief interlude of *neon burning* then occurs with reactions such as:

$$^{20}\text{Ne} + \gamma \rightarrow ^4\text{He} + ^{16}\text{O}$$

$$^4\text{He} + ^{20}\text{Ne} \rightarrow ^{24}\text{Mg} + \gamma$$

The relative rates of these and related processes are shown in Figure 12.13.

### 12.6.4 Synthesis of Nuclei with A < 60

Eventually, the helium of the star will be exhausted, leading to further gravitational collapse with a temperature increase to $6 \times 10^8$–$2 \times 10^9$ K ($kT \sim 100$–$200$ keV). At this point the fusion reactions of the "α-cluster" nuclei are possible. For example, carbon and oxygen burning occurs in charged particle reactions such as:

$$^{12}C + {}^{12}C \rightarrow {}^{20}Ne + {}^{4}He$$
$$^{12}C + {}^{12}C \rightarrow {}^{23}Na + p$$
$$^{12}C + {}^{12}C \rightarrow {}^{23}Mg + n$$
$$^{12}C + {}^{12}C \rightarrow {}^{24}Mg + \gamma$$
$$^{16}O + {}^{16}O \rightarrow {}^{24}Mg + 2\,{}^{4}He$$
$$^{16}O + {}^{16}O \rightarrow {}^{28}Si + {}^{4}He$$
$$^{16}O + {}^{16}O \rightarrow {}^{31}P + p$$
$$^{16}O + {}^{16}O \rightarrow {}^{31}S + n$$
$$^{16}O + {}^{16}O \rightarrow {}^{32}S + \gamma$$

with the production of $^{28}$Si and $^{32}$S being the most important branches of the oxygen-burning reactions. Further rises in temperature up to $5 \times 10^9$ K result in a series of silicon-burning reactions involving an equilibrium between photodisintegration and radiative capture processes such as:

$$^{28}Si + \gamma \rightarrow {}^{24}Mg + {}^{4}He$$
$$^{4}He + {}^{28}Si \rightarrow {}^{32}S + \gamma$$

The nuclei up to $A \sim 60$ are produced in these equilibrium processes. In such equilibrium processes, the final yields of various nuclei are directly related to their nuclear stability with the more stable nuclei having higher yields. So one observes greater yields of even–even nuclei than odd $A$ nuclei (due to the pairing term in the mass formula) and even $N$ isotopes are more abundant than odd $A$ isotopes of an element.

The relative time scales of the various reactions leading to nuclei with $A < 60$ are shown in Table 12.2. Note these time scales are inversely proportional to the reaction rates.

**TABLE 12.2 Time Scales of Nucleosynthetic Reactions in a 1 Solar Mass Star**

| Reaction | Time |
|---|---|
| H burning | $6 \times 10^9$ y |
| He burning | $0.5 \times 10^6$ y |
| C burning | 200 y |
| Ne burning | 1 y |
| O burning | Few months |
| Si burning | Days |

## 12.6.5 Synthesis of Nuclei with $A > 60$

The binding energy per nucleon curve peaks at $A \sim 60$ and decreases as $A$ increases beyond 60. This indicates that fusion reactions using charged particles are not energetically favored to make these nuclei. However, another possible nuclear reaction is neutron capture, that is, (n, $\gamma$). These reactions have no Coulomb barriers to inhibit them, and the rates are then governed by the Maxwell–Boltzmann distribution of velocities in a hot gas and the availability of free neutrons. If $\sigma(n, \gamma) \sim 1/v$, then the reaction rate $N_n \langle \sigma v \rangle$ is largely governed by $N_n$, the neutron density. We can identify two types of neutron capture processes for nucleosynthesis. The first of these is *slow neutron capture*, the so-called s process, where the time scale of the neutron capture process, $\tau_{\text{reaction}} \gg \tau_\beta$, where $\tau_\beta$ is the $\beta^-$-decay lifetime. In this process, each neutron capture proceeds in competition with $\beta^-$ decay. For example, consider the stable nucleus $^{56}$Fe. If it captures a neutron, the following reactions can occur:

$$^{56}\text{Fe} + \text{n} \longrightarrow {}^{57}\text{Fe (stable)} + \gamma$$

$$^{57}\text{Fe} + \text{n} \longrightarrow {}^{58}\text{Fe (stable)} + \gamma$$

$$^{58}\text{Fe} + \text{n} \longrightarrow {}^{59}\text{Fe}\ (t_{1/2} = 44.5\,\text{d}) + \gamma$$

Then, 44.5 d $^{59}$Fe will undergo $\beta^-$ decay before another neutron is captured, that is,

$$^{59}\text{Fe} \longrightarrow {}^{59}\text{Co (stable)} + \beta^- + \bar{\nu}_e$$

and further captures will start with $^{59}$Co. The mean times of neutron capture reactions $\tau_{\text{reaction}} = \ln 2/\text{rate} = \ln 2/N_n \langle \sigma v \rangle$. If $N_n \sim 10^{11}/\text{m}^3$, $\sigma = 0.1$ barn, $E_n \sim 50$ keV, then $\tau \sim 10^5$ y. Then, one will see neutron capture by all stable nuclei and many of the long-lived nuclei. A typical s-process path of nucleosynthesis for the nuclei with $Z = 45$–60 is shown in Figure 12.14. The production of nuclei follows a zig-zag path through the chart of nuclides, with increases in mass when a neutron is captured and increases in atomic number when $\beta$ decay precedes the next neutron capture.

The s process terminates at $^{209}$Bi because of the cyclic sequence:

$$^{209}\text{Bi}(n, \gamma)^{210}\text{Bi} \xrightarrow{\beta^-} {}^{210}\text{Po} \xrightarrow{\alpha} {}^{206}\text{Pb} \xrightarrow{3(n,\gamma)} {}^{209}\text{Pb} \xrightarrow{\beta^-} {}^{209}\text{Bi}$$

The source of the neutrons for the s process is ($\alpha$, n) reactions on n-rich nuclei such as $^{13}$C or $^{21}$Ne, with the latter being most important. In population II and I stars, one can get side reactions such as:

$$^{20}\text{Ne}(p, \gamma)^{21}\text{Na}$$

$$^{21}\text{Na} \longrightarrow {}^{21}\text{Ne} + e^+ + \nu_e$$

that produce the target nuclei for the ($\alpha$, n) reactions.

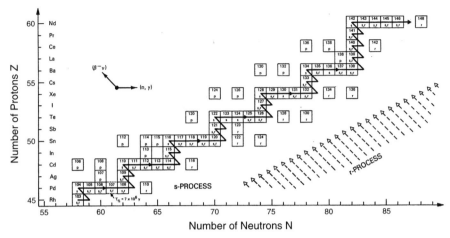

**Figure 12.14** Section of the chart of nuclides showing the s-process path. (From C. E. Rolfs and W. S. Rodney, *Cauldrons in the Cosmos*, Chicago University Press, Chicago, 1988.)

For the slow neutron capture process, there is an equilibrium between the production and loss of adjacent nuclei. Stable nuclei are only destroyed by neutron capture. For such nuclei, we can write for the rate of change of a nucleus with mass number $A$:

$$\frac{dN_A}{dt} = \sigma_{A-1} N_{A-1} - \sigma_A N_A$$

where $\sigma_i$ and $N_i$ are the capture cross sections and number of nuclei (abundance) for nucleus $i$, respectively. At equilibrium,

$$\frac{dN_A}{dt} = 0$$

Thus

$$\sigma_{A-1} N_{A-1} = \sigma_A N_A$$

This relationship between the abundances of neighboring stable nuclei is a signature for the s process.

If the time scale of neutron capture reactions is very much less than $\beta^-$-decay lifetimes, then *rapid neutron capture* or the *r process* occurs. For r-process nucleosynthesis, one needs large neutron densities, $\sim 10^{28}/\text{m}^3$, which lead to capture times of the order of fractions of a second. The astrophysical environment where such processes can occur is now thought to be in supernovas. In the r process, a large number of sequential captures will occur until the process is terminated by neutron emission or, in the case of the heavy elements, fission or $\beta$-delayed fission. The lighter "seed" nuclei capture neutrons until they reach the point where $\beta^-$-decay lifetimes have

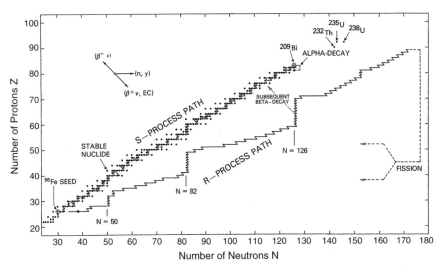

**Figure 12.15** Neutron capture paths are shown for the s and r processes. (From C. E. Rolfs and W. S. Rodney, *Cauldrons in the Cosmos*, Chicago University Press, Chicago, 1988.)

decreased and $\beta^-$ decay will compete with neutron capture. The r process is responsible for the synthesis of all nuclei with $A > 209$ and many lower mass nuclei. In a plot of abundances vs. mass number $A$ (Fig. 12.4) one sees two peaks in the abundance distributions near the magic neutron numbers ($N = 50, 82, 126$). The lower $A$ peak is due to the r process, which reaches the magic number of neutrons at a lower $Z$ value than the s process. The peaks occur because of the relative stability of $N = 50$, 82, 126 nuclei against neutron capture. A typical r-process path is shown in Figure 12.15. Notice that the path climbs up in atomic number along the neutron magic numbers. The nuclei in each zig-zag region are the places of maxima in the isotopic yields after decay.

Another important process leading to the synthesis of some proton-rich nuclei with $70 < A < 200$ is the so-called *p process*. The p process consists of a series of photonuclear reactions $(\gamma, p)$, $(\gamma, \alpha)$, $(\gamma, n)$ on "seed" nuclei from the s or r processes that produce these nuclei. (Originally, it was believed that proton capture processes during supernovas were responsible for these nuclei, but it was found that the proton densities are too small to explain the observed abundances.) The temperature during a supernova explosion is $\sim 3 \times 10^9$ K, producing blackbody radiation that can cause photonuclear reactions. The p-process contribution to the abundances of most elements is very small, but there are some nuclei ($^{190}$Pt, $^{168}$Yb) that seem to have been exclusively made by this process. The relative importance of s, r, and p processes in nucleosynthesis in a given region is shown in Figure 12.16.

A process that is sometimes related to the p process is the *rp process*, the *rapid proton capture* process. This process makes proton-rich nuclei with $Z = 7 - 26$. This process involves a set of $(p, \gamma)$ and $\beta^+$ decays that populate the p-rich nuclei. The process starts as a "breakout" from the CNO cycle, that is, a side

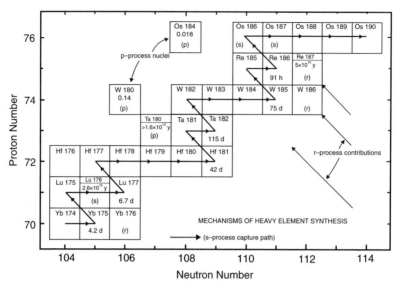

**Figure 12.16** Typical portion of the heavy element chart of the nuclides showing the relative importance of s, r, and p processes in nucleosynthesis. [From J. W. Truran, "Nucleosynthesis" in *Ann. Rev. Nucl. Part. Sci.*, **34**, 53 (1984). Copyright © 1984 by Annual Reviews, Inc. Reprinted by permission of Annual Reviews, Inc.]

chain of the CNO cycle that produces p-rich nuclei such as $^{21}$Na and $^{19}$Ne. These "seed" nuclei can form the basis for further proton captures, leading to the nucleosynthetic path shown in Figure 12.17. The rp process creates a small number of nuclei with $A < 100$. The process follows a path analogous to the r process but on the proton-rich side of stability. At present, the source of the protons for this process are certain binary stars. Note this process while close to the line of β stability, approaches the proton drip line as the nuclei become heavier.

## 12.7 SOLAR NEUTRINO PROBLEM

### 12.7.1 Introduction

Many of the nuclear reactions that provide the energy of the stars also result in the emission of neutrinos. Because of the small absorption cross sections for neutrinos interacting with matter ($\sigma_{abs} \sim 10^{-44}$ cm$^2$), these neutrinos are not generally absorbed in the sun and other stars. (This loss of neutrinos corresponds to a loss of ~2% of the energy of our sun.) Because of this, the neutrinos are a window into the stellar interior. The small absorption cross sections also make neutrinos difficult to detect, with almost all neutrinos passing through planet Earth without interacting.

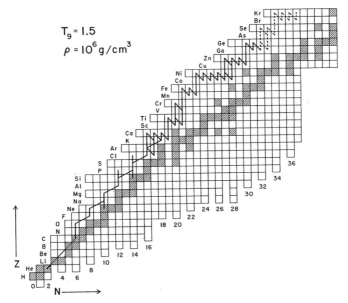

**Figure 12.17** Position of the rp process relative to the line of β stability.

Recently, a good deal of attention has been given to the "solar neutrino problem" and its solution. The 2002 Nobel Prize in physics was awarded to Ray Davis and Masatoshi Koshita for their pioneering work on this problem. Of special interest is the important role of nuclear and radiochemistry in this work as Davis is a nuclear chemist. The definition and solution of this problem is thought to be one of the major scientific advances of recent years.

### 12.7.2 Expected Solar Neutrino Sources, Energies, and Fluxes

The sun is a major source of neutrinos reaching the surface of Earth due to its close proximity. The sun emits $\sim 1.8 \times 10^{38}$ neutrinos/second, which, after an $\sim$8-min transport time, reach the surface of Earth at a rate of $6.4 \times 10^{10}$ neutrinos/s/cm². The predictions of the standard solar model for the neutrino fluxes at the surface of Earth due to various nuclear reactions are shown in Table 12.3.

The predicted energy distributions are shown in Figure 12.18. Each nuclear reaction has a characteristic neutrino energy distribution.

The source labeled "pp" in the Table 12.3 and Figure 12.18 refers to the reaction

$$p + p \longrightarrow d + e^+ + \nu_e$$

and is the most important reaction, producing one neutrino for each $^4$He nucleus made. The "pep" source is the reaction

$$p + p + e^- \longrightarrow d + \nu_e$$

**TABLE 12.3  Predicted Solar Neutrino Fluxes (Bahcall and Pena-Garay)**

| Source | Flux (particles/s/cm$^2$) |
|---|---|
| pp | $5.94 \times 10^{10}$ |
| pep | $1.40 \times 10^{8}$ |
| hep | $7.88 \times 10^{3}$ |
| $^7$Be | $4.86 \times 10^{7}$ |
| $^8$B | $5.82 \times 10^{6}$ |
| $^{13}$N | $5.71 \times 10^{8}$ |
| $^{15}$O | $5.03 \times 10^{8}$ |
| $^{17}$F | $5.91 \times 10^{6}$ |

which produces monoenergetic neutrinos, while the source "hep" refers to the reaction

$$p + {}^3He \rightarrow {}^4He + e^+ + \nu_e$$

This latter reaction produces the highest energy neutrinos with a maximum energy of 18.77 MeV due to the high reaction $Q$ value. The intensity of this source is about

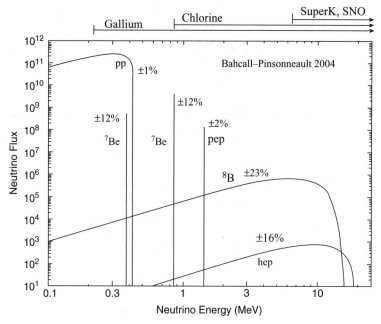

**Figure 12.18** Log–log plot of predicted neutrino fluxes from various solar nuclear reactions. The energy regions to which the neutrino detectors are sensitive are shown at the top. [From Bahcall (from Bahcall website).]

$10^7$ less than the pp source. The "$^7$Be" source refers to the pp chain electron capture decay reaction

$$e^- + {}^7\text{Be} \longrightarrow {}^7\text{Li} + \nu_e$$

This reaction produces two groups of neutrinos, one in which the ground state of $^7$Li is populated (90% abundance) and one in which the 0.477-MeV excited state is populated (10% abundance). The source "$^8$B" refers to the positron decay

$$^8\text{B} \longrightarrow {}^8\text{Be}^* + e^+ + \nu_e$$

in which the first excited state of $^8$Be (at 3.040 MeV) is populated. The weak sources "$^{13}$N," "$^{15}$O," and "$^{17}$F" refer to $\beta^+$ decays that occur in the CNO cycle, that is,

$$^{13}\text{N} \longrightarrow {}^{13}\text{C} + \beta^+ + \nu_e$$
$$^{15}\text{O} \longrightarrow {}^{15}\text{N} + \beta^+ + \nu_e$$
$$^{17}\text{F} \longrightarrow {}^{17}\text{O} + \beta^+ + \nu_e$$

### 12.7.3 Detection of Neutrinos

As indicated above, the detection of these weakly interacting neutrinos is difficult because of the low absorption cross sections. Two classes of detectors have been used to overcome this obstacle, radiochemical detectors and Cerenkov detectors. Radiochemical detectors rely on detecting the products of neutrino-induced nuclear reactions, whereas the Cerenkov detectors observe the scattering of neutrinos. The most famous radiochemical detector is that constructed by Davis and co-workers in the Homestake Gold Mine in South Dakota. In a cavern about 1500 m below the surface of Earth, a massive detector, consisting of 100,000 gal of a cleaning fluid, $C_2Cl_4$, was mounted. The cleaning fluid weighed 610 tons and corresponded to the volume of 10 railway tanker cars. The nuclear reaction occurring in the detector was

$$\nu_e + {}^{37}\text{Cl} \longrightarrow {}^{37}\text{Ar} + e^-$$

The $^{37}$Ar product nucleus decays by electron capture with a 35-d half-life. After the cleaning fluid has been exposed to solar neutrinos for a period of time, the individual $^{37}$Ar product nuclei are flushed from the detector with a stream of He gas and put into a proportional counter where the 2.8-keV Auger electrons from the EC decay are detected. The detection reaction has a threshold of 0.813 MeV, making it sensitive to the $^8$B, hep, pep, and $^7$Be (ground-state decay) neutrinos with the $^8$B being the most important. Typically ~3 atoms of $^{37}$Ar are produced per week and must be isolated from the $10^{30}$ atoms of cleaning fluid in the tank, a radiochemical *tour de force*.

The detector was placed deep underground to shield against cosmic ray background reactions.

The Davis or Cl detector was the detector used to define the "solar neutrino problem," and another type of radiochemical detector, the SAGE/GALLEX detectors, was used to further define the problem. These detectors, GALLEX in Italy and SAGE in Russia, are based on the reaction

$$\nu_e + {}^{71}\text{Ga} \longrightarrow {}^{71}\text{Ge} + e^-$$

These detectors have a threshold of 0.232 MeV and can be used to directly detect the dominant pp neutrinos from the sun. The gallium is present as a solution of $GaCl_3$. The $^{71}$Ge is collected by sweeping the detector solution with $N_2$ and converting the Ge to $GeH_4$ before counting. These detectors utilize 30–100 tons of gallium and contain a significant fraction of the world's yearly gallium production.

The Cerenkov detectors involve the scattering of neutrinos by charged particles, causing the charged particles to emit Cerenkov radiation that can be detected by scintillation detectors. The first of these detectors were placed in a mine at Kamioka, Japan. The largest of the detectors at Kamioka is called Super Kamiokande and consists of 50,000 tons of high-purity water. The detection reaction in this case is a scattering reaction

$$\nu + e^- \longrightarrow \nu + e^-$$

and the detection threshold is about 8 MeV, allowing one to observe the $^8$B neutrinos.

The Sudbury Neutrino Observatory (SNO) detector, located at Sudbury, Ontario, Canada, consists of 1000 tons of heavy water ($D_2O$) mounted ~2 km below the surface in the Sudbury Nickel Mine. In addition to neutrino–electron scattering, this detector can also utilize nuclear reactions involving deuterium:

$$\nu_e + d \longrightarrow 2p + e^-$$
$$\nu + d \longrightarrow n + p + \nu$$

where, in the latter reaction, the reaction occurs for all types of neutrinos, $\nu_e$, $\nu_\mu$, and $\nu_\tau$. The former reaction is sensitive to electron neutrinos only. These different types of reactions can be exploited to look for neutrino oscillations (see below). In the latter reaction, the emitted neutron is detected by an (n,γ) reaction in which the γ ray is detected by scintillation detectors. (The heavy water of the detector is surrounded by 7000 tons of ordinary water to shield against neutrons from radioactivity in the rock walls of the mine.) This detector also poses radiochemical challenges as the water purity must be such that there are <10 U or Th atoms per $10^{15}$ water molecules.

## 12.7.4 Solar Neutrino Problem

The solar neutrino "problem" was identified by the first results of Davis et al. using the Cl detector at the Homestake Mine. Davis et al. observed only about one-third of the expected solar neutrino flux as predicted by standard models of the sun, which assume 98.5% of the energy is from the pp chain and 1.5% of the energy is from the CNO cycle. The final result of the Cl detector experiment is that the observed solar neutrino flux is 2.1 ± 0.3 SNU compared to the predicted 7.9 ± 2.4 SNU, where the solar neutrino unit (SNU) is defined as $10^{-36}$ neutrino captures/second/target atom. The GALLEX and SAGE detectors subsequently reported solar neutrino fluxes of 77 ± 10 SNU and 69 ± 13 SNU, which are to be compared to the standard solar model prediction of 127 SNU for the neutrinos detected by these reactions.

Such large discrepancies indicated that clearly either the models of the sun were wrong or something fundamental was wrong in our ideas of the nuclear physics involved.

## 12.7.5 Solution of the Problem—Neutrino Oscillations

The solution to the solar neutrino problem is that something is wrong with our ideas of the fundamental structure of matter, the so-called standard model. This difficulty takes the form of "neutrino oscillations" as the solution to the solar neutrino

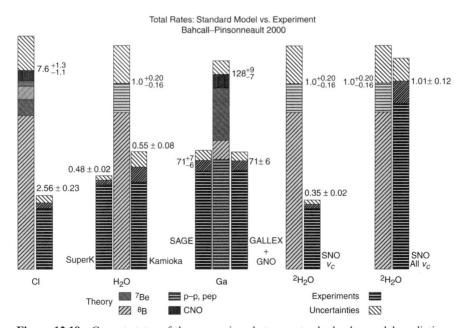

**Figure 12.19** Current status of the comparison between standard solar model predictions and experimental measurements. [From Bahcall (from Bahcall website http://www.sns.ias.edu/~jnb).] (Figure also appears in color figure section.)

problem. (The standard model predicts that the three types of neutrinos are massless and that, once created, they retain their identity for all time.) The basic idea is that as neutrinos come from the sun, they "oscillate," that is, they change from being electron neutrinos to muon neutrinos and back, and so on. This oscillation is possible if the neutrinos have mass and there is a mass difference between the electron and muon neutrinos and that they have mass. These neutrino oscillations are enhanced by neutrino–electron interactions in the sun. It is believed that the mass$_{\tau\text{neutrino}}$ > mass$_{\mu\text{neutrino}}$ > mass$_{\text{electron neutrino}}$. The current upper limits on these masses are

$$m(v_e) < 2.2 \, \text{eV}$$
$$m(v_\mu) < 170 \, \text{keV}$$
$$m(v_\tau) < 15.5 \, \text{MeV}$$

The direct observational evidence for the occurrence of neutrino oscillations came from observations with the Cerenkov detectors. The SNO detector found one-third the expected number of electron neutrinos coming from the sun in agreement with previous work with the radiochemical detectors. The Super Kamiokande detector, which is primarily sensitive to electron neutrinos, but has some sensitivity to other neutrino types found about one-half the neutrino flux predicted by the standard

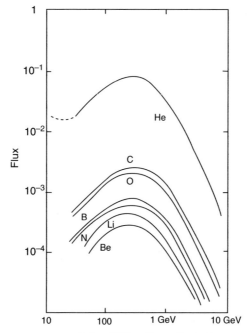

**Figure 12.20** Energy spectrum of the GCR. (From J. Audouze and S. Vauclair, *An Introduction to Nuclear Astrophysics*, Reidel, Dordrecht, 1980.)

solar models. If all neutrino types behaved similarly, the SNO and Super Kamiokande detectors should have detected the same fraction of neutrinos. Further experiments with the SNO detector operating in a mode to simultaneously detect all types of neutrinos found neutrino fluxes in agreement with the solar models. This situation is summarized in Figure 12.19.

## 12.8 SYNTHESIS OF Li, Be, AND B

Big Bang nucleosynthesis is responsible for the synthesis of hydrogen and helium and some of the $^7$Li. (Stellar nucleosynthesis in main sequence stars transforms about 7% of the hydrogen into $^4$He.) However, neither stellar nucleosynthesis or Big Bang nucleosynthesis can produce the observed abundances of Li, Be, and B. Consequently, the abundances of Li, Be, and B are suppressed by a factor of $10^6$ relative to the abundances of the neighboring elements (Fig. 12.2).

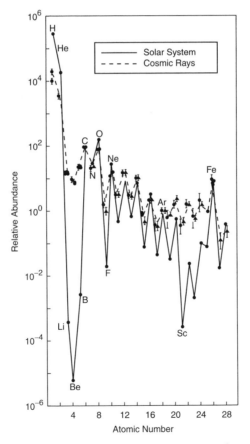

**Figure 12.21** Relative elemental abundances in the solar system and cosmic rays. (From C. E. Rolfs and N. S. Rodney, *Cauldrons in the Cosmos*, Chicago University Press, Chicago, 1988.)

The extremely low abundance is the result of two factors, the relative fragility of the isotopes of Li, Be, and B and the high binding energy of $^4$He, which makes the isotopes of Li, Be, and B unstable with respect to decay/reactions that lead to $^4$He. For example, the nuclei $^6$Li, $^7$Li, $^9$Be, $^{11}$B, and $^{10}$B are destroyed by stellar proton irradiations at temperatures of 2.0, 2.5, 3.5, 5.0, and 5.3 × 10$^6$ K, respectively. Thus, these nuclei cannot survive the stellar environment. (Only the rapid cooling following the Big Bang allows the survival of the products of primordial nucleosynthesis.)

Li, Be, and B are believed to be produced in spallation reactions in which the interstellar $^{12}$C and $^{16}$O interact with protons in the galactic cosmic rays (GCR). These reactions are high-energy reactions with thresholds of 10–20 MeV. The energy spectrum of the GCR is shown in Figure 12.20.

Typical cross sections for these spallation reactions are ~1–100 mb for $E_p > 0.1$ GeV. The time scale of the irradiation is ~$10^{10}$ y. The product nuclei are not subject to high temperatures after synthesis and can survive. A further testimonial to this mechanism is the relative abundances of the elements in the GCR relative to the solar abundances (Fig. 12.21), which shows the enhanced yields of Li, Be, and B in the GCR. This pattern is similar to the yield distributions of the fragments from the reactions of high-energy projectiles.

## PROBLEMS

1. Assume an absorption cross section of $10^{-44}$ cm$^2$ for solar neutrinos interacting with matter. Calculate the probability of a neutrino interacting as it passes through Earth.

2. What is the most probable kinetic energy of a proton in the interior of the sun ($T = 1.5 \times 10^7$ K). What fraction of these protons has an energy greater than 0.5 MeV?

3. If we want to study the reaction of $^4$He with $^{16}$O under stellar conditions, what laboratory energy would we use for the $^4$He?

4. If Earth was a neutron star, what would be its radius and density?

5. If the interior temperature of the sun is $1.5 \times 10^7$ K, what is the peak energy of the p + $^{14}$N → $^{15}$O + γ reaction.

6. Which nucleosynthetic processes are responsible for the following nuclei: $^7$Li, $^{12}$C, $^{20}$Ne, $^{56}$Fe, $^{84}$Sr, $^{96}$Zr, $^{114}$Sn, $^{124}$Sn, $^{209}$Bi, $^{238}$U.

7. Outline how you would construct a radiochemical neutrino detector based upon $^{115}$In.

8. Estimate the Coulomb barrier height for the following pairs of nuclei: (a) p + p, (b) $^{16}$O + $^{16}$O, and (c) $^{28}$Si + $^{28}$Si.

9. Calculate the rate of fusion reactions in the sun. Be sure to correct for the energy loss due to neutrino emission.

10. Assuming the sun will continue to shine at its present rate, calculate how long the sun will shine.

11. From the data given on the Davis detector, and the assumption that the $^{37}$Ar production rate is 0.5 atoms/day, calculate the neutrino capture rate in SNU. Assume the effective cross section for the $^8$B neutrinos is $10^{-42}$ cm$^2$.

12. Calculate the evolution of the n : p ratio in the primordial universe from the information given as the temporal dependence of the temperature.

13. Make an estimate of the neutron-to-proton ratio in the center of the sun if the only source of neutrons is thermal equilibrium of the weak interactions.

14. Using the information on the r-process and the s-process paths in Figures 12.14 and 12.15, make estimates of the average atomic numbers of the nuclei in the peaks for $N = Z$ in the mass abundance curves. Do the masses of these nuclei correspond to the peaks in Figure 12.4?

## REFERENCES

Anders, E. and N. Grevasse. *Geochim. Cosmochem. Acta* **53**, 197 (1989).

Burbidge, E. M., G. R. Burbidge, W. A. Fowler, and F. Hoyle. *Rev. Mod. Phys.* **29**, 547 (1957).

Cameron, A. G. W. Atomic Energy of Canada, CRL-41.

Rolfs, C. E. and W. S. Rodney. *Cauldrons in the Cosmos*, Chicago University Press, Chicago, 1988.

Truran, J. W. *Ann. Rev. Nucl. Part. Sci.* **34**, 53 (1984).

Weinberg, S. *The First Three Minutes*, Basic, New York, 1977.

Wong, S. S. M. *Introductory Nuclear Physics*, 2nd ed., Wiley, New York, 1998.

## BIBLIOGRAPHY

Audouze, J. and S. Vauclair. *An Introduction to Nuclear Astrophysics*, Reidel, Dordrecht, 1980.

Bahcall J. N. (http://www.sns.ias.edu/~jnb); a comprehensive account of the solar neutrino problem and its solution is found at the website.

Bahcall, J. N. and C. Pena-Garay. *New J. Phys.* **6**, 63 (2004).

Barnes, C. A., D. D. Clayton, and D. N. Schramm, Eds. *Essays in Nuclear Astrophysics*, Cambridge University Press, Cambridge, 1982.

Choppin, G. R., J. O. Liljenzin, and J. Rydberg. *Radiochemistry and Nuclear Chemistry*, Butterworth-Heinemann, Woburn, 2002.

Cottingham, W. N. and D. A. Greenwood. *An Introduction to Nuclear Physics*, Cambridge University Press, Cambridge, 1986.

Clayton, D. D. *Principles of Stellar Evolution and Nucleosynthesis*, Chicago University Press, Chicago, 1983.

Dunlap, R. *An Introduction to the Physics of Nuclei and Particles*, Thomson, Toronto, 2004.

Ehmann, W. D. and D. E. Vance. *Radiochemistry and Nuclear Methods of Analysis*, Wiley, New York, 1991.

Filippone, B. W. *Ann. Rev. Nucl. Part. Sci.* **36**, 717 (1986).

Friedlander, G., J. W. Kennedy, E. S. Macias, and J. M. Miller. *Nuclear and Radiochemistry*, 3rd ed., Wiley, New York, 1981.

Heyde, K. *Basic Ideas and Concepts in Nuclear Physics*, IOP, Bristol, 1999.

Hodgson, P. E., E. Gadioli, and E. Gadioli-Erba. *Introductory Nuclear Physics*, Clarendon, Oxford, 1997.

Kappeler, F., F.-K. Thielemann, and M. Wiescher. *Ann. Rev. Nucl. Part. Sci.* **48**, 175 (1998).

Krane, K. S. *Introductory Nuclear Physics*, Wiley, New York, 1988.

*Nuclear Physics: The Core of Matter, The Fuel of Stars*, NAS, Washington, 1999.

Reeves, H. *Rev. Mod. Phys.* **66**, 193 (1994).

Viola, V. E. LiBeB Nucleosynthesis and Clues to the Chemical Evolution of the Universe, in O. Manuel, Ed., *Origin of Elements in the Solar System: Implications of Post-1957 Observations*, Kluwer, New York, 2000.

Williams, W. S. C. *Nuclear and Particle Physics*, Clarendon, Oxford, 1991.

# CHAPTER 13

# ANALYTICAL APPLICATIONS OF NUCLEAR REACTIONS

As mentioned previously (Chapter 4), one of the compelling reasons to use nuclear analytical methods is their high sensitivity. The radiation from the decay or excitation of a single nucleus can be readily detected. Even when one has to have the intervening step of a nuclear reaction to produce or excite the decaying species, one still has the ability to detect very small quantities of material. This chapter deals with those nuclear analytical methods [activation analysis, particle-induced X-ray emission (PIXE), Rutherford backscattering (RBS)] in which a nuclear reaction is the necessary first step in the analysis procedure. The techniques to be discussed are known for their sensitivity, the ability to do nondestructive analysis of a large number of samples, sometimes quickly, and the ability to analyze the surfaces of materials. All these techniques are *elemental analysis techniques* and do not, in general, give information about the chemical form of the element, any attached ligands, and the like. This lack of *speciation* information is a drawback of these methods.

## 13.1 ACTIVATION ANALYSIS

### 13.1.1 Basic Description of Method

Activation analysis is an analytical technique that allows one to determine the amount of a given element X contained in some material Y. The basic steps in the activation technique are as follows:

*Modern Nuclear Chemistry*, by W.D. Loveland, D.J. Morrissey, and G.T. Seaborg
Copyright © 2006 John Wiley & Sons, Inc.

**366** ANALYTICAL APPLICATIONS OF NUCLEAR REACTIONS

1. Irradiate Y with a source of ionizing radiation so that X will change into X*, a radioactive isotope of X.
2. Using chemical or instrumental techniques, "isolate" X and X* from all other elements in Y (not necessarily quantitatively) and measure the activity of X*. Chemical "isolation" of the activity of interest is performed simply by separating it chemically from all other activities. Instrumental "isolation" of the activity of interest involves the detection of radiation that can uniquely identify the nuclide in question.
3. Calculate the amount of X present.

These basic steps are shown schematically for neutron activation analysis in Figure 13.1.

How does one calculate the amount of X present, knowing the activity of X produced in the irradiation? It can be shown that

$$A_{X^*} = N_X \sigma \phi (1 - e^{-\lambda_{X^*} t_i}) e^{-\lambda_{X^*} t_d} \qquad (13.1)$$

where $A_{X^*}$ is the activity of X* present at a time $t_d$ after the end of the bombardment, $N_X$ is the number of X nuclei present initially, $\sigma$ is the nuclear reaction cross section, $\phi$ is the flux of activating particles, $t_i$ is the length of the irradiation, and $\lambda_{X^*}$ is the decay constant of X*. From this equation one could calculate $N_X$ from $A_{X^*}$, knowing all the other variables. (The above equation for $A_{X^*}$ is valid for "thin targets," that is, samples that absorb <5% of the flux of activating particles.)

This method of analysis is called *absolute activation analysis* and is done rarely. The reasons for this are the need for detailed knowledge of the flux and energy of the bombarding particles in the sample, the compounding of the uncertainties of our knowledge of cross sections, decay branching ratios, and the like in the final results. A simpler technique is to irradiate and count a known amount of pure X under the same conditions used for the mixture of X in Y. Then

$$\text{Mass X in Y} = (\text{known mass X}) \left( \frac{\text{activity of X* in Y}}{\text{activity of X* in pure X}} \right) \qquad (13.2)$$

This is known as the *comparator technique* and is the most widely used method of activation analysis. It depends on irradiating and counting standards of known amounts of pure material using the same conditions as the samples being analyzed.

### 13.1.2 Advantages and Disadvantages of Activation Analysis

Since we know that $A = \varepsilon \lambda N$ where $A$ is the measured radioactivity, $\lambda$ is the decay constant, $N$ is the number of radioactive nuclei present, and $\varepsilon$ is a constant representing the detection efficiency, we know that just a few radioactive nuclei need to be present to give measurable activities. Use of activation analysis can lead to measurement of elemental abundances of the order of $10^{-6}$–$10^{-12}$ g. The actual detection

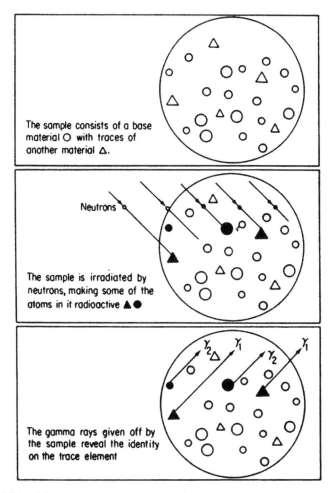

**Figure 13.1** Schematic representation of activation analysis. [From Corliss (1963).]

sensitivities for activation analysis of various elements, as practiced by a commercial activation analysis service, are shown in Figure 13.2. One can detect microgram levels of over two-thirds of the elements using activation analysis.

Although the high sensitivity of activation analysis is perhaps its most striking advantage, there are a number of other favorable aspects as well. Activation analysis is basically a multielemental technique. Many elements in the sample will become radioactive during the irradiation; and if each of these elements can be "isolated" chemically or instrumentally, their abundances may be determined simultaneously. Activation analysis can be a nondestructive method of analysis. Numerous tests have shown that with careful experimental manipulation, activation analysis is an accurate ($\sim 1\%$ accuracy) and precise ($\sim 5\%$ precision) method of measuring elemental concentrations.

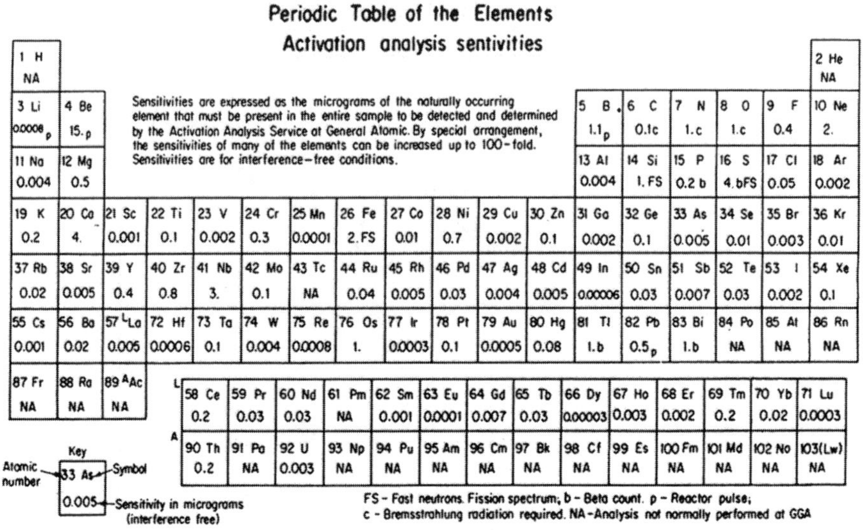

**Figure 13.2** Table of activation analysis sensitivities as offered by General Atomic Company, San Diego, California.

Activation analysis is not without its drawbacks, however. Among them are the need to use expensive equipment and irradiation facilities, the inability to determine the chemical state of the elements in question, the need to work with significant levels of radioactivity, with their attendant radiation safety and legal problems, the long times needed to complete some analyses, and complex analysis sometimes needed to unscramble the γ-ray spectra in a given experiment.

The ultimate test of the utility of activation analysis as an analytical technique is whether there are competitive technologies that have the advantages of activation analysis with fewer drawbacks. One candidate for this designation is inductively coupled plasma–mass spectroscopy (ICP–MS).

The detection limits in ICP–MS are certainly equal to those achieved by activation analysis. In addition, ICP–MS apparatus is frequently connected to ordinary chemical separation apparatus, such as liquid chromatography (LC), thus allowing a sensitive determination of both the amount and chemical species present for both metals and nonmetals.

In recent years, there has been increasing use of ICP–MS techniques to replace those of activation analysis, although there still are a large number of applications of activation analysis each year, especially in the geological sciences.

### 13.1.3 Practical Considerations in Activation Analysis

To better understand the practical details of how activation analysis may be applied to a given problem in elemental analysis, let us consider the various aspects of a typical activation analysis problem. To make our discussion more concrete, let us consider a specific problem, the measurement of the aluminum content of rocks and meteorites (Loveland et al., 1969). The choice of this problem as an example was dictated by its pedagogic simplicity and the fact that conventional chemical analyses of aluminum in rocks are known to be inaccurate for low aluminum concentrations and, in general, not very precise.

The first step in an activation analysis procedure is *sample preparation*. The unknown and known samples (sometimes referred to as the unknown and standard samples) should have the same size, composition, and homogeneity insofar as possible to ensure that any attenuation of the incoming radiation, or the sample radiation before counting, or any count rate-dependent effects, are exactly the same. In practice, this step is accomplished by making sure that the unknown sample and known sample have the same physical volume, are irradiated in a homogenous flux, and are counted under exactly the same conditions (geometry, detector, etc.). Preirradiation treatment of the sample should be kept to a minimum so as to lessen the possibility of sample contamination. The standards are either aqueous solutions of the elements in question or multielemental standard reference materials whose composition is certified by a national or international agency (IAEA, U.S. National Institute of Standards and Technology, etc.).

The second step in an activation analysis concerns the *choice of nuclear reaction* to change X into X*, plus the *irradiation facility* in which the reaction will be carried out. In addition, the length of irradiation and decay prior to counting must be chosen so the produced X* activity is enhanced relative to all other activities produced. Most activation analysis is done with thermal neutrons produced in nuclear reactors for the following reasons:

1. Many elements have high cross sections for the absorption of thermal neutrons in (n, γ) reactions.
2. Copious fluxes of thermal neutrons ($\phi \sim 10^{12}$ n/cm$^2$/s) are available in nuclear reactors.
3. Neutrons penetrate matter easily, and therefore there are few problems related to attenuation of the neutron flux in the sample.
4. The major elements, carbon, nitrogen, and oxygen, are scarcely activated by thermal neutrons, making detection of other elements easier.

Although most activation analysis is done with reactor thermal neutrons, several other nuclear reactions and irradiation facilities can be used. Spontaneous fission of $^{252}$Cf furnishes 3.8 neutrons per fission, and fluxes of up to $10^9$ n/cm$^2$/s are available from $^{252}$Cf isotopic neutron sources. Cockroft–Walton accelerators can be used to accelerate deuterons to energies of $\sim$150 keV, and then, using the $^3$H(d, n) reaction, $\sim$14 MeV neutrons can be produced (*fast neutrons*). Typical neutron generators of this type give fluxes of $\sim 10^9$ n/cm$^2$/s of 14-MeV neutrons. These fast neutrons are useful for activating the light elements, such as silicon, nitrogen, fluorine, and oxygen, via (n, p) or (n, α) reactions, leading to sensitivities of 50–200 ppm and, thus, is complementary to slow neutron activation analysis.

Charged particle or photon-induced reactions can also be used for activation. The typical charged particles used are protons, deuterons, $^3$He, and α particles. Charged particle activation analysis (CPAA) is frequently complementary to neutron activation analysis (NAA). NAA has poor sensitivity for the lighter elements, whereas CPAA has good sensitivity. Because of the limited penetrating power of charged particles in matter, CPAA either requires a thin sample or is used for surface analysis. This attenuation of the primary radiation by the sample puts especially stringent requirements on sample preparation.

Activation by photons (PAA) usually takes place via the (γ, n) reaction, although other reactions such as (γ, p), (γ, α), and the like are possible. Of special interest is the determination of lead by PAA with a detection limit of ~0.5 μg. [Lead is very hard to detect using NAA (Fig. 13.2).] Photon sources are usually electron accelerators, which produce high-energy photons through the bremsstrahlung process when the electrons strike a heavy-metal target.

For the sample problem of determining the Al content of rocks, the activating nuclear reaction was chosen to be $^{27}$Al(n, γ)$^{28}$Al, with the irradiation source being a nuclear reactor. The $^{28}$Al decays with a 2.2-min half-life and emits a β$^-$ particle and a high-energy (1.78 MeV) γ ray.

Even if you have chosen to irradiate a sample with thermal neutrons from a nuclear reactor, you may be surprised to learn that several other neutron energies may be present and cause reactions. For the popular TRIGA design of reactor, only ~25% of the neutrons at a typical irradiation position are "thermal" neutrons ($0 < E_n < 0.05$ eV). The rest of the neutrons have higher energies, with neutrons with $0.05$ eV$< E_n < 0.1$ MeV being called *epithermal* neutrons and neutrons with $0.1 < E_n < 15$ MeV being called fast neutrons. The capture cross sections for epithermal neutrons frequently involve resonance capture (Chapters 10 and 11) and can involve very large cross sections ($>10^4$ barns). Usually one uses epithermal neutrons as the activating particle when one wants to avoid interfering activities in the sample due to thermal neutron capture. For example, suppose a sample has a large content of sodium. Sodium is easily activated via the $^{23}$Na (n, γ) reaction giving rise to copious quantities of 15-h $^{24}$Na in the sample, which may interfere with the detection and measurement of other activities. How do we get rid of this sodium? We can surround our sample with a metallic cadmium cover (~0.1 cm thick). Cadmium has a very large capture cross section for neutrons in the energy region below 1.0 eV and effectively "cuts off" or removes these neutrons. The resulting neutron flux in the sample consists of the higher energy (epithermal) neutrons. Frequently, one measures a "Cd ratio" for activation of a specific element to get some idea of how much of the produced activity is due to epithermal activation. This Cd ratio, $R$, is defined as:

$$R = \frac{(\text{activity})_{\text{no Cd cover}}}{(\text{activity})_{\text{with Cd cover}}} \quad (13.3)$$

Typical values of $R$ range from 2 to 1000 depending on the reactor irradiation position. Epithermal activation is advantageous for Ag, As, Au, Ba, Cs, Ga, In, Mo, Pt, Rb, Sb, Se, Sr, Tb, Th, Tm, U, W, Zn, and Zr among others.

Once a nuclear reaction and an irradiation facility have been selected, the possibility of *interfering reactions* must be carefully considered. This term means that quite often, although X will change to X* during the irradiation, some other element Z may also change to X* during the irradiation. Thus, the activity of X* is proportional to the abundances of Z and X in the sample, not just X. This effect is referred to as *an interfering reaction or interference*, and a correction must be made for it. In the case of the aluminum analysis, there is a very important interference—namely the occurrence of the $^{28}$Si(n, p)$^{28}$Al reaction whereby silicon in the rock is converted into $^{28}$Al by reactions involving fast neutrons present in any reactor (along with the desired thermal neutrons). Thus, the measured $^{28}$Al activity will be due to the activation of $^{27}$Al and $^{28}$Si. By irradiating a known amount of silicon and counting it, and from the well-known Si abundances of rocks, a correction for the $^{28}$Al produced by the $^{28}$Si(n, p)$^{28}$Al reaction can be calculated. Other possible interferences are the fission of any uranium in the sample, or the occurrence of two nuclides that emit γ rays that have similar energies that cannot be resolved.

The final decision concerning irradiation conditions involves the determination of the flux and irradiation duration. A rough rule is that the longer one irradiates the sample and the longer one lets the sample decay before counting, the greater the activity of the long-lived species relative to the short-lived species. One must keep in mind the saturation properties of irradiations are such that it rarely pays to irradiate any material for a time corresponding to more than two half-lives of the desired activity. (In the Al analysis, a sample irradiation time of 1.0 min and a neutron flux of $5 \times 10^{10}$ n/cm$^2$/s were used.)

Frequently, multiple irradiations of a sample are made. The first irradiation is short (minutes) to determine the short-lived radioisotopes (of Ag, Al, Ba, Br, Ca, Cl, Co, Cu, Dy, F, I, In, K, Mg, Mn, Na, Se, Sb, Si, Sr, Ti, U, and V), and the subsequent irradiations (hours) are to determine the intermediate (As, Au, Br, Cd, Ga, Ge, Hg, Ho, K, La, Mo, Na, Pd, Sb, Sm, U, W, and Zn) or long-lived (Ag, Ce, Cr, Cs, Co, Eu, Fe, Hf, Hg, Lu, Nd, Ni, Rb, Sb, Sc, Se, Sn, Sr, Ta, Tb, Th, Tm, Yb, Zn, and Zr) radionuclides. In the long irradiations, it is common to let the sample "decay" for several days to get rid of the 15-h $^{24}$Na.

The next major step in any activation analysis procedure is the selection of a method of isolating the activity of interest, X*, to measure it. Two methods of isolating X* are commonly used—*instrumental activation analysis* (*IAA*) and *radiochemical activation analysis* (*RAA*). In instrumental activation analysis, the characteristic energies of the γ rays emitted by the radionuclides in the activated sample are used to identify them, and the corresponding photopeak areas give a measure of the activities. Instrumental activation analysis is nondestructive, allowing further use of the sample. Furthermore, it permits the use of short-lived activities to identify various elements that might not be possible if a lengthy chemical separation would precede the counting. Also, instrumental activation analysis (IAA) lends itself to automation and reduces the time spent per sample in the analysis. The use of Ge semiconductor detectors with excellent energy resolution has made IAA the preferred method of activation analysis.

Although most investigators prefer to use IAA, in some situations radiochemistry must be done prior to counting the sample, to isolate the activity of interest.

An example of the need for radiochemistry is the determination of trace elements in biological materials, such as blood, which have a very high sodium content. Large quantities of $^{24}$Na are produced via the $^{23}$Na(n, $\gamma$)$^{24}$Na reaction, and they tend to "mask" the trace element activities in the blood by creating a large Compton background in the region where the photopeaks of other trace-element activities are found (see the discussion in Chapter 18 on $\gamma$-ray detectors). One solution to this problem is to separate the sodium chemically from the irradiated blood (using ion exchange with hydrated antimony pentoxide) and then to instrumentally analyze the purified blood. This example does illustrate a feature of modern radiochemical activation analysis—that of not completely separating the element of interest but of making a group separation of a relatively small number of activities and further resolving these activities by $\gamma$-ray spectroscopy.

All of our discussions up to now have focused on detecting the $\gamma$ rays from the decaying activation products. There is another approach that has been used in some cases. This approach is called prompt $\gamma$-ray activation analysis (PGAA) in which one detects the prompt $\gamma$ radiation emitted during the activating reaction. For neutron activation via the (n, $\gamma$) reaction, one detects the $\gamma$ rays emitted during the neutron capture. Such analyses must be carried out with beams of activating particles (such as neutrons) and usually involves detecting high-energy ($>$5 MeV) $\gamma$ rays. Because of these constraints, this rapid analysis method is restricted usually to the determination of the major elemental constituents of the sample.

### 13.1.4 Applications of Activation Analysis

The applications of activation analysis are almost innumerable. In the physical sciences, activation analysis is used in trace-element analysis of semiconductor materials, metals, meteorites, lunar samples, and terrestrial rocks. In most cases, the multielemental analysis feature of activation analysis is used to measure the concentrations of several trace-elements simultaneously. From these detailed studies of trace-element abundance patterns, one has been able to deduce information about the thermal and chemical history of the Earth, moon, Mars, and meteorites, as well as the source or age of an object.

The use of activation analysis in criminal investigations (*forensic activation analysis*) is also well established. The basic idea here is to match the trace-element distributions found in bullets, paint, oil, and so on found at the scene of a crime with the trace-element distributions in objects found with criminal suspects. Such identification is rapid and nondestructive (allowing the actual evidence to be presented in court). Moreover, the probability of its correctness can be ascertained quantitatively. Other prominent examples of the use of forensic activation analysis involve confirmation of the notion that Napoleon was poisoned (by finding significant amounts of arsenic in hair from his head) and the finding that the activation analysis of the wipe samples taken from a suspect's hand can reveal not only if he or she has fired a gun recently but also the type of gun and ammunition used.

Applications of activation analysis in the environmental sciences are routine. Determinations of the trace element content of urban atmospheres, lakes, streams, and similar areas have been used to trace the flow of pollutants in various ecosystems. In addition, a few of the trace elements whose abundances have been measured by activation analysis have turned out to be biologically significant by themselves. The classic example is mercury and the significant mercury concentration in fish and other foodstuffs revealed by activation analysis. A particular combination of activation analysis and radiotracer methods has found important applications in the environmental sciences. This combination involves the use of stable isotopes instead of radioactive isotopes as tracers in various systems, with activation analysis of the samples collected after tracer dispersal being used to measure the tracer concentrations. Such a technique avoids the need to introduce radioactive materials into a system (such as the environment with its subsequent health and legal complications) and yet retains the selectivity and sensitivity of radiation measurements. The stable isotopes are called *stable activable tracers*. Kruger (1971) has described their use.

In summary, activation analysis is a multielemental, nondestructive, very accurate method of analysis. The best-case sensitivities are picograms per gram with an irregular variation from element to element. It is best suited for the analysis of solid samples and can be "tuned" using changes in irradiation conditions, particles, and the like, and postirradiation sample treatment. Disadvantages are the long analysis times, the need for access to an irradiation facility (usually a reactor), the need to handle radioactivity, the labor-intensive nature of sample counting, and the inability to get speciation information.

## 13.2 PARTICLE-INDUCED X-RAY EMISSION

Particle-induced X-ray emission (PIXE) is an analytical technique based upon observing fluorescent X-rays. As such, it really is not a nuclear technique since it involves an atomic process, X-ray emission. But the atomic electron shell vacancies that are filled when the X-ray is emitted are created using particle-accelerator beams and one uses typical semiconductor radiation detectors, Si (Li) detectors, to detect the X-rays.

The essential features of a PIXE setup are shown schematically in Figure 13.3. A beam of charged particles from an accelerator, typically 2- to 4-MeV protons, impinges on a thin sample in a vacuum chamber. The protons collide with the electrons in the material and some eject inner shell electrons from the atoms in the sample. A Faraday cup is used to collect the charge deposited by the incident protons, and this is integrated electronically to give the beam current. The sample is typically a thin, uniform deposit of the material to be analyzed on a thin backing material. The characteristic X-rays from the sample are detected with a Si (Li) detector. A typical spectrum is shown in Figure 13.4. The spectrum consists of discrete X-ray peaks superimposed on a continuous background of bremsstrahlung. One can see the $K_\alpha$ and $K_\beta$ lines of the lighter elements (from the filling of the K shell vacancies) and the L lines of the heaviest elements. The peaks corresponding to a given element are integrated to give peak areas and the amounts of that element obtained either from a knowledge of the absolute ionization cross

**374** ANALYTICAL APPLICATIONS OF NUCLEAR REACTIONS

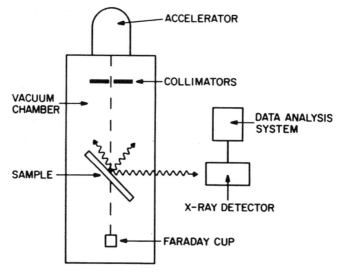

**Figure 13.3** Schematic diagram of a PIXE setup. [Reproduced from Ehmann and Vance (1991).]

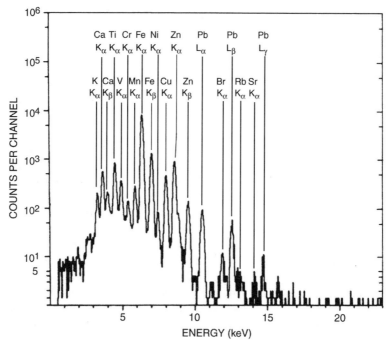

**Figure 13.4** PIXE spectrum of a rainwater sample. [From Johansson and Johansson (1976).]

sections ($\sim 1-10^4$ barns), fluorescence yields (0.1–0.9), beam current, and geometry or by comparison to the results obtained from a thin elemental standard. The term *fluorescence yield* refers to the fraction of the electron vacancies filled by X-ray emission vs. the ejection of Auger electrons.

Typical detection limits for various elements in a biological sample are shown in Figure 13.5. Typically, PIXE has sensitivity at the parts per million level for many elements. About 25% of the applications of PIXE are in biology and medicine. The light-element matrices lead to smaller continuous backgrounds, and many trace and toxic elements are easily detected by PIXE. (There are no "holes" in detection limits as there are in activation analysis as all the elements emit some X-rays.) Considerable attention has been and must be devoted to the preparation of thin, representative samples. Note that PIXE is only sensitive to the *elemental* composition of the sample and not to the *isotopic* composition.

One of the most successful applications of PIXE has been in the analysis of air pollution particulate matter. Atmospheric particulate matter is typically collected by impaction on a filter paper, which provides an ideal thin sample for PIXE analysis. Another aspect of PIXE that is very important for the analysis of aerosol samples is the ability to analyze a large number of samples in a short time. PIXE analyses typically take less than a minute, and the entire irradiation, counting, sample changing, and analysis procedure can be automated.

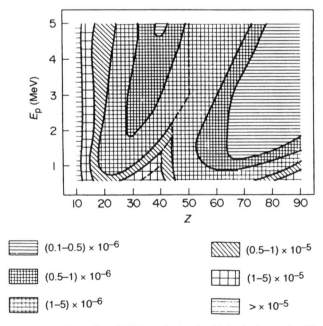

**Figure 13.5** Detection limits in a PIXE analysis of a biological sample. [From Ishii and Morita (1990).]

An important variant on PIXE is *micro-PIXE*. By using a proton beam whose spatial dimension is ~0.5 μm (rather than the usual 10 mm), one can determine the trace-element content of a small portion of the sample, giving one a "trace-element microscope." This application is important in probing samples of medical interest. A related technique is used in the *electron microprobe* where the ionization is caused by electron impact.

## 13.3 RUTHERFORD BACKSCATTERING (RBS)

One of the earliest experiments in nuclear physics was Rutherford's demonstration of large-angle scattering of α particles by gold nuclei. This experiment established the existence of a small nucleus within the atom (Chapter 10). The force acting in this process, called Rutherford scattering, is the repulsive Coulomb force between the positively charged nuclei. A schematic diagram of the phenomena is shown in Figure 13.6.

Rutherford scattering is an elastic event, that is, no excitation of either the projectile or target nuclei occurs. However, due to conservation of energy and momentum in the interaction, the kinetic energy of the backscattered ion is less than that of the incident ion. The relation between these energies is the *kinematic factor*, $K$, which is given by the expression

$$K = \frac{E_1}{E_0} = \left[ \frac{(M_2^2 - M_1^2 \sin^2 \theta)^{1/2} + M_1 \cos \theta}{M_1 + M_2} \right]^2 \quad (13.4)$$

where $M_1$ and $M_2$ are the masses of the incident and target atoms, respectively, and $\theta$ is the angle between the direction of the incident and scattered ions. Note the relative shift in energy in the collision depends only on the masses of the ions and the angle of the detector. If one measures the scattering angle and the energy shift, one can calculate the mass (identity) of the scattering atom. The largest change in energy

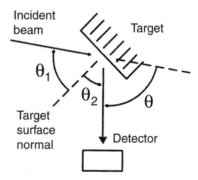

**Figure 13.6** Schematic diagram of Rutherford backscattering. [From Rauhala (1994).]

occurs for $\theta = 180°$ where

$$K = \left(\frac{M_2 - M_1}{M_2 + M_1}\right)^2 \tag{13.5}$$

A geometry that allows detection of the scattered α particles at very large angles is usually selected.

The probability or cross section for Rutherford scattering (Chapter 10) is given (Segre, 1977) as:

$$\frac{d\sigma}{d\Omega} = \left(\frac{Z_1 Z_2 e^2}{4E}\right)^2 \frac{4}{\sin^4 \theta} \frac{[\cos \theta + (1 - x^2 \sin^2 \theta)^{1/2}]^2}{(1 - x^2 \sin^2 \theta)^{1/2}} \tag{13.6}$$

where $x = M_1/M_2$, $e^2$ is the square of the electronic charge, and $E$ is the energy of the incident ion. Note the probability of scattering goes as $(Z_1 Z_2)^2$ and as $1/E^2$. If this were all that went into Rutherford backscattering, we would expect a spectrum of backscattered particles that consisted of a peak for each element in the sample with a relative height (area) $\propto Z^2$. The elemental abundances could be calculated using the relation

$$N = \frac{D}{F \int \frac{d\sigma}{d\Omega} d\Omega} \tag{13.7}$$

where $N$ is the number of target atoms, $D$ is the number of detected events, and $F$ is the incident ion flux. This is the situation if one has a very thin film as the target material or if one scatters particles from the surface of a thick sample.

In reality, the situation is usually more complicated because the incident ions lose energy as they penetrate into the sample, thus continuously changing the probability of scattering and the energies of the scattered particles. The resulting spectrum for scattering from a single element at varying depths is shown in Figure 13.7, where the

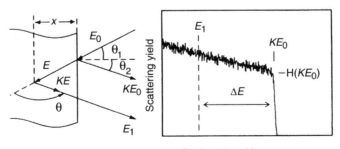

**Figure 13.7** Energy depth scale in Rutherford backscattering. [From Rauhala (1994).]

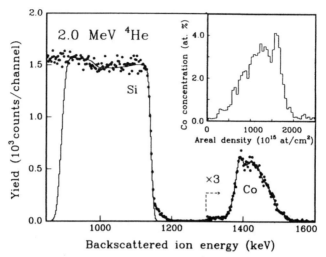

**Figure 13.8** Rutherford backscattering for 2.0 MeV $^4$He ions incident on a Si (Co) sample. Dots represent the experimental data while the solid line is a simulated spectrum. Scattering angle $\Theta = 170°$, with $\theta_1 = \theta_2 = 5°$. [From Saarilahti and Rauhala (1992).]

incident ion energy is $E_0$, the energy of ions scattered from the surface is $KE_0$, and the energy of ions scattered from a depth $x$ is $E_1$. In this situation, the energy loss in traversing (into and back out of) a foil of thickness $N_x$ is

$$\Delta E = KE_0 - E = [\varepsilon]_{BS} N_x \tag{13.8}$$

$$[\varepsilon]_{BS} = \frac{K}{\cos\theta_1}\varepsilon_{in} + \frac{1}{\cos\theta_2}\varepsilon_{out} \tag{13.9}$$

where $\varepsilon_{in}$ and $\varepsilon_{out}$ are the energy-dependent stopping cross sections (Ziegler) on the inward and outward paths of the ion.

Rutherford backscattering is an important method for determining the composition and structure of surfaces and thin films. In Figure 13.8, we show the results of an RBS measurement with 2.0-MeV $^4$He incident on an Si surface with a Co impurity that was diffused into the bulk material. One can clearly detect the Co and its depth profile.

Another important application of this technique has been to determine the elemental composition of the lunar and Martian surfaces. Turkevich et al. (1969) constructed a rugged device to measure the backscattering of $\alpha$ particles from the lunar surface, which flew on three Surveyor missions in 1967–68 and yielded the first complete and accurate analysis of the lunar surface. The $\alpha$ particles came from a radioactive source ($^{242}$Cm) that was part of the instrument package. The results of these experiments, which showed an unexpected and comparatively high abundance of Ti, were confirmed by laboratory analysis of lunar samples gathered in the Apollo missions. Since then, this technique has been used to study Martian rocks and soil.

## PROBLEMS

1. For each of the following analyses, indicate what role, if any, activation analysis could or should play. Be sure to clearly state the reasons for your choice.

   (a) Determination of the oxygen content of steel, (b) verification of the authenticity of ancient paintings, (c) determination of the radionuclides present in fallout from nuclear weapons testing, (d) determination of the extent to which radionuclides leaking from nuclear waste storage facilities contaminate the water of nearby streams, and (e) determination of lithium impurities in thin films of GaAs.

2. (a) Calculate the activity (in microcuries) of $^{49}$Ca produced when 2.7 g of CaO are irradiated in a flux of $3 \times 10^{12}$ n/cm$^2$-s for 10 min. (b) Repeat this calculation for the situation when the bombarding particle is 21 MeV deuterons, and the deuteron beam current is 10 μA. Assume the (d, p) cross section is 50 mbarns.

3. Using the Chart of the Nuclides as a guide, estimate the sensitivity (minimum quantity that can be detected) of neutron activation analysis for europium using a thermal neutron flux of $3 \times 10^{12}$ n/cm$^2$-s. Assume no irradiation may last more than 1 h and the minimum detectable activity is 10 dpm.

4. For the following analyses, indicate whether radiochemical neutron activation analysis would be preferred to instrumental neutron activation analysis. If radiochemistry is indicated, briefly sketch the separation procedures to be used: (a) the determination of ppm levels of Mo in flathead minnows, (b) the determination of the trace-element content of agricultural field-burning particulate matter, (c) the use of stable activable tracers to determine flow patterns in an ocean estuary, and (d) the determination of Dy in pine needles.

5. Consider you want to trace the deposition of particulate matter using the stable activable tracer In. The dilution factor between the point of release and the point of sampling is $10^6$. Assume the samples that are collected are activated in a thermal neutron flux of $3 \times 10^{12}$ n/cm$^2$-s for 10 min. Further assume a 1% efficiency for detecting the emitted photons. Determine the minimum amount of In that must be released to ensure the uncertainty in the measured sample concentrations is 5%.

6. Consider the following results obtained by neutron activation analysis of lake water samples for their Mn content. Assume the sample volumes are 1 L.

   | Sample | EOB Activity (cp5m) |
   |---|---|
   | 1 | 1204 |
   | 2 | 1275 |
   | 3 | 940 |
   | 4 | 1350 |
   | 10 mg Mn standard | 5000 |

   What is the Mn content of the lake water and its uncertainty?

7. Two thin 1-mg samples of dysprosium are irradiated and counted in a similar manner, except for the use of a Cd cover foil on one sample. A Cd ratio of 7 is measured, with the bare foil saturation activity of $1 \times 10^4$ dpm. Calculate the thermal neutron flux at the irradiation position in the reactor.

8. Devise an activation analysis scheme for determining the concentration of nitrogen in a sample of plant material. Assume the analysis must be nondestructive and rapid. Suggest an appropriate reaction, irradiation, and counting conditions, and indicate possible interferences in your analysis.

9. Compute the "advantage factor" for using a reactor pulse to produce 20-s $^{46}Sc^m$ compared to the activity produced by steady-state irradiation. Assume the reactor is of the TRIGA type and produces a 15-ms 3000-MW pulse with a peak instantaneous flux of $21 \times 10^{15}$ n/cm$^2$-s. Assume steady-state operation is at 1 MW.

10. Imagine you wish to detect ppm levels of Al in a matrix containing iron, calcium, and silicon. Assume you have access to a modern nuclear reactor. Describe an activation analysis procedure to do this analysis. Be sure to describe the irradiation conditions, any pre- or postirradiation chemistry, and the counting strategy. Indicate how you would deal with any interferences in the analysis.

## REFERENCES

Corliss, W. R. *Neutron Activation Analysis*, USAEC, 1963.

Ehmann, W. D. and D. E. Vance. *Radiochemistry and Nuclear Methods of Analysis*, Wiley, New York, 1991.

Ishii, K. and S. Morita. *Int. J. PIXE* **1**, 1 (1990).

Johansson, S. A. E. and T. B. Johansson. *Nucl. Instr. Meth.* **137**, 473 (1976).

Kruger, P. *Principles of Activation Analysis*, Wiley, New York, 1971.

Loveland, W., R. A. Schmitt, and D. E. Fisher. *Geochim. Cosmochim. Acta* **33**, 375 (1969).

Rauhala, E. In Z. B. Alfassi, Ed., Chemical Analysis by Nuclear Methods, Wiley, Chichester, 1994.

Saarilahti, J. and E. Rauhala. *Nucl. Instr. Meth. Phys. Res. B* **B64**, 734 (1992).

Segre, E. *Nuclei and Particles*, 2nd ed., Benjamin, Reading, 1977.

Turkevich, A., E. F. Franzgrote, and J. H. Patterson. *Science* **165**, 277 (1969).

## BIBLIOGRAPHY

Alfassi, Z. B., Ed. *Chemical Analysis by Nuclear Methods*, Wiley, Chichester, 1994. A series of essays on various aspects of nuclear analytical chemistry. Most of them are quite good.

Brune, D., B. Forkman, and B. Persson. *Nuclear Analytical Chemistry*, Chartwell-Bratt, London, 1984.

de Soete, D., R. Gijbels, and J. Hoste. *Neutron Activation Analysis*, Wiley, New York, 1974. An encyclopedic work.

Hughes, D. J. *Pile Neutron Research*, Addison-Wesley, Cambridge, 1953. The bible (old testament) of reactor neutron physics.

Johansson, S. A. E. and J. L. Campbell. *PIXE: A Novel Technique for Elemental Analysis*, Wiley, Chicester, 1988.

Johansson, S. A. E., J. L. Campbell, and K.-G. Malmqvist. *PIXE*, Wiley, New York, 1995.

Ziegler, J. F., P. J. Scanlon, W. A. Lanford, and J. L. Duggan, Eds. *Ion Beam Analysis*, North-Holland, Amsterdam, 1990.

# CHAPTER 14

# REACTORS AND ACCELERATORS

Radioactive decay is the only nuclear reaction that commonly takes place on Earth. The reasons that other nuclear reactions do not normally occur on Earth are simple. Nuclear reactions that are induced by protons or heavier charged particles all have large activation barriers and require energetic charged particles that are only present in space and the highest regions of the atmosphere. On the other hand, nuclear reactions induced by neutrons do not have an activation barrier, but neutrons are unstable, decaying by $\beta^-$ decay into protons with a half-life of approximately 10 min. Thus, neutrons cannot be stored very long and have to be produced in other nuclear reactions before they can be used in subsequent nuclear reactions.

Protons and all nuclei are positively charged and strongly repel one another through the Coulomb force. Colliding nuclei must have kinetic energies that are far in excess of normal thermal energies to reach distances that are short enough for the nuclear force to be effective ($\sim 1$ fm). We must accelerate one of the particles until it has sufficient kinetic energy to get over the Coulomb barrier for the nuclei to react.

The fact that neutrons can be absorbed by nuclei without overcoming a threshold ($l = 0$ or s-wave reactions) makes neutrons extremely effective nuclear "reactants." Neutron-induced reactions are the energy source for present-day commercial nuclear power (fission reactors) while charged-particle-induced reactions remain under study as power sources (fusion reactors). In this chapter we will consider the general features of nuclear fission reactors, following by the general features

---

*Modern Nuclear Chemistry*, by W.D. Loveland, D.J. Morrissey, and G.T. Seaborg
Copyright © 2006 John Wiley & Sons, Inc.

# 384 REACTORS AND ACCELERATORS

of charged particle accelerators, magnetic spectrometers, and the production of beams of radioactive nuclei.

**Example Problem** A very simple and potentially useful fusion reaction combines two deuterium nuclei forming $^3$He and a neutron.

$$^2H + {}^2H \longrightarrow {}^3He + {}^1n + Q$$

Estimate the Coulomb barrier for this reaction and the temperature of deuterium gas that would give such an average energy.

**Solution** The Coulomb barrier, discussed in Chapter 10, is

$$V_{\text{Coulomb}} = \frac{z_1 z_2 e^2}{r} = \frac{(1)(1) 1.439 \text{ MeV-fm}}{1.93 + 1.93 \text{ fm}} = 0.373 \text{ MeV}$$

Setting the kinetic energy equal to the Coulomb barrier as would be appropriate when all the particles are moving (in a gas), and using the thermal energy of an ideal gas:

$$V_{\text{Coulomb}} = \text{KE} = \tfrac{1}{2} m v^2 = \tfrac{3}{2} kT$$

$$T = \frac{2 V_{\text{Coulomb}}}{3k} = \frac{2(0.373 \text{ MeV}) 1.602 \times 10^{-13} \text{ J/MeV}}{3(1.38 \times 10^{-23} \text{ J/K})}$$

$$T = 3 \times 10^9 \text{ K}$$

This extraordinary temperature is characteristic of the interiors of stars and not of terrestrial objects.

## 14.1 NUCLEAR REACTORS

### 14.1.1 Neutron-Induced Reactions

Nuclear fission reactors ("nuclear power reactors") are devices that use controlled neutron-induced fission to generate energy. While a complete description of the design of these devices is beyond the scope of this book, there are certain basic principles related to nuclear reactors that are worth studying and that can be described and understood with a moderate effort.

Let us begin by reminding ourselves about the energy dependence of the cross section for neutron-induced reactions. In Figure 14.1, we show the (n, f) cross section for $^{235}$U and $^{238}$U as a function of neutron energy. In examining Figure 14.1, we see that the highest cross section for fission of $^{235}$U occurs at very low energies, so-called thermal energies where $E_n < 1$ eV. As discussed in Chapter 10, the cross section varies approximately as $1/v$ for these neutron energies. Thermal neutrons are neutrons that have come into thermal equilibrium with the surroundings.

## 14.1 NUCLEAR REACTORS

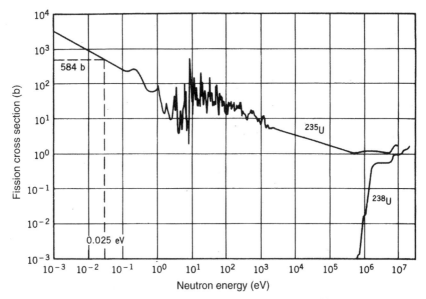

**Figure 14.1** Neutron-induced fission cross section for $^{235}$U and $^{238}$U as function of the neutron energy, $E_n$. (From D. T. Hughes and R. B. Schwartz, *Neutron Cross Sections*, 2nd ed., Brookhaven National Laboratory Report 325, 1958.)

**Example Problem** What is the velocity and de Broglie wavelength of a thermal neutron?

**Solution** The Maxwell–Boltzmann velocity distribution for the random motion of a thermally equilibrated neutron gas is

$$n(v) = 4\pi v^2 \left[\frac{m}{2\pi kT}\right]^{3/2} e^{-mv^2/2kT}$$

where we have normalized the function so that

$$\int_0^\infty n(v)\, dv = 1$$

The most probable velocity is

$$v_{mp} = (2kT/m)^{1/2}$$

If $T = 20°C$, then

$$v_{mp} = \left(\frac{2(1.38 \times 10^{-23} \text{J/K})293\text{K}}{1.675 \times 10^{-27}\text{kg}}\right)^{1/2}$$

$$v_{mp} = 2200 \text{ m/s}$$

This velocity, 2200 m/s, is taken as the characteristic velocity of thermal neutrons, and cross sections for neutrons of velocity 2200 m/s ($E_n = 1/2mv^2 = 0.0253$ eV) are referred to as "thermal" cross sections. The wavelength is

$$\lambda_{\text{deBroglie}} = \frac{h}{p} = \frac{6.626 \times 10^{-34} \text{ J/s}}{(1.675 \times 10^{-27} \text{ kg})(2200 \text{ m/s})}$$

$$\lambda_{\text{deBroglie}} = 1.80 \times 10^{-10} \text{ m}$$

Notice the de Broglie wavelength of thermal neutrons is much larger than the size of a typical nucleus ($r \sim 1\text{–}10 \times 10^{-15}$ m) and similar to the size of a typical atom. Reaction cross sections for thermal neutrons exceed the geometrical area of the nucleus.

Two other important features of Figure 14.1 deserve further comment. The first of these features is the large difference between the excitation functions for (n, f) reactions with $^{235}$U and $^{238}$U. We can understand this difference by noting the $Q$ values for neutron capture by these nuclides are

$$^{235}\text{U} + \text{n} \longrightarrow {}^{236}\text{U} + 6.54 \text{ MeV}$$

$$^{238}\text{U} + \text{n} \longrightarrow {}^{239}\text{U} + 4.80 \text{ MeV}$$

Note the $Q$ value for the n + $^{235}$U reaction is 1.7 MeV larger. The reaction with the lighter isotope converts an even–odd nucleus into an even–even nucleus, where as the other reaction creates an even–odd product. Roughly, we would then expect the $Q$ values for these reactions to differ by twice the neutron pairing energy. Since the fission barriers for $^{235}$U and $^{238}$U are about the same ($B_\text{f} \sim 6.2$ MeV), capture of neutrons of any energy will cause $^{235}$U to fission while it takes $\sim 1.4$ MeV neutrons to cause $^{238}$U to fission. The "thermally fissionable" nuclei are thus all even–odd nuclei where the energy release in neutron capture is greater than the fission barrier. The most important of these nuclei from a practical point of view are the "big three," $^{233}$U, $^{235}$U, and $^{239}$Pu.

The other feature of Figure 14.1 worthy of comment is the different regions of neutron energies and the associated cross sections. Neutrons with energies <1 eV exhibit $1/v$ cross-section behavior, and this region is referred to as the "thermal" region. Epithermal neutrons have energies from 1 to 100 eV, and their reactions are characterized by large resonances in the cross section. In the neutron energy region from 100 eV to 1 MeV, the energy levels of the excited states in the nuclei overlap and there are no discrete resonances. Neutrons with energies greater than 1 MeV are referred to as "fast" neutrons and they can cause $^{238}$U to fission.

As discussed in Chapter 10, there are other reaction mechanisms besides *fission* when neutrons interact with heavy nuclei. They include: (a) *elastic scattering* where $Q = 0$ and kinetic energy is conserved. However, the target nucleus recoils in each event and the elastically scattered neutron loses some energy. (For a collision of a neutron of energy $E$ with a nucleus containing $A$ nucleons, its energy after the collision will be $(A^2 + 1)/(A + 1)^2\, E$. Note the maximum neutron energy loss will occur when $A = 1$ when the neutron energy will be halved in each collision. (b) *Inelastic scattering* where the neutron gives up some of its energy leaving the struck nucleus in an excited state. The threshold energy for this process in the cm system will be the energy of the first excited state of the struck nucleus. For $^{235}$U and $^{238}$U, these energies are 14 and 44 keV, respectively. (c) *Radiative capture*, that is, the (n, γ) reaction in which part of the energy released by the capture of the neutron is carried away by the emitted photon. The total cross section, $\sigma_{total}$, is the sum of the cross sections for these processes, that is,

$$\sigma_{total} = \sigma_{elas} + \sigma_{inel} + \sigma_{n,\gamma} + \sigma_f$$

The distance neutrons travel between interactions, the mean free path $\lambda$ is given as:

$$\lambda = 1/\sigma_{total}\rho$$

where $\rho$ is the number density of nuclei. For uranium metal, $\rho = 4.8 \times 10^{28}/m^3$, and, if we assume $\sigma_{total} = 7$ barns, then $\lambda = 0.03$ m. If the average neutron energy is 2 MeV, the time between interactions will be $\sim 10^{-8}$ s. This mean free path will also constrain the size of a self-sustaining assembly of fissionable material.

### 14.1.2 Neutron-Induced Fission

Let us review some aspects of fission discussed in Chapter 11. Consider the case of the thermal neutron-induced fission of $^{235}$U, that is,

$$^{1}_{0}n + ^{235}_{92}U \rightarrow ^{236}_{92}U \rightarrow ^{A}_{Z}X + ^{236-A}_{92-Z}Y$$

The two fission fragments $X$ and $Y$ will have a total kinetic energy of $\sim 168$ MeV due to their mutual Coulomb repulsion at scission with the lighter fragment carrying away the larger energy due to conservation of momentum. The most probable mass split will be an asymmetric one with $A_{heavy}/A_{light} \sim 1.3 - 1.4$. Following scission, the deformed fragments will contract to a more spherical shape, heating up in the process. They will get rid of this excess energy by the emission of neutrons, emitting $\sim 2.5$ neutrons per fission event. These neutrons will have a Watt spectrum, that is, a broad peak centered below 1 MeV with an energy distribution of the form:

$$N(E) = 0.453 e^{-1.036E} \sinh[(2.29E)^{0.5}]$$

The total energy carried away by these neutrons is $\sim 5$ MeV. In competition with the last stages of neutron emission and when the excitation energies of the fission fragments are less than the neutron binding energy, the fragments will deexcite by γ-ray emission (the "prompt" γ rays) with the energy carried away by these γ rays being ~8 MeV. Following prompt γ-ray emission, one will be left with neutron-rich fragments that will decay by $\beta^-$ and γ-ray emission toward stability. About 8 and 7 MeV will be emitted in the form of $\beta^-$ particles and γ rays, respectively. About 12 MeV will appear in the form of electron antineutrinos emitted in $\beta^-$ decay; but these neutrinos will escape any practical assembly. During this $\beta^-$ and γ decay, in a small number of decays, the residual nucleus following $\beta^-$ decay is excited to an energy greater than the neutron binding energy. Such nuclei will decay by emission of neutrons (the so-called delayed neutrons) on a time scale of seconds/minutes rather than the time scale of prompt neutron emission ($\sim 10^{-15}$ s). While the energy carried away by these delayed neutrons is insignificant in the fission energy balance, these neutrons are very important in controlling the chain reaction in reactors

For the thermal neutron-induced fission of $^{235}$U, the total recoverable energy release is ~195 MeV/fission while it is ~202 MeV for the fission of $^{239}$Pu. These energy releases can be transformed into practical units by noting that 200 MeV $\cong 3.2 \times 10^{-11}$ joules. One gram of $^{235}$U contains about $3 \times 10^{21}$ atoms, corresponding to an energy release of $\sim 3.2 \times 10^{-11} \times 3 \times 10^{21} = 1$ MW/d. (The burning of one ton ($10^6$ g) of coal releases about 0.36 MW/d, so that 1 g $^{235}$U has an energy content of about $3 \times 10^6$ more energy than 1 g of coal.)

### 14.1.3 Neutron Inventory

A reactor designer must pay special attention to the inventory of neutrons in the reactor. Each fission event in the thermal neutron-induced fission of $^{235}$U produces ~2.5 fast neutrons. Upon examining the data of Figure 14.1, we conclude that the energy of these neutrons should be reduced to thermal energies to induce further fissions. Thus, we place lumps of the uranium fuel in a *moderator* that slows down the fission neutrons to thermal energies. Previously, we have demonstrated that neutrons are most rapidly slowed by elastic collisions with light nuclei. So the ideal moderator will contain light nuclei whose neutron capture cross sections are low. Graphite is an appropriate material as is $^2$H$_2$O or Be. Ordinary hydrogen has an (n, γ) cross section of 0.33 barn for thermal neutrons, making it unsuitable as a moderator unless the fuel is enriched in $^{235}$U. The number of collisions necessary to thermalize the fast neutrons from fission is 14.5 for $^1$H, 92 for $^{12}$C, and 1730 for $^{235}$U.

Of the fast neutrons produced in fission, some of them will be "moderated" to thermal energies and will induce other fission reactions while others will be "lost." The ratio of the number of neutrons in the next generation to that in the previous generation is called the *multiplication factor k*. If the value of $k$ is less than 1, then the reactor is *subcritical* and the fission process is not self-sustaining. If the value of $k$ is greater than 1, then the number of fissions will accelerate with time and the reactor is *supercritical*. The goal of reactor operation is to maintain the system in a *critical* state with $k$ exactly equal to 1. The extreme upper limit for the multiplication factor would correspond to the mean number of neutrons per fission (~2.5 for $^{235}$U(n, f)) if each neutron produces a secondary fission.

## 14.1 NUCLEAR REACTORS

This scenario is impossible to attain and, in fact, the neutron inventory must be carefully monitored in order to maintain a critical reactor.

Given that the number of neutrons emitted per fission event $v = 2.5$ for the fission of $^{235}$U, one would think that designing a system with $k = 1$ would be easy; however, there are many ways in which neutrons can be lost. First of all, the core of the reactor that contains the fuel must be finite in size. Therefore, there will be a limit or edge of the core from which some neutrons can escape. Neutrons can be "reflected" back into the core by a layer of material such as graphite (low-absorption cross section and higher mass) that surrounds the core, but the reflection is not complete.

A second unavoidable source of neutron loss occurs in the fuel itself. Consider for the moment a hypothetical reactor core made of uranium metal, There are two neutron absorption reactions with uranium nuclei, $^{235}$U(n, $\gamma$)$^{236}$U, $\sigma_{th} = 98.3$ barns and $^{238}$U(n, $\gamma$)$^{239}$U, $\sigma_{th} = 2.7$ barns that compete with the $^{235}$U(n, f), ($\sigma_{th} = 583$ barns) reaction. As an aside, one should remember that these cross sections for radiative capture and fission, like all nuclear reactions, are energy dependent. For the present discussion we will concentrate on thermal-energy ($E_n = 0.0253$ eV) cross sections. Let us define a parameter $\eta$ as the average number of fission neutrons per thermal neutron *absorbed* in the fuel. For a pure $^{235}$U sample,

$$\frac{\eta}{v} = \frac{\sigma_f}{\sigma_a} = \frac{1}{1 + \alpha}$$

where $\alpha = \sigma_\gamma/\sigma_f$ and $\sigma_a = \sigma_\gamma + \sigma_f$. For a material such as uranium metal that contains both $^{235}$U and $^{238}$U

$$\frac{\eta}{v} = \frac{x\sigma_f(235)}{x\sigma_a(235) + (1-x)\sigma_a(238)}$$

where $x$ is the atom fraction (mole fraction) $^{235}$U. For $^{235}$U in natural uranium, $\eta = 1.3$. If the $^{235}$U content of the fuel is greater than the natural abundance, $\eta$ will be larger.

The multiplication factor for an infinite sized reactor core is given by the *four-factor formula*:

$$K = \eta f p \varepsilon$$

where $p$ is the fraction of fission neutrons that are thermalized (without being captured), $f$ is the fraction of thermal neutrons that are captured in the fuel, and $\varepsilon$ is the so-called fast fission enhancement factor that expresses the fact that some fission events are due to fast neutrons. (For a typical thermal reactor, $\eta = 1.65$, $p = 0.87, f = 0.71$, and $\varepsilon = 1.02$. Because of a leakage of 4% of the neutrons in a finite reactor, $k = 1.00$.)

For safe operation of the reactor, $k$ must be exactly unity. That is difficult to achieve in practice. In fact, if the mean time between generations of neutrons is $\tau$, and the multiplication factor is $k$, and $N$ is the number of neutrons at time $t$, then there will be $kN$ neutrons at $t+\tau$, $k^2N$ neutrons at $t+2\tau$, and so forth. This can be expressed as

$$dN = (kN - N)\frac{dt}{\tau}$$

or

$$N(t) = N_0 e^{(k-1)t/\tau}$$

Suppose $k = 1.01$ and $\tau = 10^{-3}$ s, then $N(1\text{ s}) = 22{,}000 \times N_0$, a dangerous increase. The neutron inventory is regulated by inserting control rods of a neutron absorbing material such as Cd in the fissioning assembly. But such regulation cannot take place on a 1-s time scale. Fortunately, about 0.65% of the fission neutrons are "delayed neutrons" emitted on a time scale of seconds/minutes. The resulting average time constant for the "prompt + delayed" neutrons $\tau$ is $\sim 0.1$ s, instead of $10^{-3}$ s, which allows control of the reactor.

As mentioned above, reactor control is achieved using control rods containing $^{113}$Cd ($\sigma_{n,\gamma} = 20{,}000$ barns) or $^{10}$B ($\sigma_{n,\alpha} = 3800$ barns). Another important aspect of reactor control is the fact that certain fission products have very high neutron capture cross sections and thus depress the neutron inventory. Foremost among these nuclei, known as *poisons*, are $^{135}$Xe ($\sigma_{n,\gamma} = 2.65 \times 10^6$ barns) and $^{149}$Sm ($\sigma_{n,\gamma} = 4.1 \times 10^4$ barns). These nuclei lower both $f$ and $k$. In an ordinary reactor, the amount of these nuclei is regulated by their decay and their destruction by neutron capture, although their presence does affect the neutron inventory.

### 14.1.4 Light Water Reactors

A large number of light-water-cooled nuclear reactors have been constructed around the world. All of these reactors rely on the thermal fission of enriched uranium and on normal or "light" water for neutron moderation and heat transfer. These machines fall into two major categories of research reactors and power reactors. We have described the principles that underlie the construction and operation of these machines but the implementation is different. All nuclear reactors can be categorized by the thermal generating power of the core, usually given in megawatts. Power reactors are also categorized by the electrical generating power of the plant in giga- or megawatts. Note that the ratio of the electrical power to the thermal power of a power plant is the efficiency of the conversion process (a number always less than 1, see below).

Generally speaking, the research reactors are small, on the order of 1 MW of thermal power, and are optimized to provide intense neutron fluxes for the irradiation of samples. These reactors are fueled with a few kilograms of enriched uranium (20–90% $^{235}$U) in fuel rods that are clad with a zirconium alloy or with aluminum. The entire core assembly of a 1-MW research reactor is on the order of 1 m$^3$ and can produce an internal neutron flux on the order of $10^{13}$/cm$^2$-s. The two largest research reactors reach internal fluxes of $1 \times 10^{15}$/cm$^2$-s (ILL, Grenoble, France) and $3 \times 10^{15}$/cm$^2$-s (HFIR, Oak Ridge, Tennessee). The small reactors are usually submerged in a pool of water and are cooled by convection. The water layer is designed to be deep enough to provide sufficient radiation shielding so that a person can look through the water and observe the operating reactor. The heat generated by the core of a research reactor is dissipated in the large pool of water.

Nuclear power reactors are generally much larger, on the order of 2 GW of thermal power, and are designed to produce electricity by the adiabatic expansion of steam in a turbine. There are two competing designs in western countries for nuclear power reactors that differ in the primary cooling loop. In one case the water is allowed to boil (boiling water reactor, BWR) and in the other design superheated water is held in the liquid phase under pressure (pressurized water reactor, PWR) (Fig. 14.2). We will briefly consider the two types of power reactors.

Boiling water reactors are characterized by having only two coolant loops. The water in the primary coolant loop circulates through the reactor core and boils at approximately 1 atm pressure and is heated to approximately 300°C. The steam is passed to a turbine system to generate electricity, is condensed, and is cycled back to the core. A second coolant loop is used to maintain a constant output temperature at the exit of the turbines; this loop removes the so-called waste heat at the end of the thermodynamic cycle. Such cooling loops are commonly included in machines that use adiabatic expansion to do work; for example, radiators are connected to gasoline engines in cars. The waste-heat loop in a nuclear power plant is usually an external, open loop. The waste heat is released in the atmosphere in large evaporative-cooling towers or released into rivers, lakes, or the ocean. The primary coolant is also the neutron moderator and is subject to intense irradiation in the core. It will contain radioactivities from impurities extracted from the walls, and so forth, and as a result the turbines will become contaminated. Thus, the important feature of the BWR design with the primary coolant circulating through the turbines necessitates placing them inside the containment shielding.

In the other design, PWRs have two closed loops of water circulating in the plant plus a third, external loop to remove the waste heat. Water is pumped through the reactor core in the primary coolant loop to moderate the neutrons and to remove the heat from the core as in the BWR. However, the reactor vessel is pressurized so that the water does not boil. Steam is necessary to run the turbines, so the primary loop transfers the heat to a secondary loop. The water in the secondary loop is allowed to boil, producing steam that is isolated from both the core and the outside. The water in the primary loop usually contains boron (as boric acid $H_3BO_3 \sim 0.025$ M) to control the reactivity of the reactor. The steam in the secondary loop is allowed to expand and cool through a set of turbines as in the BWR; the cold steam condenses and is returned to the primary heat exchanger. A third loop of water is used to maintain the low-temperature end of the expansion near room temperature and remove the waste heat.

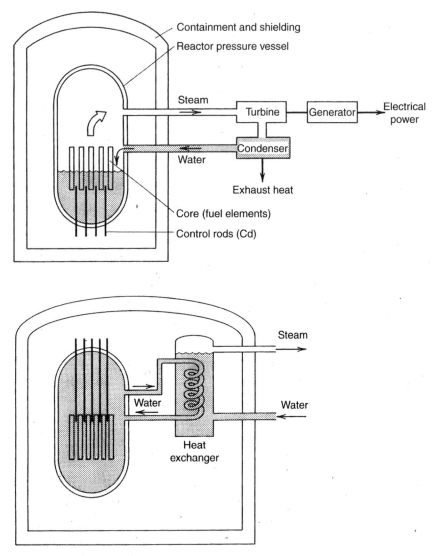

**Figure 14.2** Schematic diagram of boiling water (top) and pressurized-water reactors (bottom). (From Krane, 1988.)

The PWR is more expensive to build because the reactor vessel must be stronger to withstand the higher water pressure, and there is a secondary coolant loop with pumps and so on. The BWR, while less expensive to build, is more complicated to service since the turbines are part of the primary coolant loop. The details of the core design are different as well. Approximately twice as many PWRs have been constructed as BWRs.

## 14.1 NUCLEAR REACTORS

A limit on the efficiency of the electrical energy conduction can be obtained by applying the second law of thermodynamics to the secondary loop. The maximum thermal efficiency, $\varepsilon_{th}$, is given in terms of the input and output heats:

$$\varepsilon_{th} = (q_{in} - q_{out})/q_{in}$$

Note that the output heat is the waste heat. In the limit that the machine operates in a Carnot cycle, which can be characterized by constant temperatures at the input and output, then the maximum efficiency is

$$\varepsilon_{carnot} = (T_{in} - T_{out})/T_{in}$$

The output temperature is given by the ambient temperature of the waste-heat loop and can be taken to be 30°C for purposes of estimation. The input temperature of the steam is limited by physical constraints on the reactor primary cooling loop to be about 300°C. Therefore, the maximum Carnot efficiency is approximately $\varepsilon_{carnot} = (573\ \text{K} - 303\ \text{K})/573\ \text{K} = 0.47$, whereas the actual efficiency is typically $\varepsilon_{elec} = 0.35$ when measured as electrical power outside the plant to total thermal power in the core. For comparison, a coal-powered plant might have values of $\varepsilon_{carnot} = 0.65$, $\varepsilon_{elec} = 0.5$ due to higher steam temperatures

*Side Neutron Reactions in Water* One of the interesting side reactions that occurs in water-moderated nuclear reactors is the (n, p) reaction on $^{16}$O, which occurs with a cross section of 0.017 mbarns. The $^{16}$N product rapidly decays back to $^{16}$O with a half-life of 7.13 s; thus, the net reaction can be called a catalysis of the neutron β decay.

$$^{16}\text{O} + {^1}\text{n} \rightarrow {^{16}}\text{N} + {^1}\text{H} + Q^{\mu}_{rxn}$$
$$^{16}\text{N} \rightarrow {^{16}}\text{O} + \beta^- + \bar{\nu} + Q_\beta$$
$$\text{NET} \quad {^1}\text{n} \rightarrow {^1}\text{H} + \beta^- + \bar{\nu} + Q$$

Make an estimate of the equilibrium activity per liter of cooling water due to $^{16}$N in a reactor that has an internal flux of $10^{13}$ neutrons/cm$^2$-s.

Recall that the equilibrium activity, also called the saturation activity, occurs when the rate of production is equal to the rate of decay and requires that the sample be irradiated for more than three half-lives, or ~22 s for $^{16}$N. Also, 1 mbarn is $1 \times 10^{-27}$ cm$^2$, thus,

Activity = rate of production = rate of decay = $N_0 \sigma \phi$

$$A = N_0 \left(10^{13}\ \frac{1}{\text{cm}^2\text{s}}\right)(0.017 \times 10^{-27}\text{cm}^2)$$

$$A = \left(\frac{1\ \text{kg}\ N_A \times 0.9976^{16}\ \text{O atoms/moleH}_2\text{O}}{L\quad 0.0180\ \text{kg/moleH}_2\text{O}}\right) 0.017 \times 10^{-14}/\text{s}$$

$$A = 3.34 \times 10^{25}/L \times 0.017 \times 10^{-14}/\text{s} = 5.68 \times 10^9\ \text{Bq/L}$$

At this point it is appropriate to mention the two most significant accidents that have occurred at nuclear power plants. In both cases the seriousness of the accidents were dramatically increased by human error. In both cases the difficulties were caused by chemical reactions and not by nuclear fission.

An accident occurred at the Three-Mile Island PWR in Pennsylvania in 1979 in which the water stopped flowing due to a mechanical failure in the primary coolant loop. Subsequent actions by the operators caused the water level in the core to drop, uncovering the upper part of the fuel rods. The nuclear fission process rapidly ceased due to the loss of the water moderator, but the fuel continued to generate heat due to the decay of fission products from prior operation. This residual decay heat is a general feature of all nuclear reactors. Parts of fuel rods melted, which indicates that the local temperature reached 3000°C. As part of the accident, contaminated water from the primary coolant loop was released inside the containment building and soaked into the concrete. The noble gas fission products and a fraction of the iodine fission products were released to the environment. The difficulty of melted fuel notwithstanding, the extreme heating of the zirconium alloy that is used to clad the fuel opened the door to an exothermic chemical reaction with steam that produces hydrogen:

$$Zr(s) + 2H_2O \longrightarrow 4ZrO_2(s) + 2H_2(g)$$

An important concern during the accident was the potential chemical explosion of this hydrogen gas with oxygen inside the containment building. The cleanup process necessary inside the building continued for many years, and the perception that nuclear power is somehow very dangerous has not subsided after more than 25 years.

A much more serious accident occurred at the Chernobyl power station near Kiev in 1986 that was entirely the result of human error. This reactor relied on a large amount of graphite to moderate the neutrons with water-filled tubes to remove the heat and generate steam. This general reactor design, which also needs large amounts of uranium, was used in the United States to produce plutonium during the Cold War era but was not used for power generation. The accident in Chernobyl occurred when the operators manually removed the control rods from the reactor during a "test." The chain reaction accelerated due to the core design and the system became very hot. The cooling water was suddenly vaporized and caused the core to explode. The nuclear fission stopped due to the loss of the moderator, but the graphite was ignited and continued to burn for some time, spewing radioactivity into the air. Approximately 10% of the graphite and large fractions of the radioactive fission products were volatilized, the fraction depending on their chemical nature, all of which was spread across western Europe by the wind. The burning facility was too dangerous to approach, and the fire was extinguished by dropping sand, clay, lead, and boron onto the fire from helicopters. What was left of the reactor was buried in concrete and a massive cleanup was necessary. The inherent difficulties in the design of this reactor continue to exist in numerous other reactor facilities in the former Soviet Union.

## 14.1.5 The Oklo Phenomenon

We should not leave our discussion of nuclear reactors without mentioning "the Oklo phenomenon." In 1972, French scientists analyzing uranium ore from the Oklo uranium mine in Gabon found ore that was depleted in $^{235}$U. Further investigation showed the presence of high abundances of certain Nd isotopes, which are formed as fission products. The relative isotopic abundances of these isotopes were very different from natural abundance patterns. The conclusion was that a natural uranium chain reaction had occurred ~1.8 billion years ago.

At that time, the isotopic abundance of $^{235}$U would have been different than today, due to the differing half-lives of $^{235}$U and $^{238}$U. At $t = 1.9 \times 10^9$ y ago, the isotopic abundance of $^{235}$U was ~3%, a number characteristic of the fuels of nuclear reactors. Water apparently entered the ore deposit, acted as a moderator, and initiated chain reactions. These chain reactions appear to have lasted for ~$10^6$ y, ebbing and flowing as the water boiled away and returned. The power level was ≤ 10 kW. Some attention has been paid to the fact that these fission product deposits remained stable for more than $10^9$ y, possibly supporting the notion of geologic storage of nuclear waste.

## 14.2 NEUTRON SOURCES

Occasionally, one may need to use a radionuclide neutron source. For example, in geological applications, one may need to have a portable neutron source. Radionuclide neutron sources are generally based on either the $(\alpha, n)$ reaction or spontaneous fission. Older $(\alpha, n)$ sources utilized the 5.3-MeV $\alpha$ aprticles from 138-d $^{210}$Po, but modern sources utilize $^{238}$Pu, $^{241}$Am, or $^{242}$Cm. The $\alpha$ particles emitted from these nuclei interact with Be to produce neutrons via the $(\alpha, n)$ reaction. The resulting neutron spectrum is broad because the alpha particles can react with the Be at any point in the stopping process, usually ranging from a few MeV to greater than 10 MeV. The yield of these sources and the accompanying $\gamma$ radiation is shown in Table 14.1.

**TABLE 14.1 Properties of Radionuclide Neutron Sources**

| Radionuclide | $t_{1/2}$ | Neutron Yield (n/Ci) | $\gamma$-Radiation (mr/h at 1 m/Ci) |
|---|---|---|---|
| $^{210}$Po | 138 d | $2.5 \times 10^6$ | — |
| $^{238}$Pu | 87.8 y | $2.2 \times 10^6$ | <1 |
| $^{241}$Am | 433 y | $2.2 \times 10^6$ | <2.5 |
| $^{242}$Cm | 163 d | $2.5 \times 10^6$ | 2.5 |
| $^{252}$Cf | 2.65 y | $4.3 \times 10^9$ | 300 |

*Source*: From Brune, 1984, Kruger, 1971.

$^{252}$Cf acts as the basis of a radionuclide neutron source because 3.2% of its decays are by spontaneous fission, yielding 3.76 neutrons per fission. The neutron emission rate/Ci of material is quite high and these sources have found widespread use.

## 14.3 NEUTRON GENERATORS

Commercial neutron generators are compact charged particle accelerators designed to produce a beam of neutrons by an appropriate nuclear reaction. The most commonly used nuclear reactions are

$$^{2}\text{H}(d, n) \quad Q = 3.25 \text{ MeV}$$

$$^{3}\text{H}(d, n) \quad Q = 17.6 \text{ MeV}$$

$$^{7}\text{Li}(p, n) \quad Q = -1.646 \text{ MeV}$$

$$^{9}\text{Be}(d, n) \quad Q = 3.79 \text{ MeV}$$

The most common sources are based on the $^{3}$H(d, n) reaction. Deuterons are accelerated to ~150 keV with currents ~2.5 mA and strike a tritium target. They produce ~$2 \times 10^{11}$ of 14-MeV neutrons/s under these conditions. The neutrons produced are widely used in fast neutron activation analysis for the determination of light elements. The tritium targets are typically metals such as Ti, which have been loaded with titanium tritide. The accelerators are usually small Cockcroft–Walton machines or small sealed-tube devices where the ion source and accelerator structure are combined to produce a less expensive device with neutron yields ~$10^{8}$/s.

The $^{7}$Li(p, n) reaction is used commonly to produce approximately monoenergetic fast neutrons. The protons are accelerated to an energy of a few MeV by a small van de Graaff accelerator and strike a cooled rotating Li target. Thick target neutron yields are $> 10^{9}$ n/s-μA. The energy of the neutrons can be obtained from the $Q$ value equation (Chapter 10), which can be expressed (for 0° neutrons) as

$$8T_\text{n} - 6T_\text{p} - 2(T_\text{n}T_\text{p})^{1/2} = -11.522$$

where $T_\text{n}$ and $T_\text{p}$ are the kinetic energies of the neutron and proton in MeV.

The ultimate in accelerator-based neutron sources are the spallation neutron sources. Neutrons are generated by the spallation reaction, which occurs when high-energy (~1 GeV) protons interact with heavy nuclei, such as mercury, releasing 20–30 neutrons/reacting proton. The proton beams (and the resulting neutrons) are pulsed allowing the use of time-of-flight techniques to measure the energies of the neutrons. Expressed as a thermal neutron flux, yields of $10^{17}$ n/cm$^2$-s are possible, exceeding the neutron yields of reactors by orders of magnitude. These high-intensity neutron fluxes can be used for neutron scattering experiments in materials science and biology.

## 14.4 ACCELERATORS

As we have already indicated, charged particles must be accelerated to kinetic energies on the order of millions of electron volts (MeV) in order to overcome the Coulomb repulsion of another nucleus and induce a nuclear reaction. The Coulomb potential grows with the inverse of the separation between the two ions:

$$V_{\text{coul}} = q_1 q_2 e^2 / r$$

The Coulomb barrier is defined as the value of this potential energy for the two bare nuclei at a separation that corresponds to the sum of their radii, or

$$V_B = z_1 z_2 e^2 / (r_1 + r_2)$$

The kinetic energy necessary for a moving projectile with mass number $A_1$ to react with a stationary target with mass number $A_2$ is $\text{KE}_{\text{threshold}} = V_B(A_1 + A_2)/A_2$ being larger due to conservation of momentum.

High kinetic energies can be obtained by producing an ion with charge $q$ on a high-voltage platform held at a static potential $E$ and simply allowing the ion to move (fall) toward ground potential. The kinetic energy gain will be equal to the loss of potential energy, thus, $\text{KE} = qE$. The earliest accelerators were exactly of this type, but physical breakdown of the insulating materials limit the maximum electrostatic potential. Modern accelerators use electrode structures with alternating electric fields to accelerate charged ions. We will consider the general features of accelerators for heavy charged particles. (The acceleration of electrons is a special case due to the relative ease of production, the very large charge-to-mass $(m/q)$ ratio, and the fact that energetic electrons are relativistic and travel with essentially the same velocity, $v \approx c$.) We will start with consideration of ion sources and then consider the various machines for accelerating charged particles roughly in order of increasing final energy.

### 14.4.1 Ion Sources

All accelerators operate by the manipulation of charged ions in vacuum. Such particles do not exist naturally and must be produced in ion sources. Positive ions of all chemical elements can be produced, in principle, by ionization of atoms already in the vapor phase. The difficulty of producing an ion depends dramatically on the chemical species. A few elements have a exothermic electron affinity and can be produced as singly charged negative ions. As a result only a few accelerators utilize negative ions. Here we will consider three classes of ion sources that can produce positively charged ions.

The simplest ion sources create positive ions by bombardment of the residual gas inside a tube by electrons emitted from a hot filament (Fig. 14.3). The electrons can be accelerated to a few hundred volts, and the electron impact on atoms and small

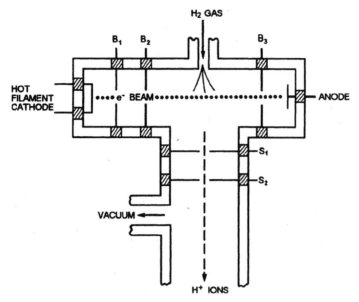

**Figure 14.3** Schematic diagram of a hot filament ion source. (From Choppin, Rydberg, and Liljenzin, 2002.)

molecules can easily create singly charged positive ions. Filament ion sources work well for producing hydrogen and helium ion beams and other gaseous elements that can be bled into the ionization region. These sources can have long lifetimes as the electrodes are not directly damaged by the ionization process, and the ions are produced from the feed material.

The energy of the electron beam can be increased in sources that are based on the features of a Penning ion gauge (PIG) that was developed to measure low pressures. A high-voltage electron arc is created between two electrodes and is confined by an external magnetic field. The arc vaporizes the electrode material and ionizes any residual gas in its path. A PIG source can produce electrons on the order of a few thousand volts and can remove substantial numbers of electrons from the top-row elements. These sources have erratic and relatively short lifetimes (10 h or less). The electrodes are worn out as they are vaporized by the arc, which also tends to produce metal coating on insulators leading to short circuits.

Very highly charged ions are produced by modern ion sources that rely on magnetically confined plasmas such as the electron-cyclotron resonance ion source (ECRIS or ECR) and the electron-beam ion source (EBIS). An ECRIS uses the superposition of axially symmetric magnetic fields with an electric field from end-cap electrodes to trap electrons in a magnetic bottle. The electrons are forced to oscillate with radiofrequency (RF) radiation that corresponds to the oscillation in the magnetic field (see discussion below about cyclotrons). The electrons move in the plasma for a long time and collide with the residual gas, creating positive ions and more electrons. The positive ions drift toward the extraction electrode and do not absorb the RF power that is tuned to the $m/q$ ratio of the electron. The plasma can remove even the innermost electrons from second-row elements.

All of these ion sources emit beams of positive ions at relatively low velocities, the ions drift or are pulled out from the ionization region with relatively small electrostatic potentials (U ~ 20 kV). These beams of charged particles can be focused and transported in vacuum to the main accelerating machines.

### 14.4.2 Electrostatic Machines

An ion source that is held at a large and stable positive electrostatic potential, $V$, will accelerate positive ions to a kinetic energy of KE = $qV$. The maximum potential is limited by the ability to sustain the high voltage without breaking down the intervening dielectric material (sparking). Consideration of the formula for the electric field $E$ at the surface of a sphere with a radius $r$, carrying a total charge $Q$, surrounded by a medium with a dielectric constant $\varepsilon$, which is $E = Q/\varepsilon r^2$, leads to several common features of electrostatic accelerators. The high-voltage terminal should be as large a sphere as possible without any sharp points (the ends of sharp points have small radii) and the terminal should be surrounded by a material with a large dielectric constant. A large, carefully prepared terminal can be held at a maximum voltage of $\approx +750$ kV in dry air. The kinetic energies of ions from such systems are (only) sufficient to induce nuclear reactions among the lightest elements and are often used to generate neutrons via the d + t $\rightarrow \alpha$ + n reaction.

The breakdown voltages of various gases were studied as a function of pressure in the 19th century by Paschen, and he showed, not surprisingly, that pure gases have higher breakdown potentials than air. Thus, higher electrostatic potentials can be maintained by insulating the platform with an inert gas such as $N_2$ or $SF_6$ and by increasing the gas pressure. An important distinguishing feature of electrostatic accelerators is that the beam is emitted "continuously" from the ion source and is literally a direct current (DC) beam of particles This feature can be good or bad depending on the application, but a DC beam can be chopped, switched, or pulsed to produce an alternating beam (AC) beam.

The underlying principle for production of high voltages is that the leakage current should be as small as possible, microamps or less, and thus the dissipated power remains low. Notice that a current of 1 μA from a 1 MV platform will have a power drain of 1 watt. The techniques developed for the production of the very high voltages necessary for electrostatic accelerators fall into two categories: *direct* and *mechanical production.* Direct production relies on electronic circuits, whereas mechanical production relies on the mechanical transportation of charge to the platform. A voltage multiplication circuit for the direct production of very high voltages was developed by Cockcroft and Walton. A schematic drawing of the circuit is shown in Figure 14.4.

An alternating current is applied to the transformer at the bottom of the circuit, which is rectified and multiplied by the stack to produce the high voltage. The resulting voltage depends on the number of elements in the stack and on the input voltage. The high voltage is not precisely constant in that it has a small variation or ripple in

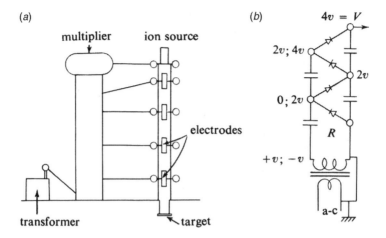

**Figure 14.4** Schematic diagram of a Cockcroft–Walton accelerator system on the left and the electronic circuit used to provide the high voltage. (From Segre, 1977.)

proportion to the input frequency. The Cockcroft–Walton design continues to be used in small machines to generate neutrons via the d + t reaction and to provide the bias voltages for ion sources for large accelerator complexes.

An important feature of all electrostatic machines is that the beam is accelerated down to ground potential by a series of electrodes at intermediate potentials. The electrodes provide a weak focusing effect that causes the beam particles to move toward the center of the tube. The focusing is due to the cylindrically symmetric shape of the electric field combined with the fact that the particle spends more time in the focusing region (first half of the gap where the field lines move toward the center) than in the defocusing region (second half of the gap) due to acceleration. Note that the Cockcroft–Walton circuit can also provide these potentials at the intermediate stages in the stacked circuit.

*Mechanical production* of high voltages for ion acceleration is the basis of the class of machines called Van de Graaff accelerators. These machines rely on the principle from electrostatics that if a charge is placed inside a hollow conducting sphere then the charge will migrate to the outer surface regardless of the amount of charge already on the sphere. Thus, a Van de Graaff uses the mechanical transportation of positive charges on an insulting belt from ground potential up to a contact on the inside of a large hollow electrode, as indicated in Figure 14.5.

The mechanical generation of a few thousand volts is relatively easy and many, many variations have been developed. You may be familiar with classroom demonstrations that use a small Van de Graaff machine to charge up a metal sphere (radius 10 cm) and make a person's hair stand on end. Larger generators that can produce moderately high potentials, a few MV, have been constructed on a large scale and have been used in spectacular displays of artificial lighting. The electrostatic potential can be written in terms of the capacitance with respect to ground of the terminal,

# COLOR PLATES

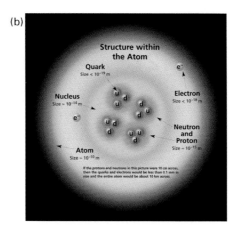

**Figure 1.4** Two artist's conceptions of the standard model. (*a*) From New York Times, 22 September, 1998. Reprinted by permission of the New York Times. (*b*) From "Nuclear Science", Contemporary Physics Education Project (CPEP), LBNL.

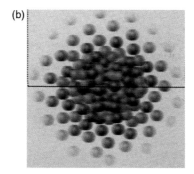

**Figure 2.10** Nuclear density distribution: (*a*) in a schematic view and (*b*) in an artist's conception from R. Mackinfosh, J. Al-Khalili, B. Jonson and T. Pena, Nucleus: A Trip to the Heart of Matter. Copyright © 2001 by The Johns Hopkins University Press, 2001; reprinted by permission of Johns Hopkins.

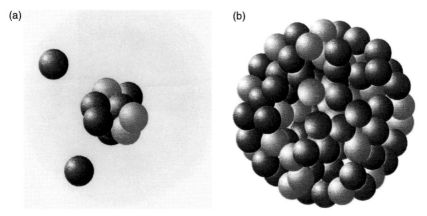

**Figure 2.12** Schematic representation of the relative sizes of the halo nucleus $^{11}$Li and $^{208}$Pb.

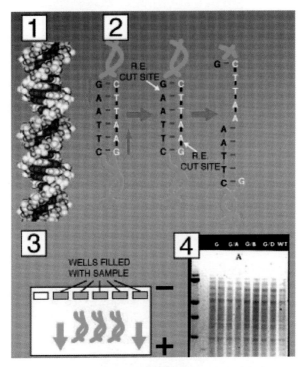

**Figure 4.2** Schematic view of DNA fingerprinting.

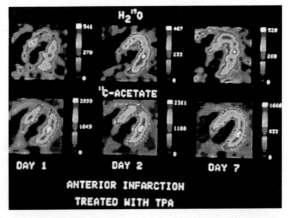

**Figure 4.6** PET pictures of the heart of a patient with acute myocardial infarction treated with a thrombolytic agent. Top row shows scans after administration of water containing $^{15}$O to trace blood flow. Bottom row shows tomograms obtained after administration of acetate containing $^{11}$C to trace the heart's metabolism, that is, its rate of oxygen use. The defects are clearly visible on day 1, both in the impaired blood flow (top left) and the impaired metabolic use of oxygen (bottom left). Recovery of blood circulation has taken place on day 2 and is maintained.

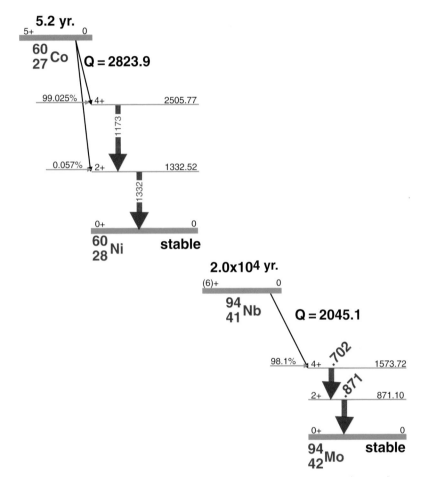

**Figure 6.8** (*a*) Energy level diagram showing the first (lowest energy) $2^+$ and $4^+$ states in $^{60}$Ni. The high-spin ground state, $5^+$, of $^{60}$Co β decays primarily to the $4^+$ state and initiates a well-known γ-ray cascade to the $2^+$ state and then the $0^+$ ground state. (*b*) For comparison, the energy level diagram showing the first (lowest energy) $2^+$ and $4^+$ states in $^{94}$Mo. The high-spin ground state, $6^+$, of $^{94}$Nb also primarily feeds the $4^+$ state initiating a γ-ray cascade.

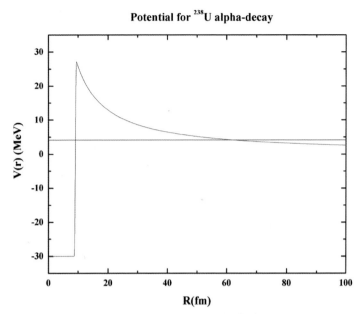

**Figure 7.5** A (reasonably accurate) one-dimensional potential energy diagram for $^{238}$U indicating the energy and calculated distances for $\alpha$ decay into $^{234}$Th. Fermi energy $\approx 30$ MeV, Coulomb barrier $\approx 28$ MeV at 9.3 fm, $Q_\alpha$ 4.2 MeV, distance of closest approach 62 fm.

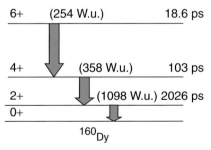

**Figure 9.4** Schematic diagram of the ground-state rotational band transitions for $^{160}$Dy.

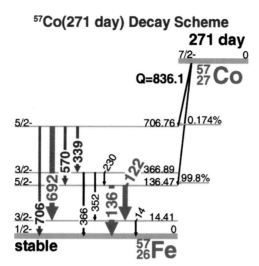

**Figure 9.10** Energy level diagram of two members of the $A = 57$ mass chain. $^{57}$Co decays to excited states of $^{57}$Fe, which result in the M1 transition from the $\frac{3}{2}^-$ state at 14.41 keV to the $\frac{1}{2}^-$ ground state.

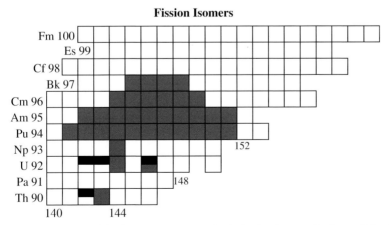

**Figure 11.6** Position of the known spontaneously fissioning isomers in the nuclide chart.

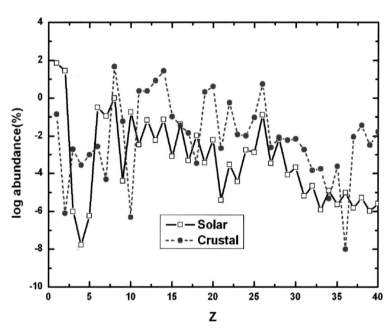

**Figure 12.1** Abundances of the first 40 elements as a percentage by mass of Earth's crust (filled squares) and in the solar system (open circles). Data from the *CRC Handbook of Chemistry and Physics*, 75th ed., 1994.

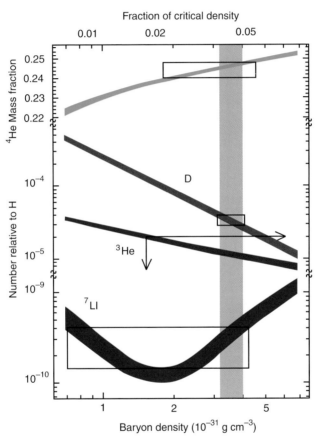

**Figure 12.6** Variation of the relative abundances of the Big Bang nuclei and the $^4$He mass fraction versus the baryon density. The boxes indicate the measured data and an estimate of the uncertainty. The curves indicate the dependence of the yield on the baryon density in the Big Bang; the vertical bar indicates the region of concordance.

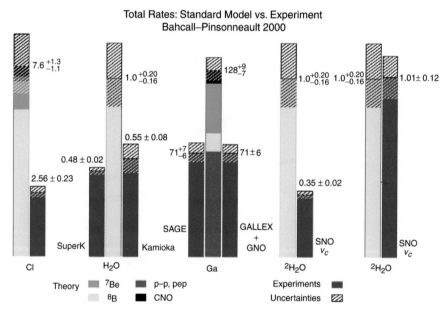

**Figure 12.19** Current status of the comparison between standard solar model predictions and experimental measurements. [From Bahcall (from Bahcall website http://www.sns.ias.edu/~jnb).]

## Projectile Fragmentation

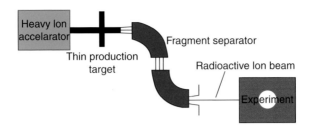

## ISOL

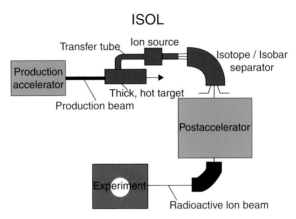

**Figure 14.13** Schematic view of the PF and ISOL techniques for generating radioactive beams.

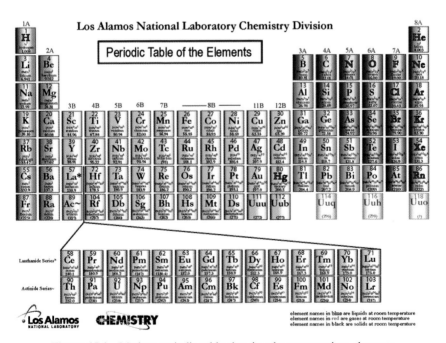

**Figure 15.1** Modern periodic table showing the transuranium elements.

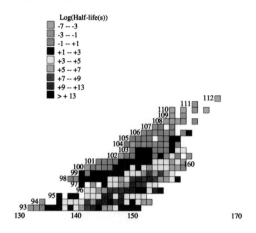

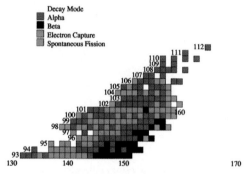

**Figure 15.2** (*a*) Half-lives of the known transuranium nuclei plotted as a function of $Z$ and $N$. (*b*) Dominant decay modes of the nuclei shown in (*a*).

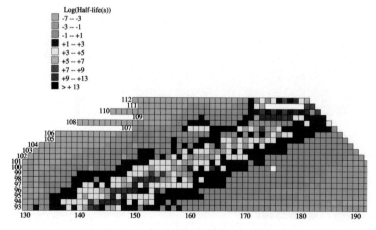

**Figure 15.3** Predicted (Möller et al., 1997) half-lives of the transuranium nuclei with $Z \leq 112$.

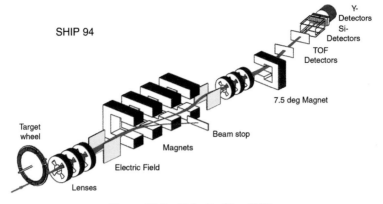

**Figure 15.7** Velocity filter SHIP

**Figure 15.9** Glenn Seaborg points out seaborgium in the periodic table.

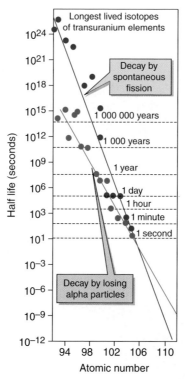

**Figure 15.10** Half-lives of the longest-lived isotope of each element versus atomic number $Z$ circa 1970.

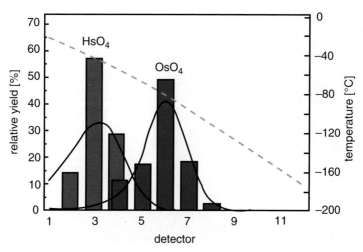

**Figure 15.17** Thermochromatogram of $HsO_4$ and $OsO_4$. From GSI Nachrichten 1/02.

**Figure 16.4** Ammonium diuranate (yellowcake) after solvent extraction. (Photo courtesy of UIC.)

**Vitrified Waste**
Radioactive material is immobilized in a glass matrix and is only released as the glass is dissolved.

**Overpack (Steel Container)**
The vitrified HLW is encapsulated in an overpack (steel container) to prevent contact with groundwater during the time when its radioactivity and heat generation are high.

**Buffer (compacted clay)**
The buffer is mainly bentonite clay compacted to high density so as to have low permeability, which slows the movement of dissolved radioactive waste. The buffer is also designed to protect the overpack

**Figure 16.13** Schematic diagram of the final steps in putting vitrified waste into a geologic repository.

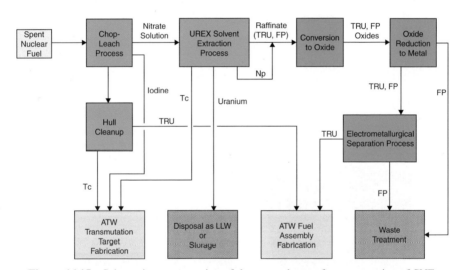

**Figure 16.15** Schematic representation of the new schemes for reprocessing of SNF.

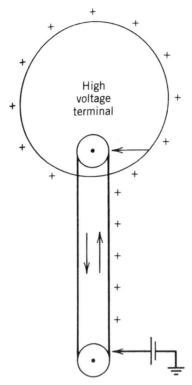

**Figure 14.5** A highly schematic view of the important components in a Van de Graaff accelerator. Positive ions created by a corona discharge near ground potential are swept by a moving belt to a similar corona contact attached to the inside of the high-voltage terminal. The positive ions then evenly distribute themselves on the surface of the terminal (Krane, 1988).

$V = Q/C$ and is limited by the designed breakdown at intermediate points along the insulating support column that provide fixed potentials for focusing the ion beam or by leakage in the surrounding gas. In practice, a Van de Graaff system reaches an equilibrium in which the added charge just compensates for charge leaking from the terminal. Rapid discharges or sparks need to be avoided because they can damage the components. An important feature of Van de Graaff high-voltage generators is that the terminal voltage can be extremely stable and ripple free.

To be used as an accelerator, a source of positive ions must be mounted *inside* the high-voltage terminal, and an insulating vacuum tube is needed to allow the ions to be accelerated to ground potential. The positive ions will be accelerated toward ground potential as in the Cockcroft–Walton machines. The terminal, accelerating column, and charging system is usually placed inside a pressurized chamber that is filled with an insulating gas such as pure $N_2$ or $SF_6$ at several atmospheres. The constraints that come from placing the ion source inside the terminal generally limit the charge on the ions to be one or two plus and thus directly limit the energy of the accelerated beam. Even though technically challenging, many so-called single-ended Van de Graaffs were produced in the 1960s and early 1970s and used in detailed studies of low-energy nuclear reactions. Note that the beam itself provides a drain on the terminal voltage, and its intensity is therefore limited to be on the order of microamperes or less. A number of advances in the technology of the construction of the belt system and of the vacuum tube and electrode structure have been made so that a modern accelerator terminal can routinely sustain 25 MV. For example, the high-voltage generator for the accelerator at Oak Ridge National Laboratory is located inside a 100-ft high, 33-ft diameter pressure vessel that is filled with $SF_6$ at a pressure of approximately 75 psig ($P_{total} \sim 6$ atm).

An important improvement of the single-ended Van de Graaff accelerators came through the replacement of the positive ion source inside the high-voltage terminal with an external negative ion source (Fig 14.6).

The negative ions are accelerated toward the high potential in one vacuum column, and they strike a very thin foil or a layer of gas placed at the center of the terminal. Electrons are readily stripped from the energetic negative ions, and the positive ions are then accelerated away from the high-voltage terminal in a second accelerating column. The final kinetic energy of the ions is the combination of the initial acceleration of the negative ion plus the final acceleration of the (multiply charged) positive ion. The charge state of the positive ion is usually much larger than one, so that the final beam energy is more than twice the terminal voltage. These devices are called tandem Van de Graaff accelerators or simply tandems. Even though there are serious limitations on the number of chemical elements that can be used to produce negative ions, the significantly higher energy beams from tandems and the relocation of the ion source away from the terminal have outweighed the limitations. Tandem accelerators have been used extensively to study low-energy nuclear reactions, particularly direct reactions induced

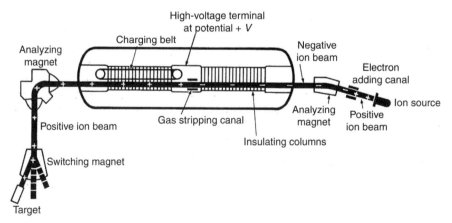

**Figure 14.6** Two-stage tandem Van de Graaff accelerator. [From R. J. Van de Graaff, *Nucl. Instr. Meth.* **8**, 195 (1960).]

by the lightest ions and fusion reactions induced by elements in the top row of the periodic table. The kinetic energy of the beam can be very precisely controlled and is very stable; however, the total energy is still limited by the terminal voltage. Attaining significantly higher kinetic energies requires a booster accelerator based on the electric fields from alternating sources of current.

### 14.4.3 Linear Accelerators

The production of very high-energy beams that are necessary for the production and study of new and exotic subnucleonic particles would require acceleration from a high-voltage platform at potentials that are unattainable in a steady state. However, one can imagine that a group of particles can be accelerated in small steps along a series of electrodes if the potential on the electrodes is synchronized with the motion of the particles (Fig. 14.7)

For example, positively charged particles will be repelled by positive electrodes and attracted to negatively charged electrodes, gaining kinetic energy as they cross the gap. (From a simple standpoint each pair of electrodes acts like an instantaneous electrostatic terminal and ground.) The particles will slowly gain energy as they synchronously cross each successive gap.

The synchronization of the arrival of the particles at the electrode gaps with an accelerating electric field can be done with an alternating electric field, as a given electrode should have the polarity that will first attract a set of particles toward it and then later repel the same particles after they have passed through the electrode. The potential difference would follow a sine function: $V = V_0 \sin(\omega t)$, where $V_0$ is the peak voltage and $\omega$ is the frequency. The key ingredient in such a simple linear accelerator is to have the electrodes be hollow tubes and have the beam particles pass through the center of each tube. The beam will be accelerated across the gap between the electrodes by the electric field, but the beam will drift while in the field-free region inside the tubular electrodes (hence the electrodes are called drift tubes). The alternating power supply can change the polarity while the beam pulse is inside a drift tube. Notice that the amount of time that the pulse of beam particles spends inside the drift tubes must be held constant to provide a uniform acceleration. The particles should reach the next gap at a time $T = 2\pi/\omega$. Thus, the length of each drift tube must increase along the path of the beam in proportion to the velocity, $L = v_i T/2$ where $v_i$ is the velocity inside the $i$th drift tube. The physical dimensions of the drift tubes will have a finite acceptance range in velocity, and so linear accelerators (linacs) have to be designed for specific velocity ranges. In practice, linacs are used as energy boosters that accept beams, usually from an electrostatic injector, started with significant initial velocities.

The phasing of the arrival of the particles with respect to the RF has two effects, one good and one bad. At first glance, one might think that the arrival of the particles should coincide with the maximum accelerating voltage. However, this point in time coincides with the top of the sine wave, and particles that arrive slightly earlier (faster particles) or slightly later (slower particles) will receive lower accelerations, and the bunch will spread out in velocity, arriving at the next electrode at different

**404** REACTORS AND ACCELERATORS

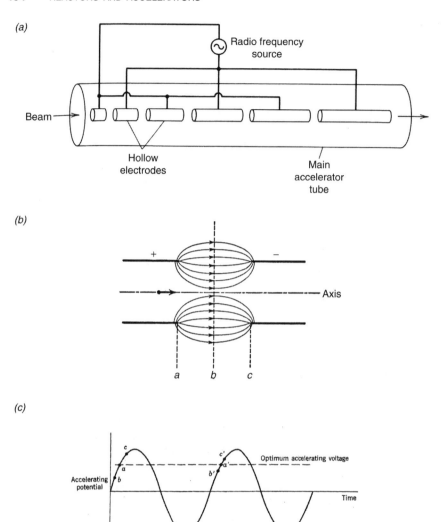

**Figure 14.7** (a) Basic design of a linear accelerator. (b) Electric field in the gap between two drift tables. (c) Phase stability in a linear accelerator. (From Krane, 1988.)

times. If the center of the pulse arrives at the electrode gap while the voltage is increasing somewhat linearly, then the faster particles will receive a lower acceleration than that applied to the average particles and the slower particles will receive a larger acceleration. These small differences in accelerations will tend to minimize the velocity distribution of the beam pulse. This phase stability is a good feature of linacs that work with nonrelativistic particles. Tuning the arrival of the beam pulse to coincide with the increasing portion of the accelerating field has the bad feature that the field across the gap sensed by a moving beam particle is asymmetric. A symmetric field provides the weak focusing in the accelerating columns of electrostatic machines, mentioned above. When the field is asymmetric, then the particle will not be focused toward the center of the tube. Phase stability is more important than the weak focusing effect, and linacs include additional components inside the drift tubes to focus the beam.

A classical linac (the Wideroe design) with fixed drift tubes connected to an external oscillator has a rather limited velocity acceptance and therefore would be used in specific applications. A much more flexible linac design (the Alvarez design) relies on creating either a standing electromagnetic wave in a resonant cavity or a traveling wave in a waveguide. Most booster accelerators used to accelerate heavy ions (nuclides more massive than helium) utilize resonant cavities to provide the accelerating voltages. Various shapes have been used to create the accelerating gaps and the drift tube regions. Early designs used copper surfaces to reflect the power, but significant power was still dissipated in the walls. Recent designs have used superconducting niobium or lead surfaces that have a much lower dissipation. Recall that the velocity of all relativistic particles is essentially constant. Thus, accelerating structures for the highest energy particles generally rely on standing wave cavities with constant drift tube lengths, $L = cT/2$. Electrons become relativistic at comparatively low energies (recall that $m_e = 0.511$ MeV), and so electron accelerators have simpler designs than heavy-ion accelerators.

The number of drift tubes in a linac has to be relatively large because the acceleration per gap is modest. A typical value of the effective accelerating electric field in a superconducting cavity is $1-2$ MV/$q$ per meter of cavity. Booster accelerators for heavy-ion beams can be 50 m long, and the electron linac at Stanford (SLAC) is 2 miles long. Linacs have the obvious difficulty that the drift tubes have to increase in length as the velocity of the particle grows. For example, the drift tube length for a relativistic particle is inversely proportional to the AC frequency: $L = c(2\pi/2\omega) = c\pi/\omega$. A typical value for the frequency in such an accelerator is 300 MHz, so that $L \sim 3$ m. The accelerator has to lie in a straight line, and thus space, alignment, and construction costs are important concerns. Notice that a given pulse of particles will only pass through the accelerating structures one time. The beam from a linac will arrive in pulses that follow the time structure of the oscillations applied to the accelerating gaps. The frequency is usually in the megahertz region (radio wave region of the spectrum) so that the pulses are usually separated by tens of nanoseconds. The time structure on this scale is usually called the beam microstructure. From a practical standpoint, a beam with such a small time separation between pulses appears to be continuous. The beam from a linac with standing-wave cavities is often pulsed on the millisecond time scale in order to allow time to dissipate heat in the walls, such pulsing makes up the macrostructure of the beam and has a large effect on experimental measurements. The fraction of time that the macrostructure of a linac is "on" is called the duty factor. The operating principles of linear accelerators were established by 1930, but the compact design of cyclotrons, which use a magnetic field to "reuse" the accelerating electrodes, overtook the linac development.

### 14.4.4 Cyclotrons, Synchrotrons, and Rings

It is well known that when a charged particle moves through a magnetic field it will experience a force that acts perpendicular to the direction of motion. Thus, a moving charged particle can be made to move in a circular orbit by placing it in a suitably large and uniform magnetic field. Essentially all of the accelerators that have been developed to produce very high-energy beams use a magnetic field to cause the particles to circle through accelerating structures. The first device to rely on a magnetic field was the cyclotron invented in 1929 by E. O. Lawrence.

The original cyclotron had a pancake-shaped vacuum chamber that was placed between the north and south poles of an electromagnet (Fig. 14.8). Two large electrodes were placed inside the vacuum chamber to provide one acceleration gap and two drift regions. Each electrode was a hollow cavity in the shape of the letter "D" and was fitted, back-to-back, inside the vacuum chamber with the acceleration gap between the straight sides of the two D's. The ions to be accelerated, in the first case $H_2^+$, were created at the center of the circle and were accelerated across the gap and enter the drift space. Thus, moving, the particles experience the Lorentz force, move on a circular path, and return to the accelerating gap! As the particles drift around, the phase of the accelerating voltage is switched to the opposite polarity as in a linac. When the particles arrive at the gap, they are accelerated again, gain energy, and move into the drift region.

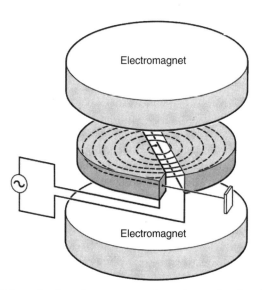

**Figure 14.8** Schematic view of components of a cyclotron. A pulse of beam particles starts in the center of the machine and is accelerated across the gap, circles through the drift space inside the D-electrode, and is accelerated again when it returns to the gap. Eventually the beam reaches the edge of the machine and can be extracted along a tangent. [From Krane, 1988.]

For many years, filament and PIG ion sources were placed in the center of the cyclotrons. However, the development of more complicated and powerful ion sources such as the ECR sources required more space than was available at the center of the machines. Present-day cyclotrons have external ion sources and a low-energy beam from an electrostatic injector is threaded into the center via an axial channel and inflector or via a radial channel in separated-sector machines (described below).

The developers of the cyclotron recognized that the frequency of the circular motion of the particle is constant. Consider the Lorentz force acting on a moving charged particle, $F_{\text{lorentz}} = B \times qv$, where $B$ is the (vector) magnetic field, $q$ is the charge, and $v$ is the (vector) velocity. The cross product follows the "right-hand rule" so that the force is perpendicular to the motion. The radius of the motion, $r$, can be found by setting the Lorentz force equal to the mass times the centripetal acceleration:

$$F_{\text{Lorentz}} = Bqv = mv^2/r$$

Solving for the radius,

$$r = \frac{mv}{Bq}$$

The time, $t_{\text{cyc}}$, that it will take a particle to complete one orbit is the circumference divided by the velocity:

$$t_{\text{cyc}} = \frac{2\pi r}{v} = \frac{2\pi mv}{Bqv} = \frac{2\pi m}{Bq}$$

and is independent of the velocity of the ion for a given value of the magnetic field. This constant time for a charged particle to orbit in a magnetic field is usually stated as a frequency called the cyclotron resonance frequency of that particle. As an aside, the circular or cyclotron motion forms the basis of mass measurements in ion-cyclotron-resonance (ICR) mass spectrometers. As long as the particles are nonrelativistic, all the beam particles in a cyclotron will drift through the D's in the same amount of time and arrive at the accelerating gaps in phase, that is, the orbits are isochronous. Notice also that we can use the concept of phase stability from linear accelerators to maintain the bunch structure of the initial beam pulse. The orbital radii will increase as the velocity of the particle increases at each gap so that the particles will appear to spiral out from the center of the cyclotron. The maximum velocity will occur when the particles reach the maximum physical radius, $\rho_{\text{max}}$, of the vacuum chamber and D's. Solving for the maximum kinetic energy, $T_{\text{max}}$, assuming a nonrelativistic beam:

$$T_{\text{max}} = \frac{1}{2} m v_{\text{max}}^2 = \frac{(Bq\rho_{\text{max}})^2}{2m} = \frac{(B\rho_{\text{max}})^2}{2}\left(\frac{q^2}{m}\right)$$

Notice that the maximum kinetic energy depends on the two machine parameters $B$ and $\rho_{max}$ times a ratio of the square of the charge to the mass of the beam particles. The first term in the expression for the maximum beam energy, $(B\rho_{max})^2/2$, is often called the $K$ value of the cyclotron and is given in units of MeV. The oscillation frequency of the D's can be tuned over a limited range (e.g., 15–30 MHz) in present-day cyclotrons to provide beams with different values of $(q/m)$ at various energies. Small cyclotrons used to produce specific isotopes for radiopharmaceuticals have $K \sim 30$ MeV and can provide 30-MeV protons ($q^2/m = 1$). The highest energy cyclotron in the world has $K = 1200$ MeV and provides heavy ions up to $E/A = 200$ MeV ($q^2/m \sim m/4$) and is limited by the vertical focusing and the relativistic mass increase of the beam.

The beam has to make many revolutions in a cyclotron in order to be accelerated up to the full energy. For example, a typical accelerating gap might have a potential difference of 100 kV so that a proton beam with total energy of 30 MeV that crosses the gap twice per revolution still requires on the order of $n_{turns} \sim E_{total}/(qV2) = 30/(1 \times 0.100 \times 2) = 150$ turns. It is important that vertical focusing be applied to the beam so that it remains, at least on average, on the central plane of the cyclotron. There are two main techniques for vertical focusing in cyclotrons. A weak focusing occurs in the simple case in which the magnetic field of the cyclotron runs between flat, uniform, and finite faces (pole faces) of an electromagnet. The magnetic field between flat pole faces will only be exactly perpendicular in the center and will increasingly bow out as one moves toward the edge. The curved shape of the magnetic field will provide a weak restoring force for particles that leave the median plane. A stronger focusing effect can be produced by dividing the flat pole face into sections that are higher (hills) and lower (valleys). As indicated in Figure 14.9, the average magnetic field should remain the same as that obtained with a flat pole face, but the local magnetic field is higher between the hills and lower between the valleys. The magnetic field will bow out from the hill region

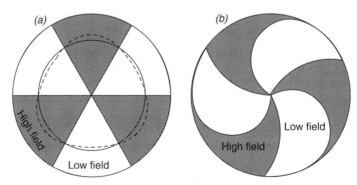

**Figure 14.9** Top view of the sectors, or the hills and valleys, in a sector-focused cyclotrons. In (a) the concept of strong vertical focusing at the transition regions, edges, of the sectors is shown for straight sectors. A larger focusing effect is obtained by spiraling the sectors as shown in (b). Note that the particles travel counterclockwise in this figure.

into the valley region along the path of the beam particles and provide a restoring force toward the median plane each time a particle crosses the transition. The vertical focusing can be increased by using a spiraled pole sector rather than a straight sector. Such cyclotrons are usually called sector-focused cyclotrons and are extensively used to provide moderate and high-energy beams of heavy ions. Another technique to produce strong focusing is to make the hill regions from separate wedge-shaped electromagnets and to leave the valley regions open. There has to be a common vacuum chamber for the particles to circulate and to house the D's. The complexity of running four large individual magnets is compensated by the very large variation in the magnetic field between the hills and valleys (called the flutter) leading to a large vertical focusing effect, and it is easier to install and service all the auxiliary equipment needed to inject, accelerate, and extract the beam. Such machines are called separated-sector cyclotrons.

An important limitation on the maximum beam energy available from cyclotrons comes from the relativistic increase of the mass of a particle with velocity. We have seen elsewhere that $m = m_0/[1 - (v/c)^2]^{1/2}$; thus, $m = 1.02\, m_0$ for $v = 0.2c$. Notice that the ratio $B/m$ occurs in all of the cyclotron equations given above. Thus, there are two classes of isochronous cyclotrons, low-energy machines in which the mass increase is small enough to be ignored, for example, $v/c < 0.2$, and medium-energy machines in which the increasing mass is compensated by increasing the strength of the magnetic field with radius. The field is usually increased by adding extra, concentric coils to the magnet pole pieces that are called trim coils. The field could also be increased by decreasing the gap between the pole faces. The field lines in a cyclotron with a magnetic field that increases with radius will bow in toward the middle (opposite to that described above) a feature that will produce a weak vertical defocusing effect! Thus, fixed frequency cyclotrons cannot produce extremely high-energy beams.

A large number of cyclotrons have been constructed over the years. For the most part, they have been spiral sector-focused machines with trim coils. The $K$ values range from 30 to 150 MeV for electromagnets with normal conducting (or resistive) coils. A number of sector-focused cyclotrons have been constructed with superconducting coils based on designs developed at Michigan State University that have $K$ values of 500, and the largest machine has $K = 1200$. Several large separated-sector, normally conducting cyclotrons were constructed at the end of the 20th century at GANIL in France ($K = 440$) and at RIKEN in Japan ($K = 540$). A project is underway at RIKEN to design and build a separated-sector machine using superconducting coils. The beams from cyclotrons will have a microstructure similar to that from a linac and appear to be continuous in most applications. A cyclotron beam will not have a macrostructure unless one is applied to the beam from the ion source for specific experimental reasons. The acceleration of the beam from a cyclotron can be rapidly stopped by simply shifting the relative phases of the D's.

A number of attempts were made to develop cyclotrons that could accelerate protons up to energies on the order of a GeV (the proton rest mass). For example, the resonant frequency of the cyclotron could be decreased in proportion to the mass increase. Such a frequency-modulated (FM) cyclotron (*synchrocyclotron*) could accelerate a single pulse of particles up to high-energy but would have to be reset to start the next pulse and thus would have a low-duty factor. In addition, the size of the magnet becomes extremely large and costly. The largest cyclotron magnet ever constructed was a 184-in. diameter machine at LBL in Berkeley, originally designed to provide 100-MeV protons, but completed in 1946 as a synchrocyclotron.

The successful acceleration of protons and heavier nuclei to relativistic energies was realized through the compensation of the increasing mass of the particle with an increasing magnetic field. The early machines were fixed-frequency cyclotrons in which the acceleration process was synchronized with a changing magnetic field produced by a very large electromagnet. The mechanical design was changed later to be just a ring of individual magnets. Simple geometry indicates that the set of individual magnets necessary to construct a ring requires much, much less iron that that of a large cyclotron magnet. The *synchrotron* design has proven to be extremely robust and is used in all of the machines built to produce the highest energy nuclear beams.

Synchrotrons use the concept that the particles are to be confined to move in a circular orbit with essentially a constant radius, that is, a ring, regardless of the energy of the particles. Thus, synchrotrons are pulsed machines that operate on a cycle in which a modest energy beam is injected into the ring, the beam is accelerated, the high-energy beam is dumped out, and the ring is returned to the injection state. A low magnetic field is necessary to confine the low-energy particles at injection. After a sufficiently large number of particles have been fed into the ring, one or more accelerating structures (originally drift tubes, now resonant cavities) are turned on and the beam begins to gain energy. The energy gain per turn is usually low ($\sim$100 kV) and during the acceleration process the magnetic field is ramped up toward the maximum value that the magnets can provide. If the synchrotron has to accept nonrelativistic particles, then the revolution frequency of the particles will increase as the velocity increases (as the radius of the orbit is constant) so that the frequency of the accelerating structure has to increase as well. Thus, both the magnetic field and the accelerator have to be synchronized with the energy of the particles. The particles can be extracted from the ring providing a single macrocycle beam pulse. The magnetic field is then returned to the initial, low value. The highest energy synchrotrons accept particles that are already relativistic (from prior acceleration in booster synchrotrons) and the revolution frequency remains constant. The time necessary for a single macrocycle is on the order of seconds and is dictated by the maximum rate of change of the magnetic field. Modern rapid-cycling synchrotrons run at 1 or 2 Hz, while the original machines from the 1960s typically ran at 1–5 Hz.

The principle of phase stability is used in synchrotrons to maintain a narrow energy distribution of the beam bunches during acceleration. The problems associated with vertical focusing in cyclotrons will be present in synchrotrons. The original machines relied on weak focusing in magnets with flat pole faces. A variation of the strong focusing obtained with hills and valleys in sector-focused cyclotrons can be obtained in synchrotrons. Rather than hills and valleys, though, the ring is divided into sectors in which the gaps between the pole pieces of the magnets are wedge shaped. The thinner side of the wedge is alternately on the inside or the outside of the ring. As in the cyclotron, the average field is set for an isochronous orbit. When the beam circles around the ring, it encounters a vertical focusing region (thin edge inside) and a vertical defocusing region (thin edge outside), which produces a net vertical restoring force. Dipole magnets that have pole pieces that are shaped (tilted) to provide focusing are called combined-function magnets as they are meant to perform two tasks, and synchrotrons that use this version of strong focusing are called alternating-gradient machines. Modern synchrotrons do not use combined-function magnets but rather use dipole magnets to bend the beam and quadrupole doublet magnets (discussed below) to focus the beam in straight sections between the dipoles. Such independent function magnets are easier to construct and allow more flexible tuning.

Synchrotrons are used to accelerate protons and heavy nuclei to the highest energies, presently 0.95 TeV protons in the TEVATRON at Fermilab and 100 GeV/nucleon heavy ions (including $^{197}$Au nuclei at almost 20 TeV) in RHIC at Brookhaven National Lab (BNL). The Large Hadron Collider (LHC) under construction at CERN will collide 7 TeV protons and collide heavy ions with a total collision energy in excess of 1250 TeV. The maximum energy of the beam is proportional to $B\rho_{max}$ as in a cyclotron, and the value for the ring is usually given in tesla-meters, or T-m. Even after extensive development, the maximum field strengths in large electromagnets is on the order of a few tesla, however, the radius of the ring of magnets is only limited by money. For example, the booster or intermediate energy synchrotron for the RHIC system is approximately 20 m radius and provides 1 GeV protons. The main ring of the AGS (alternating gradient synchrotron) at BNL is a 100-T-m system consisting of 240 combined function magnets in a radius of approximately 85 m. The main rings of the RHIC system are 839.5 T-m and contain 1740 separated function superconducting magnets in a 3.834-km circumference.

A similar but bigger and very versatile combination of big accelerators has been established at CERN in Europe. The complex includes several high-energy injectors for protons, electrons, and heavy ions, a booster synchrotron, the PS, that can accelerate pulses of these ions (e.g., 26-GeV protons), a high-energy synchrotron, the SPS, that can also accelerate the ions (e.g., 400-GeV protons, 170-GeV/nucleon heavy ions in a 1.1-km radius ring), and the largest ring system, LEP, for electrons and positrons (90-GeV electrons in 4.3-km radius). The LEP machine is being converted into the large hadron collider, LHC, that will be able to accelerate and collide all of these particles (e.g., 7-TeV protons in the LEP tunnel).

There are two important features of nuclear collisions that we have not considered in our discussion of particle accelerators. First is conservation of momentum in the collision dictates that a large fraction of the energy that goes into a collision in the laboratory between a moving particle and a resting target nucleus will go into kinematic motion of the products and will not be available for excitation of the products or for new particle production. Perhaps more surprising is the second feature that after all the effort to produce a beam of high-energy particles, the great majority of the beam particles will pass through the target material, interacting with the electrons and slowing down, and not collide with another nucleus. Both of these problems can be resolved if we create counterrotating beams of particles that are circulating at constant orbits in synchrotron rings held at their maximum magnetic fields. The beams can be forced to cross one another at specific points or interaction regions. Circular systems of magnets have also been developed without accelerating structures called storage rings. The net momentum of particles that collide head-on is zero in the laboratory frame so that all of the energy is available in principle for excitation of the products. It is much more cost effective to build a storage ring for a synchrotron system and gain a factor of 2 in energy than to double the radius of the synchrotron. If the counterrotating particles miss during one crossing, as is the most likely event, then they simply continue on their orbit and literally come back around for another try. These features, particularly the higher available energy, are part of all of the modern high-energy accelerator complexes. All of the modern high-energy synchrotrons mentioned above include storage rings and rely on colliding counterrotating beams, for example, the Tevatron collides protons and antiprotons, LEP-collided electrons and positrons, and RHIC collides heavy-ion beams from separate rings.

## 14.5 CHARGED PARTICLE BEAM TRANSPORT AND ANALYSIS

The goal of accelerating particles is to induce reactions with target nuclei. Most nuclear targets are pure elemental foils and the earliest experiments were performed by placing a metal foil in the path of the beam at the end of the acceleration process. For example, a foil could be placed at the largest radius of a cyclotron or at the end of a linear accelerator. As we have just discussed, these reactions can be made to occur between counterrotating beams in storage rings. It is very difficult to perform experiments directly in the accelerator for a number of reasons including the high radiation environment caused by beam loss during acceleration and physical constraints on the available space. Thus, beam transport techniques were developed to bring fully accelerated beams to remote and shielded vaults. These beam handling techniques are directly analogous to optical techniques using glass prisms and lenses to transport beams of photons.

In optical systems the light rays are diffracted when they make a transition between two media with different indices of refraction. Prisms use converging flat surfaces to chromatically disperse the light, lenses use curved surfaces (spherical lenses are most common) to focus or defocus the light rays. Charged particle beams are similarly effected by magnetic and electric fields. A beam of charged particles will be deflected as it travels through a uniform field created between two surfaces and can be focused or defocused as it travels through the radially increasing field created by two, four, six, or more surfaces. The fields are usually labeled as dipole, quadrupole, hexapole, and so on. The multipole fields can be created by electrostatic plates or electromagnet poles. The forces acting on the moving particles are different in electrostatic and magnetic systems. Magnetic systems have the most widespread applications primarily for technical reasons as superconducting wire technology provides a means to supply extremely large magnetic fields in compact devices, whereas the maximum attainable electric fields are small by comparison.

We have already seen that a moving charged particle will experience a force perpendicular to its direction of motion when it is moving in a magnetic field. This Lorentz force causes the particle to curve with a radius, $\rho = Bq/mv$ that depends on the charge-to-mass ratio $(q/m)$ of the ion. Thus, a simple magnetic dipole can be used to change the direction of a beam of particles. For example, consider the path of a beam that enters the magnetic field in the region between the poles of a wedge-shaped magnet, as indicated in Figure 14.10. Such magnets are commonly called magnetic sectors or sector magnets and are characterized by their magnetic field, $B$, their bend radius, $\rho$, and their bend angle, $\theta$. The beam will follow the path from $O$ to $E$ along an arc with the radius $\rho$ and turn through an angle $\theta$. The path of the beam need not be perpendicular to the straight edges of the sector. These angles are labeled $\alpha_1$ and $\alpha_2$ in Figure 14.10. If all the particles are on exactly parallel trajectories when they enter the magnetic field, they will all turn through the angle $\theta$ and emerge in a parallel bunch. If on the other hand, the particles enter the magnetic field in a diverging bunch, then the sector will focus the beam in the horizontal and vertical directions with focal lengths given by:

$$f_{\text{horiz}} = \frac{\rho}{\sin(\theta)[1 - \tan(\alpha_1)\tan(\alpha_2)] - \cos(\theta)[\tan(\alpha_1) + \tan(\alpha_2)]}$$

and

$$f_{\text{vert}} = \frac{\rho}{[\tan(\alpha_1) + \tan(\alpha_2)] - \theta[\tan(\alpha_1)\tan(\alpha_2)]}$$

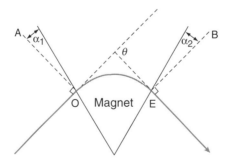

**Figure 14.10** Schematic view of the bending of a charged particle beam in a wedge-shaped dipole magnet with straight edges. The beam will be bent through an angle $\theta$; the angles $\alpha_1$ and $\alpha_2$ are used to describe the angles of incidence relative to the straight edges and are important in determining the weak focusing due to the fringing field. [From Harvey, 1964.]

Notice that the focal lengths are different. If the beam of particles enters and exits along a normal to each face, $\alpha_1 = \alpha_2 = 0$, the horizontal focal length becomes $f_{\text{horiz}} = \rho/\sin(\theta)$ while all vertical focusing is removed, $f_{\text{vert}} = \infty$. Thus, if a perfectly round beam is deflected by a dipole magnet, it will often lose its symmetry due to the different focal lengths of the dipole magnet. Subsequent ion-optical focusing elements have to be set to compensate for this difference to produce a round beam at the target position. It is important to remember that the bending radius, $\rho$, is proportional to the charge-to-momentum ratio of the beam $(q/mv)$ so that the field strengths of all of the magnets will have to be set for each beam.

Magnetic focusing of a charged particle beam can be produced by a colinear solenoidal field or by a pair of magnetic quadrupoles. In both cases the magnets create a fringing field or fields that cause diverging particles to be returned to the optical axis. If we define the $z$ axis along the path of the beam, a single solenoid will focus in both the $x$ and $y$ directions. A magnetic quadrupole can be constructed by arranging four equal strength poles on a circle at 90° from one another with polarities that alternate between north and south. An example of a quadrupole magnet is shown in Figure 14.11.

If we examine the magnetic field from such a quadrupolar device, we will find that the field along the central axis is exactly zero and increases linearly to a maximum value at each pole face. (The pole tips should be hyperbolic surfaces to conform to the shape of the magnetic field, although cylindrical pole tips are often used for ease of manufacture.) Quadrupole magnets are thus characterized by the gradient of the magnetic field, $dB/dr$, where $r$ is a radial coordinate and

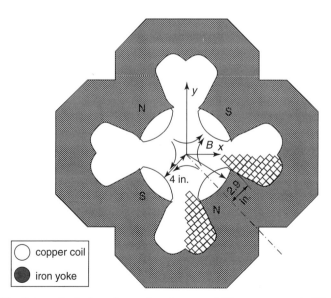

**Figure 14.11** Schematic design of a quadrupole magnet. The arrangement of the poles will provide $x$ focusing in the usual right-handed coordinate system with the $z$ axis emerging from the page. [From O. Chumberlin, *Ann. Rev. Nucl. Sci.* **10**, 161 (1960).]

## 14.5 CHARGED PARTICLE BEAM TRANSPORT AND ANALYSIS

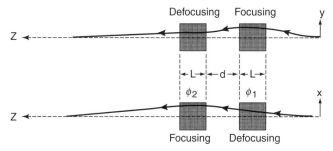

**Figure 14.12** Net focusing effect on a diverging charged particle beam from the combination of a converging and diverging pair of magnetic quadrupole lenses. The key feature of the doublet is that the particles move along paths that are closer to the optical axis in the defocusing element of the pair. (From Harvey, 1964.)

their length, $l$, along the beam direction. A particle that moves along the central axis stays in a region of no magnetic field, while particles that are off-axis will encounter an increasing magnetic field that acts like the fringing field at the edge of a dipole magnet. Notice, however, that the fringing fields will focus off-axis particles in one direction ($x$ in Fig. 14.11) and defocus them in the orthogonal direction ($y$ in Fig. 14.11). If we define the constant $k^2 = (q/mv)dB/dr$, then the focal length of one quadrupole with a length $l$ (called a singlet) is

$$f_{\text{quad}} = \frac{1}{k \sin(kl)}$$

At first glance the fact that a singlet is both focusing and defocusing at the same time might imply that uniform focusing of a beam is not possible. However, it is known from light optics that the combination of a converging lens with a diverging lens has a net focusing effect as is indicated in Figure 14.12. Therefore, focusing magnet packages are most often made by combining an $x$-focusing quadrupole with a $y$-focusing quadrupole of equal lengths into a "doublet." Three quadrupoles ($y, x, y$) with lengths ($l, 2l, l$) are sometimes combined into a "triplets," which provide more flexibility in manipulating the beam shape.

Modern accelerator complexes rely on a large combination of magnetic dipoles and quadrupoles to transport fully accelerated beams large distances without significant loss. Solenoidal magnets are only used occasionally when point-to-point focii are needed as quadrupoles are more efficient and allow the two coordinates to be tuned independently. Higher order magnetic multipoles, sextupoles, and octupoles are used to correct aberations in high-resolution applications such as spectrometers and fragment separators.

**Example Problem** A dipole magnet deflects charged particle beams through an angle of 22.5° with a radius of 2.0 m. For ease of construction the magnet has rectangular pole pieces 0.5 × 1.5 m long. The beam enters normally at the center of one of the 0.5-m faces and exits at an angle from the opposite 0.5-m edge. What are the focal lengths of this magnet?

**Solution** We need to evaluate the angles $\alpha_1$ and $\alpha_2$ in order to solve this problem. From the problem definition we know that $\alpha_1 = 0$. A little geometry will show that in this case the exit angle is equal to the bend angle:

$$f_{horiz} = \frac{\rho}{\sin(\theta)[1 - \tan(\alpha_1)\tan(\alpha_2)] - \cos(\theta)[\tan(\alpha_1) + \tan(\alpha_2)]}$$

$$= \frac{2.0 \text{ m}}{\sin(22.5°)[1 - \tan(0)\tan(22.5°)] - \cos(22.5°)[\tan(0) + \tan(22.5°)]}$$

$$= \frac{2.0 \text{ m}}{\sin(22.5°)(1 - 0) + \cos(22.5°)[\tan(22.5°)]}$$

$$= \frac{2.0 \text{ m}}{0.383 + 0.924(0.414)}$$

$$= 2.613 \text{ m}$$

For the vertical focal length we have

$$f_{vert} = \frac{h}{[\tan(\alpha_1) + \tan(\alpha_2)] - \theta[\tan(\alpha_1)\tan(\alpha_2)]}$$

$$= \frac{2.0 \text{ m}}{[\tan(0) + \tan(22.5°)] - 22.5°(2\pi/180°)[\tan(0)\tan(22.5°)]}$$

$$= \frac{2.0 \text{ m}}{\tan(22.5°) - 0}$$

$$= \frac{2.0 \text{ m}}{0.414} = 4.828 \text{ m}$$

Such different focal lengths, a weaker focus in vertical compared to horizontal, as in this situation are a common occurrence in beam transport systems.

The transport and control of charged particle beams with electrostatic elements has the nice feature that the equations of motion do not depend on the mass of the particle. The force felt by a charged particle in an electric field is simply $F_e = qE$ where $q$ is the electric charge and $E$ is the electric field strength. When a positively charged particle enters the region between two parallel plates with a separation $d$ at a voltage $V$, the particle will feel the electric force pulling it towards the electrode at the lower potential, $F_e = qV/d$, and will move on a circular orbit with a radius, $\rho = mv^2 (d/qV)$. Such a device can thus change the direction of the incident beam. (The beam will undergo a weak focusing from the fringing field at the entrance and exit of the device as discussed above.) Strong focusing of a beam of charged particles can be produced by an Einzel lens (cylindrical focusing from a set of three ring electrodes) and by a quadrupolar arrangement of electrodes.

An electrostatic mirror can be produced by an electrode at a potential energy that is greater than the kinetic energy divided by the charge of the particle. The bending and focusing power of electrostatic systems are limited by the maximum electric fields that can be applied across the electrodes. Extensive electrostatic systems have been constructed for the transport of low-energy beams, KE ≤ 50 keV, for example, beams extracted from ion sources are usually transported with electrostatic elements.

## 14.6 RADIOACTIVE ION BEAMS

Unstable nuclei with modestly short half-lives have been produced and separated as very low energy beams for some time. Recently, techniques have been developed to provide much more energetic beams of nuclei with half-lives as short as a few milliseconds with sufficient energy to induce secondary nuclear reactions. The production techniques usually rely on the creation of exotic nuclei in high-energy reactions followed by the collection and separation of a specific exotic nucleus. The physical techniques differ in that the products are either the residues of target (nearly at rest in the laboratory) or of projectile nuclei (moving with nearly the beam velocity), but in both reactions a large nucleus is fragmented into its components. The difference in the initial velocity of the product has large consequences for the physical separation techniques and reacceleration but no consequences for the reaction mechanism. Thus, the same residues can and have been produced for study in each rest frame. Very energetic proton beams ($E_{lab} \sim m_0 c^2$) from synchrotrons were used extensively in the 1960s, and the process of target fragmentation was rapidly exploited for the production of exotic nuclei. The residual nuclei left in the target after a beam pulse were thermalized and then ionized for separation. This technique is usually called the ISOL technique (Fig. 14.13). This was an acronym for isotope separator online. Today it may be more precise to associate ISOL with ion source online to more clearly distinguish these devices from in-flight devices used to separate projectile fragments. The beam interacts with a target, usually a refractory metal, that is heated to several thousand degrees. The target itself can be quite thick, even thick enough to stop the beam, but it should be thin to allow rapid release of the reaction products. This apparent paradox can be solved by using stacks of thin metal foils. The reaction products come to thermal equilibrium in the target matrix and, depending on their chemical nature, can diffuse out of the matrix. Many techniques have been used to ionize the hot atoms including surface ionization, electron beam or plasma ionization, and resonant laser ionization. Notice that the reaction mechanism creates a broad range of products, most of which diffuse out of the target at some rate. The ionization process is a second chemical process that creates singly charged ions. After being ionized, the reaction products can be readily extracted from the target system and accelerated. The chemical selectivity provides a means to select the most exotic nuclei that are only weakly produced in the primary reaction. The target ion-source combination is placed on an electrostatic platform to provide very low

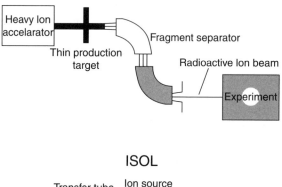

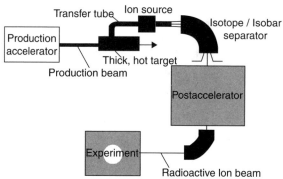

**Figure 14.13** Schematic view of the PF and ISOL techniques for generating radioactive beams. (Figure also appears in color figure section.)

energy ions (≤50 keV, total) for decay studies. A large number of facilities were operated over the years, the most successful being the ISOLDE facility at CERN that uses pulses of 1-GeV protons from the CERN PS synchrotron to irradiate targets such as Nb, Ta, and U. The ISOLDE facility has a large electrostatic beam-handling system with many experimental stations.

A number of projects are underway around the world to accelerate exotic beams to energies sufficient to induce secondary nuclear reactions. These first-generation facilities all rely on using existing accelerators and in some cases existing experimental equipment. A facility at Louvain-la-Neuve in Belgium is producing radioactive beams of light nuclei created in direct reactions induced by a 30-MeV proton beam from a cyclotron developed to produce radioisotopes for medicine. Neutral gaseous products, atoms, and small molecules are pumped from the target into an ECR ion source to be ionized and then the ions are transferred into a K110 cyclotron. The HRIBF project at Oak Ridge relies on using one of the first sector-focused cyclotrons ($K = 100$) to provide radioactive ions with an intense proton beam. In this case the K100 cyclotron produces nuclei by simple reactions such as $^{70}$Ge(p, 2n)$^{69}$As in an ion source that produces $1^+$ ions. The positive ions have to be converted into negative ions before injection into a 25-MV tandem. The tandem thus places some limits on the chemical elements that can be utilized. The ISAC facility in Vancouver, B.C. uses a 500 MeV proton beam from the TRIUMF separated sector cyclotron to irradiate heated targets. The emitted ions are then accelerated in a superconducting linear accelerator. The hallmark of all of these facilities is that they have an accelerator that provides an intense beam to produce the activities (the driver) and another accelerator for the secondary beam. Table 14.2 presents a list of ISOL facilities that is meant to indicate the variety of approaches that are being used.

The advent in the 1970s of synchrotrons that were capable of delivering beams of all elements with very high energies allowed the production of the same exotic nuclei observed as target residues with sufficient kinetic energies to allow rapid physical separation and identification. This technique is usually called the "projectile fragmentation technique" but more correctly might be referred to as in-flight separation since other reaction mechanisms besides projectile fragmentation can be used to produce the nuclei (Fig. 14.13). The beams of exotic nuclei are not stopped in this technique, and even very short-lived nuclei can be studied and used to induce secondary reactions. In the mid-1970s it was shown that up to 1% of a primary $^{12}$C beam could be converted into $^{11}$C ions and separated for implantation into biomedical samples. Nuclear physics experiments to produce exotic nuclei and calibrate space-flight instruments using similar techniques based on magnetic rigidity were also pioneered with beam-line elements at LBL (Berkeley, California) and then dramatically extended by using degraders (the LISE spectrometer at GANIL (Caen, France)). This technique has been further extended in four second-generation devices distributed around the world and third-generation devices at the NSCL (Michigan State) and at RIKEN (Japan).

Fast-ion beams of exotic nuclei are separated from the primary beam and from the other reaction products by a combination of separated function magnetic bending dipoles and focusing quadrupole doublets acting on the distribution of ions emerging from the target at high velocities ($\beta \sim 0.5c$). Achromatic magnetic systems are used, where achromatic means that the position and angle of ions at the end of the device (called the focal plane) does not depend on the ion's momentum. Such achromatic magnetic systems are generally most useful for efficient

**TABLE 14.2  Examples of ISOL Fragmentation Facilities**

| Device | Accelerator System | Reaction[a] | Country |
|---|---|---|---|
| ARENAS | Cyclotron/ECR/Cyclotron | (p, n), (p, 2p), etc. | Belgium |
| HRIBF | Cyclotron/1+//1−/TANDEM | (p, xn) | United States |
| ISAC | Synchrotron/ECR/LINAC | TF | Canada |
| REX-ISOLDE | Large Synchrotron/1+//EBIS/LINAC | TF | Europe (CERN) |
| SPIRAL | HI-Cyclotrons/ECR/Cyclotron | PF | France |

[a]Class of reactions: low-energy fusion and direct reactions (p, xn); target fragmentation (TF); or projectile fragmentation (PF).

separation at the highest energies because they can collect a large fraction of all the produced fragments and focus them to a small spot. Achromatic systems have the additional advantage that the final spot size is kept small even when the momentum acceptance is large. The key elements in these devices are an initial bend for momentum-to-charge ratio selection, an energy loss degrader for atomic number separation also called a "wedge," and a second bend for momentum-to-charge ratio selection of a specific ion.

An aperture or a slit is used at the intermediate position to limit the momentum acceptance of the device. Since the fragmentation mechanism produces all the nuclei with nearly the same velocity, the initial magnetic rigidity ($B\rho = mv/q$) and thus momentum-to-charge ratio selection is equivalent to separation according to the mass-to-charge ratio of the products. Even so, projectile fragmentation reactions produce many different ions that have the same mass-to-charge ratio, for example, the fragmentation of an $^{18}$O beam can formally produce five ions with $m/q = 3$: $^3$H, $^6$He, $^9$Li, $^{12}$Be, and $^{15}$B. An energy degrader is inserted into the beam at the intermediate dispersive image in order to remove the ions that have the same initial mass-to-charge ratio as the fragment of interest but different atomic numbers. Recall that all ions will lose some kinetic energy in the degrader, and the relative amount will depend on their atomic number. The ions, therefore, will exit the foil with different magnetic rigidities. The contaminants can then be spatially dispersed at the focal plane by an additional bend. This Z-dependent separation is proportional to the degrader thickness and to the ratio of the magnetic rigidity of the second half of the system to that of the first half. A schematic diagram of the A1900 separator operating at the NSCL is shown in Figure 14.14.

There are projectile fragmentation separators presently operating in France, Germany, Japan, Russia, and the United States. Other similar devices are in the planning stages or are under construction in several laboratories. A comparison of the various parameters that describe these fragment separators is given in Table 14.3. The LISE separator has been operated for more than 15 years and has provided

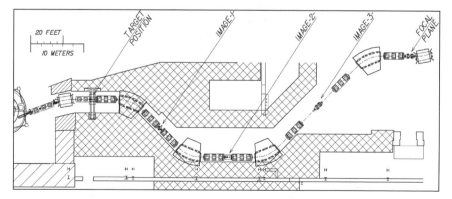

**Figure 14.14** Schematic diagram of the A1900, a new projectile fragmentation separator recently constructed at Michigan State University.

**TABLE 14.3 Comparison of Fragment Separators**

| Device | $\Omega$ (msr) | $\Delta p/p$ (%) | $B\rho$ (T − m) | Resolving Power[a] | Length (m) | Facility |
|---|---|---|---|---|---|---|
| A1200 | 0.8–4.3 | 3.0 | 5.4 | 700–1500 | 22 | NSCL (U.S.) |
| A1900 | 8.0 | 4.5 | 6.0 | ~2900 | 35 | NSCL (U.S.) |
| COMBAS | 6.4 | 20 | 4.5 | 4360 | 14.5 | JINR (Russia) |
| LISE3 | 1.0 | 5.0 | 3.2 | 800 | 18. | GANIL (France) |
| FRS | 0.7–2.5 | 2.0 | 9–18 | 240–1500 | 73 | GSI (Germany) |
| RIPS | 5.0 | 6.0 | 5.76 | 1500 | 21 | RIKEN (Japan) |
| RCNP | 1.1 | 8.0 | 3.2 | 2000 | 14 | RCNP (Japan) |

[a]Mass-to-charge resolution, see the text.

beams for a variety of experiments. The RIPS device has the largest solid angle and momentum acceptance of second-generation devices.

The more recently constructed COMBAS device at the JINR at Dubna, Russia has a significantly larger acceptance and is based on using combined function magnets. The A1200 (now retired) and the A1900(MSU) (shown schematically in Fig. 14.14), RCNP(Osaka), and FRS(GSI) separators are positioned at the beginning of the beam distribution system to allow delivery of radioactive beams to any experimental area.

Besides the obvious dependence of the RNB intensity on the intensity of the primary beam, the secondary beam rate is also directly related to the relative separator acceptance. The GSI device has been designed for very high kinetic energies, where the fragmentation angular cone and relative energy spread are relatively small. This allows the physical acceptance of the device to be smaller but yet gives "full acceptance" of individual products. Large solid angle and large momentum acceptance are especially important if the device is to be used to separate light ions at 50–200 MeV/nucleon, the energy region in which most separators are operating. Note that the larger the physical acceptance of the separator, the lower the primary beam energy at which 100% collection efficiency is attained.

## 14.7 NUCLEAR WEAPONS

While a full discussion of nuclear weapons is beyond the scope of this book, some comments about the operating principles of such devices and their connection to reactors and accelerators are desirable.

The techniques used to produce a "nuclear explosion" (i.e., an essentially instantaneous, self-perpetuating nuclear chain reaction) are very complex. A nuclear explosion must utilize a high-energy neutron spectrum (fast neutrons, that is, neutrons with energies >1 MeV). This results basically from the fact that, for an explosion to take place, the nuclear chain reaction must be very rapid—of the order of microseconds. Each generation in the chain reaction must occur within about 0.01 μs (a "shake") or less. The energy release takes place over many generations, although 99.9% of the energy release occurs within the last seven generations, that is, in a time of the order of 0.1 μs. The rapid time scale of this reaction requires the use of fast neutrons. The process by which a neutron is degraded in energy is time consuming and largely eliminates the possibility of an explosion. This also explains why power reactors that operate with a slow or thermal neutron spectrum cannot undergo a nuclear explosion, even if the worst accident is imagined. In the case of reactors that operate with higher energy neutrons, a nuclear explosion is also precluded because of the geometrical arrangement of the fissionable material and the rearrangement of this material if an accident occurs.

The explosive ingredients of fission weapons are limited, in practice, to $^{239}$Pu and $^{235}$U because these are the only nuclides that are reasonably long-lived, capable of being produced in significant quantities, and also capable of undergoing fission with neutrons of all energies, from essentially zero (thermal) to the higher energies of the secondary neutrons emitted in fission. Other nuclides, for example, $^{238}$U or $^{232}$Th, can undergo fission with some of these higher energy neutrons but not with those of lower energy. It is not possible to produce a self-sustaining chain reaction with these nuclides since an insufficient fraction of the neutrons produced in the fission reaction has an appropriate energy to induce, and hence perpetuate, the fission reaction.

Fission weapons currently use $^{239}$Pu or highly enriched $^{235}$U (usually greater than 90%), although, in principle, enrichments as low as 10% are usable. Fission weapons utilizing $^{239}$Pu have higher yield-to-weight ratios and can be made with smaller sizes and weights. One problem in plutonium-based weapons is the presence of $^{240}$Pu, whose high spontaneous fission rate can present problems with the preinitiation of the weapon. Preinitiation of the weapon is defined as the initiation of the nuclear chain reaction before the desired degree of supercriticality (see below) is achieved. The neutrons emitted during the spontaneous fission of $^{240}$Pu can cause such a preinitiation, which will decrease the yield of the weapon and increase the uncertainty in that yield. To prevent this preinitiation, weapons-grade plutonium contains less than 7% $^{240}$Pu while ordinary reactor-grade plutonium may contain more than 19% $^{240}$Pu. The $^{240}$Pu content of plutonium can be regulated by controlling the time $^{238}$U is left in the reactor for generating $^{239}$Pu. Many U.S. fission weapons contain both $^{239}$Pu and $^{235}$U as a trade-off between the higher efficiency of using $^{239}$Pu and the greater availability of $^{235}$U. (About 43 metric tons of Pu are in U.S. nuclear weapons.) If the conditions are such that the neutrons are lost at a faster rate than they are formed by fission, the chain reaction is not self-sustaining. The escape of neutrons occurs at the exterior of the $^{239}$Pu (or $^{235}$U) mass undergoing fission, and thus the rate of loss by escape will be determined by the surface area. On the other hand, the fission process, which results in the formation of more neutrons, takes place throughout the whole of the material; the rate of growth of neutron population is therefore dependent upon the mass. If the quantity of $^{239}$Pu (or $^{235}$U) is small, that is, if the ratio of the surface area to the volume is large, the proportion of neutrons lost by escape to those producing fissions will be so great that the propagation of a nuclear fission chain, and hence the production of an explosion, will not be possible. But as the size of the piece of $^{239}$Pu (or $^{235}$U) is increased and the relative loss of neutrons is thereby decreased, a point is reached at which the chain reaction can become self-sustaining. This is referred to as the "critical mass" of the fissionable material.

The critical mass of a bare sphere of normal density $^{235}$U metal has been reported to be 52 kg while the same number reported for certain phases of plutonium metal is about 10 kg. However, the critical mass may be lowered in a number of ways. Use of a reflector can lower the critical mass by a factor of 2–3. Compression of the material to increase its density will also lower the value of the critical mass, with the critical mass being approximately proportional to the inverse square of the density. Most nuclear weapons employ only a fraction of the critical mass (at normal density). Because of the presence of stray neutrons in the atmosphere or the possibility of their being generated in various ways, a quantity of $^{239}$Pu (or $^{235}$U) exceeding the critical mass would be likely to melt or possibly explode. It is necessary, therefore, that before detonation a nuclear weapon should contain no single piece of fissionable material that is as large as the critical mass for the given condition. In order to produce an explosion, the material must then be made supercritical, that is, made to exceed the critical mass, in a time so short as to completely preclude a sub-explosive change in the configuration, such as by melting.

Two general methods have been described for bringing about a nuclear explosion, that is to say, for quickly converting a subcritical system into a supercritical one. In the first procedure, two or more pieces of fissionable material, each less than a critical mass, are brought together very rapidly in the presence of neutrons to form one piece that exceeds the critical mass. This may be achieved in some kind of gun-barrel device in which a high explosive is used to blow one subcritical piece of fissionable material from the breech end of the gun into another subcritical piece firmly held in the muzzle end. The first nuclear weapons had a mass of $^{235}$U in the form of a sphere with a plug removed from its center. The plug was then fired into the center of the sphere creating a supercritical assembly. This technique is largely of historical interest.

The second method makes use of the fact that when a subcritical quantity of an appropriate isotope, that is, $^{239}$Pu (or $^{235}$U), is strongly compressed, it can become critical or supercritical. The reason for this is that compressing the fissionable material, that is, increasing its density increases the rate of production of neutrons by fission relative to the rate of loss by escape. The surface area (or neutron escape area) is decreased, while the mass (upon which the rate of propagation of fission depends) remains constant. A self-sustaining chain reaction may then become possible with the same mass that was subcritical in the uncompressed state.

In a fission weapon, the compression may be achieved by encompassing the subcritical material with a shell of chemical high explosives, which is imploded by means of a number of external detonators, so that a uniform inwardly directed "implosion" wave is produced. The implosion wave creates overpressures of millions of pounds per square inch in the core of the weapon, increasing the density by a factor of 2. A simple estimate may be made to show that the resulting assembly should have a size of 10 cm, the mean free path of a fast neutron in $^{235}$U or $^{239}$Pu. The implosion technique is used in modern nuclear weapons.

In both methods, high-density, heavy metals are used to surround the fissionable material, thereby reducing or preventing the escape of neutrons from the reacting assembly. To contain the fissionable material and ensure that a large enough fraction of the nuclei undergo fission before the expansion of the exploding material causes subcriticality, the fissile material is surrounded by a heavy-metal case that acts as a tamper (and a neutron reflector).

In a thermonuclear or hydrogen bomb, a significant fraction of the energy release occurs by nuclear fusion rather than nuclear fission. The hydrogen isotopes, $^2$H (deuterium, D) and $^3$H (tritium, T), can be made to fuse, as:

$$^2\text{H} + {}^3\text{H} \longrightarrow {}^4\text{He} + \text{n} + 17 \text{ MeV}$$

To initiate such a D–T fusion reaction requires temperatures of 10–100 million degrees. Relatively large amounts of deuterium/tritium and/or lithium deuteride can be heated to such temperatures by a fission explosion where the temperature may be $\sim 10^8$ K. (Tritium is generated in situ by the neutron bombardment of $^6$Li during the fusion reaction by the reaction $^6\text{Li} + \text{n} \to {}^3\text{H} + {}^4\text{He} + \text{n} + 17$ MeV, thus making the overall fusion reaction $^6\text{Li} + {}^2\text{H} \to 2\,{}^4\text{He} + 21.78$ MeV).

The energy release can be enhanced further by using the high-energy neutrons released in the fusion reactions to induce fission in the abundant isotope, $^{238}$U. Thus, we have fission–fusion and fission–fusion–fission weapons, which can give rise to explosions of much greater energy than those from simple fission weapons.

In a typical modern multistage thermonuclear weapon, the radiation from a fission explosion is used to transfer energy and compress a physically separate component containing the fusion material. The fissile material is referred to as the primary stage while the fusion material is called the secondary stage. A third stage can be added in which the fast neutrons from the fusion reaction are used to initiate the fission of $^{238}$U. In modern multistage thermonuclear weapons, comparable energy release is said to come from fission and fusion reactions.

A published schematic diagram of the operation of a modern multistage thermonuclear weapon is shown in Figure 14.15. The fission stage is similar to the implosion weapon used over Nagasaki but is only 12 in. in diameter. The chemical explosives are arranged in a soccer ball configuration with 20 hexagons and 12 pentagons forming a sphere. Detonator wires are attached to each face.

In this example, the fusion reaction must take place before the expanding fireball of the exploding fission trigger destroys the fusion materials (i.e., in a time scale of less than 100 shakes). This is accomplished through the use of $x$ and $\gamma$ radiation to transmit the energy of the fission reaction. The $x$ and $\gamma$ radiation travels about 100 times faster than the exploding debris from the fission reaction to the fusion assembly. As shown in Figure 14.15, the thermonuclear weapon in this example is a 3 to 4-ft-long cylinder with an 18-in. diameter with the fission stage located

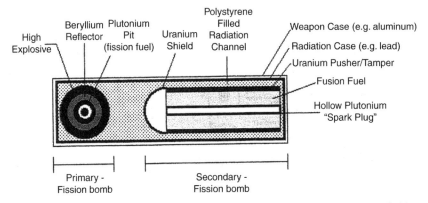

**Figure 14.15** A schematic diagram of a hydrogen bomb. Adapted from Carey Sublette, Nuclear Weapon Archive.

near one end and the fusion stage near the other. The $x$ or $\gamma$ radiation is directed to a tamper of polystyrene foam, which surrounds the fusion assembly. The radiation energy is absorbed by the polystyrene foam, which is transformed into a highly energized plasma, which compresses the fusion fuel assembly.

The "neutron bomb" or "enhanced radiation" weapon is a thermonuclear weapon in which the energy release in the form of heat and blast is minimized, and the lethal effects of the high-energy neutrons generated in fusion are maximized. This is reported to be done by the elimination of the $^{238}$U components of the weapon. The suggested net effect of this is that the instantaneously incapacitating radius (dose of 8000 rad) of a neutron bomb is about the same as a fission weapon with 10 times the yield. The instantaneously incapacitating radius for a one-kiloton (kT) neutron bomb is thus about 690 m.

Nuclear weapon yields are measured in units of kilotons of TNT (1 kT of TNT = $10^{12}$ calories = the explosive energy release from 60 g of fissile material). The energy release is mostly in the form of pressure and heat with a smaller amount (~15%) released in the form of radiation. The first nuclear explosive device, which was detonated at Alamogordo, New Mexico, had a yield of about 20 kT as did the Fat Man bomb dropped over Nagasaki, Japan (both fueled by $^{239}$Pu). The Little Boy bomb dropped over Hiroshima, Japan, had a yield of 12–15 kT (fueled by $^{235}$U). The efficiency of the plutonium-based devices was about 17%, while the uranium-based device had an efficiency of about 1.3%. For a 20-kT weapon, the radiation dose at 500 m from the center is estimated to be ~70 Gy and drops to ~4 Gy at 1.1 km. The smallest nuclear weapons have been reported to have weight that is about 0.5% of the Fat Man bomb (10,800 lb) and a total size of 25–30 in. in length and 10–12 in. in diameter, with explosive yields about 0.25 kT. Modern thermonuclear weapons with yields above 100 kT have yield/weight ratios of 1–3 kT/kg, which is far from the theoretical maximum of 80 kT/kg.

## PROBLEMS

1. A Cockcroft–Walton accelerator produces 400-keV protons. What is the maximum energy of the neutrons that can be produced with this accelerator using the $d + T$ reaction?

2. Given a reactor that contains 11 kg of $^{235}$U and operates at a power level of 1 MWe, what is the antineutrino flux 15 m from the core?

3. Verify the statement that the reactor poison $^{135}$Xe reaches a maximum $\sim 10$ h after shutdown of a high flux ($>10^{14}$ n/cm$^2$s) reactor.

4. Given the reactor of Problem 2, how long can it run before it uses up 10% of its fuel?

5. If natural uranium in dissolved in D$_2$O at a concentration of 0.4 g/g D$_2$O, calculate $k_\infty$ and the radius of a critical mass.

6. Given a reactor where the time between production and absorption of neutrons is 1 ms and the power level is 1 MWe, calculate the number of free neutrons in the reactor during operation.

7. Calculate the number of collisions needed to reduce a neutron's energy from 1 MeV to 0.025 eV in H$_2$O, D$_2$O, and C. Calculate the neutron mean free path in each case.

8. Given a 1-g source of $^{252}$Cf, calculate the neutron flux 1 m from the source and the heat produced in the source.

9. For a 1000-MWe nuclear reactor fueled with a fuel containing 5% $^{235}$U, calculate the uranium use in a year of full-time operation.

10. Describe and compare electron cooling and stochastic cooling.

11. Using the Web as an information source, compare the current Michigan State University superconducting cyclotrons (K500, K1200) and the Uppsala University GWI synchrocyclotron with a "classical" cyclotron.

12. Describe the duty cycle of a typical Van de Graaff accelerator, a linac, and a synchrotron.

13. For an accelerator with a radius equal to that of Earth, and a magnetic field of 40 kG, calculate the maximum energy of the protons that could be accelerated.

14. A cyclotron accelerates $\alpha$ particles to an energy of 42 MeV. What is the $K$ of the cyclotron? What is the maximum energy deuteron beam that it can produce?

15. A cyclotron has a diameter of 60 in. with a frequency of 10.75 MHz. Calculate the maximum energy of the proton beam and maximum field strength needed.

16. Explain the meaning of the term "phase stability" in regard to a linac and as used with a synchrocyclotron.

17. To study the structure of the proton, what energy particles do you need?
18. Consider the following situations: (a) a 20-GeV proton collides with a target electron, (b) a 20-GeV electron collides with a target proton, and (c) a 10-GeV proton collides with a 10-GeV electron. In each case, what is the cm energy available to create new particles?

## REFERENCES

Burcham, W. E. and M. Jobes. *Nuclear and Particle Physics*, Longman, Burnt Mill, 1995.

Harvey, B. G. *Introduction to Nuclear Physics and Chemistry*, 2nd ed., Prentice-Hall, Englewood Cliffs, NJ, 1964.

Krane, K. S. *Introductory Nuclear Physics*, Wiley, New York, 1988.

Segre, E. *Nuclei and Particles*, 2nd ed., Benjamin, Reading, MA, 1977.

## BIBLIOGRAPHY

Brune, D., B. Forkman, and B. Persson. *Nuclear Analytical Chemistry*, Chartwell-Bratt, Sweden, 1984.

Choppin, G., J. O. Liljenzin, and J. Rydberg. *Radiochemistry and Nuclear Chemistry*, 3rd ed., Butterworth-Heinemann, Oxford, UK, 2001.

Cochran, T. B., W. M. Arkin, and M. M. Hoenig. *Nuclear Weapons Databook*, Vol I, Ballinger, Cambridge, 1984.

Das, A. and T. Ferbel. *Introduction to Nuclear and Particle Physics*, Wiley, New York, 1994.

Draganic, I. G., Z. D. Draganic, and J. P. Adloff. *Radiation and Radioactivity on Earth and Beyond*, CRC Press, Boca Raton, FL, 1993.

Enge, H. A. *Introduction to Nuclear Physics*, Addison-Wesley, Reading, 1966.

Frauenfelder, H., and E. M. Henley. *Subatomic Physics*, 2nd ed., Prentice-Hall, Englewood Cliffs, NJ, 1991.

Friedlander, G., J. W. Kennedy, E. S. Macias, and J. M. Miller. *Nuclear and Radiochemistry*, 3rd ed., Wiley, New York, 1981.

Kruger, P. *Principles of Activation Analysis*, Wiley, New York, 1971.

LeFort, M. *Nuclear Chemistry*, van Nostrand, Princeton, NJ, 1968.

Perkins, D. H. *Introduction to High Energy Physics*, 4th ed., Cambridge University Press, Cambridge, 2000.

Seaborg, G. T. and W. Loveland. *The Elements Beyond Uranium*, Wiley, New York, 1990.

# CHAPTER 15

# THE TRANSURANIUM ELEMENTS

## 15.1 INTRODUCTION

The chemical elements are the building blocks of nature. All substances are combinations of these elements. There are (as of 2005) 113 known chemical elements with the heaviest naturally occurring element being uranium ($Z = 92$). The 22 heaviest chemical elements, the transuranium elements, are manmade. The story of their synthesis, their properties, their impact on chemistry and physics, and their importance to society is fascinating. This story is of particular importance to nuclear chemistry because most of our knowledge of these elements and their properties comes from the work of nuclear chemists, and such work continues to be a major area of nuclear chemical research. One of us (GTS) has been intimately involved in the discovery and characterization of these transuranium elements.

In this chapter, we will discuss how to make these elements, their chemical properties, and their presence in the environment. The current list of transuranium elements is shown in Table 15.1 with a modern view of their place in the periodic table being shown in Figure 15.1.

## 15.2 LIMITS OF STABILITY

There are about 260 transuranium nuclei known as of 2005. All these nuclei are unstable, with half-lives ranging from $\sim 10^{-9}$s to $10^{17}$s (Fig. 15.2a). The longest-lived nuclei are those with lower values of Z. As Z increases, the lifetimes

---

*Modern Nuclear Chemistry*, by W.D. Loveland, D.J. Morrissey, and G.T. Seaborg
Copyright © 2006 John Wiley & Sons, Inc.

## TABLE 15.1 Transuranium Elements[a]

| Atomic Number | Element | Symbol |
|---|---|---|
| 93 | Neptunium | Np |
| 94 | Plutonium | Pu |
| 95 | Americium | Am |
| 96 | Curium | Cm |
| 97 | Berkelium | Bk |
| 98 | Californium | Cf |
| 99 | Einsteinium | Es |
| 100 | Fermium | Fm |
| 101 | Mendelevium | Md |
| 102 | Nobelium | No |
| 103 | Lawrencium | Lr |
| 104 | Rutherfordium | Rf |
| 105 | Dubnium | Db |
| 106 | Seaborgium | Sg |
| 107 | Bohrium | Bh |
| 108 | Hassium | Hs |
| 109 | Meitnerium | Mt |
| 110 | Darmstadtium | Ds |
| 111 | Roentgenium | Rg |
| 112 | | |
| 113 | | |
| 114 | | |
| 115 | | |
| 116 | | |
| 117 | | |
| 118 | | |

[a]Elements 112–118 have not been named (as of 2005).

become shorter, with the lifetimes of the heaviest elements being tiny fractions of a second. For elements with $Z \geq 107$, there are few, if any, nuclei whose lifetimes lend themselves to chemical studies.

All transuranium nuclei are unstable with respect to $\alpha$ decay, meaning that $Q_\alpha$ is positive for all these nuclei. In addition, nuclei with neutron/proton ratios differing from that of nuclei along the valley of $\beta$ stability will also decay by either $\beta^-$ decay or $\beta^+$/EC decay. For most heavy nuclei, EC decay dominates over $\beta^+$ decay, and consequently the neutron-deficient heavy nuclei decay by EC decay. As the atomic number of these nuclei increases, the importance of decay by spontaneous fission (SF) increases. In Figure 15.2b, we show the dominant decay mode for each of the transuranium nuclei. (A subtle bias occurs in Fig. 15.2b in that spontaneous fission is, in general, not an acceptable way to characterize a nucleus, due to the lack of a definitive way of establishing the Z, A of the fissioning system. Consequently, the heaviest *known* nuclei are $\alpha$ emitters.) As we discussed in Chapter 11, the upper bound of the periodic table is given by spontaneous fission. At some value of (Z, A) the spontaneous fission half-life becomes so short as to prevent observation ($t_{1/2} < 10^{-9} - 10^{-6}$ s). Many transuranium nuclei decay by a combination of EC, $\alpha$ decay, and SF with the branching ratios for each mode depending on the (Z, A) of the nucleus.

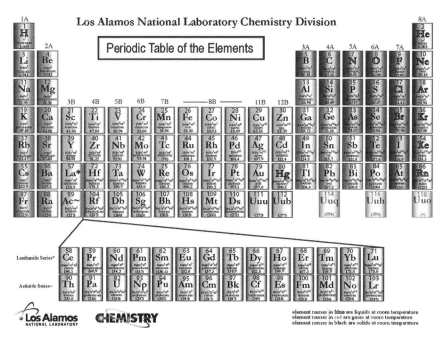

**Figure 15.1** Modern periodic table showing the transuranium elements. (Figure also appears in color figure section.)

What about the breadth of the distribution of heavy nuclei? What are the limits on $N/Z$? As in the lighter nuclei, the limits are set by the proton dripline ($S_p = 0$) and the neutron dripline ($S_n = 0$). For a typical heavy nucleus, nobelium, the proton dripline is at $N \sim 132$, while the neutron dripline is at $N \sim 236$ (Möller et al., 1997). The range of known nobelium isotopes goes from $N = 147$ to $N = 160$. Thus, it is unlikely that one will be limited by the neutron dripline for heavy nuclei, while the proton dripline may be reachable with some effort. In Figure 15.3, we show the calculated (Möller et al., 1997) changes in the half-lives of the heavy nuclei, as they become more neutron rich. Comparison of Figures 15.2a and 15.3 shows that the predicted half-lives increase by orders of magnitude as the neutron number increases modestly from those currently observed. This effect motivated recent work to make more neutron-rich heavy nuclei to study their chemistry and atomic physics. The underlying science behind these trends is that increasing $N/Z$ decreases $Z^2/A$ (reducing SF decay), decreases $Q_\alpha$ (reducing the α-decay probability), and this consequently leads to nuclei that decay by $\beta^-$ decay, a slower process.

## 15.3 ELEMENT SYNTHESIS

The synthesis of a new element involves more than just colliding two nuclei whose atomic numbers are such that they sum to those of a previously unobserved nuclide.

**432** THE TRANSURANIUM ELEMENTS

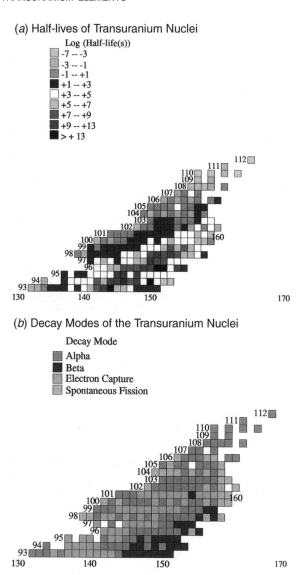

**Figure 15.2** (a) Half-lives of the known transuranium nuclei plotted as a function of Z and N. (b) Dominant decay modes of the nuclei shown in (a). (Figure also appears in color figure section.)

Heavy nuclei are, in general, quite fissionable. If they are made with significant excitation, they will decay by fission, leaving no easily identifiable heavy residue of their formation. So one must balance carefully the factors governing the "production" of a new nucleus with those factors governing its "survival". The "production factors" determine the yield of the primary reaction products while the "survival factors" determine how the primary product nuclei deexcite by particle emission, allowing them to survive or those nuclei that deexcite by fission, destroying them. Among

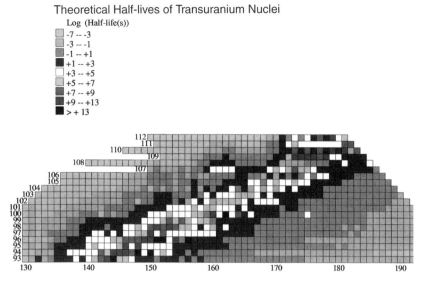

**Figure 15.3** Predicted (Möller et al., 1997) half-lives of the transuranium nuclei with $Z \leq 112$. (Figure also appears in color figure section.)

the production factors are items such as the "starting material," the target nuclei, which must be available in sufficient quantity and suitable form. We must have enough transmuting projectile nuclei also. Note that the $N/Z$ ratio of light projectile nuclei is typically lower than that of heavy nuclei and places a limit on the fusion process. The transmutation reaction must occur with adequate probability to ensure a good yield of the product nucleus in a form suitable for further study. Equally important is that the product nuclei be produced with excitation energy and angular momentum distributions such that the product nuclei will deexcite by particle or photon emission rather than the disastrous fission process. The competition between particle emission and fission as deexcitation paths depends on excitation energy, angular momentum, and the intrinsic stability of the product nucleus, which is related to the atomic and mass numbers of the product.

Nuclear synthesis is similar in some ways to inorganic or organic chemical syntheses with the synthetic chemist or physicist having to understand the reactions involved and the structure and stability of the intermediate species. While, in principle, the outcome of any synthesis reaction is calculable; in practice such calculations are, for the most part, very difficult. Instead, the cleverness of the scientists involved, their manipulative skills, and the instrumentation available for their use often determine the success of many synthetic efforts.

The synthesis reactions used to "discover" the transuranium elements are given in Table 15.2. All these reactions are complete fusion reactions in which the reacting nuclei fuse, equilibrate, and deexcite in a manner independent of their mode of formation. Other production reactions involving a partial capture of the projectile nucleus are also possible.

**TABLE 15.2 Summary of Transuranium Element Synthesis**

| Atomic Number | Name and Symbol | Synthesis Reaction | Half-Life |
|---|---|---|---|
| 93 | Neptunium (Np) | $^{238}\text{U} + \text{n} \rightarrow {}^{239}\text{U} + \gamma$<br>$^{239}\text{U} \xrightarrow{\beta^-} {}^{239}\text{Np}$ | 2.35 d |
| 94 | Plutonium (Pu) | $^{238}\text{U} + {}^{2}\text{H} \rightarrow {}^{238}\text{Np} + 2\text{n}$<br>$^{238}\text{Np} \xrightarrow{\beta^-} {}^{238}\text{Pu}$<br>$^{239}\text{Pu} + \text{n} \rightarrow {}^{240}\text{Pu} + \gamma$ | 86.4 y |
| 95 | Americium (Am) | $^{240}\text{Pu} + \text{n} \rightarrow {}^{240}\text{Pu} + \gamma$<br>$^{241}\text{Pu} \xrightarrow{\beta^-} {}^{241}\text{Am}$ | 433 y |
| 96 | Curium (Cm) | $^{239}\text{Pu} + {}^{4}\text{He} \rightarrow {}^{242}\text{Cm} + \text{n}$ | 162.5 d |
| 97 | Berkelium (Bk) | $^{241}\text{Am} + {}^{4}\text{He} \rightarrow {}^{243}\text{Bk} + 2\text{n}$ | 4.5 h |
| 98 | Californium (Cf) | $^{242}\text{Cm} + {}^{4}\text{He} \rightarrow {}^{245}\text{Cf} + \text{n}$ | 44 min |
| 99 | Einsteinium (Es) | "Mike" thermonuclear explosion (leading to $^{253}\text{Es}$) | 20 d |
| 100 | Fermium (Fm) | "Mike" thermonuclear explosion (leading to $^{255}\text{Fm}$) | 20 h |
| 101 | Mendelevium (Md) | $^{253}\text{Es} + {}^{4}\text{He} \rightarrow {}^{256}\text{Md} + \text{n}$ | 76 min |
| 102 | Nobelium (No) | $^{244}\text{Cm} + {}^{12}\text{C} \rightarrow {}^{252}\text{No} + 4\text{n}$ | 2.3 s |
| 103 | Lawrencium (Lr) | $^{250,251,252}\text{Cf} + {}^{11}\text{B} \rightarrow {}^{258}\text{Lr} + 3\text{–}5\text{n}$<br>$^{250,251,252}\text{Cf} + {}^{10}\text{B} \rightarrow {}^{258}\text{Lr} + 2\text{–}4\text{n}$ | 4.3 s |
| 104 | Rutherfordium (Rf) | $^{249}\text{Cf} + {}^{12}\text{C} \rightarrow {}^{257}\text{Rf} + 4\text{n}$<br>$+ {}^{13}\text{C} \rightarrow {}^{259}\text{Rf} + 3\text{n}$ | 3.4 s<br>3.8 s |
| 105 | Dubnium (Db) | $^{249}\text{Cf} + {}^{15}\text{N} \rightarrow {}^{260}\text{Db} + 4\text{n}$ | 1.5 s |
| 106 | Seaborgium (Sg) | $^{249}\text{Cf} + {}^{18}\text{O} \rightarrow {}^{263}\text{Sg} + 4\text{n}$ | 0.9 s |
| 107 | Bohrium (Bh) | $^{209}\text{Bi} + {}^{54}\text{Cr} \rightarrow {}^{262}\text{Bh} + \text{n}$ | 102 ms |
| 108 | Hassium (Hs) | $^{208}\text{Pb} + {}^{58}\text{Fe} \rightarrow {}^{265}\text{Hs} + \text{n}$ | 1.8 ms |
| 109 | Meitnerium (Mt) | $^{209}\text{Bi} + {}^{58}\text{Fe} \rightarrow {}^{266}\text{Mt} + \text{n}$ | 3.4 ms |
| 110 | Darmstadtium (Ds) | $^{209}\text{Bi} + {}^{59}\text{Co} \rightarrow {}^{267}\text{Ds} + \text{n}$<br>$^{208}\text{Pb} + {}^{62}\text{Ni} \rightarrow {}^{269}\text{Ds} + \text{n}$<br>$^{208}\text{Pb} + {}^{64}\text{Ni} \rightarrow {}^{271}\text{Ds} + \text{n}$<br>$^{244}\text{Pu} + {}^{34}\text{S} \rightarrow {}^{273}\text{Ds} + 5\text{n}$ | 4 μs<br>170 μs<br>56 ms<br>118 ms |
| 111 | Roentgenium (Rg) | $^{209}\text{Bi} + {}^{64}\text{Ni} \rightarrow {}^{272}\text{Rg} + \text{n}$ | 1.5 ms |
| 112 | — | $^{208}\text{Pb} + {}^{70}\text{Zn} \rightarrow {}^{277}112 + \text{n}$ | 240 μs |
| 113 | — | $^{209}\text{Bi} + {}^{70}\text{Zn} \rightarrow {}^{278}113 + \text{n}$ | 340 μs |

The cross section for production of a heavy evaporation residue, $\sigma_{\text{EVR}}$, by a complete fusion reaction can be written as:

$$\sigma_{\text{EVR}} = (\text{fusion probability})(\text{survival probability})$$

where the fusion probability refers to the probability of forming a completely fused system in the reaction, and the survival probability refers to the probability that the excited complete fusion product will deexcite by particle emission rather than fission, which destroys the nucleus. Recent synthesis reactions for heavy nuclei are divided into "cold" or "hot" fusion. Cold fusion reactions involve a heavier projectile (Ar–Kr) interacting with a Pb or Bi nucleus, where the excitation energy of the completely fused system is low (~13 MeV), giving high survival probabilities. Unfortunately, the fusion probability in such systems is low. Hot fusion reactions involve the use of lighter projectiles ($^{11}$B–$^{48}$Ca) interacting with actinide nuclei, giving a high fusion probability, but a high excitation energy ($E^* \sim 30$–$50$ MeV) with a resulting low survival probability.

The reactions in Table 15.2 can be divided into four classes: the neutron-induced reactions ($Z = 93, 95, 99, 100$), the light-charged particle-induced reactions ($Z = 94, 96$–$98, 101$), the hot fusion reactions ($Z = 102$–$106$), and the cold fusion reactions ($Z = 107$–$113$). In the neutron-induced reactions used to make the transuranium nuclei, the capture of a neutron does not create a new element, but the subsequent $\beta^-$ decays do similar to the astrophysical s-process. Light-charged particle reactions with exotic actinide target nuclei allow one to increase the atomic number of the product one or two units from the target nucleus. To make the heaviest elements, one needs to add several protons to the target nucleus by a reaction with a heavy ion. Such hot fusion reactions with actinide target nuclei lead to highly excited intermediate species that decay mostly by fission but occasionally by emitting neutrons, thus producing new heavy nuclei. However, as the atomic number of the product nuclei increases, so does the probability of fission leading to very poor survival probabilities for the putative new species. The nuclear physicist Yuri Oganessian pointed out that a way around this problem was to fuse heavier projectile nuclei with nuclei in the lead–bismuth region. Because of the special stability of the lead–bismuth nuclei, the resulting fused species would be formed "cold" and could, with some reasonable probability, decay by only emitting a single neutron.

In Figure 15.4, we show current measurements (filled squares) of the cross sections for cold fusion reactions as a function of the atomic number $Z$ of the completely fused system. (The cold fusion point at $Z = 118$ is an upper limit.) Also shown (as open circles) are the cross sections for hot fusion reactions. Clearly, future efforts will have to focus on experiments at the 0.1- to 1.0-pb cross-section level or lower. Current technology for cold fusion reaction studies would require ~12 days to observe one event at a cross-section level of 1 pb. Similarly, a cross section of 1 pb in a hot fusion reaction would require ~6–19 days to observe one event. From examining the data in Figure 15.4, it would also appear that hot fusion reactions might be the reactions of choice in pursuing future research in this area.

*Aside on Element Synthesis Calculations* The reactions used to synthesize heavy nuclei are, quite often, very improbable reactions, representing minor branches to the main reaction. Their probability of occurrence with respect to the main synthesis

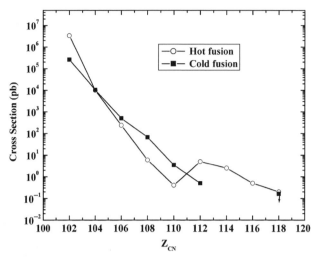

**Figure 15.4** Plot of the observed cross sections for the production of heavy elements by cold and hot fusion reactions.

reaction is frequently less than $10^{-6}$. As such, it is intrinsically difficult to accurately describe these reactions from a theoretical point of view. Instead, workers in this field have frequently resorted to semiempirical prescriptions to guide their efforts.

To give one a feel for the magnitude of the quantities involved, we outline below a very simple schematic method for estimating heavy-element production cross sections. It is intended to show the relevant factors and should not be taken too seriously, except to indicate the order of magnitude of a particular formation cross section.

The physicist Peter Armbruster has made an empirical systematic description of the probability of fusion of two heavy nuclei at energies near the reaction barrier. These systematics are shown in Figure 15.5. To use this graph, one picks values

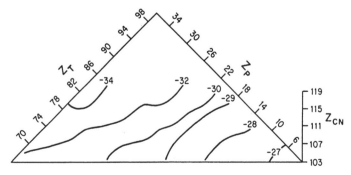

**Figure 15.5** Plot of the contours of $\log_{10} \sigma_{\text{fus}}$ (where $\sigma_{\text{fus}}$ is the $s$-wave fusion cross section at the interaction barrier).

of the atomic number of projectile and target nuclei and reads off the expected value for the cross section for producing a completely fused species. The excitation energy of the completely fused species can then be read from Figure 15.6, which is based upon the nuclear masses of Peter Möller, J. Rayford Nix, and Karl-Ludvig Kratz. Taking as a rough rule of thumb, for each 10 MeV of excitation energy, the survival probability of the fused system drops by a factor of $10^2$, one can then compute the cross section for producing a given species.

For example, the successful synthesis of $^{265}$Hs($^{265}$108) involved the reaction

$$^{58}\text{Fe} + ^{208}\text{Pb} \rightarrow ^{265}\text{Hs} + n$$

From Figure 15.5, one predicts the fusion cross section to be $10^{-32}$ cm$^2$, while Figure 15.6 would suggest an excitation energy of ~20 MeV. Thus, one would roughly estimate the overall cross section for producing $^{265}$Hs to be

$$(10^{-32})(10^{-2})^2 \approx 10^{-36} \text{cm}^2$$

(The measured cross section was $2 \times 10^{-35}$ cm$^2$.)

*Aside on the Detection of Heavy Atoms* The detection of atoms of a new element has always focused on measuring the atomic number of the new species and showing that it is different from all known values of Z. Unambiguous methods for establishing the atomic number include chemical separations, measurement of the X-ray spectrum accompanying a nuclear decay process, or establishment of a genetic relationship between the unknown new nucleus and some known nuclide. As the quest for new elements focuses on still heavier species, the probability of producing the new elements has decreased, and one has had to devote increasing attention to

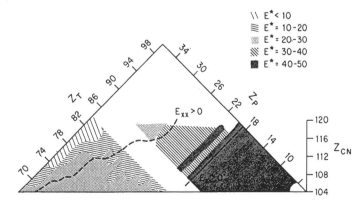

**Figure 15.6** Plot of the excitation energy of the completely fused species formed from a given target–projectile combination. Reactions are assumed to take place at the interaction barrier.

the problem of detecting a few atoms of a new species amid a background of many orders of magnitude more of other atoms. Thus, modern attempts to make new heavy-element atoms usually involve some kind of "separator."

An example of a modern separator is the velocity filter SHIP (Fig. 15.7) at the GSI in Darmstadt, Germany. In this separator, nuclear reaction products (from the target wheel) undergo different deflections (in crossed electric and magnetic fields) according to their electric and magnetic rigidity which will depend on whether they are fission fragments, scattered beam particles, or the desired heavy-element residues. The efficiency of the separator is $\sim$50% for heavy-element residues, while transfer products and scattered beam nuclei are rejected by factors of $10^{14}$ and $10^{11}$, respectively. The heavy recoil atoms are implanted in the silicon detectors. Their implantation energy and position are correlated with any subsequent decays of the nuclei to establish genetic relationships to known nuclei.

## 15.4 HISTORY OF TRANSURANIUM ELEMENT DISCOVERY

The first scientific attempts to prepare the elements beyond uranium were performed by Enrico Fermi, Emilio Segre, and co-workers in Rome in 1934, shortly after the existence of the neutron was discovered. This group of investigators irradiated uranium with slow neutrons and found several radioactive products, which were thought to be due to new elements. However, detailed chemical studies by Otto Hahn and Fritz Strassman in Berlin showed these species were isotopes of the known elements created by the fission of uranium into two approximately equal parts (see Chap. 11). This discovery of nuclear fission in December of 1938 was thus a by-product of man's quest for the transuranium elements.

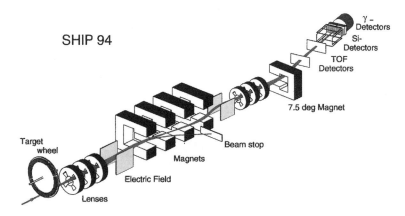

**Figure 15.7** Velocity filter SHIP. (Figure also appears in color figure section.)

## 15.4 HISTORY OF TRANSURANIUM ELEMENT DISCOVERY

With poetic justice, the actual discovery of the first transuranium element came as part of an experiment to study the nuclear fission process. Edwin McMillan, working at the University of California at Berkeley in the spring of 1939, was trying to measure the energies of the two recoiling fragments from the neutron-induced fission of uranium. He placed a thin layer of uranium oxide on one piece of paper. Next to this he stacked very thin sheets of cigarette paper to stop and collect the uranium fission fragments. During his studies he found there was another radioactive product of the reaction—one that did not recoil enough to escape the uranium layer, as did the fission products. He suspected that this product was formed by the capture of a neutron by the more abundant isotope of uranium, $^{238}_{92}U$. McMillan and Philip Abelson, who joined him in this research, showed in 1940, by chemical means, this product is an isotope of element 93, $^{239}_{93}Np$, formed in the following sequence:

$$^{238}_{92}U + ^{1}_{0}n \longrightarrow ^{239}_{92}U + \gamma$$

$$^{239}_{92}U(t_{1/2} = 23.5\,\text{min}) \xrightarrow{\beta^-} ^{239}_{93}Np(t_{1/2} = 2.36\,\text{d})$$

Neptunium, the element beyond uranium, was named after the planet Neptune because this planet is beyond the planet Uranus for which uranium is named.

Plutonium (named after the planet Pluto, following the pattern used in naming neptunium) was the second transuranium element to be discovered. By bombarding uranium with charged particles, in particular, deuterons ($^2H$), using the 60-in. cyclotron at the University of California at Berkeley, Glenn T. Seaborg, Edwin McMillan, Joseph W. Kennedy, and Arthur C. Wahl succeeded in preparing a new isotope of neptunium, $^{238}Np$, which decayed by $\beta^-$ emission to $^{238}Pu$, that is,

$$^{238}_{92}U + ^{2}_{1}H \longrightarrow ^{238}_{93}Np + 2n$$

$$^{238}_{93}Np(t_{1/2} = 2.12\,\text{d}) \xrightarrow{\beta^-} ^{238}_{94}Pu(t_{1/2} = 87.7\,\text{y})$$

Early in 1941, $^{239}Pu$, the most important isotope of plutonium was discovered by Kennedy, Segre, Wahl, and Seaborg. $^{239}Pu$ was produced by the decay of $^{239}Np$, which in turn was produced by the irradiation of $^{238}U$ by neutrons, using the reaction discovered by McMillan

$$^{238}_{92}U + ^{1}_{0}n \longrightarrow ^{239}_{92}U + \gamma$$

$$^{239}_{92}U(t_{1/2} = 23.5\,\text{min}) \xrightarrow{\beta^-} ^{239}_{93}Np(t_{1/2} = 2.35\,\text{d}) \xrightarrow{\beta^-} ^{239}_{94}Pu(t_{1/2} = 24{,}110\,\text{y})$$

This isotope, $^{239}Pu$, was shown to have a cross section for thermal neutron-induced fission that exceeded that of $^{235}U$, a property that made it important for nuclear weapons, considering that it could be prepared by chemical separation as compared to isotopic separation that was necessary for $^{235}U$.

The next transuranium elements to be discovered, americium and curium (Am and Cm; $Z = 95$ and 96, respectively), represent an important milestone in chemistry, the recognition of a new group of elements in the periodic table, the actinides. According to the periodic table of the early 20th century, one expected americium and curium to be eka-iridium and eka-platinum, that is, to have chemical properties similar to iridium and platinum. In 1944, Seaborg conceived the idea that all the known elements heavier than actinium ($Z = 89$) had been misplaced in the periodic table. He postulated that the elements heavier than actinium might form a second series similar to the lanthanide elements (Fig. 15.1), called the actinide series. This series would end in element 103 (Lr) and, analogous to the lanthanides, would show a common oxidation state of +3.

Once this redox property was understood, the use of a proper chemical procedure led quickly to the identification of an isotope of a new element. Thus, a new α-emitting nuclide, now known to be $^{242}_{96}Cm$ (half-life 162.9 d), was identified by Seaborg, Albert Ghiorso, and Ralph James in the summer of 1944 in the bombardment of $^{239}_{94}Pu$ with 32-MeV helium ions:

$$^{239}_{94}Pu + ^{4}_{2}He \rightarrow ^{242}_{96}Cm + ^{1}_{0}n$$

The bombardment took place in the Berkeley 60-in. cyclotron, after which the target material was shipped to the Metallurgical Laboratory at Chicago for chemical separation and identification. A crucial step in the identification of the α-emitting nuclide as an isotope of element 96, $^{242}_{96}Cm$, was the identification of the known $^{238}_{94}Pu$ as the α-decay daughter of the new nuclide.

The identification of an isotope of element 95, by Seaborg, Ghiorso, James, and Leon Morgan in late 1944 and early 1945, followed the identification of this isotope of element 96 ($^{242}Cm$) as a result of the bombardment of $^{239}_{94}Pu$ with neutrons in a nuclear reactor. The production reactions, involving multiple neutron capture by plutonium, are

$$^{239}_{94}Pu + ^{1}_{0}n \rightarrow ^{240}_{94}Pu + \gamma$$

$$^{240}_{94}Pu + ^{1}_{0}n \rightarrow ^{241}_{94}Pu + \gamma$$

$$^{241}_{94}Pu(t_{1/2} = 14.4\,y) \xrightarrow{\beta^-} ^{241}_{95}Am(t_{1/2} = 432.7\,y)$$

$$^{241}_{95}Am + ^{1}_{0}n \rightarrow ^{242}_{95}Am + \gamma$$

$$^{242}_{95}Am(t_{1/2} = 16.0\,h) \xrightarrow{\beta^-} ^{242}_{96}Cm$$

The years after World War II led to the discovery of elements 97–103 and the completion of the actinide series. While the story of the discovery of each of these elements is fascinating, we shall, in the interests of brevity, refer the reader elsewhere (see References) for detailed accounts of most of these discoveries. As an example of the techniques involved, we shall discuss the discovery of element 101 (mendelevium).

## 15.4 HISTORY OF TRANSURANIUM ELEMENT DISCOVERY

The discovery of mendelevium was one of the most dramatic in the sequence of transuranium element syntheses. It marked the first time in which a new element was produced and identified one atom at a time. By 1955, scientists at Berkeley had prepared an equilibrium amount of about $10^9$ atoms of $^{253}_{99}\text{Es}$ by neutron irradiation of plutonium in the Materials Testing Reactor in Idaho. As the result of a "back of the envelope" calculation done by Ghiorso during an airplane flight, they thought it might be possible to prepare element 101 using the reaction

$$^{253}_{99}\text{Es} + ^{4}_{2}\text{He} \longrightarrow ^{256}_{101}\text{Md} + ^{1}_{0}\text{n}$$

The amount of element 101 expected to be produced in an experiment can be calculated using the formula

$$N_{101} = \frac{N_{\text{Es}}\sigma\phi(1 - e^{-\lambda t})}{\lambda}$$

where $N_{101}$ and $N_{\text{Es}}$ are the number of element 101 atoms produced and the number of $^{253}_{99}\text{Es}$ target atoms, respectively, $\sigma$ is the reaction cross section (estimated to be $\sim 10^{-27}$ cm$^2$), $\phi$ is the helium ion flux ($\cong 10^{14}$ particles/s), $\lambda$ is the decay constant of $^{256}_{101}\text{Md}$ (estimated to be $\approx 10^{-4}$ s$^{-1}$), and $t$ is the length of each bombardment ($\approx 10^4$ s).

$$N_{101} = \frac{(10^9)(10^{-27})(10^{14})(1 - e^{-(10^{-4})(10^4)})}{10^{-4}} \approx 1 \text{ atom}$$

Thus, the production of only one atom of element 101 per experiment could be expected!

Adding immensely to the complexity of the experiment was the absolute necessity for the chemical separation of the one atom of element 101 from the $10^9$ atoms of einsteinium in the target and its ultimate, complete chemical identification by separation with the ion exchange method. This separation and identification would presumably have to take place in a period of hours, or perhaps even one hour or less, since the half-life of new nuclide was unknown. Furthermore, the target material had a 20-d half-life, and one needed a nondestructive technique allowing reuse of the target material.

The definitive experiments were performed in a memorable, all-night session, on February 18, 1955. To increase the number of events that might be observed at one time, three successive 3-h bombardments were made, and, in turn, their transmutation products were quickly and completely separated by the ion exchange method. Some of the nuclide $^{253}_{99}\text{Es}$ was present in each case so, together with the $^{246}_{98}\text{Cf}$ produced from $^{246}_{96}\text{Cm}$ also present in the target [via the $^{244}\text{Cm}$ ($^4$He, 2n) reaction], it was possible to define the positions in which the elements came off the column used to contain the ion exchange resin. Five spontaneous fission counters then were used to count simultaneously the corresponding drops of solution from the three runs.

A total of five spontaneous fission counts was observed in the element 101 position, while a total of eight spontaneous fission counts were also observed in the element 100 position. No such counts were observed in any other position. The original data are presented in Figure 15.8.

The synthesis of the transactinides is noteworthy from a chemical and a nuclear viewpoint. From the chemical point of view, rutherfordium ($Z = 104$) is important as an example of the first transactinide element. From Figure 15.1, we would expect rutherfordium to behave as a Group 4 (IVB) element, such as hafnium or zirconium, but not like the heavy actinides. Its solution chemistry, as deduced from chromatography experiments, is different from that of the actinides and resembles that of zirconium and hafnium. More recently, detailed gas chromatography has shown important deviations from expected periodic table trends and relativistic quantum chemical calculations.

The work on the discovery and identification of elements 104–106 was controversial and contentious due, in part, to the difficulty of the experiments. Looking back now, the following series of experiments clearly identified these elements.

Ghiorso et al. (1969) produced isotopes of element 104 in experiments at Berkeley in 1969. The nuclear reactions involved were

$$^{249}_{98}\text{Cf} + ^{12}_{6}\text{C} \rightarrow ^{257}_{104}\text{Rf}(t_{1/2} \approx 3.8\,\text{s}) + 4^1_0\text{n}$$

$$^{249}_{98}\text{Cf} + ^{13}_{6}\text{C} \rightarrow ^{259}_{104}\text{Rf}(t_{1/2} \approx 3.4\,\text{s}) + 3^1_0\text{n}$$

The atomic numbers of these isotopes were identified by detecting the known No daughters of these nuclei. The group suggested the name of rutherfordium (chemical symbol Rf) for element 104 in honor of Lord Ernest Rutherford.

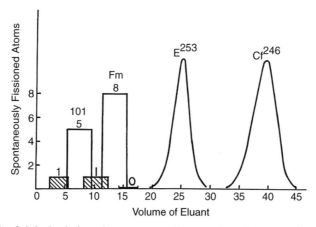

**Figure 15.8** Original elution data corresponding to the discovery of mendelevium, February 18, 1955. The curves for einsteinium-253 (given the old symbol $E^{253}$) and californium-246 are for α-particle emission. (Dowex 50 ion exchange resin was used, and the eluting agent was ammonium α-hydroxyisobutyrate.)

## 15.4 HISTORY OF TRANSURANIUM ELEMENT DISCOVERY

Contemporaneously with the Berkeley experiments, Zvara et al. (1969, 1970), working at Dubna, produced $3.2 \pm 0.8$ s $^{259}$104 by the $^{242}_{94}$Pu($^{22}_{10}$Ne, 5n) reaction. The chloride of this spontaneously fissioning activity was shown using gas chromatography to be slightly less volatile than Hf, but more volatile than the actinides. An international group of reviewers (Barber et al., 1992) has determined that the Berkeley and Dubna groups should share the credit for the discovery of element 104 and has suggested the name of rutherfordium for element 104.

In 1970, Ghiorso et al. (1970) reported the observation of an isotope of element 105 produced in the reaction

$$^{249}_{98}\text{Cf} + ^{15}_{7}\text{N} \longrightarrow ^{260}_{105}\text{Db}(t_{1/2} = 1.5\,\text{s}) + 4^{1}_{0}\text{n}$$

The Z and A of this isotope were established by correlations between the parent $^{260}$Db and its daughter $^{256}$Lr. They suggested the name of hahnium (chemical symbol Ha) for this element in honor of the German radiochemist Otto Hahn, co-discoverer of fission.

In a series of experiments occurring at a similar time, Druin et al. (1971) identified a mother–daughter pair from the decay of $^{260,261}$Db formed in the reaction $^{243}$Am($^{22}$Ne, 4–5n). A name of nielsbohrium (chemical symbol Ns) was suggested for this element in honor of the Danish physicist Niels Bohr by the Russian group. An international group (Barber et al., 1992) has suggested that credit for this discovery also be shared, and subsequently the name of dubnium (chemical symbol Db) was assigned to this element.

Element 106 was first synthesized by Ghiorso et al. (1974) at Berkeley in 1974 using the reaction

$$^{249}_{98}\text{Cf} + ^{18}_{8}\text{O} \longrightarrow ^{263}106(t_{1/2} = 0.8\,\text{s}) + 4^{1}_{0}\text{n}$$

The nuclide was identified by genetic links to its daughters $^{259}$Rf and $^{255}$No. This synthesis was reconfirmed in 1993 by Gregorich et al. (1994). Element 106 has been named seaborgium (symbol Sg) after one of the authors of this book. Glenn, the co-discoverer of plutonium and nine other transuranium elements, said upon this occasion (Fig. 15.9): "It is the greatest honor ever bestowed upon me—even better, I think, than winning the Nobel Prize."

In 1981, G. Münzenberg et al. (1981), working in Darmstadt at the velocity filter SHIP, identified the isotope $^{262}$Bh produced in the "cold fusion" reaction:

$$^{209}_{83}\text{Bi} + ^{54}_{24}\text{Cr} \longrightarrow ^{262}_{107}\text{Bh}(t_{1/2} = 102\,\text{ms}) + ^{1}_{0}\text{n}$$

This nuclide was identified by genetic links to its daughters $^{258}$Db, $^{254}$Lr, $^{250}$Md, $^{250}$Fm, and $^{246}$Cf. The cross section reported for this reaction was ~200 pb (approximately 1/5,000,000 of the production cross section assumed in the discovery of Md). This element was named bohrium (chemical symbol Bh) in honor of Niels Bohr.

**Figure 15.9** Glenn Seaborg points out seaborgium in the periodic table. (Figure also appears in color figure section.)

In 1984, Münzenberg et al. (1987), working at Darmstadt, produced three atoms of $^{265}$Hs using the cold fusion reaction $^{208}$Pb($^{58}$Fe, 1n). $^{265}$Hs was identified by genetic links to its daughters $^{261}$Sg and $^{257}$Rf. The half-life of this nucleus was ∼1.8 ms, and it decayed by α-particle emission. The production cross section was ∼20 pb. At a similar time, Oganessian et al. (1984) reported the production of $^{263-265}$Hs in the reactions $^{209}$Bi($^{55}$Mn, n), $^{206}$Pb($^{58}$Fe, n), $^{207}$Pb($^{58}$Fe, n), and $^{208}$Pb($^{58}$Fe, n). This group reported observation of spontaneous fission and α decays of the granddaughter and great-great-great-granddaughters. Because of this weaker identification, credit for this discovery (Barber et al., 1992) was assigned to the Darmstadt group, which suggested the name of hassium (chemical symbol Hs) in honor of the region of Germany, Hesse, in which the work was done.

In 1982, Münzenberg et al. (1982, 1984) reported the observation of one atom of element 109 formed in the reaction $^{209}$Bi($^{58}$Fe, n). The decay of this atom was weakly correlated to its daughters $^{262}$Bh, $^{258}$Db, and $^{258}$Rf. The production cross section was 10 pb. This discovery was confirmed by the later observation of more atoms at Darmstadt and the observed production of the descendant $^{246}$Cf at Dubna. The discoverers suggested the name of meitnerium (chemical symbol Mt) in honor of Lise Meitner.

In 1991, Ghiorso et al. (1995a, 1995b) studied the reaction of $^{209}$Bi with $^{59}$Co. They found one event that they associated with the production of $^{267}$110 ($^{267}$Ds). The evaporation residue formed decayed by the emission of an 11.6-MeV α particle 4 μs after implantation. Their evidence for the formation of $^{267}$110 was weakened by the inability (due to malfunctioning electronics) to detect the decay of the daughter $^{263}$Hs, although the decay of other members of the decay chain was observed. Further work was not pursued due to the closure of the accelerator involved. This observation was probably correct, although the evidence presented is not strong enough to justify the claim of element discovery.

There is no doubt that Hofmann et al. (1995a), working in Darmstadt in 1994, observed the production of several atoms (nominally four—see below) of $^{269}$Ds in the reaction $^{208}$Pb($^{62}$Ni, n). This nuclide was identified by genetic links to its daughters $^{265}$Hs, $^{261}$Sg, $^{257}$Rf, and $^{253}$No. The neutron-rich isotope of element 110, $^{271}$Ds, was produced later (1998) (Hofmann, 1998) using the reaction $^{208}$Pb($^{64}$Ni, n). Nine atoms were observed and identified. This latter reaction was used by workers in Berkeley (Ginter, 2003) and RIKEN (K. Morita et al., Nucl. Phys. *A734*, 101 (2004).) to confirm the discovery of element 110. The very n-rich isotope of element 110, $^{273}$Ds, was observed by Lazarev et al. (1996) using the reaction $^{244}$Pu($^{34}$S, 5n). The name darmstadtium (chemical symbol Ds) has been given to this element.

An unfortunate footnote to this chapter in the history of the discovery of the heaviest elements is the revelation (Hofmann et al., 2002) that one of the decay chains reported by Hofmann et al. (1995a) for $^{269}$Ds was "spuriously created," the result of human error or scientific misconduct. As disturbing as this finding is, it should not detract from the other correctly identified decay chains.

The first production of an isotope of element 111, $^{272}$111 ($t_{1/2} \sim$ 1.5 ms) was by Hofmann et al. (1995b) in 1994 using the reaction $^{209}$Bi($^{64}$Ni, n). Three decay chains were observed in 1994. Subsequently, three more decay chains corresponding to the decay of $^{272}$111 were reported (Hofmann et al., 2002) in 2002. The decay of $^{272}$111 is convincingly linked to its daughters $^{268}$Mt, $^{264}$Bh, $^{260}$Db, and $^{256}$Lr. This discovery was confirmed by experiments at Berkeley (Folden et al., 2004) and RIKEN (Morita, 2004). The name roentgenium (Rg) has been suggested for this element.

In 1995, Hofmann et al. (1996) reported the formation of $^{277}$112 in the reaction $^{208}$Pb($^{70}$Zn, n). Two decay chains were reported connecting $^{277}$112 to its daughters $^{273}$Ds, $^{209}$Hs, $^{265}$Sg, $^{261}$Rf, and $^{257}$No. Unfortunately, one of these decay chains, the first one "observed," is now known (Hofmann et al., 2002) to also have been "spuriously created" by human error or scientific misconduct. A subsequent experiment (Hofmann et al., 2002) in 2000 resulted in the observation of an additional decay chain for $^{277}$112. These results have been confirmed by experiments at RIKEN. It is now widely believed (S. Hofmann, private communication) that the same individual was responsible for both false reports, that is, one of the $^{269}$Ds decay chains and one of the $^{277}$112 decay chains along with false reports (see below) of the synthesis of $^{293}$118 and its decay products.

In 2004, Morita et al. reported the synthesis of a single atom of element 113 in the $^{70}$Zn + $^{209}$Bi reaction. The production cross section was 55 fb, a remarkably small value requiring several months of bombardment.

In 1999, Oganessian et al. (1999a, 1999b, 2000) reported the successful synthesis of five atoms of element 114 using the $^{48}$Ca + $^{242,244}$Pu reaction. The long half-lives associated with these atoms (~s-min) represent the possible approach to the long sought "island of stability" of superheavy nuclei near $Z = 114$ and $N = 184$. (All previous heavy nuclei with $Z \geq 110$ have decayed with millisecond half-lives.) The experimental group, consisting of scientists from Dubna and the U.S. Lawrence Livermore National Laboratory reported the following observations:

$$^{48}_{20}\text{Ca} + ^{244}_{94}\text{Pu} \longrightarrow {}^{289}114(t_{1/2} \approx 21\,\text{s}) + 3^1_0\text{n}$$

$$^{48}_{20}\text{Ca} + ^{244}_{94}\text{Pu} \longrightarrow {}^{288}114(t_{1/2} \approx 1.9\,\text{s}) + 4^1_0\text{n}$$

$$^{48}_{20}\text{Ca} + ^{242}_{94}\text{Pu} \longrightarrow {}^{287}114(t_{1/2} \approx 5.4\,\text{s}) + 3^1_0\text{n}$$

All these nuclei were observed to decay by a sequence of emitted α particles with the decay chains ending in spontaneous fission. Because these nuclei are very n-rich, their descendents have not been characterized before. Other laboratories have not confirmed these exciting results.

In an extension of this work, Oganessian et al. (2001) reported the successful synthesis of three atoms of element 116 using the reaction

$$^{48}_{20}\text{Ca} + ^{248}_{96}\text{Cm} \longrightarrow {}^{292}116(t_{1/2} \approx 0.05\,\text{s}) + 4^1_0\text{n}$$

Each observed decay sequence involved the observation of three energetic α-particle decays followed by a spontaneous fission.

In an unpublished report in 2002, Oganessian et al. (http://159.93.28.88/flnr/index.html) reported the synthesis of two atoms of element 118, using the reaction $^{249}$Cf($^{48}$Ca, 3n) $^{294}$118 (Oganessian et al., Phys. Rev. *C64*, 021604 (2004)). In 2004, Oganessian et al. reported the successful synthesis of element 115 (and by α-decay, its daughter element 113) in the reaction $^{245}$Am($^{48}$Ca, xn)$^{291-x}$115. These events are still awaiting confirmation.

In the midst of all the exciting advances in heavy-element science in the period from 1994 to 2002, there was a dark chapter, the element 118 fiasco that occurred in Berkeley. In 1999, Ninov et al. (1999) reported the successful synthesis of three atoms of element 118 using the reaction $^{208}$Pb($^{86}$Kr, n) $^{293}$118. The evidence was stunning, consisting of three decay chains involving highly correlated high-energy α-particle decays after the implantation of a putative $^{293}$118. The reported production cross section was 2 pb, a number later revised to 7 pb. The result was quite unexpected because the empirical systematics of cold fusion cross sections (Fig. 15.4) would have predicted femtobarn cross sections for this reaction.

Other laboratories were not able to reproduce this work (Hofmann and Munzenberg, 2000; Stodel et al., 2001; Morimoto, 2001), and eventually the

Berkeley group reported their inability to reproduce the original observation (Gregorich et al., Eur. J. Phys. *A18*, 633 (2003)). A subsequent investigation (Gilchriese et al., 2003) revealed the original data had been fabricated by one individual, who was later connected to similar instances of fraud at Darmstadt in the work with elements 110 and 112 mentioned above. From these episodes, one learns that "science works," fraud will be found, and the traditional method of independent confirmation of important findings is reaffirmed.

## 15.5 SUPERHEAVY ELEMENTS

Up to 1970, it was thought that the practical limit of the periodic table would be reached at about element 108. By extrapolating the experimental data on heavy-element half-lives, we concluded that the half-lives of the longest-lived isotopes of the heavy elements beyond about element 108 would be so short ($<10^{-6}$ s) due to spontaneous fission decay that we could not produce and study them (Fig. 15.10). However, in the late 1960s and early 1970s, nuclear theorists, using techniques developed by Vilen Strutinsky and Wladyslaw Swiatecki, predicted

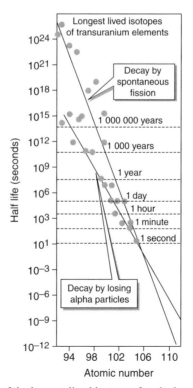

**Figure 15.10** Half-lives of the longest-lived isotope of each element versus atomic number Z circa 1970. (Figure also appears in color figure section.)

that special stability against fission would be associated with proton number $Z = 114$ and neutron number $N = 184$. These *superheavy elements* were predicted to have half-lives of the order of the age of the universe. They were predicted to form an "island" of stability separated from the "peninsula" of known nuclei (Fig. 15.11a).

We now know these predictions were wrong, in part. While we believe there are a group of "superheavy" nuclei whose half-lives are relatively long compared to lower $Z$ elements, we do not believe they form an island of stability. Rather, we picture them as a continuation of the peninsula of known nuclei (Fig. 15.11b). We also believe that their half-lives are short compared to geologic time scales. Therefore, they do not exist in nature. The most stable of the superheavy nuclei, those with $Z = 112$, $N \sim 184$, are predicted to decay by $\alpha$-particle emission with half-lives of $\sim 20$ days.

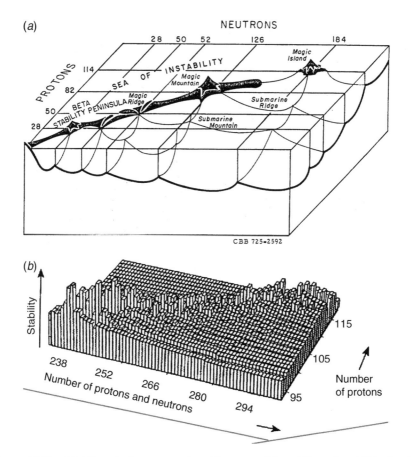

**Figure 15.11** (a) Allegorical representation of heavy nuclide stability, circa 1975, showing the superheavy island. (b) Modern plot of the predicted half-lives of the heaviest nuclei. Note there is a peninsula connecting the known and superheavy nuclei.

The principal problem associated with the superheavy nuclei is not their possible existence (which is considered relatively certain) but rather how to make them. Literally hundreds of synthesis reactions have been proposed, and several have been tried. Up to recently, all have failed because either the formation of the fused system is too improbable or its excitation energy is too large, resulting in too small a probability for formation of the product nuclei. For example, the most widely studied synthesis reaction is the $^{48}$Ca + $^{248}$Cm reaction. Using Figures 15.5 and 15.6 as guides, we roughly estimate the fusion cross section to be $3 \times 10^{-32}$ cm$^2$, the excitation energy $\sim$30 MeV (survival probability $10^{-6}$), and a predicted formation cross section of $10^{-38}$ cm$^2$. With modern technology, this would correspond to one atom per month. This production rate is at the limit of modern experiment technology.

However, as indicated in Section 15.4, observations of the synthesis of elements 114 and 116 (and possibly element 118) have been reported with low ($\sim$pb) cross sections. In addition, instead of following the trends shown in Figure 15.10, the half-lives of the longest-lived known isotopes of elements 106–112 are reported to be approximately milliseconds to seconds, an enhancement of orders of magnitude in their lifetimes. At present, none of these observations have been verified by independent measurements.

Some have taken the viewpoint that, without the special stability associated with nuclear shell structure, elements as light as $Z = 106$–108 would have negligibly short half-lives. The mere existence of these nuclei with millisecond half-lives is said to be a demonstration that we have already made superheavy nuclei, according to this view. The shell stabilization of these nuclei, which are deformed, is due to the special stability of the $N = 162$ configuration in deformed nuclei. (The "traditional" superheavy nuclei with $Z \sim 114, N = 184$ were calculated to have spherical shapes.)

Our best theoretical predictions of the expected half-lives of elements 110–120 are shown in Figure 15.12. Clearly one expects new regions of very heavy nuclei with half-lives that are substantially longer than those observed to date, but how to make them remains to be seen.

## 15.6 CHEMISTRY OF THE TRANSURANIUM ELEMENTS

The chemical behavior of the transuranium elements is interesting because of its complexity and the insights offered into the chemistry of the lighter elements. The placing of these manmade elements into the periodic table (Fig. 15.1) represents one of the few significant alterations of the original periodic table of Mendelyeev. Since so little is known about the chemistry of the transactinide elements, one has the unique opportunity to test periodic table predictions of chemical behavior before the relevant experiments are done.

The actinide and known transactinide elements are transition elements, that is, they have partially filled f or d electronic orbitals. As such, they are metals. Like other transition metals, most of them are sufficiently electropositive to dissolve in mineral acids. However, there is an important distinction that separates the actinide

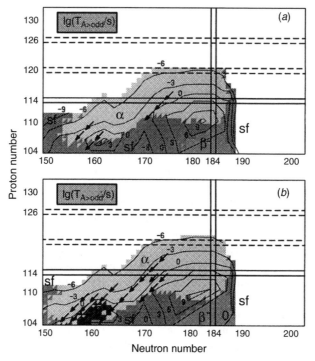

**Figure 15.12** Predicted half-lives of the transmeiterium nuclei (Möller et al., 1997).

elements from the other transition elements, including the transactinide elements. The partially filled d orbitals of most transition elements extend out to the boundary of the atoms and are influenced greatly by (or can influence) the chemical environment of the atom or ion. Thus, the chemical properties of elements with partially filled d orbitals are highly complex and seem to vary somewhat irregularly as one passes from element to element. But the 5f orbitals of the actinides are better screened from the chemical environment of the atom or ion by the higher lying s and p shell electrons, and thus there is a greater similarity in chemical properties among the actinides compared to the other transition elements. (Correspondingly, the 4f orbitals of the lanthanides are even better screened than the 5f actinide orbitals, and the chemical behavior of the lanthanides is even more homologous.) The greater extension of the 5f orbitals (relative to the 7s and 7p orbitals) compared to the lanthanide 4f orbitals allows some covalency in actinide bonding not seen with the lanthanides.

As the atomic number Z of the nucleus increases, the electrons become more tightly bound. As their binding energy increases, so does their velocity. For electrons in the 1s shell, the average velocity is roughly Z atomic units. (The speed of light, $c$, is 137.035 au.) Thus, for $Z = 90$, the velocity is about $90c/137$ or $0.66c$. The Schrödinger equation is no longer appropriate in this case and one must use the fully relativistic treatment of Dirac.

## 15.6 CHEMISTRY OF THE TRANSURANIUM ELEMENTS

The solution of the Dirac equation for the hydrogenlike atoms leads to wave functions that are products of radial and angular factors similar to the solutions of the Schrödinger equation that give the familiar atomic orbitals. The angular factors are shown in Figure 15.13. Each state is specified by four quantum numbers whose meaning is slightly different than for the Schrödinger equation. They are $n$, the principal quantum number with values of $1, 2, 3, \ldots, l$, the azimuthal quantum number with values $0, 1, 2, \ldots, n-1$, denoted by s, p, d, f, g; $j$, the angular momentum quantum number with values of $l \pm \frac{1}{2}$ (usually denoted as a subscript to $l$) and $m$, the magnetic quantum number, taking on half-integer values from $-j$ to $+j$. Thus, the three p, the five d, and the seven f orbitals are no longer degenerate and split into one $p_{1/2}$, and two $p_{3/2}$, two $d_{3/2}$, and three $d_{5/2}$, three $f_{5/2}$, and four $f_{7/2}$ levels, with the occupancy of each being $2j + 1$. This is called the "spin–orbit" splitting. The orbital shapes are given by $j$ and $m$ with orbitals of the same $j$ and $m$ having the same shape. What is surprising to the traditional chemist in Figure 15.13 is that the p orbitals are shaped like a sphere ($p_{1/2}$), a toroid or doughnut ($p_{3/2}, m = \frac{3}{2}$), and a dog-bone ($p_{3/2}, m = \frac{1}{2}$). One notes that the state with the highest $m$ value for a given $j$ value always has a doughnut-shaped distribution, while the lowest value of $m$ corresponds to a distribution stretched along the z axis with no nodes. States of intermediate $m$ are multilobed toroids.

The effect of using relativistic rather than nonrelativistic quantum mechanics to obtain the predicted atomic orbitals is threefold: (a) a contraction and stabilization of the $s_{1/2}$ and $p_{1/2}$ shells, (b) the splitting of the energy levels due to the spin–orbit coupling, and (c) an expansion (and destabilization) of the outer d and all f shells. These effects are of approximately equal magnitude and all increase as $Z^2$. In Figure 15.14, we show the magnitude of these effects for uranium. The chemical

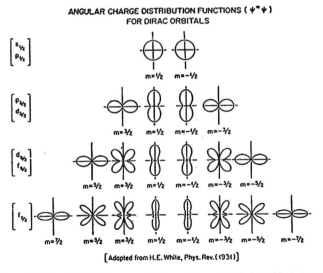

**Figure 15.13** Pictorial representation of the relativistic orbital shapes.

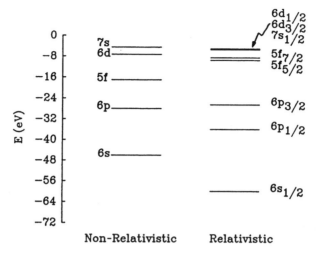

**Figure 15.14** Comparison of the predicted (nonrelativistic and relativistic) energies of the valence electronic levels for the uranium atom.

consequences of these effects have been well documented for elements such as gold where, for example, the yellow color is due to relativistic effects. (Nonrelativistic quantum mechanics predicts gold and silver to have similar colors.)

The ionic radii of the $M^{3+}$ and $M^{4+}$ ions of the actinides decrease with increasing positive charge of the nucleus (*the actinide contraction*) (Fig. 15.15). This contraction is due to the successive addition of electrons in an inner f shell where the incomplete screening of the nuclear charge by the added f electron leads to a contraction of the outer valence orbital. Because the ionic radii of ions of the same oxidation state are generally similar (Fig. 15.15), the ionic compounds of the actinides are isostructural.

The comparable energies of the 5f, 6d, 7s, and 7p orbitals and their spatial overlap will lead to bonding involving any or all of them. Thus, complex formation is an important part of actinide chemistry. The most stable oxidation states of the actinides and their solution chemistry will depend on the ligands present also because of the small differences between the energy of the electronic levels relative to chemical bond energies.

The known oxidation states of the actinide elements are shown in Table 15.3. The lower oxidation states are stabilized by acid while the higher oxidation states are more stable in basic solutions. In solution, the 2+, 3+, and 4+ species are present as metal cations, while the higher oxidation states are present as oxo-cations, $MO_2^+$, and $MO_2^{2+}$. The most common oxidation state is +3 for the transplutonium elements like the lanthanide elements. Relativistic quantum mechanics predicts the ground state of lawrencium (Lr) to be $5f^{14}7s^27p_{1/2}$ and not $5f^{14}7s^26d^1$. This might lead to a stable +1 oxidation state for lawrencium, but experiments designed to look for this state have not observed it. An upper limit for the reduction potential of $E° < 0.44$ V for the $Lr^{3+}/Lr^{1+}$ half-reaction has been determined.

## 15.6 CHEMISTRY OF THE TRANSURANIUM ELEMENTS

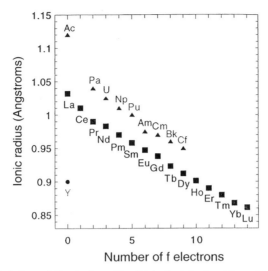

**Figure 15.15** Variation of the ionic radii of trivalent lanthanide and actinide ions with increasing Z.

The redox chemistry of the actinide elements, especially plutonium, is complex (Katz et al., 1980). Disproportionation reactions are especially important for the +4 and +5 oxidation states. Some of the equilibria are kinetically slow and irreversible. All transuranium elements undergo extensive hydrolysis with the +4 cations reacting most readily due to their large charge/radius ratio. Pu (IV) hydrolyzes extensively in acid solution and forms polymers. The polymers are of colloidal dimensions and are a serious problem in nuclear fuel reprocessing.

Hydrolysis is actually a special type of complex ion formation. The large positive charge associated with transuranium cations that leads to hydrolysis is also the

**TABLE 15.3 Oxidation States of the Actinide Elements**[a]

| Atomic Number: | 89 | 90 | 91 | 92 | 93 | 94 | 95 | 96 | 97 | 98 | 99 | 100 | 101 | 102 | 103 |
|---|---|---|---|---|---|---|---|---|---|---|---|---|---|---|---|
| Element: | Ac | Th | Pa | U | Np | Pu | Am | Cm | Bk | Cf | Es | Fm | Md | No | Lr |
| Oxidation States | | | | | | | | | | | | | 1? | | |
| | | | | | | (2) | (2) | | (2) | (2) | 2 | 2 | 2 | |
| | $\underline{3}$ | (3) | (3) | 3 | 3 | 3 | $\underline{3}$ | $\underline{3}$ | 3 | $\underline{3}$ | $\underline{3}$ | $\underline{3}$ | $\underline{3}$ | $\underline{3}$ | $\underline{3}$ |
| | | $\underline{4}$ | 4 | 4 | 4 | $\underline{4}$ | 4 | $\underline{4}$ | $\underline{4}$ | 4 | $\underline{4}$ | (4) | 4? | | |
| | | | 5 | 5 | $\underline{5}$ | 5 | 5 | 5? | | | 5? | | | | |
| | | | | 6 | 6 | 6 | 6 | 6? | | | | | | | |
| | | | | | 7 | (7) | 7? | | | | | | | | |

[a] The most common oxidation states are underlined, unstable oxidation states are shown in parentheses. Question marks indicate species that have been claimed but not substantiated.
Source: From Katz et al. (1980).

driving force for the interaction of nucleophiles with the transuranium cations. Water is only one example of a nucleophilic ligand. Other nucleophilic ligands present in solution may replace water molecules directly bound to the metal cation to form inner sphere complexes or alternatively, they may displace water molecules only from the outer hydrate shell to form outer sphere complexes.

Because of the competition between water and other ligands for positions in the inner coordination sphere of the central transuranium atom, it is not surprising that the stability of the complexes formed with a given ligand decreases, in the order $M^{4+} > MO_2^{2+} > M^{3+} > MO_2^+$. (Note that the strength of complexation does not depend simply on the net cation charge but rather the charge density seen by the anion or ligand as it approaches the metal. In the case of $MO_2^{2+}$, the effective charge is about 3.3 rather than 2.) Although there is some variation within the given cation types, the general order of complexing power of different anions is $F^- > NO_3^- > Cl^- > ClO_4^-$ for singly charged anions and $CO_3^{2-} > C_2O_4^{2-} > SO_4^{2-}$ for doubly charged anions.

The actinide cations are "hard acids," that is, their binding to ligands is described in terms of electrostatic interactions, and they prefer to interact with hard bases such as oxygen or fluorine rather than softer bases such as nitrogen or sulfur. The actinide cations do form complexes with the soft bases but only in nonaqueous solvents.

As typical hard acids, the stabilities of the actinide complexes are due to favorable entropy effects. The enthalpy terms are either endothermic or very weakly exothermic and are of little importance in determining the overall position of the equilibrium in complex formation.

The formation of complexes could be thought to be a three-step process:

$$M(aq) + X(aq) \longleftrightarrow [M(H_2O)_nX](aq) \longleftrightarrow [M(H_2O)X](aq) \longleftrightarrow MX(aq)$$

The first step is diffusion controlled while in the second step an "outer sphere" complex is formed with at least one water molecule intervening between the ligand and the metal atom. In the third or rate-determining step, a direct connection between the metal and ligand is established with the formation of an "inner sphere" complex. The process could terminate after the second step if the ligand cannot displace the water. Actinides form inner and outer sphere complexes, although in most cases the stronger inner sphere complexes are formed. The halide, nitrate, sulfonate, and trichloroacetate ligands form outer sphere complexes of the trivalent actinides while fluoride, iodate, sulfate, and acetate form inner sphere complexes.

It should be noted that the study of the chemistry of the elements with $Z > 100$ is very difficult. These elements have short half-lives and the typical production rates are about one atom/experiment. The experiments must be carried out hundreds of times, and the results summed to produce statistically meaningful results.

The elements Lr–112 are expected (nonrelativistically) to be d-block elements because they are expected to involve the filling of the 6d orbital. However, relativistic calculations have shown that rutherfordium prefers a 6d 7p electron configuration rather than the $6d^2$ configuration expected for nonrelativistic behavior and a simple extrapolation of periodic table trends. This prediction also implies that RfCl$_4$ should be more covalently bonded than its homologs $HfCl_4$ and $ZrCl_4$. In particular, the calculations show $RfCl_4$ to be more volatile than $HfCl_4$, which is more volatile than $ZrCl_4$ with bond dissociation energies in the order $RfCl_4 > ZrCl_4 > HfCl_4$. (The periodic table extrapolations would predict the volatility sequence $ZrCl_4 > HfCl_4 > RfCl_4$.)

The first aqueous chemistry of rutherfordium showed that it eluted from liquid chromatography columns as a $4+$ ion, consistent with its position in the periodic table as a d-block element rather than a trivalent actinide. Gas chromatography of the rutherfordium halides has shown the volatility sequence $ZrCl_4 > RfCl_4 > HfCl_4$ with a similar sequence for the tetrabromides (Fig. 15.16). Thus, rutherfordium does not follow the expected periodic table trend nor is its behavior in accord with relativistic calculations.

The aqueous chemistry of dubnium has also shown unexpected trends. Dubnium does not behave like its homolog tantalum in aqueous solutions but is similar to niobium or the pseudo-Group 5 element, protactinium, under certain conditions. For example, dubnium did not extract from methylisobutylketone under conditions where tantalum is extracted but niobium is not. The extraction of dubnium, niobium, tantalum, and protactinium from 6 M HCl solutions by amines agreed with relativistic calculations. Gas-phase thermochromatography of $NbBr_5$, $TaBr_5$, and $DbBr_5$ shows $NbBr_5$ and $TaBr_5$ to behave similarly, while $DbBr_5$ is less volatile. Just the

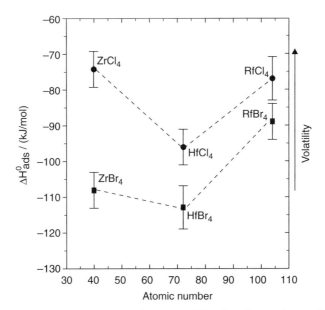

**Figure 15.16** Adsorption enthalpies, $\Delta H_a$, on $SiO_2$ for Group 4 tetrachlorides and tetrabromides. [From K. E. Gregorich, In Radiochemistry of Rutherfordium and Hahnium, Proc. "The Robert A. Welch Foundation. 41st Conference on Chemical Research—The Transactinide Elements," Houston, Texas, Oct. 27–28 (1997), p. 95.]

opposite trend was predicted by relativistic calculations. Thus, the chemistry of dubnium and rutherfordium deviates significantly from periodic table trends, a fact that is partly explained by relativistic calculations.

The study of the chemistry of seaborgium is remarkable for its technical difficulty as well as the insight offered. In an experiment carried out over a 2-year period, 15 atoms of seaborgium were identified. From this experiment, one concluded that the volatility sequence $MoO_2Cl_2 > WO_2Cl_2 \approx SgO_2Cl_2$ was followed. This observation agreed with both the extrapolations of periodic table trends and relativistic calculations. In an aqueous chemistry experiment, three atoms of seaborgium were detected, showing seaborgium to have a hexavalent character expected of a Group 6 element. The most stable oxidation state of seaborgium is +6, and, like its homologs molybdenum and tungsten, seaborgium forms neutral or anionic oxo- or oxohalide compounds.

To study the chemistry of elements 107 (Bh) and 108 (Hs) one must be able to produce isotopes of these elements where half-lives are long enough for chemical studies. $^{269}$Hs is reported (Schadel, 2002) to have $t_{1/2} \sim 14$ s, $^{270}$Hs has $t_{1/2} = 2-7$ s, and $^{267}$Bh is reported to have a half-life of $\sim 17$ s. Because of the small probability of producing these nuclei, methods for chemical study must be very sensitive. Among the projected methods of study, liquid–liquid extraction and gas-phase thermochemistry are thought to be the most viable.

Thermochromatographic measurements have indicated that Bh is less volatile than Re, which is less volatile than Tc, in agreement with periodic table trends.

The chemistry of hassium has been studied recently using the formation of chemically stable, volatile $HsO_4$, a property of Group 10 (VIII) elements. Thermochromatography has shown $HsO_4$ to be less volatile than $OsO_4$, a result in agreement with some relativistic predictions but not others (Fig. 15.17).

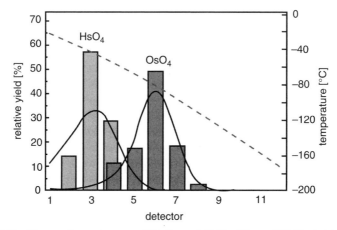

**Figure 15.17** Thermochromatogram of $HsO_4$ and $OsO_4$. [From GSI Nachrichten 1/02.] (Figure also appears in color figure section.)

## 15.7 ENVIRONMENTAL CHEMISTRY OF THE TRANSURANIUM ELEMENTS

With the large annual production of neptunium, plutonium, and the higher actinides in the nuclear power industry, there has been increasing concern about the possible release of these elements to the environment. This concern has been heightened by the nuclear reactor accidents at Three Mile Island and Chernobyl. Coupled with the prospect of cleaning up the detritus of the nuclear weapons programs of the major nations and the general lack of a publicly acceptable method of long-term disposal of nuclear waste, there is considerable interest in the environmental chemistry of the transuranium elements.

Plutonium is clearly the most significant transuranium element in the environment. The plutonium present in the environment is due primarily to atmospheric testing of nuclear weapons, secondarily to the disintegration upon reentry of satellites equipped with $^{238}$Pu power sources, and, lastly, to the processing of irradiated fuel and fuel fabrication in the nuclear power industry and the plutonium production program. Some major radionuclide releases are summarized in Table 15.4. During the period from 1950 to 1963, about 4.2 tons of plutonium (mostly a mixture of $^{239}$Pu and $^{240}$Pu) was injected into the atmosphere as a result of nuclear weapons testing. Because of the high temperatures involved, most of this plutonium was thought to be in the form of a refractory oxide. Most of this plutonium has been redeposited on Earth with concentrations being highest at the midlatitudes. Of the 350,000 Ci of $^{238}$Pu and $^{239}$Pu originally injected into the atmosphere, about 1000 Ci remained in 1989. Approximately $9.7 \times 10^6$ Ci of $^{241}$Pu were also injected into the atmosphere during weapons testing. When this completely decays ($t_{1/2}$ of $^{241}$Pu is 14.4 y), a total of $\sim 3.4 \times 10^5$ Ci of $^{241}$Am will be formed. There is an additional $\sim 1.4$ tons of plutonium deposited in the ground (1989) due to surface and subsurface nuclear weapons testing. Approximately 16,000 Ci of $^{238}$Pu were

TABLE 15.4 Events Leading to Large Injections of Radionuclides into the Atmosphere

| Source | Country | Time | Radioactivity (Bq)[a] | Important Nuclides |
|---|---|---|---|---|
| Hiroshima & Nagasaki | Japan | 1945 | $4 \times 10^{16}$ | Fiss. prod. Actinides |
| Atmospheric weapons tests | USA USSR | 1963 | $2 \times 10^{20}$ | Fiss. prod. Actinides |
| Windscale | UK | 1957 | $1 \times 10^{15}$ | $^{131}$I |
| Chelyabinsk (Kysthym) | USSR | 1957 | $8 \times 10^{16}$ | Fiss. prod. $^{90}$Sr, $^{137}$Cs |
| Three Mile Island | USA | 1979 | $1 \times 10^{12}$ | Noble gases, $^{131}$I |
| Chernobyl | USSR | 1986 | $2 \times 10^{18}$ | $^{137}$Cs |

[a]1 becquerel (Bq) is one disintegration/s.
Source: From G. Choppin, J. O. Liljinzin, and J. Rydberg, *Radiochemistry and Nuclear Chemistry*, Pergamon, London, 1994.

injected into the atmosphere when a satellite containing an isotopic power source disintegrated over the Indian Ocean in 1964. The Chernobyl accident caused the release of ~800 Ci $^{238}$Pu, ~700 Ci of $^{239}$Pu and ~1000 Ci of $^{240}$Pu, representing ~3% of the reactor core inventory. This activity was dispersed over large areas of the former Soviet Union and Europe. The amount of plutonium in the environment due to fuel reprocessing is small.

Over 99% of the plutonium released to the environment ends up in the soil and in sediments. The global average concentration of plutonium in soils is $5 \times 10^{-4}$ to $2 \times 10^{-2}$ pCi/g dry weight, with most of the plutonium being near the soil surface. The concentrations of plutonium in natural waters are quite low, with an average concentration being $\sim 10^{-4}$ pCi/L that is, $\sim 10^{-18}$ M. (Greater than 96% of any plutonium released to an aquatic ecosystem ends up in the sediments. In these sediments, there is some translocation of the plutonium to the sediment surface due to the activities of benthic biota.) Less than 1% (and perhaps closer to 0.1%) of all the plutonium in the environment ends up in the biota. The concentrations of plutonium in vegetation range from $10^{-5}$ to 2%, with concentrations in litter and animals ranging from $10^{-4}$ to 3% and $10^{-8}$ to 1%, respectively. *None of these concentrations has been observed to cause any discernible effect.*

Despite the extremely low concentrations of the transuranium elements in water, most of the environmental chemistry of these elements has been focused on their behavior in the aquatic environment. One notes that the neutrality of natural water (pH = 5–9) results in extensive hydrolysis of the highly charged ions except for Pu(V) and a very low solubility. In addition, natural waters contain organics as well as micro- and macroscopic concentrations of various inorganic species such as metals and anions that can compete with, complex, or react with the transuranium species. The final concentrations of the actinide elements in the environment are thus the result of a complex set of competing chemical reactions such as hydrolysis, complexation, redox reactions, and colloid formation. As a consequence, the aqueous environmental chemistry of the transuranium elements is significantly different from their ordinary solution chemistry in the laboratory.

In natural waters, hydrolysis is the primary factor affecting concentration. The tendency to hydrolyze follows the relative effective charge of the ions. This is known to be

$$An^{4+} > AnO_2^{2+} > An^{3+} > AnO_2^+$$

(where An represents an actinide element).

The hydrolysis reaction can be written as

$$x\,An^{m+} + y\,OH^- \leftrightarrows An_x(OH)_y$$

The hydrolysis products can be monomeric, polynuclear, or colloidal.

A number of strongly complexing inorganic anions are present in natural waters, such as $HCO_3^-/CO_3^{2-}$, $Cl^-$, $SO_4^{2-}$, $PO_4^{3-}$, and so on. The complexation order of these anions is

$$CO_3^{2-} > SO_4^{2-} > PO_4^{3-} > Cl^- > \cdots$$

Also present in many natural waters are humic/fulvic acid, citric acid, and the like. These organics also can complex actinides. In Figure 15.18, we show the relative stability constants for the first complexation reaction of various ligands with actinides of different oxidation states. Clearly, the carbonate and humate ions along with hydrolysis dominate the chemistry. The tetravalent actinide ions will tend toward hydrolysis reactions or carbonate complexation rather than humate/fulvate formation.

The aquatic solution chemistry of the actinides is also influenced by pH and redox potential ($E_h$). The approximate ranges of pH and $E_h$ for natural waters are shown in Figure 15.19. The pH varies from 4 to 9.5 and $E_h$ from $-300$ to $+500$ mV. In these

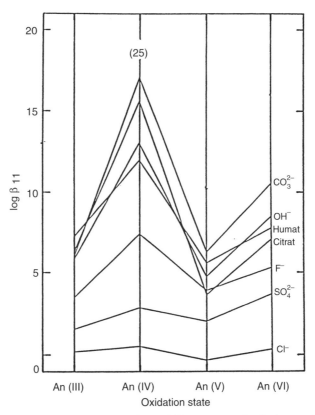

**Figure 15.18** Comparison of complexation stability constants for the interaction of various ligands with different actinide oxidation states (Kim, 1986).

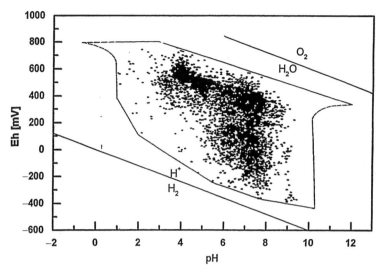

**Figure 15.19** A Pourbaix diagram showing the ranges of pH and $E_h$ values in natural waters.

pH and $E_h$ ranges, neptunium and plutonium can be present in several different oxidation states while americium and curium will be trivalent. In the oxidizing environment of surface waters, Np(V), Pu(IV), Pu(V), and Am(III) will be the dominant species, while in the reducing environment of deep groundwater, other species may be present (Fig. 15.20).

Colloids are always present in natural waters containing the transuranium elements. (Colloids are defined as particles with sizes ranging from 1 to 450 nm. These particles form stable suspensions in natural waters.) Colloids of the transuranium elements can be formed by hydrolysis of transuranium ions, or by the sorption of transuranium elements on the "naturally occurring colloids." The naturally occurring colloids include such species as metal hydroxides, silicate polymers, organics (such as humates), and the like. The mobility of the transuranium elements in an aquifer is determined largely by the mobility of its pseudocolloids, that is, those colloidal species formed by the adsorption of the transuranium ions upon the naturally occurring colloids.

The speciation of the transuranium elements in waters is thus a complex function of hydrolysis, colloid formation, redox reactions, and complexation with available ligands. The solubility (mobility) is, thus, highly dependent on the particular aquatic environment and its characteristics.

However, bearing in mind these caveats, we can make certain generalizations about the behavior of the actinide elements in natural waters. Americium and curium remain in the +3 oxidation state over the natural range of environmental conditions. For plutonium, Pu(III) is unstable to oxidation at environmental acidities, and so the other three states are observed with the dominant oxidation state in natural waters being Pu(V). [Humic materials cause a slow reduction of

## 15.7 ENVIRONMENTAL CHEMISTRY OF THE TRANSURANIUM ELEMENTS

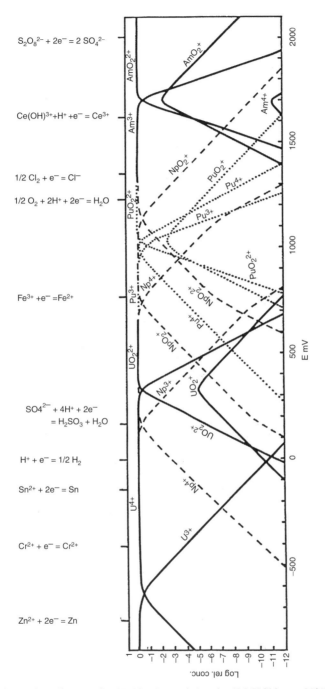

**Figure 15.20** Redox diagram for U, Np, Pu and Am in 1M $HClO_4$ at 25°C (Choppin, Liljenzin, and Rydberg, 2002).

Pu(V) to Pu(IV) so that Pu(IV) is found to be important in waters containing significant amounts of organic material.] Under reducing conditions, neptunium should be present as Np(IV) and behave like Pu(IV); under oxidizing conditions, $NpO_2^+$ will be the stable species. In marine waters, Pu(IV) and the transplutonium elements will tend to undergo hydrolysis to form insoluble hydroxides and oxides. However, these elements can also form strong complexes with inorganic anions ($OH^-, CO_3^{2-}, HPO_4^{2-}, F^-$, and $SO_4^{2-}$) and organic complexing agents that may be present. The speciation and solubility of these elements are largely determined by hydrolysis and formation of carbonate, fluoride, and phosphate complexes. Stable soluble species include Pu(V, VI) and Np(V), although under most conditions the actinides will form insoluble species that concentrate in the sediments.

Pu(IV), which forms highly charged polymers, strongly sorbs to soils and sediments. Other actinide III and IV oxidation states also bind by ion exchange to clays. The uptake of these species by solids is in the same sequence as the order of hydrolysis: Pu > Am(III) > U(VI) > Np(V). The uptake of these actinides by plants appears to be in the reverse order of hydrolysis: Np(V) > U(VI) > Am(III) > Pu(IV), with plants showing little ability to assimilate the immobile hydrolyzed species. The further concentration of these species in the food chain with subsequent deposit in humans appears to be minor. Of the ~4 tons of plutonium released to the environment in atmospheric testing of nuclear weapons, the total amount fixed in the world population is less than 1 g [of this amount, most (99.9%) was inhaled rather than ingested].

## PROBLEMS

1. Predict the aqueous solution chemistry of element 114. What is the expected oxidation state? By extrapolating periodic table trends, estimate the first ionization potential of element 114 [see Nash and Bursten (1999)].

2. Suppose you want to synthesize the nucleus $^{271}$Mt using the $^{37}$Cl + $^{238}$U reaction. Estimate the production cross section for this reaction. What is the expected half-life of this nucleus? What is the expected decay mode?

3. What is the expected relative population of the +4, +5, and +6 oxidation states of Pu in seawater ($E_h = -90$ mV)?

4. The reported discoveries of elements 114 and 116 do not connect the observed decays to the region of known nuclei. Devise an experimental program to make this connection or to establish the Z of the 114 and 116 nuclei.

5. It has been proposed to do a Stern–Gerlach experiment with Lr atoms to better understand the electron configuration of Lr. (a) Describe the original Stern–Gerlach experiment and what it demonstrated about the quantization of electron spin. Why was an inhomogeneous magnetic field used instead of a uniform field? (b) Using nonrelativistic quantum mechanics, one predicts the electron configuration of Lr to be [Rn] $5f^{14}6d^17s^2$. Using relativistic

quantum mechanics, the predicted ground-state electron configuration is [Rn] $5f^{14}7p_{1/2}7s^2$. Predict the expected pattern of spots one would observe when one passes a beam of neutral Lr atoms through an inhomogeneous magnetic field assuming (a) the nonrelativistic prediction is correct and (b) assuming the relativistic prediction is correct. (c) More sophisticated relativistic calculations predict a spacing of ~1 eV between the ground state [Rn] $5f^{14}7p_{1/2}7s^2$ electron configuration and an electronic excited state with the configuration [Rn] $5f^{14}6d^17s^2$. If the temperature of an assembly of Lr atoms is 300 K, predict the relative population of the two configurations.

6. In the discussion of Figures 15.5 and 15.6, a rough rule of estimating the survival probability as $(E^*/10) \times 10^{-2}$ was given. Using the equations presented in Chapter 11, calculate $\Gamma_n/\Gamma_f$ (the survival probability) for $^{254}$No with $B_f = 7.1$ MeV and an excitation energy of 50 MeV. Compare this estimate to the rough rule.

7. Consider the reaction $^{48}$Ca + $^{206}$Pb → $^{252}$No + 2n. Assume the energy of the $^{48}$Ca projectile in the lab system was 217 MeV. Using the equations in Chapters 10 and 11, calculate the formation cross section for this reaction. Compare your calculation with the rough estimate made using Figures 15.5 and 15.6.

8. Estimate the total decay power (*W*) produced in a sample of 10 g of $^{238}$Pu.

## REFERENCES

Barber, R. C., N. N. Greenwood, A. Z. Hrynkiewics, Y. P. Jeanin, M. Lefort, M. Sakai, I. I. Uleuhla, A. H. Wapstra, and D. H. Wilkinson. *Prog. Part. Nucl. Phys.* **29**, 453 (1992).

Druin, V. A., A. G. Demin, Yu. P. Kharitonov, G. N. Akap'ev, V. I. Rud, G. Y. Sung-Chin-Yang, L. P. Chelnokov, and K. A. Gavrilov. *Yad. Fiz.* **13**, 251 (1971).

Folden, C. M. et al. *Phys. Rev. Lett.* **93**, 212702 (2004).

Ghiorso, A., M. Nurmia, J. Harris, K. Eskola, and P. Eskola. *Phys. Rev. Lett.* **22**, 1317 (1969).

Ghiorso, A., M. Nurmia, K. Eskola, J. Harris, and P. Eskola. *Phys. Rev. Lett.* **24**, 1498 (1970).

Ghiorso, A., J. M. Nitschke, J. R. Alonso, C. T. Alonso, M. Nurmia, G. T. Seaborg, E. K. Hulet, and R. W. Lougheed. *Phys. Rev. Lett.* **33**, 1490 (1974).

Ghiorso, A., et al. *Nucl. Phys.* **A583**, 861 (1995a).

Ghiorso, A., et al. *Phys. Rev.* **C51**, R2293 (1995b).

Gilchriese, M., A. Sessler, G. Trilling, and R. Vogt. Lawrence Berkeley National Laboratory Report, LBNL-51773 (2003).

Ginter, T. N., et al. *Phys. Rev. C* **67**, 064609 (2003).

Gregorich, K. E., M. R. Lane, M. F. Mohar, D. M. Lee, C. D. Kacher, E. R. Sylwester, and D. C. Hoffman. *Phys. Rev. Lett.* **72**, 1423 (1994).

Gregorich, K. E., et al. *Eur. Phys. J. A* **18**, 633 (2003).

Hofmann, S. *Rep. Prog. Phys.* **61**, 639 (1998).

Hofmann, S., and G. Münzenberg. *Rev. Mod. Phys.* **72**, 733 (2000).

Hoffmann, S., et al. *Z. Phys. A* **350**, 277 (1995a).
Hofmann, S., et al. *Z. Phys. A* **350**, 281 (1995b).
Hofmann, S., et al. *Z. Phys. A* **354**, 229 (1996).
Hofmann, S., et al. *Eur. Phys. J. A* **14**, 147 (2002).
Katz, J. J., G. T. Seaborg, and L. Morss. *Chemistry of the Actinide Elements*, 2nd ed., Chapman and Hall, London, 1980.
Kim, J. I. Chemical Behaviour of Transuranic Elements in Natural Aquatic Systems, in *Handbook on the Physics and Chemistry of the Actinides*, Vol. 4, A. J. Freeman and C. Keller, Eds., North-Holland, Amsterdam, 1986, pp. 413–456.
Lazarev, Y. A., et al. *Phys. Rev. C* **54**, 620 (1996).
Möller, P., J. R. Nix, and K. L. Kratz. *At. Data Nucl. Data Tables* **66**, 131 (1997).
Morimoto, K. *AIP Conf. Series*, **561**, 351 (2001).
Münzenberg, G., et al. *Z. Phys. A* **300**, 107 (1981).
Münzenberg, G., et al. *Z. Phys. A* **309**, 89 (1982).
Münzenberg, G., et al. *Z. Phys. A* **315**, 145 (1984).
Münzenberg, G., et al. *Z. Phys. A* **317**, 235 (1984); G. Münzenberg, et al., *Z. Phys. A* **328**, 49 (1987).
Nash, C. S. and B. E. Bursten. *J. Phys. Chem. A* **103**, 402 (1999).
Ninov, V., et al. *Phys. Rev. Lett.* **83**, 1101 (1999).
Oganessian, Y. T., et al. *Z. Phys. A* **319**, 215 (1984).
Oganessian, Y. T., et al. *Phys. Rev. Lett.* **83**, 3154 (1999a).
Oganessian, Y. T., et al. *Nature* **400**, 242–245 (1999b).
Oganessian, Y. T., et al. *Phys. Rev. C* **62**, 041604 (2000).
Oganessian, Y. T., et al. *Phys. Rev. C* **63**, 011301 (2001).
Schadel, M. *J. Nucl. Radiochem. Sci.* **3**, 113 (2002).
Stodel, C., et al. *AIP Conf. Series* **561**, 311 (2001).
Zvara, I., Yu. T. Chuburicov, R. Tsaletka, and M. R. Shalaevskii. *Radiokhimiya* **11**, 163 (1969).
Zvara, I., Yu. T. Chuburkov, V. Z. Belov, G. V. Buklanov, B. B. Zakhvataev, T. S. Zvarova, O. D. Maslov, R. Caletka, and M. R. Shalaevsky. *J. Inorg. Nucl. Chem.* **32**, 1885 (1970).

## BIBLIOGRAPHY

Seaborg, G. T., and W. Loveland. *The Elements Beyond Uranium*, Wiley, New York, 1990, and the many references contained therein. The material in this chapter represents a condensation of this book.

Seaborg, G. T., and E. G. Valens. *Elements of the Universe*, Dutton, New York, 1958.

Schädel, M., ed. *The Chemistry of Superheavy Elements*, Kluwer, Dordrecht, 2003.

*The Transactinide Elements*. Proceedings of the 41st Robert Welch Conference on Chemical Research, Welch, Houston, 1998. The best modern summary of many aspects of the study of the heaviest chemical elements.

# CHAPTER 16

# NUCLEAR REACTOR CHEMISTRY

One of the most important applications of nuclear and radiochemistry is in the area of nuclear power. Chemistry and chemical processes are intimately involved in reactor operation, the preparation and processing of reactor fuel, and the storage and ultimate disposal of radioactive waste. In this chapter, we shall examine some of the most important chemistry associated with nuclear power.

## 16.1 INTRODUCTION

The basic principles behind nuclear reactors and the current design of light-water reactors are discussed in Chapter 14. The two types of light-water reactors, the pressurized water reactor (PWR) and the boiling water reactor (BWR) are shown schematically in Figure 14.2. In these light-water reactors, the reactor fuel is enriched to $\sim 3\%$ in $^{235}$U, and ordinary water is used as the coolant and moderator. In a PWR, the reactor core is encased in a steel pressure vessel filled with water under a pressure of $\sim 150$ bar or greater. The water does not boil due to this high pressure even though the temperature of the water is as high as 350°C. It passes through a heat exchanger where its heat is used to boil water in the secondary coolant loop. The steam is dried and used to drive a turbine, then condensed and returned to the boiler. The water in the primary loop usually contains boron (as 0.025 M boric acid, $H_3BO_3$) to control the reactivity of the reactor.

---

*Modern Nuclear Chemistry*, by W.D. Loveland, D.J. Morrissey, and G.T. Seaborg
Copyright © 2006 John Wiley & Sons, Inc.

In a BWR, the water passing through the reactor core is allowed to boil because it is at a lower pressure, ~75 bar. The steam produced is passed through a turbine, condensed, and returned to the reactor.

In PWRs, the fuel is $UO_2$, enriched typically to 3.3% $^{235}U$; while for BWRs, the fuel is $UO_2$, enriched to 2.6%. (Natural uranium is 0.72% $^{235}U$). The fuel elements are clad in Zircaloy, a zirconium alloy that includes tin, iron, chromium, and nickel that prevents fission product release and protects them against corrosion by the coolant. The control rod material in BWRs is $B_4C$, while PWRs have Ag–In–Cd or Hf control materials.

The nuclear fuel cycle is a set of steps in the processing of the reactor's fissile materials that begins with the mining of uranium and extends through the final disposition of the waste from the reactor. These steps are referred to as a cycle because it is possible that the material taken from the reactor after use can be recycled. A schematic diagram of the nuclear fuel cycle is shown in Figure 16.1.

The diagram in Figure 16.1 shows two possible paths for this cycle, that is, with and without fuel reprocessing. The majority of reactors in the world and all U.S. reactors operate with a once-through cycle without reprocessing. Some countries, particularly France, do fuel reprocessing with reuse of the plutonium from spent fuel. The portions of the cycle, that precede the introduction of the fissile material into the reactor are referred to as the *front end* of the cycle, while the *back end* includes those steps that occur after the removal of the fuel from the reactor. The details of this cycle and the chemistry involved are discussed below.

## 16.2 FISSION PRODUCT CHEMISTRY

In the chemistry of the fuel cycle and reactor operations, one must deal with the chemical properties of the actinide elements, particularly uranium and plutonium and those of the fission products. In this section, we focus on the fission products and their chemistry. In Figures 16.2 and 16.3, we show the chemical composition and associated fission product activities in irradiated fuel. The fission products include the elements from zinc to dysprosium, with all periodic table groups being represented.

The inert gases (Group 18) are represented by isotopes of Kr and Xe. These isotopes are generally short-lived and will decay before fuel reprocessing. As inert gases, they are unreactive and consequently they are isolated using cryogenic techniques.

The alkali metals (Group 1) are represented by Rb and Cs where long-lived $^{137}Cs$ ($t_{1/2} = 30$ y) is the most important nuclide. Their solution chemistry is that of the generally soluble +1 cation. The alkaline earths (Group 2) are represented by the high-yield nuclides $^{140}Ba$, $^{90}Sr$, and $^{91}Sr$. These nuclides can be separated using ion exchange or solvent extraction or gravimetric techniques. The 28-y $^{90}Sr$ is an important radiation hazard in aged spent fuel, while 12.8-d $^{140}Ba$ frequently determines the shielding requirements during fuel for the 10- to 100-d cooling period.

## 16.2 FISSION PRODUCT CHEMISTRY

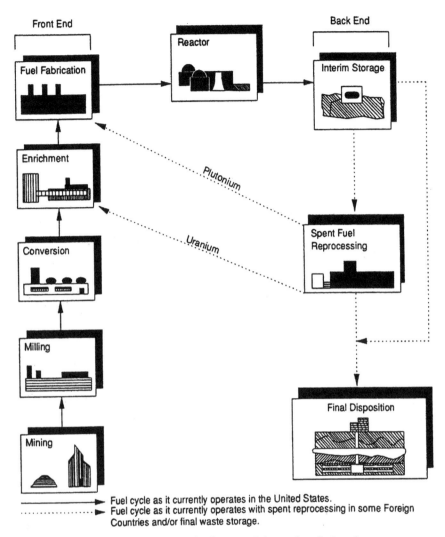

**Figure 16.1** Schematic diagram of the nuclear fuel cycle.

Group 3 contains Y and the lanthanides. These elements are chemically similar. They can be separated from one another by ion exchange, while their separation from U and Pu can be done using solvent extraction with tri-butyl phosphate (TBP). These elements have soluble nitrates, chlorides, and sulfates while their fluorides and hydroxides are insoluble. $LaF_3$ is frequently used as a carrier for this group. They form stable complexes with strong chelating agents such as DPTA, EDTA, and the like. The Group 4 element in fission product mixtures is zirconium whose chemistry is that of the +4 oxidation state. The principal nuclide of interest is

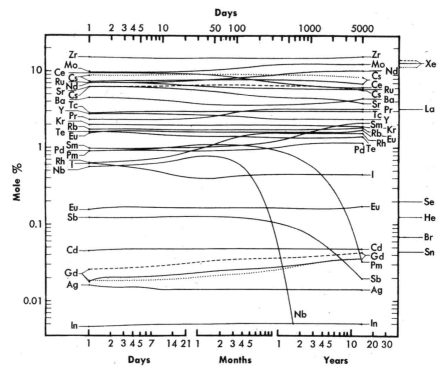

**Figure 16.2** Chemical composition of the fission products in irradiated fuel as a function of decay time after a 2-month irradiation. [From J. Prawitz and J. Rydberg, *Acta. Chem. Scand.* **12**, 393 (1958).]

the 63-d $^{95}$Zr and its 35-d daughter, $^{95}$Nb. The chemistry of Zr can be tricky as it readily forms complexes and does form colloids, which can lead to poor separation factors.

The Group 5 element niobium occurs as a decay product, $^{95}$Nb, of $^{95}$Zr. Like its zirconium parent, $^{95}$Nb forms colloids and is readily hydrolyzed. The principal Group 6 fission product is $^{99}$Mo ($t_{1/2} = 67$ h), which is important at short cooling times because of its high yield but is insignificant in aged fuel. The most important Group 7 fission product is the long-lived ($2.1 \times 10^5$ y) $^{99}$Tc. Its chemistry is that of the pertechnate ion $TcO_4^-$. The elements of Groups 8, 9, and 10 are important because of their activity, and in the case of ruthenium and its multiple oxidation states and the slow interconversion kinetics, the chemistry can be troublesome. $RuO_4$ can be volatilized leading to the loss of Ru in radiochemical procedures. The chemistry of the Group 11 element, Ag, is straightforward (+1 cation, forming insoluble compounds) and the Groups 12 and 13 elements, Zn, Ga, Cd, and In have low yields and small activities. The Groups 14 and 15 elements, Ge, As, Sn, and Sb also have low activities in aged fuel. The Group 16 element, Te, is present in the form of 30- to 100-d activities, but most interest is focused on

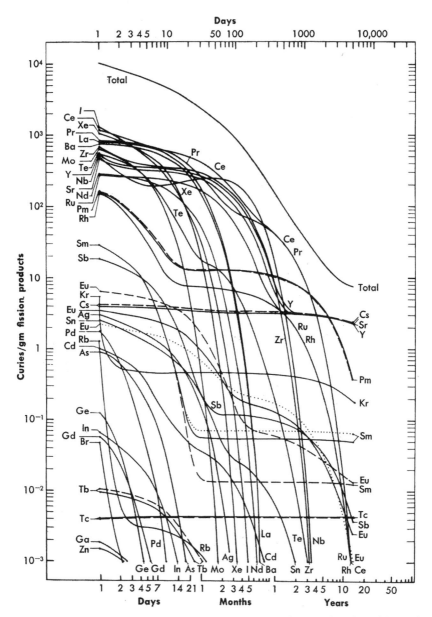

**Figure 16.3** Principal fission product activities in irradiated fuel as a function of decay time after a 2-month irradiation. [From J. Prawitz and J. Rydberg, *Acta. Chem. Scand.* **12**, 385 (1958).]

78-h $^{132}$Te, which decays to short-lived $^{132}$I, which is volatile and can be released in rapid processing of fuel. The halogens, Br and I, are not important in fuel reprocessing due to their short half-lives but can be important in reactor operation and accidents due to their volatility.

## 16.3 RADIOCHEMISTRY OF URANIUM

### 16.3.1 Uranium Isotopes

Natural uranium is 99.274 atom % $^{238}$U, 0.7205 atom % $^{235}$U, and 0.0056 atom % $^{234}$U. The 234/238 ratio is exactly the ratio of their half-lives as expected for nuclei in secular equilibrium. The isotope $^{233}$U is produced by neutron capture on $^{232}$Th, followed by $\beta^-$ decay. $^{232}$U is a short-lived ($t_{1/2} = 72$ y) nuclide that is a contaminant in $^{233}$U samples (from fast neutron reactions). The daughters of $^{232}$U are hard $\gamma$-ray emitters that make working with $^{232}$U containing samples difficult. $^{236,237,239}$U are produced by neutron captures on $^{235}$U and $^{238}$U. $^{236}$U is long-lived but $^{237,239}$U are short-lived and decay to $^{237}$Np and $^{239}$Pu, respectively.

### 16.3.2 Metallic Uranium

Metallic uranium can exist in three different solid phases with differing densities, depending on temperature. At room temperature, the $\alpha$ phase is observed with a density of 19.07 g/cm$^3$ and a melting point of 1132°C. Metallic uranium is a very reactive metal that is silvery in color. (Frequently, a surface oxide layer makes metallic uranium look black.) Uranium powder is pyrophoric. When uranium metal is cut or scratched in the laboratory, a shower of sparks is sometimes observed due to the creation of small particles that ignite. Uranium metal with an oxide coating will burn at 700°C to form U$_3$O$_8$. Uranium reacts with hot water to produce UO$_2$ and UH$_3$. In reactors, uranium is alloyed with zirconium to resist corrosion and radiation damage. Metallic uranium can be produced by the reduction of UF$_4$, for example, by reaction with magnesium

$$UF_4 + 2Mg \longrightarrow 2MgF_2 + U$$

### 16.3.3 Uranium Compounds

Uranium exists in the +3, +4, +5, and +6 oxidation states. The +5 state disproportionates to the +4 and +6 states and is of little importance. Trivalent uranium reduces water and therefore there is no stable aqueous chemistry of U$^{3+}$ although compounds do exist.

The most important uranium compounds are the oxides. UO$_2$ is the compound used in reactor fuel. It is a stable refractory material that is brown-black in color and is nonreactive with H$_2$O. It has density of 10.97 g/cm$^3$ and can be prepared by the reduction of UO$_3$ with hydrogen. U$_3$O$_8$ (UO$_2 \cdot$ 2UO$_3$) is a green-black solid that occurs in the mineral pitchblende. It has a density of 8.38 g/cm$^3$, is soluble in HNO$_3$, and can be prepared by oxidizing UO$_2$ or reducing UO$_3$. UO$_3$ is a yellow-orange solid ("orange oxide") and is important as an intermediate in the production of UO$_2$ or UF$_6$.

Uranium hydride, UH$_3$, is a reactive black powder. It is a powerful reducing agent and is pyrophoric. A mixture of uranium and zirconium hydrides is used as the fuel for the TRIGA research reactors. The resulting very large negative temperature coefficient of reactivity provides a large safety margin for reactor operation.

Uranium halides exist in the +3 oxidation state ($UF_3$, $UCl_3$, $UBR_3$, $UI_3$), the +4 oxidation state ($UF_4$, $UCl_4$, $UBr_4$, and $UI_4$), the +5 oxidation state ($UF_5$, $UCl_5$), and the +6 oxidation state ($UF_6$, $UCl_6$).

$UF_4$ ("green salt") is an intermediate in the production of U and $UF_6$. It can be made by reacting $UO_2$ with excess HF as

$$UO_2 + 4HF \Longleftrightarrow UF_4 + 2H_2O$$

or, for laboratory use, by the reaction

$$U^{4+} + 4F^- \longrightarrow UF_4 \downarrow$$

(Uranium tetrafluoride precipitated from aqueous solutions exists as $UF_4 \cdot 5H_2O$, and it is difficult to remove the waters of hydration so that the dry reaction above is preferred.) $UF_4$ is frequently used to make accelerator targets of uranium by vacuum volatilization.

$UF_6$ ("hex") is the only readily available uranium compound that is volatile at room temperature. It is a colorless solid that is used in the uranium enrichment process. It sublimes at room temperature without melting. $UF_6$ is rapidly hydrolyzed by water and is a fluorinating agent. This latter property means that one must carefully choose the materials to contain $UF_6$.

### 16.3.4 Uranium Solution Chemistry

The solution chemistry of uranium is that of the +4 and +6 oxidation states, that is, $U^{4+}$ and $UO_2^{2+}$. The formal reduction potential of uranium in aqueous solution (i.e., 1 M $HClO_4$) is

$$UO_2^{2+} \xrightarrow{+0.063} UO_2^+ \xrightarrow{+0.58} U^{4+} \xrightarrow{-0.031} U^{3+} \xrightarrow{-1.70} U$$
$$\underbrace{\phantom{UO_2^{2+} \xrightarrow{+0.063} UO_2^+ \xrightarrow{+0.58}}}_{+0.32}$$

The U(IV) chemistry is similar to that of $Th^{4+}$, except for the difference in the charge/radius ratio of the ions. $U^{4+}$ solutions are green in color, stable, and slowly oxidized by air to $UO_2^{2+}$. Solutions of $U^{4+}$ are generally prepared by reduction of solutions of the uranyl ($UO_2^{2+}$) ion. U(IV) forms complexes with many anions ($C_2O_4^{2-}$, $C_2H_3O_2^-$, $CO_3^{2-}$, $Cl^-$, and $NO_3^-$). The chlorides and bromides of U(IV) are soluble while the fluorides and hydroxides are insoluble. In aqueous solution, U(IV) hydrolyzes via the reaction,

$$U^{4+} + H_2O \longrightarrow U(OH)^{3+} + H^+ \quad K = 0.027$$

The U(VI) can be prepared by dissolving $UO_3$ in acid or U metal in $HNO_3$. Solutions of the uranyl ion show a characteristic yellow-green color and are very stable. U(VI) shows complex solution equilbria due to the occurrence of hydrolysis, which leads to $[(UO_2)_2OH]^{3+}$, $[(UO_2)_2(OH)_2]^{2+}$, and $[(UO_2)_3(OH)_4]^{2+}$ mixtures. Because of hydrolysis reactions, aqueous solutions of uranyl salts are slightly acidic. Addition of base to uranyl solutions results in precipitation beginning at a pH between 4 and 7.

Uranyl ions form complexes in solutions with most anions. Uranyl sulfate and carbonate complexes are especially strong and are used in extracting uranium from its ores. Of great practical importance are the complexes of the uranyl ions with nitrate that are soluble in organic liquids such as alcohols, ethers, ketones, and esters. One of the most important of these reactions is that involving the extraction of uranyl nitrate into TBP (the Purex process):

$$UO_2^{2+}(aq) + 2NO_3^-(aq) + 2TBP(org) \Longleftrightarrow UO_2(NO_3)_2 \cdot 2TBP(org)$$

Neglecting activity coefficients, the coefficient for the distribution of uranium between the organic and aqueous phases is written as

$$D = \frac{[UO_2(NO_3)_2 \cdot 2TBP]_{org}}{[UO_2^{2+}]_{aq}} = K(NO_3)_{aq}^2 (TBP)_{org}^2$$

where $K$ is the equilibrium constant ($K \sim 15\text{–}60$). One can use the concentration of a salting agent such as $NO_3^-$ to control the extraction process. (In the Purex process, these salting agents ultimately leave the cycle with the fission products, contributing to the problems of dealing with these wastes.)

## 16.4 NUCLEAR FUEL CYCLE—THE FRONT END

The nuclear fuel cycle (Fig. 16.1) begins with the mining of uranium ore. Uranium is by no means rare. Its overall abundance in Earth's crust is $\sim 4$ ppm (which is more abundant than Ag, Hg, Bi, or Cd). There are $\sim 10^4$ tonnes of uranium in Earth's crust. The problem is one of concentration in that most uranium deposits contain <0.001% uranium.

### 16.4.1 Mining and Milling

Uranium ore can be classified as high grade (1–4% U), medium grade (0.1–0.5% U), and low grade (<0.1% U). In the high-grade deposits, in Zaire and Canada, uranium is found as pitchblende or uranite, materials of general composition $xUO_2 \cdot yUO_3$ where $0 < y/x < 2$. The medium-grade ores are found in places such as the Colorado plateau of the United States, where uranium is incorporated in carnotite ($K_2O \cdot 2UO_3 \cdot V_2O_5 \cdot xH_2O$) or autunite ($CaO \cdot 2UO_3 \cdot P_2O_5 \cdot xH_2O$). Low-grade sources include the gold ore residues of South Africa, seawater where the nominal uranium concentration of 3 ppb corresponds to a reservoir of $\sim 10^{10}$ tonnes of uranium, or the fertilizer by-products of the phosphate fields of Florida and Idaho, Tennessee shale, or the lignites of Wyoming and the Dakotas. The average uranium content of the ores used in the nuclear fuel cycle in the United States in recent years is ~0.24%.

After mining, the uranium must be concentrated before further operations are carried out. This is done in the mills, which are located near the mines. Here the uranium ore content is increased from a few tenths of a percent (in the ore) to 85–95% (in a semirefined concentrate known as "yellowcake"), while eliminating other elements that are present (the "tailings").

While very high grade pitchblende deposits can be concentrated by physical techniques involving their specific gravity, most concentration of uranium involves chemical leaching from the ore. The ore is prepared for leaching by crushing and roasting (to destroy organic material). It is then subjected to chemical leaching followed by a separation of the liquids and solids, followed by concentration/purification and final product recovery.

The normal choice (~80% of all U.S. ores) for chemical leaching is acid leaching with sulfuric acid (due to its low cost and great availability). For the ores that have high carbonate content, an alkaline leach with $Na_2CO_3$ is used to prevent high acid concentration.

The problem in acid leaching is that only U(VI) dissolves in $H_2SO_4$. Any U(IV) present must be oxidized to U(VI) prior to leaching. The chemical steps can be summarized as

$$FeS_2, Fe + H_2SO_4 \rightarrow Fe^{2+} + H_2 + SO_4^{2-}$$
$$2Fe^{2+} + MnO_2 + 4H^+ \rightarrow 2Fe^{3+} + Mn^{2+} + 2H_2O$$

or

$$6Fe^{2+} + NaClO_3 + 6H^+ \rightarrow 6Fe^{3+} + NaCl + 3H_2O$$

leading to

$$UO_2 + 2Fe^{3+} \rightarrow UO_2^{2+} + 2Fe^{2+}$$
$$UO_3 + H_2SO_4 \rightarrow UO_2^{2+} + SO_4^{2-} + H_2O$$
$$UO_2^{2+} + 2SO_4^{2-} \Longleftrightarrow [UO_2(SO_4)_2]^{2-} \Longleftrightarrow [UO_2(SO_4)_3]^{4-}$$

The final product thus appears in three chemical forms, $UO_2^{2+}$, $[UO_2(SO_4)_2]^{2-}$, and $[UO_2(SO_4)_3]^{4-}$.

In alkaline leaching of high carbonate materials, one takes advantage of the unique solubility of $[UO_2(CO_3)_3]^{4-}$. One begins with finely divided material and must also deal with the oxidation of any U(IV) that is present. The basic reaction (pun intended) is

$$UO_2 + (\text{oxidant}, Ox) + 3CO_3^{2-} \rightarrow [UO_2(CO_3)_3]^{4-} + Ox^{2-}$$

After leaching, a concentration/purification is done to get rid of other materials leached from the ore. This can be done by ion exchange or solvent extraction. In the ion exchange method, three steps are employed: (a) the absorption of uranium from the leach liquor onto the resin, (b) the selective elution of uranium from the resin, and (c) the regeneration of the resin. *Anion exchange* is the preferred method of ion exchange with the relevant chemical reactions for acid leach being

$$[UO_2(SO_4)_3]^{4-} + 4RX \iff R_4[UO_2(SO_4)_3] + 4X^-$$

and for alkaline leach

$$[UO_2(CO_3)_3]^{4-} + 4RX \iff R_4[UO_2(CO_3)_3] + 4X^-$$

The eluant is usually 1 M $NO_3^-$ in the form of $NH_4NO_3$. The physical method for carrying out the extraction can involve: (a) a fixed resin bed, (b) a "resin-in-pulp" technique where resin in baskets is passed through a stream of pulp or slurry from the leach process, or (c) a moving bed of resin.

Concentration/purification by solvent extraction usually involves four steps: (a) *extraction* of uranium from the leach liquor in a solvent, (b) *scrubbing* to remove impurities from the solvent, (c) *stripping* to remove uranium from the solvent, and (d) regeneration of the solvent. The solvent phase in solvent extraction will contain the *extractant* that complexes uranium to make it soluble in the organic phase, a *diluent*, an inexpensive material to dilute the extractant, and a *modifier* to improve the solubility of the extractant in the diluent. Typical extractants are amines with isodecanol acting as a modifier to improve the amine solubility in a diluent such as kerosene. The typical chemistry of the extraction would involve the reactions

$$2R_3N(org) + H_2SO_4 \rightarrow (R_3NH)_2SO_4(org)$$

$$2(R_3NH)_2SO_4(org) + [UO_2(SO_4)_3]^{4-}(aq) \rightarrow (R_3NH)_4[UO_2(SO_4)_3](org)$$
$$+ 2SO_4^{2-}(aq)$$

with the stripping reactions being

$$(R_3NH)_4[UO_2(SO_4)_3](org) + 4NH_4OH \rightarrow 4R_3N(org) + 4H_2O + [UO_2(SO_4)_2]^{2-}$$
$$+ 4NH_4^+ + SO_4^{2-}$$

Following solvent extraction, uranium is precipitated from the solution by the addition of gaseous ammonia with the yellowcake product (Fig. 16.4) being collected, packaged in 55-gal drums, and shipped to a refinery for further purification and conversion to $UF_6$ for enrichment.

Alkaline leach mills will use NaOH for final product recovery (in the form of sodium uranate, $Na_2U_2O_7$). Often these uranium ore concentrates (yellowcake) are transformed into $U_3O_8$ by drying at 200°C to remove water and calcining, that is, heating until decomposition.

**Figure 16.4** Ammonium diuranate (yellowcake) after solvent extraction. (Photo courtesy of UIC.) (Figure also appears in color figure section.)

### 16.4.2 Refining and Chemical Conversion

Following milling, the yellowcake is shipped for *refining* and *conversion*. In most refineries, the uranium ore concentrates are purified by solvent extraction and then converted to $UF_6$ for enrichment. (A schematic diagram for this process is shown in Fig. 16.5.)

The yellowcake is dissolved in nitric acid. Uranyl nitrate is separated from metallic impurities in the dissolver by solvent extraction with TBP in some organic hydrocarbon, such as hexane. The organic extract is scrubbed with 1 M nitric acid, and uranium is stripped from the organic phase by back-extraction with 0.01 M $HNO_3$. The uranyl nitrate–hexahydrate, $UO_2(NO_3)_2 \cdot 6H_2O$ or UNH is converted to $UO_3$ (orange oxide) in two steps, a concentration by evaporation and a denitrification by heating. $UO_3$ is reduced to $UO_2$ (brown oxide) using hydrogen via the overall reaction

$$UO_3 + H_2 \longrightarrow UO_2 + H_2O$$

$UO_2$ is hydrofluorinated via the reaction

$$UO_2 + 4HF \longrightarrow UF_4 + 2H_2O$$

using anhydrous HF gas to produce green salt, $UF_4$. This green salt can be burned in $F_2$ to produce $UF_6$. $UF_4$ can also be reduced to metallic uranium.

### 16.4.3 Enrichment

As discussed earlier, natural uranium is 0.72 atom % $^{235}U$, and the fuel used in light-water reactors is typically 3% $^{235}U$. This means the refined uranium must be enriched in the lighter 235 isotope prior to fuel fabrication. This can be done by a

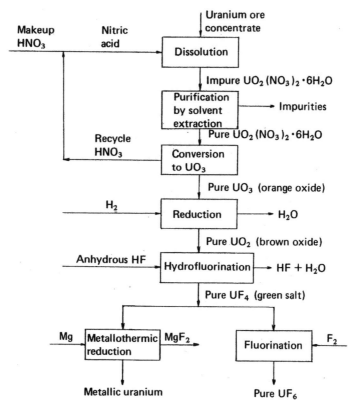

**Figure 16.5** Schematic diagram of the refining and conversion of uranium ore concentrates. (From Benedict et al., 1981.)

variety of physical approaches that take advantage of the small mass difference between $^{235}$U and $^{238}$U. We shall discuss four of these methods, gaseous diffusion, gas centrifuges, electromagnetic separation, and laser enrichment.

In *gaseous diffusion*, one takes advantage of the fact that in a gas, the lighter molecules have a higher velocity. If we assume the average kinetic energy of all gas molecules at a given temperature is the same, then we can write

$$\text{Average kinetic energy} = mv^2/2$$

For $^{235}$UF$_6$ and $^{238}$UF$_6$, we have

$$m_{235}v_{235}^2 = m_{238}v_{238}^2$$

$$\frac{v_{235}}{v_{238}} = \left(\frac{m_{238}}{m_{235}}\right)^{1/2} = \left(\frac{314}{311}\right)^{1/2} = 1.0043$$

If we send a stream of $UF_6$ gas into a vessel with porous walls (Fig. 16.6), then the lighter 235 molecules will pass through the pores slightly more frequently (due to the greater number of impacts on the walls per time). The maximum separation factor is 1.0043, but, in practice, this is not achieved and the less enriched part of the gas stream is recycled. Typically, if one starts with natural uranium (0.72% $^{235}U$) and with tails depleted to 0.3%, about 1200 enrichment stages are needed to get a 4% enrichment. In practice, membranes are used with several million 10- to 100-nm pores/$cm^2$, and the process must be carried out at elevated temperatures, dealing with $UF_6$, a strong fluorinating agent that decomposes upon contact with water. Most enrichment is done using this technique.

In *gas centrifuges* (Fig. 16.7), one takes advantage of the fact that the centrifugal force will push the heavier $^{238}UF_6$ molecules to the wall of the centrifuge, and the gas near the center will be enriched in $^{235}UF_6$. The gas flow pattern allows the heavier gas to be collected near the top and the lighter gas near the bottom. Separation is more efficient than in the gaseous diffusion method, requiring only about 10 stages to enrich $^{235}U$ from 0.72 to 3% with a 0.2% tail.

In *electromagnetic separation*, one uses the fact that when ions of differing mass with the same charge state are accelerated through a potential difference, their kinetic energy is the same, and the radius of curvature in a magnetic field is proportional to $\sqrt{mass}$. Thus, the material to be separated is ionized, accelerated, and passed through an analyzing magnet that bends the isotopes into separate beams, depositing them onto collectors for each isotope. The technology is straightforward, but the cost is high for ionizing macroscopic amounts of material, and so this technique has largely been abandoned, although it was used recently in Iraq in an attempt to obtain enriched uranium for nuclear weapons.

In *laser enrichment*, one takes advantage of the fact that the atomic energy levels of different isotopes differ slightly (called the isotope shift). This is due to the differing reduced electron masses for the different isotopes and the change in the overlap between the wave functions of the inner electrons and the nucleus, due to isotopic

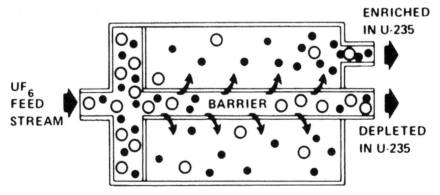

**Figure 16.6** Schematic diagram of the operation of a gaseous diffusion cell. (From Leuze, 1981.)

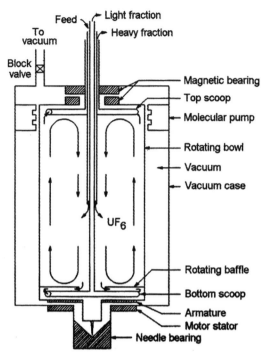

**Figure 16.7** Schematic diagram of a gas centrifuge. (From Leuze, 1981.)

differences in nuclear radii. Lasers can be tuned to excite $^{235}$U atoms, but not $^{238}$U atoms, to higher energy levels. The excited atoms are then ionized with another laser beam. The resulting ionized $^{235}$U atoms can be separated electromagnetically. To date, the feasibility of this technique has been demonstrated, but there are no commercial applications.

### 16.4.4 Fuel Fabrication

The enriched $UF_6$ is converted into $UO_2$ at the fuel fabrication plants. The $UF_6$ is reacted with water to produce a solution of $UO_2F_2$ and HF:

$$UF_6(g) + 2H_2O \longrightarrow UO_2F_2 + 4HF$$

Ammonium hydroxide is added to the uranyl fluoride solution to quantitatively precipitate ammonium diuranate

$$2UO_2F_2 + 6NH_4OH \longrightarrow (NH_4)_2U_2O_7 + NH_4F + 3H_2O$$

This product is collected, calcined in air to produce $U_3O_8$, and heated with hydrogen to make $UO_2$ powder. The $UO_2$ powder is pressed into pellets, which are sintered, ground to size, and loaded into zircalloy tubing filled with helium. The tubes are sealed and assembled into fuel bundles.

## 16.5 NUCLEAR FUEL CYCLE—THE BACK END

At one point in the history of nuclear power, the concept of reprocessing spent reactor fuel to recover its plutonium content to fuel other reactors was considered central to reactor development. The idea of an energy source that could generate its own fuel was very appealing. But, as outlined earlier, most fuel is not reprocessed/recycled but used in a "once-through" manner. The reasons for this are complex, that is, no shortage in near-term uranium supply, low uranium prices, some technical problems in reprocessing, and a concern that reprocessing would make plutonium too readily available for use in weaponry. Nonetheless there are operating plants for reprocessing reactor fuel, and the overall fate of spent fuel continues to be of great concern and interest.

### 16.5.1  Properties of Spent Fuel

Periodically, a portion of the fuel in a nuclear reactor is removed and replaced with fresh fuel. In the past, the average lifetime of fuel in the reactor was 3 years with one-third of the fuel being removed each year. More recently, attempts are being made to extend fuel lifetimes.

Initially, the radioactivity levels of the irradiated fuel are very high (Fig. 16.8). Chemically, the $^{235}$U content has been reduced from its initial 3 to 1% while the

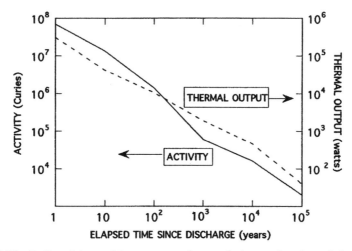

**Figure 16.8**  Radioactivity and heat output of spent fuel as a function of time. (From Bodansky, 1996.)

$^{238}$U content has been reduced from 97 to 94% (Fig. 16.9). $^{239}$Pu and other Pu isotopes are produced along with the fission products.

The original plan for this spent fuel was to store it for about 150 d and then transfer it to other facilities for disposal as waste or reprocessing. In the United States, this transfer has not occurred, and most of the fuel has remained in cooling ponds at the reactor sites, for times that have exceeded 20 y. In this case, a closer look at the properties of irradiated fuel is justified.

In Figure 16.8, we show the activity and heat output for spent fuel, beginning one year after removal from the reactor. (During the first year, the activity drops to 1% of its initial value and drops another factor of 5 in the first 10-y period.) The "waste disposal problem" begins at the 10-y point. Interpreted in this light, the dominant activities over the long term in unseparated waste are $^{90}$Sr, $^{137}$Cs, $^{241}$Pu, $^{241}$Am (from the decay of $^{241}$Pu), $^{239,240}$Pu, $^{99}$Tc, and $^{237}$Np. The activity level of the waste decreases faster than the heat output due to the shorter half-lives of the

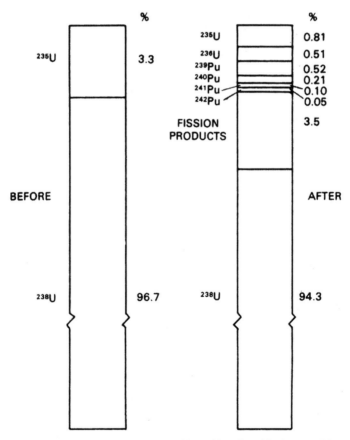

**Figure 16.9** Changes in the chemical composition of irradiated fuel. (From Murray, 2003.)

β emitters, which do not contribute as much to the heat output as the longer-lived α emitters. The heat output of the waste is sufficiently large to require care in waste storage.

### 16.5.2 Fuel Reprocessing

Fuel reprocessing has three objectives: (a) to recover U or Pu from the spent fuel for reuse as a nuclear reactor fuel or to render the waste less hazardous, (b) to remove fission products from the actinides to lessen short-term radioactivity problems and in the case of recycle of the actinides, to remove reactor poisons, and (c) to convert the radioactive waste into a safe form for storage. Fuel reprocessing was/is important in the production of plutonium for weapons use.

The Purex process is used for almost all fuel reprocessing today. Irradiated $UO_2$ fuel is dissolved in $HNO_3$ with the uranium being oxidized to $UO_2(NO_3)_2$ and the plutonium oxidized to $Pu(NO_3)_4$. A solution of TBP in a high-boiling hydrocarbon, such as $n$-dodecane, is used to selectively extract the hexavalent $UO_2(NO_3)_2$ and the tetravalent $Pu(NO_3)_4$ from the other actinides and fission products in the aqueous phase. The overall reactions are

$$UO_2^{2+}(aq) + 2NO_3^-(aq) + 2TBP(org) \longrightarrow UO_2(NO_3)_2 \cdot 2TBP(org)$$

or

$$Pu^{4+}(aq) + 4NO_3^-(aq) + 2TBP(org) \longrightarrow Pu(NO_3)_4(TBP)_2(org)$$

[These equilibria can be shifted to the right (i.e., improved extraction) by increasing the TBP concentration in the organic phase or increasing the $[NO_3^-](aq)$.] In a second step, the TBP solution is treated with a dilute nitric acid solution of a reducing agent, such as ferrous sulfamate or U(IV), which reduces the plutonium to a trivalent state but leaves the uranium in a hexavalent state. Plutonium will then transfer to the aqueous phase, leaving uranium in the organic phase. The uranium is stripped from the organic phase.

The only fission fragments that extract during the Purex process are Zr, Ru, Nb, and Tc, with the most troublesome being Zr and Ru. Zr forms a number of complex species with the most important being $[Zr(NO_3)_4 \cdot 2TBP]$. The formation of this complex is inhibited by the addition of $F^-$ whereby

$$Zr(NO_3)_4 \cdot 2TBP + 6F^- \iff ZrF_6^{2-} + 4NO_3^- + 2TBP$$

An overall schematic view of the Purex process is shown in Figure 16.10. Detailed descriptions of the process can be found in Benedict and co-workers (1981) and Wymer and Vondra (1981). We shall briefly summarize the important steps. The first step is to prepare the irradiated fuel for dissolution by mechanically chopping it into small pieces (~1–5 cm). This opening of the cladding causes the

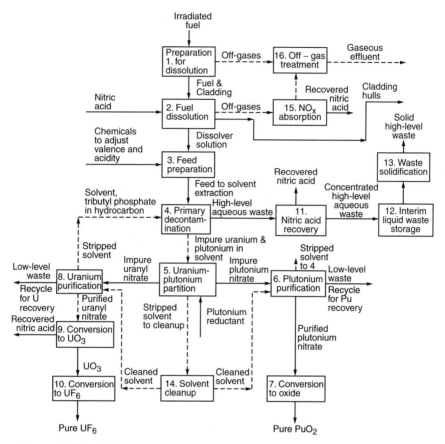

**Figure 16.10** Schematic diagram of the Purex process. (From Benedict et al., 1981.)

release of ~10% of the Kr and Xe fission products as well as some $^3$H and volatile fission products. These off-gases are combined with those from the dissolution step.

In the dissolution step, the fuel pieces are dissolved in near boiling 10 M HNO$_3$. This step, which takes a few hours, dissolves the uranium, plutonium, and fission products, leaving the cladding to be recovered. The Kr and Xe are recovered from the off-gas of steam, air, and NO$_x$. The chemical reactions for the dissolution of uranium involve processes like

$$3UO_2 + 8HNO_3 \rightarrow 3UO_2(NO_3)_2 + 2NO + 4H_2O$$

and

$$UO_2 + 4HNO_3 \rightarrow UO_2(NO_3)_2 + 2NO_2 + 4H_2O$$

The plutonium is oxidized to Pu(IV) and Pu(VI), while the neptunium ends up in the pentavalent or hexavalent states. Small amounts of plutonium and fission products may not dissolve, and they can be leached with acid solutions containing the oxidant $Ce^{4+}$.

The off-gas treatment involves primarily iodine, krypton, and xenon. There are a variety of processes for capturing the iodine and disposing of it. Kr and Xe are captured by either cryogenic techniques or selective absorption, such as absorption in chlorofluoromethane. Most of the off-gas volume is due to Xe ($\sim$800 L/Mg fuel) with the activity being mostly 10.7-y $^{85}$Kr ($\sim$11,000 Ci/Mg fuel).

The dissolver solution is treated with chemicals to adjust the acidity, valence, and concentrations of the species involved. The $HNO_3$ concentrations are $\sim$2–3 M, the $UO_2(NO_3)_2$ concentrations are $\sim$1–2 M, and the Pu is stabilized as Pu(IV) using $N_2O_4$ or hydroxylamine. In these and subsequent manipulations of these solutions, attention must be given to criticality control. This is done by regulating the solution geometry, the concentrations of fissile materials, and by the addition of neutron absorbers such as Gd.

The primary separation of plutonium and uranium from the fission products involves a solvent extraction with 30 vol % TBP at room temperature. The activity levels in this separation are quite high ($\sim$1700 Ci/L for the fission products) and the aqueous waste, which contains 99$^+$% of the fission products, is a high-level waste. Am and Cm are not extracted and Np is partially extracted. Because of the high radiation levels, there are radiolysis problems with TPB, leading to solvent degradation. Primary products of the radiolysis of TBP are the dibutyl- and monobutylphosphoric acids along with phosphoric acid. These degradation products are removed in the solvent purification steps.

Following decontamination of the uranium/plutonium from the fission products, the plutonium is separated from the uranium. This is done by reducing the Pu(IV) to nonextractable Pu(III), leaving uranium in the hexavalent state. In the older Purex plants, this was done using $Fe^{2+}$ while the newer plants add $U^{4+}$. The plutonium thus ends up in an aqueous phase while the uranium remains in the organic phase.

Uranium is back-extracted (and thus removed from the organic phase) with 0.01 M $HNO_3$. It is purified by a series of solvent extraction cycles until the Pu/U ratio is $<10^{-8}$ and the total $\beta\gamma$ activity is less than twice that of aged natural uranium.

## 16.6 RADIOACTIVE WASTE DISPOSAL

Radioactive waste management began with the advent of nuclear energy and has been studied since then, with the expenditure of billions of dollars. Despite this Herculean effort, great uncertainty remains about when and how, many aspects of waste disposal, especially high-level waste, will be understood and dealt with effectively.

### 16.6.1 Classification of Radioactive Waste

The simplest way to classify radioactive waste is by its physical state, that is, whether it is a gas, liquid, or solid. Gaseous waste arises from gas evolution, during nuclear fuel reprocessing and by activation of air during reactor operation. The principal gaseous activation product is 1.8-h $^{41}$Ar, which is usually dispersed into the atmosphere from a stack whose height ensures safe ground-level concentrations of the released gas.

The off-gases from fuel reprocessing are the largest contributors to the gaseous waste. The fission products krypton and xenon escape when the fuel elements dissolve. Molecular iodine and ruthenium tetraoxide can also be released. Iodine and ruthenium are removed from the waste stream by trapping. Radio xenon has mostly decayed after a cooling time of about one year, while $^{85}$Kr is trapped cryogenically. While small quantities of $^{14}$C are formed in reactors, the release of this $^{14}$C as $CO_2$ is an important component ($\sim \frac{1}{2}$) of the public dose due to the fuel cycle.

The most important liquid wastes are the high-level effluents, containing fission products from fuel reprocessing. They contain >99% of the fission products in the fuel with small quantities of U and Pu. Medium-level liquid waste has an activity of $\sim 4$ GBq/L and results from various steps in fuel reprocessing. Low-level (<0.1 GBq/m$^3$) waste is treated or concentrated. Liquid organic waste is usually incinerated or chemically destroyed.

Solid waste comes from the mining and milling of uranium ore and the sludge from spent fuel storage. It also includes contaminated equipment and structures. High-level solid waste includes the hulls from the dissolving of spent fuel, ion exchange resin, and the like.

Radioactive waste may also be classified as to origin (defense or commercial waste), the material present (transuranium waste, spent fuel), or the level of radioactivity present (high, medium, low). The principal categories of waste using this classification scheme are: *high-level waste* (HLW) resulting from spent fuel reprocessing and consisting mainly of fission products and a small portion (<0.5%) of the original U and Pu; *spent nuclear fuel* (SNF), which is irradiated fuel that has not been reprocessed; *transuranic waste* (TRU), which is α-emitting waste with >100 nCi/g of the transuranium nuclei; *low-level waste* (LLW), which is waste with small amounts of radioactivity (non-TRU) arising from a variety of sources; and *mill tailings*, a special type of LLW that contains α-emitting radionuclides resulting from uranium mills. This finely ground sandy material contains U daughters. Waste may contain *hazardous* waste (i.e., carcinogens, flammable materials, etc.) as well as *radioactive* waste. Such waste is referred to as *mixed* waste.

### 16.6.2 Amounts and Associated Hazards

The volume and activity of this waste, as of 1996, is summarized in Table 16.1.

The largest volume, but lowest activity, is the mill tailings, which have a specific activity of $\sim 1.8$ nCi/g, mostly due to the $^{238}$U decay series daughters ($^{230}$Th, $^{226}$Ra, $^{222}$Rn, $^{210}$Pb, $^{210}$Bi, $^{210}$Po, etc.). Of special concern is gaseous $^{222}$Rn. Most of the activity associated with radioactive waste is in the unprocessed spent fuel. The military high-level waste tends to have larger volumes than the commercial HLW because the latter has been compacted. Most of the commercial spent fuel is

**TABLE 16.1 1996 U.S. Radioactive Waste Inventory**

| Type/Origin | Volume ($10^3 m^3$) | Activity (MCi) |
|---|---|---|
| Military | | |
| HLW | 347 | 853 |
| LLW | 3,474 | 16 |
| SNF (MTHM)[a] | 2,483 | 9,980 |
| TRU | 238 | 1.3 |
| Mill tailings | 28,000 | 0.08 |
| Commercial | | |
| HLW | 2 | 26 |
| LLW | 1,782 | 8 |
| SNF (MTIHM)[a] | 34,252 | 12,537 |
| Mill tailings | 11,870 | 0.03 |

[a] By convention, the quantity of spent fuel is reported in mass units (MTHM, or MTIHM, metric tons of heavy metal or metric tons of initial heavy metal).
*Source*: Integrated Data Base Report-1996, DOE/RW-0006, Rev. 13, U.S. Dept. of Energy, Washington, DC, 1997.

stored at the reactor sites while most of the military/USDOE waste is stored at a few major production facilities.

One measure of the hazard associated with this waste is the *water dilution volume* ($m^3$). The water dilution volume is the volume of water needed to dilute a radionuclide to its maximum permissible concentration in water. A plot of the water dilution volume (WDV) for spent fuel is shown in Figure 16.11.

Though the activity of the spent fuel falls by more than a factor of 1000 in the first thousand years (Fig. 16.3), the WDV falls more slowly. This is due to the hazards posed by the long-lived α emitters in the spent fuel. For times greater than 500 years, the actinide radio toxicity prevails. On a time scale of $10^4$–$10^6$ years, the WDV values approach those of the original ore used to make the reactor fuel. (There are limitations to this measure of hazard because of how radionuclides enter the biosphere and are concentrated.)

### 16.6.3 Storage and Disposal of Nuclear Waste

What does one do with the radioactive waste described in the previous section? Clearly, the most important component of the waste is the spent fuel. Currently, most spent fuel assemblies are held in cooling ponds at the reactor sites, although one cannot do this indefinitely. In a few reactor sites, dry storage of the spent fuel is used. The fuel rods are transferred to special casks when the heat output and activity are such that air cooling will suffice.

Because a permanent disposal strategy has proven illusive due to technical and political considerations, plans have been made for interim storage facilities where

**486** NUCLEAR REACTOR CHEMISTRY

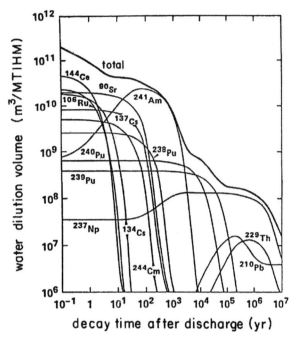

**Figure 16.11** Water dilution volume for radionuclides in PWR spent fuel. (From National Research Council, *A Study of the Isolation System for Geologic Disposal of Radioactive Waste*, NAS, Washington, 1983.)

the fuel is stored in a retrievable manner until a permanent storage facility is developed.

The favored method for permanent storage of radioactive waste is deep geologic repositories. This option is the only option for unprocessed spent fuel assemblies and for most HLW. (An alternative, supplemental strategy discussed below is to remove some of the actinides in the HLW by chemical separations prior to geologic storage.)

In general terms, the goal of long-term waste storage is to isolate the radioactive waste from humans and the environment. The prevailing design strategy for waste repositories is that of multiple barriers (Fig. 16.12).

The first barrier is the form of the waste, which will immobilize the radioactive materials. The waste form should not be damaged by heat or radiation nor be attacked by groundwater. The waste is placed in a steel canister, which is resistant to leaching. The canister is surrounded by packing materials that prevent radioactivity from escaping, and the entire repository is backfilled with a material that absorbs or resists chemical intrusion. The final barrier is the host medium that separates the repository from the surrounding area.

The host medium can be bedded salt, salt domes, granite, basalt, or volcanic tuff. Each medium has advantages and disadvantages with regard to resistance to water intrusion, site availability, and political considerations. Each medium can work and the most important factor is how the local site is designed. An additional factor is the

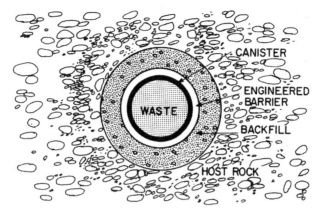

**Figure 16.12** Schematic representation of the multiple barrier waste disposal strategy. (From Murray, 2003.)

position of the repository relative to the groundwater table, with most repositories being below the water table. The United States has chosen the Yucca Mountain region in Nevada, near the Nevada nuclear weapons test site, as its location for permanent geologic storage. The Yucca Mountain area features welded volcanic tuff as the host medium. Welded volcanic tuff is a material of low permeability, and the Yucca Mountain site is above the water table.

The biggest concern with respect to radionuclides in a waste repository is their movement in the groundwater. Attention is focused on the horizontal motion of the groundwater not the upward motion, as the repositories are several hundred meters below the surface. The repositories and their boundaries and locations are such that the biosphere is 10–100 km away from the center of the repository. The velocity of the groundwater in typical geologic media proposed for repositories ranges from 0.01 to 10 m/y. The velocity of radionuclides in groundwater is smaller than the velocity of the groundwater due to sorption phenomena. This sorption can be expressed by a retardation factor $R$ where $R$ is the ratio of the groundwater velocity to the average radionuclide velocity. $R$ can be expressed by an approximation as

$$R \approx 1 + 10 K_d$$

where $K_d$ is the ratio of the radionuclide concentration in the rock of the repository to that in the groundwater. Retardation factors for radionuclides in geologic media of interest as repository sites range from 1 to 3000 and depend strongly on the ion being sorbed (Saling and Fentimann, 2001). For $^{239}$Pu, $R = 200$ for volcanic tuff. The Yucca Mountain groundwater velocities are $\sim 0.025$ m/y, meaning that $^{239}$Pu will decay before it migrates 6 km. A similar conclusion can be reached for the transplutonium nuclei and $^{90}$Sr. From this standpoint, special concern is given to $^{99}$Tc and $^{237}$Np, whose long half-lives can allow geologic transport.

***Spent Fuel*** The largest single radioactive waste disposal problem is the spent fuel from military and commercial reactors. As discussed earlier, the spent fuel from commercial reactors is stored in water ponds at the reactor sites. The spent fuel storage facility consists of a cooling and cleanup system for the water along with equipment to safely transfer the fuel rods from the reactor to the storage area. A typical pool will have a volume of ~400,000 gal. The water will contain ~2000 ppm boron that acts as a neutron absorber and will be maintained at a temperature of <70°C.

The long-term fate of the spent fuel is geologic storage. The spent fuel assemblies are packaged in canisters with a stabilizing material (powder or sand) in the canister. The fuel canisters are placed in the geologic storage site and covered by a backfill to impede water movement.

***High-Level Waste*** The high-level waste (HLW) consists primarily of liquid waste from fuel reprocessing. It contains all the fission products from the spent fuel along with all the neptunium and transplutonium nuclei and less than 1% of the uranium and plutonium. HLW is intensely radioactive with a high heat output (Fig. 16.8). The hazard potential of this waste remains high for much longer than the time scale of nuclear energy use. In addition to its radiological hazard, it is very corrosive, being up to 7 M in $HNO_3$ and containing ~250 g/L salt. This waste is ultimately to be stored in geologic repositories after solidification of the liquid waste. The volume of the solidified waste is modest in that a commercial nuclear power reactor plant running for 1 GW-year will produce about 2 $m^3$ of solidified waste after reprocessing.

The liquid waste is stored for at least 6 y prior to solidification to reduce the decay heat (Fig. 16.8) by a factor of 10 or more. The first U.S. military fuel reprocessing wastes were stored as neutralized waste in mild steel tanks at the Hanford reservation in eastern Washington. These steel-lined, reinforced-concrete tanks were 500,000–1,000,000 gal in capacity with provisions for removal of waste heat and radiolysis products. Corrosion of several tanks occurred with the release of waste. Fortunately, the soil around these tanks retarded nuclide transport. A better (and more expensive) design for storage tanks was implemented at the Savannah River site in South Carolina consisting of a second steel tank inside of a Hanford-style tank. The storage of acid waste in these tanks has not encountered the corrosion problems seen with the Hanford tanks.

The solidification of waste takes place in two steps, a calcining step and an incorporation of the calcined material into borosilicate glass. Calcining can be done in various ways but primarily involves the removal of volatile products like water or $NO_3^-$ and conversion of all species in the HLW to solid stable oxides. The oxides are then mixed with $SiO_2$, $B_2O_3$, and the like to make a borosilicate glass, which is then prepared for geologic storage (Fig. 16.13).

***Transuranic Waste*** Transuranic waste (TRU) results from fuel reprocessing and fuel fabrication facilities, the production of nuclear weapons, and the decommissioning of nuclear reactors or fuel cycle facilities. TRU includes clothing,

**Figure 16.13** Schematic diagram of the final steps in putting vitrified waste into a geologic repository. (Figure also appears in color figure section.)

equipment, and the like from reprocessing facilities along with plutonium and other transplutonium elements removed in fuel reprocessing. Because of the relatively low activity levels of this waste, the primary processing steps prior to storage involve volume reduction. In the United States, this waste is stored in an interim storage facility, the Waste Isolation Pilot Plant (WIPP), an underground salt-bed facility near Carlsbad, New Mexico.

***Low-Level Waste*** Low-level waste (LLW) consists of contaminated dry trash, paper, plastics, protective clothing, organic liquids such as liquid scintillation samples, and the like. LLW is produced by any facility that handles radioactive materials such as nuclear power plants, medical facilities, colleges, and so forth. In the United States, commercial LLW is sent to one of three disposal sites (Barnwell, South Carolina, Richland, Washington, and Clive, Utah). Due to the limited size of these sites (and similar disposal sites through the world) and steeply escalating costs for waste disposal, the primary goal of LLW treatment prior to disposal is volume reduction, either by incineration or compaction, followed by immobilization. For noncombustible solids, volume reduction can be achieved by mechanical disassembly, crushing, melting, or dissolution. For noncombustible liquids, evaporation, calcination, filtration, or concentration on ion exchange resins are used for volume reduction. Combustible material is oxidized to ashes. Mechanical techniques of volume reduction reduce volume by 5–10 times, while combustion reduces volume by 50–100 times. The products of volume reduction are immobilized using absorbents (vermiculite, clay, etc.), cement, or salt matrices.

**Mill Tailings**  The tailings from uranium mining and milling contain all the daughters of uranium present in the original ore. The mill tailings thus have $\sim$70% of the original activity of the ore. This activity decays with the half-life of $^{230}$Th, $\sim 8 \times 10^4$ y. The radiological hazard is mostly in the $^{226}$Ra and its daughter, $^{222}$Rn. The inert gas $^{222}$Rn can escape from the tailings and can create a radiological inhalation risk. Other radionuclides, such as $^{230}$Th or $^{226}$Ra, can be leached from the tailings by water, being transferred to the biosphere. The tailings themselves are finely divided sandy material that can be dispersed by wind. Remediation of sites containing mill tailings involves covering the tailings by a clay or earth overburden that is 3–8 m thick. This overburden reduces Rn release, minimizes leaching, and prevents wind dispersal.

**Partitioning of Waste**  As discussed earlier, for a variety of reasons, the nuclear fuel cycle is operated in a "once-through" mode in the United States and for the majority of reactors worldwide. The long-lived transuranium nuclides are thus mixed with the shorter-lived fission products in the HLW and SNF. As concern about the lack of long-term, widely accepted programs to deal with this waste mounts, attention is being focused on *partitioning* (chemically separating) the transuranium nuclei from the waste and *transmuting* it to nuclides of shorter half-life by irradiation in high neutron fluxes. These waste destruction steps also can be incorporated into advanced design reactor systems or advanced fuel cycle designs. Other targets of partitioning are troublesome fission products such as $^{99}$Tc, $^{129}$I, $^{90}$Sr, and $^{134,137}$Cs.

At the outset, one should comment that there is a division (Benedict et al., 1981; Bodansky, 1996) among scientists as to whether the gains in reduced radiotoxicity of the waste offset the additional cost and risk of further treatment and handling of the waste. Removal of the actinides from a waste repository gives a small reduction in risk because the actinides migrate so slowly through the repository. Several additional problems would be created by partitioning and transmutation such as: a greater volume of wastes, the need for large, expensive processing facilities, an increase in the neutron flux coming from the waste, necessitating increased shielding, and the possibility of additional releases of radioactivity to the environment. A U.S. study has suggested 99.9% removal of the actinides and 95% removal of Tc and I is a minimum requirement to justify the effort.

Nonetheless, the United States and other nations have been developing additional chemical separation processes to bring about these goals. Strontium and cesium can be removed from HLW by extraction with crown ethers, such as ditertiarybutyl-dicyclohexanone-18-crown-6, which can also extract $^{99}$Tc. In the United States,

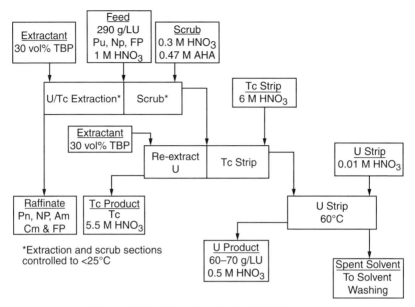

**Figure 16.14** Flowsheet for the UREX process.

a variation of the Purex process is being developed to reprocess spent fuel. This modification of the Purex process is such that only U and Tc are extracted from the fission products and TRU nuclei (Np, Pu, Am, Cm). It is called the UREX process (uranium extraction). It is intended to recover >99.9% of the U, >95% of the Tc and leave >99.9% of the TRU nuclei in the acid waste. All chemicals used in the process are converted to gases in subsequent processing to minimize waste. A Purex separation is carried out using 30% TBP in kerosene working on a 1 M $HNO_3$ solution of spent fuel and then acetohydroxamic acid (AHA), an analog of hydroxylamine, is used to reduce Np(VI) to nonextractible Np(V) and to complex Pu(IV) and Np(IV), preventing their extraction. Figures 16.14 and 16.15 show the flowsheet for the UREX process and its place in the total, new proposed treatment of spent fuel. Other work on pyrochemical processes involving electrochemical reduction steps in molten salt baths is also underway.

***Transmutation*** Transmutation is the term used in connection with waste management to describe a nuclear process in which long-lived nuclides are changed into shorter-lived nuclides by nuclear reactions. Possible sources of bombarding particles are neutrons from reactors or accelerators that have been configured to produce large neutron fluxes, by reactions, such as spallation. Fast neutrons are preferred as the bombarding particle because some of the relevant isotopes of the actinides Np, Am, and Cm are not thermally fissionable. $^{129}$I and $^{99}$Tc, on the other hand, are readily destroyed in thermal fluxes. Transmutation can also be used to destroy plutonium from dismantled nuclear weapons. Of course, the

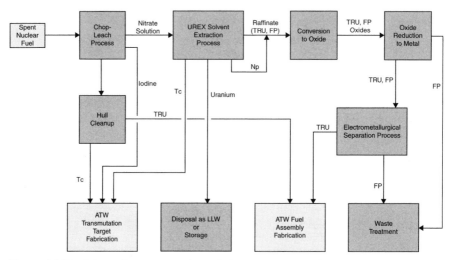

**Figure 16.15** Schematic representation of the new schemes for reprocessing of SNF. (Figure also appears in color figure section.)

resulting fission or reaction products from transmutation are radioactive and do constitute radioactive waste, albeit with shorter half-lives.

Current attention is centered on methods using charged particle accelerators for transmutation. In a typical scheme, an accelerator is used to produce 10- to 200-mA beams of 1-GeV protons. The proton beam strikes a liquid lead or lead–bismuth eutectic target giving rise to 30–40 fast neutrons/proton, via spallation. These fast (predominantly 1- to 10-MeV neutrons) are slowed down in a graphite moderator surrounding the Pb target. Some designs produce fast (75–225 keV) neutrons in the moderator while other designs thermalize the neutrons. The actinides and fission products to be transmuted are dissolved in molten salts or other media allowing high heat transfer in channels passing through the moderator. The neutron flux in the channels is $\sim 10^{15}-10^{16}$ n/cm$^2$-s.

Typical heat production in the moderator-fuel blanket is $\sim$750–1500 MW. The excess heat is used to generate electricity that helps to pay for the operation of the facility. The transmuted material will have $\sim$20% of the original plutonium and minor actinides of the input material and will contain significant fission product activities. This transmuted material can be put into geologic storage, reducing the long-term hazard of the repository material. The overall feasibility of this accelerator transmutation of waste (ATW) has not been established yet.

## 16.7 CHEMISTRY OF OPERATING REACTORS

A complex set of chemical processes occurs during the operation of a nuclear reactor. Up to now, we have concentrated our attention on the chemical, physical, and nuclear processes occurring in the fuel and their consequences. Now we turn our attention to the coolant, the moderator, and the reactor materials and the changes in them due to the hostile chemical environment. The reactor environment has coolant temperatures up to 350°C, pressures of ~75–150 atm, and intense neutron and $\gamma$ radiation. Further complications are introduced by the two phases present in BWRs, and the use of chemical shims like B to control the reactivity of PWRs. Both reactor types exhibit potentially serious possibilities of corrosion. (One PWR in Ohio was closed for several years to repair the steel pressure vessel that was corroded by boric acid.)

### 16.7.1 Radiation Chemistry of Coolants

About 2% of the total neutron and $\gamma$-ray energy released in a nuclear reactor is deposited in the cooling water which leads to decomposition of the water itself. The situation is of special concern in BWRs where the neutron and $\gamma$-ray fluxes near the core are $\sim 10^9 \, R/h$ of neutrons and $10^8 \, R/h$ of photons. (There is a hydrogen gas overpressure in PWRs suppressing to some extent water decomposition or oxygen gas production.) Due to radiolysis of the coolant water, this energy deposit produces an oxidizing environment with 100–300 ppb oxygen and hydrogen peroxide, with lesser concentrations in the vapor phase. The primary process is

$$H_2O \longrightarrow H^\bullet + OH^\bullet$$

These radicals react to form $H_2$ or $H_2O_2$ by

$$H^\bullet + H^\bullet \longrightarrow H_2$$
$$H^\bullet + OH^\bullet \longrightarrow H_2O$$
$$OH^\bullet + OH^\bullet \longrightarrow H_2O_2$$

Molecular $O_2$ is generated by

$$H_2O_2 + OH^\bullet \longrightarrow HO_2^\bullet$$
$$HO_2^\bullet + OH^\bullet \longrightarrow H_2O + O_2$$
$$HO_2^\bullet + HO_2^\bullet \longrightarrow H_2O_2 + O_2$$
$$H_2O_2 + HO_2^\bullet \longrightarrow H_2O + {}^\bullet OH + O_2$$

and destroyed by

$$O_2 + H^\bullet \longrightarrow HO_2^\bullet$$

### 16.7.2 Corrosion

The oxygen and radiolysis products attack the outer layers of the stainless steel or nickel-based alloys used in the reactor structure, forming a thin oxide layer on these components. Corrosion products are released from this thin oxide layer by the cooling water and become activated as they pass near the reactor core. These products can deposit on fuel surfaces, in coolant channels, and on reactor materials. These deposits are referred to as *crud* (Chalk River unidentified deposits). In PWRs the chemical composition of the crud is $Ni_xFe_yO_4$ where $x/y = 0.25$ and $x+y = 3$. In PWRs, the crud that also contains boron compounds causes a phenomenon known as *axial offset anomaly* (AOA), which is an unexpected deviation from predictions of the core axial power distribution during operation. Because it is not well understood, AOA limits the operating power of several PWRs, at great cost to the utilities.

Attempts are made to minimize corrosion by controlling the chemistry of the coolant water. Adding $^7$LiOH raises the pH to 8. One can use oxygen scavengers such as hydrazine to reduce the oxygen concentration.

### 16.7.3 Coolant Activities

When corrosion products are deposited on the fuel surfaces, they are activated by neutron capture. Some of the most prominent of these activities are $^{55}$Fe, $^{63}$Ni, $^{60}$Co, $^{54}$Mn, $^{58}$Co, and $^{59}$Fe. These radionuclides will then be found in the reactor coolant.

Fission products can be released from defects in the fuel rods or from tramp uranium on the fuel cladding. Of special importance are the volatile fission products $^{131-135}$I ($^{89}$Kr, $^{137,138}$Xe in BWR steam). Cations include the Sr and Cs isotopes, which are present along with $^{129}$I and $^{99}$Tc. One can use the ratio of short/long-lived isotopes such as the ratio of $^{133}$I/$^{131}$I to determine the source of the fission product release, by assuming the short-lived species can only result from tramp fuel or large cracks in the fuel assembly.

Impurities in the water and water activation products also contribute to the radioactivity of the coolant water. Tritium is produced as a low yield ($\sim 0.01\%$) fission product that can diffuse out of the fuel, by activation of boron or $^6$Li impurities in PWRs. $^{24}$Na and $^{38}$Cl are produced by neutron activation of water impurities. In BWRs, the primary source of radiation fields in the coolant and steam systems during normal operations is 7.1s $^{16}$N. This nuclide is produced by $^{16}$O(n, p)$^{16}$N reactions from fast neutrons interacting with the coolant water. This $^{16}$N activity can exist as $NO_2^-$, $NO_3^-$ in the coolant and $NH_4^+$ in the steam.

### PROBLEMS

1. Define or describe the following terms or concepts: (a) crud, (b) axial offset anomaly, (c) accelerator transmutation of waste, (d) UREX process, (e) LLW, (f) TRU, (g) HLW, (h) SNF, (i) Purex process, (j) back end of the fuel cycle, and (k) pyroprocessing.

2. Define or describe the following terms or concepts: (a) yellowcake, (b) orange oxide, and (c) green salt.

3. Since $UO_2$ can be converted directly to $UF_6$, why is it first converted to $UF_4$ and then to $UF_6$?

4. Write balanced chemical equations for three different methods to produce metallic uranium.

5. Nitric acid readily dissolves $UO_2$. Why doesn't hydrochloric acid?

6. Given a $G$ value of 1.59 for the production of hydrogen gas by the irradiation of TBP with 1-MeV electrons, calculate the rate of hydrogen gas evolution in a liter of TBP irradiated for 1 h at a dose rate of 200 W/L. Assume STP conditions.

7. Draw a flowsheet for the Purex process like Figure 16.10. Estimate the relative volumes of all streams in the process using data from the references cited in the text.

8. Discuss the disposal of the following examples of radioactive waste:

   a. Water solutions containing 1 mCi of $^3H$ from a research lab
   b. Gas escaping from the dissolution of 1 kg of irradiated reactor fuel
   c. Ion exchange resin used to purify the cooling water of a 1-MW research reactor

9. An amount of 500 g of natural uranium is irradiated in a neutron flux of $10^{13}$ n/cm$^2$-s for 1 y. What is the heat output of this material after cooling for 1 week? One month? One year?

10. What is the theoretical maximum separation factor for separation by gaseous diffusion of $^3He$ and $^4He$?

11. If a gaseous diffusion plant produces uranium with a 235/238 ratio of 5, what is the expected 234/235 ratio in the resulting material?

12. Discuss quantitatively the relative merits of using LiOH, NaOH, or KOH for pH control in reactor coolant water.

13. In reactors based on a Th fuel cycle, $^{233}Pa$ is produced. Discuss the radiochemistry of this radionuclide.

## REFERENCES

Benedict, M., T. H. Pigford, and H. W. Levy. *Nuclear Chemical Engineering*, 2nd ed., McGraw-Hill, New York, 1981.

Bodansky, D. *Nuclear Energy—Principles, Practices and Prospects*, AIP. Woodbury, NY, 1996.

Murray, R. L. *Understanding Radioactive Waste*, 5th ed., BMI, Columbus, 2003.

Saling, J. H. and A. W. Fentimann, Eds. *Radioactive Waste Management*, 2nd ed., Taylor & Francis, New York, 2001.

Wymer R. G. and B. L. Vondra. *Light Water Reactor Nuclear Fuel Cycle*, CRC, Boca Raton, FL, 1981.

## BIBLIOGRAPHY

Bowman, C. D. Accelerator-Driven Systems for Nuclear Waste Transmutation, *Ann. Rev. Nucl. Part. Sci.* **48**, 505 (1998).

Choppin, G., J. O. Liljenzin, and J. Rydberg. *Radiochemistry and Nuclear Chemistry*, 3rd ed., Butterworth-Heinemann, Woburn, 2002.

Eisenbud, M. *Environmental Radioactivity*, 3rd ed., Academic, Orlando, 1987.

Glasstone, S. *Sourcebook on Atomic Energy*, 3rd ed., Van Nostrand, New York, 1967.

Lin, C. C. *Radiochemistry in Nuclear Power Reactors, NAS-NS-3119*, National Academy Press, Washington, 1996.

National Research Council, *A Study of the Isolation System for Geologic Disposal of Radioactive Wastes*, National Academy Press, Washington, 1983.

Peterson, S. and R. G. Wymer. *Chemistry in Nuclear Technology*, Addison-Wesley, Reading, MA, 1963.

Roberts, L. E. J. Radioactive Waste Management, *Ann. Rev. Nucl. Part. Sci.* **40**, 79 (1990).

# CHAPTER 17

# INTERACTION OF RADIATION WITH MATTER

## 17.1 INTRODUCTION

At this point we have described nuclear transitions and reactions that produce various forms of nuclear rad\iation. The radiation propagates out from the originating nucleus and interacts with other matter along its path. These interactions with external matter allow us to observe the radiation, and its effects, and to determine the nature of the transition inside the nucleus. The interaction of radiation with matter is also the cause of chemical, physical, and biological changes that concern the public at large. We will specifically address the operating principles of radiation detectors in the next chapter, but first we will consider the fundamental interactions of nuclear radiation with matter.

It should be clear that radiation (with the exception of neutrons) primarily interacts with bound electrons. For example, a silicon atom contains 14 electrons in a sphere with a radius of 0.12 nm that presents a geometrical cross section of $4.5 \times 10^{-20}$ m$^2$. The nucleus at the center of the sphere has a radius of 3.6 fm with a geometrical cross section of $4.1 \times 10^{-29}$ m$^2$. The geometrical probability to strike the electrons in an atom is something like 9 orders of magnitude higher than that to strike the nucleus. A scattered electron leaves the original atom and creates an ion pair. The interaction of a single particle of nuclear radiation can lead to tens or hundreds of thousands of ion pairs, and so nuclear radiation is generally called ionizing radiation.

*Modern Nuclear Chemistry*, by W.D. Loveland, D.J. Morrissey, and G.T. Seaborg
Copyright © 2006 John Wiley & Sons, Inc.

From the starting point that we should consider the interaction of the radiation with electrons, we can divide nuclear radiation into four classes as indicated in Table 17.1. The overall scattering of the particle from an electron will be dominated by the mass and the charge on the particle. Thus, massive charged particles will tend to scatter the small mass electrons widely without losing much energy, while the collision of an electron with another electron will lead to energy sharing, and a photon can be scattered or even absorbed by a single bound electron. Neutrons only interact very weakly with electrons through their small magnetic moment and predominately interact with nuclei. (As will be discussed later, neutrons are very penetrating and difficult to detect due to the small probability of striking a nucleus.) As a general rule, all of the interactions end up in releasing energetic electrons. The heavy charged particles and recoiling atoms scatter electrons. The electrons interact with matter to create moving charged particles, while neutrons create recoiling nuclei and photons create moving electrons.

We will discuss the interaction of each class of radiation starting with the simplest, those charged particles that are more massive than electrons. We will consider what happens as they pass through various types of matter. The radiation generally penetrates through many, many atomic layers, so we can generally assume that the atoms are randomly distributed in space. This assumption is certainly true for liquids and gases, but we usually think of solids as having a regular crystal or lattice structure. The solid material that we encounter in everyday life certainly has a lattice structure on the microscopic scale, but nearly all materials are polycrystalline on a larger scale. Thus, the types of radiation that we will generally consider will cross many crystal boundaries in normal materials. The exceptions are single crystals of silicon, germanium, or other special materials that are used in semiconductor-based radiation detectors that rely on their special electronic properties. We have to be aware of the orientation of the crystal axes in these devices relative to the propagation direction of the radiation.

Also before starting the discussion, we should define a quantity called the areal density. We have said that ionizing radiation will pass through a significant amount of material and sometimes will even pass completely through an object. We can imagine that the solid piece of material can be compressed into a thin, two-dimensional sheet as far as the radiation is concerned. The origins of this concept lie in early experiments with α rays in which very thin sheets of metals were necessary. We can easily measure the length $x$ and breadth $y$ of such thin foils, but it is very difficult to measure the thickness $z$ without destroying the foil. On the other hand, we can determine the mass $m$ of the foil rather easily and then

**TABLE 17.1 Particle Classes for Interaction with Matter**

| Charged Particles | Uncharged Particles |
|---|---|
| Protons, heavy ions | Neutrons |
| Electrons | γ rays |

calculate the thickness using the density of the material. For example, if the foil is made from a pure chemical element, so that the density is known, then:

$$\text{Linear thickness} \equiv z = \frac{\left(\dfrac{m}{x \times y}\right)}{\rho} = \frac{\text{areal density}}{\text{density}}$$

where $\rho$ is the density (and has dimensions of mass per unit volume, of course). The measured quantity of mass per unit area is often used to characterize thin foils of pure materials. This quantity is called the areal density. Typical dimensions for materials that we might use in experiments are $mg/cm^2$. Thus, if the density of a typical metal is in the range of 5–10 $g/cm^3$, the thickness of a typical foil would be on the order of fractions of a millimeter.

**Example Problem** A ream of a certain type of standard letter-sized paper (in the United States) was found to have a mass of 2.26 kg. What is the areal density of one sheet of this paper?

**Solution** Recall that a ream of paper contains 500 sheets, so that the areal density of one sheet is simply:

$A = $ total mass/surface area

$A = 2260\,\text{g}/[(500 \times 8.5\,\text{in.} \times 11\,\text{in.}) \times (2.54\,\text{cm/in.})^2]$

$\quad \times 1000\,\text{mg/g} = 7.5\,\text{mg/cm}^2$

## 17.2 HEAVY CHARGED PARTICLES ($A \geq 1$)

We can imagine the progress of any ionizing radiation through material as a series of straight-line segments between scattering events. The scattering events primarily involve electrons. The total path is thus made up of these line segments, and the overall trajectory of the particle in the material will depend on the kinematics of these scattering events. Elastic scattering is, of course, governed by the conservation of momentum and energy, so we should expect that the mass of the particle will play a large role in determining the overall features of the trajectory of the particle.

The tracks of a few heavy charged particles that have stopped in a photographic film are shown in Figure 17.1. The photographic film, called a nuclear emulsion, is sensitive to the ionization that is caused by the charged particles as they move. Normally, visible photons "expose" photographic film by creating photoelectrons, and the ionization is converted into an image through the development process. All of the heavy charged ions that we have to consider have positive charges. As can be seen in Figure 17.1, energetic charged ions move through material on essentially straight trajectories, giving up or losing kinetic energy through collisions with the atomic electrons of the material. Only rarely by comparison is an ion scattered by the Coulomb potential of a nucleus, and even more rarely does a nuclear reaction take place. Nuclear reactions are excluded when the initial kinetic energy of the

**500** INTERACTION OF RADIATION WITH MATTER

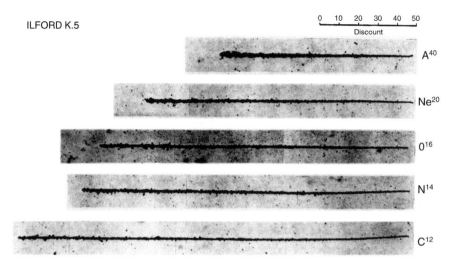

**Figure 17.1** Trails of ionization left by heavy charged particles (initial energy of 10 MeV/nucleon) as they penetrate through a photographic plate (nuclear emulsion). The ions interact with the atomic electrons in the emulsion creating ion pairs to "expose" the emulsion and the tracks become visible after the film is developed. Notice the straight-line tracks. (From Knoll, 2000.)

heavy charged particle is lower than the Coulomb barrier (as discussed in Chapter 10). Thus, the ions interact with an extremely large number of electrons, and we can examine the average behavior of the ions as they pass through material.

The rate at which charged particles lose energy as they travel through a given material is called the *stopping power* of that material. The stopping power is made up from two parts, the electronic stopping power due to the interaction with the atomic electrons of the material and the nuclear stopping power. Thus

$$-\frac{dE}{dx} = S_{\text{electronic}} + S_{\text{nuclear}} \approx S_{\text{electronic}} \quad (17.1)$$

because the electronic stopping power is always much larger than the nuclear stopping power. (We should note that some authors use the term "nuclear stopping" not for nuclear processes but rather for the atomic scattering processes of neutralized cores at the very end of their range.) Notice that the minus sign on the rate to indicate that the ions are losing kinetic energy. The nuclear stopping power is not zero, of course, because we know that nuclear reactions do take place even if they are rare. The stopping powers are functions of the mass, charge, and velocity of the ion, the atomic number, and density of the medium.

Niels Bohr suggested that the energy loss rate could be estimated in a very simple picture as the series of impulses delivered to individual electrons by the ion. Imagine an ion moving on a straight-line trajectory past an electron (see Fig. 17.2). A net impulse to the electron will occur in the direction perpendicular to the trajectory

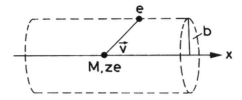

**Figure 17.2** Trajectory of a moving ion past an electron. (From Leo, 1987.)

of the ion because any impulse due to the approaching ion will be cancelled by that of the receding ion. It can be shown that the energy gained by a single electron and thus lost by an ion in one encounter depends on the impact parameter, $b$, as:

$$\Delta E(b) = \frac{2q^2 e^4}{m_e v^2 b^2} \quad (17.2)$$

where $q$ is the charge of the ion (sometimes equal to the atomic number), $v$ is the velocity of the ion, and $m_e$ is the mass of the electron. This expression can be converted to a differential expression by multiplying by the electron number density, $N_e$, times the volume element:

$$-dE(b) = \Delta E(b) N_e \, dV = \Delta E(b) N_e 2\pi \, db \, dx \quad (17.3)$$

by using the cylindrical coordinates of impact parameter and taking $x$ along the ion's path. This expression should not be integrated from $b = 0$ to $b = \infty$ but only over the range $b_{\min}$ to $b_{\max}$ that are appropriate to the initial assumptions with regard to the ion and the electron, so that:

$$-\frac{dE}{dx} = \frac{4\pi q^2 e^4}{m_e v^2} N_e \ln \frac{b_{\max}}{b_{\min}} \quad (17.4)$$

The minimum impact parameter will correspond to those collisions in which the maximum amount of kinetic energy is transferred to the electron. Due to conservation of momentum, the maximum electron energy is $W_{\max} = (\tfrac{1}{2}) m_e (2\gamma v)^2$ where we have included the relativistic factor $\gamma$ due to the low mass of the electron. Recall that

$$\gamma = \sqrt{\frac{1}{1-\beta^2}} \quad \text{and} \quad \beta = \frac{v}{c}$$

Thus, substituting into the expression for the energy loss at a given impact parameter:

$$\Delta E(b_{\min}) = \frac{2q^2 e^4}{m_e v^2 b_{\min}^2} = 2\gamma^2 m_e v^2 \quad (17.5)$$

we find that

$$b_{min} = \frac{qe^2}{\gamma m_e v^2} \quad (17.6)$$

The maximum impact parameter has to be estimated from different considerations. The basis of this process is that the ion rapidly moves past the electron and delivers a sharp impulse to the electron. The electrons are bound in atoms and thus are orbiting with their own characteristic frequencies or time scales. Thus, the time for the ion to cross the atom should be less than the average time for an electron orbit; otherwise the collision will not be adiabatic or "rapid." The time for the ion to move past can be estimated as the ratio of the impact parameter to the ion's velocity, the average orbital time for an electron will clearly depend on the chemical element, as there will be an average radius and velocity, thus

$$\frac{b_{max}}{\gamma v} \leq \frac{R_e}{v_e} = f(Z) \quad (17.7)$$

where $f(Z)$ is a function of the atomic number of the stopping material. So that we can combine these two limits into the expression for the stopping power or energy loss rate to get Bohr's classical formula:

$$\left(-\frac{dE}{dx}\right)_{Bohr} = \frac{4\pi q^2 e^4}{m_e v^2} N_e \ln \frac{\gamma^2 m_e v^3 f(Z)}{qe^2} \quad (17.8)$$

This expression has been superseded by the expression derived by Bethe and Bloch based on momentum transfer in a quantum mechanically correct formalism. Their expression with the expanded form of the electron number density is

$$\left(-\frac{dE}{dx}\right)_{Bethe-Bloch} = 4\pi N_A r_e^2 m_e c^2 \rho \frac{Zq^2}{A\beta^2} \left[\ln\left(\frac{W_{max}}{I}\right) - \beta^2\right] \quad (17.9)$$

where $N_A$ is Avogadro's number, $r_e$ is the classical radius of the electron, $\rho$ is the density of the stopping medium with atomic number, $Z$, mass number, $A$, and ionization potential, $I$. Finally, $W_{max}$ is the maximum energy transfer, encountered above. The structure is very similar to the classical formula, as should be expected, but it includes an extra term in the logarithm. Various formulas are available to give the average variation of the ionization potential for the chemical elements. For example, the expressions:

$$\begin{aligned} I/Z &= (12 + 7Z^{-1})\,\text{eV} & Z < 13 \\ &= (9.76 + 58.8Z^{-1.19})\,\text{eV} & Z \geq 13 \end{aligned} \quad (17.10)$$

are the results from one empirical fitting of the data, but one should realize that the variation could be quite complicated due to the filling of the atomic shells.

**Example Problem** Evaluate the stopping power of beryllium metal for $^{18}O^{8+}$ ions with a kinetic energy of 540 MeV ($E/A = 30$ MeV) using the Bethe–Bloch formula.

**Solution** Finding some necessary constants, the density of beryllium metal is 1.85 g/cm$^3$ and $Z = 4$ so that the ionization potential can be estimated as:

$$I/Z = (12 + 7Z^1)\,\text{eV} = 13.75\,\text{eV}$$
$$I = 55\,\text{eV}$$

The values of $\beta$ and $\gamma$ for the ion can be obtained from the relativistic expressions (derived elsewhere):

$$\beta = [1 - (m_0^2/(m_0c^2 + (E/A))]^{1/2}$$
$$\beta = [1 - (931.5/(931.5 + 30)]^{1/2} = 0.1766$$
$$\gamma = [1/(1-\beta^2)^{1/2}] = 1.01598$$

The value of $W_{max}$ is can be evaluated as:

$$W_{max} = 2m_ec^2(\gamma\beta)^2$$
$$= 2 \times 0.511\,\text{MeV}(1.01598 \times 0.1766)^2 = 0.03291\,\text{MeV}$$

Finally, the expression with the constants evaluated is

$$\left(-\frac{dE}{dx}\right)_{\text{Bethe-Bloch}} = 0.3071\,\frac{\text{MeV}\cdot\text{cm}^2}{\text{g}}\,\rho\,\frac{Zq^2}{A\beta^2}\left[\ln\left(\frac{W_{max}}{I}\right) - \beta^2\right]$$

$$\left(-\frac{dE}{dx}\right)_{\text{Bethe-Bloch}} = 0.3071\,\frac{\text{MeV}\cdot\text{cm}^2}{\text{g}}\,(1.85\,\text{g/cm}^3)$$

$$\times \frac{8^2}{9(0.1766)^2}\left[\ln\left(\frac{0.03291\,\text{MeV}}{55 \times 10^{-6}\,\text{MeV}}\right) - 0.1766^2\right]$$

$$= 518.0\,\text{MeV} \times [\ln(598.4) - 0.03119] = 3.3\,\text{GeV/cm}$$

Notice that the answer indicates that an ion with only 540 MeV of kinetic energy will lose all its energy and stop before it travels a fraction of a centimeter. As discussed later, we need to integrate this expression to determine the predicted range.

The modern form of the stopping power includes two corrections. The first correction applies at high energies at which polarization of electrons by the electric field of the moving ion tends to shield distant electrons; this correction depends on the electron density; it is subtractive and given the symbol $\delta$. The second correction applies at low energies when the collisions are no longer adiabatic, similar to the limit applied by Bohr. This correction is termed the shell correction as it depends on the orbital velocities of the electrons. It is also a subtractive term and given the symbol C. If we evaluate all the constants, then the modern form is

$$\left(-\frac{dE}{dx}\right)_{\text{Bethe-Bloch}} = 0.3071 \frac{\text{MeV} \cdot \text{cm}^2}{\text{g}} \rho \frac{Zq^2}{A\beta^2} \left[\ln\left(\frac{W_{\max}}{I}\right) - \beta^2 - \frac{\delta}{2} - \frac{C}{Z}\right]$$

(17.11)

which has the dimensions of MeV/cm when the usual form of the density in g/cm³ is used. The actual evaluation of this function is complicated due to the detailed variation of the ionization potential and the two correction terms. The reader is referred to more detailed discussions (Leo, 1987) for actual formulas for the correction factors. In practice, several computer codes and detailed tables of the stopping powers are available. In addition, some authors divide through by the density, $\rho$, and report the mass stopping power:

$$\frac{-1}{\rho}\frac{dE}{dx}$$

with dimensions of MeV-cm²/g, which is convenient for combining materials.

If we look at the form of these equations for the stopping power, we would see that they all have a part that depends on the moving ion and another part that depends on the stopping medium. If we concentrate on the part that depends on the ion, we find that

$$-\frac{dE}{dx} \propto \frac{q^2}{v^2} \ln(\gamma^2 v^2) g(Z)$$

(17.12)

in which we can convert the factors of $v^2$ into kinetic energy, $E$, by suitably applying factors of $\frac{1}{2}m_{\text{ion}}$. The function $g(Z)$ collects all the variation on the absorbing medium. The revised expression shows that the energy loss rate will be proportional to the mass of the ion:

$$-\frac{dE}{dx} \propto \frac{Aq^2}{2E} \ln\left(\frac{\gamma^2 2E}{A}\right) g(Z)$$

(17.13)

and inversely proportional to the kinetic energy. At low ion velocities $(E/A < 10 \text{ MeV}/A)$, the $\ln(\gamma^2 2E/A)$ term is approximately constant and

$$-\frac{dE}{dx} \propto \frac{Aq^2}{E}$$

(17.14)

Thus, a more energetic ion will tend to lose energy at a lower rate than a less energetic ion. Be careful to note that we have ignored the relativistic terms, $\gamma^2$ and $\beta^2$, in the parentheses, which produce a minimum in the complete function near $\beta \sim 0.96$ and a small rise at higher velocities. (Particles with $\beta \sim 0.96$ are called *minimum ionizing particles*.) The proportionality of the stopping power on the mass and square of the charge of the ion for a given kinetic energy provides the basis for a very effective particle identification technique using thin silicon semiconductor detectors, as discussed in Chapter 18.

We are now in a position to examine the slowing down of a charged particle as it penetrates into material. Kinetic energy is lost through scattering electrons away from the essentially straight-line path of the ion. If the initial kinetic energy of the ion is a few MeV/A or higher, the rate at which kinetic energy is dissipated slowly increases as the ion penetrates into the material. For example, the stopping power of beryllium metal for a very energetic $^{40}$Ar ion is shown in Figure 17.3 along with the residual energy of the ion. (The thickness scale can be converted into a linear distance by dividing by the density in appropriate units.) Notice that the stopping power is relatively constant over most of the ion's path. The kinetic energy of the ion uniformly decreases as it moves through the material. However, two changes occur as the velocity of the ion approaches the Bohr velocity of the atomic electrons,

$$v_{\text{Bohr}} = \frac{Zh}{m_e n a_0} = 5.51 \times 10^{-3} \frac{Zc}{n}$$

The energy loss rate begins to increase dramatically as $\beta \to 0$, but, more importantly, the charge state on the ion starts to decrease as the ion captures orbital electrons causing the rate to drop. As indicated in Figure 17.3, the ion rapidly loses energy at the end of its range and stops rather suddenly. The energy loss for an $\alpha$ particle near the end of its range is shown in Figure 17.4. The resulting

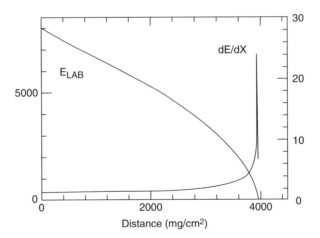

**Figure 17.3** Energy loss rate as a function of thickness for a $^{40}$Ar projectile in beryllium metal is shown on the scale to the right for an ion that enters the foil at the very high energy of 8 GeV. The remaining kinetic energy of the ion is shown on the left scale. Note that the ion penetrates approximately 21 mm into the metal.

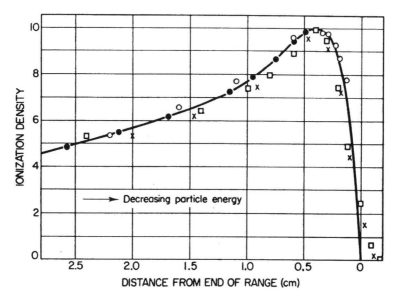

**Figure 17.4** Density of ionization along the path of an α particle stopping in air is shown. The Bragg peak in the ionization density is evident. [From M. G. Hollaway and M. S. Livingston, *Phys. Rev.* **54**, 18 (1938).]

peak in the energy loss function just before the end of the charged particle's range is called the *Bragg peak*. The fact that charged particles deliver a significant fraction of their kinetic energy at the end of their range makes charged particles useful for radiation therapy.

All of these expressions for the stopping power only apply to pure chemical elements. The stopping power of a compound or any complicated mixture will depend on the overall density and the relative numbers of electrons from each chemical element. Recognizing that the ionizing radiation *will* usually move through macroscopic distances, we can use an averaging procedure called *Bragg's rule*. The average mass stopping power is

$$\left(\frac{1}{\rho}\frac{dE}{dx}\right)_{total} = \frac{w_1}{\rho_1}\left(\frac{dE}{dx}\right)_1 + \frac{w_2}{\rho_2}\left(\frac{dE}{dx}\right)_2 + \frac{w_3}{\rho_3}\left(\frac{dE}{dx}\right)_3 + \cdots \quad (17.15)$$

where $w_1$, $\rho_1$, and so forth refer to the fraction by mass of element 1 in the entire mixture and its elemental density. The sum ranges over all the elements in the mixture. Thus, if the mixture was a pure compound, then we would combine the numbers of each element in the molecular formula. If the mixture had several components, then we would combine the masses of each element from all the components and so on to get an overall mass stopping power.

## 17.2 HEAVY CHARGED PARTICLES ($A \geq 1$)

**Example Problem** What is the rate of energy loss of an 8-MeV $\alpha$ particle in air? Assume air is 21% oxygen and 79% nitrogen:

**Solution**

$$\left(\frac{1}{\rho}\frac{dE}{dx}\right)_{total} = \sum_i w_i \frac{1}{\rho_i}\left(\frac{dE}{dx}\right)_i$$

where $\rho$ is the total density, and $\rho_i$ the density of the $i$th element. For oxygen:

$$\left(\frac{1}{\rho}\frac{-dE}{dx}\right)_{oxygen} = 0.3070\,\frac{\text{MeV}\cdot\text{cm}^2}{\text{g}}\,\frac{Zq^2}{A\beta^2}\left[\ln\left(\frac{w_{max}}{I}\right) - \beta^2\right]$$

neglecting any correction terms

$$I/Z = (12 + 7Z^{-1})\,\text{eV} = 12.875$$

$$I = 103\,\text{eV}$$

$$\beta = \left[1 - \frac{m_0 c^2}{m_0 c^2 + E/A}\right]^{1/2} = 0.04629$$

$$\gamma = [1/(1-\beta^2)]^{1/2} = 1.0011$$

$$w_{max} = 2 \times 0.511(\gamma\beta)^2 = 0.002194\,\text{MeV}$$

$$\left(\frac{1}{\rho}\frac{-dE}{dx}\right)_{oxygen} = 0.3070\,\frac{2^2}{16(0.04629)^2}\left[\ln\left(\frac{0.002194}{103 \times 10^{-6}}\right) - (0.04629)^2\right]$$

$$\left(\frac{-1}{\rho}\frac{dE}{dx}\right)_{oxygen} = 109.4\,\text{MeV/g/cm}^2$$

For nitrogen

$$I/Z = (12 + 7Z^{-1})\,\text{eV} = 13$$

$$I = 91\,\text{eV}$$

$$\left(\frac{-1}{\rho}\frac{dE}{dx}\right)_{nitrogen} = 0.3070\,\frac{2^2}{14(0.04629)^2}\left[\ln\frac{0.002194}{91 \times 10^{-6}} - (0.04629)^2\right]$$

$$= 130.2\,\text{MeV/g/cm}^2$$

For air

$$\left(\frac{-1}{\rho}\frac{dE}{dx}\right)_{air} = 0.21\left(\frac{-1}{\rho}\frac{dE}{dx}\right)_{oxygen} + 0.79\left(\frac{-1}{\rho}\frac{dE}{dx}\right)_{nitrogen}$$

$$= 0.21(109.4) + 0.79(130.2) = 125.8\,\text{MeV/g/cm}^2$$

One of the implications of the stopping power formulas for heavy charged particles is that all particles of a given type will follow the same energy loss pattern in a given material. More specifically, the example shown in Figures 17.3 started with 8 GeV $^{40}$Ar ions in Be. However, these curves display the expected result for *all* $^{40}$Ar ions with kinetic energies less than 8 GeV. The energy loss rate for an ion with 4 GeV or even 4 MeV can be read off the graph by finding the position at which the residual energy of the ion is equal to the required energy. This may seem a trivial point, but it has the more subtle meaning that all the ions *will* follow *exactly* the same energy loss pattern, within the limits of the statistical process, if we ignore nuclear Coulomb scattering. Formally, we can write that the amount of kinetic *energy lost*, $\Delta E$, in a finite thickness, $\Delta x$, of material is

$$\Delta E(\Delta x) = \left(\frac{dE}{dx}\right)\Delta x \qquad (17.16)$$

and the statistical variation in the energy lost $\delta \Delta E$ would be evidenced as a width in the measured value that is called the amount of *energy straggling*. In a colloquial expression, the ions are said to *straggle* through the material, and the width of the energy distribution is due to this straggling. Thus, as ions pass through matter, the spread in their energies increases (Fig. 17.5). This can be represented mathematically as a Gaussian distribution:

$$\frac{N(E)\,dE}{N} = \frac{1}{\alpha\pi^{1/2}}\exp\left[-\frac{(E-\overline{E})^2}{\alpha^2}\right] \qquad (17.17)$$

where the straggling parameter $\alpha$, which is the half-width at $(1/e)$th height, is given by the expression

$$\alpha^2 = 4\pi q^2 e^4 N_e x_0 \left[1 + \frac{kI}{m_e v^2}\ln\left(\frac{2m_e v^2}{I}\right)\right] \qquad (17.18)$$

where $k$ is a constant (about $\frac{4}{3}$) and the thickness of the absorber the ion has penetrated is $\rho x_0$.

The range or distance that a heavy charged particle will travel in a material can be obtained by integrating the energy loss rate *along the path* of the ion. In the approximation that the ion follows a straight-line trajectory, then the range for a given kinetic energy, $R(T)$, would be given by the integral:

$$R(T) = \int_0^T -\left(\frac{dE}{dx}\right)^{-1} dE \qquad (17.19)$$

where the function $dE/dx$ is the appropriate function for the ion in the material. There are two difficulties in applying this simple integral, the ions will suffer a different number of collisions with atomic electrons, and, more importantly, the

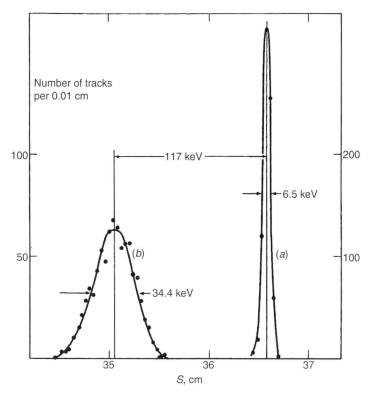

**Figure 17.5** Energy spectrum of 3-MeV protons: (*a*) before and (*b*) after passing through a 3.3-mg/cm$^2$ gold foil. [From L. P. Nielson, *Dan. Mat. Fys. Medd.* **33** (6), 1961.]

ions will undergo some scattering from the Coulomb fields of the atomic nuclei. The *multiple Coulomb scattering* leads to an effect that the ion's trajectory is not straight but rather is made up from a series of straight line segments. Thus, the apparent range or the projection of the range onto the initial velocity vector of the ion will not be a single value but rather will consist of a statistical distribution of values. Thus, the distribution of ranges is due to *range straggling*. It is important to note that size of range straggling will grow as an ion penetrates into material because it will literally add-up, in contrast to the energy straggling mentioned above. The range of an ion and its fluctuations are integral quantities, whereas the energy loss rate and its fluctuations are differential quantities. It is still true that Coulomb and nuclear scattering are relatively rare, so that the range straggling for typical ion energies in metals is on the order of a few percent of the range. The qualitative features of the range distribution and the attenuation curve for a typical heavy charged particle are shown in Figure 17.6. Heavy charged particles penetrate uniformly into matter with essentially no attenuation in intensity until they are nearly at rest; at this point the intensity of moving ions rapidly drops to zero.

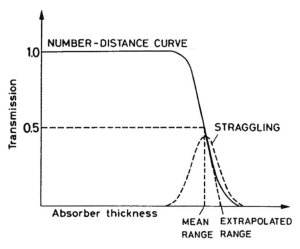

**Figure 17.6** Intensity distribution or attenuation curve is shown as a function of absorber thickness for a typical energetic heavy ion penetrating into a metal. The effect of range straggling is indicated by the Gaussian distribution of ranges. (From Leo, 1987.)

The calculated range–energy curves for some low-mass charged particles in silicon are shown in Figure 17.7. We can see from the integral form of the range as a function of initial kinetic energy, given above, that $R \propto aE^b$. The exponent should be of order 2 at low energies where the energy loss rate is dominated by the $1/\beta^2$ or $1/E$ term. The range–energy relationships are very useful in determining the kinetic energies of particles by measuring the attenuation curves. More recently, range–energy relationships are used to identify charged particles that are detected in silicon semiconductor telescopes as they emerge from nuclear reactions. The scaling rules that apply to the stopping power for different ions in a given medium can be extended to the range–energy relationship. For example, given the range of ion 1 at an initial kinetic energy $T_1$, the range of ion 2 with a different mass, charge, and kinetic energy is

$$R_2(T_2) = \frac{M_2 \, q_1^2}{M_1 \, q_2^2} R_1\left(T_2 \frac{M_1}{M_2}\right) \qquad (17.20)$$

Notice that we have to scale the range of the known, first particle at the appropriate energy of the unknown, second particle.

As a final, more practical, point about the stopping power and ranges of charged particles, we should consider the best method to calculate the amount of energy deposited in a thin foil. Clearly, the ion will slow down as it passes through the material so that the energy loss rate will change as the particle passes through the foil. Thus, we should use the average energy loss rate but notice that the function is not linear, so that we will need a technique to determine the average. Two cases can be identified: thin foils in which the initial, average, and final energy

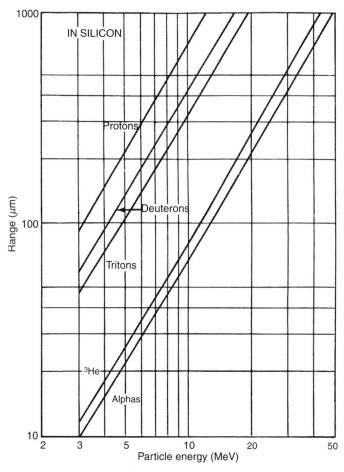

**Figure 17.7** Range–energy curves for some charged particles in silicon. Note the data has the form $R = aE^b$ with a similar exponent for all ions. (From Knoll, 2000.)

loss rates are nearly the same and thick foils in which the particle undergoes a substantial energy loss. In the former example of a thin foil, we can use the expression written above that:

$$\Delta E(\Delta x) = \left(\frac{dE}{dx}\right)\Delta x$$

and we should verify that the final rate is approximately equal to the initial rate:

$$\left(\frac{dE}{dx}\right)_{\text{initial}} = \left(\frac{dE}{dx}\right)_{\text{final}}$$

If the energy loss rates are not substantially different, then we can use the initial rate to obtain the average in a successive approximation procedure. For the case of substantial slowing, we can recall that the range relations come from the integration of $dE/dx$ and thus provide the average energy loss rate that we should use. The technique relies on determining the ranges of ions in graphs or tables of ranges as follows: Imagine that an incident particle with an energy $E_0$ passes through some material with thickness $t$. These are the "known quantities." The particle will emerge from the foil with an energy $E_1$, which we would like to determine. We can find the total range of the ion in the material from tables, $R_0$. The particles that emerge from the foil will have a residual range equal to $R_0 - t$. We can then use the range table or graph to determine $E_1$ that corresponds to the range $R_1 = R_0 - t$. The slowing down and averaging of the energy loss rate will be contained in the range function and do not have to be explicitly evaluated.

**Example Problem** Imagine that a beam of $^{40}$Ar ions at 400 MeV (10 MeV/A) is incident on a 18.5-mg/cm$^2$ beryllium foil (0.1 mm thick). Do the ions pass through the foil, and, if they do, what is their residual kinetic energy?

**Solution** Using a standard reference for stopping power, the tables of Northcliffe and Schilling (1970), we find for these ions that $dE/dx = 9.597$ MeV-cm$^2$/mg in beryllium. Thus, for our first estimate of the energy lost:

$$\Delta E \sim (dE/dx)_{\text{initial}} \Delta x = 9.567 \times 18.5 = 177.5 \text{ MeV}$$

giving a residual energy of $400 - 177.5 = 222$ MeV, which is almost half the initial kinetic energy. The ions will pass through the foil, but this estimate of energy loss is probably too low. Recall that the ions lose more energy per distance traveled as they slow down. Checking, we see that the energy loss rate for these ions at 178 MeV is substantially larger, that is, $dE/dx = 15.3$-MeV-cm$^2$/mg. Thus, this is not a "thin" foil for these ions.

We can use the range technique with information in the same table, for $^{40}$Ar ions with $E_0 = 400$ MeV:

$$R(400 \text{ MeV}) = 28.278 \text{ mg/cm}^2$$
$$R(E_1) = 28.278 - 18.5 = 9.8 \text{ mg/cm}^2$$

This range lies between the tabulated values, and, by linear interpolation between the range values for $E = 160$ and 200 MeV, one finds that $E_1 \sim 185$ MeV. Thus, just using the initial energy loss rate gives a substantial error.

For the practicing nuclear chemist, range–energy tables or relationships are among the most commonly used tools. The largest collection of data on stopping powers and ranges of ions in matter is that of Ziegler and Biersack[1] in the form of the computer programs SRIM/TRIM. Subsets of these tables exist for

---

[1] Check out http://www.srim.org. This website and the references cited therein represent the largest and most widely used compilation about the stopping of energetic ions in matter. The computer programs SRIM and TRIM found these are used widely to estimate stopping powers, ranges, and straggling.

low-energy heavy ions interacting with matter (Northcliffe and Schilling, 1970), α particles interacting with matter (Williamson et al., 1966) and for energetic heavy ions (Hubert et al., 1990).

For the most commonly encountered heavy charged particles, the α particles from radioactive decay, some semiempirical range–energy rules are used. For the range of α particles in air, $R_{air}$, we have

$$R_{air} \text{ (cm)} = [0.005 E_\alpha \text{ (MeV)} + 0.285] E_\alpha^{3/2} \text{ (MeV)} \tag{17.21}$$

or

$$R_{air} \text{ (mg/cm}^2\text{)} = 0.40 E_\alpha^{3/2} \text{ (MeV)} \tag{17.22}$$

so that the range of a 7-MeV α particle in air is about 5.9 cm. For a pure element with $10 < Z < 15$, we have

$$\frac{R_Z}{R_{air}} = 0.90 + 0.0275 Z + (0.06 - 0.0086 Z) \log_{10}\left(\frac{E_\alpha}{4}\right) \tag{17.23}$$

where $R_Z$ is the range in a pure element of atomic number Z expressed in mg/cm², $R_{air}$ is the range in air in mg/cm², and $E_\alpha$ is the α-particle energy in MeV. [For $Z < 10$, substitute 1.00 for the term $(0.09 + 0.0275 Z)$. For $Z > 15$, replace the term $R_Z$ by $(R_Z - 0.005 Z)$.] For compounds or mixtures, the range in the compound or mixture, $R_C$ in mg/cm², is given as

$$\frac{1}{R_C} = \sum_i \frac{p_i}{R_i} \tag{17.24}$$

where $p_i$ is the weight fraction of the ith element in the mixture or compound and $R_i$ is the range of an α particle of this energy in the ith element.

**Example Problem**  What is the range of an 8-MeV α particle in air?

***Solution***

$$R_{air} \text{ (cm)} = (0.005 E_\alpha \text{ (MeV)} + 0.285) E_\alpha^{3/2} \text{ (MeV)}$$
$$= (0.005 \cdot 8 + 0.285) 8^{3/2} = 7.4 \text{ cm}$$

What is the range of this same α particle in Al?

$$\frac{R_Z}{R_{air}} = 0.90 + 0.0275 Z + (0.06 - 0.0086 Z) \log_{10} \frac{E_\alpha}{4}$$

$$R_{air} \text{ (mg/cm}^2\text{)} = 0.40 E_\alpha^{3/2} = 0.0510 \text{ mg/cm}^2$$

$$R_{Al} = 0.0510 \left[ 0.90 + 0.0275 \cdot 13 + (0.06 - 0.0086 \cdot 13) \log_{10} \frac{8}{4} \right]$$

$$= 0.065 \text{ mg/cm}^2$$

Application of these formulas show that the ranges of decay α particles in solids are very short. A sheet of paper will stop the α particles from most radioactive sources. α-Emitting nuclei are not external radiation hazards, but, because of their high LET, they do represent significant inhalation or ingestion hazards.

## 17.3 ELECTRONS

The passage of energetic electrons through matter is similar to that of heavy charged particles in that the Coulomb interaction plays a dominant role. However, three clear differences can be easily seen: The incident electrons are generally relativistic particles (notice that 1 MeV of kinetic energy corresponds to nearly twice the rest mass of an electron, 0.511 MeV); the scattering is predominantly between identical particles and repulsive; and the interactions with nuclei are attractive, and the direction of the electron can be dramatically changed, even reversed, in a collision with a heavy nucleus. A fourth difference that is not so obvious is that a fraction of the kinetic energy is lost through the *radiative* process of bremsstrahlung. Bremsstrahlung (the German word can be literally translated as "braking radiation") is a general process in which electromagnetic radiation is emitted whenever a charged particle undergoes a substantial acceleration. The scattering of electrons, particularly to large angles, corresponds to a classical acceleration that creates/requires the emission of bremsstrahlung. By comparison, very few heavy charged particles undergo large accelerations as they slow down in material.

Summarizing this overview of the possible interaction mechanisms for fast electrons in material, we find that the rate of energy loss in a material is

$$-\left(\frac{dE}{dx}\right)_{\text{electron}} = S_{\text{electronic}} + S_{\text{radiative}} \tag{17.25}$$

where the electronic stopping power is similar to the electronic interaction between charged particles and electrons discussed above, whereas the radiative stopping power is specific to electrons. The electronic stopping power for electrons is written as:

$$S_{\text{electronic}} = -\left(\frac{dE}{dx}\right)_{\text{electronic}} = \frac{2\pi Z e^4 \rho_N}{m_e v^2} \left[ \ln\left(\frac{m_e v^2 E}{2I^2(1-\beta^2)}\right) \right.$$
$$\left. - \ln 2\left(2\sqrt{1-\beta^2} - 1 + \beta^2\right) + (1-\beta^2) + \frac{1}{8}\left(1 - \sqrt{1-\beta^2}\right)^2 \right]$$
$$\tag{17.26}$$

using the same definitions as for the stopping power of heavy charged particles. On the other hand, the expression for the radiative stopping power is

$$S_{\text{radiative}} = -\left(\frac{dE}{dx}\right)_{\text{radiative}} = \frac{(Z+1)Ze^4\rho_N E}{137\, m_e^2 c^4}\left[4\ln\left(\frac{2E}{m_e c^2}\right) - \frac{4}{3}\right] \tag{17.27}$$

and given that $\beta \approx 1$ contains an extra factor of $Z^*E$ in the term in front of the parenthesis. With some algebra and for a typical electron energy, one can show that the ratio of the two contributions to the stopping power depends on the atomic number of the material, $Z$, and the electron kinetic energy $E$:

$$\frac{S_{\text{radiative}}}{S_{\text{electronic}}} \approx \frac{ZE}{800\,\text{MeV}} \tag{17.28}$$

which indicates that the radiative contribution is only significant for large atomic numbers ($Z \sim 80-90$) and high electron energies ($E \sim 10-100$ MeV). Typical $\beta$ particles from radioactive sources are emitted with only 1–10 MeV of kinetic energy, often much less, and the radiative contribution to the stopping power is very small. The bremsstrahlung spectrum is smooth and continuous, ranging from zero energy up to the electron energy due to the random distribution of electron scattering angles.

The important feature that the electrons are occasionally scattered to large angles during the penetration of material causes each electron to follow a tortuous path. A beam of electrons will not have a fixed range in the sense of a beam of heavy charged particles. In fact, given the identical nature of the particles involved in the scattering process, obtaining the range distribution is problematic. Moreover, the primary or a secondary electron can be scattered backwards and emitted from the material. A schematic plot of the range distribution is shown in Figure 17.8 for a monoenergetic source of electrons. The falloff of the intensity with penetration depth starts immediately as the electron enters the material and then gradually approaches zero. This figure emphasizes the fact that the concept of range cannot be applied in a simple way to energetic electrons. As an approximation, the electron range is taken as the extrapolation of the linear portion of the attenuation curve to zero. As shown in Figure 17.9 the product of this definition of the range times the density of a material is a smooth function of incident electron energy for a wide range of materials. Such behavior comes from the fact that the range has a strong dependence on the electron density through the electronic stopping and a weak dependence on the atomic number of the material through the ionization potential and radiative stopping.

Many related measurements have been made of the range distribution of electrons emitted in $\beta^-$ decay. These studies were particularly important before

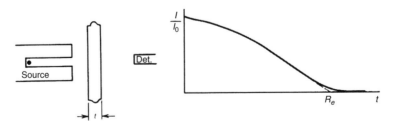

**Figure 17.8** Schematic attenuation curve for an energetic electron in a solid material. (From Knoll, 2000.)

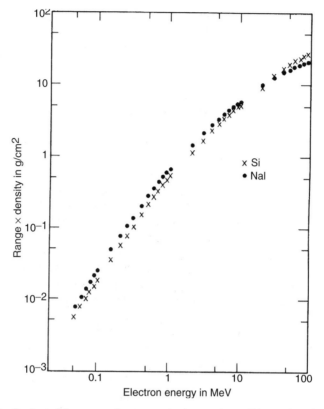

**Figure 17.9** Product of the range of an energetic electron in a solid material with the density as a function of the incident kinetic energy is shown. (From Knoll, 2000.)

solid-state detectors were available. The measurements have shown that the combination of the Fermi energy distribution of electrons from the decay with the sloping range distribution leads to an approximately exponential attenuation of the β-decay electrons. This can be expressed as

$$N_t = N_0 e^{-\mu t} \tag{17.29}$$

where $N_t$ is the number of β particles transmitted through a thickness $t$. The absorption coefficient $\mu$ can be related to the endpoint energy $E_{max}$ of the β spectrum as:

$$\mu(\text{m}^2/\text{kg}) = 1.7 E_{max}^{-1.14} \tag{17.30}$$

where $E_{max}$ is given in MeV.

**Example Problem** What fraction of the β particles emitted by $^{32}$P ($E_\beta^{max} = 1.71$ MeV) will be stopped by a sheet of Al of thickness 1 mm?

***Solution*** The mass absorption coefficient is $1.7(1.71)^{-1.14} = 0.922 \text{ m}^2/\text{kg}$. The fraction of these β particles that are transmitted is $N_t/N_0 = e^{-\mu t} = \exp[-0.922 \text{ m}^2/\text{kg} \times (1 \times 10^{-3} \text{ m}) \times (2.7 \times 10^3 \text{ kg/m}^3)] = \exp[-2.489] = 0.083$. The fraction absorbed $= 1.0 - 0.083 = 0.917$.

In the distant past, measurements of the attenuation of the β spectrum from a newly discovered isotope were used to identify the energy of the β decay. Recently, the attenuation of strong sources has been used to monitor the thickness of materials during manufacturing processes. Notice that the monitoring can be continuous, nondestructive, and a physical probe does not need to "touch" the material being measured.

The backscattering of energetic electrons from materials is a feature that is extremely rare for heavy charged particles. Backscattering is primarily due to multiple interactions with (heavy) nuclei that significantly alter the direction of the incident electron that is enhanced by the fact that two energetic electrons can be created when the incident electron scatters from an atomic electron. The coefficient of backscattering is used to quantify the probability that an electron will emerge "backwards" from a surface that is irradiated with electrons. The coefficient is a function of the energy of the incident electron and the atomic number of the absorber. Formally, the coefficient is the fraction of time that an electron is emitted from the surface of a material following the entry of an energetic electron. The coefficient is approximately 0.5 for $E < 1$ MeV electrons in gold, approximately 0.3, 0.04, and 0.1 for copper, aluminum, and carbon, respectively. It falls below 0.1 for $E = 10$ MeV electrons in gold, below 0.05 for copper, and to near zero in aluminum and carbon.

The bremsstrahlung radiation from electron beams has important practical applications even though it is a small contribution to the stopping power. Bremsstrahlung forms the basis for operation of X-ray tubes and other "controllable" high-fluence sources of radiation. Such devices collide an electron beam with an energy of the order of 10–50 keV with a large electrode, usually made out of a heavy element like tungsten or tantalum. The electrons penetrate the electrode and the bulk of their kinetic energy is lost through electron scattering and eventually creates heat. However, a small fraction of the incident energy is converted into electromagnetic energy in the X-ray region. This is called "thick-target" bremsstrahlung because the incident electrons are completely stopped inside the material. If we assume that the bremsstrahlung is independent of electron energy, then the fraction of the electron energy would be $f_{\text{rad}} = 0.0014ZE$ from the expression above. The observed fraction is about a factor of 2 lower, which would be consistent with simply taking the average energy of the electron in the material to be one-half the initial energy. We should note that some of the energetic electrons can create inner shell atomic vacancies in the atoms that make up the lattice. These vacancies will be filled by discrete K and L X-ray transitions. These sharp lines will add to the continuous bremsstrahlung spectrum and will depend, of course, on the atomic number of the material.

Another mechanism for electron energy loss in matter is the emission of Cerenkov radiation. When a beam of fast moving charged particles with a velocity

$v$ near the speed of light $c$ enters another medium with index of refraction $n$, the particle velocity will exceed the speed of light in the new medium (which is $c/n$). The electron radiates the "excess" energy in the form of a blue-white light call Cerenkov radiation. This light is localized in a cone of half-angle $\theta$ around the direction of motion of the electron such that $\cos\theta = c/nv$.

## 17.4 ELECTROMAGNETIC RADIATION

As photons move through material, they only interact or "scatter" in localized or discrete interactions, and they do not interact at long distances, that is, they are not subject to the Coulomb or nuclear forces. This behavior is in clear contrast to the long-range interactions felt by charged particles. Thus, as a beam of photons propagates through any material, the *intensity* of the beam will decrease as the photons that interact are removed, but the *energy* of all the noninteracting photons will remain constant. The photons will interact in ways that predominantly release fast moving electrons, low-energy photons will interact only once and give rise to a single primary electron, energetic photons can interact several times and give rise to a few primary electrons. The most energetic photons can create a matter–antimatter pair of electrons that induce a cascade of secondary electrons.

The energy of the noninteracting photons remains constant so that the probability that a photon will interact in a fixed thickness of material will also remain constant regardless of the photon energy. This leads immediately to an exponential attenuation of electromagnetic radiation that is called the Beer–Lambert law. The law was applied to the absorption of visible light but applies to all electromagnetic radiation. The derivation of the exponential attenuation law is similar to the derivation of the exponential decay law of radioactive nuclei and will not be repeated here. (The analogy is that the probability of radioactive decay is constant in a given time interval.)

The general expression for the attenuation of photons is

$$I = I_0 e^{-\mu x} \quad (17.31)$$

where $I$ and $I_0$ are transmitted and incident intensities, respectively, $x$ is the thickness, and $\mu$ is an energy-dependent *total linear absorption coefficient* that depends on the nature of the material. The mean free path of the photons in the material is simply $\lambda = 1/\mu$, and we can define a half-thickness as $x_{1/2} = \ln(2)/\mu$ in analogy to the radioactive half-life. The exponential nature of the attenuation means that the intensity of the transmitted radiation does not go to zero, although it can be made arbitrarily small. The mass attenuation coefficient is obtained by dividing the linear attenuation coefficient by the density of the material, $\mu/\rho$. The mass attenuation coefficient is independent of the physical state of the absorber and represents the fact that the fundamental interactions can be expressed in terms of cross sections per atom. Extensive tabulations and figures, such as Figure 17.10, are available for the mass attenuation coefficients of photons with energies in the range of 0.01 to 10 MeV.

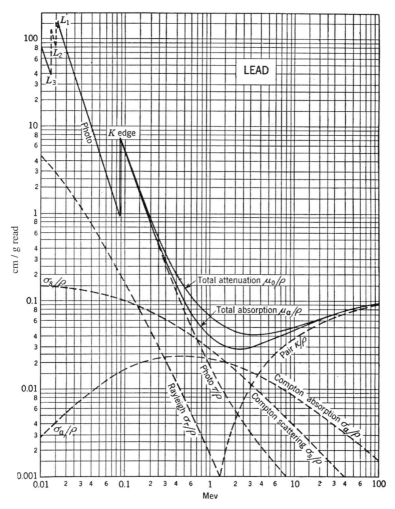

**Figure 17.10** Mass attenuation coefficient for energetic photons in lead. (From Evans, 1955.)

**Example Problem** Estimate the fraction of 1.0 MeV photons that will be transmitted through a lead absorber that is 5 cm thick (the thickness of "lead bricks" commonly used in radiation shields).

**Solution** The transmitted fraction is simply

$$f = \frac{I}{I_0} = e^{-\mu x}$$

where $x$ is 5 cm and $\mu$ can be obtained from the mass attenuation coefficient in Figure 17.10. Reading the value from Figure 17.8, $\mu_0/\rho = 0.07 \text{ cm}^2/\text{g}$ for the total *attenuation*. $\mu_0$ is equal to $\mu_0/\rho * \rho$, of course. Thus:

$$f = \frac{I}{I_0} = e^{-(\mu_0/\rho)\rho x}$$
$$= e^{-(0.07\,\text{cm}^2/\text{g})*11.35\,\text{g/cm}^3*5\,\text{cm}} = e^{-0.795*5} = 0.019$$

Approximately 2% will be transmitted. Notice that the half-thickness for these photons in lead, $x_{1/2} = \ln 2/\mu$ is 0.87 cm.

Concentrating on photon energies that are associated with the differences between nuclear energy levels, those in the region from 10 keV to 10 MeV, we find that only three types of interactions play a role in attenuating a photon beam. These mechanisms are shown schematically in Figure 17.11. Each photon that interacts with an atom via any one of the mechanisms will be lost from beam. The type of interaction is random, but their relative probabilities depend on the photon energy. *Photoelectric absorption* dominates at low energies, $E_\gamma < 0.1$ MeV in the heaviest elements, while *Compton scattering* is most important at intermediate energies, whereas *pair production* has an absolute threshold at 1.022 MeV and only is important for the highest energy photons. We will consider each of these processes in turn.

### 17.4.1 Photoelectric Effect

The photoelectric effect was originally described by Albert Einstein and helped to establish the quantized nature of light. The photoelectric effect has many extremely

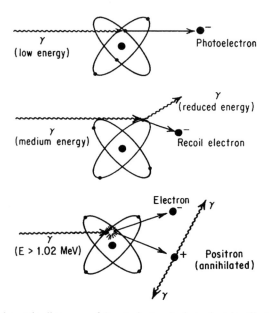

**Figure 17.11** Schematic diagrams of (top to bottom) photoelectric effect, Compton effect, and pair production.

important applications, for example, the detection of visible light by photocells and the photovoltaic conversion of sunlight. The photoelectric effect converts a single photon into a single free electron. When the photon interacts with a bound electron, the photon can be completely absorbed and the electron emerges with a kinetic energy (KE) that corresponds to the photon energy, $h\nu$, minus the electron binding energy (BE), $KE_{e-} = h\nu - BE$ (see Fig. 17.11). Photocells use a semiconductor like silicon for the absorbing material, and the electrons released by visible light have relatively small kinetic energies and are collected as a photocurrent. In the present application of the absorption of a nuclear photon, a fast electron is usually created in the bulk of a solid medium because the binding energy is often small compared to the photon energy. The fast electron goes on to lose its kinetic energy by scattering through the material (as discussed above). Conservation of momentum requires that the electron be bound in an atom (that could be in a lattice) that recoils.

The cross section or probability of the photoelectric effect is on the order of the square of the atomic size for photons in the keV region and decreases rapidly with increasing photon energy. The cross section also has a strong dependence on the atomic number of the absorbing material, as there is a sharp increase in the cross section at each threshold for the emission of bound electrons. As an example, the heavy-element lead ($Z = 82$) has K, L, and M (principal quantum numbers, $n = 1, 2,$ and 3) binding energies of approximately 88, 15, and 3 keV, respectively, which provide strong photoelectric absorption for photons in this energy region. The sharp increase in the photoelectric cross section can be seen as the sharp peaks on the left in Figure 17.10. The overall probability for photoelectric absorption follows the very rough expression:

$$\sigma_{\text{photoelectric}} \propto Z^n / E_\gamma^{7/2} \qquad (17.32)$$

where the exponent $n$ is between 4 and 5. This expression only includes the dramatic effects of the electron binding energies in an overall way and is not meant to replace the measured values.

We should note that the photoelectric effect often leaves an inner shell vacancy in the atom that previously contained the "ejected" electron. This vacancy will be filled by an atomic transition, called fluorescence, and generally produces an X-ray photon. In an interesting twist of fate, the X-ray photon will have an energy that is just below the sharp rise in the attenuation coefficient due to conservation of momentum and can often escape from the absorber. Recall that the direction of the fluorescence photon will be uncorrelated with the direction of the incident photon and a fraction will be emitted "backwards" from the absorber. The absorber will thus emit its own characteristic X-rays when it is irradiated with high-energy photons.

In γ-ray spectroscopy lead shields are commonly used. This can result in the production of Pb X-rays that can interfere with the measurement of low-energy photons. Lining the Pb shields with layers of Al and Cu that absorb the Pb X-rays and other subsequent radiation ameliorates these problems.

## 17.4.2 Compton Scattering

If the energy of the incident photon exceeds the typical binding energies of the innermost atomic electrons, the probability of photoelectric absorption drops below the probability that the photon will scatter from an electron leading to a scattered electron and a lower energy photon. This process is called *Compton scattering*. A schematic diagram of this process is shown in Figure 17.12.

From the conservation of momentum in the $x$ direction we have

$$p = p' \cos \theta + p_e \cos \phi \tag{17.33}$$

From conservation of momentum in the $y$ direction, we have

$$0 = -p' \sin \theta + p_e \sin \phi \tag{17.34}$$

Assuming the collision is elastic, conservation of energy gives us

$$E = E' + T_e \tag{17.35}$$

Combining these equations (see Problems) and utilizing the relativistically correct expressions for the energy and momentum of the electron

$$p_e^2 = \frac{1}{c^2}[T_e(T_e + 2m_e c^2)] \tag{17.36}$$

we get

$$\lambda' - \lambda = \frac{h}{m_e c}(1 - \cos \theta) \tag{17.37}$$

where $\lambda'$ and $\lambda$ are the wavelengths of the scattered and incident γ rays. The quantity $h/m_e c$ is called the Compton wavelength of the electron and is equal to 2426 fm.

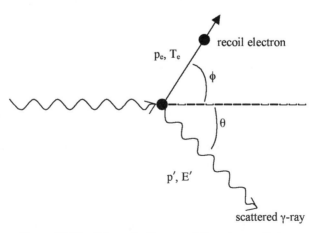

**Figure 17.12** Schematic diagram of Compton scattering.

**TABLE 17.2 Compton Scattering at Fixed Angles**

| θγ (degrees) | Emitted Photon Energy ($h\nu'$) | Electron Kinetic Energy |
|---|---|---|
| 0 | $h\nu$ | $\sim 0$ |
| 90 | $h\nu \left( \dfrac{mc^2}{mc^2 + h\nu} \right)$ | $\sim h\nu - mc^2$ |
| 180 | $h\nu \left( \dfrac{mc^2}{mc^2 + 2h\nu} \right)$ | $\sim h\nu - mc^2/2$ |

Note that the shift in γ-ray energy is independent of the incident energy. The expressions for the energies of the scattered photon and electron are given in Table 17.2.

It is clear that the minimum energy of the scattered γ ray occurs when $\theta = 180°$ ($\cos \theta = -1$). In this case, we have

$$E_{\gamma^1}^{\min} = \frac{m_e c^2}{2} \left( \frac{1}{1 + \dfrac{m_e c^2}{2 E_\gamma}} \right) \approx 255 \text{ keV} \qquad (17.38)$$

In this case, the electron energy, $T_e$ will be maximum and $T_e = E_\gamma - 0.255$ MeV. If we consider all scattering angles θ, then the distribution of scattered electron kinetic energies is as shown in Figure 17.13. The sharp peak at $E_\gamma - 0.255$ MeV is called the *Compton edge*. The minimum energy photon, $E \approx 225$ keV, will be a noticeable

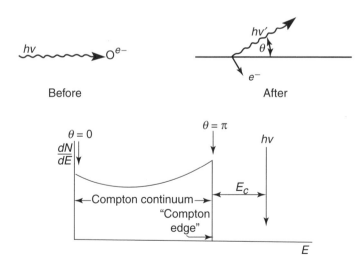

**Figure 17.13** Top: schematic version of Compton scattering. Bottom: schematic variation of the distribution of electron kinetic energies produced in Compton scattering. (From Knoll, 2000.)

component in γ-ray spectra resulting from the interaction of a photon from a radioactive source with the lead shield surrounding the detector, resulting in a backscattered photon ($E \sim 225$ keV), which strikes the radiation detector (the "backscatter peak").

The Compton scattering cross section per electron of the stopping material is independent of Z, and thus the cross section per atom goes as Z. For energies about 0.5 MeV, it varies roughly as $1/E_\gamma$.

### 17.4.3 Pair Production

Whenever the energy of the initial photon exceeds the rest mass of two electrons, 1.022 MeV, the process of pair production is possible. During the process of pair production, the initial photon interacts with the Coulomb field of a nucleus and is converted into an electron and a positron, a matter–antimatter pair, that shares the initial energy of the photon (see Fig. 17.11). Conservation of energy and momentum in the Coulomb field cause the pair of electrons to move forward along the initial direction of the photon with a small opening angle. The pair of particles will interact with the electrons and nuclei in the remaining material as described above.

The process of bremsstrahlung observed in electron stopping is closely related to the process of pair production. From a schematic standpoint, in the first case, a moving electron interacts with the Coulomb field of an atom, making a transition between two energy states and a (bremsstrahlung) photon is emitted. In the second case, a photon is destroyed by interaction with an atomic Coulomb field and a pair of electrons is created. The probability of pair production has an absolute threshold of 1.022 MeV, that is, this process cannot take place if the photon has a lower energy. The cross section increases relatively rapidly and saturates above $\sim 10$ MeV, as indicated in Figure 17.10. The variation of the pair-production cross section with photon energy is complicated, but the cross section depends on the square of the atomic number of the absorber. For large photon energies, $\sigma_{\text{pair}} \propto Z^2 \ln(E\gamma/mc^2)$. Pair production is the predominant attenuation process for high-energy photons.

Pair production has a threshold energy of 1.022 MeV because two particles are created, one electron and one positron. Thus, some energy is "stored in" or "used to create" the mass of the pair. Notice the total electric charge is conserved because the electron charge is $-1e$ and the positron charge is $+1e$. One of the unique features of this process is that the energy that went into the creation of the two particles will be "released" when the positron comes to rest and annihilates with an electron. The annihilation process is

$$e^+ + e^- \longrightarrow \gamma + \gamma$$

in which the two γ rays have exactly the same energy, $mc^2 = 0.511$ MeV, and are emitted at 180°, or back to back. The axis along which the two γ rays are emitted will be random with respect to the initial direction of the incident photon because the positron will undergo a slowing down process involving multiple scattering with atoms and atomic electrons. In the final phase of the process, the positron captures a single electron and forms a neutral species called positronium. Therefore, the characteristic annihilation radiation (photons with an energy of 0.511 MeV) can escape from the absorber whenever pair production can occur.

In summary, photons pass through material until they interact individually with the atoms or nuclei in the material. Depending on the energy of the photons the interaction will be predominately pair production (high energy), Compton scattering, or photoelectric absorption (low energy). The relative importance of these processes is summarized in Figure 17.14 as a function of atomic number and photon energy. In each interaction, the photon ionizes the material, creating one or two fast moving electrons and leaving a positive ion. Pair production gives two fast moving electrons—one positive, one negative, Compton scattering gives one fast moving electron and a lower energy photon, and the photoelectric effect gives one fast moving electron. The fast electrons have a much higher rate of ionization than the photons, and the general features of the interactions of these electrons with the material have been described above. The "path" of the scattered photon will be erratic when Compton scattering is the predominant process. γ Rays from nuclear decay processes tend to have energies on the order of 1 MeV. From Figure 17.14 we expect these photons to interact via Compton scattering in all materials. The first interaction will give an electron and a lower energy photon. The interaction probability of the secondary photon will usually be higher than that of the primary photon and will often result in a photoelectric absorption. Recall also that, depending on the energy of the primary photon, the absorber will weakly emit lower energy photons such as annihilation radiation, backscatter radiation, or fluorescence X-rays.

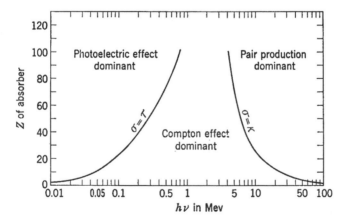

**Figure 17.14** Summary of the relative importance of the three mechanisms by which photons interact with matter. The curves indicate the locations in the atomic number–photon energy plane at which the cross section for Compton scattering is equal to that for photoelectric absorption, left side, or is equal to that for pair production, right side.

## 17.5 NEUTRONS

We will now briefly consider the propagation of neutrons through material. Neutrons are the most penetrating radiation for the simple reason that their only significant interaction is with nuclei via the strong force. (Neutrons only have a very small interaction with electrons through their magnetic dipole moment that can be ignored.) As we discussed at the beginning of this chapter, nuclei are very much smaller than atoms, and so the probability that a fast neutron will interact with (strike) a nucleus is very, very small. On the other hand, neutrons cause significant radiation damage because all of their interactions cause nuclear recoil and many lead to nuclear transmutations.

A neutron will move through material along a straight line with a constant energy until it encounters a nucleus and induces a nuclear reaction. Thus, neutron attenuation follows an exponential law similar to that for photons. Written in terms of an energy-dependent attenuation length, $\mu_E$ we have

$$I = I_0 e^{-\mu_E x} \qquad (17.39)$$

where $x$ is a linear dimension and $I_0$ is the incident intensity. The attenuation length is the inverse of the mean free path, $\lambda$, and is proportional to the total nuclear reaction cross section:

$$\mu_E = \frac{1}{\lambda_E} = N_0 \sigma_{\text{Total}}(E) \qquad (17.40)$$

where $N_0$ is a constant that gives the total number of nuclei per unit volume in the material. The total nuclear reaction cross section is a characteristic of each isotope in the absorbing material and has the dimensions of an area. If we have a monoisotopic element such as gold or bismuth, then we will only have to account for the energy dependence of the neutron. If the material contains several isotopes such as silver ($^{107}$Ag and $^{109}$Ag), nickel (five isotopes), or is a compound NaF (one isotope of each element), and so forth, then the effective cross section will be the *number-weighted* cross section:

$$\sigma_{\text{average}} = f_1 \, \sigma_{\text{Total}}(E)_1 + f_2 \, \sigma_{\text{Total}}(E)_2 + f_3 \, \sigma_{\text{Total}}(E)_3 + \cdots \qquad (17.41)$$

where the constants, $f_i$, are the fraction by number of each isotope in the sample.

**Example Problem** Calculate the average thermal neutron capture cross section and the mean free path for LiF, a solid crystalline material at room temperature with a density of 2.635 g/cm$^3$ and a molar mass of 25.94 g/mol. Lithium has two stable isotopes $^6$Li (7.5%) and $^7$Li (92.5%) with thermal neutron capture cross sections of $\sigma_{\text{thermal}} = 39$ mb and 45 mb, respectively. Fluorine is monoisotopic, $^{19}$F, with $\sigma_{\text{thermal}} = 9.6$ mb.

***Solution***

$$\sigma_{\text{average}} = f_1 \, \sigma_{\text{Total}}(E)_1 + f_2 \, \sigma_{\text{Total}}(E)_2 + f_3 \, \sigma_{\text{Total}}(E)_3$$

Notice that half the atoms are fluorine and half the atoms are lithium, but the lithium atoms are split unevenly between $A=6$ and $A=7$. The fractions of each isotope must reflect this distribution.

$$\sigma_{\text{average}} = 0.075 \times 0.5 \times 39\,\text{mb} + 0.925 \times 0.5 \times 45\,\text{mb} + 1.0 \times 0.5 \times 9.6\,\text{mb}$$
$$\sigma_{\text{average}} = 27.1\,\text{mb}$$

Rearranging the equation above relating the mean free path and the total reaction cross one has

$$\lambda_{\text{thermal}} = \frac{1}{N_0 \sigma_{\text{thermal}}}$$

$$N_0 = \frac{N_A \rho}{\text{molar mass}} = \frac{6.022 \times 10^{23}/\text{mol} \times 2.634\,\text{g/cm}^3}{25.94\,\text{g/mol}}$$

$$N_0 = 6.11 \times 10^{22}\,\text{cm}^{-3}$$

And finally for the mean free path:

$$\lambda_{\text{thermal}} = \frac{1}{6.11 \times 10^{22}\,\text{cm}^{-3} \times 27.1\,\text{mb} \times 1 \times 10^{-27}\,\text{cm}^2/\text{mb}}$$
$$\lambda_{\text{thermal}} = 604\,\text{cm}$$

Thus, the average thermal neutron travels more than 6 m in solid LiF before undergoing a nuclear capture reaction! Note the total reaction cross section will be larger and the neutrons will most likely scatter before being captured.

Neutrons can interact with matter via a number of different reactions, depending on their energy. The following are among the most important of these reactions:

1. Elastic scattering, $A(n, n)A$, which is the principal interaction mechanism for neutrons.
2. Inelastic scattering, $A(n, n')A^*$, where the product nucleus $A^*$ is left in an excited state. To undergo inelastic scattering, the incident neutron must have sufficient energy to excite the product nucleus, generally about 1 MeV or more.
3. Radiative capture, $A(n, \gamma)A + 1$. As discussed earlier, this cross section shows a $1/v$ energy dependence, and this process is important for low-energy neutrons.
4. Fission, $A(n, f)$, which is most likely at thermal energies but occurs at all energies where the neutron binding energy exceeds the fission barrier height for fissile nuclei.

5. Knockout reactions, such as (n, p), (n, α), (n, t), and so forth, which are maximum for neutrons of eV–keV energy but occur at higher energies.

The total neutron interaction cross section, $\sigma_{total}(E)$, is the sum of the various reaction cross sections:

$$\sigma_{total} = \sigma_{elastic} + \sigma_{inelastic} + \sigma_{capture} + \cdots \tag{17.42}$$

One of the technologically most important interactions of neutrons with matter is their loss of energy ("slowing down") by a series of elastic collisions. Let us consider the case where particle 1 of mass $m_1$, speed $v_1$, collides with particle 2, mass $m_2$, at rest. After the collision, the particles will have speeds $v'_1$ and $v'_2$ in the lab system.

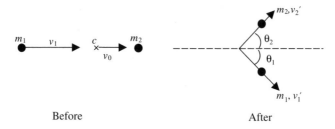

Before                    After

In the center of mass (cm) system, we have

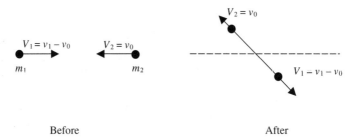

Before                    After

After the collision, the relationship between the cm and the lab systems is

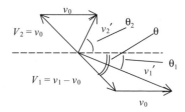

$$v_0 = \frac{M_1}{M_1 + M_2} v_1 \equiv \text{velocity of cm} \tag{17.43}$$

## 17.5 NEUTRONS

If we designate the kinetic energy of particle 1 after the collision at $T'_1$, we have

$$T'_1 = \frac{1}{2}m_1(v'_1)^2 \tag{17.44}$$

$$T'_1 = \frac{1}{2}m_1(V_1 + v_0)^2 = \frac{1}{2}m_1(V_1^2 + v_0^2 + 2V_1 v_0 \cos\theta) \tag{17.45}$$

Thus, $T'_1$ will have a maximum value for $\theta = 0°$ and a minimum value for $\theta = 180°$. We have

$$T'_1(\max) = \frac{1}{2}m_1(V_1^2 + v_0^2 + 2V_1 v_0) = \frac{1}{2}m_1 v_1^2 = T_1$$

$$T'_1(\min) = \frac{1}{2}m_1(V_1^2 + v_0^2 - 2V_1 v_0) = \frac{1}{2}m_1(V_1 - v_0)^2$$

$$= \frac{1}{2}m_1(v_1 - 2v_0)^2 \tag{17.46}$$

$$= T_1\left(\frac{m_1 - m_2}{m_1 + m_2}\right)^2$$

For the special case where particle 1 is a neutron and particle 2 is a proton, $m_1 \approx m_2$:

$$T'_1(\min) = 0$$
$$T'_1(\max) = T_1$$

It follows that

$$T'_1 = \frac{T_1(1 + \cos\theta)}{2}$$
$$\theta_1 + \theta_2 = 90° \tag{17.47}$$
$$v_1 = v_2 = v_0$$

If we assume the angular distribution of the scattered neutrons is isotropic in the cm system, then the probability of a neutron scattering into and solid angle $d\Omega$, $P(d\Omega)$, is a constant given by

$$P(d\Omega) = \frac{d\Omega}{4\pi} \tag{17.48}$$

where the solid angle $d\Omega$ is given in steradians. Substituting in for $d\Omega$, assuming spherical symmetry, we have

$$P(d\Omega) = \frac{2\pi \sin\theta \, d\theta}{4\pi}$$
$$= \frac{1}{2}\sin\theta \, d\theta \tag{17.49}$$

When a neutron is scattered into an angular interval ($\theta$ to $\theta + d\theta$), its energy is changed from $T_1$ to the interval ($T_1'$ to $T_1' + dT_1'$). Here

$$dT_1' = -m_1 V_1 v_0 \sin\theta \, d\theta \tag{17.50}$$

Thus, we have for the probability of scattering into an energy interval $dT_1'$:

$$\begin{aligned} P(dT_1') &= P(d\Omega) \\ &= \frac{1}{2}\sin\theta \, d\theta \\ &= \frac{dT_1'}{2m_1 V_1 v_0} \end{aligned} \tag{17.51}$$

We have an equal probability of scattering into each energy interval. For neutrons scattering off hydrogen

$$T_1'(\min) = 0$$

$$T_1'(\text{average}) = \frac{T_1}{2}$$

After $n$ collisions

$$T_1'(\text{average}) \approx \left(\frac{1}{2}\right)^n T_1 \tag{17.52}$$

Thus, to reduce a 1-MeV neutron to thermal energies ($\sim \frac{1}{40}$ eV) would require about 25 collisions.

## 17.6 RADIATION EXPOSURE AND DOSIMETRY

Up to this point we have taken a very microscopic view of the propagation of beams of particles through material. We have described the degradation of the intensity and the energies of the beams in terms of individual interactions. Now we will take a more macroscopic view from the standpoint of the absorber.

As we have seen, the passage of radiation through material causes ionization of the atoms and molecules. The creation of free electrons, recoiling positive ions, and in some cases transmuted nuclei can disrupt the chemical structure of the material. It is important to note that the effect of the passage of a single particle through a macroscopic object will usually cause a minimal overall effect due to the enormous number of atoms present in the object. Special devices are necessary to observe the ionization from individual particles as described in the next chapter on radiation detectors. Physical changes in an everyday object will be observed when it has been exposed to large amounts of radiation. At the same time we should realize that certain materials will be relatively immune to the ionization caused by photons and fast electrons, whereas neutron irradiation of the same material generally will have a substantial effect. For example, a metal lattice is characterized by delocalized electrons, and the local ionization caused by a Compton scattering or photoelectric absorption will be quickly neutralized (or repaired). Neutron absorption by a metal will generally lead to $\beta$ decay and the transmutation of one atom into the neighboring element. On the other hand, local ionization created by photons and electrons in insulating materials like glasses will persist for a long time. Similarly, lattice defects caused by atomic recoil have to be removed by annealing, but atomic recoil has little effect in liquids (and none in gases). Thus, we can see that the effects of radiation on a material will depend in great detail on the type and amount of radiation, and on the physical and chemical nature of the material being irradiated.

We have seen that the neutral forms of radiation, photons and neutrons, are very penetrating and can pass through layers of material without interacting. In these cases, we need to distinguish between the amount of *radiation exposure* and the amount of *energy absorbed* by the material. Photons such as X-rays, bremsstrahlung, and $\gamma$ rays play an important role in nuclear medicine, but they are not strongly absorbed by tissue. The exposure to these photons is not equal to the dose. In the case of highly ionizing radiation such as charged particles, the exposure will correspond to the absorbed energy except for very thin materials that allow the particle to escape. In order to characterize radiation effects, we need to know the amount of energy absorbed by the material, which is called the *absorbed dose*.

The unit of radiation exposure is the roentgen (R). It is a historical unit of the exposure and characterizes the radiation incident on an absorbing material without regard to the character of the absorber. The unit was formalized in 1928 as "The amount of radiation which produces one electrostatic unit of ions (esu), either positive or negative, per cubic centimeter of air at standard temperature and pressure." Translated in modern units:

$$1 \text{ roentgen} = 2.58 \times 10^{-4} \text{coulomb/kg-air} = 0.3336 \text{ nC/cm}^3 \text{ at STP}$$

This value corresponds to an absorbed energy of approximately 8.8 mJ/kg using the effective ionization energy of 34 eV per ion pair in air. The roentgen is most often used to describe the intensity of a photon source such as a medical X-ray machine or other irradiator. The exposure should be measured at some distance from the source so that the radiation field is uniform compared to the dimensions of the detector. The detector is usually an ion chamber filled with dry air that is sensitive to pico-coulombs of charge.

As studies of the effects of exposure to all types of radiation went on, it became clear that these effects were correlated with the amount of absorbed energy, which is generally less than the exposure. In 1962 the "rad" was formally defined as a special unit of energy called the "radiation absorbed dose" with a value of 100 ergs per gram of absorbing material. The rad is a convenient physical standard that correlates well with chemical and biological effects of radiation, whereas the roentgen defined in terms of an air ionization measurement was left for exposure. More recently, the gray (Gy) was introduced as the SI unit for the absorbed dose:

$$1\,\text{Gy} = 1\,\text{J/kg} = 100\,\text{rad} = 6.24 \times 10^{12}\,\text{MeV/kg}$$

so that one centigray is exactly one rad. One gray corresponds to a relatively large amount of energy to be absorbed from a radiation source per unit mass. For example, 1 Gy in water is 18 mJ/mol. Modern dosimeters routinely measure doses at the few millirad (mr) level or few tens of µGy.

The acronym "kerma" for "kinetic energy released in absorbing material" has been used to conceptually connect the energy deposited by ionizing radiation with the radiation field. It is defined to include the kinetic energy, which is locally absorbed from products of interaction with the particular medium such as Compton electrons, photoelectrons, and pair production while excluding the energy, which is not locally absorbed, from Compton-scattered photons, characteristic fluorescence radiation, and annihilation photons. The kerma is defined as:

$$K = \phi E \mu_x / \rho$$

where $\phi$ is the particle fluence (m$^{-2}$), $E$ is the energy of the radiation, $\mu_x$ is a linear energy attenuation coefficient (energy/m), and $\rho$ is the density (kg/m$^3$). The dimensions of the kerma are thus J/kg or gray. The concept of kerma in air is very close to the practical definition of the roentgen as a unit of exposure, times a factor for the amount of energy necessary to create an ion pair. However, bremsstrahlung photons are lost in the secondary-particle equilibrium condition assumed in the definition of the roentgen, but they represent a small effect under most conditions.

Just as the effect of radiation on a specific material depends on the dose or amount of absorbed energy in contrast to the exposure, the effect of radiation on biological systems depends on the energy density and not just the energy. One can imagine that a biological system could survive the formation of a single ion pair and the following chemical transformations from a single photoelectric event. However, if a large number of direct ionization events take place in a small volume due to the passage of a heavy charged particle through some biological material, the resulting chemical changes could be profound. The important parameter is called the "linear energy transfer," or LET, which is very close to the specific energy loss, $-dE/dx$, discussed earlier in this chapter. The value of LET for a given particle is smaller than the value of $-dE/dx$ because the LET does not include the radiative energy loss term, as the bremsstrahlung radiation is not absorbed "locally." Recall that the radiative energy loss term was only significant for high-energy electrons. Typical values of LET for photons and fast electrons are a few MeV/mm but are one or two orders of magnitude larger for heavy charged particles.

**TABLE 17.3 Radiation Weighting Factors for Various Radiations**

|  | γ | β | Proton (>2 MeV) | α | Fast n (2–20 MeV) | Thermal n |
|---|---|---|---|---|---|---|
| Radiation weighting factor | 1 | 1 | 5 | 20 | 10 | 5 |

The concept of dose equivalent has developed over time to quantify the more damaging effects of high LET radiation. The original definition of the absorbed dose in rads was multiplied by a quality factor, $Q > 1$. The quality factor increased with increasing LET. The historical unit for dose equivalent is called a rem for "roentgen equivalent man," and measurements of dose equivalents to biological systems, especially people, are most commonly reported in millirem (mrem). With the more recent SI dose unit of gray, a new SI unit of dose equivalent (or as it is now called, equivalent dose) was introduced called the sievert (Sv). The sievert and rem are different by a factor of 100 in the same way as the rad and gray:

$$1\,\text{Sv} = 100\,\text{rem} \quad \text{or} \quad 10\,\mu\text{Sv} = 1\,\text{mrem}$$

The equivalent dose in Sv = absorbed dose in grays × $w_R$ (radiation weighting factor, formerly the quality factor). The absorbed dose for low LET radiation, β and γ rays, is taken as having a radiation weighting factor of unity, $w_R = 1$. The radiation weighting factor has been defined to increase in proportion to the log of the LET. Thus, the radiation weighting factor for α particles in tissue is about 20. The factor for neutrons takes an intermediate value due to the high probability for scattering protons in tissue. A listing of radiation weighting factor values for various types of radiation is shown in Table 17.3.

Notice that the dose has a strict definition of energy per unit mass of the absorber and, in principle, can be measured for a given radiation at a certain energy in a specific material. The equivalent dose is a relative unit in that a radiation weighting factor is applied to a measured quantity. The dose can be measured from ionization in an electronic radiation detector; the equivalent dose must take into account the type of radiation causing the ionization.

## PROBLEMS

1. Calculate $dE/dx$ for a 10-MeV α particle interacting with aluminum.

2. At what kinetic energy does an electron have the same energy loss as a 6-MeV α particle interacting with aluminum?

3. Calculate $dE/dx$ for an 8-MeV α particle interacting with Mylar.

4. It has been said very approximately that α particles and protons having the same speed have approximately the same range in matter. Why is this false? Which has the longer range and why?

5. Verify that the minimum ionization for heavy charged particles takes place at $\beta \sim 0.96$.

6. Thin nickel foil is used to slow down monoenergetic 10-MeV protons. What is the maximum thickness that can be used if one wants the straggling to be <1% of the mean transmitted energy? What is the mean transmitted energy in this case?

7. Calculate the thickness of aluminum foil needed to degrade a beam of 10 MeV/nucleon $^{12}$C ions to 3 MeV/nucleon.

8. Calculate the energy loss of a 6-MeV α particle in passing through 50 μg/cm² of natural nickel?

9. Assuming no energy losses occur, calculate the heating of a 500-μg/cm² foil of $^{208}$Pb when bombarded with 1 particle microampere of $^{86}$Kr$^{19+}$ ions.

10. Devise a way, using measurements of $dE/dx$ and $E$, to build a particle identification system.

11. Calculate the range in aluminum of: (a) an 80-MeV $^{80}$Br ion, (b) a 12-MeV α particle, and (c) a 1-MeV electron.

12. A GM counter window is made of mica, NaAl$_3$Si$_3$O$_{10}$(OH)$_2$, with a thickness of 2 mg/cm². (a) What is the minimum energy $\beta^-$ particle that can penetrate this window? (b) What is the minimum energy α particle that can penetrate this window?

13. Repeat the calculation outlined in Problem 12 for skin of thickness 1 mm with an average density of 1 g/cm³. Assume skin is 65% O, 18% C, 10% H, and 7% N.

14. Particles of kinetic energy 400 MeV are incident on a medium of index of refraction of 1.888. One observes Cerenkov radiation with an opening angle of $\theta = \cos^{-1}(0.55)$. What are the particles?

15. How far does a $^{32}$P $\beta^-$ particle ($E_{\max} = 1.7$ MeV) penetrate in P-10 counter gas?

16. Suppose you have a sample that contains radionuclides that emit 1 MeV β particles and 1 MeV γ rays. Devise an attenuation technique that would allow you to count the γ rays without interference from the β particles.

17. What is the fractional attenuation of a beam of 1 MeV photons in 2.5 cm of Pb?

18. Prove that a photon with $E_\gamma > 1.022$ MeV cannot undergo pair production in free space.

19. Lead is thought to be a "better" absorber of photons than aluminum. At what γ-ray energies is the mass absorption coefficient of lead greater than that of aluminum? Why?

**20.** A 1-MeV photon undergoes Compton scattering through angles of $0°, 90°$, and $180°$. What is the energy of the scattered photon in each case?

**21.** What is the mean free path of a 0.1-, a 1.0-, and a 3.0-MeV photon in NaI?

**22.** Calculate the mean free path of a 200-keV photon in water.

**23.** How much lead shielding will it take to reduce the radiation exposure level to $<10$ mrem/h 1 ft from a 5-mCi $^{60}$Co source?

**24.** Prove the scattering angle is $90°$ for $A + A$ elastic scattering.

**25.** Consider a particle with mass $m_1$ scattering elastically from a particle (at rest) with mass $m_2$. If $m_1 > m_2$, show that the scattering angle cannot exceed $\sin^{-1}(m_2/m_1)$.

**26.** In graphite, how many collisions are necessary to reduce the kinetic energy of a 1-MeV neutron to thermal energies? What is the approximate time scale for this process?

## REFERENCES

Evans, R. D. *The Atomic Nucleus*, McGraw-Hill, New York, 1955.

Hubert, F., R. Bimbot, and H. Gauvin. *At. Data and Nucl. Data Tables*, **46**, 1 (1990).

Knoll, G. F. *Radiation Detection and Measurement*, 3rd ed., Wiley, New York, 2000.

Leo, W. R. *Techniques for Nuclear and Particle Physics Experiments*, Springer, Berlin, 1987.

Northcliffe, L. C. and R. P. Schilling. *At. Data and Nucl. Data Tables*, **A7**, 233 (1970).

Williamson, G. F., J. P. Boujot, and J. Picard. CEA—R03A2, 1966.

## BIBLIOGRAPHY

Harvey, B. G. *Introduction to Nuclear Physics and Chemistry*, 2nd ed., Prentice-Hall, Englewood Cliffs, NJ, 1969. The same information as Meyerhof from a chemist's perspective.

Lieser, K. H. *Nuclear and Radiochemistry: Fundamentals and Applications*, VCH, New York, 1997. Covers a number of the practical aspects of the subject that are important to radiochemists.

Marmier, P. and E. Sheldon. *Physics of Nuclei and Particles*, Vol. I, Academic, New York, 1969. Detailed treatment of many of the important concepts. Still a useful reference.

Meyerhof, W. E. *Elements of Nuclear Physics*, McGraw-Hill, New York, 1967. A simplified treatment that captures the essential details.

Tsoulfanidis, N. *Measurement and Detection of Nuclear Radiation*, 2nd ed., Taylor and Francis, Washington, DC, 1983. Many detailed numerical examples.

# CHAPTER 18

# RADIATION DETECTORS

A fundamental feature of nuclear processes is that the energy released is generally larger than the binding energies of atomic electrons. Any emitted particles will have sufficient energy to ionize atoms. Nuclear radiation is called *ionizing radiation*, therefore, and detecting this ionization allows us to observe nuclear processes. Radiations that interact with matter via the electromagnetic force, that is, electrons, charged particles, and photons, can directly ionize or excite atoms. These radiations are readily detected. Neutrons interact with nuclei only via the nuclear force and are detected through indirect or secondary ionization processes.

We should note that though the energy released in nuclear processes is several (even many) orders of magnitude larger than atomic binding energies, the total number of ion pairs that can be created when radiation interacts with matter is small on a macroscopic scale. For example, typical electron binding energies are about 10 eV. If the total energy available from a 1-MeV nuclear decay was completely converted into electron/ion pairs, then the total number of pairs would be $\sim 10^5$ corresponding to a charge of $\sim 10^{-14}$ C. Even this estimate of the charge created is optimistic because it is unlikely that all of the energy will create ion pairs. (The "effective" ionization energy of most gases is about 35 eV/ion pair because some ion pairs recombine.)

To measure the radiation, the primary ionization must be preserved and not be lost to recombination or scavenging by electronegative atoms. Metals are not useful for creating radiation detectors, therefore. On the other hand, the created ions must be mobile so that they can be collected. This rules out insulating materials

---

*Modern Nuclear Chemistry*, by W.D. Loveland, D.J. Morrissey, and G.T. Seaborg
Copyright © 2006 John Wiley & Sons, Inc.

in most cases. The small electrical signals must be amplified to be observed, and so electronic instrumentation plays a role in modern nuclear chemistry.

Although the various types of radiation detectors differ in many respects, several common criteria are used to evaluate the performance of any detector type. The criteria used for this purpose are as follows:

1. *Sensitivity of Detector* What types of radiation will the detector detect? For example, solid scintillation detectors are normally not used to detect $\alpha$ particles from radioactive decay because the $\alpha$ particles cannot penetrate the detector covering.
2. *Energy Resolution of Detector* Will the detector measure the energy of the radiation striking it, and if so, how precisely does it do this? If two $\gamma$ rays of energies 1.10 and 1.15 MeV strike the detector, can it distinguish between them?
3. *Time Resolution of Detector or Its Pulse-Resolving Time* How high a counting rate will be measured by the detector without error? How accurately and precisely can one measure the time of arrival of a particle at the detector?
4. *Detector Efficiency* If 100 $\gamma$ rays strike a detector, exactly how many will be detected? Each detector discussed here will be evaluated using these basic criteria.

In this chapter we will consider the techniques developed to detect and quantitatively measure how much ionization and/or excitation is caused by different nuclear radiations. As all radiation creates ionization and/or excitation, we will separate the discussion of detection methods according to the general techniques used to collect and amplify the results of the interaction of the primary radiation with matter rather than by the type of radiation. These detection methods can be classified as: (a) collection of the ionization produced in a gas or solid, (b) detection of secondary electronic excitation in a solid or liquid scintillator, or (c) detection of specific chemical changes induced in sensitive emulsions.

A brief summary of these detector types is as follows:

**Gas Ionization** Several detector types take advantage of the ionizing effect of radiation on gases. The ion pairs so produced can be separately collected. When a potential gradient is applied between the two electrodes in a gas-filled ion chamber, the positively charged molecules move to the cathode and the negative ions (electrons) move swiftly to the anode, thereby creating a measurable pulse. Such pulses can be readily measured by the associated devices as individual events or integrated current.

**Ionization in a Solid (Semiconductor Detectors)** In a semiconductor radiation detector, incident radiation interacts with the detector material, a semiconductor such as Si or Ge, to create hole–electron pairs. These hole–electron pairs are collected by charged electrodes with the electrons migrating to the positive electrode and the holes to the negative electrode, thereby creating an electrical pulse. Such pulses contain information on the type, energy, time of arrival, and number of particles arriving per unit time. The important features of semiconductor detectors are their superior energy resolution due to a lower ionization potential and compact size.

***Solid Scintillators*** Some of the energy of ionizing radiation can be transferred to fluor molecules (i.e., compounds that can produce fluorescence) in a crystalline solid. The absorbed energy causes excitation of orbital electrons in the fluor. Deexcitation causes the emission of the absorbed energy as electromagnetic radiation in the visible or near-ultraviolet region (*scintillations*). Observing these weak scintillations visually under certain circumstances is possible (see Deeper Look), but visual observation is normally not a feasible detection method. Instead, a photomultiplier tube close to the solid fluor is employed. In the photomultiplier, the photons are converted to photoelectrons, which are greatly amplified by secondary electron emission through a series of electrodes (dynodes) to cause a sizable electrical pulse. Thus, the original excitation energy is transformed into a measurable pulse.

***Liquid Scintillators*** This detection mechanism is quite similar in principle to the preceding one. Here, however, the radioactive sample and the fluor are the solute in a liquid medium, usually a nonpolar solvent. The energy of nuclear radiation first excites the solvent molecules. This excitation energy eventually appears as photons emitted from the fluor following an intermediate transfer stage. The photons are detected by means of a photomultiplier arrangement.

***Nuclear Emulsions*** The process involved here is a chemical one. Ionizing radiation from a sample interacts with the silver halide grains in a photographic emulsion to cause a chemical reaction. Subsequent development of the film produces an image and so permits a semiquantitative estimate of the radiation coming from the sample.

*Deeper Look—Visual Detection of Radiation* There are very few cases in which nuclear radiation can be directly observed by humans. The eye is not sensitive to photons in the X-ray and γ-ray regions. The energy per decay is small so that only large sources generate enough energy that they warm up and glow. An exception is the Cerenkov radiation emitted by very energetic electrons from the β decay of fission products and Compton-scattered electrons from γ decay of these fragments in nuclear reactors. During operation, the reactor produces very neutron-rich fission products that rapidly emit very penetrating β particles and Compton-scattered electrons. These particles can leave the fuel rods and enter the reactor coolant. The energetic electrons are relativistic and travel with velocities near the speed of light. However, because the speed of light is lower in liquid water than in a vacuum, the electrons emit characteristic blue photons—called Cerenkov radiation—as they adjust their speed downward. Large power reactors do not have viewing ports, but the cores of research reactors are usually visible. The cores are surrounded by an eerie blue glow when the reactors are operating.

## 18.1 DETECTORS BASED ON IONIZATION

Many detectors have been developed to collect and amplify the primary ionization created by nuclear particles. In principle, the careful measurement of this ionization provides the most information about the particle and its energy. The devices with the highest resolution are these detectors based upon ionization. Broadly speaking, ionization-based detectors have the common feature that the incident radiation creates ion pairs in an active volume of the device. An electric field is applied to the active volume to separate the charge pairs and sweep the ions to the electrodes.

Ionization-based detectors have mostly used gases as the active medium. Very few devices use liquids because extremely pure materials are needed to preserve the primary ionization. Gas-filled detectors are easy to construct and operate, but the density of the stopping material is low. The effective ionization potential is large, typically $\sim 20$ eV. Semiconductors are $\sim 10^3$ times denser than gases and have lower ionization potentials, $\sim 2$ eV, but producing large volumes of suitably pure material is expensive. Liquids also have high densities, however, successful devices have only been made with liquefied rare gases, liquid argon, and xenon. The impurity level has limited these devices.

### 18.1.1 Gas Ionization Detectors

As an energetic charged particle passes through a gas, its electrostatic field will dislodge orbital electrons from atoms sufficiently close to its path. In each case, the negatively charged electron dislodged and the more massive positive ion comprising the remainder of the atom form an *ion pair*. The minimum energy (in electron volts) required for such ion pair formation in a given gas is called the *ionization potential*. This value differs markedly for different gases and is dependent on the type and energy of the charged particle. A more meaningful value is the average energy lost by the particle in producing one ion pair, which is nearly independent of particle energy and type (and is about 35 eV).

The *rate of energy loss* will depend on the energy and type of charged particle as discussed in Chapter 17. The $\alpha$ particles will create intense ionization ($-10^4-10^5$ ion pairs/cm of path length), whereas $\beta^-$ particles will produce $10^2-10^3$ ion pairs/cm, and the passage of $\gamma$ rays will result in $1-10$ ion pairs/cm.

***Ion Chambers*** How can we use this primary ionization to produce a detectable signal? The first class of devices to be discussed is the *pulse-type ion chambers*. A sketch of such a device (a parallel-plate ion chamber) is shown in Figure 18.1. Note that one electrode has been connected to the negative terminal of the voltage source, making it the cathode, while the other electrode acts as the anode.

If a 3.5-MeV $\alpha$ particle traverses the chamber, intense ionization will occur along its short path. Since about 35 eV are expended, on the average, in forming an ion pair in air, the 3.5-MeV $\alpha$ particle could form approximately $1 \times 10^5$ such ion pairs before dissipating all its energy. Because of the potential on the chamber electrodes, these ions migrate rapidly to the respective electrodes. The less massive electrons move

## 18.1 DETECTORS BASED ON IONIZATION

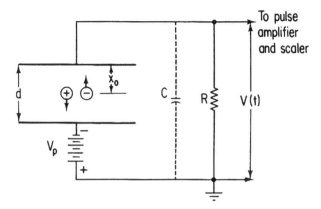

**Figure 18.1** Schematic representation of a parallel-plate ionization chamber in which one ion pair has been formed. $V$ is the voltage source, $R$ denotes resistor, and $C$ denotes the capacitor. (From O'Kelley, 1962.)

very quickly to the anode and produce a rapid buildup of charge there. (Because the positive ions move about 1000 times slower than the electrons, their effect can be neglected for the moment.) The time for collection of the electron charge is about 0.1 to 1 μs depending on the volume of the chamber and the potential gradient. The magnitude of this charge due to the electrons can be calculated as follows:

$$\text{One } e^- \text{ charge} = 1.6 \times 10^{-19} \text{ C}$$

$$10^5 \ e^- \times 1.6 \times 10^{-19} \text{ C/}e^- = 1.6 \times 10^{-14} \text{ C} \tag{18.1}$$

The collected charge flows through the external circuit as a surge, or pulse. If a 20-pF capacitor is used, the potential of the pulse, $V(t)$, is found as follows:

$$V = \frac{Q}{C} \tag{18.2}$$

where $Q$ is the charge (coulombs) and $C$ is the capacitance (farads). For this case, $V = 0.0008$ V. The precise measurement of such small pulses is difficult and sensitive low-noise electronic modules are needed to measure the signals accurately. Note that in these ion chambers, *there is no amplification of the primary ionization.*

The discussion up to now has not been completely accurate in that the effect of the positive ions on charge collection has been totally neglected. In practice, the positive ions are troublesome. Although they move very slowly to the cathode, as they move, they induce a charge on the negative electrode. If no correction is made for this induced charge, the size of the output pulse will depend on the position of the particle track in the chamber volume. A simple method for eliminating this positive ion induction is the addition of a grid to the ionization chamber, as shown in Figure 18.2.

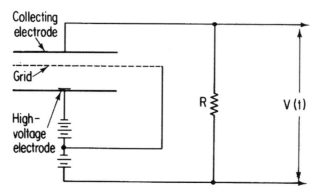

**Figure 18.2** Schematic diagram of a gridded ionization chamber. (From O'Kelley, 1962.)

The grid is charged positively with respect to the cathode, but it is less positive than the anode or collecting electrode. The grid acts to shield the collecting electrode from the effects of the positive ions and accelerates electrons toward the anode. An "internal sample" can be placed on the cathode or high-voltage electrode.

In many applications, instead of recording pulses from each particle that strikes an ionization chamber, the charge from several events is integrated or added. The total current from the chamber is then measured as a function of time. These devices are generally useful for high radiation field measurements. For example, if one 3.5-MeV $\alpha$ particle produces $10^5$ ion pairs, if we have $10^7$ particles/s entering the chamber, we will produce $10^{12}$ ion pairs/s, producing a current of $10^{-7}$ A, which can be readily detected.

The electronic signals from the passage of individual particles through an ion chamber can be accurately measured, along with the energy deposited in the gas as a function of position inside the volume. The *rate of ionization* is a characteristic of the nature of the radiation as discussed in Chapter 17. For example, devices with multiple or segmented anodes have been constructed to take samples of the rate of ionization (Fig. 18.3). Bragg curve counters determine the relative ionization along the path of the particle by measuring the time distribution of ions as they arrive at the anode. These detectors require sophisticated electronic readout to measure the ionization collected as a function of time which encodes the position at which a particle passed through the gas volume. Therefore, Bragg curve detectors and segmented anode ionization chambers are usually only used to detect charged particles from nuclear reactions.

The most sophisticated gas ionization detector is the *time projection chamber* (*TPC*). It is a large, gas-filled cylinder with a negative high-voltage electrode at its center and an external magnetic field. Electrons produced when ionizing radiation passes through the chamber drift toward the end of the cylinder under the influence of axial magnetic and electric fields. The location where the electrons hit the end of the chamber is measured by a set of anode wires. The arrival time of the electrons and the total charge deposited are also recorded. The original particle trajectory through the chamber can be reconstructed from this information.

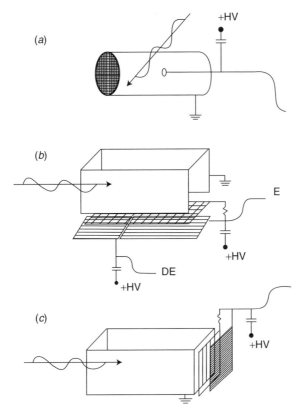

**Figure 18.3** Schematic diagram of an ion chamber that drifts the ionization perpendicular to the particle's path is shown. In this case the anode is segmented and the relative rate of ionization along the path can be determined. The device also contains a Frisch grid between the anode and chamber to improve the pulse–shape response of the device. (c) The schematic version of a detector that drifts the ionization along the particle's path, called a Bragg counter, is shown. The time distribution of the output signal will contain information on the relative rate of ionization all along the particle's path.

Gas-filled ionization counters that collect the primary electrons on a wire, as opposed to a plate, can internally amplify the initial ionization. The cylindrical electric field can be very large near thin wire anodes ($\sim 50$ μm), causing the primary electrons to be accelerated past the point at which they create a secondary ionization cascade (Fig. 18.4). The secondary ions so formed are accelerated by the prevailing potential gradient, thereby producing still more ionization. Thus, from a few primary ion pairs, a geometrical increase results in a veritable torrent of negative ions moving toward the chamber electrodes. The process described is known as *gas amplification*; the flood of ions produced is termed the *Townsend avalanche*, in honor of the discoverer of this phenomenon. Because of gas amplification, most of the electrons are collected at the anode within a microsecond or less from the entrance of a single

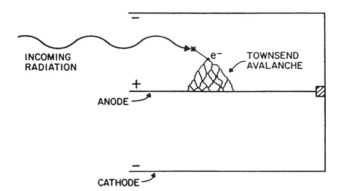

**Figure 18.4** Schematic illustration of the Townsend avalanche in a gas ionization device. The avalanche occurs very close to the wire in reality. (From Knoll, 2000.)

charged particle into the chamber. A strong pulse is thereby formed, fed into the external circuit, and is directly measured after only low amplification.

As the potential gradient between the electrodes of the ionization chamber is further increased, the number of electrons, mostly secondary, reaching the anode rises sharply for a given original ionization event from a charged particle. Eventually, a potential will be reached at which the chamber undergoes continuous discharge and is no longer usable as a detector. There are two distinct potential regions between the ion chamber region (no amplification) and continuous discharge that are useful for gas ionization devices. They are called the proportional region and Geiger–Müller region.

**Proportional Counters** In detectors operating in the *proportional region*, the number of ions that form an output pulse is very much greater than, yet proportional to, the number formed by the initial ionization. Gas amplification factors of about $10^3$–$10^4$ are generally obtained. The amplification factor is primarily dependent on the composition of the chamber filling gas and the potential gradient. At a given potential, the amplification factor is the same for all ionizing events. Consequently, if an $\alpha$ particle traversing the ionization chamber causes $10^5$ primary ion pairs, with an amplification factor of $10^3$, a charge equivalent to $10^8$ electrons would be collected at the anode. An incident $\beta$ particle, on the other hand, producing only $10^3$ ion pairs, would, after amplification by the factor of $10^3$, result in a collected charge equivalent to only $10^6$ electrons.

As with simple ionization chambers, then, it is possible to differentiate between $\alpha$ and $\beta$ particles in the proportional region based on pulse size. This is one advantage of operating a detector in the region. Because the amplification factor in the proportional region is so heavily dependent on the applied potential, highly stable high-voltage supplies are necessary.

The avalanche of electrons in proportional detectors is collected only on part of the anode wire. Furthermore, only a small fraction of the gas volume of the ionization chamber is involved in the formation of ions. These factors result in a very short *dead time*, that is, the interval during which ion pairs from a previous ionization event are being collected and the chamber is rendered unresponsive to a new ionizing particle. Ionization chambers operating in the proportional region are thus inactivated for only 1–2 µs following each ionization event. Dead times as low as 0.2–0.5 µs can be achieved, but if a proportional counter is used for spectroscopy purposes, the average time between pulses should be ~100 µs or greater due to the slower operation of the external amplifiers and other electronics.

Some practical designs for proportional counters are shown in Figure 18.5. In the cylindrical detector, a very thin window of split mica or Mylar plastic covers one end of the tube. It can be so thin (down to 150 µg/cm$^3$) that the absorption of α particles by the window is not extensive. An even more efficient arrangement is found with the hemispherical detector, where the radioactive sample can be introduced directly into the detector chamber. In the hemispherical detector, one detects ~50% of all the

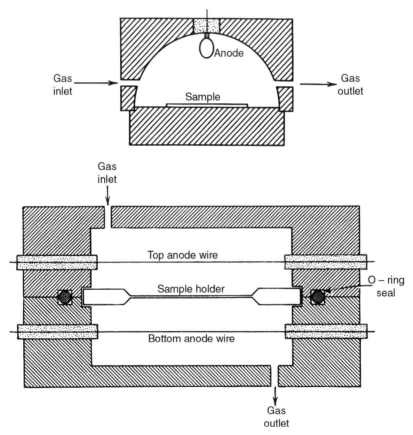

**Figure 18.5** (*a*) Diagram of a 2π gas flow proportional counter. The sample is introduced by sliding out the bottom of the chamber. (*b*) A 4π gas flow proportional counter for absolute counting. (From Wang et al., 1975.)

particles emitted by the source. Such windowless detectors are widely used for α- and weak β-particle counting.

With either ultrathin end-window or windowless detectors, a certain amount of air leaks into the counting chamber. Both the oxygen and the water vapor of the air reduce the detection efficiency because they scavenge the electrons creating slow moving negative ions. Detectors of this variety, therefore, must be purged with an appropriate counting gas before counting is started and must be continually flushed at a lower flow rate during the counting operation. Consequently, such chambers are often called *gas flow detectors*. The operating potential of the chamber is determined, largely, by the gases used for this purpose. Argon, methane, a 90% argon–10% methane mixture, known as P-10 gas, or a 4% isobutane–96% helium mixture, known as Q-gas, are some commonly used counting gases.

Other high-purity gases and gas mixtures are used to fill the detectors. Often argon is used for its relatively high density, but fluorocarbons like $CF_4$ and $C_2F_6$ and hydrocarbons like isooctane and isobutane are also used in devices designed to detect charged particles.

The electronic instrumentation necessary for the operation of the proportional counter is shown in Figure 18.6. Pulses from the detector pass through a preamplifier and amplifier, where they are shaped and amplified. Emerging from the amplifier, the pulses go to a discriminator. The discriminator is set so as not to trip on noise pulses but rather to trip on radiation pulses of any larger size. The number of discriminator pulses produced is recorded by the scaler.

When the count rate of a sample emitting both α and β particles is determined over the voltage range of a proportional detector and the data are plotted, the results are as seen in Figure 18.7. The *characteristic curve* for a proportional detector exhibits two plateaus. The plateau at the lower voltage represents α radiation alone because, in this voltage range, only the α particles, with their much greater specific ionization, produce pulses large enough to trigger the discriminator. Not only may the α particles thus be counted separately from accompanying β radiation at this potential, but also the background radiation counting rate (primarily cosmic rays and γ rays) is extremely low, on the order of a few counts per hour, as well.

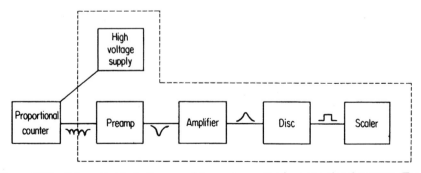

**Figure 18.6** Schematic block diagram of the components of a proportional counter. (From Wang et al., 1975.)

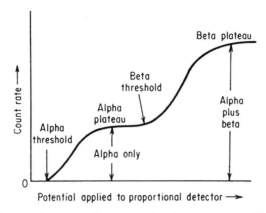

**Figure 18.7** Characteristic curve for a proportional counter. (From Wang et al., 1975.)

As the potential gradient in the chamber is increased, the amplification factor becomes correspondingly greater. Eventually the primary ions produced by even the most energetic β particles are amplified sufficiently to produce pulses large enough to be recorded. This point represents the *β threshold*. Further increases in potential gradient allow even the pulses from the weaker β particles to be registered. The *β plateau* in the operating voltage has now been reached. The count rate here actually represents α plus β radiation. A good proportional counter has a β plateau slope of less than 0.2% per 100 V. The efficiency of proportional detectors for γ radiation is so low that they are seldom used for γ counting. Often, in discussions of the proportional counter, one forgets to mention that the proportional counter is an excellent spectrometer (i.e., an energy-measuring instrument) for low-energy radiation, such as X-rays.

At still higher potential gradients, the gas amplification factor may reach $10^8$. Now even a weak β particle or γ ray can create sufficient ion pairs to completely saturate the available "ion space" in the chamber. Consequently, the size of the charge collected on the anode no longer depends on the number of primary ions produced, and, thus, it is no longer possible to distinguish between the various types of radiation. This potential level is called the *Geiger–Müller region*, after the German physicists who first investigated it. Ionization chambers operated in this potential region are commonly called Geiger–Müller (G–M) detectors. Since the maximal gas amplification is realized in this region, the size of the output pulse from the detector will remain the same over a considerable voltage range until continuous discharge occurs. This fact makes it possible to use a less expensive high-voltage supply than that required for proportional detectors.

Use of a very high amplification factor in the Geiger–Müller region is not without problems. One is the longer dead time of the chamber. Following the passage of an ionizing particle through a detector, an electron avalanche occurs along the entire anode wire, resulting in a cylindrical sheath of positive ions around the anode. The number of such positive ions per pulse will be one or more orders of magnitude greater than that in chambers operated in the proportional region. To be neutralized, the positive ions must migrate to the cathode wall. Being much more massive than the electrons, these ions move at a slower velocity in the electrical field. During this migration, the chamber is unresponsive to any new ionizing particles passing through it. Thus, the dead time of a detector operated in the Geiger–Müller region is from 100 to 300 μs or more.

A correction for "dead time loss" can be made from the value of the dead time, τ. If the true counting rate is $n$ ($\tau = 0$), and $m$ is the measured rate, we have

$$n = \frac{m}{1 - m\tau} \tag{18.3}$$

So, if we measure a counting rate of 1000 cps with a dead time of 250 μs, the true counting rate is

$$N = \frac{1000}{1 - (1000 \times 250 \times 10^{-6})} = 1333 \text{ cps}$$

with the measured counting rate being low by 33%.

Another important problem is the perpetuated chamber ionization resulting from complications associated with the discharge of positive ions at the cathode wall of a detector. As a result, we need a means of terminating or *quenching* the perpetual ionization in the detector. This is done by introducing polyatomic organic compounds or halogen gases into the counter gas. These molecules absorb energy by collisions and dissociate into neutral species, which stops the discharge. Since they are destroyed during quenching, the lifetime of a G–M tube is typically limited to $10^8 - 10^9$ pulses.

One of us (GTS), as a graduate student in 1936, was involved with the discovery of the quenching phenomenon in Geiger–Müller counters. He and a fellow graduate student, David C. Grahame, were plagued with erratic Geiger–Müller counter behavior until they discovered the beneficial effect of water vapor, which was introduced by accident into the argon gas in their counter. They found that reliable behavior also followed the admixture of small amounts of other gases, such as ammonia and natural gas. They did not publish a description of this discovery and were quite interested to read in 1937 the publication by Trost [*Z. Phys.* **105**, 399 (1937)] of his observation that ethyl alcohol had a similar quenching effect leading to reliable operations of such counters.

### 18.1.2 Semiconductor Detectors (Solid-State Ionization Chambers)

As mentioned at the beginning of this section, the primary ionization must be collected to make a direct measurement of the energy of nuclear radiation. Condensed phases have higher densities than gases and so provide more efficient stopping of the radiation per unit length. However, metals allow rapid recombination of the electron/positive ion pairs and insulators inhibit the collection of the charge. Therefore, only semiconductors have been used extensively for radiation detectors. Metals and insulators (like concrete) are used extensively in radiation shielding and some transparent inorganic crystals have a special sensitivity to radiation, which is discussed below.

Silicon and germanium are the most common semiconductors used to construct "solid-state ionization chambers." These materials must be extremely pure to observe the primary ionization ($\sim 10^5$ electrons) and, as we will see below, germanium devices must be cooled to reduce the thermal noise to observe the signals. The properties of small-scale devices based on Group III/Group V materials, for example, GaAs, have been studied, but no large-scale applications have been made. The size and shape of the available semiconductors have grown over time but are still severely limited by production techniques and the availability of high-purity material.

Early solid-state devices relied on observing the ionization in intrinsic semiconductors. Early devices were impractical due to the requirement of extremely pure material. Modern devices are based on semiconductor junction diodes. These diodes have a rectifying junction that only allows the flow of current in one direction. Incident radiation creates ionization inside the bulk of the diode and creates a pulse of current in the opposite direction to the normal current flow through a diode that is straightforward to detect.

To understand how semiconductor radiation detectors operate, it is necessary to review a little of the basic chemistry of semiconductors. Consider a typical Group IV element, such as Si or Ge. It will crystallize in the diamond lattice structure, as shown in Figure 18.8. Each silicon atom is bound by four electron pair bonds to adjacent silicon atoms. The electrons are not free to migrate through the crystal, and therefore pure silicon is a poor conductor of electricity. The electron energy levels of silicon are shown in Figure 18.9. The electron energy levels of the valence electrons are so close together that they form a nearly continuous "band" of energies, known as the *valence band*. In pure silicon, there is a region of energies above the valence band in which there are no allowed energy levels. This energy region is called the *forbidden gap* and corresponds to $\sim 1.08$ eV for silicon. Just above the forbidden gap is the *conduction band*, another band of energies that

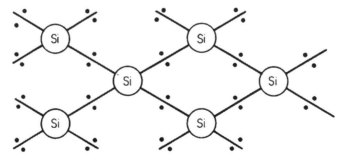

**Figure 18.8** Schematic view of the crystal lattice of Si. The dots represent electron pair bonds between the Si atoms. (From Wang et al., 1975.)

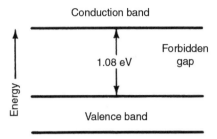

**Figure 18.9** Schematic diagram of the energy levels of crystalline silicon. (From Wang et al., 1975.)

allows free electron migration through the crystal, that is, the conduction of electricity. Suppose we replace a silicon atom in the silicon lattice with a Group V atom, such as phosphorus. Then we will have the situation depicted in Figure 18.10. Phosphorus has five valence electrons. After forming four electron pair bonds to the adjacent silicon atoms, there is one electron left over. This leftover electron will be very loosely bound to the phosphorus atom and will be easily removed to conduct electricity through the crystal.

In terms of our diagrams of the crystalline–electron energy levels, we have the situation shown in Figure 18.11. The "extra" phosphorus electron occupies a "donor level" very close to the conduction band and is easily promoted into this conduction band. Silicon containing Group V impurities, such as phosphorus, is called *n-type silicon* because the species that carries charge through the crystal is negative.

What happens when a Group III atom, like boron, replaces an atom in the silicon lattice? The situation is shown in Figure 18.12. Boron has three valence electrons and can form electron pair bonds with three of its neighbors. It has no electron to pair up with the electron on the fourth silicon atom. We say we have an electron *hole* in the silicon lattice.

In terms of our energy level diagrams, we have the situation illustrated in Figure 18.13. The hole occupies an energy level very close to the valence band (an *acceptor level*) and can be easily promoted into the valence band. (Promotion

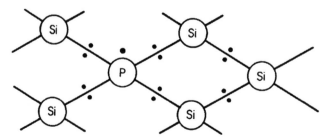

**Figure 18.10** Schematic diagram of a typical n-type impurity in a silicon crystal lattice. (From Wang et al., 1975.)

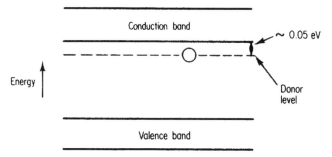

**Figure 18.11** Schematic diagram of the energy levels of crystalline silicon with a donor impurity. (From Wang et al., 1975.)

of a hole into the valence band simply means that an electron in the valence band and a hole in the acceptor level switch places, so that a hole is created in the valence band.)

We must realize (unphysical as it may sound) that a hole in the valence band can conduct electricity as well as an electron in the conduction band. How does this work? Consider Figure 18.14. Imagine that electron 1 moves to fill hole 0. This step creates a hole at position 1. Electron 2 moves to fill this hole, leaving a hole at position 2. Electron 3 fills the hole at position 2, leaving a hole at position 3, and so forth. Thus, as the hole moves to the right in Figure 18.14, negative charge is moving toward the left. Since electricity is the movement of charge, the motion of the hole corresponds to the flow of electricity. Silicon containing Group III impurities is said to be *p-type silicon* because of the positive charge carriers (the holes).

A silicon–semiconductor–radiation detector of a layer of p-type silicon in contact with a layer of n-type Si is shown in Figure 18.15. What happens when this *p-n junction* is created? The electrons from the n-type silicon will migrate across the junction and fill the holes in the p-type silicon to create an area around the p-n junction in which there is no excess of holes or electrons. (We say that a *depletion region* has been formed around the junction.) Imagine that we apply a positive voltage to the n-type material and a negative voltage to the p-type material (the junction is said to be *reverse biased*). The electrons will be "pulled farther

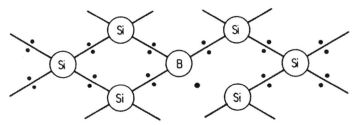

**Figure 18.12** Schematic diagram of a silicon crystal lattice with a p-type impurity in it. (From Wang et al., 1975.)

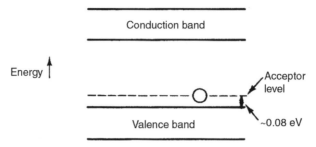

**Figure 18.13** Schematic diagram of the energy levels of silicon with a p-type impurity. (From Wang et al., 1975.)

away" from the junction by the positive voltage on the n-type material, thus creating a much thicker depletion region around the p-n junction. The exact thickness of the depletion region, $d$, is given by

$$d \propto (\rho V)^2 \qquad (18.4)$$

where $\rho$ is the resistivity of the silicon and $V$ is the magnitude of the applied reverse-bias voltage. Note that the depletion of the depletion region can be varied at will by changing the voltage applied to the detector.

The depletion region acts as the sensitive volume of the detector. The passage of ionizing radiation through this region will create holes in the valence band and electrons in the conduction band. The electrons will migrate to the positive charge on the $n$ side, while the holes will migrate to the negative voltage on the $p$ side, thereby creating an electrical pulse at the output of the device.

The *average* energy necessary to create a hole–electron pair in silicon is ~3.6 eV. [This average energy is about three times the forbidden gap energy (~1.1 eV) because most electrons are promoted from deep in the valence band to high in the conduction band.] The energy required to create a hole–electron pair is independent of particle charge and mass, thus causing semiconductor detector response to be independent of particle type. If we remember that the average

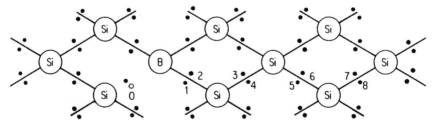

**Figure 18.14** Schematic diagram of a silicon crystal lattice with a p-type impurity boron, at one lattice point. The hole is labeled 0, while the electrons are denoted as 1, 2, etc. (From Wang et al., 1975.)

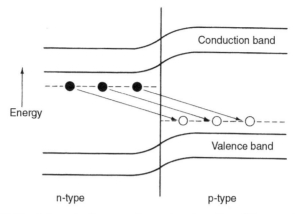

**Figure 18.15** Schematic diagram of a p-n junction. (From Wang et al., 1975.)

energy to create an ion–electron pair in a gas ionization device was $\sim$35 eV, then we see that, for the same energy deposit in the detector, we get $\sim 35/3.6 \approx 10$ times more charged pairs. If we note that the energy resolution of a detector, $\Delta E/E$, is proportional to $N^{-1/2}$ where $N$ is the number of charge pairs formed, we can see that the energy resolution of a semiconductor is approximately $10^{1/2} = 3.2$ times better than the energy resolution of a gas ionization detector. [Furthermore, as we will see later, the average $\gamma$-ray energy deposit required to liberate one photoelectron at the cathode of a photomultiplier tube is $\approx$1000 eV, then we can say that the resolution of a semiconductor detector is $(1000/3.6)^2 \approx 17$ times better than that of a scintillation detector.] More detailed considerations show that the observed resolution can be smaller than the estimate based on the statistics of ion pair formation due to correlations between processes giving rise to ion pair formation.

For some semiconductor detectors, germanium is used instead of silicon for the detector material. The reasons for this substitution are as follows: (a) The average energy needed to create a hole–electron pair in germanium is 2.9 eV rather than the 3.6 eV necessary for Si. Thus, the energy resolution for germanium should be $(3.6/2.9)^2 = 1.1$ times better than silicon. (b) The atomic number of germanium (32) is much higher than that of silicon (14), leading to increased probability of $\gamma$-ray interaction with the detector material. Consequently, germanium is preferred to silicon for $\gamma$-ray detection. The forbidden gap is so small, however, for germanium (0.66 eV) that room temperature thermal excitation leads to the formation of hole–electron pairs in the solid. Therefore, germanium detectors must be operated at liquid nitrogen temperature (77 K) to prevent this thermal electron noise from overwhelming the small signals from the primary ionization.

The silicon-based solid-state detectors fall into three general categories: surface barrier devices, PIN diodes, and Si(Li) (pronounced "silly") devices. These detectors are used to measure short-ranged radiation: charged particles in the first two cases and low-energy $\gamma$ rays and X-rays in the third case. The detector consists of a thin layer of silicon material (often ~200 μm thick, but thicknesses from 5 μm up to 5 mm are available). An electric field (typically ~V/μm) is applied in the direction opposite to the "normal" flow of current through the diode. Radiation creates electron–hole pairs that are swept to the electrodes by the electric field and induce a current signal. These signals are amplified in an external circuit.

*Silicon-surface barrier* (SSB) detectors consist of a thin cylindrical piece of high resistivity ($10^3$ Ω cm) pure n-type silicon with a thin gold contact on one side and an aluminum contact on the other (Fig. 18.16). The gold contact is a thin layer through which the radiation enters the silicon. Just under the gold is an oxide layer that forms the semiconductor junction (or barrier). The gold layer is sensitive to physical wear, and the oxide layer can be depleted by extended exposure to vacuum. The oxide layer is also very sensitive to organic molecules but can be reconstructed with proper treatment. During use these detectors must be shielded from visible light as electron–hole pairs can be created by photons that enter the silicon through the thin gold contact. The gold and oxide layers are also thin to reduce the amount of kinetic energy lost by the particle before it enters the active silicon region. These layers make up a dead layer that can be significant in α spectroscopy. Recently, SSBs with very thin and uniform dead layers have been created with ion-implanted junctions. A thin layer of boron ions is implanted near the surface of n-type silicon to form the junction.

"Ruggedized" detectors are available in which the radiation enters the silicon through the thicker and light-tight aluminum contact. The bulk of the material is p-type silicon and a negative bias is applied to the gold contact so that the entry

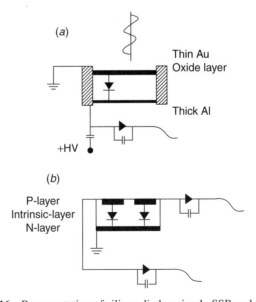

**Figure 18.16** Representation of silicon diodes, simple SSB and segmented.

window can remain at ground potential. Another design of a "rugged" surface barrier detector replaces the gold contact with a thicker nickel contact. The nickel is resilient enough to be wiped clean.

Silicon surface barrier detectors have found widespread application in α-particle spectroscopy and in nuclear reaction studies. These detectors can be used in stacks to identify particles uniquely. Consider the situation in which a penetrating ion with a total kinetic energy (KE) passes through a thin SSB detector and is stopped in a second thicker SSB detector. Such a stack of detectors is called a silicon detector telescope and provides two signals, the energy lost in the thinner detector, $\Delta E$, and the remainder, $KE - \Delta E$. The rate of energy loss for a charged particle is approximately given by the simple expression:

$$\frac{dE}{dX} \sim \frac{mZ^2}{KE} \tag{18.5}$$

where $m$ is the ion's mass, $Z$ its nuclear charge, for a given kinetic energy. When the first detector is thin, $dE/dX$ is approximately constant and $\Delta E \cong (dE/dX) \Delta X$. Thus, $\Delta E$ will be a hyperbolic function of KE for each ion with a different value of $mZ^2$. Therefore, the components of a mixture of penetrating charged particles can be identified by their relative values of ionization. Stacks of SSBs with several thin detectors are used when unambiguous particle identification through redundant measurements is necessary. Variations in the thickness of the silicon in the manufacturing process place a practical limit on the particle identification. Thickness variations as small as 1 μm are currently achievable.

The typical resolution of a single detector is ~20 keV but depends on the detector geometry, in particular, on the detector capacitance. Notice that SSB detectors have parallel electrodes separated by a thin dielectric; the capacitance of such an object will increase with increasing area and with decreasing thickness. Thus, thin large-area devices will have the largest capacitance and thus the poorest resolution.

*Silicon PIN diodes* are a more recent class of detectors that have become available, in large part, due to the growth of the semiconductor industry. These devices are made up with a p-type layer on one side of an intrinsic silicon wafer and an n-type layer on the opposite, therefore a p-I-n sandwich. The detectors are available in a much larger range of sizes and shapes than surface-barrier detectors. For example, 25-cm$^2$ devices with single or multiple specially shaped contacts are routinely available. The contacts on the front and back of PIN diodes can have different shapes and sizes. Consider a 5-cm × 5-cm rectangular wafer that has two horizontal electrical contacts on the front and two vertical contacts on the back. The divided contacts define four quadrants of silicon that are electrically separate. A particle that enters the detector will generate two signals, one on the front and another on the back that uniquely identifies the quadrant of silicon. Rectangular devices are often used with 16 stripes on the front and 16 stripes on the back and provide very accurate position measurements in nuclear reaction studies.

All solid-state detectors are damaged by long exposures to charged particles. The threshold dose for observable damage (in particles/cm$^2$) is $\sim 10^8$ for fission fragments, $10^9$ for $\alpha$ particles, $10^{12}$ for fast neutrons, and $10^{13}$ for electrons. Notice that the particles generally come to rest in the silicon and stay there. The lattice will be disrupted and poisoned by the presence of many stopped particles. If the particles have the same energy (same range), then all of the particles will stop in a very narrow band of the silicon and can create a dead layer inside the silicon.

*Lithium-drifted silicon detectors* have been developed for measurements of $\beta$ particles. Electrons are more penetrating than heavier charged particles and so these devices, usually called Si(Li) detectors have to be much thicker than heavy-ion detectors. Si(Li) detectors are commonly 5 mm thick. Such large volumes of very pure silicon are not readily available; thus, the technique of drifting lithium ions into the bulk material to compensate for internal lattice defects in p-type silicon has been developed. A layer of lithium metal is applied to the surface and some atoms diffuse into the bulk silicon. The lithium atoms readily donate an electron into the conduction band and become ions. A bias can be applied to the silicon that causes the lithium ions to migrate from the surface through the lattice. The migrating ions will be trapped by negative impurities in the lattice, thus "compensating" for the effect of the impurity. The lithium ions retain their high mobility in the lattice and the detectors have to be stored with a small retaining bias if they are stored for long periods at room temperature.

Si(Li) detectors are favored over Ge(Li) detectors for $\beta$ detection because of their low $\gamma$ sensitivity and their lower (by $\sim\frac{1}{3}$ to $\sim\frac{1}{2}$) backscattering. The energy resolution of Si(Li) detectors for electrons is $\sim 1$–$2$ keV for electron energies up to 1000 keV. The detection efficiency of Si(Li) detectors for $\beta$ particles ranges from one-half that of a gas counter for a low-energy $\beta$ emitter like $^{14}$C to greater than that of a gas counter for an energetic $\beta$ emitter such as $^{32}$P. The background of these detectors is exceptionally low because of their small size for a given stopping power, and they do not require any peripheral gas supply, and the like. Very good energy resolution for X-ray detection is possible. A resolution of 180 eV for the 5.9-keV Mn K$_\alpha$ X-ray have been obtained with Si(Li) detectors, whereas the best energy resolution available from a scintillation detector is about 1000 eV.

All of the silicon detectors can be cooled to reduce the thermal noise that produces a background under all the induced signals. The thermal noise is created by random fluctuations that promote an electron across the bandgap into the conduction band resulting in an electron–hole pair. The number of promoted electrons will be proportional to a Boltzmann function containing the bandgap $\Delta$ and the temperature $T$:

$$N_{\text{thermal}} \propto e^{\Delta/kT} \qquad (18.6)$$

For practical reasons, silicon detectors are usually cooled from room temperature down to approximately $-20°$C; cooling below $-60°$C is not useful because the system noise becomes dominated by the external electronic circuit. Temperatures below $-20°$C are not used also because the internal physical stresses from differences in the thermal expansion coefficients of the construction materials become important. Specially prepared detectors are recommended before cooling to the lower temperatures.

We can estimate the factor by which the thermal noise will be reduced with the Boltzmann expression:

$$\frac{N_{\text{thermal}}(T=-20°)}{N_{\text{thermal}}(T=+25°)} \sim e^{253/298} = 2.3 \quad (18.7)$$

Cooled silicon detectors are particularly useful in experiments in which the measured particles are expected to cause significant damage to the crystal lattice during the experiment. If the detector is not cooled, the thermal noise will dramatically change during the measurement and the detector resolution will decrease with time.

*Germanium detectors* have the highest resolution of any direct ionization devices. This is due to the small bandgap of germanium of 0.73 eV (at 80 K) and effective ionization potential of 2.95 eV that allows the creation of many ion pairs for a given amount of radiation. The bandgap is also small enough that the number of electron–hole pairs created by thermal fluctuations causes a very significant electronic noise. The noise is reduced by enclosing the germanium and the first stage of the amplification circuit in a cryostat and cooling both to liquid nitrogen temperature.

Other things being equal, the size of the signals produced in a germanium diode compared with a silicon diode should be larger by the inverse ratio of the effective ionization potentials, $3.76/2.95 \sim 1.27$ (at 80 K). However, the thermal noise will be larger in proportion to a Boltzmann exponential distribution with the bandgaps, so at the same temperature the noise in the germanium will be larger by the factor:

$$\frac{e^{-\Delta_{\text{Ge}}/kT}}{e^{-\Delta_{\text{Si}}/kT}} = e^{1.16/0.73} = 4.9 \quad (18.8)$$

These facts would appear to favor the use of silicon detectors strongly. However, the "stopping power" of matter for photons is much lower that than for charged particles giving photons long penetration depths in all materials. Moreover, the probability of a photoelectric interaction with an atom, which contributes significantly to the absorption of the full energy of photons, increases in proportion to $Z^5$. This makes high-Z materials more effective total absorbers of photons. Also from the practical standpoint, manufacturing techniques have been developed to produce very high purity germanium crystals that are much larger than silicon crystals.

Germanium detectors are used almost exclusively to detect $\gamma$ radiation. Energetic photons can easily penetrate the cryostat, and the high resolution of germanium detectors are well suited to the very precise energies of the $\gamma$ rays emitted by the deexcitation of nuclear levels. There are two main classes of germanium detectors, those that use lithium compensated material, called Ge(Li) (pronounced like *jelly*), which have been supplanted by intrinsic germanium, also called high-purity germanium.

Ge(Li) detectors are similar in principle to the Si(Li) detectors described above. Large ingots (~1 L) of p-type germanium material are prepared in a relatively pure state. Lithium metal is applied to the surface, some atoms diffuse into the lattice, donate their electron, and the ions are subsequently drifted through the material. The mobility of lithium ions is much higher in germanium than in silicon. This allows very effective compensation of the impurities in large volumes of germanium but also allows the rapid loss of compensation by the lattice at room temperature. The hallmark of Ge(Li) detectors is that the germanium crystals must be kept at liquid nitrogen temperatures for their entire useful lifetime. If the detectors are allowed to warm up, even for a very short time, the lithium compensation is lost to some extent and their high resolution is degraded. The crystals can be redrifted but the manufacturer must do this.

The shape of the germanium crystals is generally cylindrical with the lithium applied to the outer surface and drifted in toward the center. The lithium contact produces an n-type region on the surface of the crystal, the bulk becomes intrinsic through compensation, and a small p-type region is deliberately left undrifted to produce a P-I-N diode structure. The detector thus has coaxial p- and n-type "electrodes" and is sometimes called a coax detector. A reverse bias is applied to the germanium diode, as to the silicon detectors, and the small current pulses from the primary ionization events are collected and amplified. The requirement that Ge(Li) detectors must be kept very cold during their entire life span has spurred the development of germanium purification techniques. Large volumes of intrinsic germanium material can now be produced in sufficient quantity to produce PIN diode detectors without lithium drifting. These devices can be stored at room temperature and only need be cooled when they are used. The n-type region is prepared by lithium diffusion without drifting and an extremely thin p-type region is prepared by implanting boron ions. The crystals are generally cylindrical with coaxial electrodes. Other shapes, such as thin planar detectors for low-energy γ rays and X-rays and crystals with hollow wells with large geometrical efficiency are available.

For γ-ray detection, the detector of choice is the Ge detector because of its higher Z. The most spectacular feature of the Ge detector is the superior energy resolution. An energy resolution of 1.75 keV for the 1332-keV γ ray of $^{60}$Co is routinely obtained [compared with the typical 90–100 keV for a 3 × 3 in. NaI(Tl) detectors]. What this means in terms of the ability to resolve γ-ray spectra is shown in Figure 18.17.

## 18.2 SCINTILLATION DETECTORS

In a scintillation detector, a fraction of the energy deposited by the primary radiation in the detector is converted to light that, in turn, is converted into an electrical signal. Conceptually, the process can be divided into the scintillation process itself (energy → light), the collection and conversion of the light into electrons

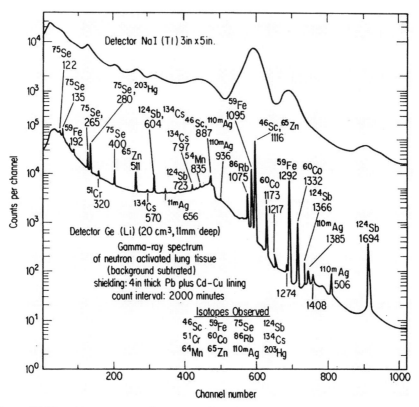

**Figure 18.17** γ-Ray spectrum of neutron-activated lung tissue as measured with a 3 × 3 in. NaI(Tl) detector and a Ge detector. (From Cooper, 1971.)

and the multiplication of the electrons to make a macroscopic signal. We will divide our discussion similarly.

As discussed in Chapter 17, as radiation interacts with matter, it will lose energy by ionizing or exciting matter. As we have seen, only a few materials have the right properties to allow the collection of the primary ionization from nuclear radiation. If the ionization is not preserved and collected, the electron/positive ion pairs are expected to recombine eventually. During this recombination, the energy used to separate the charges will be reemitted to the surroundings, very often as lattice vibrations and heat. Occasionally, a triplet electronic excited state is populated, and the energy from such states is released as visible photons. This emission process is well known as atomic or molecular fluorescence and is called *scintillation* when it is caused by exciting radiation. (We will use the term "visible light" loosely in our discussion. The wavelengths of the fluorescent photons from excited electronic states are characteristic of the material and range from UV to red.) These visible, secondary photons can be easily detected and amplified with photomultiplier tubes.

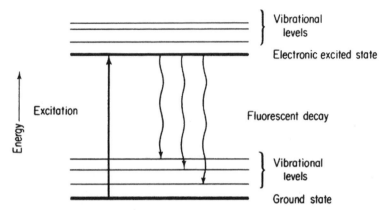

**Figure 18.18** Schematic view of the scintillation mechanism in organic crystals. (From Wang et al., 1975.)

The details of the scintillation process are complicated and depend very much on the molecular structure of the scintillator. In organic crystals, the molecules of the organic solid are excited from their ground states to their electronic excited states (see Fig. 18.18). The decay of these states by the emission of photons occurs in about $10^{-8}$ s (fluorescence). Some of the initial energy absorbed by the molecule is dissipated as lattice vibrations before or after the decay by photon emission. As a result, the crystal will generally transmit its own fluorescent radiation without absorption.

There are three common types of organic scintillator. The first type is a pure crystalline material, such as anthracene. The second type, *the liquid scintillator*, is the solution of an organic scintillator in an organic liquid, such as a solution of *p*-terphenyl in toluene ($\approx$3 g solute/L solution). The third type is the solution of an organic scintillator, such as *p*-terphenyl, in a solid plastic, such as polystyrene.

All these organic scintillators are characterized by short fluorescence lifetimes, that is, 2–3 ns. This allows their use in high-count rate situations or for fast time measurements. The light output is modest, being 10–50% of that of NaI(Tl) (see below). Because of the low Z of the organic scintillators, they are primarily used in the detection of heavy charged particles or electrons. The plastic scintillators are easily machined into a variety of shapes and/or made as thin films. Energy resolutions of about 10–14% are possible. Liquid scintillators are used to assay low-energy $\beta$ emitters, like $^{14}$C or $^{3}$H. The radioactive material is dissolved or suspended in the scintillator solution. Another application involves the use of liquid scintillators for large volume (several m$^3$) detectors where the liquid scintillator has been "loaded" with a neutron absorbing material, such as gadolinium. The gadolinium captures neutrons producing $e^-$ and $\gamma$ rays that are detected by the scintillator.

The scintillation process in inorganic scintillators differs from that in organic scintillators. Consider the structure of an ionic crystal, as shown in Figure 18.19. When an energetic electron passes through the crystal, it may raise valence electrons from the valence band to the conduction band. The electron vacancy in the valence

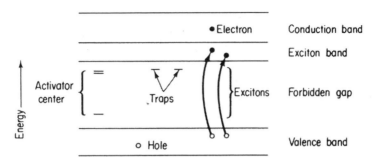

**Figure 18.19** Electronic energy levels in an ionic crystal. (From Wang et al., 1975.)

band resulting from this ionization is called a *hole*, in the valence band. The electron in the conduction band and the hole in the valence band can migrate independently through the crystal.

Alternatively, another process called *excitation* can occur by which a valence band electron is excited to an energy level lower than the conduction band. The electron remains bound to the hole in the valence band. This neutral electron–hole pair is called an *exciton*, and it can move through the crystal also. Associated with the exciton is a band of energy levels called the exciton band (see Fig. 18.19).

The presence of lattice defects and/or intentionally placed impurities in the alkali halide crystal will cause the formation of local energy levels in the forbidden gap, called traps or *activator centers*. Figure 18.19 shows the energy levels of an alkali halide crystal, including the activator centers and traps. (Atomic thallium is a common activator for alkali halide crystals.)

Excitons, holes, and electrons produced by the interaction of radiation with the crystal wander through the crystal until they are trapped at an activator center or trap. Migration of an exciton in a crystal may be thought of as a 6- to 8-eV excited iodide ion, $I^{-*}$, transferring its energy to an adjacent stable $I^-$, which, in turn, becomes excited. Thus, energy may be transferred from $I^-$ to $I^-$ in the crystal lattice to final capture by either an activator center or crystal impurity. By exciton capture or hole–electron capture, the activator centers are raised from their ground state to an excited state. The deexcitation of this activator center by emission of light occurs in a time about 0.3 μs. Hence the energy deposited by the radiation in the scintillator is emitted as light by the activator center (Tl). The amount of light emitted by the entire crystal is directly proportional to the amount of energy deposited in the crystal by the incident radiation. The fraction of the deposited energy converted into fluorescence photons is small, about 10%.

Thallium-activated sodium iodide [NaI(Tl)] is the most widely used inorganic scintillator. This material is used extensively to detect γ rays because it is relatively inexpensive, has a high stopping power for photons, and is rugged and easy to use. The fluorescence light output has a relatively slow decay time of almost 230 ns, limiting the count rate in such detectors. The energy resolution of NaI(Tl) detectors is rarely better than 6% for the 1332-keV γ ray of $^{60}$Co (as compared to the 0.13% typically seen with Ge detectors). NaI(Tl) detectors are very efficient for detecting γ radiation (with typical detection efficiencies of 1–10%). The efficiency of a 3-in. diameter right cylinder that is 3 in. long is the reference standard for γ-ray detectors. Other inorganic scintillators of note are bismuth germanate ($Bi_4Ge_3O_{12}$, BGO) which is a high Z, high-density material. Its low-light output [10–20% that of NaI(Tl)] limits its use to situations where a high efficiency (with poorer resolution) is needed. Barium fluoride ($BaF_2$) is a high Z material with a fast light output ($\tau < 1$ ns) with reduced light output. It is used in situations where its high density, high Z, and fast timing are important.

How is the light emitted by the scintillator converted into an electrical signal? To answer this question, let us consider a schematic diagram of a typical scintillation detector (Fig. 18.20). The photons of visible light emitted by the activator centers,

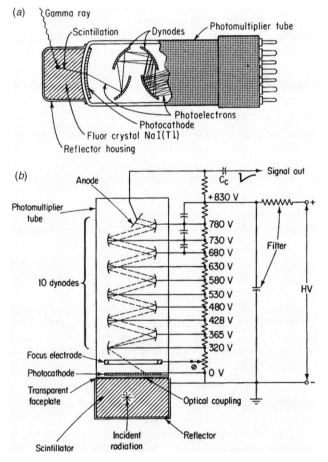

**Figure 18.20** (a) Cutaway diagram of a typical solid scintillation detector. (b) Diagram of a scintillation detector illustrating schematically the way in which light from the scintillator is transmitted to a photomultiplier tube. A typical wiring diagram is shown for the 10-stage photomultiplier operated with a positive high-voltage supply. (From O'Kelley, 1962.)

such as Tl$^+$ ions in a NaI(Tl) crystal, pass through the transparent fluor substance and out through a clear window to impinge on an adjacent photocathode. The typical photocathode is composed of a thin, photosensitive layer (commonly a cesium–antimony alloy) on the inner surface of the end of the photomultiplier tube. Here impinging photons, particularly those having wave lengths between 3000 and 6000 A, are absorbed, with a consequent emission of photoelectrons. The number of photoelectrons ejected is slightly less than, but directly proportional to, the number of incident photons. Such a burst of photoelectrons resulting from a single γ-ray interaction in the crystal is still far too weak to be registered directly.

Amplification occurs by means of a series of electrodes, called *dynodes*, spaced along the length of the photomultiplier tube (see Fig. 18.20). Each dynode is maintained at a higher potential (usually about 50 V higher) than the preceding one. The photoelectrons emitted from the photocathode are focused by an electric field to hit the first dynode. Striking the dynode surface, they cause the secondary emission of a larger number of electrons. This new burst of electrons is guided by the potential gradient to the second dynode in the series, where a still larger number of electrons are dislodged. This electron-multiplying process continues at each dynode until at last the collecting anode is struck by $10^5$–$10^6$ electrons for each original photoelectron ejected from the photocathode. Thus, the size of the output pulse from the photomultiplier is directly related to the quantity of energy dissipated by the incident γ-ray photon in the fluor.

To present a quantitative example of the energy conversions involved in scintillation detection, we trace the results of the interaction of a single 1.17-MeV γ ray from $^{60}$Co with a thallium-activated sodium iodide crystal [NaI(Tl)]:

1. If 20% of the energy of the γ ray results in exciton production in a fluor crystal, and if it is assumed that 7 eV is needed to produce an exciton, then approximately 33,000 excitons could result from this γ-ray photon.
2. Assuming that only 10% of the excitation events result in the production of photons of visible light seen by the adjacent photocathode, this would mean that about 3300 photons would reach the photocathode.
3. This number of photons striking a photocathode with a conversion efficiency of 10% would eject approximately 330 photoelectrons.
4. The successive dynodes of a photomultiplier operated at an overall gain of $10^6$ could then amplify this quantity of photoelectrons so that $\sim 3.3 \times 10^8$ electrons would be collected at the photomultiplier anode, or a charge of $\sim 5 \times 10^{-11}$ C.
5. This charge could then be transformed by a preamplifier circuit with a capacitance of 30 pF into an output pulse of 1.8 V. A pulse of this size would be capable of directly triggering a scaler.

$^{60}$Co emits two γ rays per disintegration. The other γ ray has an energy of 1.33 MeV. Following the preceding calculations, this 1.33-MeV γ ray would result in an output pulse of about 2.05 V from the detector. The statistical uncertainty in the pulse heights would be determined by the step in the chain with the greatest uncertainty (Step 3). In this case, we would expect a resolution of ≈6% so that the two pulse heights would be barely resolved.

## 18.3 NUCLEAR TRACK DETECTORS

The passage of highly ionizing radiation through an insulating solid leaves a wake of destruction in the material. In covalently bonded materials, the chemical structure of the material along the track can be significantly and permanently changed by the passage of a single energetic ion. Certain polymeric (plastic) materials and the mineral mica (a form of silicon dioxide) are particularly sensitive to such radiation damage. The original radiation damage remains localized on the molecular scale but is not visible without enhancement. However, the track can be expanded by chemical etching from the molecular scale (nanometers) up to the microscopic scale (micrometers).

Nuclear track detectors are very simple and very efficient detectors of rare events that produce highly ionizing radiation. Carefully prepared and scanned track detectors have been used to identify individual rare decays. The detectors are integrating in that the damage caused by a track is not spontaneously repaired. The drawback to track detectors is that the tracks are small and can only be observed with a microscope. In the past, scanning by eye was extremely labor intensive and prone to error. Modern computer-controlled scanning has improved the speed and reliability of the analysis. Plastic track detectors that are sensitive to $\alpha$ particles are used extensively in commercial radon detectors.

Chemical etching of the material takes place on all surfaces that are exposed to the etching solution. The exposed surfaces of the material are eroded along with the material along the track. Therefore, the rate of etching has to be carefully controlled to get the maximum amount of information from the track. Notice that etching of a uniform track will generally form a circular cone because the material will be more easily removed from the surface than from deep along the track. Mica tracks are diamond-shaped due to the lattice structure as opposed to being circular.

Nuclear emulsions are closely related "track detectors" that trace their origins to the original discovery of radiation by Becquerel. Nuclear emulsions are very fine-grained photographic film. The film is "exposed" by the passage of radiation through it, and the grains of AgCl are activated by the ionization. The film is developed, and, with careful handling and microscopic observation, the track or path of individual particles can be traced. Occasionally, a particle interacts with a nucleus in the emulsion, creating many fragments or particles, and the tracks of the reaction products can be traced. The emulsion is also sensitive to the rate of ionization and the nature of the particle in each track can often be determined. On the other hand, most people are familiar with the shadow images of skeletal features taken with X-rays. The X-rays are absorbed and scattered more efficiently by the heavy elements in bones (essentially calcium) than by the light elements in soft tissue (carbon, oxygen, hydrogen) and create a shadow. The grains in the emulsion are then exposed by the transmitted X-rays and are developed to form the negative image.

## 18.4 NUCLEAR ELECTRONICS AND DATA COLLECTION

As we have seen, essentially all of the nuclear radiation detectors produce electronic pulses in response to the interaction of some ionizing radiation. These signals are processed by standardized *nuclear instrumentation modules* (NIM) electronics to count the number of pulses or to more fully analyze the size or even the shape of the signal. In addition, computer-based electronics in the CAMAC (computer automated measurement and control) system are used to measure the time relationships of pulses, the pulse heights, and the signal shapes. The signals are recorded and stored by computers for later analysis. An important feature of scientific studies with radioactivities and with nuclear beams is that the data must be collected as rapidly as possible usually during a very limited time. A radioactive source will decay away after being produced and cannot be "stopped" because the scientist is not ready to use it. Similarly, the nuclear reactions induced by particle beams take place in a very short time and must be recorded when they occur. Then after a set of events has been collected "on-line," the data are analyzed "off-line."

We will give a very brief overview of the kinds of modules used. CAMAC and NIM electronics fall into three broad categories: *linear electronics* that maintain a linear relationship to the size of the initial signal, *logic circuits* that provide only a standard (or single sized) pulse indicating that a given logical condition was met, and *data acquisition modules* to measure the signals and record the data. One should realize that with modern high-density electronics the functions that we will describe can correspond to a single electronic module or may be condensed into a single integrated circuit. Therefore, we will only describe the *functions* performed by the electronic modules and not specific equipment.

The output of most detectors is an electrical pulse that carries information about the energy deposited in the detector, the time of the interaction, and the like. Linear electronics are described as modules that preserve and extract information about the energy deposit in the detector from the detector signal. An overview of these modules and their function is given in Table 18.1. A typical pulse-height analysis system is shown in Figure 18.21. The signal from the detector is given a preliminary amplification and shaping by a *preamp* before being sent through a coaxial cable to a *linear amplifier*. This is done to prevent noise in the cable from destroying the tiny detector signal. In the amplifier the signal is further amplified and shaped before analysis. The height of the pulse is related to the energy deposited in the detector. The analog-to-digital converter (ADC) converts the signal from the amplifier into digital data (a number of standard pulses) thus measuring its size. The ADC could be contained on a plug-in card in a personal computer (used to measure the distribution of pulses from a single detector monitoring a radioactive source), or it might be one of many identical ADC units in a CAMAC

**TABLE 18.1 Summary of Common Pulse-Processing Modules**

| | Input | Output |
|---|---|---|
| *Linear input– linear output* | | |
| Preamplifier | Linear charge pulse from detector | Linear tail pulse |
| Linear amplifier | Linear tail pulse | Amplified and shaped linear pulse |
| Biased amplifier | Shaped linear pulse | Linear pulse proportional to amplitude of input pulse that lies above input bias level |
| Pulse stretcher | Fast linear pulse | Conventional shaped linear pulse of amplitude equal to input pulse |
| Sum amplifier | Two or more shaped linear pulses | Shaped linear pulse with amplitude equal to the sum of the coincident input pulses |
| Delay | Fast linear or shaped linear pulse | Identical pulse after a fixed time delay |
| Linear gate | Shaped linear pulse and gate pulse | Linear pulse identical to linear input if gate pulse is supplied in time overlap |
| *Linear input– logic output* | | |
| Integral discriminator | Shaped linear pulse | Logic pulse if input amplitude exceeds discriminator level |
| Single-channel analyzer | Shaped linear pulse | Logic pulse if input amplitude lies within acceptance window |
| Time pickoff | Fast linear or shaped linear pulse | Logic pulse synchronized with some feature of input pulse |
| *Logic input– linear output* | | |
| Time to amplitude converter | Logic start and stop pulses separated by a time $\Delta t$ | Shaped linear pulse with amplitude proportional to $\Delta t$ |
| *Logic input– linear output* | | |
| Coincdence | Logic pulses at two or more inputs | Logic pulse if pulses appear at all inputs within a time interval $\Delta t$ |
| Anticoincidence | Logic pulses at two inputs | Logic pulse only if pulse appears at one input without pulse at second input within time $\Delta t$ |
| Scaler | Logic pulses | One logic pulse for $N$ input pulses |

module (used to record the signals from many detectors monitoring nuclear collisions simultaneously).

Logic modules are used to monitor the counting rate of single detectors and the relative times at which radiation is detected. A fast signal derived from the detector itself, the preamplifier, or from a timing-filter amplifier is sent to a *discriminator*.

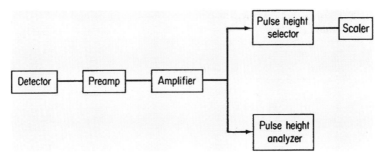

**Figure 18.21** Schematic diagram of a simple pulse height analysis system for nuclear spectroscopy. (From Wang et al., 1975.)

The discriminator produces an output pulse with a fixed shape (generally square) and size when the input signal crosses a reference. Discriminators usually have multiple identical output signals. The logic pulses can be sent to a scaler that simply counts the number of pulses, to a count rate meter to monitor radiation rates or doses, and to a time-to-amplitude converter (TAC) to measure the relative times of arrival of two or more logic signals.

## 18.5 NUCLEAR STATISTICS

Radioactive decay is a random process. The number of nuclei in a sample of radioactive material that decay in any time period is not a fixed number but will differ, usually, for various time periods. This point can be readily shown by making repeated measurements of the activity of a long-lived radionuclide, each for the same time duration. The results of such an experiment might be as shown in Table 18.2. Note that in these measurements there is a large range of activity values with a clustering near the center of the range. We can plot these data (Table 18.2) as a distribution function, by "binning" the data (Fig. 18.22). We can now ask ourselves if we can understand this distribution function. Statisticians have given us mathematical models that describe these and other similar distribution functions. As a background for our discussion of how to extract the maximum amount of information from these data, let us consider some of these models.

The most general model to describe radioactive decay is the *binomial distribution*. For a process that has two outcomes (success or failure, decay or no decay), we can write for the distribution function $P(x)$

$$P(x) = \frac{n!}{(n-x)!x!} p^x (1-p)^{n-x} \tag{18.9}$$

where $n$ is the number of trials where each trial has a probability of success $p$ and $P(x)$ is the predicted probability of getting $x$ successes. Applying this distribution to radioactivity, $P(x)$ might be taken as the probability of getting $x$ counts in a given time interval and $p = \lambda \Delta t$ where $\Delta t$ is a time short compared with the

**TABLE 18.2  Typical Sequence of Counts of a Long-Lived Sample ($^{170}$Tm)**[a]

| Measurement Number | cp0.1 m | $x_i - x_m$ | $(x_i - x_m)^2$ |
|---|---|---|---|
| 1  | 1880 | −18 | 324 |
| 2  | 1887 | −11 | 121 |
| 3  | 1915 | 17  | 289 |
| 4  | 1851 | −47 | 2209 |
| 5  | 1874 | −24 | 576 |
| 6  | 1853 | −45 | 2025 |
| 7  | 1931 | 33  | 10899 |
| 8  | 1886 | −32 | 1024 |
| 9  | 1980 | 82  | 6724 |
| 10 | 1893 | −5  | 25 |
| 11 | 1976 | 78  | 6084 |
| 12 | 1876 | −22 | 484 |
| 13 | 1901 | 3   | 9 |
| 14 | 1979 | 81  | 6561 |
| 15 | 1836 | −62 | 3844 |
| 16 | 1832 | −66 | 4536 |
| 17 | 1930 | 32  | 1024 |
| 18 | 1917 | 19  | 361 |
| 19 | 1899 | 1   | 1 |
| 20 | 1890 | −8  | 64 |

[a]We are indebted to Prof. R. A. Schmitt for providing these data.

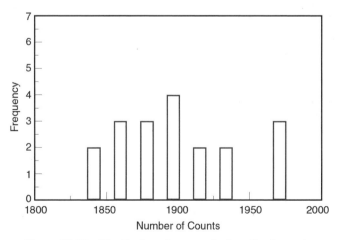

**Figure 18.22**  Distribution of counts of a long-lived sample.

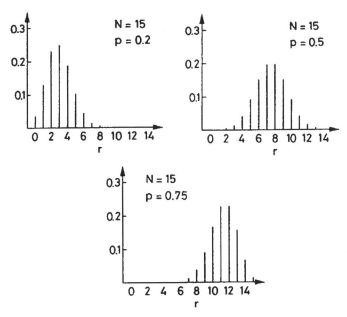

**Figure 18.23** Binomial distributions for various values of $N$ and $p$. (From Leo, 1987.)

measurement time and the half-life. Note that $x$ and $n$ are both integers. Typical binomial distribution functions are shown in Figure 18.23.

The binomial distribution function is cumbersome and a simplification can be made. If the probability of success $p$ is small ($p \ll 1$) (the measurement time is very short compared with the half-life), we can approximate the binomial distribution by the Poisson distribution. The Poisson distribution is written as

$$P(x) = (x_m)^x \exp(-x_m)/x! \tag{18.10}$$

where

$$x_m = pn \tag{18.11}$$

Thus, we have a simplified distribution characterized by one parameter, $x_m$ compared to two parameters in the binomial distribution. The Poisson distribution is an asymmetric distribution as shown in Figure 18.24. Besides being a more tractable function to use, the Poisson distribution has certain important properties that we will use in analyzing radioactivity data. Let us consider a parameter, the *variance*, $\sigma^2$, which expresses something about the width of the distribution of values about the mean, $x_m$.

For a set of $N$ measurements, we can calculate $\sigma^2$ as:

$$\sigma^2 = \frac{\sum_{i=1}^{N}(x_i - x_m)^2}{N - 1} \tag{18.12}$$

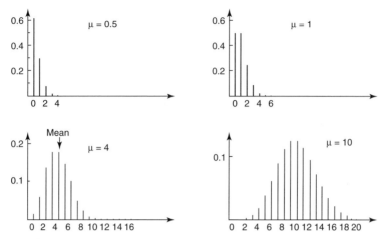

**Figure 18.24** Poisson distribution for various values of $x_m$. (From Leo, 1987.)

(For the data of Table 18.2, $x_m = 1898$, $\sigma = 44.2$.) For a binomial distribution

$$\sigma^2 = np(1-p) \tag{18.13}$$

which is cumbersome to use. But, for a Poisson distribution, we can show that

$$\sigma^2 = x_m \tag{18.14}$$

$$\sigma = (x_m)^{1/2} \tag{18.15}$$

Applying these equations to the data of Table 18.2, we get $\sigma_{\text{Pois}} = 43.6$. This illustrates the important point that these distribution functions are models, not physical laws, and when they are applied to finite data sets, their predictions may deviate from observation.

The Poisson distribution can be applied also to describe the action of detectors. For example, suppose the interaction of a $\gamma$-ray photon with an inefficient scintillator produced, on average, 3.3 photoelectrons from the photocathode. The probability of producing no photoelectrons (not seeing the event) is given by the Poisson distribution as:

$$P(0) = \exp(-3.3) = 3.7\% \tag{18.16}$$

Thus, 3.7% of the events will be missed due to "statistical fluctuations."

A further simplification of the parent binomial distribution occurs when the number of successes is relatively large, that is, we get more than about 30 counts in a measurement. Then, the binomial distribution can be represented as a *normal* or *Gaussian distribution*. Here we write

$$P(x) = \frac{1}{\sqrt{2\pi x_m}} \exp\left(-\frac{(x-x_m)^2}{2x_m}\right) \tag{18.17}$$

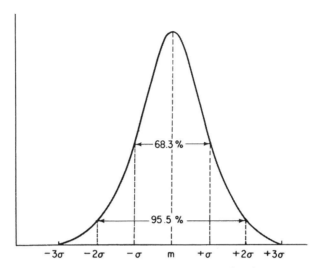

**Figure 18.25** Plot of the normal distribution function showing the mean $x_m$ and $1\sigma$ and $2\sigma$ points. (From Wang et al., 1975.)

This analytical approximation is symmetric (Fig. 18.25). As shown in Figure 18.25, 68.3% of the measured values lie within $\pm 1\sigma$ of the mean, $x_m$. Furthermore 95.5% of all measurements lie within $\pm 2\sigma$ of the mean and 99.7% lie within $\pm 3\sigma$ of the mean. The full width at half maximum (FWHM) is $2.35\sigma$.

Thus, for a single measurement of a count rate of 100, we would estimate that $\sigma = 10$. We could say, with a 68.3% chance of being correct that the true rate was between $100 - 10 = 90$ and $100 + 10 = 110$. With 95.5% certainty, we could say the true rate lies between 80 and 120. Generalizing, we can quote the results of a measurement as $x \pm n\sigma$ where $n$ is related to the probability that an infinite number of measurements would give a value within the quoted range. For $n = 0.6745, 1, 1.6449, 1.96, 2, 2.5758, 3$, the "confidence limits" are 50, 68.3, 90, 95, 95.5, 99, and 99.7%, respectively. Commonly, people will quote the results of a measurement as $x \pm \sigma$. One should remember that doing so means one will be wrong 31.7% of the time, that is, the mean count rate will be outside $x \pm \sigma$. If this risk is not acceptable, one should pick a greater confidence level, that is, $2\sigma$, $3\sigma$, and so forth.

Another distribution function of interest relates to the distribution of time intervals between successive counts. We know the average time between counts is (1/count rate). The distribution of time intervals is given by the *interval distribution*. This distribution (applicable to all random events) states that for a process with an average time between events $t_m$, the probability of getting a time $t$ between successive events is

$$I(t) = \frac{1}{t_m}\exp\left(\frac{-t}{t_m}\right)dt \qquad (18.18)$$

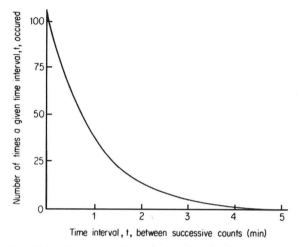

**Figure 18.26** Plot of the frequency of occurrence of a time interval $t$ between successive counts, where the average rate is 1 cpm. (From Wang et al., 1975.)

For radioactive decay

$$t_m = \frac{1}{\lambda} \qquad (18.19)$$

This distribution function is shown in Figure 18.26. Note the most probable time between events is zero. Random events (counts, natural disasters, etc.) occur in "bunches."

Let us summarize how we describe the statistical uncertainty in measurements of radioactivity. If we measure the activity of a sample (+background) as 64 counts in 1 min, then we estimate

$$S + B = 64 \text{ cpm}$$

with an uncertainty $\sigma_{S+B}$

$$\sigma_{S+B} = 8 \text{ cpm}$$

What if a second measurement with no sample showed a background of 10 counts in 1 min? We would then estimate

$$B = 10 \text{ cpm}$$

$$\sigma_B = (10)^{1/2} = 3.2 \text{ cpm}$$

What is the net sample rate and its uncertainty? This raises the general question of calculating the uncertainty in the result of some mathematical operations on an uncertain number. If we consider two independently determined numbers and their uncertainties (standard deviations), $A \pm \sigma_a$, $B \pm \sigma_b$, we can write down, as

## 18.5 NUCLEAR STATISTICS

**TABLE 18.3  Uncertainties for Some Common Operations**

| Operation | Answer | Uncertainty |
|---|---|---|
| Addition | $A + B$ | $(\sigma_A^2 + \sigma_B^2)^{1/2}$ |
| Subtraction | $A - B$ | $(\sigma_A^2 + \sigma_B^2)^{1/2}$ |
| Multiplication | $A \times B$ | $A \times B[(\sigma_A/A)^2 + (\sigma_B/B)^2]^{1/2}$ |
| Division | $A/B$ | $A/B[(\sigma_A/A)^2 + (\sigma_B/B)^2]^{1/2}$ |

shown in Table 18.3, some rules for the uncertainty in the result of some common mathematical operations.

As an example of the use of the relations outlined in Table 18.3, we would calculate that for our sample and background counting case,

$$\text{Net rate} = (\text{sample} + \text{background}) - (\text{background})$$
$$= 64 - 10 = 54 \text{ cpm}$$
$$\text{Uncertainty in net rate} = (8^2 + 3.2^2)^{1/2} = 8.6 \text{ cpm}$$

Up to now we have carefully restricted our discussion of nuclear statistics to cases where 1-min counts were taken. If the number of counts recorded in 1 min was $x$, then the counting *rate* has been quoted as $x \pm (x)^{1/2}$ cpm. Suppose, however, that we recorded 160 counts in 5 min. What would be the standard deviation of the average counting rate (in cpm)? The best estimate of the mean number of counts in the 5-min period would by $160 \pm (160)^{1/2}$ that is, $160 \pm 13$ counts. The average rate would be $160/5 \pm 13/5 = 32 \pm 3$ cpm. In general, therefore, the rate $R$ is given as:

$$R = \frac{\text{number of counts recorded}}{\text{measurement time}} = \frac{x}{t}$$

The standard deviation of the rate, $\sigma_R$, is

$$\sigma_R = \frac{(x)^{1/2}}{t} = \frac{(R*t)^{1/2}}{t} = \left(\frac{R}{t}\right)^{1/2} \qquad (18.20)$$

Thus, for the preceding example we could have calculated directly that

$$\sigma_R = \left(\frac{R}{t}\right)^{1/2} = \left(\frac{32}{5}\right)^{1/2} = 3$$

Often we wish to compute the average of two numbers, $x_1$ and $x_2$, both of which have an uncertainty denoted by their standard deviations $\sigma_1$ and $\sigma_2$, respectively. The best average of these two numbers is not the simple average but the weighted average $x_m$

given by

$$x_m = \left(\frac{x_1}{\sigma_1^2} + \frac{x_2}{\sigma_2^2}\right) \bigg/ \left(\frac{1}{\sigma_1^2} + \frac{1}{\sigma_2^2}\right)$$

$$x_m = \frac{x_1 + wx_2}{1 + w}$$

where

$$w = \left(\frac{\sigma_1}{\sigma_2}\right)^2$$

In short, each number is weighted by the inverse of its standard deviation squared. For the weighted average of $N$ values, $x_i$, with standard deviation, $\sigma_i$, we have

$$x_m = \frac{\sum_{i=1}^{N}(x_i/\sigma_i^2)}{\sum_{i=1}^{N}(1/\sigma_i^2)}$$

The uncertainty or standard deviation of $x$ is given by

$$\sigma_{x_m} = \left(\frac{1}{\sum_{i=1}^{N}(1/\sigma_i^2)}\right)^{1/2}$$

For example, suppose that we make two independent measurements of an activity, obtaining results of $35 \pm 10$ cpm and $46 \pm 2$ cpm. The weighted average of the two measurements is

$$w = (10/2)^2 = 25$$

$$x_m = \frac{[(35)(1 + (25)(46))]}{1 + 25} \cong 46 \text{ cpm}$$

The standard deviation of the weighted average is

$$\sigma_x = \left\{\frac{[100 + (25)^2(4)]}{26^2}\right\}^{1/2} \cong 2.0$$

Thus, we would say that the average rate was $46 \pm 2$ cpm.

### 18.5.1 Rejection of Abnormal Data

In our discussions so far, we have only considered the uncertainty in the experimental data due to the randomness of radioactive decay. But there may also be systematic error that contributes to the overall uncertainty in the data. As a result, when we make repeated measurements of a sample activity under seemingly identical situations, we will find occasionally one measurement that differs from the others by a large amount. If included in the average, this abnormal observation may cause significant error. When are we justified in rejecting such data? One criterion for rejecting such data is to reject suspected values that deviate from the mean by more than $2\sigma$ or $3\sigma$. The probabilities of occurrence of such deviations are 4.5 and 0.27%, respectively.

What about the question of whether a detector or counting system is working properly? For example, the data in Table 18.2 do not exactly match a Poisson or normal distribution. Was the counting system malfunctioning? One parameter that we can calculate that will help us answer such questions is $\chi^2$ (chi squared). Formally,

$$\chi^2 = \frac{\sum_{i=1}^{N}(x_i - x_m)^2}{x_m}$$

For the data in Table 18.2, we calculate that $\chi^2 = 37194/1898 = 19.60$. Figure 18.27 shows the properties of the $\chi^2$ distribution in terms of $p$, the probability that a random sample of $N$ values from a Poisson distribution. For the data of Table 18.2, $\chi^2/\nu = 1.03$, which is acceptable. To be suspicious of the data $\chi^2/\nu$ should have been $>1.7$ or $<0.6$.

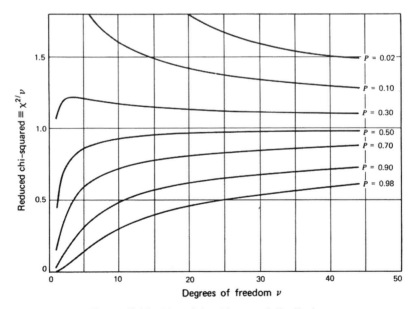

**Figure 18.27** Plot of the chi-squared distribution.

## 18.5.2  Setting Upper Limits When No Counts are Observed

Suppose your experiment failed to detect a type of decay you were seeking: What can you say about its occurrence? The simplest answer is what is termed the *one-event upper limit*. You assume that you had detected one event, and calculate the resulting decay rate, cross section, and the like, taking into account detection efficiencies, solid angles, and so forth.

A more sophisticated answer can be obtained by using the properties of a Poisson distribution. The probability of observing $n$ events if the mean value is $\mu$ is given as:

$$p\left(\frac{n}{\mu}\right) = \frac{\mu^n}{n!} = e^{-\mu}$$

The probability of observing 0 events in a time period $T$ for a process with mean rate $\lambda$ is

$$p(0/\lambda T) = e^{-\lambda T}$$

It can be shown that the upper limit on the rate (when zero counts are observed), $\lambda_0$, is given by

$$\lambda_0 = -\frac{1}{T}\ln(1 - CL)$$

where CL is the confidence limit you wish to attach to your upper limit. (If you want to quote an upper limit with 95% confidence, then CL $= 0.95$.)

**Example**  Suppose you are attempting to observe a process that should produce 1 decay per week, and you observe no counts (with a 100% efficient detector) in 4 weeks. At the 95% confidence level, the upper limit on the decay rate is

$$\lambda_0 = -\tfrac{1}{4}\ln(1 - 0.95) = 0.75/\text{week}$$

## PROBLEMS

1. What detector would you chose to detect the following? Please give your reasoning. (a) 0.1 µCi $^{32}$P? (b) A mixture of 0.1 µCi $^3$H and 0.5 µCi $^{14}$C. (c) A mixture of 10 nCi each of $^{235}$U and $^{238}$U. (d) 50 Bq $^{60}$Co.

2. An advertisement for a high-purity Ge detector quoted its relative efficiency as being 200%. Is that possible? Please explain.

3. How would you detect 10-MeV neutrons?

4. How many counts would you have to collect to have 90% confidence in the quoted counting rate?

5. How would you detect the individual β particles, γ rays, and delayed neutrons from a fission product mixture?

6. What is the maximum allowed counting rate with a scintillation detector with a 0.25-μs dead time if you can only tolerate a 3% counting loss?

7. Given five measurements of the counting rate of a long-lived sample that are 2712, 2753, 2708, 2739, and 2726. Do a chi-squared test to see if these data are consistent with a Poisson distribution.

8. How many counts do you need to collect to have a 2% uncertainty in your result?

9. Sketch a diagram of the nuclear instruments you would need to make logical AND, OR, NAND, and NOR circuits.

10. If a sample has an average counting rate of 16 counts/h and you read out the data every hour for each of 1000 h, how many times will you get 2, 8, 16, and 32 counts?

## REFERENCES

Cooper, J. A. Applied Ge(Li) Gamma Ray Spectroscopy, BNWL-SA-3603, 1971.
Knoll, G. F. *Radiation Detection and Measurement*, 3rd ed., Wiley, New York, 2000.
Leo, W. R. *Techniques for Nuclear and Particle Physics Experiments*, Springer, Berlin, 1987.
O'Kelley, G. D. *Detection and Measurement of Nuclear Radiation*, NAS-NS-3105, 1962.

## BIBLIOGRAPHY

Birks, J. B. *Scintillation Counters*, McGraw-Hill, New York, 1953. The classic work on scintillators.
Evans, R. D. *The Atomic Nucleus*, McGraw-Hill, New York, 1955. Noted for its discussion of statistics.
Kleinknecht, K. *Detectors for Particle Radiation*, 2nd ed., Cambridge, 1998. An excellent discussion from a high-energy physics perspective.
Tsoulfanidis, N. *Measurement and Detection of Radiation*, 2nd ed., Taylor and Francis, Washington, DC, 1995.
Wang, C. H., D. L. Willis, and W. D. Loveland. *Radiotracer Methodology in the Biological, Environmental and Physical Sciences*, Prentice-Hall, Englewood Cliffs, NJ, 1975. Noted for its discussion of liquid scintillation counting, this book served as a model for much of the discussion of this chapter.
Wilkinson, D. H. *Ionization Chambers and Counters*, Cambridge University Press, Cambridge, UK, 1950. The classic discussion of ionization chambers, proportional counters, and Geiger–Müller counters.

# CHAPTER 19

# RADIOCHEMICAL TECHNIQUES

Radiochemistry is defined as "the chemical study of radioactive elements, both natural and artificial, and their use in the study of chemical processes" (*Random House Dictionary*, 1984). Operationally, radiochemistry is defined by the activities of radiochemists, that is, (a) nuclear analytical methods, (b) the application of radionuclides in areas outside of chemistry, such as medicine, (c) the physics and chemistry of the radioelements, (d) the physics and chemistry of high-activity-level matter, and (e) radiotracer studies. We have dealt with several of these topics in Chapters 4, 13, 15, and 16. In this chapter, we will discuss the basic principles behind radiochemical techniques and some details of their application.

Because of the small amounts of material involved, the presence of radioactivity that implies certain regulatory and safety concerns and the frequent need to deal with short-lived nuclei, these techniques are not the same as ordinary chemical techniques. Specialized techniques have evolved from the early part of the 20th century when chemistry was a principal tool in identifying the basic nature of radioactive decay through the extensive use of chemistry in the Manhattan Project in World War II to the present, "high-tech" character of many radiochemical manipulations. These techniques are quite important for they are often the key to a successful experiment, even though they may get scant mention in descriptions of the experiment. Often the successful application of these techniques requires careful, painstaking attention to detail, frequent practice to develop the necessary manipulative skills, and a thorough knowledge of the underlying scientific principles.

*Modern Nuclear Chemistry*, by W.D. Loveland, D.J. Morrissey, and G.T. Seaborg
Copyright © 2006 John Wiley & Sons, Inc.

In addition to the discussions of these topics in textbooks such as this, there are excellent textbooks that focus primarily on radiochemistry (Adloff and Guillamont, 1984; Lieser, 1997; Ehmann and Vance, 1991).

## 19.1 UNIQUE ASPECTS OF RADIOCHEMISTRY

Radiochemistry involves the application of the basic ideas of inorganic, organic, physical, and analytical chemistry to the manipulation of radioactive material. However, the need to manipulate radioactive materials imposes some special constraints (and features) upon these endeavors. The first of these features is the number of atoms involved and the solution concentrations. The range of activity levels in radiochemical procedures ranges from pCi to MCi. For the sake of discussion, let us assume an activity level, $D$, typical of radiotracer experiments of 1 $\mu$Ci ($= 3.7 \times 10^4$ dis/s $= 3.7 \times 10^4$ Bq), of a nucleus with mass number $A \sim 100$. If we assume a half-life for this radionuclide of 3 d, the number of nuclei present can be calculated from the equation

$$N = \frac{D}{\lambda} = \frac{(1\ \mu Ci)(3.7 \times 10^4 dps/\mu Ci)(3\ d)(24\ h/d)(3600\ s/h)}{\ln 2}$$

where $\lambda$ is the decay constant of the nuclide ($= \ln 2/t_{1/2}$). Then

$$N \sim 1.4 \times 10^{10}\ \text{atoms}$$
$$\text{Mass of sample} = 2.3 \times 10^{-12}\ \text{g}$$

This tiny quantity of material, if prepared as an aqueous solution of volume 1 L, would have a concentration of $10^{-14}$ mol/L. This simple calculation demonstrates a number of the important features of radiochemistry, that is, (a) the manipulation of samples involving infinitesimal quantities of material, (b) the power of nuclear analytical techniques (since 1 $\mu$Ci is a significant, easily detectable quantity of radioactivity), and (c) in an extension of the calculation, since the decay of a single atom might occur by $\alpha$-particle emission (with 100% detection efficiency), the ability to do chemistry one atom at a time.

The small number of atoms involved in some radiochemical procedures can alter the expected behavior. Although time-dependent processes obeying first-order kinetics are not changed by changes in concentration, the same is not true of second-order kinetics. For example, at $10^{-2}$ M, isotopic exchange between U(IV) and U(VI) has a lifetime of $\sim$2 h, whereas at $10^{-10}$ M, the same lifetime is $\sim$400 d. Another example is Np(V), which is unstable with respect to disproportionation and yet $\mu$Ci/L solutions of NpO$_2^+$ are stable. The extreme dilution in some solutions can mean that equilibrium is not reached due to kinetic limitations. Fallout plutonium, present in the aqueous environment at concentrations of $10^{-18}$–$10^{-17}$ M, has not reached equilibrium in over 40 y.

In addition to the limitations posed by kinetics or thermodynamics, there are certain practical problems associated with very low solution concentrations. An important problem is the adsorption of tracer levels of radioactivity on the surfaces of laboratory glassware. Glass has an ion exchange capacity of $10^{-10}$ mol/cm$^2$ along with a similar number of chemisorption sites. A 100-mL beaker can thus absorb $\sim 10^{-8}$ mol, which is significant if the concentration of the tracer is $\leq 10^{-6}$ M. One suppresses this absorption by having high [H$^+$] (thus blocking adsorption sites), by treating glass surfaces with nonadsorbing silicone coatings, or by the use of holdback carriers (see below).

Conventional analytical techniques generally operate at the part per million or higher levels. Some techniques such as laser photo acoustic spectroscopy are capable of measuring phenomena at the $10^{-8}$–$10^{-6}$ mol/L level. The most sensitive conventional analytical techniques, time-resolved laser-induced fluorescence, and ICP-MS are capable of measuring concentrations at the part per trillion level, that is, 1 part in $10^{12}$, but rarely does one see detection sensitivities at the single atom level as routinely found in some radioanalytical techniques. While techniques such as ICP-MS are replacing the use of neutron activation analysis in the routine measurement of part per billion concentrations, there can be no doubt about the unique sensitivity associated with radioanalytical methods.

Along with the unique sensitivity and small quantities of material associated with radiochemistry, there is the need to comply with the regulations governing the safe use and handling of radioactive material. This task is a primary focus in the design and execution of radiochemical experiments and is often a significant factor in the cost of the experiment. Because so many of these rules are site specific, they are not treated in this chapter.

There are some chemical effects that accompany high specific activities that are unique to radiochemistry and are worth noting. Foremost among these are the chemical changes accompanying radioactive decay. The interaction of ionizing radiation from a radioactive source with air can result in the generation of ozone and the nitrous oxides, which can lead to corrosion problems. Sources containing Ra or Rn produced from the decay of heavier elements, such as U, will emanate Rn gas as the decay product of Ra. The decay products of gaseous Rn are particulates that deposit on nearby surfaces, such as the interior of the lungs, leading to contamination problems. In high-activity aqueous solutions, one can make various species such as the solvated electron, $e^-_{aq}$, hydroxyl radicals, OH$^\bullet$, as well as the solvated proton, H$_3$O$^+$. The hydroxyl radical, OH$^\bullet$, is a strong oxidizing agent with

$$OH^\bullet + e^- \longrightarrow OH^- \qquad E_0 = 2.8 \text{ V}$$

while the solvated electron, $e^-_{aq}$, is a strong reducing agent

$$e^-_{aq} + H^+ \longrightarrow \frac{1}{2} H_2 \qquad E_0 = 2.77 \text{ V}$$

Solutions involving high activity levels will change their redox properties as a function of time. For example, all the atoms in a 100-Bq/mL ($10^{-7}$ mol/L) solution of $^{239}$Pu will undergo a redox change in a period of one year. In general, it is hard to keep high specific activity solutions stable. Reagents, column materials, and the like can suffer radiation damage also. In radiotracer studies, the self-decomposition (radiolysis) of $^3$H or $^{14}$C-labeled compounds can lead to a variable concentrations and variable number of products.

Many of these effects of radioactive decay can be treated quantitatively using "$G$ values." Historically, the $G$ value was defined as the number of molecules or species decomposed or formed per 100 eV of absorbed energy. A newer (SI) definition of the $G$ value is the number of moles of molecules or species formed or decomposed per Joule of energy absorbed. (Note that 1 mol/J $= 9.76 \times 10^6$ molecules/100 eV.) The $G$ values depend on the radiation and the medium being irradiated and its physical state. Table 19.1 shows some typical $G$ values for the irradiation of neutral liquid water.

The actual final products of radiolysis are the result of a complex set of chemical reactions. Detailed quantitative estimates of product yields are, therefore, more complicated and beyond the scope of this book. The reader is referred to other textbooks that discuss how these estimates are made (Woods et al., 1999; Mozumder, 1999).

Radioactive decay also causes chemical transmutations. The daughter nucleus in $\alpha$ or $\beta$ decay is a different chemical element than the mother nucleus, but it is in the same chemical environment as the mother nucleus. Change of oxidation state or bonding is a possibility.

In $\alpha$ decay, one expects all chemical bonds to the decaying atom to be broken as the recoil energy of the daughter nucleus exceeds chemical bond energies. Surprisingly, the oxidation state of the daughter nucleus is frequently that of the parent nucleus after all electronic and atomic rearrangements have taken place. (An obvious exception is when the daughter cannot exhibit the parent's oxidation state such as the $\alpha$ decay of U(VI) as $UO_2^{2+}$, where the thorium daughter does not exhibit the +6 oxidation state.)

In $\beta^-$ decay, especially for low-energy $\beta^-$ emitters like $^{14}$C or $^3$H, the effects on chemical bonding are modest. So if we have

$$^{14}CH_4 \longrightarrow {}^{14}NH_4^+$$

the $\beta^-$ decay can be considered an oxidizing process. In fact, $\beta^-$ decay (of $^{83}SeO_4^{2-}$, and $^{242}AmO_2^+$) was used successfully to prepare new higher oxidation states (of $^{83}BrO_4^-$ and $^{242}CmO_2^{2+}$) of some elements. In electron capture or internal conversion decay, there are massive rearrangements of the atomic electrons, which makes these considerations more complicated.

**TABLE 19.1 Product Yields ($\mu$mol/J) in Irradiated Neutral Water**

| Radiation | $G(-H_2O)$ | $G(H_2)$ | $G(H_2O_2)$ | $G(e_{aq})$ | $G(H^\bullet)$ | $G(^\bullet OH)$ |
|---|---|---|---|---|---|---|
| $\gamma$ and fast electrons | 0.43 | 0.047 | 0.073 | 0.062 | 0.28 | 0.0027 |
| 12-MeV $\alpha$ particles | 0.29 | 0.12 | 0.11 | 0.028 | 0.056 | 0.007 |

## 19.1 UNIQUE ASPECTS OF RADIOCHEMISTRY

Some tracers (usually cations) in solution behave as colloids rather than true solutions. Such species are termed radiocolloids and are aggregates of $10^3$–$10^7$ atoms, with a size of the aggregate in the range 0.1–500 nm. They are quite often formed during hydrolysis, especially of the actinides in high oxidation states. One can differentiate between real radiocolloids and pseudocolloids, in which a radionuclide is sorbed on an existing colloid, such as humic acid or $Fe(OH)_3$. Formation of real colloids can be prevented by using solutions of low pH or by addition of complexing agents. The chemical behavior of these radiocolloids is difficult to predict, as the systems are not at equilibrium.

There are certain unique features to the chemical separations used in radiochemistry compared to those in ordinary analytical chemistry that are worth noting. First of all, high yields are not necessarily needed, provided the yields of the separations can be measured. Emphasis is placed on radioactive purity, expressed as decontamination factors rather than chemical purity. Chemical purity is usually expressed as the ratio of the number of moles (molecules) of interest in the sample after separation to the number of *all* the moles (molecules) in the sample. Radioactive purity is usually expressed as the ratio of the activity of interest to that of all the activities in the sample. The decontamination factor is defined as the ratio of the radioactive purity after the separation to that prior to the separation. Decontamination factors of $10^5$–$10^7$ are routinely achieved with higher values possible. In the event that the radionuclide(s) of interest are short-lived, then the time required for the separation is of paramount importance, as it does no good to have a very pure sample in which most of the desired activity has decayed during the separation.

As indicated above, frequently the amount of material involved in a radiochemical procedure is quite small. To obviate some of the difficulties associated with this, a weighable amount ($\sim$mg) of inactive material, the *carrier*, is added to the procedure at an early stage. It is essential that this carrier and the radionuclide (tracer) be in the same chemical form. This is achieved usually by subjecting the carrier + tracer system to one or more redox cycles prior to initiating any chemical separations to ensure that the carrier and tracers are in the same oxidation state.

Carriers frequently are stable isotopes of the radionuclide of interest, but they need not be. Nonisotopic carriers are used in a variety of situations. *Scavengers* are nonisotopic carriers used in precipitations that carry/incorporate other radionuclides into their precipitates indiscriminately. For example, the precipitation of $Fe(OH)_3$ frequently carries, quantitatively, many other cations that are absorbed on the surface of the gelatinous precipitate. Such scavengers are frequently used in chemical separations by precipitation in which a radionuclide is put in a soluble oxidation state, a scavenging precipitation is used to remove radioactive impurities, and then the nuclide is oxidized/reduced to an oxidation state where it can be precipitated. In such scavenging precipitations, *holdback carriers* are introduced to dilute the radionuclide atoms by inactive atoms and thus prevent them from being scavenged.

It is certainly possible, although usually more difficult, to do *carrier-free* radiochemistry in which one works with the radionuclides in their low, tracer-level concentrations. Such carrier-free radiochemistry is used when the presence of the additional mass of carrier atoms would lead to problems of sample thickness ($\alpha$ emitters), biological side effects (radiopharmaceuticals), or where high specific activities are needed (synthesis of labeled compounds). (Formally, *specific activity* is the activity per mass unit, such as mCi/mg or $\mu$Ci/$\mu$mol, etc.)

## 19.2 AVAILABILITY OF RADIOACTIVE MATERIAL

To do radiochemistry, one needs radioactive materials. As indicated in Chapter 3, radionuclides may be classified as primordial (remnants of nucleosynthesis), cosmogenic (being continuously generated by the action of cosmic rays with the upper atmosphere) or anthropogenic (made by humans). Most of the radionuclides used in radiochemistry are of the latter type, that is, made artificially in response to perceived needs. In Table 4.1, we summarized the commonly used radionuclides and their method of preparation. As indicated in this table, a large number of these nuclides can be made by neutron irradiation in a nuclear reactor using (n,$\gamma$) reactions. Such nuclei are, of course, not carrier-free, are largely $\beta^-$ emitters and have low specific activities. Charged-particle-induced reactions, using cyclotrons, are used to synthesize neutron-deficient nuclei, that decay by EC or $\beta^+$ emission. The short-lived nuclei used in PET or other procedures in nuclear medicine fall into this category.

The transuranium nuclei are a special class of radionuclides, being made by both reactor irradiation and production in charged particle accelerators. In Table 19.2, we summarize the properties and available amounts for research in the United States by qualified individuals. One should also note that while large quantities of $^{239}$Pu are available, it is classified as a special nuclear material because of its use in weaponry, and very strict regulations govern the possession and use of this nuclide (along with $^{233}$U and $^{235}$U).

## 19.3 TARGETRY

As indicated above, a combination of reactor and cyclotron irradiations is used to prepare most radionuclides. While many of these radionuclides are available commercially, some are not. In addition, nuclear structure, nuclear reactions, and heavy-element research require accelerator or reactor irradiations to produce short-lived nuclei or to study the dynamics of nuclear collisions, and so on. One of the frequent chores of radiochemists is the preparation of accelerator targets and samples for reactor irradiation. It is this chore that we address in this section.

The first question to be addressed in preparing accelerator targets or samples for irradiation is the question of impurities and/or other chemical constituents of the sample. For neutron irradiation, one generally prefers metals or nonactivable inorganic salts. The salts usually include nitrates, sulfates, and the like but not halides (especially chlorides due to the activation of chlorine) nor sodium nor potassium salts. In general, one avoids materials that undergo radiolysis, although it is

**TABLE 19.2 Availability of Transuranium Element Materials**

| Nuclide | $t_{1/2}$ | Decay Mode | Amounts Available | Specific Activity (dpm/μg) |
|---|---|---|---|---|
| $^{237}$Np | $2.14 \times 10^6$ y | α, SF($10^{-10}$%) | kg | 1565.0 |
| $^{238}$Pu | 87.7 y | α, SF($10^{-7}$%) | kg | $3.8 \times 10^7$ |
| $^{239}$Pu | $2.41 \times 10^4$ y | α, SF($10^{-4}$%) | kg | $1.38 \times 10^5$ |
| $^{240}$Pu | $6.56 \times 10^3$ y | α, SF($10^{-6}$%) | 10–50 g | $5.04 \times 10^6$ |
| $^{241}$Pu | 14.4 y | β, α($10^{-3}$%) | 1–10 g | $2.29 \times 10^8$ |
| $^{242}$Pu | $3.76 \times 10^5$ y | α, SF($10^{-3}$%) | 100 g | $8.73 \times 10^3$ |
| $^{244}$Pu | $8.00 \times 10^7$ y | α, SF(0.1%) | 10–100 mg | 39.1 |
| $^{241}$Am | 433 y | α, SF($10^{-10}$%) | kg | $7.6 \times 10^6$ |
| $^{243}$Am | $7.38 \times 10^3$ y | α, SF($10^{-8}$%) | 10–100 g | $4.4 \times 10^5$ |
| $^{242}$Cm | 162.9 d | α, SF($10^{-5}$%) | 100 g | $7.4 \times 10^9$ |
| $^{243}$Cm | 28.5 y | α, (0.2%) | 10–100 mg | $1.15 \times 10^8$ |
| $^{244}$Cm | 18.1 y | α, SF($10^{-4}$%) | 10–100 g | $1.80 \times 10^8$ |
| $^{248}$Cm | $3.40 \times 10^5$ y | α, SF(8.3%) | 10–100 mg | $9.4 \times 10^3$ |
| $^{249}$Bk | 320 d | β, α($10^{-3}$%), SF($10^{-8}$%) | 10–50 mg | $3.6 \times 10^9$ |
| $^{249}$Cf | 350.6 y | α, SF($10^{-7}$%) | 1–10 mg | $9.1 \times 10^6$ |
| $^{250}$Cf | 13.1 y | α, SF(0.08%) | 10 mg | $2.4 \times 10^8$ |
| $^{252}$Cf | 2.6 y | α, SF(3.1%) | 10–1000 mg | $1.2 \times 10^9$ |
| $^{254}$Cf | 60.5 d | SF, α(0.3%) | μg | $1.9 \times 10^{10}$ |
| $^{253}$Es | 20.4 d | α, SF($10^{-5}$%) | 1–10 mg | $5.6 \times 10^{10}$ |
| $^{254}$Es | 276 d | α | 1–5 μg | $4.1 \times 10^9$ |
| $^{257}$Fm | 100.5 d | α, SF(0.2%) | 1 pg | $1.1 \times 10^{10}$ |

possible, with suitable precautions, to irradiate materials, such as gasoline, oil, and other flammable materials in reactors. Liquid samples can be irradiated in reactors easily, but one must generally pay attention to pressure buildup in the irradiation container due to radiolytic decomposition of water. Unless purged with nitrogen, water will contain dissolved argon, which will activate to form $^{41}$Ar, and the radioactive atoms will be released upon opening the irradiation container. For irradiations of a few hours in moderate flux ($\sim 10^{12}$ n/cm$^2$ s) reactors, the samples to be irradiated may be heat sealed in polyethylene vials, usually using double encapsulation. For long irradiations or higher fluxes ($\sim 10^{15}$ n/cm$^2$s) encapsulation in quartz is needed. One never uses Pyrex or other boron-containing glasses due to their high cross sections for neutron absorption. When irradiating larger samples, one must pay attention to self-shielding in the samples. For example, the flux reduction in a 0.5-mm Au foil is about 27% due to self-absorption. However, in irradiations of most liquid samples or geological samples, these self-absorption corrections can be neglected.

Preparation of the targets for charged particle irradiations requires more effort due to the large rate of energy loss of charged particles in matter. In general, material to be irradiated must be in vacuum, thus making the irradiation of liquids and gases more difficult. Solids must be in the form of thin foils or deposits on thin backing material. Typical backing materials are carbon, aluminum, beryllium, and titanium. The typical measure of thickness of accelerator targets is in units of areal density (mass/unit area, i.e., $mg/cm^2$). The thickness expressed in units of areal density ($mg/cm^2$) is the linear thickness (cm) multiplied by the density ($mg/cm^3$). One can weigh very thin samples and determine their area and thus their areal density. Typical thicknesses of accelerator targets are $\sim$0.1–5 $mg/cm^2$ but depend, of course, on the rate of energy loss of the irradiating ion in passing through the target material. Target backings are frequently 10–100 $\mu g/cm^2$. Such thicknesses qualify as being "thin," that is, easily breakable and require special preparation techniques.

Because of the high rate of energy loss of charged particles in matter, one must pay attention to cooling the targets or in some way dissipating the energy deposited in the target material as the beam nuclei pass through it. For example, consider the irradiation of a 0.5-$mg/cm^2$ $^{208}$Pb target by 450-MeV $^{86}$Kr ions. Each Kr ion passing through the target deposits $\sim$8.1 MeV in the target. If the Kr beam intensity is 1 particle-$\mu$A (6.24 × $10^{12}$ ions/s), then the rate of energy deposit in the target is $\sim 5 \times 10^{13}$ MeV/s $\approx$ 8.1 J/s. If the foil has an area of 2 $cm^2$, it would have a mass of 1 g. The specific heat of Pb is 0.130 J/g/°C. Thus, in the absence of any cooling, the temperature of the target would rise 8.1°/s, and the foil would soon melt. Since the foil will generally be in vacuum, without further intervention, it would only cool radiatively, which will not suffice. Heat transfer from such a foil can occur by clamping it to the front of a cooled block (remembering now that the entire beam energy will be dumped into the cooling block). Alternately, a jet of a light gas such as helium can be used to cool the backside of the foil or the particle beam can be spread over a larger area foil thus reducing the temperature increase.

Over the years, a number of specialized techniques have been developed for the preparation of accelerator targets. These techniques are also used in the preparation of thin sources for counting, such as those used in $\alpha$ or $\beta$ spectroscopy. The first and simplest technique for depositing a target material on a backing foil is by *evaporation* of a solution of the desired material on the foil. Generally, this is a poor choice as the solute tends to deposit at the edges of the dried droplet, leading to variations of up to a factor of 100 in thickness over the area of the deposit. Uniformity can be improved by using a spreading agent such as insulin to coat the surface of the backing material prior to evaporation.

A method that was widely used in the past is *electrospraying*. A solution of the nuclide to be deposited is prepared in a volatile, nonconducting liquid like acetone, alcohol, and the like. A capillary is drawn out to a fine point such that no liquid can escape under normal conditions and filled with the solution. A fine wire is threaded though the capillary to within a few millimeters of the tip. A high voltage (3–10 kV) is applied between the wire in the tube and the backing material on which the deposit is to be made. One gets a spray of charged drops that are collected on the backing material, placed $\sim$1 cm from the capillary. The volatile solvent evaporates leaving a uniform film. The deposit can be calcined.

Another widely used technique for preparing thin deposits on a backing material is *electrodeposition*. Two types of electrodeposition are commonly used: (a) the direct deposit of a metal on a cathodic surface by reduction or (b) precipitation of a cationic species in an insoluble form on an electrode. This latter technique is widely used to deposit actinides and lanthanides. A 10- to 100-μL aqueous solution of the actinide or lanthanide is mixed with ~15 mL of isopropyl alcohol and placed in a plating cell (Fig. 19.1). The inorganic material forms a positively charged complex in which the inorganic molecule is surrounded by a cluster of solvent molecules. A high voltage (~600 V) is applied between a rotating anode and the cathodic backing material. The positively charged complex is attracted to the cathode of the cell. The lanthanides/actinides precipitate as hydrous oxides near the cathode, which is a region of high pH. The alcohol is withdrawn from the cell and the deposit is dried and calcined. This technique is called *molecular plating* because the film is not that of the metal but some molecular form of it. Deposit thicknesses are restricted to $<0.5$ mg/cm$^2$ but the deposition is rapid and quantitative and allows the use of active metals such as Al as backing foils (Parker and Slatis, 1966).

*Vacuum deposition* is a well-established technique for making very uniform deposits of non-refractory materials on a backing material. In Figure 19.2, we show a typical simple evaporation apparatus.

The material to be evaporated is placed in a sample holder. Frequently, these sample holders are indented strips of W, Ta, or Mo, or wire baskets of the same metals, or carbon crucibles. These sample holders can be heated resistively by passing a large current through them, thus melting and then volatilizing the material. Alternatively, the sample holder can be bombarded by low-energy electrons to heat the sample. The entire process takes place in vacuum. Under reduced pressure, most

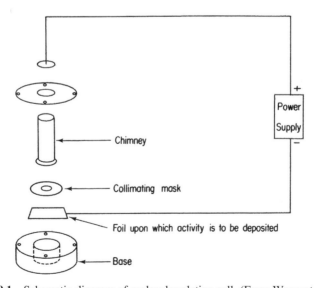

**Figure 19.1** Schematic diagram of molecular plating cell. (From Wang et al., 1975.)

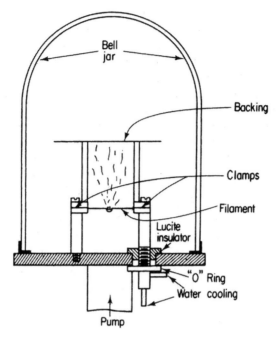

**Figure 19.2** Simple schematic diagram of a vacuum deposition apparatus. (From Wang et al., 1975.)

materials melt readily and then evaporate. The substrate on which the vapors from the heated sample condense is placed some distance from the source of evaporating material. The area of the deposit may be defined by collimators. The deposits produced by vacuum evaporation are very uniform, but the process is not efficient, with <1% of the sample material being deposited in typical applications. Self-supporting deposits can be prepared if the substrate on which the vapors condense is coated with a release agent prior to evaporation.

The thicknesses of these thin targets can be measured using a variety of techniques. The simplest and most reliable method is weighing. One weighs a known area of target material and computes its areal density. Such a technique will give the average deposit thickness but no information about its uniformity or composition. The uniformity of targets or thin foils can be measured with $\alpha$-particle thickness gauges. In such devices, a collimated beam of low-energy $\alpha$ particles passes through the foil whose thickness is to be measured. Changes in count rate are noted as the beam scans over the area of the target or foil. To get absolute thicknesses from such devices, calibration with foils of known thickness is needed. This technique works best when the energy of the $\alpha$ particle is so low as to barely pass through the foil. In this case, small changes in thickness are magnified in the observed count rate. X-ray fluorescence or neutron activation analysis can be used to determine the elemental composition of the target or source material.

Approximate values of the target thicknesses can also be obtained by noting the energy loss of monoenergetic ions as they pass through a foil. The thicknesses are calculated from the observed $dE/dx$ and empirical stopping power relationships.

## 19.4 MEASURING BEAM INTENSITY AND FLUXES

Measurement of charged particle beam intensities is largely done using physical methods, although some older radiochemical methods are used occasionally. The most common techniques to measure the intensity of a charged particle beam is through the use of a *Faraday cup*. The beam is stopped in an electrically isolated section of beam pipe referred to as a Faraday cup (Fig. 19.3).

The collected charge is measured with an electrometer that functions as a current integrator. The beam intensity is just the current divided by the charge on each ion. Care must be taken regarding the loss of secondary electrons. (The beam will liberate secondary electrons in the material in which it stops.) If these electrons escape, their positive ion partners will add to the positive charges of the stopped ions, thus causing an overestimate of the beam current. The Faraday cup is thus made as a long cylinder to inhibit electron escape geometrically, and a magnetic field is applied to the cup along with a suppressing voltage to further prevent electron loss.

When the energy of the charged particle beam is too large to easily stop the beam in a Faraday cup, the beam intensity is frequently monitored by a *secondary ionization chamber*. These ion chambers have thin entrance and exit windows and measure the differential energy loss when the beam traverses them. They must be calibrated to give absolute beam intensities. If the charged particle beam intensity is very low ($<10^6$ particles/s), then individual particles can be counted in a plastic scintillator detector mounted on a photomultiplier tube.

When performing irradiations with neutrons or high-energy protons, it is common to measure the beam intensity using a *monitor reaction*. A thin foil of a

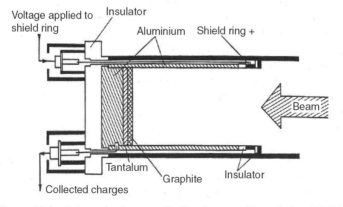

**Figure 19.3** Schematic diagram of a Faraday cup. (From Lefort, 1968.)

pure element is placed in the irradiating flux near the target and irradiated simultaneously with the target. Then both the reaction products from the target foil and the monitor foil are collected and counted. The flux is calculated using a known cross section for the monitor reaction.

Assuming that the monitor and target foils are exposed to the same irradiating flux, we have, for the activity of the monitor and target foils, $A_{\text{mon}}$ and $A_{\text{tgt}}$, respectively, at the end of the irradiation:

$$A_{\text{tgt}} = N_r \sigma_r \phi (1 - e^{-\lambda_r t})$$

$$A_{\text{mon}} = N_m \sigma_m \phi (1 - e^{-\lambda_m t})$$

where $N_i$, $\sigma_i$, $\lambda_i$ are the number of target atoms, cross section, and product decay constant, respectively, for the $i$th reaction. The cross section for the reaction of interest, $\sigma_r$, is

$$\sigma_r = \frac{A_{\text{tgt}}}{N_r \phi (1 - e^{-\lambda_r t})}$$

Substituting for $\phi$ in terms of the monitor reaction, we get

$$\sigma_r = \frac{A_{\text{tgt}}}{A_{\text{mon}}} \cdot \frac{N_r}{N_{\text{mon}}} \cdot \frac{(1 - e^{-\lambda_m t})}{(1 - e^{-\lambda_r t})} \cdot \sigma_{\text{mon}}$$

If the irradiation is long enough to produce saturation activities in both the target and monitor foils, we have

$$\sigma_r = \frac{A_{\text{tgt}}}{A_{\text{mon}}} \cdot \frac{N_r}{N_{\text{mon}}} \cdot \sigma_{\text{mon}}$$

For high-energy protons, typical monitor reactions are $^{27}\text{Al} \rightarrow {}^{24}\text{Na}$, $^{27}\text{Al} \rightarrow {}^{22}\text{Na}$, $^{12}\text{C} \rightarrow {}^{7}\text{Be}$, $^{27}\text{Al} \rightarrow {}^{18}\text{F}$, $^{197}\text{Au} \rightarrow {}^{149}\text{Tb}$, and $^{12}\text{C}(\text{p},\text{pn}){}^{11}\text{C}$ where the arrows indicate a complex set of reaction paths leading from the initial nucleus to the product nucleus. Care must be taken in the case of reactions producing $^{24}\text{Na}$ or $^{11}\text{C}$ to correct for secondary neutron-induced reactions that produce these nuclides. In high-energy reactions, the loss of recoils from the monitor or target foils can be corrected for by irradiating a stack of three identical foils and only counting the center foil. The forward-going recoils from the first foil enter the second foil and compensate for its forward recoil loss. The backward recoils from the third foil enter the second foil and compensate for its backward recoil loss.

The measurement of neutron fluxes by foil activation is more complicated because the neutrons are not monoenergetic and the monitor cross sections are energy dependent. The simplest case is monitoring slow neutron fluxes. Radiative capture (n,γ) reactions have their largest cross sections at thermal energies and are thus used in slow neutron monitors. Typical slow neutron activation detectors are Mn, Co, Cu, Ag, In, Dy, and Au. Each of these elements has one or more odd $A$ isotopes with a large thermal (n,γ) cross section, ~1–2000 barns. The (n,γ) reaction products have half-lives ranging from minutes to hours. The activation cross sections generally vary as $1/v$, although some nuclides have resonances in the capture cross sections for neutrons with energies between 1 and 1000 eV. A correction for such resonance capture can be made by irradiating the monitor foils with and without a Cd cover. The (n,$\gamma$) cross section for Cd below 0.4 eV is very large and is small for energies above this, and thus very few low-energy neutrons penetrate the Cd cover. Irradiation of a foil without a Cd cover will cause reactions induced by both thermal and resonance neutrons, while the Cd-covered foil will just respond to resonance neutrons.

One can also use so-called threshold monitor detectors where the activating reaction has an energy threshold, such as the (n,$\alpha$), (n,p), and (n,2n) reactions. By exposing a set of threshold detectors (involving different reactions with different thresholds) to a neutron flux, one can determine the relative amounts of different energy groups in the neutron spectrum. Further information about the use of activation detectors to measure neutron fluxes can be found in the textbooks by Knoll (2000) and Tsoulfanidis (1995).

## 19.5 RECOILS, EVAPORATION RESIDUES (EVRs), AND HEAVY RESIDUES

In a nuclear reaction, the momentum transfer to the struck nucleus is not negligible. If an $A = 100$ nucleus fuses completely with a 100-MeV $\alpha$-particle projectile, the kinetic energy of the completely fused system is $\sim 4$ MeV. A similar fusion of an $A = 100$ nucleus with a 100-MeV $^{16}$O projectile will give the completely fused system an energy of 13.8 MeV. These energies are extremely large compared to chemical bond energies. Depending upon the position in the target foil where the nuclear reaction takes place, some or all of these recoiling nuclei may escape from the target foil. These recoil nuclei, which are usually radioactive, can be collected or studied using physical or radiochemical techniques. In reactor irradiations, these recoils produce "contamination" on the surface of irradiation containers.

When these heavy recoil nuclei are the result of a complete fusion of the projectile and target nuclei, they are usually called *evaporation residues* because they result from a deexcitation of the primary complete fusion product by particle evaporation (emission). In intermediate energy and relativistic nuclear collisions, the momentum transfer to the target nucleus is much less, and the energy of the recoiling nucleus is $\sim 5-100$ keV/nucleon. Such recoils are usually called *heavy residues*.[1]

---

[1]It should be noted that in this discussion we are tacitly assuming "normal" reaction kinematics with the lighter nucleus being the projectile (that is in motion) and the heavier collision partner being at rest in the laboratory system. In reactions studied using *inverse kinematics* with a heavier projectile striking a lighter target nucleus, the momentum of the recoiling heavy nucleus is approximately the same as that of the projectile nucleus. In inverse kinematics reactions, the energies of the EVRs or heavy residues are large and their spatial and energy distribution is compressed accordingly. Collection of these recoils is relatively easy, but high resolution is needed because of the spatial and energy compression.

There are a variety of ways to collect the recoiling heavy products of a nuclear reaction. One radiochemical technique is the so-called thick target–thick catcher method. Here a target foil whose thickness exceeds the average range of the recoils is surrounded by catcher foils of C or Al or some other material whose thickness exceeds the range of the recoiling product nucleus, which will not lead to production of the nuclide of interest. The average range of the recoiling product, $\langle R \rangle$, (which can be related to its total kinetic energy) is given as

$$\langle R \rangle = \frac{N_c W}{N_c + N_w}$$

where $N_c$ is the number of recoils that escape from the target, $N_w$ is the number that remains in the target, and $W$ is the thickness of the target. The fraction of product nuclei that recoil into the forward catcher foil, $F$, and the fraction that recoil into the backward catcher foil, $B$, can be used to deduce something about the relative velocity imparted to the recoiling nucleus by the initial projectile–target interaction, $v$, and the isotropic velocity kick, $V$, given to it by successive momentum kicks by sequential particle emission. Formally, we define the quantity $\eta$ where $\eta = v/V$. Also $\eta$ can be related to $F$ and $B$:

$$\eta = \frac{(F/B)^{1/2} - 1}{(F/B)^{1/2} + 1}$$

So the thick target–thick catcher method can lead to crude information about the kinematics of the nuclear reaction under study (Harvey, 1960). This technique can be used to advantage in the study of intermediate energy and relativistic nuclear collisions where the energy of the heavy residues is low ($\sim$10–100 keV/nucleon). In this case, most of the residues stop in the target foil and cannot be studied any other way.[1]

For the study of recoils in low-energy and some intermediate-energy reactions, one can use a thin target ($<0.5$ mg/cm$^2$). The energy loss of the recoils in emerging from these targets is negligible or calculable and tolerable. With thin target irradiations, one can stop the recoiling nuclei in a catcher foil, which can be counted, perhaps after intervening chemical separations to isolate the products of interest. The "catcher foil" can take the form of a tape or rotating wheel that can rapidly transport the activity to a remote, low background location for counting. Alternatively, the catcher foil can take the form of a stack of thin foils that stops the products. These foils, upon disassembly and counting, can be used to construct a differential range distribution for products of interest. Catcher foils or stacks of foils can be mounted at various angles with respect to the incident beam and can be used to measure product angular or energy distributions. These catcher foil techniques are now only used to study reactions with very low cross sections where their use provides a high detection sensitivity.

The catcher foil can take the form of a jet of rapidly moving gas, a helium jet. The atoms produced in a nuclear reaction recoil out of a thin target and are stopped in $\sim$1 atm of helium gas in the target chamber. The gas contains an aerosol, typically an alkali halide or more recently, carbon clusters, to which charged reaction product recoils attach themselves via van der Waals forces. The helium gas (and the aerosol particles) escapes through a small orifice to a vacuum chamber, with the gas achieving sonic velocity. The gas and aerosol can be transported for substantial distances in thin capillary tubes. The aerosol particles (and the attached atoms) are collected by allowing the gas stream to strike a collector surface. The resulting deposit can be counted directly or dissolved for further chemical processing. The aerosol loaded gas stream (jet) can also be used to transport the atoms through a thin capillary a distance of several meters in a few seconds. If the aerosol particles are carbon clusters, they can be burned in oxygen, freeing the transportal activity. If the carrier gas is extremely pure helium gas, then the residues will remain ionized and can be collected using electrostatic devices. A number of systems have been developed and used to collect reaction products as ions without involving aerosols in the carrier gas.

The principal limitation of the isolation devices discussed previously (tapes, jets, etc.) is that the reaction product must be stopped and mechanically transported to radiation detectors before product identification can occur. This restricts their use to studies of nuclei whose $t_{\frac{1}{2}} > 1$ ms. For detection and identification of species whose $t_{\frac{1}{2}} < 1$ ms, one employs instruments based upon direct magnetic and/or electrostatic deflection of target recoils. The most spectacularly successful of these devices is the velocity filter SHIP (Fig. 15.7). Evaporation residues produced in compound nucleus reactions emerge from the target and pass through a thin carbon foil that has the effect of equilibrating the ionic charge distribution of the residues. The ions then pass through two filter stages consisting of electric deflectors, dipole magnets, and a quadruple triplet for focusing. The solid angle of acceptance of the separator is 2.7 msr with a separation time for the reaction products of $\sim 2$ µs with a total efficiency of collecting evaporation residues of $\sim 20\%$ for $A_{proj} > 40$. Since complete fusion evaporation residues have very different velocities and angular distributions than targetlike transfer and deep-inelastic products (a factor of $\sim 2$ difference in velocity between transfer products and evaporation residues) and beam nuclei, the separator with its $\pm 5\%$ velocity acceptance range and narrow angular acceptance very effectively separates the evaporation residues from the other reaction products and the beam. Following separation, the residues pass through a large-area time-of-flight detector and are stopped in an array of position-sensitive detectors. From their time of flight and the energy deposited as they stop in the position-sensitive detectors, a rough estimate of their mass may be obtained ($dA/A \sim 0.01$). The final genetic identification of the residues is made by recording the time correlations between the original position signals from the detectors and subsequent decay signals from the same location (due to $\alpha$ or spontaneous fission decay) and/or signals from $\gamma$ or X-ray detectors placed next to the position-sensitive detector. This device for studying heavy-element evaporation residues was used in the discovery of elements 107–112 (Chapter 15).

An alternate method of recoil collection and separation from other reaction products is the gas-filled separator. Unlike the vacuum separators SHIP at GSI or the Fragment Mass Analyzer at ANL that can only accept a limited range of charge-to-mass ratios, a gas-filled separator is a magnetic separator that is filled with a low pressure (~1 torr) gas (usually H or He). Evaporation residues emerging from the target undergo charge changing collisions in the gas and quickly equilibrate to a common average charge state. The change in recoil charge states with gas pressure in a gas-filled magnet is shown in Figure 19.4. At a pressure of ~1 torr, all recoils have a common average charge state. This charge state will differ from the charge of the projectile nuclei or other reaction products. This charge equilibration allows a very efficient collection of the recoils but with a loss of the detailed mass selection obtained in vacuum separators like SHIP that use both electric and magnetic fields for separation. The appropriate magnetic rigidity to collect a given product nuclide with a gas-filled separator can be roughly described by the simple formula

$$B\rho = \frac{mv}{q_{ave}} = \frac{mv}{(v/v_0)eZ^{1/3}} = \frac{0.0227A}{Z^{1/3}} Tm$$

where $m$, $Z$, and $v$ are the mass, atomic number, and velocity of the recoiling ion, respectively, and $v_0$ is the Bohr velocity, $2.18 \times 10^6$ m/s. [In reality, $q_{ave}$ is a sensitive function of the atomic structure of the recoiling ion and the gas (Ghiorso et al., 1988; Oganessian et al., 2001).] The magnetic rigidity is proportional to the recoil velocity so that recoils of the same charge and velocity are focused. The primary

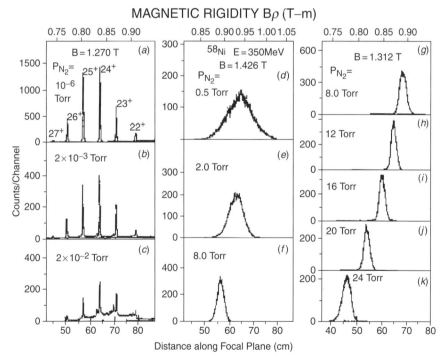

**Figure 19.4** Changes in focal plane distributions in a gas-filled magnet for the reaction involving 350-MeV $^{58}$Ni ions. (From Paul, 1989.)

beam is separated from the recoils right after the target by a dipole bending magnet followed by two quadrupole magnets that focus the beam onto the focal plane. For the study of heavy-element production reactions, collection efficiencies of 25–50% can be achieved.

## 19.6 RADIOCHEMICAL SEPARATION TECHNIQUES

In the study of nuclear reactions, nuclear structure, and the heaviest elements, one frequently needs to *chemically* separate the nuclide(s) of interest from other radioactive species that are present. This is done by performing radiochemical separations that involve the conventional separation techniques of analytical chemistry adapted to the special needs of radiochemistry. For example, radiochemical purity is generally more important than chemical purity. When dealing with short-lived nuclides, speed may be more important than yield or purity. The high cost of radioactive waste disposal may require unusual waste minimization steps. As noted earlier, radiochemical separations need not be quantitative. One only needs to know the yield. Because of the availability of modern high-resolution counting equipment, such as Ge γ-ray spectrometers, modern radiochemical separations frequently are designed only to reduce the level of radioactive impurities in the sample rather than producing a pure sample. [The counting instrumentation is used to "isolate" the nuclide(s) of interest from other nuclides.] Thus, modern procedures sometimes are similar to qualitative analysis schemes, breaking products into chemically similar groups and using instrumentation to further separate the group members. A recent review summarizes some newer developments of relevance to radiochemistry (Bond et al., 1999).

### 19.6.1 Precipitation

The oldest, most well-established chemical separation technique is *precipitation*. Because the amount of the radionuclide present may be very small, *carriers* are frequently used. The carrier is added in macroscopic quantities and ensures the radioactive species will be part of a kinetic and thermodynamic equilibrium system. Recovery of the carrier also serves as a measure of the yield of the separation. It is important that there is an isotopic exchange between the carrier and the radionuclide. There is the related phenomenon of *co-precipitation* wherein the radionuclide is incorporated into or adsorbed on the surface of a precipitate that does not involve an isotope of the radionuclide or isomorphously replaces one of the elements in the precipitate. Examples of this behavior are the sorption of radionuclides by $Fe(OH)_3$ or the co-precipitation of the actinides with $LaF_3$. Separation by precipitation is largely restricted to laboratory procedures and apart from the bismuth phosphate process used in World War II to purify Pu, has little commercial application.

As a practical matter, precipitation is usually carried out in hot, dilute aqueous solutions to allow the slow formation of large crystals. The pH of the solution is chosen to minimize colloid formation. After precipitation, the precipitate is washed carefully to remove impurities, dissolved, and reprecipitated to cause further purification. The precipitate is collected by filtration (Fig. 19.5).

The filter paper is supported by a glass frit clamped between two glass tubes. The precipitate is washed finally with acetone or alcohol to dry it. The precipitate is chosen to have a known stoichiometry to allow calculation of the yield of the separation and should not absorb water or $CO_2$ so that an accurate weight can be obtained. (The filter paper used in the filtration must be treated with all the reagents beforehand, dried, and weighed so that any material loss in filtration is minimized.)

### 19.6.2 Solvent Extraction

Separation by liquid–liquid extraction (solvent extraction) has played an important role in radiochemical separations. Ether extraction of uranium was used in early weapons development, and the use of tributyl phosphate (TBP) as an extractant for U and Pu was recognized in 1946, resulting in the commercial PUREX process for reprocessing spent reactor fuel (see Chapter 16). In recent years, there has been a good deal of development of solvent extraction processes for the removal of transuranic elements, or $^{90}$Sr, and $^{137}$Cs from acidic high-level waste. Laboratory demonstrations of the TRUEX process that uses the neutral extractant CMPO [octyl(phenyl)-*N*,*N*-diisobutylcarbamoyl-methylphosphine oxide] to separate the transuranium elements from acidic high-level waste have been successful. More recently crown ethers have been used as specific extractants for Sr and Cs.

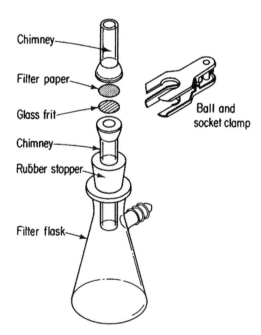

**Figure 19.5** Schematic diagram of a filtration apparatus used in radiochemistry.

In solvent extraction, the species to be separated is transferred between two immiscible or partially miscible phases, such as water and a nonpolar organic phase. To achieve sufficient solubility in the organic phase, the species must be in the form of a neutral, nonhydrated species. The transfer between phases is achieved by selectively complexing the species of interest causing its solubility in water to decrease with a concomitant increase in its solubility in the organic phase.

A hydrated metal ion ($M^{Z+}$) will always prefer the aqueous phase to the organic phase. To get the metal ion to extract, some or all of the inner hydration sphere must be removed. The resulting complex must be electrically neutral and organophilic, that is, have an organic "surface" that interacts with the organic solvent. This can be done by:

1. Forming a neutral complex $MA_Z$ by coordination with organic anions $A^-$.
2. Replacing water in the inner coordination sphere by large organic molecules B such that one forms $MB_N^{Z+}$, which is extracted into the organic phase as an ion association complex $(MB_N)^{Z+}L_x^{Z-}$.
3. Forming metal complexes of form $ML_N^{Z-N}$ with ligands (L) such that they combine with large organic cations $RB^+$ to form ion pair complexes $(RB^+)_{N-Z}(ML_N)^{N-Z}$.

*The extracting agents are thus divided into three classes: polydentate organic anions $A^-$, neutral organic molecules B, or large organic cations $RB^+$.*

*Polydentate organic anions*, which form chelates (ring structures of four to seven atoms) are important extracting agents. Among these are the β-diketonates, such as acetylacetonate, the pyrazolones, benzoylacetonate, and thenoyltrifluoroacetone (TTA), with the extraction increasing strongly through this sequence. Representing the organic chelating agent as HA, the overall reaction involved in the chelate extraction of a metal ion, $M^{n+}$, is

$$M^{n+}(aq) + nHA(o) \longleftrightarrow MA_n(o) + nH^+(aq)$$

When an aqueous solution containing extractable metal ions is brought into contact with an organic phase containing chelating agent, the chelating agent dissolves in the water phase, ionizes, complexes the metal ion, and the metal chelate dissolves in the organic phase. The low solubility of the metal complexes and their slow rates of formation limit the industrial use of this type of anionic extraction.

However, a number of organophosphorus compounds are efficient extractants as they and their complexes are very soluble in organic solvents. The most important of these are monobasic diethylhexylphosphoric acid (HDEHP) and dibutylphosphoric acid (HDBP). The actinide $MO_2^{2+}$ ions are very effectively extracted by these reagents as are the actinide (IV) ions.

Among the *neutral extractants*, alcohols, ethers, and ketones have been used extensively. The most famous example of these is the extraction of uranyl nitrate into diethyl ether, the process used in the Manhattan Project to purify the uranium used in the first reactors. In one of the early large-scale processes (the Redox process) to recover uranium and plutonium from irradiated fuel, methyl isobutyl ketone was used to extract the actinides as nitrates.

The most widely used neutral extractants, however, are the organophosphorus compounds, of which the ester, TBP, is the most important. TBP forms complexes with the actinide elements thorium, uranium, neptunium, and plutonium by bonding to the central metal atom via the phosphoryl oxygen in the structure

$$(C_4H_9O)_3P^+O^-$$

The overall reactions are

$$MO_2^{2+}(aq) + 2NO_3^-(aq) + 2TBP(o) \leftrightarrow MO_2(NO_3)_2 \cdot 2TBP(o)$$

or

$$M^{4+}(aq) + 4NO_3^-(aq) + 2TBP(o) \leftrightarrow M(NO_3)_4(TBP)_2(o)$$

These equilibria can be shifted to the right, increasing the degree of extraction by increasing the concentration of uncombined TBP in the organic phase or by increasing the concentration of [$NO_3^-$(aq)]. The latter increase is achieved by adding a salting agent such as $HNO_3$ or $Al(NO_3)_3$. These extraction equilibria are the basis of the PUREX process, used almost exclusively in all modern reprocessing of spent nuclear fuel.

A third group of extractants (the *cationic extractants*) are the amines, especially the tertiary or quarternary amines. These strong bases form complexes with actinide metal cations. The efficiency of the extraction is improved when the alkyl groups have long carbon chains, such as trioctylamine or triisooctylamine. The extraction is conventionally thought of as a "liquid anion exchange" in that the reaction for metal extraction can be written as an anion exchange, that is,

$$xRB^+L^-(o) + ML_n^{-x} \leftrightarrow (RB^+)_xML_n^x(o) + xL^-$$

where $ML_n^{-x}$ is the metal anion complex being extracted and $RB^+$ is the ammonium salt of the amine. Hexavalent and tetravalent actinides are efficiently extracted using this technique while trivalent actinides are not well extracted under ordinary conditions.

As a practical matter, the distribution ratio $D$ is defined as

$$D = [M]_{org}/[M]_{aq}$$

where $[M]_i$ is the metal ion concentration in the $i$th phase. The relevant equilibria, for example, to describe the extraction in systems of lipophilic acidic chelating agents are

$$HL_{org} \rightleftarrows HL_{aq}$$
$$HL_{aq} \rightleftarrows H^+ + L^-_{aq}$$
$$M^{3+}_{aq} + 3HL_{org} \rightleftarrows (ML_3)_{org} + 3H^+_{aq}$$

where $K_e$ is the equilibrium constant for the last reaction. The distribution coefficient $D$ is given as:

$$D = [ML_3]_{org}/[M^{3+}]_{aq} = K_e[HL]^3_{org}/[H^+]^3_{aq}$$

If one introduces a water-soluble complexing agent into the system, the $[M^{3+}]_{aq}$ becomes $[M^{3+}] + [MX^{2+}] + [MX_2^+] + \ldots$ and the measured distribution ratio will include these species as well. The separation factor between two ions, $S$, is given by the ratio of their distribution coefficients:

$$S = D_A/D_B$$

Thus, the most effective separations will involve cases where the target ion interacts strongly with the extractant but is less strongly complexed by the aqueous ligand X. The percent extraction is given by:

$$\% \text{ extraction} = \frac{100D}{D + \left(\frac{V_{aq}}{V_{org}}\right)}$$

where $V_i$ is the volume of the $i$th phase.

### 19.6.3 Ion Exchange

Ion exchange is one of the most popular radiochemical separation techniques due to its high selectivity and the ability to perform separations rapidly. In ion exchange, a solution containing the ions to be separated is brought into contact with a synthetic organic resin containing specific functional groups that selectively bind the ions in question. In a later step the ions of interest can be removed from the resin by elution with another suitable solution that differs from the initial solution. Typically, the solution containing the ions is run through a column packed with resin beads. The resins are typically crosslinked polystyrenes with attached functional groups. Most cation exchangers (such as Dowex 50) contain free sulfonic acid groups, $SO_3H$, where the cation displaces the hydrogen ion. Anion exchangers (such as Dowex 1) contain quaternary amine groups, such as $CH_2N(CH_3)_3Cl$ where the anion replaces the chloride ion. The resin particles have diameters of 0.08–0.16 mm and exchange capacities of 3–5 meq per gram of dry resin.

It is common to absorb a group of ions on the column material and then selectively elute them. Complexing agents, which form complexes of varying solubility with the absorbed ions, are used as eluants. A competition occurs between the complexing agent and the resin for each ion, and each ion will be exchanged between the resin and the complexing agent several times as it moves down the column. This is akin to a distillation process. The rates at which the different ions move down the column vary, causing a spatial separation between "bands" of different ions. The ions can be collected separately in successive eluant fractions (see Fig. 19.6).

The most widely cited application of ion exchange techniques is the separation of the rare earths or actinides from one another. This is done with cation exchange using a complexing agent of α-hydroxyisobutyric acid ("α-but"). The order of elution of the ions from a cation exchange column is generally in order of the radii of the hydrated ions with the largest hydrated ions leaving first; thus lawrencium elutes first and americium last among the tri-positive actinide ions (see Fig. 19.6). In the case of the data of Figure 19.6, the separation between adjacent cations and the order of elution is derived from the comparative stability of the aqueous actinide or lanthanide complexes with α-hydroxyisobutyrate. As shown in Figure 19.6, there is a strikingly analogous behavior in the elution of the actinides

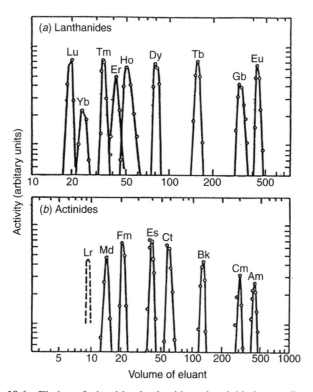

**Figure 19.6** Elution of tripositive lanthanide and actinide ions on Dowex-50.

and lanthanides that allowed chemists to prove the identity of new elements in the discovery of elements 97–102 (Bk–No). For cation exchange, the strength of absorption goes as $M^{4+} > M^{3+} > MO_2^{2+} > M^{2+} > MO_2^+$.

The anion exchange behavior of various elements has been extensively studied. For example, consider the system of Dowex 1 resin and an HCl eluant. Typical distribution ratios for various elements as a function of [Cl$^-$] are shown in Figure 19.7. Note that groups 1, 2, and 3 are not absorbed on the column. One usually sees a rise in the distribution coefficient $D$ until a maximum is reached, and then $D$ decreases gradually with further increases in [Cl$^-$]. The maximum occurs when the number of ligands bonding to the metal atom equals the initial charge on the ion. The decrease in $D$ with further increases in eluant concentration is due to free anions from the eluant competing with the metal complexes for ion exchange resin sites. Figure 19.7 or similar data can be used to plan separations. For example, to separate Ni(II) and Co(II), one needs simply to pass a 12 M HCl solution of the elements through a Dowex 1 column. The Co(II) sticks to the column while the Ni(II) is not absorbed. A mixture of Mn(II), Co(II), Cu(II), Fe(III), and Zn(II) can be separated by being placed on a Dowex 1 column using 12 M HCl, followed by elutions with 6 M HCl (Mn), 4 M HCl (Co), 2.5 M HCl (Cu), 0.5 M HCl (Fe), and 0.005 M HCl (Zn).

In addition to the organic ion exchange resins, some inorganic ion exchanges, such as the zeolites, have been used. Inorganic ion exchangers are used in situations where heat and radiation might preclude the use of organic resins, although the establishment of equlibria may be slow.

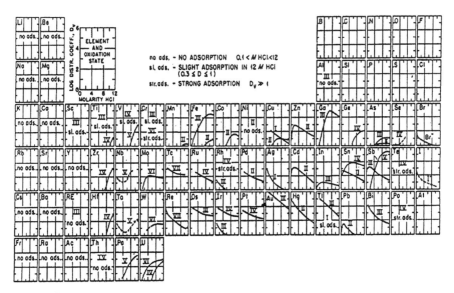

**Figure 19.7** Elution of elements from anion exchange resin. (From Kraus and Nelson, 1956.)

Newer developments have emphasized the preparation of more selective resins. Among these are the chelating resins (such as Chelex-100) that contain functional groups that chelate metal ions. Typical functional groups include iminodiacetic acids, 8-hydroxyquinoline, or macrocyclic units such as the crown ethers, calixarenes, or cryptands. The bifunctional chelating ion exchange material, Diphonix resin—a substituted diphosphonic acid resin, shows promise in treating radioactive waste. Important newer resins include those with immobilized phosphorus ligands (Bond et al., 1999).

### 19.6.4 Extraction Chromatography

Extraction chromatography is an analytical separation technique that is closely related to solvent extraction. Extraction chromatography is a form of solvent extraction where one of the liquid phases is made stationary by adsorption on a solid support. The other liquid phase is mobile. Either the aqueous or the organic phase can be made stationary. Extraction chromatography has the selectivity of solvent extraction and the multistage character of a chromatographic process. It is generally used for laboratory-scale experiments, although some attempts have been made to use it in larger scale operations. The common applications involve the adsorption of an organic extractant onto a variety of inorganic substrates such as silica or alumina or organic substrates such as cellulose or styrene-divinyl benzene copolymers. When the stationary phase is organic, the technique is referred to as reversed-phase high-performance liquid chromatography. The stationary phase is used in a column just as in ion exchange chromatography. High-pressure pumps are usually used to force the liquid phase through these columns, just as in conventional high-performance liquid chromatography.

The same extracting agents as used in solvent extraction can be used in extraction chromatography. Early applications of extraction chromatography have employed various traditional extractants such as the acidic organophosphorus compounds [di-(2-ethylhexyl) phosphoric acid, HDEHP] or TBP as extractants for the actinide elements. Recent advances have led to a variety of new solvent exchange extractants such as the crown ethers, cryptands, or bifunctional organophosphorus compounds. A particularly successful application is the selective sorption of actinides on TRU resins, involving solutions of carbamolymethyl-phosphoryl (CMPO) compounds in TBP sorbed on Amberlite XAD-7. This resin has found a number of applications in the isolation and subsequent determination of the actinides in complex matrices.

### 19.6.5 Rapid Radiochemical Separations

Many of the separation techniques we have described take hours to perform. Many interesting nuclei, such as the heavier actinides, the transactinides, or the light nuclei used in PET studies, have much shorter half-lives. Thus, we will briefly review the principles of rapid radiochemical separations (procedures that take seconds to minutes) and refer the reader to Herrman and Trautman (1982), Meyer and Henry (1979), Schädel et al. (1988), and Trautman (1995) for details.

In most chemical separation procedures, the goal is to selectively transfer the species of interest from one phase to another, leaving behind any unwanted species. The phase-to-phase transfer is rapid, but the procedures to place the species in the proper form for transfer to occur are slow. The goal of rapid radiochemical separations is to speed up existing chemical procedures or to use new, very fast chemical transformations.

Two procedures are commonly used for rapid radiochemical separations, the batch approach and the continuous approach. In the batch approach, the desired activities are produced in a short irradiation, separated and counted with the procedure being repeated many times to reduce the statistical uncertainty in the data. In the continuous approach, the production of the active species is carried out continuously, and the species is isolated and counted as produced.

One of the most widely used techniques for rapid chemical separation is that of gas chromatography, which has been developed for use with the transuranium elements by Zvara and co-workers (1972). In gas chromatography, volatile elements or compounds are separated from one another by their differences in distribution between a mobile gas phase and a stationary solid phase. Thermochromatography involves passing a gas through a column whose temperature decreases continuously with distance from the entrance. Thus, the less volatile species condense on the column walls first with the more volatile species depositing last. Measurement of the migration times, the deposition temperature, the temperature gradient in the column, and so on can allow one to deduce the molar enthalpy of absorption of the compound on the column material. This physical quantity can be compared to quantum chemical calculations of this quantity to gain insight into the bonding properties of the element in question. This technique was used to show the chemical properties (Schädel, 1997; Eichler, 2000; Düllmann et al., 2002) of the transactinides Rf–Hs and their behavior relative to their chemical congeners.

Another rapid chemical separation technique is separation by volatilization. There are a variety of volatile compounds that can be released from an irradiated material upon dissolution that can, with proper conditions, serve to rapidly chemically separate the elements involved. Examples of such volatile species include $I_2$, At, $GeCl_4$, $AsCl_3$, $SeCl_4$, $OsO_4$, $RuO_4$, $Re_2O_7$, $Tc_2O_7$, and so on. Separation by volatilization has largely been used for the elements forming volatile hydrides, As, Se, Sn, Sb, and Te.

## 19.7 LOW-LEVEL MEASUREMENT TECHNIQUES

### 19.7.1 Introduction

One of the areas in which the skills of radiochemists are used is the area of low-level chemistry and low-level counting. Areas as diverse as the detection of solar neutrinos or the study of environmental radioactivity involve low-level techniques. For example, despite concentration of the radiotracers of interest during sampling procedures in environmental studies, quite often one is left with a sample containing a small quantity <10 ppm of radioactivity that must be assayed. Such essays are referred to as *low-level techniques*. Let us begin our discussion of low-level techniques by considering any chemical manipulations of the sample that must be made prior to counting it. Understandably, the fact of having activity levels <10 dpm puts severe restrictions on the nature of *low-level chemistry*. Among the requirements for low-level chemistry are a small constant blank, high chemical yields for all procedures, high radiochemical purity for all reagents employed, and the ability to place the sample in suitable chemical form for counting.

### 19.7.2 Blanks

The *blank* in low-level chemistry is defined as the contribution of the added reagents and other sample constituents to the activity being measured. The blank is determined by performing the chemical procedures without the radioactive sample being analyzed. Care must be taken to ensure that the blank is properly measured and includes all possible contributions to the activity that would be encountered in a real system. For example, in the determination of fission product $^{144}$Ce in seawater, the blank must be determined for each new bottle of reagents used due to the high variability of the $^{144}$Ce content in chemical reagents.

Clearly, one of the most effective ways of dealing with a blank correction is to reduce it to its lowest level. Among the factors contributing to the blank correction that can be reduced with care is *radiochemical contamination of analytical reagents used in chemical procedures*. DeVoe (1961) and Sugihara (1961) have written extensive review articles on this subject, and their work should be consulted for detailed information. Typical contamination of most reagents is in the range of $\sim$10–100 ppm/g reagent, although individual reagents may contain activity levels of >10,000 ppm/g. Some especially troublesome reagents are rare earths (Ce salts in particular), chlorine or sulfur-containing reagents that may contain $^{32}$P contamination, cesium salts (which may contain $^{40}$K or $^{87}$Rb), and potassium salts and other obvious offenders. Precipitating agents, such as tetraphenylborates and chloroplatinates are also particularly pernicious with regards to contamination problems.

Airborne contamination is another possible contribution to the blank correction. Here one is chiefly concerned with sample contamination with the daughters of $^{222}$Rn, which have half-lives in the 30- to 40-min range. Steps that can be taken to avoid this problem include eschewing the use of suction filtration in chemical procedures, prefiltering of room air, and use of radon traps.

Further lowering of the blank correction occurs when *nonisotopic carriers* in chemical procedures are used to replace inert carriers of the element of interest when it is difficult to obtain the inert carrier in a contamination-free condition. Obviously, only clean glassware should be used, reagents should not be reused, and the laboratory should be kept in an immaculate condition. Separations that have high chemical yields and high radiochemical purity reduce the blanks.

### 19.7.3 Low-Level Counting—General Principles

Once the low-level radioactive sample has been collected and any chemical procedures performed prior to counting, it is ready for counting. Because of the extremely small disintegration rates encountered, special techniques, called low-level counting, must be used to assay the sample. We shall survey some of the highlights of this area, which has been the subject of many articles and monographs (Knoll, 2000).

What are the most important characteristics the detector must possess for low-level counting? The first general characteristic is one of stability. Low-level counting frequently requires counting periods of long duration; hence counter stability is quite important. If the sample count rate $S$ (source + background) is only slightly larger than the background rate $B$, one's detector should be picked to maximize the ratio $S^2/B$—that is, low background and high efficiency. If the sample count rate is large with respect to background, one need only maximize $S$—that is, one chooses a high-efficiency detector.

### 19.7.4 Low-Level Counting—Details

For *low-level α-particle counting*, the choice is generally between the gas-filled ionization chamber and a semiconductor detector system. The former can have a counting efficiency of approximately 50% and a background of approximately 3–4 cpm; the latter has a background rate of approximately 0.5 cph and a detection efficiency that approaches 50%. The semiconductor detector is usually the detector of choice, although large sample sizes may be better assayed with gas-filled ionization chambers. Background radiation is primarily due to α-particle emitting impurities in the counter, counter support material, and so on, plus the occurrence of cosmic-ray-induced (n,α) reactions. Because of its Ra content, Al is not used in constructing α spectrometers.

Low-level "soft" radiation counting has its own techniques. The term *soft radiation counting* refers to detecting EC and low-energy $\beta^-$ emitters where the self-absorption of the radiation in the sample is important. To solve this problem, one typically tries to incorporate the radionuclide to be counted into the detector itself. One typical method of assay is liquid scintillation counting, which is used to assay samples whose activity is greater than 10 pCi. Typical liquid scintillation counter backgrounds can be as high as approximately 100 cpm, whereas special counters have been built with background rates of approximately 10 cpm or less. Liquid scintillation counting is a speedy, simple method of low-level counting. Another technique that has been used to count low-level soft radiation samples involves the use of gas-filled proportional counter. The sample to be counted is converted to gaseous form and added to the counter gas at a concentration of approximately 0.05 mol % or less. This method of low-level counting, although tedious and time-consuming, allows one to assay samples whose activity is less than 0.5 pCi. Typical counter backgrounds are $\sim$1–2 cpm with 100% counting efficiency for energies as low as $\sim$10 eV.

The counting of tritium in water is a special problem about which much has been written. Current methods for assay of tritium in water have a range of 0.1–5000 TU, where a tritium unit (TU) has the value of 7.2 dpm/L. The most desirable feature of a tritium measuring system is that it be capable of measuring a large number of samples rapidly, simply, and cheaply as possible with an uncertainty of $\sim \pm 10\%$ or better. It is generally more important to assay 100 samples with an uncertainty of $\pm 10\%$ than to assay 10 samples with an uncertainty of $\pm 3\%$.

The basic feature of low-level $\beta^-$ counting that distinguishes it from ordinary $\beta^-$ counting is the use of an anticoincidence shield around the main $\beta^-$ counters. An anticoincidence shield is a single detector, or array of detectors, that surrounds the primary detector. The output of the anticoincidence detector is fed to an anticoincidence circuit along with the output of the primary detector. When nuclear radiation passes through both detectors simultaneously, as in the case of a highly penetrating cosmic ray striking both detectors, no output results from the anticoincidence circuit. When the anticoincidence circuit receives a signal *only* from the primary detector, an output signal results. The net effect is that the anticoincidence shield detector "guards" or shields against exterior radiation background radiation entering the primary detector. Typical ring assemblies reduce the background rate in the primary counter by a factor of $\sim 50$. A well-designed guard ring will allow several different types of central counter to be inserted into it. Low background $\beta^-$ counters constructed of especially pure materials with anticoincidence shields have exhibited background rates of $\sim 1$ cph with efficiencies of approximately 50%.

Low-level counting of $\gamma$-ray emitters using solid scintillation counters is an extensively used technique. The most important aspect of low-level solid scintillation counting is to decrease the counter background. Typical contributions to a solid scintillation counter's background rate from various sources are shown in Table 19.3.

Here four factors are seen as the major contributors to the detector background rate. They are (a) the cosmic ray shield, (b) the atmosphere surrounding the detector, (c) the detector itself, and (d) the cosmic rays. For the cosmic ray shield about the detector, it is advisable to use old or virgin lead, that is, lead that was purified over 100 years ago, thus allowing any $^{210}$Pb present to decay. One should expect $\sim 1$ cpm/g shield material. Iron can also be used in constructing the detector

**TABLE 19.3 Components of a NI(Tl) Scintillation Counter Background**

| | |
|---|---|
| Outside shield | 29.200 cpm |
| Inside shield | |
|    Cosmic ray mesons | 116.4 cpm |
|    Cosmic ray neutrons | 19.4 cpm |
|    $^{222}$Rn daughters | 25.9 cpm |
|    $^{40}$K | 8.6 cpm |
|    Remaining background | 33.1 cpm |
| Total | 203.4 cpm |

*Source*: Stenberg and Olsson (1968).

shield, but care must be taken to ensure that the iron or steel is pre-1945 in origin. (Iron processed in the post-1945 period has $^{60}$Co contamination due to the use of $^{60}$Co in the blast furnace operation.) Mercury is a very good, easily purified shield material but it is quite expensive. "Graded" shields consisting of an outer thick layer of Pb lined with Cd, which in turn is coated with Cu, are used to reduce the production of X-rays in the Pb shield material. The main portion of the atmospheric contribution to the detector background is due to radon and its daughters. Particularly troublesome in this regard is the fact that atmospheric radon concentrations can fluctuate by a factor of 40 during the course of a day. Once again the problem is best handled by filtering the room air, rapid air turnover, and the use of inert atmospheres ($N_2$ from evaporation of liquid nitrogen) inside counting assemblies. A NaI detector will contain some $^{40}$K impurity, which will contribute to the background. The detector housing is also a potential contributor to the background. Copper appears to be the best material for detector housing, with aluminum being the least preferred, for it can be expected to contain $\sim 10^{-13}$ Ci Ra/g Al. In reducing the cosmic ray component of the background, one tries to stop the "soft" cosmic rays (electrons, X, and $\gamma$ rays) in the detector shield while using an anticoincidence system to stop the "hard' component of the cosmic radiation (mesons, etc.).

At first one might think that germanium detectors with their low detection efficiencies would have little use in low-level counting. However, such is not the case because of the complexity of a radionuclide mixture found in environmental samples, the virtual impossibility of drawing significant conclusions from NaI detector spectra due to poor energy resolution. Clearly, there are many cases in which the very good energy resolution of the germanium detector is a necessity. Furthermore, recent developments in detector fabrication techniques allow the production of germanium detectors with detector efficiencies equal to that of standard (3 in. × 3 in.) NaI detectors.

A number of special techniques have evolved to increase the detection sensitivity in $\gamma$-ray counting. One of the most important is the suppression of the Compton scattering events in the $\gamma$-ray spectrum by the use of anticoincidence annulus around the central $\gamma$-ray detector. The idea behind a Compton suppression spectrometer is that most events in which the incident photon undergoes one or more Compton scattering events in the central detector will result in partial energy deposition in the detector with a low-energy photon escaping the detector.

Suppression is accomplished by setting up an anticoincidence between the central detector signal and any signal coming from the annulus. A photon that is Compton scattered from the central detector will probably give rise to a signal from the annulus. Thus, such events will not be accepted. Other events, such as photoelectric events in the central detector will not produce signals from the annulus and will be counted. This reduction in number of Compton scattering events in the $\gamma$-ray spectrum leads to a more easily interpreted spectrum since the peak to Compton ratios are much higher. Typically, the use of such annuli reduces the number of Compton events 10-fold. More sophisticated designs have been used to further improve the rejection of Compton scattering events.

## 19.7.5 Limits of Detection

Suppose you have performed a low-level experiment and you wish to state your results in a statistically meaningful manner. You wish to answer such questions as "Is there a result/signal/event? What is the chance it will be detected with my apparatus? How big is it?" Currie (1968) has provided answers to these questions by defining three different limits of detection:

- The critical level $L_C$, the signal level above which an observed instrument response may be reliably recognized as "detected."
- The detection limit $L_D$, the true net signal that may be expected a priori to lead to detection.
- The determination limit $L_Q$, the signal level above which a quantitative measurement can be performed with a stated relative uncertainty.

Operationally, the recipes for calculating these limits are as follows:

|  | $L_C$ | $L_D$ | $L_Q$ |
|---|---|---|---|
| Paired observations | $2.33\sigma_B$ | $4.65\sigma_B$ | $14.1\sigma_B$ |
| "Well-known" blank | $1.64\sigma_B$ | $3.29\sigma_B$ | $10\sigma_B$ |

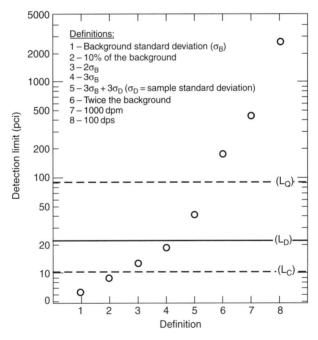

**Figure 19.8** Various measures of detection limits for low-level counting. (From Currie, 1968.)

For example, if the background under a photopeak of interest in a γ-ray spectrum was 100 cpm, then $\sigma_B = \sqrt{100} = 10$ cpm. According to the recipes given above, one would need to detect a photopeak area of 23 cpm to say that a nuclide was present and would need to detect at least 141 cpm to measure the amount of nuclide present. One would need a count rate of 47 cpm to ensure, before making the measurement, that the nuclide in question could be detected. The relationship of these limits to other measures used to describe low-level counting is shown in Figure 19.8.

## PROBLEMS

1. A beam of 1 particle-μA of $^{48}$Ca$^{10+}$ ions is incident on an Al foil that is 5 mg/cm² thick. (a) Estimate the energy deposit/s in the foil. (b) If the foil has an area of 4 cm² and it is mounted in a vacuum with no cooling, how long will it take until the foil reaches the melting point of Al (660°C)? Assume the specific heat of Al is independent of temperature and is 0.25 cal/deg/g.

2. Au foils are to be used as flux monitors in a nuclear reactor. What is the maximum thickness that can be used if the self-shielding corrections are to be less than 10%?

3. The reaction $^{27}$Al(p,3pn)$^{24}$Na is to be used to measure the proton flux in an irradiation with 300-GeV protons. The cross section for this reaction is known to be 10.1 mbarn. The flux monitor is 5.0 mg/cm². One measures a $^{24}$Na counting rate (background corrected) of 10,000 cpm (in a 2% efficient detector) 24 h after a 0.5-h irradiation. What was the average proton flux during the irradiation?

4. Imagine you have for your use a nuclear reactor with a flux of $10^{13}$ n/cm²/s in its irradiation facility and a cyclotron with a beam of 10.5-MeV protons, 21-MeV deuterons, and 42-MeV α particles. What would be the best way to prepare: (a) $^{140}$Ba, (b) carrier-free $^{99}$Tcm, (c) carrier-free $^{144}$Ce, (d) $^{237}$Np, (e) $^{254}$Es, and (f) carrier free $^{18}$F for incorporation into glucose. Outline the target and reaction to be used, how the target would be prepared and any chemical separations to be done following irradiation.

5. What value of $B\rho$ would you use in a gas-filled separator if you wanted to separate $^{254}$No produced in the reaction of 215 MeV $^{48}$Ca with $^{208}$Pb. You may assume $q_{\text{ave}}$ for $^{254}$No is 1 torr He gas is 17+.

6. As a radiochemist, you have been asked to prepare the following accelerator targets for your research group. They are: (a) $^{226}$Ra, (b) $^{208}$Pb, (c) $^{238}$U, (d) $^{244}$Pu, (e) $^{90}$Zr, and (f) $^{124}$Sn. Assume all targets are to be 0.5 mg/cm². For each target, outline the best method of target preparation, the backing foil used, if any, and the reasons for your choice. Describe any anticipated problems in each procedure.

## REFERENCES

Adloff, J. P. and R. Guillamont. *Fundamentals of Radiochemistry*, CRC Press, Boca Raton, FL, 1984.

Bond, A. H., M. L. Dietz, and R. D., Rogers, Eds. *Metal Ion Separations and Preconcentration*, ACS Symposium Series 716, ACS, Washington, DC, 1999.

Currie, L. A. *Anal. Chem.* **40**, 586 (1968).

DeVoe, J. R. Radioactive Contamination of Materials Used in Scientific Research, *NASNRC* 895 (1961).

Düllmann, Ch. E. et al. *GSI Scientific Report 2001*; GSI, Darmstadt, 2002; p. 179.

Ehmann, W. D. and D. E. Vance. *Radiochemistry and Nuclear Methods of Analysis*, Wiley, New York, 1991.

Eichler, R. et al. *Nature* **407**, 63 (2000).

Ghiorso, A. et al. *Nucl. Instru. Meth.* **A269**, 192 (1988).

Harvey, B. G. Recoil Techniques in Nuclear Reactions and Fission Studies, *Ann. Rev. Nucl. Sci.* **10**, 235 (1960).

Herrman, G. and N. Trautman. *Annu. Rev. Nucl. Part. Sci.* **32**, 117 (1982).

Knoll, G. F. *Radiation Detection and Measurement*, 3rd ed., Wiley, New York, 2000.

Kraus, K. and D. Nelson. Paper 837, Geneva Conference, Vol. 7, 1956.

LeFort, M. *Nuclear Chemistry*, Van Nostrand, Princeton, 1968.

Lieser, K. *Nuclear and Radiochemistry: Fundamentals and Applications*, VCH, New York, 1997.

Meyer, R. A. and E. A. Henry. *Proc. Workshop Nucl. Spectrosc. Fission Products*, Grenoble 1979, Bristol, London, 1979, pp. 59–103.

Mozumder, A. *Fundamentals of Radiation Chemistry*, Academic, New York, 1999.

Oganessian, Y. T. et al. *Phys. Rev. C* **64**, 064309 (2001).

Parker, W. C. and H. Slatis. Sample and Window Techniques, in K. Siegbahn, Ed., *Alpha-, Beta- and Gamma-Ray Spectroscopy*, Vol. 1, North Holland, Amsterdam, 1966, pp. 379–408.

Paul, M. *Nucl. Instru. Meth.* **A277**, 418 (1989).

*Random House College Dictionary, Revised Edition*, Random House, New York, 1984.

Schädel, M. et al. *Nucl. Instr. Meth. Phys. Res.* **A 264**, 308 (1988).

Schädel, M. et al. *Nature* **388**, 55 (1997).

Stenberg, A. and I. U. Olsson. *Nucl. Instr. Meth.* **61**, 125 (1968).

Sugihara, T. T. Low Level Radiochemical Separations, *NAS-NRC* 3103.

Trautmann, N. *Radiochimica Acta* **70/71**, 237 (1995).

Tsoulfanidis, N. *Measurement and Detection of Radiation*, 2nd ed., Taylor and Francis, Washington, DC, 1995.

Wang, C. H., D. L. Willis, and W. Loveland. *Radiotracer Methods in the Biological, Physical and Environmental Sciences*, Prentice-Hall, Englewood Cliffs, NJ, 1975.

Woods, R. J., J. W. T. Spinks, and B. Spinks. *An Introduction to Radiation Chemistry*, Wiley, New York, 1990.

Zvara, I. et al. *Sov. Radiochemistry* **14**, 115 (1972).

# BIBLIOGRAPHY

Brune, D., B. Forkman, and G. Persson. *Nuclear Analytical Chemistry*, Chartwell-Brandt, Lund, 1984.

Choppin, G. R. *Experimental Nuclear Chemistry*, Prentice-Hall, Englewood Cliffs, NJ, 1961.

Choppin, G., J.-O. Liljenzin, and J. Rydberg. *Radiochemistry and Nuclear Chemistry*, 3rd ed., Butterworth-Heinemann, Woburn, 2002.

Ehmann, W. D., and D. E. Vance. *Radiochemistry and Nuclear Methods of Analysis*, Wiley, New York, 1991.

Friedlander, G., J. W. Kennedy, J. M. Miller, and E. S. Macias. *Nuclear and Radiochemistry*, Wiley, New York, 1981.

Lieser, K. *Nuclear and Radiochemistry: Fundamentals and Applications*, VCH, New York, 1997.

Seaborg, G. T. and W. Loveland. *The Elements Beyond Uranium*, Wiley, New York, 1990.

# APPENDIX A

# FUNDAMENTAL CONSTANTS AND CONVERSION FACTORS

**TABLE A.1  Fundamental Constants**

| Quantity | Symbol (Expression) | Value | SI Units | cgs Units |
|---|---|---|---|---|
| Speed of light in vacuum | $c$ | 2.99792458 | $10^8$ m s$^{-1}$ | $10^{10}$ cm s$^{-1}$ |
| Elementary charge | $e$ | 4.8032068 | | $10^{-10}$ esu |
| | | 1.60217733 | $10^{-19}$ C | $10^{-20}$ emu |
| Planck constant | $h$ | 6.6260755 | $10^{-34}$ J s | $10^{-27}$ ergs |
| (reduced) | $\hbar$ | 1.05457266 | $10^{-34}$ J s | $10^{-27}$ ergs |
| Boltzmann constant | $k$ | 1.380658 | $10^{-23}$ J K$^{-1}$ | $10^{-16}$ erg K$^{-1}$ |
| Avogadro's number | $N_A$ | 6.0221367 | $10^{23}$ mol$^{-1}$ | $10^{23}$ mol$^{-1}$ |
| Molar gas constant | $R$ | 8.314510 | J mol$^{-1}$ K$^{-1}$ | $10^7$ erg mol$^{-1}$ K$^{-1}$ |
| Rydberg constant | $R_\infty = m_e c\, \alpha^2/2h$ | 1.0973731534 | $10^7$ m$^{-1}$ | $10^5$ cm$^{-1}$ |

*(continued)*

*Modern Nuclear Chemistry*, by W.D. Loveland, D.J. Morrissey, and G.T. Seaborg
Copyright © 2006 John Wiley & Sons, Inc.

**TABLE A.1** *Continued*

| Quantity | Symbol (Expression) | Value | SI Units | cgs Units |
|---|---|---|---|---|
| Bohr magneton | $\mu_B = eh[c]/2m_ec$ | 9.2740154 | $10^{-24}$ J T$^{-1}$ | $10^{-21}$ erg G$^{-1}$ |
| Nuclear magneton | $\mu_N = eh[c]/2m_pc$ | 5.0507866 | $10^{-27}$ J T$^{-1}$ | $10^{-24}$ erg G$^{-1}$ |
| Fine structure constant | $\alpha = [4\pi\varepsilon_0]^{-1} e^2/hc$ | 7.29735308 | | |
| | $\alpha^{-1}$ | 137.0359895 | | |
| Permittivity, free space | $\varepsilon_0$ | 8.854187817 | $10^{-12}$ C$^2$ N$^{-1}$ m$^{-2}$ | |
| Atomic mass unit | u | 1.6605402 | $10^{-27}$ kg | $10^{-24}$ g |
| Electron rest mass | $m_e$ | 9.10938974 | $10^{-31}$ kg | $10^{-28}$ g |
| Proton rest mass | $m_p$ | 1.6726231 | $10^{-27}$ kg | $10^{-24}$ g |
| Neutron rest mass | $m_n$ | 1.674929 | $10^{-27}$ kg | $10^{-24}$ g |
| Muon rest mass | $m_\mu$ | 1.8835327 | $10^{-28}$ kg | $10^{-25}$ g |
| Pion rest mass | $m_\pi^\pm$ | 2.4880187 | $10^{-28}$ kg | $10^{-25}$ g |
| | $m_\pi^0$ | 2.406120 | $10^{-28}$ kg | $10^{-25}$ g |
| Bohr radius | $a_0 = r_e/\alpha^2$ | 5.29177249 | $10^{-11}$ m | $10^{-9}$ cm |
| Compton wavelength | | | | |
| Electron | $\lambda_{c,e} = h/m_ec$ | 2.42031058 | $10^{-12}$ m | $10^{-10}$ cm |
| Proton | $\lambda_{c,p} = h/m_pc$ | 1.32141002 | $10^{-15}$ m | $10^{-13}$ cm |
| Neutron | $\lambda_{c,n} = h/m_nc$ | 1.31959110 | $10^{-15}$ m | $10^{-13}$ cm |
| Classical electron radius | $r_e = \alpha h/m_ec$ | 2.81794092 | $10^{-15}$ m | $10^{-13}$ cm |
| Magnetic dipole moment | | | | |
| Electron | $\mu_e$ | 1.001159652193 | $\mu_B$ | |
| Proton | $\mu_p$ | 2.792847386 | $\mu_N$ | |
| Neutron | $\mu_n$ | −1.91304275 | $\mu_N$ | |
| Proton gyromagnetic ratio | $\gamma_p$ | 2.67522128 | $10^8$ s$^{-1}$ T$^{-1}$ | $10^4$ s$^{-1}$ G$^{-1}$ |

**TABLE A.2 Conversion Factors and Handy Units**

| Quantity | Symbol | Value |
|---|---|---|
| Atomic mass unit | u | 931.494 MeV |
| Electron mass | $m_e$ | 0.510999 MeV |
| Proton mass | $m_p$ | 938.272 MeV |
| Neutron mass | $m_n$ | 939.566 MeV |
| Electron volt | 1 eV | $1.602177 \times 10^{-19}$ J |
| Electron volt/particle | 1 eV/k | 11604.45 K |
| Planck constant | h | $6.582122 \times 10^{-22}$ MeV·s |
|  | hc | 197.327053 MeV·fm |
|  | $(hc)^2$ | 0.389380 GeV$^2$·mb |
| Rydberg constant | $R_\infty hc$ | 13.605698 eV |
| Gas constant | R | 1.987216 cal/mol |
| 1 degree | ° | $1.7453 \times 10^{-2}$ rad |
| 1 calorie | cal | 4.184 J |
| 1 British thermal unit | Btu | 1054.4 J |
| 1 erg |  | $10^{-7}$ J |
| 1 ton (equivalent of TNT) |  | $4.184 \times 10^9$ J |
| 1 electron radius | $r_e$ | $2.8179 \times 10^{-15}$ m |
| 1 fermi |  | $10^{-15}$ m |
| 1 light year | ly | $9.4605 \times 10^{15}$ m |
| 1 parsec | pc | $3.0857 \times 10^{16}$ m |
| 1 atmosphere | atm | 101325 Pa |
| 1 torr (mm Hg, 0°C) |  | 133.32 Pa |
| 1 day | d | 86400 s |
| 1 year (365.25636 d) | y | $3.1558150 \times 10^7$ s |
| 1 Curie (Ci) |  | $3.700 \times 10^{10}$ Bq |
| 1 rad |  | $1.000 \times 10^{-2}$ Gy |
| 1 rem |  | $1.000 \times 10^{-2}$ Sv |
| 1 Roentgen (R) |  | $2.580 \times 10^{-4}$ C/kg |

# APPENDIX B

# NUCLEAR WALLET CARDS

What follows is the 2005 version of the Nuclear Wallet Cards

---

*Modern Nuclear Chemistry*, by W.D. Loveland, D.J. Morrisscy, and G.T. Seaborg
Copyright © 2006 John Wiley & Sons, Inc.

## INTRODUCTION

This is an updated edition of the 2000 booklet of the same name[†]. The 2000 edition is being archived for the US, DOE nuclear material inventory control (see p. vii).

This booklet presents selected properties of all known nuclides and their known isomeric states. Properties of ionized atoms are not included.

The data given here are taken mostly from the adopted properties of the various nuclides as given in the *Evaluated Nuclear Structure Data File* (ENSDF)[1]. The data in ENSDF are based on experimental results and are published in *Nuclear Data Sheets*[2] for A>20 and in *Nuclear Physics*[3] for A≤20. For nuclides for which either there are no data in ENSDF or those data that have since been superseded, the half-life and the decay modes are taken either from recent literature[4] or from other sources, *e.g.*, [5].

For other references, experimental data, and information on the data measurements, please refer to the original evaluations [1–3]. The data were updated to **January 15, 2005**.

[†]The first *Nuclear Wallet Cards* was produced by F. Ajzenberg-Selove and C. L. Busch in 1971. The Isotopes Project, Lawrence Berkeley National Laboratory, produced the next edition in 1979 based on the *Table of Isotopes*, 7th edition (1978)[9]. The subsequent editions, the third in 1985, the fourth in 1990, the fifth in 1995, and the sixth in 2000 were produced by J.K. Tuli, NNDC. In 2004, **Nuclear Wallet Cards for Radioactive Nuclides** aimed at Homeland Security personnel was produced by J.K. Tuli, NNDC (see p. vii).

## Explanation of Table

### Column 1, Nuclide (Z, El, A):

Nuclides are listed in order of increasing atomic number (Z), and are suborbered by increasing mass number (A). All isotopic species, as well as all isomers with half-life ≥0.1 s, and some with half-life≥1 ms which decay by SF, α or p emissions, are included. A nuclide is given even if only its mass estimate or its production cross section is available. For the latter nuclides half-life limit or an approximate value is given as estimated from systematics [5].

Isomeric states are denoted by the symbol "m" after the mass number and are given in the order of increasing excitation energy.

The $^{235}$U thermal fission products, with fractional cumulative yields≥$10^{-6}$, are *italicized* in the table. The information on fission products is taken from the ENDF/B–VI fission products file [8].

The names and symbols for elements are those adopted by the International Union of Pure and Applied Chemistry (2004). No names and symbols have as yet been adopted for Z>111.

### Column 2, Jπ:

Spin and parity assignments, without and with parentheses, are based upon strong and weak arguments, respectively. See the introductory pages of any issue of *Nuclear Data Sheets*[2] for description of strong and weak arguments for Jπ assignments.

## Explanation of Table (cont.)

### Column 3, Mass Excess, Δ:

Mass excesses, M−A, are given in MeV (from [6]) with $\Delta(^{12}C)=0$, by definition. For isomers the values are obtained by adding the excitation energy to the Δ(g.s.) values. Wherever the excitation energy is not known, the mass excess for the next lower isomer (or the g.s.) is given. The values are given to the accuracy determined by uncertainty in Δ(g.s.) (maximum of three figures after the decimal). The uncertainty is ≤9 in the last significant figure. An appended "s" denotes that the value is obtained from systematics [6].

### Column 4, T$_½$,Γ or Abundance:

The half-life and the abundance (in **bold face** from [7]) are shown followed by their units ("%" symbol in the case of abundance) which are followed by the uncertainty, in the last significant figures. For example, 8.1 s *10* means 8.1±1.0 s. For some very short–lived nuclei, level widths rather than half–lives are given. There also, the width is followed by units (*e.g.*, eV, keV, or MeV) which are followed by the uncertainty in *italics*, if known. As stated above when a limit or an approximate value is given it is based on systematics (*sy*), mostly from [5]. A '?' in this field indicates that T$_½$ is not known.

For 2β$^-$ and 2ε decay only the lowest value of their several limits (*e.g.*, for 0ν or 2ν, etc.) is given.

If a new measurement of half–life, has since become available it is presented in place of the evaluated value in ENSDF.

## Explanation of Table (cont.)

### Column 5, Decay Mode:

Decay modes are given in decreasing strength from left to right, followed by the percentage branching, if known ("w" indicates a weak branch). The percentage branching is omitted where there is no competing mode of decay or no other mode has been observed. A "?" indicates an expected but not observed mode of decay[5]. The various modes of decay are given below:

| | |
|---|---|
| β− | β$^-$ decay |
| ε | ε (electron capture), or ε+β$^+$, or β$^+$ decay |
| IT | isomeric transition (through γ or conversion–electron decay) |
| n, p, α, … | neutron, proton, alpha, … decay |
| SF | spontaneous fission |
| 2β−, 3α, … | double β$^-$ decay (β$^-$β$^-$), decay through emission of 3 α's, … |
| β−n, β−p, β−α, … | delayed n, p, α, … (emission following β$^-$ decay) |
| εp, εα, εSF, … | delayed p, α, SF, … (emission following ε or β$^+$ decay) |

## NNDC Web Services

The centerfold presents the NNDC home page on the web (*www.nndc.bnl.gov*). The greatly expanded NNDC web service offers a wealth of Nuclear Physics information which includes analysis programs, reference data, and custom-tailored retrievals from its many databases.

## DOE Standard for Nuclear Material Inventory

The 2000 edition was adopted as the standard by the the Department of Energy for the purposes of their nuclear material inventory. The 2000 edition, as well as, the current version are available through the NNDC web site, *www.nndc.bnl.gov/wallet/*.

## Homeland Security

Nuclear Wallet Cards for Radioactive Nuclides, a reference for homeland security personnel based on this booklet was published in March 2004. The booklet, although limited to radioactive nuclides, contains additional radiation information. It is available on the web as well as in printed form from NNDC.

## Acknowledgements

The appendix on $\Lambda$ hypernuclides has been prepared by R. Chrien, BNL. The author is thankful to T. Burrows, M. Herman, B. Pritychenko, D. Rochman, A. Sonzogni, and D. Winchell for many useful suggestions. The help in library matters from J. Totans, proof-reading from M. Blennau, and in all computer matters from R. Arcilla is acknowledged. The author is grateful for encouragement and support from P. Oblozinsky and C. Dunford.

This research was supported by the Office of Nuclear Physics, Office of Science, US Department of Energy.

## References

1. *Evaluated Nuclear Structure Data File* – a computer file of evaluated experimental nuclear structure data maintained by the National Nuclear Data Center, Brookhaven National Laboratory (file as of January, 2005).

2. *Nuclear Data Sheets* – Elsevier B.V., Amsterdam. Evaluations published by mass number for A = 21 to 294. See page ii of any issue for the index to A-chains. See also Energy Levels of A = 21–44 Nuclei (VII), P. M. Endt, *Nuclear Physics* A521, 1 (1990). Supplement, *Nuclear Physics* A633, 1 (1998).

3. *Nuclear Physics* – North Holland Publishing Co., Amsterdam – Evaluations for A = 3 to 20.

4. *Nuclear Science Reference File* – a bibliographic computer file of nuclear science references continually updated and maintained by the National Nuclear Data Center, Brookhaven National Laboratory. Recent literature is scanned by D. Winchell.

5. NUBASE Evaluation of Nuclear and Decay Properties, G. Audi, O. Bersillon, J. Blachot, and A.H. Wapstra, *Nuclear Physics* A729, 3 (2003).

6. The AME2003 Atomic Mass Evaluation, A. H. Wapstra, G. Audi, and C. Thibault, *Nuclear Physics* A729, 129 (2003).

## References (cont.)

7. Table of Isotopes, N. Holden, *The CRC Handook of Physics and Chemistry* (2004).

8. Evaluation and Compilation of Fission Product Yields 1993, T.R. England and B.F. Rider; Rept. LA–UR–94–3106 (1994). ENDF/B–VI evaluation; MAT #9228, Revision 1.

9. *Table of Isotopes* (1978), 7$^{th}$ edition, Editors: C.M. Lederer, V.S. Shirley, Authors: E. Browne, J.M. Dairiki, R.E. Doebler, A.A. Shihab–Eldin, J. Jardine, J.K. Tuli, and A.B. Buyrn, John Wiley, New York.

## APPENDIX B

### Nuclear Wallet Cards

| Nuclide Z El A | Jπ | Δ (MeV) | T½, Γ, or Abundance | Decay Mode |
|---|---|---|---|---|
| 0 n 1 | 1/2+ | 8.071 | 10.24 m 2 | β– |
| 1 H 1 | 1/2+ | 7.289 | **99.985% 1** | |
| 2 | 1+ | 13.136 | **0.015% 1** | |
| 3 | 1/2+ | 14.950 | 12.32 y 2 | β– |
| 4 | 2– | 25.9 | 4.6 MeV 9 | n |
| 5 | | 32.9 | 5.7 MeV 21 | n |
| 6 | (2–) | 41.9 | 1.6 MeV 4 | n |
| 7 | | 49s | $29 \times 10^{-23}$ y 7 | |
| 2 He 3 | 1/2+ | 14.931 | **0.000137% 3** | |
| 4 | 0+ | 2.425 | **99.999863% 3** | |
| 5 | 3/2– | 11.39 | 0.60 MeV 2 | α, n |
| 6 | 0+ | 17.595 | 806.7 ms 15 | β– |
| 7 | (3/2)– | 26.10 | 150 keV 20 | n |
| 8 | 0+ | 31.598 | 119.0 ms 15 | β–, β–n 16% |
| 9 | (1/2–) | 40.94 | 65 keV 37 | n |
| 10 | 0+ | 48.81 | 0.17 MeV 11 | 2n? |
| 3 Li 3 | | 29s | unstable | p? |
| 4 | 2– | 25.3 | 6.03 MeV 4 | p |
| 5 | 3/2– | 11.68 | ≈1.5 MeV 4 | α, n |
| 6 | 1+ | 14.087 | **7.59% 4** | |
| 7 | 3/2– | 14.908 | **92.41% 4** | |
| 8 | 2+ | 20.947 | 838 ms 6 | β–, β–α |
| 9 | 3/2– | 24.954 | 178.3 ms 4 | β–, β–n 50.8% |
| 10 | (1–,2–) | 33.05 | 1.2 MeV 3 | n |
| 11 | 3/2– | 40.80 | 8.59 ms 14 | β–, β–nα 0.027%, β–n |
| 12 | | 50.1s | <10 ns | n? |
| 4 Be 5 | (1/2+) | 38s | ? | p |
| 6 | 0+ | 18.375 | 92 keV 6 | p, α |
| 7 | 3/2– | 15.770 | 53.22 d 6 | ε |
| 8 | 0+ | 4.942 | 6.8 eV 17 | α |
| 9 | 3/2– | 11.348 | **100%** | |
| 10 | 0+ | 12.607 | $1.51 \times 10^6$ y 6 | β– |
| 11 | 1/2+ | 20.174 | 13.81 s 8 | β–, β–α 3.1% |
| 12 | 0+ | 25.08 | 21.49 ms 3 | β–, β–n ≤1% |
| 13 | (1/2–) | 33.25 | $2.7 \times 10^{-21}$ s 18 | n |
| 14 | 0+ | 40.0 | 4.84 ms 10 | β–, β–n 94%, β–2n 6% |
| 15 | | 49.8s | <200 ns | n? |
| 16 | 0+ | 57.7s | <200 ns | 2n? |
| 5 B 6 | | 43.6s | unstable | 2p? |
| 7 | (3/2–) | 27.87 | 1.4 MeV 2 | p, α |
| 8 | 2+ | 22.921 | 770 ms 3 | ε, εα |
| 9 | 3/2– | 12.416 | 0.54 keV 21 | p, |
| 10 | 3+ | 12.051 | **19.8% 3** | |
| 11 | 3/2– | 8.668 | **80.2% 3** | |
| 12 | 1+ | 13.369 | 20.20 ms 2 | β–, β–3α 1.58% |
| 13 | 3/2– | 16.562 | 17.33 ms 17 | β– |
| 14 | 2– | 23.66 | 12.5 ms 5 | β–, β–n 6.04% |
| 15 | | 28.97 | 9.93 ms 7 | β–, β–n 93.6%, β–2n 0.4% |

| Nuclide Z El A | Jπ | Δ (MeV) | T½, Γ, or Abundance | Decay Mode |
|---|---|---|---|---|
| 5 B 16 | 0– | 37.08 | <190 ps | n |
| 17 | (3/2–) | 43.8 | 5.08 ms 5 | β–, β–n 63%, β–2n 11%, β–3n 3.5%, β–4n 0.4% |
| 18 | (4–) | 52.3s | <26 ns | n? |
| 19 | (3/2–) | 59.4s | 2.92 ms 13 | β–, β–n 72%, β–2n 16% |
| 6 C 8 | 0+ | 35.09 | 230 keV 50 | p, α |
| 9 | (3/2–) | 28.910 | 126.5 ms 9 | ε, εp 83%, εα 17% |
| 10 | 0+ | 15.699 | 19.26 s 3 | ε |
| 11 | 3/2– | 10.650 | 20.334 m 24 | ε |
| 12 | 0+ | 0.000 | **98.89% 1** | |
| 13 | 1/2– | 3.125 | **1.11% 1** | |
| 14 | 0+ | 3.020 | 5700 y 30 | β– |
| 15 | 1/2+ | 9.873 | 2.449 s 5 | β– |
| 16 | 0+ | 13.694 | 0.747 s 8 | β–, β–n 99% |
| 17 | | 21.04 | 193 ms 13 | β–, β–n 31.5% |
| 18 | 0+ | 24.93 | 92 ms 2 | β–, β–n 61%, β– |
| 19 | | 32.42 | 49 ms 4 | β–n 61%, β– |
| 20 | 0+ | 37.6 | 14 ms +6–5 | β–, β–n 72% |
| 21 | (1/2+) | 46.0s | <30 ns | n? |
| 22 | 0+ | 53.3s | 6.1 ms +14–12 | β–, β–n 61%, β–2n >0% |
| 7 N 10 | (1–) | 38.8 | $20 \times 10^{-23}$ y 14 | p? |
| 11m | 1/2+ | 24.62 | 1.58 MeV +75–52 | p |
| 12 | 1+ | 17.338 | 11.000 ms 16 | ε |
| 13 | 1/2– | 5.345 | 9.965 m 4 | ε |
| 14 | 1+ | 2.863 | **99.634% 20** | |
| 15 | 1/2– | 0.101 | **0.366% 20** | |
| 16 | 2– | 5.684 | 7.13 s 2 | β–, β–α $1.2 \times 10^{-3}$% |
| 17 | 1/2– | 7.87 | 4.173 s 4 | β–, β–n 95.1% |
| 18 | 1– | 13.11 | 624 ms 12 | β–, β–n 14.3%, β–α 12.2% |
| 19 | | 15.86 | 271 ms 8 | β–, β–n 54.6% |
| 20 | | 21.77 | 130 ms 7 | β–, β–n 57% |
| 21 | (1/2–) | 25.25 | 85 ms 7 | β–, β–n 81% |
| 22 | | 32.0 | 18 ms 4 | β–, β–n 36%, β–2n <13% |
| 23 | | 38.4s | 14.1 ms +12–15 | β–, β–n 42%, β–2n 8% |
| 24 | | 47.5s | <52 ns | n? |
| 25 | | 56.5s | <260 ns | n? |
| 8 O 12 | 0+ | 32.05 | 0.40 MeV 25 | p |
| 13 | (3/2–) | 23.112 | 8.58 ms 5 | ε, εp≈100% |
| 14 | 0+ | 8.007 | 70.641 s 20 | ε |
| 15 | 1/2– | 2.856 | 122.24 s 16 | ε |
| 16 | 0+ | –4.737 | **99.762% 16** | |
| 17 | 5/2+ | –0.809 | **0.038% 1** | |
| 18 | 0+ | –0.781 | **0.200% 14** | |
| 19 | 5/2+ | 3.335 | 26.88 s 5 | β– |
| 20 | 0+ | 3.797 | 13.51 s 5 | β– |

| Nuclide Z El A | Jπ | Δ (MeV) | T½, Γ, or Abundance | Decay Mode |
|---|---|---|---|---|
| 8 O 21 | (5/2+) | 8.06 | 3.42 s 10 | β– |
| 22 | 0+ | 9.28 | 2.25 s 15 | β–, β–n <22% |
| 23 | | 14.6 | 82 ms 37 | β–, β–n 31% |
| 24 | 0+ | 19.1 | 65 ms 5 | β–, β–n 18% |
| 25 | (3/2+) | 27.4s | <50 ns | n |
| 26 | 0+ | 35.7s | <40 ns | n |
| 27 | | 45.0s | <260 ns | n? |
| 28 | 0+ | 53.8s | <100 ns | n? |
| 9 F 14 | (2–) | 32.7s | ? | p |
| 15 | (1/2+) | 16.8 | 1.0 MeV 2 | p |
| 16 | 0– | 10.680 | 40 keV 20 | p |
| 17 | 5/2+ | 1.952 | 64.49 s 16 | ε |
| 18 | 1+ | 0.874 | 1.8291 h 4 | ε |
| 19 | 1/2+ | –1.487 | **100%** | |
| 20 | 2+ | –0.017 | 11.07 s 6 | β– |
| 21 | 5/2+ | –0.048 | 4.158 s 20 | β– |
| 22 | (4+) | 2.79 | 4.23 s 4 | β–, β–n <1% |
| 23 | (3/2,5/2)+ | 3.33 | 2.23 s 14 | β– |
| 24 | (1,2,3) | 7.56 | 400 ms 50 | β–, β–n <5.9% |
| 25 | (5/2+) | 11.27 | 50 ms 6 | β–, β–n 14% |
| 26 | 1+ | 18.3 | 9.6 ms 8 | β–, β–n 11% |
| 27 | (5/2+) | 24.9 | 5.0 ms 2 | β–, β–n 77% |
| 28 | | 33.2s | <40 ns | n |
| 29 | (5/2) | 40.3s | 2.5 ms 4 | β–, β–n, β– |
| 30 | | 48.9s | <260 ns | n? |
| 31 | | 56.3s | >260 ns | β–?, β–n? |
| 10 Ne 16 | 0+ | 24.00 | 122 keV 37 | p |
| 17 | 1/2– | 16.46 | 109.2 ms 6 | ε, εp≈100%, εα |
| 18 | 0+ | 5.317 | 1672 ms 8 | ε |
| 19 | 1/2+ | 1.751 | 17.22 s 2 | ε |
| 20 | 0+ | –7.042 | **90.48% 3** | |
| 21 | 3/2+ | –5.732 | **0.27% 1** | |
| 22 | 0+ | –8.025 | **9.25% 3** | |
| 23 | 5/2+ | –5.154 | 37.24 s 12 | β– |
| 24 | 0+ | –5.951 | 3.38 m 2 | β– |
| 25 | (5/2+) | –2.11 | 602 ms 8 | β– |
| 26 | 0+ | 0.43 | 192 ms 6 | β–, β–n <0.2% |
| 27 | (3/2+) | 7.1 | 32 ms 2 | β–, β–n 2% |
| 28 | 0+ | 11.2 | 19 ms 3 | β–, β–n 16% |
| 29 | (3/2+) | 18.1 | 15.6 ms 5 | β–, β–n 17%, β–2n <2.9% |
| 30 | 0+ | 23.1 | 5.8 ms 2 | β–, β–n <26% |
| 31 | | 30.8s | 3.4 ms 8 | β– |
| 32 | 0+ | 37.3s | 3.5 ms 9 | β– |
| 33 | | 46.0s | <260 ns | n? |
| 34 | 0+ | 53.1s | >1.5 μs | β–?, β–n? |
| 11 Na 18 | (1–) | 24.19 | $1.3 \times 10^{-21}$ s d | p?, ε? |
| 19 | (5/2+) | 12.93 | <40 ns | p |
| 20 | 2+ | 6.848 | 447.9 ms 23 | ε, εα 20.05% |
| 21 | 3/2+ | –2.184 | 22.49 s 4 | ε |
| 22 | 3+ | –5.182 | 2.6027 y 10 | ε |

| Nuclide Z El A | Jπ | Δ (MeV) | T½, Γ, or Abundance | Decay Mode |
|---|---|---|---|---|
| 11 Na 23 | 3/2+ | –9.530 | **100%** | |
| 24 | 4+ | –8.418 | 14.951 h 3 | β– |
| 25 | 5/2+ | –9.358 | 59.1 s 6 | β– |
| 26 | 3+ | –6.862 | 1.077 s 5 | β– |
| 27 | 5/2+ | –5.517 | 301 ms 6 | β–, β–n 0.13% |
| 28 | 1+ | –0.99 | 30.5 ms 4 | β–, β–n 0.58% |
| 29 | 3/2+ | 2.66 | 44.9 ms 12 | β–, β–n 21.5% |
| 30 | 2+ | 8.36 | 48 ms 2 | β–, β–n 30%, β–1 1.17%, β–0.5 $\times 10^{-3}$% |
| 31 | 3/2+ | 12.7 | 17.0 ms 4 | β–, β–n 37%, β– 0.9% |
| 32 | | 19.1 | 13.2 ms 4 | β–, β–n 24%, β– 8% |
| 33 | | 24.9 | 8.1 ms 4 | β–, β–n 47%, β–2n 13% |
| 34 | | 32.8s | 5.5 ms 10 | β–, β–n ≈100%, β– |
| 35 | | 39.6s | 1.5 ms 5 | β–, β–n |
| 36 | | 48.0s | <260 ns | n? |
| 37 | | 55.3s | >1.5 μs | β–?, β–n? |
| 12 Mg 19 | | 33.0 | ? | 2p? |
| 20 | 0+ | 17.57 | 90.8 ms 24 | ε, εp≈2% |
| 21 | 5/2+ | 10.91 | 122 ms 3 | ε, εp 32.6%, εα <0.5% |
| 22 | 0+ | –0.397 | 3.8755 s 12 | ε |
| 23 | 3/2+ | –5.474 | 11.317 s 11 | ε |
| 24 | 0+ | –13.934 | **78.99% 4** | |
| 25 | 5/2+ | –13.193 | **10.00% 1** | |
| 26 | 0+ | –16.215 | **11.01% 3** | |
| 27 | 1/2+ | –14.587 | 9.458 m 12 | β– |
| 28 | 0+ | –15.019 | 20.915 h 9 | β– |
| 29 | 3/2+ | –10.62 | 1.30 s 12 | β– |
| 30 | 0+ | –8.911 | 335 ms 17 | β– |
| 31 | | –3.22 | 230 ms 20 | β–, β–n 1.7% |
| 32 | 0+ | –0.95 | 86 ms 5 | β–, β–n 5.5% |
| 33 | | 4.89 | 90.5 ms 16 | β–, β–n 17% |
| 34 | 0+ | 8.8 | 20 ms 10 | β–, β–n |
| 35 | (7/2+) | 16.2s | 70 ms 40 | β–, β–n 52% |
| 36 | 0+ | 21.4s | 3.9 ms 13 | β– |
| 37 | (7/2+) | 29.2s | >260 ns | β–, β–n |
| 38 | 0+ | 35.0s | >260 ns | β–? |
| 39 | | 43.6s | >260 ns | n? |
| 40 | 0+ | 50.2s | 1 ms y | β–?, β–n? |
| 13 Al 21 | (5/2+) | 26.1s | <35 ns | p |
| 22 | (3)+ | 18.18 | 59 ms 3 | ε, εp≈60%, ε 0.9%, εα 0.31% |
| 23 | 3/2+ | 6.77 | 0.47 s 3 | ε, εp≈1.1% |
| 24 | 4+ | –0.057 | 2.053 s 4 | ε, εα 0.04%, εp $1.6 \times 10^{-3}$% |
| 24m | 1+ | 0.369 | 131.3 ms 25 | IT 82%, ε 18%, εα 0.03% |
| 25 | 5/2+ | –8.916 | 7.183 s 12 | ε |
| 26 | 5+ | –12.210 | $7.17 \times 10^5$ y 24 | ε |

# APPENDIX B

## Nuclear Wallet Cards

| Nuclide Z El A | Jπ | Δ (MeV) | T½, Γ, or Abundance | Decay Mode |
|---|---|---|---|---|
| 13 Al 26m | 0+ | -11.982 | 6.3452 s 19 | ε |
| 27 | 5/2+ | -17.197 | 100% | |
| 28 | 3+ | -16.850 | 2.2414 m 12 | β- |
| 29 | 5/2+ | -18.215 | 6.56 m 6 | β- |
| 30 | 3+ | -15.87 | 3.60 s 6 | β- |
| 31 | (3/2,5/2‡) | -14.95 | 644 ms 25 | β- |
| 32 | 1+ | -11.06 | 33 ms 4 | β- |
| 33 | (5/2+) | -8.53 | 41.7 ms 2 | β-, β-n 8.5% |
| 34 | | -2.9 | 42 ms 6 | β-, β-n 27% |
| 35 | | -0.1 | 38.6 ms 4 | β-, β-n 41% |
| 36 | | 5.8 | 90 ms 40 | β-, β-n<31% |
| 37 | | 9.9 | 10.7 ms 13 | β- |
| 38 | | 16.1 | 7.6 ms 6 | β- |
| 39 | (3/2+) | 21 | 7.6 ms 16 | β- |
| 40 | | 29.3s | >260 ns | β-, β-n |
| 41 | | 35.7s | >260 ns | β- |
| 42 | | 43.7s | 1 ms sy | β-?, β-n? |
| 14 Si 22 | 0+ | 32.2s | 29 ms 2 | ε, εp 32% |
| 23 | | 23.8s | 42.3 ms 4 | ε, εp=73%, ε2p<4% |
| 24 | 0+ | 10.75 | 140 ms 8 | ε, εp 38% |
| 25 | 5/2+ | 3.82 | 220 ms 3 | ε, εp |
| 26 | 0+ | -7.145 | 2.234 s 13 | ε |
| 27 | 5/2+ | -12.384 | 4.16 s 2 | ε |
| 28 | 0+ | -21.493 | 92.230% 19 | |
| 29 | 1/2+ | -21.895 | 4.683% 8 | |
| 30 | 0+ | -24.433 | 3.087% 5 | |
| 31 | 3/2+ | -22.949 | 157.3 m 3 | β- |
| 32 | 0+ | -24.081 | 132 y 13 | β- |
| 33 | (3/2+) | -20.49 | 6.18 s 18 | β- |
| 34 | 0+ | -19.96 | 2.77 s 20 | β- |
| 35 | | -14.36 | 0.78 s 12 | β- |
| 36 | 0+ | -12.5 | 0.45 s 6 | β-, β-n<10% |
| 37 | (7/2-) | -6.6 | 90 ms 60 | β-, β-n 17% |
| 38 | 0+ | -4.1 | >1 μs | β-, β-n |
| 39 | (7/2-) | 1.9 | 47.5 ms 20 | β- |
| 40 | 0+ | 5.5 | 33.0 ms 10 | β-, β-n |
| 41 | | 14. | 20.0 ms 25 | |
| 42 | 0+ | 18.4s | 13 ms 4 | β- |
| 43 | | 26.7s | >260 ns | β-?, β-n? |
| 44 | 0+ | 32.8s | 10 ms sy | β-? |
| 15 P 24 | (1+) | 32.0s | ? | p?, ε? |
| 25 | (1/2+) | 18.9s | <30 ns | p |
| 26 | (3+) | 11.0s | 43.7 ms 6 | ε, εp |
| 27 | 1/2+ | -0.72 | 260 ms 80 | ε, εp 0.07% |
| 28 | 3+ | -7.159 | 270.3 ms 5 | ε, εp 1.3×10⁻³%, εα 8.6×10⁻⁴% |
| 29 | 1/2+ | -16.953 | 4.142 s 15 | ε |
| 30 | 1+ | -20.201 | 2.498 m 4 | ε |
| 31 | 1/2+ | -24.441 | 100% | |
| 32 | 1+ | -24.305 | 14.262 d 14 | β- |
| 33 | 1/2+ | -26.337 | 25.34 d 12 | β- |
| 34 | | -24.558 | 12.43 s 8 | β- |

| Nuclide Z El A | Jπ | Δ (MeV) | T½, Γ, or Abundance | Decay Mode |
|---|---|---|---|---|
| 15 P 35 | 1/2+ | -24.858 | 47.3 s 7 | β- |
| 36 | 4- | -20.25 | 5.6 s 3 | β- |
| 37 | | -18.99 | 2.31 s 13 | β- |
| 38 | | -14.8 | 0.64 s 14 | β-, β-n<10% |
| 39 | | -12.9 | 0.25 s 8 | β-, β-n 26% |
| 40 | (2-,3-) | -8.1 | 125 ms 25 | β-, β-n 15.8% |
| 41 | | -5.3 | 100 ms 2 | β-, β-n 30% |
| 42 | | 0.9 | 48.5 ms 15 | β-, β-n 50% |
| 43 | | 5.8 | 36.5 ms 15 | β-, β-n |
| 44 | | 12.1s | 18.5 ms 25 | β- |
| 45 | | 17.9s | >200 ns | β-? |
| 46 | | 25.5s | >200 ns | β- |
| 16 S 26 | 0+ | 26.0s | ~10 ms | 2p? |
| 27 | (5/2+) | 17.5s | 15.5 ms 15 | ε, εp 2.3%, ε2p 1.1% |
| 28 | 0+ | 4.1 | 125 ms 10 | ε, εp 21% |
| 29 | 5/2+ | -3.16 | 187 ms 4 | ε, εp 47% |
| 30 | 0+ | -14.063 | 1.178 s 5 | ε |
| 31 | 1/2+ | -19.045 | 2.572 s 13 | ε |
| 32 | 0+ | -26.016 | 95.02% 9 | |
| 33 | 3/2+ | -26.586 | 0.75% 1 | |
| 34 | 0+ | -29.932 | 4.21% 8 | |
| 35 | 3/2+ | -28.846 | 87.51 d 12 | β- |
| 36 | 0+ | -30.664 | 0.02% 1 | |
| 37 | 7/2- | -26.896 | 5.05 m 2 | β- |
| 38 | 0+ | -26.861 | 170.3 m 7 | β- |
| 39 | (3/2,5/2,7/2)- | -23.16 | 11.5 s 5 | β- |
| 40 | 0+ | -22.9 | 8.8 s 22 | β- |
| 41 | (7/2-) | -19.0 | 1.99 s 5 | β-, β-n |
| 42 | 0+ | -17.7 | 1.013 s 15 | β- |
| 43 | | -12.0 | 0.28 s 3 | β-, β-n 40% |
| 44 | 0+ | -9.1 | 100 ms 1 | β-, β-n 19% |
| 45 | | -3 | 68 ms 2 | β-, β-n 54% |
| 46 | 0+ | 0.7s | 50 ms 8 | β- |
| 47 | | 8.0s | >200 ns | β-? |
| 48 | 0+ | 13.2s | ≥200 ns | β- |
| 49 | | 22.0s | <200 ns | n |
| 17 Cl 28 | (1+) | 26.6s | ? | p? |
| 29 | (3/2+) | 13.1s | <20 ns | p |
| 30 | (3+) | 4.4s | <30 ns | p |
| 31 | | -7.07 | 150 ms 25 | ε, εp 0.7% |
| 32 | 1+ | -13.330 | 298 ms 1 | ε, εα 0.05%, εp 0.03% |
| 33 | 3/2+ | -21.003 | 2.511 s 3 | ε |
| 34 | 0+ | -24.440 | 1.5264 s 14 | ε |
| 34m | 3+ | -24.293 | 32.00 m 4 | ε 55.4%, IT 44.6% |
| 35 | 3/2+ | -29.014 | 75.77% 4 | |
| 36 | 2+ | -29.522 | 3.01×10⁵ y 2 | β- 98.1%, ε 1.9% |
| 37 | 3/2+ | -31.761 | 24.23% 4 | |
| 38 | 2- | -29.798 | 37.24 m 5 | β- |
| 38m | 5- | 29.127 | 715 ms 3 | IT |
| 39 | 3/2+ | -29.800 | 55.6 m 2 | β- |

| Nuclide Z El A | Jπ | Δ (MeV) | T½, Γ, or Abundance | Decay Mode |
|---|---|---|---|---|
| 17 Cl 40 | 2- | -27.56 | 1.35 m 2 | β- |
| 41 | (1/2+,3/2+) | -27.31 | 38.4 s 8 | β- |
| 42 | | -24.9 | 6.8 s 3 | β- |
| 43 | | -24.2 | 3.07 s 7 | β- |
| 44 | | -20.2 | 0.56 s 11 | β-, β-n<8% |
| 45 | | -18.4 | 400 ms 43 | β-, β-n 24% |
| 46 | | -14.7 | 232 ms 2 | β-, β-n 60% |
| 47 | | -10.5s | 101 ms 6 | β-, β-n>0% |
| 48 | | -4.7s | ≥200 ns | β- |
| 49 | | 0.3s | ≥170 ns | β- |
| 50 | | 7.3s | 20 ms sy | β- |
| 51 | (3/2+) | 13.5s | >200 ns | β- |
| 18 Ar 30 | 0+ | 20.1s | <20 ns | p? |
| 31 | 5/2+ | 11.3s | 15.1 ms 13 | ε, εp 69%, ε 7.6% |
| 32 | 0+ | -2.200 | 98 ms 2 | ε, εp 43% |
| 33 | 1/2+ | -9.384 | 173.0 ms 20 | ε, εp 38.7% |
| 34 | 0+ | -18.377 | 844.5 ms 34 | ε |
| 35 | 3/2+ | -23.047 | 1.775 s 4 | ε |
| 36 | 0+ | -30.232 | 0.3365% 30 | |
| 37 | 3/2+ | -30.948 | 34.95 d 4 | ε |
| 38 | 0+ | -34.715 | 0.0632% 5 | |
| 39 | 7/2- | -33.242 | 269 y 3 | β- |
| 40 | 0+ | -35.040 | 99.6003% 30 | |
| 41 | 7/2- | -33.068 | 109.61 m 4 | β- |
| 42 | 0+ | -34.423 | 32.9 y 11 | β- |
| 43 | (5/2-) | -32.010 | 5.37 m 6 | β- |
| 44 | 0+ | -32.673 | 11.87 m 5 | β- |
| 45 | | -29.771 | 21.48 s 15 | β- |
| 46 | 0+ | -29.72 | 8.4 s 6 | β- |
| 47 | (3/2-) | -25.9 | 1.23 s 3 | β-, β-n<0.002% |
| 48 | 0+ | -23.7s | 0.48 s 40 | β- |
| 49 | | -18.1s | ≥170 ns | β- |
| 50 | 0+ | -12.6s | ≥170 ns | β- |
| 51 | | -7.8s | >200 ns | β-? |
| 52 | 0+ | -3.0s | 10 ms | β- |
| 53 | (5/2-) | 4.6s | 3 ms sy | β-, β-n |
| 19 K 32 | | 20.4s | ? | p? |
| 33 | (3/2+) | 6.8s | <25 ns | p |
| 34 | (1+) | -1.5s | <25 ns | p |
| 35 | 3/2+ | -11.17 | 178 ms 8 | ε, εp 0.37% |
| 36 | 2+ | -17.426 | 342 ms 2 | ε, εp 0.05%, εα 3.4×10⁻³% |
| 37 | 3/2+ | -24.800 | 1.226 s 7 | ε |
| 38 | 3+ | -28.801 | 7.636 m 18 | ε |
| 38m | 0+ | -28.670 | 924.2 ms 3 | ε |
| 39 | 3/2+ | -33.807 | 93.2581% 44 | |
| 40 | 4- | -33.535 | 1.248×10⁹ y 3 | β- 89.28%, ε 10.72% |
| | | | 0.0117% 1 | |
| 41 | 3/2+ | -35.559 | 6.7302% 44 | |
| 42 | 2- | -35.022 | 12.321 h 25 | β- |
| 43 | 3/2+ | -36.593 | 22.3 h 1 | β- |
| 44 | 2- | -35.81 | 22.13 m 19 | β- |

| Nuclide Z El A | Jπ | Δ (MeV) | T½, Γ, or Abundance | Decay Mode |
|---|---|---|---|---|
| 19 K 45 | 3/2+ | -36.61 | 17.3 m 6 | β- |
| 46 | (2-) | -35.42 | 105 s 10 | β- |
| 47 | 1/2+ | -35.696 | 17.50 s 24 | β- |
| 48 | (2-) | -32.12 | 6.8 s 2 | β-, β-n 1.14% |
| 49 | (3/2+) | -30.32 | 1.26 s 5 | β-, β-n 86% |
| 50 | (0-,1,2-) | -25.4 | 472 ms 4 | β-, β-n 29% |
| 51 | (1/2+,3/2+) | -22.0s | 365 ms 5 | β-, β-n 47% |
| 52 | (2-) | -16.2s | 105 ms 5 | β-, β-n<64%, β- |
| 53 | (3/2+) | -12.0s | 30 ms 5 | β-, β-n<67%, β-2n 17% |
| 54 | | -5.4s | 10 ms 5 | β-, β-n>0% |
| 55 | | -0.3s | 3 ms sy | β-?, β-n? |
| 20 Ca 34 | 0+ | 13.2s | <35 ns | p |
| 35 | | 4.6s | 25.7 ms 2 | ε, εp 95.7%, ε2p 4.2% |
| 36 | 0+ | -6.44 | 102 ms 2 | ε, εp 57% |
| 37 | 3/2+ | -22.59 | 181.1 ms 10 | ε, εp 82.1% |
| 38 | 0+ | -22.059 | 440 ms 8 | ε |
| 39 | 3/2+ | -27.274 | 859.6 ms 14 | ε |
| 40 | 0+ | -34.846 | >3.0×10²¹ y | 2ε |
| | | | 96.94% 16 | |
| 41 | 7/2- | -35.138 | 1.02×10⁵ y 7 | ε |
| 42 | 0+ | -38.547 | 0.647% 23 | |
| 43 | 7/2- | -38.409 | 0.135% 10 | |
| 44 | 0+ | -41.468 | 2.09% 11 | |
| 45 | 7/2- | -40.812 | 162.61 d 9 | β- |
| 46 | 0+ | -43.135 | >0.28×10¹⁶ y | 2β- <25% |
| | | | 0.004% 3 | |
| 47 | 7/2- | -42.340 | 4.536 d 3 | β- |
| 48 | 0+ | -44.2142 | 3×10¹⁹ y 12-6 | 2β- 84%, |
| | | | 0.187% 21 | 2β-<25% |
| 49 | 3/2- | -41.289 | 8.718 m 6 | β- |
| 50 | 0+ | -39.571 | 13.9 s 6 | β-, β-n |
| 51 | (3/2-) | -35.86 | 10.0 s 8 | β-, β-n |
| 52 | 0+ | -32.5 | 4.6 s 3 | β-, β-n<2% |
| 53 | (3/2-,5/2-) | -27.9s | 90 ms 15 | β-, β-n>30% |
| 54 | 0+ | -23.9s | >300 ns | β- |
| 55 | | -18.1s | >300 ns | β-? |
| 56 | 0+ | -13.3s | 10 ms sy | β-? |
| 57 | | -7s | 5 ms sy | β-?, β-n? |
| 21 Sc 36 | | 13.9s | ? | p? |
| 37 | | 2.8s | ? | p? |
| 38 | (2-) | -4.9s | <300 ns | p |
| 39 | (7/2-) | -14.17 | <300 ns | p |
| 40 | 4- | -20.523 | 182.3 ms 7 | ε, εp 0.44%, εα 0.02% |
| 41 | 7/2- | -28.642 | 596.3 ms 17 | ε |
| 42 | 0+ | -32.121 | 681.3 ms 7 | ε |
| 42m | (7+) | -32.5 | 61.7 s 4 | ε |
| 43 | 7/2- | -36.188 | 3.891 h 12 | ε |
| 44 | 2+ | -37.816 | 3.97 h 4 | ε |
| 44m | 6+ | -37.545 | 58.61 h 19 | IT 98.8%, ε 1.2% |

## Nuclear Wallet Cards

| Nuclide Z El A | Jπ | Δ (MeV) | T½, Γ, or Abundance | Decay Mode |
|---|---|---|---|---|
| 21 Sc 45 | 7/2− | −41.068 | 100% | |
| 45m | 3/2+ | −41.056 | 318 ms 7 | IT |
| 46 | 4+ | −41.757 | 83.79 d 4 | β− |
| 46m | 1− | −41.615 | 18.75 s 4 | IT |
| 47 | 7/2− | −44.332 | 3.3492 d 6 | β− |
| 48 | 6+ | −44.496 | 43.67 h 9 | β− |
| 49 | 7/2− | −46.552 | 57.2 m 2 | β− |
| 50 | 5+ | −44.54 | 102.5 s 5 | β− |
| 50m | (2,3)+ | −44.28 | 0.35 s 4 | IT≈97.5%, β−<2.5% |
| 51 | (7/2)− | −43.22 | 12.4 s 1 | β− |
| 52 | 3(+) | −40.4 | 8.2 s 2 | β− |
| 53 | (7/2−) | −37.6s | >3 s | β− |
| 54 | (3,4+) | −34.2 | 0.36 s 6 | β− |
| 55 | (7/2−) | −29.6 | 0.115 s 15 | β−, β−n |
| 56 | (1+) | −25.3s | 35 ms 5 | β−n |
| 56 | (6+,7+) | −25.3s | 60 ms 7 | β−n |
| 57 | | −20.7s | 13 ms 4 | β− |
| 57 | (7/2−) | −20.7s | 13 ms 4 | β−n |
| 58 | (3+) | −15.2s | 12 ms 5 | β− |
| 59 | | −10.0s | 10 ms sy | β−?, β−n? |
| 60 | | −4.0s | 3 ms sy | β− |
| 22 Ti 38 | 0+ | 9.1s | <120 ns | 2p? |
| 39 | (3/2+) | 1.5s | 31 ms +6−4 | ε, εp 14% |
| 40 | 0+ | −8.9 | 53.3 ms 15 | ε, εp |
| 41 | 3/2+ | −15.7s | 80.4 ms 9 | ε, εp≈100% |
| 42 | 0+ | −25.122 | 199 ms 6 | ε |
| 43 | 7/2− | −29.321 | 509 ms 5 | ε |
| 44 | 0+ | −37.549 | 60.0 y 11 | ε |
| 45 | 7/2− | −39.006 | 184.8 ms 5 | ε |
| 46 | 0+ | −44.123 | **8.25% 3** | |
| 47 | 5/2− | −44.932 | **7.44% 2** | |
| 48 | 0+ | −48.488 | **73.72% 3** | |
| 49 | 7/2− | −48.559 | **5.41% 2** | |
| 50 | 0+ | −51.427 | **5.18% 2** | |
| 51 | 3/2− | −49.728 | 5.76 m 1 | β− |
| 52 | 0+ | −49.465 | 1.7 m 1 | β− |
| 53 | (3/2)− | −46.8 | 32.7 s 9 | β− |
| 54 | 0+ | −45.6 | 1.5 s 4 | β− |
| 55 | (3/2−) | −41.7 | 1.3 s 1 | β− |
| 56 | 0+ | −38.9 | 200 ms 5 | β−, β−n |
| 57 | | −33.5 | 60 ms 16 | β−, β−n |
| 58 | 0+ | −30.8s | 59 ms 9 | β− |
| 59 | (5/2−) | −25.2s | 30 ms 3 | β− |
| 60 | 0+ | −21.6s | 22 ms 2 | β− |
| 61 | | −15.6s | >300 ns | β−? |
| 62 | 0+ | −11.7s | 10 ms sy | β−? |
| 63 | | −5.2s | 3 ms sy | β−?, β−n? |
| 23 V 40 | | 10.3s | ? | p? |
| 41 | | −0.2s | ? | p? |
| 42 | | −8.2s | <55 ns | p |
| 43 | | −18.0s | >800 ms | β− |
| 44 | (2+) | −24.1 | 111 ms 7 | ε, εα |

## Nuclear Wallet Cards

| Nuclide Z El A | Jπ | Δ (MeV) | T½, Γ, or Abundance | Decay Mode |
|---|---|---|---|---|
| 23 V 44m | (6+) | −24.1 | 150 ms 3 | ε |
| 45 | 7/2− | −31.88 | 547 ms 6 | ε |
| 46 | 0+ | −37.073 | 422.50 ms 11 | ε |
| 47 | 3/2− | −42.002 | 32.6 m 3 | ε |
| 48 | 4+ | −44.475 | 15.9735 d 25 | ε |
| 49 | 7/2− | −47.957 | 329 d 3 | ε |
| 50 | 6+ | −49.222 | 1.4×10^17 y 4 | ε 83%, β− 17% |
| | | | **0.250% 2** | |
| 51 | 7/2− | −52.201 | **99.750% 2** | |
| 52 | 3+ | −51.441 | 3.743 m 5 | β− |
| 53 | 7/2− | −51.849 | 1.60 m 4 | β− |
| 54 | 3+ | −49.89 | 49.8 s 5 | β− |
| 55 | (7/2−) | −49.2 | 6.54 s 15 | β− |
| 56 | (1+) | −46.1 | 216 ms 4 | β− |
| 56 | 1+ | −46.1 | 216 ms 4 | β−n 0.06% |
| 57 | (3/2−) | −44.2 | 0.35 s 1 | β−, β−n 0.04% |
| 58 | (1+) | −40.2 | 185 ms 10 | β− |
| 59 | (5/2−,3/2−) | −37.1 | 75 ms 7 | β− |
| 60 | | −32.6 | 68 ms 5 | β− |
| 60m | | −32.6 | 122 ms 18 | β−, β−n |
| 61 | (3/2−) | −29.4s | 47 ms 1 | β− |
| 62 | | −24.4s | 33.5 ms 2 | β− |
| 63 | (7/2−) | −20.9s | 17 ms 3 | β− |
| 64 | | −15.4s | >150 ns | β− |
| 65 | | −11.3s | 10 ms sy | β−?, β−n? |
| 24 Cr 42 | 0+ | 6.0s | 13 ms +4−2 | ε |
| 43 | (3/2+) | −2.1s | 21.6 ms 7 | ε, εp 23%, ε 6% |
| 44 | 0+ | −13.46s | 53 ms +4−3 | ε, εp 7% |
| 45 | | −19.0 | 50 ms 6 | ε, εp 27% |
| 46 | 0+ | −29.47 | 0.26 s 6 | ε |
| 47 | 3/2− | −34.56 | 500 ms 15 | ε |
| 48 | 0+ | −42.819 | 21.56 h 3 | ε |
| 49 | 5/2− | −45.331 | 42.3 m 1 | ε |
| 50 | 0+ | −50.259 | >1.3×10^18 y | 2ε |
| | | | **4.345% 13** | |
| 51 | 7/2− | −51.449 | 27.7025 d 24 | ε |
| 52 | 0+ | −55.417 | **83.789% 18** | |
| 53 | 3/2− | −55.285 | **9.501% 17** | |
| 54 | 0+ | −56.932 | **2.365% 7** | |
| 55 | 3/2− | −55.107 | 3.497 m 3 | β− |
| 56 | 0+ | −55.281 | 5.94 m 10 | β− |
| 57 3/2−,5/2−,7/2− | | −52.524 | 21.1 s 10 | β− |
| 58 | 0+ | −51.8 | 7.0 s 3 | β− |
| 59 | (1/2−) | −47.9 | 0.46 s 5 | β− |
| 60 | 0+ | −46.5 | 0.57 s 6 | β− |
| 61 | | −42.2 | 0.27 s 4 | β− |
| 62 | 0+ | −40.4 | 209 ms 12 | β−, β−n |
| 63 | (1/2−) | −35.5s | 129 ms 2 | β− |
| 64 | 0+ | −33.2s | 43 ms 1 | β− |
| 65 | (1/2−) | −27.8s | 27 ms 3 | β−, β−n? |
| 66 | 0+ | −24.8s | 10 ms 6 | β− |
| 67 | | −19.0s | ≈50 ms | β−? |

## Nuclear Wallet Cards

| Nuclide Z El A | Jπ | Δ (MeV) | T½, Γ, or Abundance | Decay Mode |
|---|---|---|---|---|
| 25 Mn 44 | (2−) | 6.4s | <105 ns | ε, p |
| 45 | (7/2−) | −5.1s | <70 ns | p |
| 46 | [4+] | −12.4s | 34 ms +5−4 | ε, εp 22% |
| 47 | | −22.3s | 100 ms 50 | ε, εp>3.4% |
| 48 | 4+ | −29.3 | 158.1 ms 22 | ε, εp 0.28%, εα<6.0×10^−4% |
| 49 | 5/2− | −37.62 | 382 ms 7 | ε |
| 50 | 0+ | −42.627 | 283.29 ms 8 | ε |
| 50m | 5+ | −42.398 | 1.75 m 3 | ε |
| 51 | 5/2− | −48.241 | 46.2 m 1 | ε |
| 52 | 6+ | −50.705 | 5.591 d 3 | ε |
| 52m | 2+ | −50.328 | 21.1 m 2 | ε 98.25%, IT 1.75% |
| 53 | 7/2− | −54.688 | 3.74×10^6 y 4 | ε |
| 54 | 3+ | −55.555 | 312.12 d 6 | ε, β−<2.9×10^−4 |
| 55 | 5/2− | −57.711 | **100%** | |
| 56 | 3+ | −56.910 | 2.5789 h 1 | β− |
| 57 | 5/2− | −57.487 | 85.4 s 18 | β− |
| 58 | 1+ | −55.91 | 3.0 s 1 | β− |
| 58m | (4)+ | −55.82 | 65.2 s 5 | β−≈80%, IT≈20% |
| 59 | (5/2−) | −55.48 | 4.59 s 5 | β− |
| 60 | 1+ | −53.18 | 51 s 6 | β− |
| 60m | 3+ | −52.91 | 1.77 s 2 | β−88.5%, IT 11.5% |
| 61 | (5/2−) | −51.6 | 0.67 s 4 | β− |
| 62 | 1+ | −48.0 | 92 ms 13 | β− |
| 63 | (3+,4+) | −48.0 | 671 ms 5 | β−, β−n |
| 63 | (5/2−) | −46.4 | 0.29 s 2 | β− |
| 64 | | −42.6 | 89 ms 4 | β−, β−n 1.42% |
| 65 | | −40.7 | 92 ms 1 | β−, β−n 6.92% |
| 66 | | −36.3s | 64 ms 2 | β−, β−n 10.88% |
| 67 | (5/2−) | −33.4s | 47 ms 4 | β−, β−n |
| 69 | 5/2− | −25.3s | 14 ms 4 | β− |
| 26 Fe 45 | (3/2+) | 13.6s | 3.8 ms +20−8 | 2p |
| 46 | 0+ | 0.8s | 12 ms +4−3 | |
| 47 | | −6.6s | 21.8 ms 7 | ε, εp |
| 48 | 0+ | −18.16s | 44 ms 7 | ε, εp 3.6% |
| 49 | (7/2−) | −24.6s | 70 ms 3 | ε, εp≥52% |
| 50 | 0+ | −34.48 | 155 ms 11 | ε, εp≈0% |
| 51 | 5/2− | −40.22 | 305 ms 5 | ε |
| 52 | 0+ | −48.332 | 8.275 h 8 | ε |
| 52m | (12+) | −41.512 | 45.9 s 6 | ε |
| 53 | 7/2− | −50.945 | 8.51 m 2 | ε |
| 53m | 19/2− | −47.905 | 2.526 m 24 | IT |
| 54 | 0+ | −56.252 | **5.845% 35** | |
| 55 | 3/2− | −57.479 | 2.737 y 11 | ε |
| 56 | 0+ | −60.605 | **91.754% 36** | |
| 57 | 1/2− | −60.180 | **2.119% 10** | |
| 58 | 0+ | −62.153 | **0.282% 4** | |
| 59 | 3/2− | −60.663 | 44.495 d 9 | β− |
| 60 | 0+ | −61.412 | 1.5×10^6 y 3 | β− |
| 61 3/2−,5/2− | | −58.92 | 5.98 m 6 | β− |
| 62 | 0+ | −58.90 | 68 s 2 | β− |
| 63 | (5/2−) | −55.5 | 6.1 s 6 | β− |

## Nuclear Wallet Cards

| Nuclide Z El A | Jπ | Δ (MeV) | T½, Γ, or Abundance | Decay Mode |
|---|---|---|---|---|
| 26 Fe 64 | 0+ | −54.8 | 2.0 s 2 | β− |
| 65 | | −50.9 | 1.3 s 3 | β− |
| 66 | 0+ | −49.6 | 0.44 s 6 | β− |
| 67 | | −45.7 | 0.47 s 5 | β−, β−n 1.13% |
| 68 | 0+ | −43.1 | 187 ms 6 | β− |
| 69 | 1/2− | −38.4s | 109 ms 9 | β− |
| 70 | 0+ | −35.9s | 94 ms 17 | β− |
| 71 | (7/2−) | −31.0s | >150 ns | β− |
| 72 | 0+ | −28.3s | >150 ns | β− |
| 27 Co 47 | | 10.7s | ? | p? |
| 49 | | −9.6s | <35 ns | ε, p |
| 50 | (6+) | −17.2s | 44 ms 4 | ε, εp>54% |
| 51 | (7/2−) | −27.3s | >200 ns | ε |
| 52 | (6+) | −33.92s | 115 ms 23 | ε |
| 53 | (7/2−) | −42.64 | 240 ms 9 | ε |
| 53m (19/2−) | | −39.45 | 247 ms 12 | ε≈98.5%, p=1.5% |
| 54 | 0+ | −48.009 | 193.28 ms 7 | ε |
| 54m | (7)+ | −47.812 | 1.48 m 2 | ε |
| 55 | 7/2− | −54.028 | 17.53 h 3 | ε |
| 56 | 4+ | −56.039 | 77.233 d 27 | ε |
| 57 | 7/2− | −59.344 | 271.74 d 6 | ε |
| 58 | 2+ | −59.846 | 70.86 d 6 | ε |
| 58m | 5+ | −59.821 | 9.04 h 11 | IT |
| 59 | 7/2− | −62.228 | **100%** | |
| 60 | 5+ | −61.649 | 1925.28 d 14 | β− |
| 60m | 2+ | −61.590 | 10.467 m 6 | IT 99.76%, β− 0.24% |
| 61 | 7/2− | −62.898 | 1.650 h 5 | β− |
| 62 | 2+ | −61.43 | 1.50 m 4 | β− |
| 62m | 5+ | −61.41 | 13.91 m 5 | β−>99%, IT<1% |
| 63 | 7/2− | −61.84 | 27.4 s 5 | β− |
| 64 | 1+ | −59.79 | 0.30 s 3 | β− |
| 65 | (7/2−) | −59.17 | 1.20 s 6 | β− |
| 66 | (3+) | −56.1 | 0.18 s 1 | β− |
| 67 | (7/2−) | −55.1 | 0.425 s 20 | β− |
| 68 | (7−) | −51.4 | 0.199 s 21 | β− |
| 68m | (3+) | −51.4 | 1.6 s 3 | β− |
| 69 | 7/2− | −50.0 | 0.22 s 2 | β− |
| 70 | (6−) | −45.6 | 119 ms 6 | β− |
| 70 | (3−) | −45.6 | 0.50 s 18 | β− |
| 71 | | −43.9 | 79 ms 5 | β−, β−n 2.61% |
| 72 | (6−,7−) | −39.3s | 62 ms 3 | β−, β−n 4.8% |
| 73 | | −37.0s | 41 ms 4 | β− |
| 74 | 0+ | −32.2s | >150 ns | β− |
| 75 | (1/2−) | −29.5s | >150 ns | β− |
| 28 Ni 48 | 0+ | 18.4s | >0.5 µs | ε, εp? |
| 49 | | 9.0s | 12 ms +5−3 | ε, εp? |
| 50 | 0+ | −3.8s | 12 ms 3 | εp 70%, ε |
| 51 | (7/2−) | −11.4s | >200 ns | ε |
| 52 | 0+ | −22.65s | 38 ms 5 | ε, εp 17% |
| 53 | (7/2−) | −29.4s | 45 ms 15 | ε, εp≈45% |
| 54 | 0+ | −39.21 | 104 ms 7 | ε |
| 55 | 7/2− | −45.34 | 202 ms 3 | ε |

## APPENDIX B

### Nuclear Wallet Cards

| Nuclide Z El A | Jπ | Δ (MeV) | T½, Γ, or Abundance | Decay Mode |
|---|---|---|---|---|
| 28 Ni 56 | 0+ | −53.90 | 6.075 d *10* | ε |
| 57 | 3/2− | −56.082 | 35.60 h *6* | ε |
| 58 | 0+ | −60.228 | **68.077% *9*** | |
| 59 | 3/2− | −61.156 | 7.6×10⁴ y *5* | ε |
| 60 | 0+ | −64.472 | **26.223% *8*** | |
| 61 | 3/2− | −64.221 | **1.140% *1*** | |
| 62 | 0+ | −66.746 | **3.634% *2*** | |
| 63 | 1/2− | −65.513 | 100.1 y *20* | β− |
| 64 | 0+ | −67.099 | **0.926% *1*** | |
| 65 | 5/2− | −65.126 | 2.5172 h *3* | β− |
| 66 | 0+ | −66.006 | 54.6 h *3* | β− |
| 67 | (1/2)− | −63.743 | 21 s *1* | β− |
| 68 | 0+ | −63.464 | 29 s *2* | β− |
| 69 | 9/2+ | −59.979 | 11.4 s *3* | β− |
| 69m | 1/2− | −59.658 | 3.5 s *5* | β− |
| 70 | 0+ | −59.1 | 6.0 s *3* | β− |
| 71 | | −55.2 | 2.56 s *3* | β− |
| 72 | 0+ | −53.9 | 1.57 s *5* | β−, β−n |
| 73 | (9/2+) | −49.9s | 0.84 s *3* | β− |
| 74 | 0+ | −48.4s | 0.68 s *18* | β−, β−n 8.43% |
| 75 | (7/2+) | −43.9s | 0.6 s *2* | β−, β−n 8.43% |
| 76 | 0+ | −41.6s | 0.24 s +55−24 | β−, β−n |
| 77 | | −36.7s | >150 ns | β−? |
| 78 | 0+ | −34s | >150 ns | β− |
| 29 Cu 52 | (3+) | −2.6s | ? | |
| 53 | (3/2−) | −13.5s | <300 ns | ε, p |
| 54 | (3+) | −21.7s | <75 ns | p |
| 55 | 3/2− | −31.6s | >200 ns | ε |
| 56 | 4+ | −38.6s | 94 ms *3* | ε |
| 57 | 3/2− | −47.31 | 196.3 ms *7* | ε |
| 58 | 1+ | −51.662 | 3.204 s *7* | ε |
| 59 | 3/2− | −54.35 | 81.5 s *5* | ε |
| 60 | 2+ | −58.344 | 23.7 m *4* | ε |
| 61 | 3/2− | −61.984 | 3.333 h *5* | ε |
| 62 | 1+ | −62.798 | 9.67 m *3* | ε |
| 63 | 3/2− | −65.579 | **69.17% *3*** | |
| 64 | 1+ | −65.424 | 12.700 h *2* | ε 61%, β− 39% |
| 65 | 3/2− | −67.264 | **30.83% *3*** | |
| 66 | 1+ | −66.258 | 5.120 m *14* | β− |
| 67 | 3/2− | −67.319 | 61.83 h *12* | β− |
| 68 | 1+ | −65.567 | 31.1 s *15* | β− |
| 68m | (6−) | −64.845 | 3.75 m *5* | IT 84%, β− 16% |
| 69 | 3/2− | −65.736 | 2.85 m *15* | β− |
| 70 | (6−) | −62.976 | 44.5 s *2* | β− |
| 70m | (3−) | −62.875 | 33 s *2* | β− 52%, IT 48% |
| 70m | 1+ | −62.734 | 6.6 s *2* | β− 93.2%, IT 6.8% |
| 71 | (3/2−) | −62.711 | 19.5 s *16* | β− |
| 72 | (1+) | −59.783 | 6.6 s *1* | β− |
| 73 | (3/2−) | −58.987 | 4.2 s *3* | β− |
| 74 | (1+,3+) | −56.006 | 1.594 s *10* | β− |
| 75 | (3/2−) | −54.1 | 1.224 s *3* | β−, β−n 3.5% |
| 76m | | −50.976 | 0.641 s *6* | β−, β−n 3% |

| Nuclide Z El A | Jπ | Δ (MeV) | T½, Γ, or Abundance | Decay Mode |
|---|---|---|---|---|
| 29 Cu 76m | | −50.976 | 1.27 s *30* | β− |
| 77 | | −48.6s | 0.469 s *8* | β− |
| 78 | | −44.7s | 342 ms *11* | β− |
| 79 | | −42.3s | 188 ms *25* | β−, β−n 55% |
| 80 | | −36.4s | >300 ns | β− |
| 30 Zn 54 | 0+ | −6.6s | ? | 2p? |
| 55 | | −14.9s | >0.5 μs | ε, p |
| 56 | 0+ | −25.7s | >0.5 μs | ε, p |
| 57 | (7/2−) | −32.8s | 38 ms *4* | ε, εp ≥65% |
| 58 | 0+ | −42.30 | 84 ms *9* | ε |
| 59 | 3/2− | −47.26 | 182.0 ms *18* | ε, εp 0.1% |
| 60 | 0+ | −54.19 | 2.38 m *5* | ε |
| 61 | 3/2− | −56.35 | 89.1 s *2* | ε |
| 61m | 1/2− | −56.26 | <430 ms | IT |
| 61m | 3/2− | −55.93 | 0.14 s *7* | IT |
| 61m | 5/2− | −55.59 | <0.13 s | IT |
| 62 | 0+ | −61.17 | 9.186 h *13* | ε |
| 63 | 3/2− | −62.213 | 38.47 m *5* | ε |
| 64 | 0+ | −66.004 | >2.8×10¹⁶ y | 2ε |
| | | | **48.63% *60*** | |
| 65 | 5/2− | −65.912 | 243.66 d *9* | ε |
| 66 | 0+ | −68.899 | **27.90% *27*** | |
| 67 | 5/2− | −67.880 | **4.10% *13*** | |
| 68 | 0+ | −70.007 | **18.75% *51*** | |
| 69 | 1/2− | −68.418 | 56.4 m *9* | β− |
| 69m | 9/2+ | −67.979 | 13.76 h *2* | IT 99.97%, β− 0.03% |
| 70 | 0+ | −69.565 | >1.3×10¹⁶ y | 2β− |
| | | | **0.62% *3*** | |
| 71 | 1/2− | −67.33 | 2.45 m *10* | β− |
| 71m | 9/2+ | −67.17 | 3.96 h *5* | β−, IT≤ 0.05% |
| 72 | 0+ | −68.131 | 46.5 h *1* | β− |
| 73 | (1/2)− | −65.41 | 23.5 s *10* | β− |
| 73m | | −65.41 | 5.8 s *8* | IT, β− |
| 74 | 0+ | −65.71 | 95.6 s *12* | β− |
| 75 | (7/2+) | −62.47 | 10.2 s *2* | β− |
| 76 | 0+ | −62.14 | 5.7 s *3* | β− |
| 77 | (7/2+) | −58.7 | 2.08 s *5* | β− |
| 77m | (1/2−) | −57.9 | 1.05 s *10* | IT>50%, β− <50% |
| 78 | 0+ | −57.34 | 1.47 s *15* | β− |
| 79 | (9/2+) | −53.4s | 0.995 s *19* | β−, β−n 1.3% |
| 80 | 0+ | −51.8 | 0.54 s *2* | β−, β−n 1% |
| 81 | | −46.1s | 0.29 s *5* | β−, β−n 7.5% |
| 82 | 0+ | −42.5s | >150 ns | β− |
| 83 | (5/2+) | −36.3s | >150 ns | β− |
| 31 Ga 56 | | −4.7s | ? | |
| 57 | | −15.9s | ? | p? |
| 58 | | −24.0s | ? | p? |
| 59 | | −34.1s | ? | p? |
| 60 | (2+) | −40.0s | 70 ms *13* | ε 98.4%, εp 1.6%, εα<0.02% |
| 61 | 3/2− | −47.09 | 168 ms *3* | ε |
| 62 | 0+ | −52.00 | 116.18 ms *4* | ε |

| Nuclide Z El A | Jπ | Δ (MeV) | T½, Γ, or Abundance | Decay Mode |
|---|---|---|---|---|
| 31 Ga 63 | (3/2−) | −56.547 | 32.4 s *5* | ε |
| 64 | 0+ | −58.834 | 2.627 m *12* | ε |
| 65 | 3/2− | −62.657 | 15.2 m *2* | ε |
| 66 | 0+ | −63.724 | 9.49 h *7* | ε |
| 67 | 3/2− | −66.880 | 3.2623 d *15* | ε |
| 68 | 1+ | −67.086 | 67.71 m *9* | ε |
| 69 | 3/2− | −69.328 | **60.108% *9*** | |
| 70 | 1+ | −68.910 | 21.14 m *3* | β− 99.59%, ε 0.41% |
| 71 | 3/2− | −70.140 | **39.892% *9*** | |
| 72 | 3− | −68.589 | 14.095 h *3* | β− |
| 73 | 3/2− | −69.699 | 4.86 h *3* | β− |
| 74 | (3−) | −68.050 | 8.12 m *12* | β− |
| 74m | (0) | −67.990 | 9.5 s *10* | IT 75%, β−<50% |
| 75 | (3/2−) | −68.465 | 126 s *2* | β− |
| 76 | (2+,3+) | −66.297 | 32.6 s *6* | β− |
| 77 | (3/2−) | −65.992 | 13.2 s *2* | β− |
| 78 | (3+) | −63.707 | 5.09 s *5* | β− |
| 79 | (3/2−) | −62.51 | 2.847 s *3* | β−, β−n 0.09% |
| 80 | (3) | −59.1 | 1.676 s *14* | β−, β−n 0.86% |
| 81 | (5/2−) | −58.0 | 1.217 s *5* | β−, β−n 11.9% |
| 82 | (1,2,3) | −53.1s | 0.599 s *2* | β−, β−n 19.8% |
| 83 | | −44.4s | 0.308 s *1* | β−, β−n 37% |
| 84 | | −41.4s | 0.085 s *10* | β−, β−n 70% |
| 85 | (3/2−) | −40.1s | >150 ns | β− |
| 86 | | −34.4s | >150 ns | β− |
| 32 Ge 58 | 0+ | −8.4s | ? | 2p? |
| 59 | | −17.0s | ? | 2p? |
| 60 | 0+ | −27.8s | ≈30 ms | ε?, 2p? |
| 61 | (3/2−) | −33.7s | 39 ms *12* | ε, εp≈80% |
| 62 | 0+ | −42.2s | 129 ns *35* | ε |
| 63 | (3/2−) | −46.9s | 142 ms *8* | ε |
| 64 | 0+ | −54.35 | 63.7 s *25* | ε |
| 65 | (3/2−)− | −56.4 | 30.9 s *5* | ε |
| 66 | 0+ | −61.62 | 2.26 h *5* | ε |
| 67 | 1/2− | −62.658 | 18.9 m *4* | ε |
| 68 | 0+ | −66.980 | 270.95 d *16* | ε |
| 69 | 5/2− | −67.101 | 39.05 h *10* | ε |
| 70 | 0+ | −70.563 | **20.37% *18*** | |
| 71 | 1/2− | −69.908 | 11.43 d *3* | ε |
| 72 | 0+ | −72.586 | **27.31% *26*** | |
| 73 | 9/2+ | −71.298 | **7.76% *8*** | |
| 73m | 1/2− | −71.231 | 0.499 s *11* | IT |
| 74 | 0+ | −73.422 | **36.73% *15*** | |
| 75 | 1/2− | −71.856 | 82.78 m *4* | β− |
| 75m | 7/2+ | −71.717 | 47.7 s *5* | IT 99.97%, β− 0.03% |
| 76 | 0+ | −73.213 | 1.2×10²¹ y *14* | 2β− |
| | | | **7.83% *7*** | |
| 77 | 7/2+ | −71.214 | 11.30 h *1* | β− |
| 77m | 1/2− | −71.054 | 52.9 s *6* | β− 81%, IT 19% |
| 78 | 0+ | −71.862 | 88.0 m *10* | β− |
| 79 | (1/2) | −69.49 | 18.98 s *3* | β− |
| 79m | (7/2+) | −69.30 | 39.0 s *10* | β− 96%, IT 4% |

| Nuclide Z El A | Jπ | Δ (MeV) | T½, Γ, or Abundance | Decay Mode |
|---|---|---|---|---|
| 32 Ge 80 | 0+ | −69.52 | 29.5 s *4* | β− |
| 81 | (9/2+) | −66.3 | 7.6 s *6* | β− |
| 81m | (1/2+) | −65.6 | 7.6 s *6* | β− |
| 82 | 0+ | −65.6 | 4.55 s *5* | β− |
| 83 | (5/2+) | −60.9s | 1.85 s *6* | β− |
| 84 | 0+ | −58.2s | 0.947 s *11* | β−, β−n 10.8% |
| 85 | | −53.1s | 535 ms *47* | β−, β−n 14% |
| 86 | 0+ | −49.8s | >150 ns | β− |
| 87 | (5/2+) | −44.2s | ≈0.14 s | β−, β−n |
| 88 | 0+ | −40.1s | ≥300 ns | β−? |
| 89 | | −33.7s | >150 ns | β−? |
| 33 As 60 | | −6.4s | ? | p? |
| 61 | | −18.1s | ? | p? |
| 62 | | −25.0s | ? | p |
| 63 | (3/2−) | −33.8s | ? | p |
| 64 | | −39.5s | 18 ms *+43−7* | ε? |
| 65 | | −47.0s | 128 ms *16* | ε |
| 66 | | −51.5 | 95.79 ms *22* | ε |
| 67 | (5/2−) | −56.6 | 42.5 s *12* | ε |
| 68 | 3+ | −58.90 | 151.6 s *8* | ε |
| 69 | 5/2− | −63.09 | 15.2 m *2* | ε |
| 70 | 4+ | −64.34 | 52.6 m *3* | ε |
| 71 | 5/2− | −67.894 | 65.28 h *15* | ε |
| 72 | 2− | −68.230 | 26.0 h *1* | ε |
| 73 | 3/2− | −70.957 | 80.30 d *6* | ε |
| 74 | 2− | −70.860 | 17.77 d *2* | ε 66%, β− 34% |
| 75 | 3/2− | −73.032 | **100%** | |
| 76 | 2− | −72.289 | 1.0942 d *7* | β− |
| 77 | 3/2− | −73.917 | 38.83 h *5* | β− |
| 78 | 2− | −72.817 | 90.7 m *2* | β− |
| 79 | 3/2− | −73.636 | 9.01 h *1* | β− |
| 80 | 1+ | −72.16 | 15.2 s *2* | β− |
| 81 | 3/2− | −72.533 | 33.3 s *8* | β− |
| 82 | (1+) | −70.3 | 19.1 s *5* | β− |
| 82m | (5−) | −70.3 | 13.6 s *4* | β− |
| 83 | (5/2−,3/2−) | −69.9 | 13.4 s *3* | β− |
| 84 | (3−) | −66.1s | 3.24 s *26* | β−, β−n 0.28% |
| 85 | (3/2−) | −63.3s | 2.021 s *10* | β− 59.4% |
| 86 | 0+ | −59.2s | 0.945 s *8* | β−, β−n 33% |
| 87 | (3/2−) | −56.0s | 0.56 s *8* | β−, β−n 15.4% |
| 88 | | −51.3s | ≥300 ns | β−, β−n? |
| 89 | | −47.1s | ≥300 ns | β−? |
| 90 | | −41.5s | >150 ns | β−? |
| 91 | | −36.9s | >150 ns | β− |
| 92 | | −30.9s | >300 ns | β− |
| 34 Se 65 | | −32.9s | <50 ms | ε |
| 66 | 0+ | −41.7s | 33 ms *12* | ε |
| 67 | | −46.5s | 133 ms *11* | ε, εp 0.5% |
| 68 | 0+ | −54.21 | 35.5 s *7* | ε |
| 69 | (1/2−,3/2−) | −56.30 | 27.4 s *2* | ε, εp 0.05% |
| 70 | 0+ | −62.05 | 41.1 m *3* | ε |
| 71 | 5/2− | −63.12 | 4.74 m *5* | ε |

## Nuclear Wallet Cards

| Nuclide Z El A | Jπ | Δ (MeV) | T½, Γ, or Abundance | Decay Mode |
|---|---|---|---|---|
| 34 Se 72 | 0+ | −67.89 | 8.40 d 8 | ε |
| 73 | 9/2+ | −68.22 | 7.15 h 8 | ε |
| 73m | 3/2− | −68.19 | 39.8 m 13 | IT 72.6%, ε 27.4% |
| 74 | 0+ | −72.213 | 0.89% 4 | |
| 75 | 5/2+ | −72.169 | 119.779 d 4 | ε |
| 76 | 0+ | −75.252 | 9.37% 29 | |
| 77 | 1/2− | −74.600 | 7.63% 16 | |
| 77m | 7/2+ | −74.438 | 17.36 s 5 | IT |
| 78 | 0+ | −77.026 | 23.77% 28 | |
| 79 | 7/2+ | −75.918 | 2.95×10⁵ y 38 | β− |
| 79m | 1/2− | −75.822 | 3.92 m 1 | IT 99.94%, β− 0.06% |
| 80 | 0+ | −77.760 | 49.61% 41 | |
| 81 | 1/2− | −76.390 | 18.45 m 12 | β− |
| 81m | 7/2+ | −76.286 | 57.28 m 2 | IT 99.95%, β− 0.05% |
| 82 | 0+ | −77.594 | 9.1×10¹⁹ y 9 **8.73% 22** | 2β− |
| 83 | 9/2+ | −75.341 | 22.3 m 3 | β− |
| 83m | 1/2− | −75.112 | 70.1 s 4 | β− |
| 84 | 0+ | −75.95 | 3.10 m 10 | β− |
| 85 | (5/2+) | −72.43 | 31.7 s 9 | β− |
| 86 | 0+ | −70.54 | 15.3 s 9 | β− |
| 87 | (5/2+) | −66.58 | 5.50 s 12 | β−, β−n 0.2% |
| 88 | 0+ | −63.88 | 1.53 s 6 | β−, β−n 0.99% |
| 89 | (5/2+) | −59.2s | 0.41 s 4 | β−, β−n 7.8% |
| 90 | 0+ | −55.9s | >150 ns | β−? |
| 91 | | −50.3s | 0.27 s 5 | β−n 21% |
| 92 | 0+ | −46.6s | >300 ns | β− |
| 93 | (1/2+) | −40.7s | >150 ns | β− |
| 94 | 0+ | −36.8s | >150 ns | β− |
| 35 Br 67 | | −32.8s | ? | ? |
| 68 | | −38.6s | <1.2 μs | p? |
| 69 | | −46.5s | <24 ns | p |
| 70 | 0+ | −51.4s | 79.1 ms 8 | ε |
| 70m | 9+ | −49.1s | 2.2 s 2 | ε |
| 71 | (5/2)− | −57.1 | 21.4 s 6 | ε |
| 72 | 1+ | −59.02 | 78.6 s 24 | ε |
| 72m | 1− | −58.91 | 10.6 s 3 | IT=100%, ε |
| 73 | 1/2− | −63.63 | 3.4 m 2 | ε |
| 74 | (0−) | −65.31 | 25.4 m 3 | ε |
| 74m | 4(+) | −65.29 | 46 m 2 | ε |
| 75 | 3/2− | −69.14 | 96.7 m 13 | ε |
| 76 | 1− | −70.289 | 16.2 h 2 | ε |
| 76m | (4)+ | −70.186 | 1.31 s 2 | IT>99.4%, ε<0.6% |
| 77 | 3/2− | −73.235 | 57.036 h 6 | ε |
| 77m | 9/2+ | −73.129 | 4.28 m 10 | IT |
| 78 | 1+ | −73.452 | 6.46 m 4 | ε≥99.99%, β−≤0.01% |
| 79 | 3/2− | −76.068 | 50.69% 7 | |
| 79m | 9/2+ | −75.861 | 4.86 s 4 | IT |
| 80 | 1+ | −75.890 | 17.68 m 2 | β− 91.7%, ε 8.3% |
| 80m | 5− | −75.804 | 4.4205 h 8 | IT |
| 81 | 3/2− | −77.975 | 49.31% 7 | |

## Nuclear Wallet Cards

| Nuclide Z El A | Jπ | Δ (MeV) | T½, Γ, or Abundance | Decay Mode |
|---|---|---|---|---|
| 36 Kr 98 | 0+ | −44.8s | 46 ms 8 | β−, β−n 7% |
| 99 | (3/2+) | −39.5s | 40 ms 11 | β−, β−n 11% |
| 100 | 0+ | −36.2s | >150 ns | β− |
| 37 Rb 71 | | −32.3s | ? | p? |
| 72 | (3+) | −38.1s | <1.2 μs | p |
| 73 | | −46.1s | >30 ns | ε, p>0% |
| 74 | (0+) | −51.917 | 64.9 ms 5 | ε |
| 75 | (3/2−) | −57.222 | 19.0 s 12 | ε |
| 76 | 1(−) | −60.480 | 36.5 s 6 | ε, εα 3.8×10⁻⁷% |
| 77 | 3/2− | −64.825 | 3.77 m 4 | ε |
| 78 | 0+ | −66.936 | 17.66 m 8 | ε |
| 78m | 4(−) | −66.833 | 5.74 m 5 | ε 90%, IT 10% |
| 79 | 5/2+ | −70.803 | 22.9 m 5 | ε |
| 80 | 1+ | −72.173 | 33.4 s 7 | ε |
| 81 | 3/2− | −75.455 | 4.570 h 4 | ε |
| 81m | 9/2+ | −75.368 | 30.5 m 3 | IT 97.6%, ε 2.4% |
| 82 | 1+ | −76.188 | 1.273 m 2 | ε |
| 82m | 5− | −76.111 | 6.472 h 5 | ε, IT<0.33% |
| 83 | 5/2− | −79.075 | 86.2 d 1 | ε |
| 84 | 2− | −79.750 | 33.1 d 1 | ε 96.2%, β− 3.8% |
| 84m | 6− | −79.286 | 20.26 m 4 | IT |
| 85 | 5/2− | −82.167 | 72.17% 2 | |
| 86 | 2− | −82.747 | 18.642 d 18 | β− 99.99%, ε 5.2×10⁻³% |
| 86m | 6− | −82.191 | 1.017 m 3 | IT, β−<0.3% |
| 87 | 3/2− | −84.598 | 4.97×10¹⁰ y 3 **27.83% 2** | β− |
| 88 | 2− | −82.609 | 17.773 m 11 | β− |
| 89 | 3/2− | −81.713 | 15.15 m 12 | β− |
| 90 | 0− | −79.362 | 158 s 5 | β− |
| 90m | 3− | −79.255 | 258 s 4 | β− 97.4%, IT 2.6% |
| 91 | 3/2(−) | −77.745 | 58.4 s 4 | β− |
| 92 | 0− | −74.772 | 4.492 s 20 | β−, β−n 0.01% |
| 93 | 5/2− | −72.618 | 5.84 s 2 | β−, β−n 1.39% |
| 94 | 3(−) | −68.553 | 2.702 s 5 | β−, β−n 10.01% |
| 95 | 5/2− | −65.85 | 377.5 ms 8 | β−, β−n 8.73% |
| 96 | 2+ | −61.22 | 202.8 ms 33 | β−, β−n 14% |
| 97 | 3/2+ | −58.36 | 169.9 ms 7 | β−, β−n 25.1% |
| 98 | (0,1) | −54.22 | 114 ms 5 | β−, β−n 13.8%, β− 0.05% |
| 99 | (5/2+) | −50.9 | 50.3 ms 7 | β−, β−n 15.9% |
| 100 | | −46.7s | 51 ms 8 | β−, β−n 6%, β−2n 0.16% |
| 101 | (3/2+) | −43.6 | 32 ms 5 | β−, β−n 28% |
| 38 Sr 73 | | −31.7s | >25 ms | ε, εp>0% |
| 74 | 0+ | >−1.2 μs | | ε |
| 75 | (3/2−) | −46.6 | 88 ms 3 | ε, εp 5.2% |
| 76 | 0+ | −54.24 | 7.89 s 7 | ε, εp 0.34% |
| 77 | 5/2+ | −57.804 | 9.0 s 2 | ε, εp<0.25% |
| 78 | 0+ | −63.174 | 2.5 m 3 | ε |
| 79 | 3/2(−) | −65.477 | 2.25 m 10 | ε |
| 80 | 0+ | −70.308 | 106.3 m 15 | ε |

## Nuclear Wallet Cards

| Nuclide Z El A | Jπ | Δ (MeV) | T½, Γ, or Abundance | Decay Mode |
|---|---|---|---|---|
| 35 Br 82 | 5− | −77.496 | 35.282 h 7 | β− |
| 82m | 2− | −77.451 | 6.13 m 5 | IT 97.6%, β− 2.4% |
| 83 | 3/2− | −79.009 | 2.40 h 2 | β− |
| 84 | 2− | −77.80 | 31.80 m 8 | β− |
| 84m | 6− | −77.48 | 6.0 m 2 | β− |
| 85 | 3/2− | −78.61 | 2.90 m 6 | β− |
| 86 | (2−) | −75.64 | 55.1 s 4 | β− |
| 87 | 3/2− | −73.86 | 55.65 s 13 | β−, β−n 2.6% |
| 88 | (2−) | −70.73 | 16.29 s 6 | β−, β−n 6.58% |
| 89 | (3/2−,5/2−) | −68.5 1 | 4.40 s 3 | β−, β−n 13.8% |
| 90 | | −64.62 | 1.91 s 1 | β−, β−n 25.2% |
| 91 | | −61.51 | 0.541 s 5 | β−, β−n 20% |
| 92 | (2−) | −56.58 | 0.343 s 15 | β−, β−n 33.1% |
| 93 | (5/2−) | −53.0s | 102 ms 10 | β−, β−n 68% |
| 94 | | −47.8s | 70 ms 20 | β−, β−n 70% |
| 95 | (3/2+) | −43.9s | >150 ns | β− |
| 96 | | −38.6s | >150 ns | β− |
| 97 | (3/2−) | −34.7s | >150 ns | β− |
| 36 Kr 69 | | −32.4s | 32 ms 10 | ε |
| 70 | 0+ | −41.7s | 52 ms 17 | ε, εp≤1.3% |
| 71 | (5/2)− | −46.9 | 100 ms 3 | ε, εp 5.2% |
| 72 | 0+ | −53.941 | 17.1 s 2 | ε |
| 73 | 3/2− | −56.552 | 27.3 s 10 | ε, εp 0.25% |
| 74 | 0+ | −62.332 | 11.50 m 11 | ε |
| 75 | 5/2+ | −64.324 | 4.29 m 17 | ε |
| 76 | 0+ | −69.014 | 14.8 h 1 | ε |
| 77 | 5/2+ | −70.169 | 74.4 m 6 | ε |
| 78 | 0+ | −74.180 | ≥2.3×10²⁰ y **0.35% 1** | 2ε |
| 79 | 1/2− | −74.443 | 35.04 h 10 | ε |
| 79m | 7/2+ | −74.313 | 50 s 3 | IT |
| 80 | 0+ | −77.893 | 2.28% 6 | |
| 81 | 7/2+ | −77.694 | 2.29×10⁵ y 11 | ε, ε 2.5×10⁻³% |
| 81m | 1/2− | −77.503 | 13.10 s 3 | IT |
| 82 | 0+ | −80.590 | 11.58% 14 | |
| 83 | 9/2+ | −79.982 | 11.49% 6 | |
| 83m | 1/2− | −79.940 | 1.83 h 2 | IT |
| 84 | 0+ | −82.431 | 57.00% 4 | |
| 85 | 9/2+ | −81.480 | 3916.8 d 25 | β− |
| 85m | 1/2− | −81.175 | 4.480 h 8 | β− 78.6%, IT 21.4% |
| 86 | 0+ | −83.266 | 17.30% 22 | |
| 87 | 5/2+ | −80.709 | 76.3 m 5 | β− |
| 88 | 0+ | −79.69 | 2.84 h 3 | β− |
| 89 | 3/2(+) | −76.73 | 3.15 m 4 | β− |
| 90 | 0+ | −74.97 | 32.32 s 9 | β− |
| 91 | 5/2(+) | −71.31 | 8.57 s 4 | β− |
| 92 | 0+ | −68.79 | 1.840 s 8 | β−, β−n 0.03% |
| 93 | 1/2+ | −64.0 | 1.286 s 10 | β−, β−n 1.95% |
| 94 | 0+ | −61.1s | 212 ms 5 | β−, β−n 1.26% |
| 95 | 1/2 | −56.0s | 114 ms 3 | β−, β−n 2.87% |
| 96 | 0+ | −53.0s | 80 ms 8 | β−, β−n 3.8% |
| 97 | | −47.9s | 63 ms 4 | β−, β−n 8.2% |

## Nuclear Wallet Cards

| Nuclide Z El A | Jπ | Δ (MeV) | T½, Γ, or Abundance | Decay Mode |
|---|---|---|---|---|
| 38 Sr 81 | 1/2− | −71.528 | 22.3 m 4 | ε |
| 82 | 0+ | −76.008 | 25.55 d 15 | ε |
| 83 | 7/2+ | −76.80 | 32.41 h 3 | ε |
| 83m | 1/2− | −76.54 | 4.95 s 12 | IT |
| 84 | 0+ | −80.644 | 0.56% 1 | |
| 85 | 9/2+ | −81.103 | 64.84 d 2 | ε |
| 85m | 1/2− | −80.864 | 67.63 m 4 | IT 86.6%, ε 13.4% |
| 86 | 0+ | −84.524 | 9.86% 1 | |
| 87 | 9/2+ | −84.880 | 7.00% 1 | |
| 87m | 1/2− | −84.492 | 2.815 h 12 | IT 99.7%, ε 0.3% |
| 88 | 0+ | −87.922 | 82.58% 1 | |
| 89 | 5/2+ | −86.209 | 50.57 d 3 | β− |
| 90 | 0+ | −85.942 | 28.90 y 3 | β− |
| 91 | 5/2+ | −83.645 | 9.63 h 5 | β− |
| 92 | 0+ | −82.868 | 2.66 h 4 | β− |
| 93 | 5/2+ | −80.085 | 7.423 m 24 | β− |
| 94 | 0+ | −78.840 | 75.3 s 2 | β− |
| 95 | 1/2+ | −75.117 | 23.90 s 14 | β− |
| 96 | 0+ | −72.94 | 1.07 s 1 | β− |
| 97 | 1/2+ | −68.79 | 429 ms 5 | β−, β−n≤0.05% |
| 98 | 0+ | −66.65 | 0.653 s 2 | β−, β−n 0.25% |
| 99 | 3/2+ | −62.19 | 0.269 s 1 | β−, β−n 0.1% |
| 100 | 0+ | −60.2 | 202 ms 3 | β−, β−n 0.78% |
| 101 | (5/2−) | −55.4 | 118 ms 3 | β−, β−n 2.37% |
| 102 | 0+ | −53.1 | 69 ms 6 | β−, β−n 4.8% |
| 103 | | −47.6s | >150 ns | β− |
| 104 | 0+ | −44.4s | >300 ns | β− |
| 105 | | −38.6s | >150 ns | β− |
| 39 Y 76 | | −38.7s | >200 ns | ε?, p? |
| 77 | | −46.90s | ~0.06 s | ε, εp |
| 78 | (0+) | −52.5s | 50 ms 8 | ε |
| 78m | (5+) | −52.5s | 5.7 s 7 | ε |
| 79 | (5/2+) | −58.4s | 14.8 s 6 | ε, εp |
| 80 | (4−) | −61.2 | 30.1 s 5 | ε, εp |
| 80m | (1−) | −61.0 | 4.8 s 3 | IT 81%, ε 19% |
| 81 | (5/2+) | −66.82 | 70.4 s 10 | ε |
| 82 | 1+ | −68.2 | 8.30 s 20 | ε |
| 83 | 9/2+ | −72.33 | 7.08 m 6 | ε |
| 83m | 3/2− | −72.26 | 2.85 m 2 | ε 60%, IT 40% |
| 84 | 1+ | −74.16 | 4.6 s 2 | ε |
| 84m | (5−) | −74.16 | 39.5 m 8 | ε |
| 85 | (1/2)− | −77.84 | 2.68 h 5 | ε |
| 85m | 9/2+ | −77.82 | 4.86 h 13 | ε, IT<2.0×10⁻³% |
| 86 | 4− | −79.28 | 14.74 h 2 | ε |
| 86m | (8+) | −79.07 | 48 m 1 | IT 99.31%, ε 0.69% |
| 87 | 1/2− | −83.019 | 79.8 h 3 | ε |
| 87m | 9/2+ | −82.638 | 13.37 h 3 | IT 98.43%, ε 1.57% |
| 88 | 4− | −84.299 | 106.616 d 13 | ε |
| 89 | 1/2− | −87.702 | 100% | |
| 89m | 9/2+ | −86.793 | 15.28 s 17 | IT |
| 90 | 2− | −86.488 | 64.053 h 20 | β− |
| 90m | 7+ | −85.806 | 3.19 h 6 | IT, β− 1.8×10⁻³% |

APPENDIX B  **625**

## Nuclear Wallet Cards

| Nuclide Z El A | Jπ | Δ (MeV) | T½, Γ, or Abundance | Decay Mode |
|---|---|---|---|---|
| 39 Y 91 | 1/2− | −86.345 | 58.51 d 6 | β− |
| 91m | 9/2+ | −85.789 | 49.71 m 4 | IT, β−<1.5% |
| 92 | 2− | −84.813 | 3.54 h 1 | β− |
| 93 | 1/2− | −84.22 | 10.18 h 8 | β− |
| 93m | 7/2+ | −83.46 | 0.82 s 4 | IT |
| 94 | 2− | −82.349 | 18.7 m 1 | β− |
| 95 | 1/2− | −81.207 | 10.3 m 1 | β− |
| 96 | 0− | −78.35 | 5.34 s 5 | β− |
| 96m | (8)+ | −78.35 | 9.6 s 2 | β− |
| 97 | (1/2−) | −76.26 | 3.75 s 3 | β−, β−n 0.058% |
| 97m | (9/2)+ | −75.59 | 1.17 s 3 | β−>99.3%, IT<0.7%, β−n<0.08% |
| 97m | (27/2−) | −72.73 | 142 ms 8 | IT>80%, β−<20% |
| 98 | (0)− | −72.47 | 0.548 s 2 | β−, β−n 0.33% |
| 98m | (4,5) | −72.06 | 2.0 s 2 | β−>80%, IT<20%, β−n 3.4% |
| 99 | (5/2+) | −70.20 | 1.470 s 7 | β−, β−n 1.9% |
| 100 | 1−,2− | −67.29 | 735 ms 7 | β−, β−n 0.92% |
| 100m | (3,4,5) | −67.29 | 0.94 s 3 | β− |
| 101 | (5/2+) | −64.91 | 0.45 s 2 | β−, β−n 1.5% |
| 102 | | −61.89 | 0.30 s 1 | β−, β−n 4% |
| 102m | | −61.89 | 0.36 s 4 | β−, β−n 8% |
| 103 | (5/2+) | −58.9s | 0.23 s 2 | β−, β−n |
| 104 | | −54.9s | 180 ms 60 | β−, β−n? |
| 105 | | −51.4s | >300 ns | β−? |
| 106 | | −46.8s | >150 ns | β− |
| 107 | (5/2+) | −42.7s | ≈30 ms | β− |
| 108 | | −37.7s | 20 ms sy | β− |
| 40 Zr 78 | 0+ | −41.7s | >200 ns | ε?, εp? |
| 79 | | −47.4s | 56 ms 30 | ε, εp |
| 80 | 0+ | −56 | 4.6 s 6 | ε, εp |
| 81 | (3/2−) | −58.5 | 5.5 s 4 | ε, εp 0.12% |
| 82 | 0+ | −64.2s | 32 s 5 | ε |
| 83 | (1/2−) | −66.46 | 41.6 s 24 | ε, εp |
| 84 | 0+ | −71.5s | 25.9 m 7 | ε |
| 85 | 7/2+ | −73.1 | 7.86 m 4 | ε |
| 85m | (1/2−) | −72.9 | 10.9 s 3 | IT≤92%, ε>8% |
| 86 | 0+ | −77.80 | 16.5 h 1 | ε |
| 87 | (9/2)+ | −79.348 | 1.68 h 1 | ε |
| 87m | (1/2)− | −79.012 | 14.0 s 2 | IT |
| 88 | 0+ | −83.62 | 83.4 d 3 | ε |
| 89 | 9/2+ | −84.869 | 78.41 h 12 | ε |
| 89m | 1/2− | −84.281 | 4.161 m 17 | IT 93.77%, ε 6.23% |
| 90 | 0+ | −88.767 | 51.45% 40 | |
| 90m | 5− | −86.448 | 809.2 ms 20 | IT |
| 91 | 5/2+ | −87.890 | 11.22% 5 | |
| 92 | 0+ | −88.454 | 17.15% 8 | |
| 93 | 5/2+ | −87.117 | 1.53×10⁶ y 10 | β− |
| 94 | 0+ | −87.267 | 17.38% 28 | |
| 95 | 5/2+ | −85.658 | 64.032 d 6 | β− |
| 96 | 0+ | −85.443 | >3.9×10²⁰ y | 2β−; 2.80% 9 |

| Nuclide Z El A | Jπ | Δ (MeV) | T½, Γ, or Abundance | Decay Mode |
|---|---|---|---|---|
| 41 Nb 104 | (1+) | −72.2 | 4.9 s 3 | β−, β−n 0.06% |
| 104m | | −72.0 | 0.94 s 4 | β−, β−n 0.05% |
| 105 | (5/2+) | −70.85 | 2.95 s 6 | β−, β−n 1.7% |
| 106 | | −67.1s | 1.02 s 5 | β−, β−n 4.5% |
| 107 | | −64.9s | 330 ms 50 | β− |
| 108 | (2+) | −60.7s | 0.193 s 17 | β−, β−n 6.2% |
| 109 | (5/2) | −58.1s | 0.19 s 3 | β−, β−n 31% |
| 110 | | −53.6s | 0.17 s 2 | β−, β−n 40% |
| 111 | (5/2+) | −50.6s | 80. ms sy | β− |
| 112 | (2+) | −45.8s | >150 ns | β− |
| 113 | | −42.2s | 30 ms sy | β− |
| 42 Mo 83 | | −47.7s | 6 ms +30 −3 | ε |
| 84 | 0+ | −55.8s | 3.7 s +10 −8 | ε |
| 85 | (1/2−) | −59.1s | 3.2 s 2 | εp 0.14%, ε |
| 86 | 0+ | −64.6 | 19.6 s 11 | ε |
| 87 | 7/2+ | −67.7 | 14.02 s 26 | ε, εp 15% |
| 88 | 0+ | −72.70 | 8.0 m 2 | ε |
| 89 | (9/2+) | −75.00 | 2.11 m 10 | ε |
| 89m | (1/2−) | −74.62 | 190 ms 15 | IT |
| 90 | 0+ | −80.167 | 5.56 h 9 | ε |
| 91 | 9/2+ | −82.20 | 15.49 m 1 | ε |
| 92 | 0+ | −81.55 | 64.6 s 6 | ε 50%, IT 50% |
| 92 | 0+ | −86.805 | 14.84% 35 | |
| 93 | 5/2+ | −86.803 | 4.0×10³ y 8 | ε |
| 93m | 21/2+ | −84.379 | 6.85 h 7 | IT 99.88%, ε 0.12% |
| 94 | 0+ | −88.410 | 9.25% 12 | |
| 95 | 5/2+ | −87.707 | 15.92% 13 | |
| 96 | 0+ | −88.790 | 16.68% 2 | |
| 97 | 5/2+ | −87.540 | 9.55% 8 | |
| 98 | 0+ | −88.112 | 24.13% 31 | |
| 99 | 1/2+ | −85.966 | 2.7489 d 6 | β− |
| 100 | 0+ | −86.184 | 0.78×10¹⁹ y 8 | 2β−; 9.63% 3 |
| 101 | 1/2+ | −83.511 | 14.61 m 3 | β− |
| 102 | 0+ | −83.56 | 11.3 m 2 | β− |
| 103 | (3/2+) | −80.85 | 67.5 s 15 | β− |
| 104 | 0+ | −80.33 | 60 s 2 | β− |
| 105 | (5/2−) | −77.34 | 35.6 s 16 | β− |
| 106 | 0+ | −76.26 | 8.4 s 5 | β− |
| 107 | (7/2−) | −72.9 | 3.5 s 5 | β− |
| 108 | 0+ | −71.3s | 1.09 s 2 | β− |
| 109 | (7/2−) | −67.2s | 0.53 s 6 | β− |
| 110 | 0+ | −65.5s | 0.27 s 1 | β− |
| 111 | | −61.1s | 200. ms sy | β− |
| 112 | 0+ | −58.8s | >150 ns | β−? |
| 113 | | −54.1s | 100 ms sy | β− |
| 114 | 0+ | −51.3s | 80 ms sy | β− |
| 115 | | −46.3s | 60 ms sy | β−, β−n |
| 43 Tc 85 | | −47.7s | ≈0.5 s | ε? |
| 86 | (0+) | −53.2s | 54 ms 7 | ε |
| 87 | (9/2+) | −59.1s | 2.2 s 2 | ε |
| 88 | (3+) | −62.7s | 5.8 s 2 | ε |

## Nuclear Wallet Cards

| Nuclide Z El A | Jπ | Δ (MeV) | T½, Γ, or Abundance | Decay Mode |
|---|---|---|---|---|
| 40 Zr 97 | 1/2+ | −82.947 | 16.744 h 11 | β− |
| 98 | 0+ | −81.29 | 30.7 s 4 | β− |
| 99 | (1/2+) | −77.77 | 2.1 s 1 | β− |
| 100 | 0+ | −76.60 | 7.1 s 4 | β− |
| 101 | (3/2+) | −73.46 | 2.3 s 1 | β− |
| 102 | 0+ | −71.74 | 2.9 s 2 | β− |
| 103 | (5/2−) | −68.4 | 1.3 s 1 | β− |
| 104 | 0+ | −66.3s | 1.2 s 3 | β− |
| 105 | | −62.4s | 0.6 s 1 | β− |
| 106 | 0+ | −59.7s | >150 ns | β−? |
| 107 | | −55.2s | ≈150 ms | β− |
| 108 | 0+ | −52.2s | 80 ms sy | β−, β−n |
| 109 | | −47.3s | >150 ns | β−, β−n |
| 110 | 0+ | −43.9s | >150 ns | β− |
| 41 Nb 81 | | −47s | ≈0.8 s | ε?, εp?, p? |
| 82 | 0+ | −53.0s | 50 ms 5 | ε |
| 83 | (5/2+) | −59.0 | 4.1 s 3 | ε |
| 84 | 3+ | −61.9s | 9.5 s 10 | ε, εp |
| 85 | (9/2+) | −67.1 | 20.9 s 7 | ε |
| 86 | | −69.83 | 56 s 8 | ε |
| 86m | (6+) | −69.83 | 88 s 1 | ε |
| 87 | (1/2)− | −74.18 | 3.75 m 9 | ε |
| 87m | (9/2)+ | −74.18 | 2.6 m 1 | ε |
| 88 | (8+) | −76.1 | 14.55 m 6 | ε |
| 88m | (4−) | −76.1 | 7.78 m 5 | ε |
| 89 | (9/2+) | −80.65 | 2.03 h 7 | ε |
| 89m | (1/2)− | −80.62 | 66 m 2 | ε |
| 90 | 8+ | −82.656 | 14.60 h 5 | ε |
| 90m | 4− | −82.532 | 18.81 s 6 | IT |
| 91 | 9/2+ | −86.632 | 6.8×10² y 13 | ε |
| 91m | 1/2− | −86.528 | 60.86 d 22 | IT 96.6%, ε 3.4% |
| 92 | (7)+ | −86.448 | 3.47×10⁷ y 24 | ε, β−<0.05% |
| 92m | 2+ | −86.313 | 10.15 d 2 | ε |
| 93 | 9/2+ | −87.208 | 100% | |
| 93m | 1/2− | −87.177 | 16.13 y 14 | IT |
| 94 | (6)+ | −86.365 | 2.03×10⁴ y 16 | β− |
| 94m | 3+ | −86.324 | 6.263 m 4 | IT 99.5%, β− 0.5% |
| 95 | 9/2+ | −86.782 | 34.991 d 6 | β− |
| 95m | 1/2− | −86.546 | 3.61 d 3 | IT 94.4%, β− 5.6% |
| 96 | 6+ | −85.604 | 23.35 h 5 | β− |
| 97 | 9/2+ | −85.606 | 72.1 m 7 | β− |
| 97m | 1/2− | −84.863 | 58.7 s 18 | IT |
| 98 | 1+ | −83.529 | 2.86 s 6 | β− |
| 98m | (5+) | −83.445 | 51.3 m 4 | β− 99.9%, IT<0.2% |
| 99 | 9/2+ | −82.33 | 15.0 s 2 | β− |
| 99m | 1/2− | −81.96 | 2.6 m 2 | β−>96.2%, IT<3.8% |
| 100 | 1+ | −79.94 | 1.5 s 2 | β− |
| 100m | (4+,5+) | −79.44 | 2.99 s 11 | β− |
| 101 | (5/2+) | −78.94 | 7.1 s 3 | β− |
| 102m | 1+ | −76.35 | 1.3 s 2 | β− |
| 102m | | −76.35 | 4.3 s 4 | β− |
| 103 | (5/2+) | −75.32 | 1.5 s 2 | β− |

| Nuclide Z El A | Jπ | Δ (MeV) | T½, Γ, or Abundance | Decay Mode |
|---|---|---|---|---|
| 43 Tc 88m | (6+) | −62.7s | 6.4 s 8 | ε |
| 89 | (9/2+) | −67.8s | 12.8 s 9 | ε |
| 89m | (1/2−) | −67.8s | 12.9 s 8 | ε, IT<0.01% |
| 90 | 1+ | −71.2 | 8.7 s 2 | ε |
| 90 | (6+) | −70.7 | 49.2 s 4 | ε |
| 91 | (9/2+) | −76.0 | 3.14 m 2 | ε |
| 91m | (1/2−) | −75.8 | 3.3 m 1 | ε, IT<1% |
| 92 | (8+) | −78.93 | 4.25 m 15 | ε |
| 93 | 9/2+ | −83.603 | 2.75 h 5 | ε |
| 93m | 1/2− | −83.211 | 43.5 m 10 | IT 76.6%, ε 23.4% |
| 94 | 7+ | −84.154 | 293 m 1 | ε |
| 94m | (2)+ | −84.079 | 52.0 m 10 | ε, IT<0.1% |
| 95 | 9/2+ | −86.017 | 20.0 h 1 | ε |
| 95m | 1/2− | −85.978 | 61 d 2 | ε 96.12%, IT 3.88% |
| 96 | 7+ | −85.817 | 4.28 d 7 | ε |
| 96m | 4+ | −85.783 | 51.5 m 10 | IT 98%, ε 2% |
| 97 | 9/2+ | −87.220 | 4.21×10⁶ y 16 | ε |
| 97m | 1/2− | −87.123 | 91.4 d 8 | IT, ε 3.94% |
| 98 | (6)+ | −86.428 | 4.2×10⁶ y 3 | β− |
| 99 | 9/2+ | −87.323 | 2.111×10⁵ y 12 | β− |
| 99m | 1/2− | −87.180 | 6.0058 h 12 | IT, β− 3.7×10⁻³% |
| 100 | 1+ | −86.016 | 15.8 s 1 | β−, ε 1.8×10⁻³% |
| 101 | 9/2+ | −86.34 | 14.22 m 1 | β− |
| 102 | 1+ | −84.566 | 5.28 s 15 | β− |
| 102m | (4,5) | −84.566 | 4.35 m 7 | β− 98%, IT 2% |
| 103 | 5/2+ | −84.60 | 54.2 s 8 | β− |
| 104 | (3+) | −82.49 | 18.3 m 3 | β− |
| 105 | (3/2−) | −82.29 | 7.6 m 1 | β− |
| 106 | (1,2) | −79.78 | 35.6 s 6 | β− |
| 107 | (3/2−) | −79.1 | 21.2 s 2 | β− |
| 108 | (2+) | −76.0 | 5.17 s 7 | β− |
| 109 | (5/2+) | −74.54 | 0.86 s 4 | β−, β−n 0.09% |
| 110 | (2+) | −70.96 | 0.92 s 3 | β− 99.96%, β−n 0.04% |
| 111 | (7/2+,9/2+) | −69.2 | 290 ms 20 | β−, β−n 0.85% |
| 112 | | −66.0 | 0.29 s 2 | β−, β−n 1.5% |
| 113 | | −63.7s | 170 ms 20 | β−, β−n 2.1% |
| 114 | | −59.7s | 150 ms 10 | β− |
| 115 | | −57.1s | 100 ms sy | β−n |
| 116 | | −52.8s | 90 ms sy | β− |
| 117 | | −49.9s | 40 ms sy | β− |
| 118 | | −45.2s | >150 ns | β− |
| 44 Ru 87 | | −47.3s | >1.5 μs | ε? |
| 88 | 0+ | −55.6s | 1.2 s +3 −2 | ε, εp |
| 89 | | −59.5s | 1.5 s 2 | ε, εp<0.15% |
| 90 | 0+ | −65.3s | 11.7 s 9 | ε |
| 91 | (9/2+) | −68.7s | 7.9 s 4 | ε |
| 91m | (1/2−) | −68.7s | 7.6 s 8 | ε>0%, εp>0%, IT |
| 92 | 0+ | −74.4s | 3.65 m 5 | ε |
| 93 | (9/2+) | −77.27 | 59.7 s 6 | ε |
| 93m | (1/2−) | −76.53 | 10.8 s 3 | ε 78%, IT 22%, εp 0.03% |

## Nuclear Wallet Cards

| Nuclide Z El A | Jπ | Δ (MeV) | T½, Γ, or Abundance | Decay Mode |
|---|---|---|---|---|
| 44 Ru 94 | 0+ | −82.57 | 51.8 m 6 | ε |
| 95 | 5/2+ | −83.45 | 1.643 h 14 | ε |
| 96 | 0+ | −86.072 | **5.54% 14** | |
| 97 | 5/2+ | −86.112 | 2.791 d 4 | ε |
| 98 | 0+ | −88.225 | **1.87% 3** | |
| 99 | 5/2+ | −87.617 | **12.76% 14** | |
| 100 | 0+ | −89.219 | **12.60% 7** | |
| 101 | 5/2+ | −87.950 | **17.06% 2** | |
| 102 | 0+ | −89.098 | **31.55% 14** | |
| 103 | 3/2+ | −87.259 | 39.26 d 2 | β− |
| 104 | 0+ | −88.089 | **18.62% 27** | |
| 105 | 3/2+ | −85.928 | 4.44 h 2 | β− |
| 106 | 0+ | −86.322 | 373.59 d 15 | β− |
| 107 | (5/2)+ | −83.9 | 3.75 m 5 | β− |
| 108 | 0+ | −83.7 | 4.55 m 5 | β− |
| 109 | (5/2+) | −80.85 | 34.5 s 10 | β− |
| 110 | 0+ | −79.98 | 11.6 s 6 | β− |
| 111 | (5/2+) | −76.67 | 2.12 s 7 | β− |
| 112 | 0+ | −75.48 | 1.75 s 7 | β− |
| 113 | (5/2+) | −72.20 | 0.80 s 5 | β− |
| 113m | (11/2−) | −72.07 | 510 ms 30 | β− 92%, IT 8% |
| 114 | 0+ | −70.6s | 0.53 s 6 | β− |
| 115 | | −66.4 | 740 ms 80 | β−, β−n |
| 116 | 0+ | −64.4s | 400 ms sy | β−? |
| 117 | | −60.0s | 300 ms sy | β−? |
| 118 | 0+ | −57.9s | >150 ns | β−? |
| 119 | | −53.2s | >150 ns | β− |
| 120 | 0+ | −50.9s | >150 ns | β− |
| 45 Rh 89 | | −47.7s | >1.5 μs | ε |
| 90 | | −53.2s | 12 ms +9−4 | ε? |
| 90m | | −53.2s | 1.0 s +3−2 | ε? |
| 91 | (9/2+) | −59.1s | 1.47 s 22 | ε |
| 91m | (1/2−) | −59.1s | 1.46 s 11 | ε |
| 92 | (2+) | −63.4s | 0.5 s 4 | ε |
| 92 | (≥6+) | −63.4s | 4.66 s 25 | ε |
| 93 | (9/2+) | −69.2s | 11.9 s 7 | ε |
| 94 | (8+) | −72.9s | 25.8 s 2 | ε |
| 94m | (3+) | −72.9s | 70.6 s 6 | ε |
| 95 | (9/2+) | −78.3 | 5.02 m 10 | ε |
| 95m | (1/2)− | −77.8 | 1.96 m 4 | IT 88%, ε 12% |
| 96 | (6+) | −79.68 | 9.90 m 10 | ε |
| 96m | (3+) | −79.63 | 1.51 m 2 | IT 60%, ε 40% |
| 97 | 9/2+ | −82.59 | 30.7 m 6 | ε |
| 97m | 1/2− | −82.33 | 46.2 m 16 | ε 94.4%, IT 5.6% |
| 98 | (2)+ | −83.17 | 8.72 m 12 | ε |
| 98m | (5+) | −83.17 | 3.6 m 2 | IT 89%, ε 11% |
| 99 | 1/2− | −85.574 | 16.1 d 2 | ε |
| 99m | 9/2+ | −85.510 | 4.7 h 1 | ε >99.84%, IT<0.16% |
| 100 | 1− | −85.58 | 20.8 h 1 | ε |
| 100m | (5+) | −85.58 | 4.6 m 2 | IT=98.3%, ε=1.7% |
| 101 | 1/2− | −87.41 | 3.3 y 3 | ε |

| Nuclide Z El A | Jπ | Δ (MeV) | T½, Γ, or Abundance | Decay Mode |
|---|---|---|---|---|
| 45 Rh 101m | 9/2+ | −87.25 | 4.34 d 1 | ε 92.8%, IT 7.2% |
| 102 | (1−,2−) | −86.775 | 207 d 3 | ε 78%, β− 22% |
| 102m | 6(+) | −86.634 | −2.9 y | ε 99.77%, IT 0.23% |
| 103 | 1/2− | −88.022 | **100%** | |
| 103m | 7/2+ | −87.982 | 56.114 m 9 | IT |
| 104 | 1+ | −86.950 | 42.3 s 4 | β− 99.55%, ε 0.45% |
| 104m | 5+ | −86.821 | 4.34 m 3 | IT 99.87%, β− 0.13% |
| 105 | 7/2+ | −87.846 | 35.36 h 6 | β− |
| 105m | 1/2− | −87.716 | 42.9 s 3 | IT |
| 106 | 1+ | −86.362 | 29.80 s 8 | β− |
| 106m | (6)+ | −86.225 | 131 m 2 | β− |
| 107 | 7/2+ | −86.86 | 21.7 m 4 | β− |
| 108 | 1+ | −85.0 | 16.8 s 5 | β− |
| 108m | (5+) | −85.0 | 6.0 m 3 | β− |
| 109 | 7/2+ | −85.01 | 80 s 2 | β− |
| 110 | 1+ | −82.78 | 3.2 s 2 | β− |
| 110m | (24) | −82.78 | 28.5 s 15 | β− |
| 111 | (7/2+) | −82.36 | 11 s 1 | β− |
| 112 | 1+ | −79.74 | 3.45 s 37 | β− |
| 112m (4,5,6) | | −79.74 | 6.73 s 15 | β− |
| 113 | (7/2+) | −78.68 | 2.80 s 12 | β− |
| 114 | 1+ | −75.6 | 1.85 s 5 | β− |
| 114m | (4,5) | −75.6 | 1.85 s 5 | β− |
| 115 | (7/2+) | −74.21 | 0.99 s 5 | β− |
| 116 | 1+ | −70.7 | 0.68 s 6 | β− |
| 116m | (6−) | −70.6 | 0.57 s 5 | β− |
| 117 | (7/2+) | −68.9s | 0.44 s 4 | β− |
| 118 | 0+ | −65.1s | 0.30 s 6 | β− |
| 119 | | −63.2s | >150 ns | β− |
| 120 | | −59.2s | >150 ns | β−? |
| 121 | | −57.1s | >150 ns | β−? |
| 122 | | −52.9s | ≈50 ms | β−? |
| 46 Pd 91 | | −47.4s | >1 μs | ε? |
| 92 | 0+ | −55.5s | 0.7 s +4−2 | ε |
| 93 | (7/2+,9/2+) | −59.7s | 1.3 s 2 | ε, ερ 1.5% |
| 93m | | −59.7s | 9.3 s +25−17 | ε, IT |
| 94 | 0+ | −66.3s | 9.0 s 5 | ε |
| 95 | | −70.2s | 10 s sy | ε |
| 95m | (21/2+) | −68.2s | 13.3 s 3 | ε ≥91.3%, IT ≤9.7%, ερ 0.9% |
| 96 | 0+ | −76.2 | 122 s 2 | ε |
| 97 | 5/2+ | −77.8 | 3.10 m 9 | ε |
| 98 | 0+ | −81.30 | 17.7 m 3 | ε |
| 99 | (5/2)+ | −82.19 | 21.4 m 2 | ε |
| 100 | 0+ | −85.23 | 3.63 d 9 | ε |
| 101 | 5/2+ | −85.43 | 8.47 h 6 | ε |
| 102 | 0+ | −87.925 | **1.02% 1** | |
| 103 | 5/2+ | −87.479 | 16.991 d 19 | ε |
| 104 | 0+ | −89.390 | **11.14% 8** | |
| 105 | 5/2+ | −88.413 | **22.33% 8** | |
| 106 | 0+ | −89.902 | **27.33% 3** | |
| 107 | 5/2+ | −88.368 | 6.5×10⁶ y 3 | β− |

| Nuclide Z El A | Jπ | Δ (MeV) | T½, Γ, or Abundance | Decay Mode |
|---|---|---|---|---|
| 46 Pd 107m | 11/2− | −88.153 | 21.3 s 5 | IT |
| 108 | 0+ | −89.524 | **26.46% 9** | |
| 109 | 5/2+ | −87.607 | 13.7012 h 24 | β− |
| 109m | 11/2− | −87.418 | 4.696 ms 3 | IT |
| 110 | 0+ | −88.35 | **11.72% 9** | |
| 111 | 5/2+ | −86.00 | 23.4 m 2 | β− |
| 111m | 11/2− | −85.83 | 5.5 h 1 | IT 73%, β− 27% |
| 112 | 0+ | −86.34 | 21.03 h 5 | β− |
| 113 | (5/2+) | −83.69 | 93 s 5 | β− |
| 113m | | −83.69 | ≥100 s | |
| 113m | (9/2−) | −83.61 | 0.3 s 1 | IT |
| 114 | 0+ | −83.50 | 2.42 m 6 | β− |
| 115 | (5/2+) | −80.40 | 25 s 2 | β− |
| 115m | (11/2−) | −80.31 | 50 s 3 | β− 92%, IT 8% |
| 116 | 0+ | −79.96 | 11.8 s 4 | β− |
| 117 | (5/2+) | −76.53 | 4.3 s 3 | β− |
| 118 | 0+ | −75.75 | 1.9 s 1 | β− |
| 119 | | −71.6s | 0.92 s 13 | β− |
| 120 | 0+ | −70.1 | 0.5 s 1 | β− |
| 121 | | −66.3s | >150 ns | β− |
| 122 | 0+ | −64.7s | >150 ns | β−, β−n |
| 123 | | −60.0s | >150 ns | β− |
| 124 | 0+ | −58.8s | ≈0.2 s | β−? |
| 47 Ag 93 | | −46.8s | >1.5 μs | ε?, p? |
| 94 | (0+) | −53.3s | 26 ms +26−9 | ε, εp |
| 94m | (21+) | −53.3s | 0.47 s 8 | ε, εp |
| 94m | (7+) | −53.3s | 0.59 s 2 | ε, εp>0% |
| 95 | | −60.1s | 2.0 s 1 | ε, εp |
| 96 | (8+) | −64.6s | 4.40 s 6 | ε, εp 8.15% |
| 96 | (2+) | −64.6s | 6.9 s 6 | ε, εp 19% |
| 97 | 9/2+ | −70.8 | 25.9 s 4 | ε |
| 98 | (6+) | −73.06 | 47.5 s 3 | ε, εp 1.1×10⁻³% |
| 99 | (9/2+) | −76.8 | 124 s 3 | ε |
| 99m | (1/2−) | −76.3 | 10.5 s 5 | IT |
| 100 | (5)+ | −78.15 | 2.01 m 9 | ε |
| 100m | (2+) | −78.13 | 2.24 m 13 | ε, IT |
| 101 | 9/2+ | −81.2 | 11.1 m 3 | ε |
| 101m | (1/2−) | −81.0 | 3.10 s 10 | IT |
| 102 | 5+ | −82.26 | 12.9 m 3 | ε |
| 102m | 2+ | −82.26 | 7.7 m 5 | ε 51%, IT 49% |
| 103 | 7/2+ | −84.79 | 65.7 m 7 | ε |
| 103m | 1/2− | −84.66 | 5.7 s 3 | IT |
| 104 | 5+ | −85.111 | 69.2 m 10 | ε |
| 104m | 2+ | −85.104 | 33.5 m 20 | ε 99.93%, IT<0.07% |
| 105 | 1/2− | −87.07 | 41.29 d 7 | ε |
| 105m | 7/2+ | −87.04 | 7.23 m 16 | IT 99.66% |
| 106 | 1+ | −86.937 | 23.96 m 4 | ε 99.5%, β− <1% |
| 106m | 6+ | −86.847 | 8.28 d 2 | ε |
| 107 | 1/2− | −88.402 | **51.839% 8** | |
| 107m | 7/2+ | −88.309 | 44.5 s 8 | IT |
| 108 | 1+ | −87.602 | 2.37 m 1 | β− 97.15%, ε 2.85% |
| 108m | 6+ | −87.492 | 438 y 9 | ε 91.3%, IT 8.7% |

| Nuclide Z El A | Jπ | Δ (MeV) | T½, Γ, or Abundance | Decay Mode |
|---|---|---|---|---|
| 47 Ag 109 | 1/2− | −88.723 | **48.161% 8** | |
| 109m | 7/2+ | −88.635 | 38.0 s 12 | IT |
| 110 | 1+ | −87.461 | 24.6 s 2 | β− 99.7%, ε 0.3% |
| 110m | 6+ | −87.343 | 249.76 d 4 | β− 98.64%, IT 1.36% |
| 111 | 1/2− | −88.221 | 7.45 d 2 | β− |
| 111m | 7/2+ | −88.161 | 64.8 s 8 | IT 99.3%, β− 0.7% |
| 112 | 2(−) | −86.62 | 3.130 h 9 | β− |
| 113 | 1/2− | −87.03 | 5.37 h 5 | β− |
| 113m | 7/2+ | −86.99 | 68.7 s 16 | IT 64%, β− 36% |
| 114 | 1+ | −84.95 | 4.6 s 1 | β− |
| 115 | 1/2− | −84.99 | 20.0 m 5 | β− |
| 115m | 7/2+ | −84.95 | 18.0 s 7 | β− 79%, IT 21% |
| 116 | (2)− | −82.57 | 2.68 m 10 | β− |
| 116m | (5+) | −82.49 | 8.6 s 3 | β− 94%, IT 6% |
| 117 | (1/2−) | −82.27 | 72.8 s +20−7 | β− 100% |
| 117m | (7/2+) | −82.24 | 5.34 s 5 | β− 94%, IT 6% |
| 118 | 1(−) | −79.57 | 3.76 s 15 | β− |
| 118m | 4(+) | −79.44 | 2.0 s 2 | β− 59%, IT 41% |
| 119 | (7/2+) | −78.56 | 2.1 s 1 | β− |
| 119m | (1/2−) | −78.56 | 6.0 s 5 | β− |
| 120 | 3(+) | −75.65 | 1.23 s 4 | β−, β−n<3.0×10⁻³% |
| 120m | 6(−) | −75.45 | 0.40 s 3 | β−=63%, IT=37% |
| 121 | (7/2+) | −74.7 | 0.79 s 2 | β−, β−n 0.08% |
| 122 | (3+) | −71.2s | 0.529 s 13 | β−, β−n 0.19% |
| 122m | (8−) | −71.2s | 1.5 s 5 | β−, β−n |
| 123 | (7/2+) | −70.0s | 0.300 s 5 | β−, β−n>0.1% |
| 124 | | −68.4s | 0.172 s 5 | β−, β−n |
| 125 | (7/2+) | −64.8s | 166 ms 7 | β−, β−n |
| 126 | | −61.0s | 107 ms 12 | β−, β−n |
| 127 | (1/2−) | −58.9s | 79 ms 3 | β−, β−n |
| 128 | | −54.8s | 58 ms 5 | β−, β−n |
| 129 | (9/2+) | −52.5s | 46 ms +5−9 | β−, β−n |
| 129m | (1/2−) | −52.5s | >160 ms | β−, β−n |
| 130 | | −46.2s | ≈50 ms | β− |
| 48 Cd 95 | | −46.7s | 5 ms sy | ε?, εp? |
| 96 | 0+ | −56.1s | ≈1 s | ε? |
| 97 | | −60.6s | 2.8 s 6 | ε, εp |
| 98 | 0+ | −67.63 | 9.2 s 3 | ε, εp<0.03% |
| 99 | (5/2+) | −69.9s | 16 s 3 | ε, εp 0.17%, εα<1.0×10⁻⁴% |
| 100 | 0+ | −74.25 | 49.1 s 5 | ε |
| 101 | (5/2+) | −75.7 | 1.36 m 5 | ε |
| 102 | 0+ | −79.68 | 5.5 m 5 | ε |
| 103 | 5/2+ | −80.65 | 7.3 m 1 | ε |
| 104 | 0+ | −83.975 | 57.7 m 10 | ε |
| 105 | 5/2+ | −84.33 | 55.5 m 4 | ε |
| 106 | 0+ | −87.132 | ≥2.6×10¹⁷ y | 2ε |
| | | | **1.25% 6** | |
| 107 | 5/2+ | −86.985 | 6.50 h 2 | ε |
| 108 | 0+ | −89.252 | >1.0×10¹⁸ y | 2ε |
| | | | **0.89% 3** | |
| 109 | 5/2+ | −88.508 | 461.4 d 12 | ε |

# APPENDIX B

## Nuclear Wallet Cards

| Nuclide Z El A | Jπ | Δ (MeV) | T½, Γ, or Abundance | Decay Mode |
|---|---|---|---|---|
| 48 Cd 110 | 0+ | −90.353 | 12.49% 18 | |
| 111 | 1/2+ | −89.257 | 12.80% 12 | |
| 111m | 11/2− | −88.861 | 48.50 m 9 | IT |
| 112 | 0+ | −90.580 | 24.13% 21 | |
| 113 | 1/2+ | −89.049 | $7.7\times10^{15}$ y 3 | β− |
| | | | 12.22% 12 | |
| 113m | 11/2− | −88.786 | 14.1 y 5 | β− 99.86%, IT 0.14% |
| 114 | 0+ | −90.021 | $>6.4\times10^{18}$ y | 2β− |
| | | | 28.73% 42 | |
| 115 | 1/2+ | −88.090 | 53.46 h 5 | β− |
| 115m | (11/2−) | −87.910 | 44.56 d 24 | β− |
| 116 | 0+ | −88.719 | $3.1\times10^{19}$ y 4 | 2β− |
| | | | 7.49% 18 | |
| 117 | 1/2+ | −86.425 | 2.49 h 4 | β− |
| 117m | (11/2−) | −86.289 | 3.36 h 5 | β− |
| 118 | 0+ | −86.71 | 50.3 m 2 | β− |
| 119 | 3/2+ | −83.91 | 2.69 m 2 | β− |
| 119m | (11/2−) | −83.76 | 2.20 m 2 | β− |
| 120 | 0+ | −83.97 | 50.80 s 21 | β− |
| 121 | (3/2+) | −81.06 | 13.5 s 3 | β− |
| 121m | (11/2−) | −80.85 | 8.3 s 8 | β− |
| 122 | 0+ | −80.73 | 5.24 s 3 | β− |
| 123 | (3/2+) | −77.31 | 2.10 s 2 | β− |
| 123m | (11/2−) | −76.99 | 1.82 s 3 | β− ≤100%, IT |
| 124 | 0+ | −76.71 | 1.25 s 2 | β− |
| 125 | (3/2+) | −73.36 | 0.65 s 2 | β− |
| 125m | (11/2−) | −73.31 | 0.48 s 3 | β− |
| 126 | 0+ | −72.33 | 0.515 s 17 | β− |
| 127 | (3/2+) | −68.52 | 0.37 s 7 | β− |
| 128 | 0+ | −67.3 | 0.28 s 4 | β− |
| 129 | (3/2+) | −63.2 | 0.27 s 4 | β− |
| 130 | 0+ | −61.6 | 162 ms 7 | β−, β−n=3.5% |
| 131 | | −55.3 s | 68 ms 3 | β−, β−n 3.5% |
| 132 | 0+ | −50.7 s | 97 ms 10 | β−, β−n 60% |
| 49 In 97 | | −47.0 s | 5 ms sy | p?, ε? |
| 98 | | −53.9 s | 32 ms +32−11 | ε |
| 98m | | −53.9 s | 1.2 s +12−4 | ε |
| 99 | (9/2+) | −61.3 s | 3.0 s +8−7 | ε |
| 100 | (6,7)+ | −64.2 | 5.9 s 2 | ε, εp 1.6% |
| 101 | | −68.6 s | 15.1 s 3 | ε≈100%, εp |
| 102 | (6+) | −70.7 | 23.3 s 1 | ε, εp $9.3\times10^{-3}$% |
| 103 | (9/2+) | −74.60 | 65 s 7 | ε |
| 103m | (1/2−) | −73.97 | 34 s 2 | ε 67%, IT 33% |
| 104 | 5,6(+) | −76.11 | 1.80 m 3 | ε |
| 104m | (3+) | −76.01 | 15.7 s 5 | IT 80%, ε 20% |
| 105 | 9/2+ | −79.48 | 5.07 m 7 | ε |
| 105m | (1/2−) | −78.81 | 48 s 6 | IT |
| 106 | 7+ | −80.61 | 6.2 m 1 | ε |
| 106m | (3+) | −80.58 | 5.2 m 1 | ε |
| 107 | 9/2+ | −83.56 | 32.4 m 3 | ε |
| 107m | 1/2− | −82.88 | 50.4 s 6 | IT |
| 108 | 7+ | −84.116 | 58.0 m 12 | ε |

| Nuclide Z El A | Jπ | Δ (MeV) | T½, Γ, or Abundance | Decay Mode |
|---|---|---|---|---|
| 49 In 108m | 2+ | −84.086 | 39.6 m 7 | ε |
| 109 | 9/2+ | −86.489 | 4.2 h 1 | ε |
| 109m | 1/2− | −85.839 | 1.34 m 7 | IT |
| 109m | (19/2+) | −84.387 | 0.209 s 6 | IT |
| 110 | 7+ | −86.47 | 4.9 h 1 | ε |
| 110m | 2+ | −86.41 | 69.1 m 5 | ε |
| 111 | 9/2+ | −88.396 | 2.8047 d 5 | ε |
| 111m | 1/2− | −87.859 | 7.7 m 2 | IT |
| 112 | 1+ | −87.996 | 14.97 m 10 | ε 56%, β− 44% |
| 112m | 4+ | −87.840 | 20.56 m 6 | IT |
| 113 | 9/2+ | −89.370 | 4.29% 5 | |
| 113m | 1/2− | −88.978 | 99.476 m 23 | IT |
| 114 | 1+ | −88.572 | 71.9 s 1 | β− 99.5%, ε 0.5% |
| 114m | 5+ | −88.382 | 49.51 d 1 | IT 96.75%, ε 3.25% |
| 115 | 9/2+ | −89.537 | $4.41\times10^{14}$ y 25 | β− |
| | | | 95.71% 5 | |
| 115m | 1/2− | −89.200 | 4.486 h 4 | IT 95%, β− 5% |
| 116 | 1+ | −88.250 | 14.10 s 3 | β− 99.98%, ε 0.02% |
| 116m | 5+ | −88.123 | 54.29 m 17 | β− |
| 116m | 8− | −87.960 | 2.18 s 4 | IT |
| 117 | 9/2+ | −88.945 | 43.2 m 3 | β− |
| 117m | 1/2− | −88.630 | 116.2 m 3 | β− 52.9%, IT 47.1% |
| 118 | 1+ | −87.230 | 5.0 s 5 | β− |
| 118m | 5+ | −87.170 | 4.45 m 5 | β− |
| 118m | 8− | −87.030 | 8.5 s 3 | IT 98.6%, β− 1.4% |
| 119 | 9/2+ | −87.704 | 2.4 m 1 | β− |
| 119m | 1/2− | −87.393 | 18.0 m 3 | β− 94.4%, IT 5.6% |
| 120 | 1+ | −85.74 | 3.08 s 8 | β− |
| 120m | (8−) | −85.74 | 47.3 s 5 | β− |
| 120m | (5)+ | −85.67 | 46.2 s 8 | β− |
| 121 | 9/2+ | −85.84 | 23.1 s 6 | β− |
| 121m | 1/2− | −85.52 | 3.88 m 10 | β− 98.8%, IT 1.2% |
| 122 | 1+ | −83.58 | 1.5 s 3 | β− |
| 122m | 5+ | −83.54 | 10.3 s 6 | β− |
| 122m | 8− | −83.29 | 10.8 s 4 | β− |
| 123 | (9/2)+ | −83.43 | 6.17 s 5 | β− |
| 123m | (1/2)− | −83.10 | 47.4 s 4 | β− |
| 124 | 3+ | −80.88 | 3.11 s 10 | β− |
| 124m | (8−) | −80.83 | 3.7 s 2 | β− |
| 125 | 9/2+ | −80.48 | 2.36 s 4 | β− |
| 125m | 1/2− | −80.12 | 12.2 s 2 | β− |
| 126 | 3(+) | −77.81 | 1.53 s 1 | β− |
| 126m | (8−) | −77.71 | 1.64 s 5 | β− |
| 127 | (9/2+) | −76.99 | 1.09 s 1 | β−, β−n≤0.03% |
| 127m | (1/2−) | −76.52 | 3.67 s 4 | β−, β−n 0.69% |
| 128 | (3)+ | −74.36 | 0.84 s 6 | β−, β−n<0.05% |
| 128m | (8,+) | −74.02 | 0.72 s 10 | β−, β−n 0.05% |
| 129 | (9/2+) | −72.94 | 0.61 s 1 | β−, β−n 0.25% |
| 129m | (1/2−) | −72.56 | 1.23 s 3 | β−, β−n 2.5%, IT<0.3% |
| 130 | 1(−) | −69.89 | 0.29 s 2 | β−, β−n 0.93% |
| 130m | (10−) | −69.84 | 0.54 s 1 | β−, β−n 1.65% |

| Nuclide Z El A | Jπ | Δ (MeV) | T½, Γ, or Abundance | Decay Mode |
|---|---|---|---|---|
| 49 In 130 | (5+) | −69.49 | 0.54 s 1 | β−, β−n 1.65% |
| 131 | (9/2+) | −68.14 | 0.28 s 3 | β−, β−n≤2% |
| 131m | (1/2−) | −67.77 | 0.35 s 5 | β− ≥99.98%, β−n≤2%, IT≤0.02% |
| 131m | | −63.87 | 0.32 s 6 | β− >99%, IT<1%, β−n 0.03% |
| 132 | (7−) | −62.42 | 0.207 s 6 | β−, β−n 6.3% |
| 133 | (9/2+) | −57.9 s | 165 ms 3 | β−, β−n 85% |
| 134 | (4− to 7−) | −52.0 s | 140 ms 4 | β−, β−n 65% |
| 135 | | −47.2 s | 92 ms 10 | β−, β−n>0% |
| 50 Sn 99 | | −47.2 s | 5 ms sy | ε?, εp? |
| 100 | 0+ | −56.8 | 0.94 s +54−27 | ε, εp<17% |
| 101 | | −59.6 s | 3 s 1 | ε, εp |
| 102 | 0+ | −64.9 | 4.5 s 7 | ε |
| 103 | | −67.0 s | 7.0 s 6 | ε, εp |
| 104 | 0+ | −71.6 | 20.8 s 5 | ε |
| 105 | (5/2+) | −73.26 | 34 s 1 | ε, εp |
| 106 | 0+ | −77.43 | 115 s 5 | ε |
| 107 | (5/2+) | −78.58 | 2.90 m 5 | ε |
| 108 | 0+ | −82.04 | 10.30 m 8 | ε |
| 109 | 5/2(+) | −82.64 | 18.0 m 2 | ε |
| 110 | 0+ | −85.84 | 4.11 h 10 | ε |
| 111 | 7/2+ | −85.945 | 35.3 m 6 | ε |
| 112 | 0+ | −88.661 | 0.97% 1 | |
| 113 | 1/2+ | −88.333 | 115.09 d 3 | ε |
| 113m | 7/2+ | −88.256 | 21.4 m 4 | IT 91.1%, ε 8.9% |
| 114 | 0+ | −90.561 | 0.66% 1 | |
| 115 | 1/2+ | −90.036 | 0.34% 1 | |
| 116 | 0+ | −91.528 | 14.54% 9 | |
| 117 | 1/2+ | −90.400 | 7.68% 7 | |
| 117m | 11/2− | −90.085 | 13.76 d 4 | IT |
| 118 | 0+ | −91.656 | 24.22% 9 | |
| 119 | 1/2+ | −90.068 | 8.59% 4 | |
| 119m | 11/2− | −89.979 | 293.1 d 7 | IT |
| 120 | 0+ | −91.105 | 32.58% 9 | |
| 121 | 3/2+ | −89.204 | 27.03 h 4 | β− |
| 121m | 11/2− | −89.198 | 43.9 y 5 | IT 77.6%, β− 22.4% |
| 122 | 0+ | −89.945 | 4.63% 3 | |
| 123 | 11/2− | −87.821 | 129.2 d 4 | β− |
| 123m | 3/2+ | −87.796 | 40.06 m 1 | β− |
| 124 | 0+ | −88.237 | 5.79% 5 | |
| 125 | 11/2− | −85.898 | 9.64 d 3 | β− |
| 125m | 3/2+ | −85.871 | 9.52 m 5 | β− |
| 126 | 0+ | −86.02 | $2.30\times10^5$ y 14 | β− |
| 127 | (11/2−) | −83.50 | 2.10 h 4 | β− |
| 127m | (3/2+) | −83.49 | 4.13 m 3 | β− |
| 128 | 0+ | −83.33 | 59.07 m 14 | β− |
| 128m | (7−) | −81.24 | 6.5 s 5 | IT |
| 129 | (3/2+) | −80.59 | 2.23 m 4 | β− |
| 129m | (11/2−) | −80.56 | 6.9 m 1 | β−, IT<2.0×10⁻³% |
| 130 | 0+ | −80.14 | 3.72 m 7 | β− |
| 130m | (7−) | −78.19 | 1.7 m 1 | β− |

| Nuclide Z El A | Jπ | Δ (MeV) | T½, Γ, or Abundance | Decay Mode |
|---|---|---|---|---|
| 50 Sn 131 | (3/2+) | −77.31 | 56.0 s 5 | β− |
| 131m | (11/2−) | −77.07 | 58.4 s 5 | β−, IT≤4.0×10⁻⁴% |
| 132 | 0+ | −76.55 | 39.7 s 8 | β− |
| 133 | (7/2−) | −70.95 | 1.45 s 3 | β−, β−n 0.08% |
| 134 | 0+ | −66.80 | 1.050 s 11 | β−, β−n 17% |
| 135 | (7/2+) | −60.8 s | 530 ms 20 | β−, β−n 21% |
| 136 | 0+ | −56.5 s | 0.25 s 3 | β−, β−n 30% |
| 137 | | −50.3 s | 190 ms 60 | β−, β−n 58% |
| 51 Sb 103 | | −56.2 s | >1.5 μs | ε? |
| 104 | | −59.2 s | 0.44 s +15−11 | ε, εp<7%, p<1% |
| 105 | (5/2+) | −63.8 | 1.12 s 16 | ε 99%, p 1% |
| 106 | (4+) | −66.3 s | 0.6 s 2 | ε |
| 107 | (5/2+) | −70.7 s | 4.0 s 2 | ε |
| 108 | (4+) | −72.5 s | 7.4 s 3 | ε |
| 109 | (5/2+) | −76.26 | 17.3 s 5 | ε |
| 110 | (3+,4+) | −77.5 s | 23.0 s 4 | ε |
| 111 | (5/2+) | −80.89 | 75 s 1 | ε |
| 112 | 3+ | −81.60 | 51.4 s 10 | ε |
| 113 | 5/2+ | −84.42 | 6.67 m 7 | ε |
| 114 | 3+ | −84.52 | 3.49 m 3 | ε |
| 115 | 5/2+ | −87.00 | 32.1 m 3 | ε |
| 116 | 3+ | −86.821 | 15.8 m 8 | ε |
| 116m | 8− | −86.438 | 60.3 m 6 | ε |
| 117 | 5/2+ | −88.645 | 2.80 h 1 | ε, ε 1.7% |
| 118 | 1+ | −87.999 | 3.6 m 1 | ε |
| 118m | 8− | −87.749 | 5.00 h 2 | ε |
| 119 | 5/2+ | −89.477 | 38.19 h 22 | ε |
| 119m | (27/2+) | −86.636 | 0.85 s 9 | IT |
| 120 | 1+ | −88.424 | 15.89 m 4 | ε |
| 120m | (5)+ | −88.424 | 5.76 d 2 | ε |
| 121 | 5/2+ | −89.595 | 57.21% 5 | |
| 122 | 2− | −88.330 | 2.7238 d 2 | β− 97.59%, ε 2.41% |
| 122m | (8)− | −88.167 | 4.191 m 3 | IT |
| 123 | 7/2+ | −89.224 | 42.79% 5 | |
| 124 | 3− | −87.620 | 60.11 d 7 | β− |
| 124m | 5+ | −87.609 | 93 s 5 | IT 75%, β− 25% |
| 124m | (8)− | −87.584 | 20.2 m 2 | IT |
| 125 | 7/2+ | −88.256 | 2.7586 y 3 | β− |
| 126 | (8)− | −86.40 | 12.35 d 6 | β− |
| 126m | (5+) | −86.38 | 19.15 m 8 | β− 86%, IT 14% |
| 126m | 3− | −86.36 | ≈11 s | IT |
| 127 | 7/2+ | −86.700 | 3.85 d 5 | β− |
| 128 | 8− | −84.61 | 9.01 h 4 | β− |
| 128m | 5+ | −84.61 | 10.4 m 2 | β− 96.4%, IT 3.6% |
| 129 | 7/2+ | −84.63 | 4.40 h 1 | β− |
| 129m | (19/2−) | −82.78 | 17.7 m 1 | β− 85%, IT 15% |
| 130 | (8−) | −82.39 | 39.5 m 8 | β− |
| 130m | (4,5)+ | −82.29 | 6.3 m 2 | β− |
| 131 | (7/2+) | −81.99 | 23.03 m 4 | β− |
| 132 | (4)+ | −79.67 | 2.79 m 7 | β− |
| 132m | (8−) | −79.67 | 4.10 m 5 | β− |
| 133 | (7/2+) | −78.94 | 2.5 m 1 | β− |

## Nuclear Wallet Cards

| Nuclide Z El A | Jπ | Δ (MeV) | T½, Γ, or Abundance | Decay Mode |
|---|---|---|---|---|
| 51 Sb 134 | (0−) | −74.17 | 0.78 s 6 | β− |
| 134m | (7−) | −74.17 | 10.07 s 5 | β−, β−n 0.09% |
| 135 | (7/2+) | −69.7 | 1.68 s 2 | β−, β−n 22% |
| 136 | 1− | −64.9s | 0.923 s 14 | β−, β−n 16.3% |
| 137 | | −60.3s | >150 ns | β−?, β−n? |
| 138 | | −55.2s | >300 ns | β−?, β−n? |
| 139 | | −50.3s | >150 ns | β−? |
| 52 Te 105 | | −52.5s | 1 μs xy | α?, ε? |
| 106 | 0+ | −58.2 | 70 μs +20−10 | α |
| 107 | | −60.5s | 3.1 ms 1 | α 70%, ε 30% |
| 108 | 0+ | −65.7 | 2.1 s 1 | ε 51%, α 49%, εp 2.4% |
| 109 | (5/2+) | −67.61 | 4.6 s 3 | ε 96.1%, εp 9.4%, α 3.9%, εα<5.0×10⁻³% |
| 110 | 0+ | −72.28 | 18.6 s 8 | ε=100%, α<3.0×10⁻³% |
| 111 | (5/2)+ | −73.48 | 19.3 s 4 | ε, εp |
| 112 | 0+ | −77.3 | 2.0 m 2 | ε |
| 113 | (7/2)+ | −78.35 | 1.7 m 2 | ε |
| 114 | 0+ | −81.89 | 15.2 m 7 | ε |
| 115 | 7/2+ | −82.06 | 5.8 m 2 | ε |
| 115m | (1/2)+ | −82.04 | 6.7 m 4 | ε≤100%, IT |
| 116 | 0+ | −85.27 | 2.49 h 4 | ε |
| 117 | 1/2+ | −85.10 | 62 m 2 | ε, ε 25% |
| 117m | (11/2−) | −84.80 | 103 ms 3 | IT |
| 118 | 0+ | −87.72 | 6.00 d 2 | ε |
| 119 | 1/2+ | −87.184 | 16.05 h 5 | ε, ε 2.0% |
| 119m | 11/2− | −86.923 | 4.70 d 4 | ε, ε 0.41%, IT<8.0×10⁻³% |
| 120 | 0+ | −89.405 | >2.2×10¹⁶ y 0.09% 1 | 2ε |
| 121 | 1/2+ | −88.55 | 19.16 d 5 | ε |
| 121m | 11/2− | −88.26 | 154 d 7 | IT 88.6%, ε 11.4% |
| 122 | 0+ | −90.314 | 2.55% 12 | |
| 123 | 1/2+ | −89.172 | >9.2×10¹⁶ y 0.89% 3 | ε |
| 123m | 11/2− | −88.924 | 119.2 d 1 | IT |
| 124 | 0+ | −90.524 | 4.74% 14 | |
| 125 | 1/2+ | −89.022 | 7.07% 15 | |
| 125m | 11/2− | −88.877 | 57.40 d 15 | IT |
| 126 | 0+ | −90.065 | 18.84% 25 | |
| 127 | 3/2+ | −88.281 | 9.35 h 7 | β− |
| 127m | 11/2− | −88.193 | 109 d 2 | IT 97.6%, β− 2.4% |
| 128 | 0+ | −88.992 | 8.8×10¹⁸ y 4 31.74% 8 | 2β− |
| 129 | 3/2+ | −87.003 | 69.6 m 3 | β− |
| 129m | 11/2− | −86.898 | 33.6 d 1 | IT 63%, β− 37% |
| 130 | 0+ | −87.351 | >5×10²³ y 34.08% 62 | 2β− |
| 131 | 3/2+ | −85.210 | 25.0 m 1 | β− |
| 131m | 11/2− | −85.027 | 30 h 2 | β− 77.8%, IT 22.2% |

## Nuclear Wallet Cards

| Nuclide Z El A | Jπ | Δ (MeV) | T½, Γ, or Abundance | Decay Mode |
|---|---|---|---|---|
| 53 I 139 | (7/2+) | −68.84 | 2.280 s 11 | β−, β−n 10% |
| 140 | (3) | −64.3s | 0.86 s 4 | β−, β−n 9.3% |
| 141 | | −60.5s | 0.43 s 2 | β−, β−n 21.2% |
| 142 | | −55.7s | ≈0.2 s | β− |
| 143 | | −51.6s | >150 ns | β−? |
| 144 | | −46.6s | >300 ns | β−? |
| 54 Xe 110 | 0+ | −51.9 | 105 ms +35−25 | α≈64% |
| 110 | 0+ | −51.9 | ≈0.2 s | α |
| 111 | | −54.4s | 0.74 s 20 | α 0%, ε |
| 112 | 0+ | −60.0 | 2.7 s 8 | ε 99.16%, α 0.84% |
| 113 | (5/2+) | −62.09 | 2.74 s 8 | ε 0%, εp 7%, α=0.01%, εα=7.0×10⁻³% |
| 114 | 0+ | −67.09 | 10.0 s 4 | ε |
| 115 | (5/2+) | −68.66 | 18 s 4 | ε, εp 0.34%, α 3.0×10⁻⁴% |
| 116 | 0+ | −73.05 | 59 s 2 | ε |
| 117 | 5/2+ | −74.19 | 61 s 2 | ε, εp 2.9×10⁻³% |
| 118 | 0+ | −78.08 | 3.8 m 9 | ε |
| 119 | (5/2+) | −78.79 | 5.8 m 3 | ε |
| 120 | 0+ | −82.17 | 40 m 1 | ε |
| 121 | (5/2+) | −82.47 | 40.1 m 20 | ε |
| 122 | 0+ | −85.36 | 20.1 h 1 | ε |
| 123 | (1/2+) | −85.249 | 2.08 h 2 | ε |
| 124 | 0+ | −87.660 | ≥1.1×10¹⁷ y 0.095% 3 | 2ε |
| 125 | 1/2(+) | −87.192 | 16.9 h 2 | ε |
| 125m | 9/2(−) | −86.939 | 56.9 s 9 | IT |
| 126 | 0+ | −89.169 | 0.089% 1 | |
| 127 | 1/2+ | −88.321 | 36.4 d 1 | ε |
| 127m | 9/2− | −88.024 | 69.2 s 9 | IT |
| 128 | 0+ | −89.860 | 1.910% 22 | |
| 129 | 1/2+ | −88.697 | 26.40% 18 | |
| 129m | 11/2− | −88.461 | 8.88 d 2 | IT |
| 130 | 0+ | −89.882 | 4.071% 53 | |
| 131 | 3/2+ | −88.415 | 21.232% 62 | |
| 131m | 11/2− | −88.251 | 11.934 d 21 | IT |
| 132 | 0+ | −89.281 | 26.909% 68 | |
| 133 | 3/2+ | −87.644 | 5.243 d 1 | β− |
| 133m | 11/2− | −87.410 | 2.19 d 1 | IT |
| 134 | 0+ | −88.124 | >5.8×10²² y 10.436% 29 | 2β− ≥0% |
| 134m | 7− | −86.159 | 290 ms 17 | IT |
| 135 | 3/2+ | −86.417 | 9.14 h 2 | β− |
| 135m | 11/2− | −85.890 | 15.29 m 5 | IT 99.4%, β− <0.6% |
| 136 | 0+ | −86.425 | >2.4×10²¹ y 8.857% 33 | 2β− |
| 137 | 7/2− | −82.379 | 3.818 m 13 | β− |
| 138 | 0+ | −80.15 | 14.08 m 8 | β− |
| 139 | 3/2− | −75.64 | 39.68 s 14 | β− |
| 140 | 0+ | −72.99 | 13.60 s 10 | β− |
| 141 | 5/2(−) | −68.33 | 1.73 s 1 | β−, β−n 0.04% |

## Nuclear Wallet Cards

| Nuclide Z El A | Jπ | Δ (MeV) | T½, Γ, or Abundance | Decay Mode |
|---|---|---|---|---|
| 52 Te 132 | 0+ | −85.182 | 3.204 d 13 | β− |
| 133 | (3/2+) | −82.94 | 12.5 m 3 | β− |
| 133m | (11/2−) | −82.61 | 55.4 m 4 | β− 82.5%, IT 17.5% |
| 134 | 0+ | −82.56 | 41.8 m 8 | β− |
| 135 | (7/2−) | −77.83 | 19.0 s 2 | β− |
| 136 | 0+ | −74.43 | 17.63 s 8 | β−, β−n 1.31% |
| 137 | (7/2−) | −69.6 | 2.49 s 5 | β−, β−n 2.69% |
| 138 | 0+ | −65.9s | 1.4 s 4 | β−, β−n 6.3% |
| 139 | (7/2−) | −60.8s | >150 ns | β−, β−n |
| 140 | 0+ | −57.0s | >150 ns | β−?, β−n? |
| 141 | | −51.6s | >150 ns | β−?, β−n? |
| 142 | 0+ | −47.4s | >150 ns | β−? |
| 53 I 108 | (1) | −52.7s | 36 ms 6 | α 91%, ε 9%, p<1% |
| 109 | 1/2+ | −57.6 | 103 μs 5 | p |
| 110 | | −60.3s | 0.65 s 2 | ε 83%, α 17%, εp 11%, εα 1.1% |
| 111 | (5/2+) | −64.9s | 2.5 s 2 | ε 99.9%, α≈0.1% |
| 112 | | −67.1s | 3.42 s 11 | ε, α=1.2×10⁻³% |
| 113 | 5/2+ | −71.13 | 6.6 s 2 | ε, α 3.3×10⁻⁴% |
| 114 | 1+ | −72.8s | 2.1 s 2 | ε, εp |
| 114m | (7) | −72.5s | 6.2 s 5 | ε 91%, IT 9% |
| 115 | (5/2+) | −76.34 | 1.3 m 2 | ε |
| 116 | 1+ | −77.49 | 2.91 s 15 | ε |
| 117 | (5/2)+ | −80.43 | 2.22 m 4 | ε |
| 118 | 2− | −80.97 | 13.7 m 5 | ε |
| 118m | (7−) | −80.87 | 8.5 m 5 | ε<100%, IT>0% |
| 119 | 5/2+ | −83.77 | 19.1 m 4 | ε |
| 120 | 2− | −83.79 | 81.6 m 2 | ε |
| 120m | (7−) | −83.47 | 53 m 4 | ε |
| 121 | 5/2+ | −86.29 | 2.12 h 1 | ε |
| 122 | 1+ | −86.080 | 3.63 m 6 | ε |
| 123 | 5/2+ | −87.943 | 13.232 h 6 | ε |
| 124 | 2− | −87.365 | 4.1760 d 3 | ε |
| 125 | 5/2+ | −88.836 | 59.400 d 10 | ε |
| 126 | 2− | −87.910 | 12.93 d 5 | ε 52.7%, β− 47.3% |
| 127 | 5/2+ | −88.983 | 100% | |
| 128 | 1+ | −87.738 | 24.99 m 2 | β− 93.1%, ε 6.9% |
| 129 | 7/2+ | −88.503 | 1.57×10⁷ y 4 | β− |
| 130 | 5+ | −86.932 | 12.36 h 1 | β− |
| 130m | 2+ | −86.892 | 8.84 m 6 | IT 84%, β− 16% |
| 131 | 7/2+ | −87.444 | 8.02070 d 11 | β− |
| 132 | 4+ | −85.700 | 2.295 h 13 | β− |
| 132m | (8−) | −85.580 | 1.387 h 15 | IT 86%, β− 14% |
| 133 | 7/2+ | −85.887 | 20.8 h 1 | β− |
| 133m | (19/2−) | −84.252 | 9 s 2 | IT |
| 134 | (4)+ | −84.073 | 52.5 m 2 | β− |
| 134m | (8−) | −83.756 | 3.52 m 4 | IT 97.7%, β− 2.3% |
| 135 | 7/2+ | −83.790 | 6.57 h 2 | β− |
| 136 | (1−) | −79.50 | 83.4 s 10 | β− |
| 136m | (6−) | −78.86 | 46.9 s 10 | β− |
| 137 | (7/2+) | −76.50 | 24.5 s 2 | β−, β−n 6.97% |
| 138 | (2−) | −72.33 | 6.23 s 3 | β−, β−n 5.56% |

## Nuclear Wallet Cards

| Nuclide Z El A | Jπ | Δ (MeV) | T½, Γ, or Abundance | Decay Mode |
|---|---|---|---|---|
| 54 Xe 142 | 0+ | −65.5 | 1.250 s 25 | β−, β−n 0.21% |
| 143 | 5/2− | −60.4s | 0.511 s 6 | β−, β−n 1% |
| 144 | 0+ | −57.3s | 0.388 s 7 | β−, β−n 3% |
| 145 | (3/2−) | −52.1s | 188 ms 4 | β− |
| 145 | | −52.1s | 188 ms 4 | β− |
| 146 | 0+ | −48.7s | 146 ms 6 | β−, β−n 6.9% |
| 147 | | −43.3s | 0.10 s +10−5 | β−n <% |
| 55 Cs 112 | (0+,3+) | −46.3s | 0.5 ms 1 | p |
| 113 | (3/2+) | −51.7 | 16.7 μs 7 | p, α |
| 114 | (1+) | −54.5s | 0.57 s 2 | ε=100%, εp 8.7%, εα 0.19%, α 0.02% |
| 115 | | −59.7s | 1.4 s 8 | ε, εp 0.07% |
| 116 | 1+ | −62.1s | 0.70 s 4 | ε, εp>0%, εα>0% |
| 116 | 4,5,6 | −62.0s | 3.85 s 13 | ε, εp>0%, εα>0% |
| 117 | (9/2+) | −66.44 | 8.4 s 6 | ε |
| 117m | (3/2+) | −66.29 | 6.5 s 4 | ε |
| 118 | 2 | −68.41 | 14 s 2 | ε, εp<0.04%, εα<2.4×10⁻³% |
| 118m | 6,7,8 | −68.41 | 17 s 3 | ε, εp<0.04%, εα<2.4×10⁻³% |
| 119 | 9/2+ | −72.31 | 43.0 s 2 | ε |
| 119m | 3/2(+) | −72.31 | 30.4 s 1 | ε |
| 120 | 2(+) | −73.89 | 61.3 s 11 | ε, εα 2.0×10⁻⁵%, εp 7.0×10⁻⁶% |
| 120m | (7−) | −73.89 | 57 s 6 | ε |
| 121 | 3/2(+) | −77.10 | 155 s 4 | ε |
| 121m | 9/2(+) | −77.03 | 122 s 3 | ε 83%, IT 17% |
| 122 | 1+ | −78.14 | 21.18 s 19 | ε |
| 122m | 8− | −78.01 | 3.70 m 11 | ε |
| 122m | (5)− | −78.01 | 0.36 s 2 | IT |
| 123 | 1/2+ | −81.04 | 5.88 m 3 | ε |
| 123m | (11/2−) | −80.87 | 1.64 s 12 | IT |
| 124 | 1+ | −81.731 | 30.8 s 5 | ε |
| 124m | (7+) | −81.269 | 6.3 s 2 | IT |
| 125 | 1/2(+) | −84.088 | 46.7 m 1 | ε |
| 126 | 1+ | −84.34 | 1.64 m 2 | ε |
| 127 | 1/2+ | −86.240 | 6.25 h 10 | ε |
| 128 | 1+ | −85.931 | 3.66 m 2 | ε |
| 129 | 1/2+ | −87.500 | 32.06 h 6 | ε |
| 130 | 1+ | −86.900 | 29.21 m 4 | ε 98.4%, β− 1.6% |
| 130m | 5− | −86.737 | 3.46 m 6 | IT 99.84%, ε 0.16% |
| 131 | 5/2+ | −88.060 | 9.689 d 16 | ε |
| 132 | 2+ | −87.156 | 6.480 d 6 | ε 98.13%, β− 1.87% |
| 133 | 7/2+ | −88.071 | 100% | |
| 134 | 4+ | −86.891 | 2.0652 y 4 | β−, ε 3.0×10⁻⁴% |
| 134m | 8− | −86.753 | 2.912 h 2 | IT |
| 135 | 7/2+ | −87.582 | 2.3×10⁶ y 3 | β− |
| 135m | 19/2− | −85.949 | 53 m 2 | IT |
| 136 | 5+ | −86.339 | 13.04 d 3 | β− |
| 136m | 8− | −86.339 | 19 s 2 | IT>0%, β− |
| 137 | 7/2+ | −86.546 | 30.03 y 5 | β− |
| 138 | 3− | −82.887 | 33.41 m 18 | β− |

## Nuclear Wallet Cards

| Nuclide Z El A | | Jπ | Δ (MeV) | T½, Γ, or Abundance | Decay Mode |
|---|---|---|---|---|---|
| 55 Cs | 138m | 6− | −82.808 | 2.91 m 8 | IT 81%, β− 19% |
| | 139 | 7/2+ | −80.701 | 9.27 m 5 | β− |
| | 140 | 1− | −77.051 | 63.7 s 3 | β− |
| | 141 | 7/2+ | −74.48 | 24.84 s 16 | β−, β−n 0.04% |
| | 142 | 0− | −70.52 | 1.684 s 14 | β−, β−n 0.09% |
| | 143 | 3/2+ | −67.67 | 1.791 s 7 | β−, β−n 1.64% |
| | 144 | 1 | −63.27 | 0.994 s 4 | β−, β−n 3.2% |
| | 144m | (≥4) | −63.27 | <1 s | β− |
| | 145 | 3/2+ | −60.06 | 0.594 s 13 | β−, β−n 14.3% |
| | 146 | 1− | −55.62 | 0.321 s 2 | β−, β−n 14.2% |
| | 147 | (3/2+) | −52.02 | 0.235 s 3 | β−, β−n 43% |
| | 148 | | −47.3 | 146 ms 6 | β−, β−n 25.1% |
| | 149 | | −43.8s | >50 ms | β− |
| | 150 | | −39.0s | >50 ms | β−, β−n |
| | 151 | | −35.2s | >50 ms | β−?, β−n? |
| 56 Ba | 114 | 0+ | −45.9 | 0.43 s +30−15 | ε=100%, εp 20%, α 9.0×10⁻⁵%, 12−3.0×10⁻⁵% |
| | 115 | (5/2+) | −49.0s | 0.45 s 5 | ε, εp>19% |
| | 116 | 0+ | −54.6s | 1.3 s 2 | ε, εp 3% |
| | 117 | (3/2) | −57.3s | 1.75 s 7 | ε, εp>0%, εα>0% |
| | 118 | 0+ | −62.4s | 5.2 s 2 | ε |
| | 119 | (5/2+) | −64.6 | 5.4 s 3 | ε, εp<25% |
| | 120 | 0+ | −68.9 | 24 s 2 | ε |
| | 121 | 5/2(+) | −70.7 | 29.7 s 15 | ε |
| | 122 | 0+ | −74.61 | 1.95 ms 15 | ε |
| | 123 | 5/2(+) | −75.65 | 2.7 m 4 | ε |
| | 124 | 0+ | −79.09 | 11.0 m 5 | ε |
| | 125 | 1/2(+) | −79.67 | 3.5 m 4 | ε |
| | 126 | 0+ | −82.67 | 100 m 2 | ε |
| | 127 | 1/2+ | −82.82 | 12.7 m 4 | ε |
| | 127m | 7/2− | −82.74 | 1.9 s 2 | IT |
| | 128 | 0+ | −85.40 | 2.43 d 5 | ε |
| | 129 | 1/2+ | −85.06 | 2.23 h 11 | ε |
| | 129m | 7/2+ | −85.06 | 2.16 h 2 | ε≤100%, IT |
| | 130 | 0+ | −87.262 | ≥3.5×10¹⁴ y | 2ε |
| | | | | **0.106% 1** | |
| | 131 | 1/2+ | −86.684 | 11.50 d 6 | ε |
| | 131m | 9/2− | −86.497 | 14.6 m 2 | IT |
| | 132 | 0+ | −88.435 | >3.0×10²¹ y | 2ε |
| | | | | **0.101% 1** | |
| | 133 | 1/2+ | −87.553 | 3841 d 7 | ε |
| | 133m | 11/2− | −87.265 | 38.9 h 1 | IT 99.99%, ε 9.6×10⁻³% |
| | 134 | 0+ | −88.950 | **2.417% 18** | |
| | 135 | 3/2+ | −87.851 | **6.592% 12** | |
| | 135m | 11/2− | −87.582 | 28.7 h 2 | IT |
| | 136 | 0+ | −88.887 | **7.854% 24** | |
| | 136m | 7− | −86.856 | 0.3084 s 19 | IT |
| | 137 | 3/2+ | −87.721 | **11.232% 24** | |
| | 137m | 11/2− | −87.060 | 2.552 m 1 | IT |
| | 138 | 0+ | −88.262 | **71.698% 42** | |

| Nuclide Z El A | | Jπ | Δ (MeV) | T½, Γ, or Abundance | Decay Mode |
|---|---|---|---|---|---|
| 57 La | 144 | (3−) | −74.89 | 40.8 s 4 | β− |
| | 145 | (5/2+) | −72.99 | 24.8 s 20 | β− |
| | 146 | 2− | −69.12 | 6.27 s 10 | β− |
| | 146m | (6−) | −69.12 | 10.0 s 1 | β− |
| | 147 | (5/2+) | −66.85 | 4.015 s 8 | β−, β−n 0.04% |
| | 148 | (2−) | −63.13 | 1.26 s 8 | β−, β−n 0.15% |
| | 149 | (3/2,5/2) | −60.8s | 1.05 s 3 | β−, β−n 1.43% |
| | 150 | (3−) | −57.0s | 0.51 s 3 | β−, β−n 2.7% |
| | 151 | | −54.3s | >150 ns | β−? |
| | 152 | | −50.1s | >150 ns | β−? |
| | 153 | | −46.9s | >100 ns | ε |
| | 154 | | −42.4s | ≈0.1 s | β−? |
| | 155 | | −38.8s | ≈0.6 s | β−? |
| 58 Ce | 119 | | −44.0s | ≈0.2 s | ε? |
| | 120 | 0+ | −49.7s | ≈0.25 s | ε? |
| | 121 | | −52.7s | 1.1 s 1 | ε, εp≈1% |
| | 122 | 0+ | −57.8s | ≈2 s | ε?, εp? |
| | 123 | (5/2) | −60.2s | 3.8 s 2 | ε, εp>0% |
| | 124 | 0+ | −64.8s | 6 s 2 | ε |
| | 125 | (5/2+) | −66.7s | 10.2 s 4 | ε, εp |
| | 126 | 0+ | −70.82 | 51.0 s 3 | ε |
| | 127 | (5/2+) | −71.98 | 31 s 2 | ε |
| | 128 | 0+ | −75.53 | 3.93 m 2 | ε |
| | 129 | 5/2+ | −76.29 | 3.5 m 5 | ε>0% |
| | 130 | 0+ | −79.42 | 22.9 m 5 | ε |
| | 131 | (7/2+) | −79.72 | 10.2 m 3 | ε |
| | 131m | (1/2+) | −79.72 | 5.0 m 10 | ε |
| | 132 | 0+ | −82.47 | 3.51 h 11 | ε |
| | 133 | 1/2+ | −82.42 | 97 m 4 | ε |
| | 133m | 9/2− | −82.39 | 4.9 h 4 | ε |
| | 134 | 0+ | −84.84 | 3.16 d 4 | ε |
| | 135 | 1/2(+) | −84.62 | 17.7 h 3 | ε |
| | 135m | (11/2−) | −84.18 | 20 s 1 | IT |
| | 136 | 0+ | −86.47 | >0.7×10¹⁴ y | 2ε |
| | | | | **0.185% 2** | |
| | 137 | 3/2+ | −85.88 | 9.0 h 3 | ε |
| | 137m | 11/2− | −85.62 | 34.4 h 3 | IT 99.22%, ε 0.78% |
| | 138 | 0+ | −87.57 | ≥0.9×10¹⁴ y | 2ε |
| | | | | **0.251% 2** | |
| | 139 | 3/2+ | −86.952 | 137.641 d 20 | ε |
| | 139m | 11/2− | −86.198 | 54.8 s 10 | IT |
| | 140 | 0+ | −88.083 | **88.450% 18** | |
| | 141 | 7/2− | −85.440 | 32.508 d 13 | β− |
| | 142 | 0+ | −84.538 | >2.6×10¹⁷ y | 2β− |
| | | | | **11.114% 17** | |
| | 143 | 3/2− | −81.612 | 33.039 h 6 | β− |
| | 144 | 0+ | −80.437 | 284.91 d 5 | β− |
| | 145 | (3/2−) | −77.10 | 3.01 m 6 | β− |
| | 146 | 0+ | −75.68 | 13.52 m 13 | β− |
| | 147 | (5/2−) | −72.03 | 56.4 s 10 | β− |
| | 148 | 0+ | −70.39 | 56 s 1 | β− |
| | 149 | (3/2−) | −66.70 | 5.3 s 2 | β− |

## Nuclear Wallet Cards

| Nuclide Z El A | | Jπ | Δ (MeV) | T½, Γ, or Abundance | Decay Mode |
|---|---|---|---|---|---|
| 56 Ba | 139 | 7/2− | −84.914 | 83.06 m 28 | β− |
| | 140 | 0+ | −83.271 | 12.752 d 3 | β− |
| | 141 | 3/2− | −79.726 | 18.27 m 7 | β− |
| | 142 | 0+ | −77.823 | 10.6 m 2 | β−, β−n 0.09% |
| | 143 | 5/2− | −73.94 | 14.5 s 3 | β− |
| | 144 | 0+ | −71.77 | 11.5 s 2 | β−, β−n 3.6% |
| | 145 | 5/2− | −67.42 | 4.31 s 16 | β− |
| | 146 | 0+ | −65.00 | 2.22 s 7 | β− |
| | 147 | (3/2+) | −60.6s | 0.893 s 1 | β−, β−n 0.06% |
| | 148 | 0+ | −58.01 | 0.612 s 17 | β−, β−n 0.4% |
| | 149 | | −53.5s | 0.344 s 7 | β−, β−n 0.43% |
| | 150 | 0+ | −50.6s | 0.3 s | β− |
| | 151 | | −45.8s | >150 ms | β−? |
| | 152 | 0+ | −42.6s | ≈0.1 s | β−? |
| | 153 | | −37.6s | ≈0.08 s | β−? |
| 57 La | 117 | (3/2+,3/2−) | −46.5s | 23.5 ms 26 | p 93.9%, ε 6.1% |
| | 117m | (9/2+) | −46.5s | 10 ms 5 | p 97.4%, ε 2.6% |
| | 118 | | −49.6s | ≈1 s | ε? |
| | 119 | | −55.0s | ≈2 s | ε? |
| | 120m | | −57.7s | 2.8 s 2 | ε, εp>0% |
| | 121 | | −62.4s | 5.3 s 2 | ε |
| | 122 | | −64.5s | 8.6 s 5 | ε, εp |
| | 123 | | −68.7s | 17 s 3 | ε |
| | 124m low | | −70.26 | 29 s 1 | ε |
| | 124 | (7,8−) | −70.26 | <1 s | ε |
| | 125 | | −73.76 | 64.8 s 12 | ε |
| | 125m | | −73.65 | 0.4 s 2 | IT |
| | 126m 0−,1,2− | | −74.97 | <50 s 1 | ε, IT |
| | 126m | (5+) | −74.97 | 54 s 2 | ε>0% |
| | 127 | (11/2−) | −77.90 | 5.1 m 1 | ε |
| | 127m | (5/2+) | −77.88 | 3.7 m 4 | ε, IT |
| | 128 | (5+) | −78.63 | 5.18 m 14 | ε |
| | 128m | (1+,2−) | −78.63 | <1.4 m | ε |
| | 129 | 3/2+ | −81.33 | 11.6 m 2 | ε |
| | 129m | 11/2− | −81.15 | 0.56 s 5 | IT |
| | 130 | 3(+) | −81.63 | 8.7 m 1 | ε |
| | 131 | 3/2+ | −83.77 | 59 m 2 | ε |
| | 132 | 2− | −83.74 | 4.8 h 2 | ε |
| | 132m | 6− | −83.55 | 24.3 m 5 | IT 76%, ε 24% |
| | 133 | 5/2+ | −85.49 | 3.912 h 8 | ε |
| | 134 | 1+ | −85.22 | 6.45 m 16 | ε |
| | 135 | 5/2+ | −86.65 | 19.5 h 2 | ε |
| | 136 | 1+ | −86.04 | 9.87 m 3 | ε |
| | 136m | (8+) | −85.81 | 114 ms 3 | IT |
| | 137 | 7/2+ | −87.10 | 6×10⁴ y 2 | ε |
| | 138 | 5+ | −86.525 | 1.02×10¹¹ y 1 | ε 65.6%, β− 34.4% |
| | | | | **0.090% 1** | |
| | 139 | 7/2+ | −87.231 | **99.910% 1** | |
| | 140 | 3− | −84.321 | 1.6781 d 3 | β− |
| | 141 | (7/2+) | −82.938 | 3.92 h 3 | β− |
| | 142 | 2− | −80.035 | 91.1 m 5 | β− |
| | 143 | (7/2)+ | −78.19 | 14.2 m 1 | β− |

| Nuclide Z El A | | Jπ | Δ (MeV) | T½, Γ, or Abundance | Decay Mode |
|---|---|---|---|---|---|
| 58 Ce | 150 | 0+ | −64.82 | 4.0 s 6 | β− |
| | 151 | | −61.8s | 1.02 s 6 | β− |
| | 152 | 0+ | −59.1s | 1.4 s 2 | β− |
| | 153 | | −55.3s | >100 ns | β−? |
| | 154 | 0+ | −52.7s | >100 ns | β−? |
| | 155 | | −48.4s | >300 ns | β−? |
| | 156 | 0+ | −45.4s | ≈0.15 s | β−? |
| | 157 | | −40.7s | ≈0.05 s | β−? |
| 59 Pr | 121 | (3/2−) | −41.6s | 1.4 s 8 | p |
| | 122 | | −44.9s | ≈0.5 s | ε? |
| | 123 | | −50.3s | ≈0.8 s | ε? |
| | 124 | | −53.1s | 1.2 s 2 | ε, εp |
| | 125 | | −57.9s | 3.3 s 7 | ε, εp |
| | 126 | ≥4 | −60.3s | 3.14 s 22 | ε |
| | 126 | ≥4 | −60.3s | 3.14 s 22 | εp |
| | 127 | | −64.4s | 4.2 s 3 | ε |
| | 128 | 4,5,6 | −66.3s | 2.84 s 9 | ε |
| | 129 | (11/2−) | −69.77 | 32 s 3 | ε>0% |
| | 130+ | (4,5) | −71.18 | 40 s 4 | ε |
| | 131 | (3/2+) | −74.13 | 94 s 4 | ε |
| | 131m | (11/2−) | −74.13 | 5.7 s 2 | IT 96.4%, ε 3.6% |
| | 132 | (2+) | −75.21 | 1.6 m 3 | ε |
| | 133 | (3/2+) | −77.94 | 6.5 m 3 | ε |
| | 134m | (6−) | −78.51 | 11 m | ε |
| | 134 | 2− | −78.51 | 17 m 2 | ε |
| | 135 | 3/2(+) | −80.94 | 24 m 2 | ε |
| | 136 | 2+ | −81.33 | 13.1 m 1 | ε |
| | 137 | 5/2+ | −83.18 | 1.28 h 3 | ε |
| | 138 | 1+ | −82.77 | 1.45 m 5 | ε |
| | 138m | 5/2+ | −84.823 | 4.41 h 4 | ε |
| | 139 | 5/2+ | −84.695 | 3.39 m 1 | ε |
| | 140 | 1+ | −86.021 | 3.39 m 1 | ε |
| | 141 | 5/2+ | −86.021 | **100%** | |
| | 142 | 2− | −83.793 | 19.12 h 4 | β− 99.98%, ε 0.02% |
| | 142m | 5− | −83.789 | 14.6 m 5 | IT |
| | 143 | 7/2+ | −83.074 | 13.57 d 2 | β− |
| | 144 | 0− | −80.756 | 17.28 m 5 | β− |
| | 144m | 3− | −80.697 | 7.2 m 3 | IT 99.93%, β− 0.07% |
| | 145 | 7/2+ | −79.632 | 5.984 h 10 | β− |
| | 146 | 2− | −76.71 | 24.15 m 18 | β− |
| | 147 | (3/2+) | −75.45 | 13.4 m 4 | β− |
| | 148 | 1− | −72.53 | 2.29 m 2 | β− |
| | 148m | (4) | −72.44 | 2.01 m 7 | β− |
| | 149 | (5/2+) | −71.06 | 2.26 m 7 | β− |
| | 150 | (1)− | −68.30 | 6.19 s 16 | β− |
| | 151 | (3/2−) | −66.77 | 18.90 s 7 | β− |
| | 152 | 4+ | −63.8 | 3.63 s 12 | β− |
| | 153 | | −61.6 | 4.28 s 11 | β− |
| | 154 | (3+,2+) | −58.2 | 2.3 s 1 | β− |
| | 155 | | −55.8s | >300 ns | β−? |
| | 156 | | −51.9s | >300 ns | β−? |
| | 157 | | −49.0s | ≈0.3 s | β−? |

# APPENDIX B

## Nuclear Wallet Cards

| Nuclide Z El A | Jπ | Δ (MeV) | T½, Γ, or Abundance | Decay Mode |
|---|---|---|---|---|
| 59 Pr 158 | | −44.7s | ~0.2 s | β−? |
| 159 | | −41.5s | ~0.1 s | β−? |
| 60 Nd 124 | 0+ | −44.5s | 0.5 s sy | ε |
| 125 | (5/2) | −47.6s | 0.60 s 15 | ε, εp>0% |
| 126 | 0+ | −52.9s | >200 ns | ε, εp |
| 127 | | −55.4s | 1.8 s 4 | ε, εp |
| 128 | 0+ | −60.2s | 5 s | ε, εp |
| 129 | (5/2+) | −62.2s | 7 s 1 | ε, εp |
| 130 | 0+ | −66.60 | 21 s 3 | ε |
| 131 | (5/2) | −67.77 | 33 s 3 | ε, εp |
| 132 | 0+ | −71.43 | 94 s 8 | ε |
| 133 | (7/2+) | −72.33 | 70 s 10 | ε |
| 133m | (1/2)+ | −72.20 | ~70 s | ε, IT |
| 134 | 0+ | −75.65 | 8.5 m 15 | ε |
| 135 | 9/2(−) | −76.21 | 12.4 m 6 | ε |
| 135m | (1/2)+ | −76.15 | 5.5 m 5 | ε>99.97%, IT<0.03% |
| 136 | 0+ | −79.20 | 50.65 m 33 | ε |
| 137 | 1/2+ | −79.58 | 38.5 m 15 | ε |
| 137m | 11/2− | −79.06 | 1.60 s 15 | IT |
| 138 | 0+ | −82.02 | 5.04 h 9 | ε |
| 139 | 3/2+ | −81.99 | 29.7 m 5 | ε |
| 139m | 11/2− | −81.76 | 5.50 h 20 | ε 88.2%, IT 11.8% |
| 140 | 0+ | −84.25 | 3.37 d 2 | ε |
| 141 | 3/2+ | −84.198 | 2.49 h 3 | ε |
| 141m | 11/2− | −83.441 | 62.0 s 8 | IT, ε<0.05% |
| 142 | 0+ | −85.955 | 27.2% 5 | |
| 143 | 7/2− | −84.007 | 12.2% 2 | |
| 144 | 0+ | −83.753 | 2.29×10¹⁵ y 16 | α |
| | | | 23.8% 3 | |
| 145 | 7/2− | −81.437 | 8.3% 1 | |
| 146 | 0+ | −80.931 | 17.2% 3 | |
| 147 | 5/2− | −78.152 | 10.98 d 1 | β− |
| 148 | 0+ | −77.413 | 5.7% 1 | |
| 149 | 5/2− | −74.381 | 1.728 h 1 | β− |
| 150 | 0+ | −73.690 | 0.79×10¹⁹ y 2 | 2β− |
| | | | 5.6% 2 | |
| 151 | 3/2+ | −70.953 | 12.44 m 7 | β− |
| 152 | 0+ | −70.16 | 11.4 m 2 | β− |
| 153 | (3/2−) | −67.35 | 31.6 s 10 | β− |
| 154 | 0+ | −65.7 | 25.9 s 2 | β− |
| 155 | | −62.5s | 8.9 s 2 | β− |
| 156 | 0+ | −60.5 | 5.49 s 7 | β− |
| 157 | | −56.8s | >100 ns | β−? |
| 158 | 0+ | −54.4s | >50 ms | β−? |
| 159 | | −50.2s | ~0.7 s | β−? |
| 160 | 0+ | −47.4s | ~0.3 s | β−? |
| 161 | | −43.0s | ~0.2 s | β−? |
| 61 Pm 126 | | −39.6s | 0.5 s sy | ε? |
| 127 | | −45.1s | 1 s sy | ε?, p? |
| 128 | | −48.0s | 1.0 s 3 | ε, α, εp |
| 129 | (5/2−) | −52.9s | 2.4 s 9 | ε |

41

## Nuclear Wallet Cards

| Nuclide Z El A | Jπ | Δ (MeV) | T½, Γ, or Abundance | Decay Mode |
|---|---|---|---|---|
| 61 Pm 130 | (4,5,6) | −55.5s | 2.6 s 2 | ε, εp |
| 131 | (11/2) | −59.7s | 6.3 s 8 | ε |
| 132 | (3+) | −61.7s | 6.2 s 6 | ε, εp=5.0×10⁻⁵ |
| 133 | (11/2−) | −65.41 | 15 s 3 | ε |
| 134 | (2+) | −66.74 | ~5 s | ε |
| 134m | (5+) | −66.74 | 22 s 1 | ε |
| 135 | (11/2−) | −69.98 | 45 s 4 | ε |
| 135m3/2+,5/2− | | −69.98 | 49 s 3 | ε |
| 136 | (2+) | −71.20 | 47 s 2 | ε |
| 136m | (5−) | −71.20 | 107 s 6 | ε |
| 137 | 11/2− | −74.07 | 2.4 m 1 | ε |
| 138 | | −74.94 | 10 s 2 | ε |
| 138m | | −74.92 | 3.24 m 5 | ε |
| 139 | (5/2+) | −77.50 | 4.15 m 5 | ε |
| 139m | (11/2)− | −77.31 | 180 ms 20 | IT 99.94%, ε 0.06% |
| 140 | 1+ | −78.21 | 9.2 s 2 | ε |
| 140m | 8− | −78.21 | 5.95 m 5 | ε |
| 141 | 5/2+ | −80.52 | 20.90 m 5 | ε |
| 142 | 1+ | −81.16 | 40.5 s 5 | ε |
| 143 | 5/2+ | −82.966 | 265 d 7 | ε, ε<5.7×10⁻⁶% |
| 144 | 5− | −81.421 | 363 d 14 | ε |
| 145 | 5/2+ | −81.274 | 17.7 y 4 | ε, α 3×10⁻⁸ |
| 146 | 3− | −79.460 | 5.53 y 5 | ε 66%, β− 34% |
| 147 | 7/2+ | −79.048 | 2.6234 y 2 | β− |
| 148 | 1− | −76.872 | 5.368 d 2 | β− |
| 148m | 5−,6− | −76.734 | 41.29 d 11 | β− 95.8%, IT 4.2% |
| 149 | 7/2+ | −76.071 | 53.08 h 5 | β− |
| 150 | (1−) | −73.60 | 2.68 h 2 | β− |
| 151 | 5/2+ | −73.395 | 28.40 h 4 | β− |
| 152 | 1+ | −71.26 | 4.12 m 8 | β− |
| 152m | 4− | −71.11 | 7.52 m 8 | β− |
| 152m | (8) | −71.11 | 13.8 m 2 | β−≤100%, IT≥0% |
| 153 | 5/2− | −70.68 | 5.25 m 2 | β− |
| 154 | (3,4) | −68.50 | 2.68 m 7 | β− |
| 154m | (0,1) | −68.50 | 1.73 m 10 | β− |
| 155 | 5/2− | −66.97 | 41.5 s 2 | β− |
| 156 | 4− | −64.22 | 26.70 s 10 | β− |
| 157 | (5/2−) | −62.4 | 10.56 s 10 | β− |
| 158 | | −59.1 | 4.8 s 5 | β− |
| 159 | | −56.8s | 1.47 s 15 | β− |
| 160 | | −53.1s | ~2 s | β−? |
| 161 | | −50.4s | ~0.7 s | β−? |
| 162 | | −46.3s | ~0.5 s | β−? |
| 163 | | −43.1s | ~0.2 s | β−? |
| 62 Sm 128 | 0+ | −39.0s | 0.5 s sy | ε?, εp? |
| 129 | (1/2+) | −42.3s | 0.55 10 | ε, εp |
| 130 | 0+ | −47.6s | 1 s sy | ε |
| 131 | | −50.2s | 1.2 s 2 | ε, εp>0% |
| 132 | 0+ | −55.2s | 4.0 s 3 | ε, εp |
| 133 | (5/2+) | −57.1s | 3.7 s 7 | ε, εp>0% |
| 134 | 0+ | −61.5s | 9.5 s 8 | ε |
| 135 | (3/2+,5/2+) | −62.9 | 10.3 s 5 | ε, εp 0.02% |

42

## Nuclear Wallet Cards

| Nuclide Z El A | Jπ | Δ (MeV) | T½, Γ, or Abundance | Decay Mode |
|---|---|---|---|---|
| 62 Sm 136 | 0+ | −66.81 | 47 s 2 | ε |
| 137 | (9/2−) | −68.03 | 45 s 1 | ε |
| 138 | 0+ | −71.50 | 3.1 m 2 | ε |
| 139 | 1/2+ | −72.38 | 2.57 m 10 | ε |
| 139m | 11/2− | −71.92 | 10.7 s 6 | IT 93.7%, ε 6.3% |
| 140 | 0+ | −75.46 | 14.82 m 12 | ε |
| 141 | 1/2+ | −75.939 | 10.2 m 2 | ε |
| 141m | 11/2− | −75.763 | 22.6 m 2 | ε 99.69%, IT 0.31% |
| 142 | 0+ | −78.993 | 72.49 m 5 | ε |
| 143 | 3/2+ | −79.523 | 8.75 m 8 | ε |
| 143m | 11/2− | −78.769 | 66 s 2 | IT 99.76%, ε 0.24% |
| 144 | 0+ | −81.972 | 3.07% 7 | |
| 145 | 7/2− | −80.658 | 340 d 3 | ε |
| 146 | 0+ | −81.002 | 10.3×10⁷ y 5 | α |
| 147 | 7/2− | −79.272 | 1.06×10¹¹ y 2 | α |
| | | | 14.99% 18 | |
| 148 | 0+ | −79.342 | 7×10¹⁵ y 3 | α |
| | | | 11.24% 10 | |
| 149 | 7/2− | −77.142 | 13.82% 7 | |
| 150 | 0+ | −77.057 | 7.38% 1 | |
| 151 | 5/2− | −74.582 | 90 y 8 | β− |
| 152 | 0+ | −74.769 | 26.75% 16 | |
| 153 | 3/2+ | −72.566 | 46.284 h 4 | β− |
| 154 | 0+ | −72.462 | 22.75% 29 | |
| 155 | 3/2− | −70.197 | 22.3 m 2 | β− |
| 156 | 0+ | −69.370 | 9.4 h 2 | β− |
| 157 | (3/2−) | −66.73 | 8.03 m 7 | β− |
| 158 | 0+ | −65.21 | 5.30 m 3 | β− |
| 159 | 5/2− | −62.2 | 11.37 s 15 | β− |
| 160 | 0+ | −60.4s | 9.6 s 3 | β− |
| 161 | | −57.0s | 4.8 s 8 | β− |
| 162 | 0+ | −54.8s | ~2 s | β−? |
| 163 | | −50.9s | ~1 s | β−? |
| 164 | 0+ | −48.2s | ~0.5 s | β−? |
| 165 | | −43.8s | ~0.2 s | β−? |
| 63 Eu 130 | (1+) | −33.9s | 0.9 ms +5−3 | p |
| 131 | 3/2+ | −39.4s | 17.8 ms 19 | ε 87.9%, ε 12.1% |
| 132 | | −42.5s | 200 ms sy | ε |
| 133 | | −47.3s | ~1 s | ε? |
| 134 | | −49.8s | 0.5 s 2 | ε, εp>0% |
| 135 | | −54.2s | 1.5 s 2 | ε, εp |
| 136m | (7+) | −56.3s | 3.3 s 3 | ε, εp 0.09% |
| 136m | (3+) | −56.3s | 3.8 s 3 | ε, εp 0.09% |
| 137 | (11/2−) | −60.0s | 11 s 2 | ε |
| 138 | (6−) | −61.75 | 12.1 s 6 | ε |
| 139 | (11/2)− | −65.40 | 17.9 s 6 | ε |
| 140 | 1+ | −66.99 | 1.51 s 2 | ε |
| 140m | (5−) | −66.80 | 125 ms 2 | IT, ε<1% |
| 141 | 5/2+ | −69.93 | 40.7 s 7 | ε |
| 141m | 11/2− | −69.83 | 2.7 s 3 | IT 87%, ε 13% |
| 142 | 1+ | −71.32 | 2.34 s 12 | ε |
| 142m | 8− | −71.32 | 1.223 m 8 | ε |

43

## Nuclear Wallet Cards

| Nuclide Z El A | Jπ | Δ (MeV) | T½, Γ, or Abundance | Decay Mode |
|---|---|---|---|---|
| 63 Eu 143 | 5/2+ | −74.24 | 2.59 m 2 | ε |
| 144 | 1 | −75.62 | 10.2 s 1 | ε |
| 145 | 5/2+ | −77.998 | 5.93 d 4 | ε |
| 146 | 4 | −77.122 | 4.61 d 3 | ε |
| 147 | 5/2+ | −77.550 | 24.1 d 6 | ε, α 2.2×10⁻³% |
| 148 | 5− | −76.30 | 54.5 d 5 | ε, α 9.4×10⁻⁷% |
| 149 | 5/2+ | −76.447 | 93.1 d 4 | ε |
| 150 | 5(−) | −74.797 | 36.9 y 9 | ε |
| 150m | 0− | −74.755 | 12.8 h 1 | β− 89%, ε 11%, IT≤5.0×10⁻⁸% |
| 151 | 5/2+ | −74.659 | 47.81% 3 | |
| 152 | 3− | −72.895 | 13.506 y 6 | ε 72.1%, β− 27.9% |
| 152m | 0− | −72.849 | 9.3116 h 13 | β− 72%, ε 28% |
| 152m | 8− | −72.747 | 96 m 1 | IT |
| 153 | 5/2+ | −73.373 | 52.19% 3 | |
| 154 | 3− | −71.744 | 8.590 y 3 | β− 99.98%, ε 0.02% |
| 154m | (8−) | −71.599 | 46.3 m 4 | IT |
| 155 | 5/2+ | −71.825 | 4.753 y 14 | β− |
| 156 | 0+ | −70.093 | 15.19 d 8 | β− |
| 157 | 5/2+ | −69.467 | 15.18 h 3 | β− |
| 158 | (1−) | −67.21 | 45.9 m 2 | β− |
| 159 | 5/2+ | −66.053 | 18.1 m 1 | β− |
| 160 | 1 | −63.4s | 38 s 4 | β− |
| 161 | | −61.8s | 26 s 3 | β− |
| 162 | | −58.6s | 10.6 s 10 | β− |
| 163 | | −56.6s | 6 s sy | β−? |
| 164 | | −53.1s | ~2 s | β−? |
| 165 | | −50.6s | ~1 s | β−? |
| 166 | | −46.5s | ~0.4 s | β−? |
| 167 | | −43.6s | ~0.2 s | β−? |
| 64 Gd 134 | 0+ | −41.6s | 0.4 s sy | ε? |
| 135 | | −44.2s | 1.1 s 2 | ε, εp~2% |
| 136 | 0+ | −49.1s | ≥200 ns | ε |
| 137 | (7/2) | −51.2s | 2.2 s 2 | ε, εp |
| 138 | 0+ | −55.8s | 4.7 s 9 | ε |
| 139 | (9/2−) | −57.5s | 5.8 s 9 | ε>0%, εp>0% |
| 139m | | −57.5s | 4.8 s 9 | ε>0%, εp>0% |
| 140 | 0+ | −61.78 | 15.8 s 4 | ε |
| 141 | 1/2+ | −63.22 | 14 s 4 | ε, εp 0.03% |
| 141m | 11/2− | −62.85 | 24.5 s 5 | ε 89%, IT 11% |
| 142 | 0+ | −66.96 | 70.2 s 6 | ε |
| 143 | (1/2) | −68.2 | 39 s 2 | ε |
| 143m (11/2−) | | −68.1 | 110.0 s 14 | ε |
| 144 | 0+ | −71.76 | 4.47 m 6 | ε |
| 145 | 1/2+ | −72.93 | 23.0 m 4 | ε |
| 145m | 11/2− | −72.18 | 85 s 3 | IT 94.3%, ε 5.7% |
| 146 | 0+ | −76.093 | 48.27 d 10 | ε |
| 147 | 7/2− | −75.363 | 38.06 h 12 | ε |
| 148 | 0+ | −76.276 | 70.9 y 10 | α |
| 149 | 7/2− | −75.133 | 9.28 d 10 | ε, α 4.3×10⁻⁴% |
| 150 | 0+ | −75.769 | 1.79×10⁶ y 8 | α |
| 151 | 7/2− | −74.195 | 124 d 1 | ε, α=8.0×10⁻⁷% |

44

## Nuclear Wallet Cards

| Nuclide Z El A | Jπ | Δ (MeV) | T½, Γ, or Abundance | Decay Mode |
|---|---|---|---|---|
| 64 Gd 152 | 0+ | −74.714 | 1.08×10^14 y 8 0.20% 1 | α |
| 153 | 3/2− | −72.890 | 240.4 d 10 | ε |
| 154 | 0+ | −73.713 | 2.18% 3 | |
| 155 | 3/2− | −72.077 | 14.80% 12 | |
| 156 | 0+ | −72.542 | 20.47% 9 | |
| 157 | 3/2− | −70.831 | 15.65% 2 | |
| 158 | 0+ | −70.697 | 24.84% 7 | |
| 159 | 3/2− | −68.568 | 18.479 h 4 | β− |
| 160 | 0+ | −67.949 | >3.1×10^19 y 21.86% 19 | 2β− |
| 161 | 5/2− | −65.513 | 3.66 m 5 | β− |
| 162 | 0+ | −64.287 | 8.4 m 2 | β− |
| 163 | (5/2−,7/2+) | −61.5s | 68 s 3 | β− |
| 164 | 0+ | −59.7s | 45 s 3 | β− |
| 165 | | −56.5s | 10.3 s 16 | β− |
| 166 | 0+ | −54.4s | ≈7 s | β− |
| 167 | | −50.7s | ≈3 s | β− ? |
| 168 | 0+ | −48.1s | ≈0.3 s | β− ? |
| 169 | | −43.9s | ≈1 s | β− ? |
| 65 Tb 135 | (7/2−) | | 0.94 ms +33−22 | p |
| 136 | | −36.0s | 0.2 s xy | p |
| 137 | | −41.0s | 0.6 s sy | p?, ε? |
| 138? | | −43.6s | ≥200 ns | ε, p |
| 139 | | −48.2s | 1.6 s 2 | ε, εp? |
| 140 | (7+) | −50.5 | 2.1 s 4 | ε, εp 0.26% |
| 141 | (5/2−) | −54.5 | 3.5 s 2 | ε |
| 141m | | −54.5 | 7.9 s 6 | ε |
| 142 | 1+ | −57.1s | 597 ms 17 | ε, εp 2.2×10^−3% |
| 142m | (5−) | −56.8s | 303 ms 17 | IT |
| 143 | (11/2−) | −60.43 | 12 s 1 | ε |
| 143m | | −60.43 | <21 s | ε |
| 144 | 1+ | −62.37 | ≈1 s | ε |
| 144m | (6−) | −61.97 | 4.25 s 15 | IT 60%, ε 34% |
| 145 | (3/2+) | −65.88 | ≈20 m | ε? |
| 145m (11/2−) | | −65.88 | 30.9 s 7 | ε |
| 146 | 1+ | −67.77 | 8 s 4 | ε |
| 146m | 5− | −67.77 | 23 s 2 | ε |
| 147 | (1/2+) | −70.75 | 1.7 h 1 | ε |
| 147m (11/2−) | | −70.70 | 1.83 m 6 | ε |
| 148 | 2− | −70.54 | 60 m 1 | ε |
| 148m | (9)+ | −70.45 | 2.20 m 5 | ε |
| 149 | 1/2+ | −71.496 | 4.118 h 25 | ε 83.3%, α 16.7% |
| 149m | 11/2− | −71.460 | 4.16 m 4 | ε 99.98%, α 0.02% |
| 150 | (2−) | −71.110 | 3.48 h 16 | ε |
| 150m | 9+ | −70.637 | 5.8 m 2 | ε |
| 151 | 1/2(+) | −71.630 | 17.609 h 1 | ε, α 9.5×10^−3% |
| 151m (11/2−) | | −71.530 | 25 s 3 | IT 93.8%, ε 6.2% |
| 152 | 2− | −70.72 | 17.5 h 1 | ε, α<7.0×10^−7% |
| 152m | 8+ | −70.22 | 4.2 m 1 | IT 78.8%, ε 21.2% |
| 153 | 5/2+ | −71.320 | 2.34 d 1 | ε |
| 154 | 0 | −70.16 | 21.5 h 4 | ε, β−<0.1% |

| Nuclide Z El A | Jπ | Δ (MeV) | T½, Γ, or Abundance | Decay Mode |
|---|---|---|---|---|
| 65 Tb 154m | 3− | −70.16 | 9.4 h 4 | ε 78.2%, IT 21.8%, β−<0.1% |
| 154m | 7− | −70.16 | 22.7 h 5 | ε 98.2%, IT 1.8% |
| 155 | 3/2+ | −71.25 | 5.32 d 6 | ε |
| 156 | 3− | −70.098 | 5.35 d 10 | ε |
| 156m | (7−) | −70.048 | 24.4 h 10 | IT |
| 156m | (0+) | −70.009 | 5.3 h 2 | IT<100%, ε>0% |
| 157 | 3/2+ | −70.771 | 71 y 7 | ε |
| 158 | 3− | −69.477 | 180 y 11 | ε 83.4%, β− 16.6% |
| 158m | 0− | −69.367 | 10.70 s 17 | IT, β−<0.6%, ε<0.01% |
| 159 | 3/2+ | −69.539 | 100% | |
| 160 | 3− | −67.843 | 72.3 d 2 | β− |
| 161 | 3/2+ | −67.468 | 6.906 d 19 | β− |
| 162 | 1− | −65.68 | 7.60 m 15 | β− |
| 163 | 3/2+ | −64.601 | 19.5 m 3 | β− |
| 164 | (5+) | −62.1 | 3.0 m 1 | β− |
| 165 | (3/2+) | −60.7s | 2.11 m 10 | β− |
| 166 | | −57.8 | 21 s 6 | β− |
| 167 | (3/2+) | −55.8s | 19.4 s 27 | β− |
| 168 | (4−) | −52.5s | 8.2 s 13 | β− |
| 169 | | −50.1s | ≈2 s | β− ? |
| 170 | | −46.3s | ≈3 s | β− ? |
| 171 | | −43.5s | ≈0.5 s | β− |
| 66 Dy 138 | 0+ | −34.9s | 200 ms xy | ε? |
| 139 | (7/2+) | −37.7s | 0.6 s 2 | ε, εp |
| 141 | (9/2−) | −45.3s | 0.9 s 2 | ε, εp |
| 142 | 0+ | −50.0s | 2.3 s 3 | ε, εp 0.06% |
| 143 | (1/2+) | −52.3s | 5.6 s 10 | ε, εp |
| 143m (11/2−) | | −52.3s | 3.0 s 3 | ε, εp |
| 144 | 0+ | −56.58 | 9.1 s 4 | ε, εp |
| 145 | (1/2+) | −58.29 | 10.5 s 15 | ε |
| 145m (11/2−) | | −58.29 | 13.6 s 10 | ε |
| 146 | 0+ | −62.55 | 29 s 3 | ε |
| 146m | (10+) | −59.62 | 150 ms 20 | IT |
| 147 | 1/2+ | −64.19 | 40 s 10 | ε, εp>0% |
| 147m | 11/2− | −64.19 | 55.7 s 7 | 65%, IT 35% |
| 148 | 0+ | −67.86 | 3.3 m 2 | ε |
| 149 | (7/2−) | −67.715 | 4.20 m 14 | ε |
| 149m (27/2−) | | −65.054 | 0.490 s 15 | IT 99.3%, ε 0.7% |
| 150 | 0+ | −69.317 | 7.17 m 5 | ε 64%, α 36% |
| 151 | 7/2(−) | −68.759 | 17.9 m 3 | ε 94.4%, α 5.6% |
| 152 | 0+ | −70.124 | 2.38 h 2 | ε 99.9%, α 0.1% |
| 153 | 7/2(−) | −69.150 | 6.4 h 1 | ε 99.99%, α 9.4×10^−3% |
| 154 | 0+ | −70.398 | 3.0×10^6 y 15 | α |
| 155 | 3/2− | −69.16 | 9.9 h 2 | ε |
| 156 | 0+ | −70.530 | 0.06% 1 | |
| 157 | 3/2− | −69.428 | 8.14 h 4 | ε |
| 158 | 0+ | −70.412 | 0.10% 1 | |
| 159 | 3/2− | −69.174 | 144.4 d 2 | ε |
| 160 | 0+ | −69.678 | 2.34% 8 | |

| Nuclide Z El A | Jπ | Δ (MeV) | T½, Γ, or Abundance | Decay Mode |
|---|---|---|---|---|
| 66 Dy 161 | 5/2+ | −68.061 | 18.91% 24 | |
| 162 | 0+ | −68.187 | 25.51% 26 | |
| 163 | 5/2− | −66.386 | 24.90% 16 | |
| 164 | 0+ | −65.973 | 28.18% 37 | |
| 165 | 7/2+ | −63.618 | 2.334 h 1 | β− |
| 165m | 1/2− | −63.510 | 1.257 m 6 | IT 97.76%, β− 2.24% |
| 166 | 0+ | −62.590 | 81.6 h 1 | β− |
| 167 | (1/2−) | −59.94 | 6.20 m 8 | β− |
| 168 | 0+ | −58.6 | 8.7 m 3 | β− |
| 169 | (5/2−) | −55.6 | 39 s 8 | β− |
| 170 | 0+ | −53.7s | ≈30 s | β− ? |
| 171 | | −50.1s | ≈6 s | β− |
| 172 | 0+ | −47.7s | ≈3 s | β− |
| 173 | | −43.8s | ≈2 s | β− ? |
| 67 Ho 140 | (6−,0−,8+) | −29.3s | 6 ms 3 | p |
| 141 | 7/2− | −34.4s | 4.1 ms 3 | ε |
| 142 | | −37.5s | ≈0.3 s | ε |
| 143 | | −42.3s | >200 ns | ε ?, εp ? |
| 144 | | −45.2s | 0.7 s 1 | ε, εp |
| 145 | | −49.2s | 2.4 s 1 | ε |
| 146 | (10+) | −51.6s | 3.6 s 3 | ε |
| 147 | (11/2−) | −55.84 | 5.8 s 4 | ε |
| 148 | (1+) | −58.0 | 2.2 s 11 | ε |
| 148m | (6−) | −58.0 | 9.59 s 15 | ε, εp 0.08% |
| 149 | (11/2−) | −61.69 | 21.1 s 2 | ε |
| 149m | (1/2+) | −61.64 | 56 s 3 | ε |
| 150 | 2− | −61.95 | 72 s 4 | ε |
| 150m | (9+) | −61.15 | 23.3 s 3 | ε |
| 151 | 11/2− | −62.63 | 35.2 s 1 | ε 78%, α 22% |
| 151m | (1/2+) | −63.59 | 47.2 s 10 | α 80%, ε 20% |
| 152 | 2− | −63.61 | 161.8 s 3 | ε 88%, α 12% |
| 152m | 9+ | −63.45 | 50.0 s 4 | ε 89.2%, α 10.8% |
| 153 | 11/2− | −65.019 | 2.01 m 3 | ε 99.95%, α 0.05% |
| 153m | 1/2+ | −64.951 | 9.3 m 5 | ε 99.82%, α 0.18% |
| 154 | 2− | −64.644 | 11.76 m 19 | ε 99.98%, α 0.02% |
| 154m | 8+ | −64.644 | 3.10 m 14 | ε, α<1.0×10^−3%, IT? |
| 155 | 5/2+ | −66.04 | 48 m 1 | ε |
| 156 | 4− | −65.35 | 56 m 1 | ε |
| 156m | (1+) | −65.30 | 9.5 s 15 | IT |
| 156m | 9+ | −65.30 | 7.8 m 3 | ε 75%, IT 25% |
| 157 | 7/2− | −66.83 | 12.6 m 2 | ε |
| 158 | 5+ | −66.19 | 11.3 m 4 | ε |
| 158m | (9+) | −66.12 | 28 m 2 | IT>81%, ε<19% |
| 158m | (9+) | −66.01 | 21.3 m 23 | ε≥93%, IT≤7% |
| 159 | 7/2− | −67.326 | 33.05 m 11 | ε |
| 159m | 1/2+ | −67.130 | 8.30 s 8 | IT |
| 160 | 5+ | −66.39 | 25.6 m 3 | ε |
| 160m | 2− | −66.33 | 5.02 h 5 | IT 73%, ε 27% |
| 160m | (9+) | −66.22 | 3 s | IT |
| 161 | 7/2− | −67.203 | 2.48 h 5 | ε |
| 161m | 1/2+ | −66.992 | 6.76 s 7 | IT |

| Nuclide Z El A | Jπ | Δ (MeV) | T½, Γ, or Abundance | Decay Mode |
|---|---|---|---|---|
| 67 Ho 162 | 1+ | −66.047 | 15.0 m 10 | ε |
| 162m | 6− | −65.941 | 67.0 m 7 | IT 62%, ε 38% |
| 163 | 7/2− | −66.384 | 4570 y 25 | ε |
| 163m | 1/2+ | −66.086 | 1.09 s 3 | IT |
| 164 | 1+ | −64.987 | 29 m 1 | ε 60%, β− 40% |
| 164m | 6− | −64.847 | 37.5 m +15−5 | IT |
| 165 | 7/2− | −64.905 | 100% | |
| 166 | 0− | −63.077 | 26.83 h 2 | β− |
| 166m | (7−) | −63.071 | 1.20×10^3 y 18 | β− |
| 167 | 7/2− | −62.287 | 3.003 h 18 | β− |
| 168 | 3+ | −60.07 | 2.99 m 7 | β− |
| 168m | (6+) | −60.01 | 132 s 4 | IT≥99.5%, β−≤0.5% |
| 169 | 7/2− | −58.80 | 4.72 m 10 | β− |
| 170 | (6+) | −56.24 | 2.76 m 5 | β− |
| 170m | (1+) | −56.12 | 43 s 2 | β− |
| 171 | (7/2−) | −54.5 | 53 s 2 | β− |
| 172 | | −51.4s | 25 s 3 | β− |
| 173 | | −49.1s | ≈10 s | β− ? |
| 174 | | −45.5s | ≈8 s | β− ? |
| 175 | | −42.8s | ≈5 s | β− ? |
| 68 Er 143 | | −31.4s | 0.2 s xy | ε ? |
| 144 | 0+ | −36.9s | ≥200 ns | ε |
| 145 | (11/2−) | −41.0s | 0.9 s 3 | ε |
| 146 | 0+ | −44.7s | 1.7 s 6 | ε, εp |
| 147 | (11/2−) | −47.0s | 2.5 s 2 | ε, εp |
| 147m | 1/2+ | −47.0s | ≈2.5 s | ε, εp>0% |
| 148 | 0+ | −51.7s | 4.6 s 2 | ε |
| 149 | (1/2−) | −53.74 | 4 s 2 | ε, εp |
| 149m (11/2−) | | −53.00 | 8.9 s 2 | ε 96.5%, IT 3.5%, εp 0.18% |
| 150 | 0+ | −57.83 | 18.5 s 7 | ε |
| 151 | (7/2−) | −58.27 | 23.5 s 13 | ε |
| 151m (27/2−) | | −55.68 | 0.58 s 2 | IT 95.3%, ε 4.7% |
| 152 | 0+ | −60.50 | 10.3 s 1 | α 90%, ε 10% |
| 153 | (7/2−) | −60.488 | 37.1 s 2 | α 53%, ε 47% |
| 154 | 0+ | −62.612 | 3.73 m 9 | ε 99.53%, α 0.47% |
| 155 | 7/2− | −62.215 | 5.3 m 3 | ε 99.98%, α 0.02% |
| 156 | 0+ | −64.21 | 19.5 m 10 | ε ≈100%, α 1.7×10^−5% |
| 157 | 3/2− | −63.42 | 18.65 m 10 | ε ≈100% |
| 158 | 0+ | −65.30 | 2.29 h 6 | ε |
| 159 | 3/2− | −64.567 | 36 m 1 | ε |
| 160 | 0+ | −66.06 | 28.58 h 9 | ε |
| 161 | 3/2− | −65.209 | 3.21 h 3 | ε |
| 162 | 0+ | −66.343 | 0.139% 5 | |
| 163 | 5/2− | −65.174 | 75.0 m 4 | ε |
| 164 | 0+ | −65.950 | 1.601% 3 | |
| 165 | 5/2− | −64.528 | 10.36 h 4 | ε |
| 166 | 0+ | −64.932 | 33.503% 36 | |
| 167 | 7/2+ | −63.297 | 22.869% 9 | |
| 167m | 1/2− | −63.089 | 2.269 s 6 | IT |
| 168 | 0+ | −62.997 | 26.978% 18 | |

## Nuclear Wallet Cards

| Nuclide Z El A | Jπ | Δ (MeV) | T½, Γ, or Abundance | Decay Mode |
|---|---|---|---|---|
| 68 Er 169 | 1/2− | −60.929 | 9.392 d 18 | β− |
| 170 | 0+ | −60.115 | 14.910% 36 | |
| 171 | 5/2− | −57.725 | 7.516 h 2 | β− |
| 172 | 0+ | −56.489 | 49.3 h 3 | β− |
| 173 | (7/2−) | −53.7s | 1.4 m 1 | β− |
| 174 | 0+ | −51.9s | 3.2 m 2 | β− |
| 175 | (9/2+) | −48.7s | 1.2 m 3 | β− |
| 176 | 0+ | −46.5s | ≈20 s | β−? |
| 177 | | −42.8s | ≈3 s | β−? |
| 69 Tm 145 | (11/2−) | −27.9s | 3.1 μs 3 | p |
| 146 | (5−) | −31.3s | 80 ms 10 | p, ε |
| 146m | (8+) | −31.1s | 200 ms 10 | p, ε |
| 147 | 11/2− | −36.4s | 0.58 s 3 | ε 85%, p 15% |
| 148m | (10+) | −39.3s | 0.7 s 2 | ε |
| 149 | (11/2−) | −44.0s | 0.9 s 2 | ε, εp 0.2% |
| 150 | (6−) | −46.6s | 2.2 s 2 | ε |
| 151 | (11/2−) | −50.78 | 4.17 s 10 | ε |
| 151m | (1/2+) | −50.78 | 6.6 s 14 | ε |
| 152 | (2)− | −51.77 | 8.0 s 10 | ε |
| 152m | (9)+ | −51.77 | 5.2 s 6 | ε |
| 153 | (11/2−) | −54.02 | 1.48 s 1 | α 91%, ε 9% |
| 153m | (1/2+) | −53.97 | 2.5 s 2 | α 92%, ε 8% |
| 154 | (2−) | −54.43 | 8.1 s 3 | α 54%, ε 46% |
| 154m | (9+) | −54.43 | 3.30 s 7 | α 58%, ε 42%, IT |
| 155 | 11/2− | −56.64 | 21.6 s 2 | ε 99.11%, α 0.89% |
| 155m | 1/2+ | −56.59 | 45 s 3 | ε >98%, α <2% |
| 156 | 2− | −56.84 | 83.8 s 18 | ε 99.94%, α 0.06% |
| 157 | 1/2+ | −58.71 | 3.63 m 9 | ε |
| 158 | 2− | −58.70 | 3.98 m 6 | ε |
| 158m | (5+) | −58.70 | ≈20 s | IT? |
| 159 | 5/2+ | −60.57 | 9.13 m 16 | ε |
| 160 | 1− | −60.30 | 9.4 m 3 | ε |
| 160m | 5 | −60.23 | 74.5 s 15 | IT 85%, ε 15% |
| 161 | 7/2+ | −61.90 | 30.2 m 8 | ε |
| 162 | 1− | −61.48 | 21.70 m 19 | ε |
| 162m | 5+ | −61.48 | 24.3 s 17 | IT 82%, ε 18% |
| 163 | 1/2+ | −62.735 | 1.810 h 5 | ε |
| 164 | 1+ | −61.89 | 2.0 m 1 | ε, ε 39% |
| 164m | 6− | −61.89 | 5.1 m 1 | IT~80%, ε~20% |
| 165 | 1/2+ | −62.936 | 30.06 h 3 | ε |
| 166 | 2+ | −61.89 | 7.70 h 3 | ε |
| 167 | 1/2+ | −62.548 | 9.25 d 2 | ε |
| 168 | 3+ | −61.318 | 93.1 d 2 | ε 99.99%, β− 0.01% |
| 169 | 1/2+ | −61.280 | 100% | |
| 170 | 1− | −59.801 | 128.6 d 3 | β− 99.87%, ε 0.13% |
| 171 | 1/2+ | −59.216 | 1.92 y 1 | β− |
| 172 | 2− | −57.380 | 63.6 h 2 | β− |
| 173 | (1/2−) | −56.259 | 8.24 h 8 | β− |
| 174 | (4)− | −53.87 | 5.4 m 1 | β− |
| 175 | (1/2+) | −52.32 | 15.2 m 5 | β− |
| 176 | (4+) | −49.4 | 1.9 m 1 | β− |
| 177m | (7/2−) | −47.5s | 90 s 6 | β−≤100% |

| Nuclide Z El A | Jπ | Δ (MeV) | T½, Γ, or Abundance | Decay Mode |
|---|---|---|---|---|
| 69 Tm 178 | | −44.1s | ≈30 s | β−? |
| 179 | | −41.6s | ≈20 s | β−? |
| 70 Yb 148 | 0+ | −30.3s | ≈0.25 s | ε? |
| 149 | (1/2+,3/2+) | −33.5s | 0.7 s 2 | ε, εp≈100% |
| 150 | 0+ | −38.7s | >200 ns | ε? |
| 151 | (1/2+) | −41.5 | 1.6 s 1 | ε, εp |
| 151m | (11/2−) | −41.5 | 1.6 s 1 | ε=100%, εp, IT? |
| 152 | 0+ | −46.3 | 3.04 s 6 | ε, εp |
| 153 | 7/2− | −47.1s | 4.2 s 2 | α 60%, ε 40% |
| 154 | 0+ | −49.93 | 0.409 s 2 | α 92.6%, ε 7.4% |
| 155 | 7/2− | −50.50 | 1.793 s 19 | α 89%, ε 11% |
| 156 | 0+ | −53.26 | 26.1 s 7 | ε 90%, α 10% |
| 157 | 7/2− | −53.44 | 38.6 s 10 | ε 99.5%, α 0.5% |
| 158 | 0+ | −56.015 | 1.49 m 13 | ε, α=2.1×10⁻³% |
| 159 | 5/2(−) | −55.84 | 1.67 m 9 | ε |
| 160 | 0+ | −58.17 | 4.8 m 2 | ε |
| 161 | 3/2− | −57.84 | 4.2 m 2 | ε |
| 162 | 0+ | −59.83 | 18.87 m 19 | ε |
| 163 | 3/2− | −59.30 | 11.05 m 35 | ε |
| 164 | 0+ | −61.02 | 75.8 m 17 | ε |
| 165 | 5/2− | −60.29 | 9.9 m 3 | ε |
| 166 | 0+ | −61.589 | 56.7 h 1 | ε |
| 167 | 5/2− | −60.594 | 17.5 m 2 | ε |
| 168 | 0+ | −61.575 | 0.13% 1 | |
| 169 | 7/2+ | −60.370 | 32.018 d 5 | ε |
| 169m | (1/2−) | −60.346 | 46 s 2 | IT |
| 170 | 0+ | −60.769 | 3.04% 15 | |
| 171 | 1/2− | −59.312 | 14.28% 57 | |
| 172 | 0+ | −59.260 | 21.83% 67 | |
| 173 | 5/2− | −57.556 | 16.13% 27 | |
| 174 | 0+ | −56.950 | 31.83% 92 | |
| 175 | (7/2−) | −54.701 | 4.185 d 1 | β− |
| 176 | 0+ | −53.494 | ≥1.6×10¹⁷ y | 2β− |
| | | | 12.76% 41 | |
| 176m | (8)− | −54.701 | 11.4 s 3 | IT≥90%, β−≤10% |
| 177 | (9/2+) | −50.989 | 1.911 h 3 | β− |
| 177m | (1/2−) | −50.658 | 6.41 s 2 | IT |
| 178 | 0+ | −49.70 | 74 m 3 | β− |
| 179 | (1/2−) | −46.4s | 8.0 m 4 | β− |
| 180 | 0+ | −44.4s | 2.4 m 5 | β− |
| 181 | | −40.8s | 1 m sy | β−? |
| 71 Lu 150 | (2+) | −24.9s | 43 ms 5 | p 69%, ε 32% |
| 151 | 11/2− | −30.2s | 80.6 ms 19 | p 63.4%, ε 36.6% |
| 152 | (5−,6−) | −33.4s | 0.7 s 1 | ε, εp 15% |
| 153 | 11/2− | −38.4 | 0.9 s 2 | α~70%, ε~30% |
| 154 | (2−) | −39.6s | ≈2 s | ε? |
| 154m | (9+) | −39.6s | 1.12 s 8 | ε=100% |
| 155 | 11/2− | −42.55 | 68 ms 1 | α 90%, ε 10% |
| 155m | 1/2+ | −42.53 | 138 ms 8 | α 76%, ε 24% |
| 156 | (2)− | −43.75 | 494 ms 12 | α=95%, ε=5% |
| 156m | 9+ | −43.75 | 198 ms 2 | α |
| 157 | (1/2+,3/2+) | −46.28 | 6.8 s 18 | α>0% |

| Nuclide Z El A | Jπ | Δ (MeV) | T½, Γ, or Abundance | Decay Mode |
|---|---|---|---|---|
| 71 Lu 157m (11/2−) | | −46.46 | 4.79 s 12 | ε 94%, α 6% |
| 158 | | −47.21 | 10.6 s 3 | ε 99.09%, α 0.7% |
| 159 | | −49.72 | 12.1 s 10 | ε, α 0.1% |
| 160 | | −50.27 | 36.1 s 3 | ε, α≤1.0×10⁻⁴% |
| 160m | | −50.27 | 40 s 1 | ε≤100%, α |
| 161 | 1/2+ | −52.56 | 77 s 2 | ε |
| 161m | (9/2−) | −52.43 | 7.3 ms 4 | IT |
| 162 | (1−) | −52.84 | 1.37 m 2 | ε≤100% |
| 162m | (4−) | −52.84 | 1.5 m | ε≤100% |
| 162m | | −52.84 | 1.9 m | ε≤100% |
| 163 | 1/2(+) | −54.79 | 3.97 m 13 | ε |
| 164 | 1 | −54.64 | 3.14 m 3 | ε |
| 165 | 1/2− | −56.44 | 10.74 m 10 | ε |
| 166 | (6−) | −56.02 | 2.65 m 10 | ε |
| 166m | (3−) | −55.99 | 1.41 m 10 | ε 58%, IT 42% |
| 166m | (0−) | −55.98 | 2.12 m 10 | ε>80%, IT<20% |
| 167 | 7/2+ | −57.50 | 51.5 m 10 | ε |
| 167m | 1/2+ | −57.50 | ≥1 m | ε, IT |
| 168 | 3+ | −57.06 | 5.5 m 1 | ε |
| 168m | 3+ | −56.84 | 6.7 m 4 | ε>95%, IT<5% |
| 169 | 7/2+ | −58.077 | 34.06 h 5 | ε |
| 169m | 1/2− | −58.048 | 160 s 10 | IT |
| 170 | 0+ | −57.31 | 2.012 d 20 | ε |
| 170m | (4−) | −57.22 | 0.67 s 10 | IT |
| 171 | 7/2+ | −57.833 | 8.24 d 3 | ε |
| 171m | 1/2− | −57.762 | 79 s 2 | IT |
| 172 | 4− | −56.741 | 6.70 d 3 | ε |
| 172m | 1− | −56.699 | 3.7 m 5 | IT |
| 173 | 7/2+ | −56.886 | 1.37 y 1 | ε |
| 174 | (1)− | −55.575 | 3.31 y 5 | ε |
| 174m | (6)− | −55.404 | 142 d 2 | IT 99.38%, ε 0.62% |
| 175 | 7/2+ | −55.171 | 97.41% 2 | |
| 176 | 7− | −53.387 | 3.76×10¹⁰ y 7 | β− |
| | | | 2.59% 2 | |
| 176m | 1− | −53.264 | 3.664 h 19 | β− 99.91%, ε 0.1% |
| 177 | 7/2+ | −52.389 | 6.6475 d 20 | β− |
| 177m | 23/2− | −51.419 | 160.44 d 6 | β− 78.6%, IT 21.4% |
| 177m | (39/2−) | −49.689 | 6 m +3 −2 | ε≤100%, IT |
| 178 | 1(+) | −50.343 | 28.4 m 2 | β− |
| 178m | (9−) | −50.223 | 23.1 m 3 | β− |
| 179 | 7/2(+) | −49.064 | 4.59 h 6 | β− |
| 180 | 5+ | −46.69 | 5.7 m 1 | β− |
| 181 | (7/2+) | −44.7s | 3.5 m 3 | β− |
| 182 | (0,1,2) | −41.9s | 2.0 m 2 | β− |
| 183 | (7/2+) | −39.5s | 58 s 4 | β− |
| 184 | (3+) | −36.4s | 20 s 3 | β− |
| 72 Hf 153 | | −27.3s | >60 ns | ε? |
| 154 | 0+ | −32.7s | 2 s 1 | ε=100%, α≈0% |
| 155 | | −34.1s | 0.89 s 12 | ε |
| 156 | 0+ | −37.9 | 23 ms 1 | ε |
| 157 | 7/2− | −38.8s | 110 ms 6 | α 86%, ε 14% |
| 158 | 0+ | −42.10 | 2.85 s 7 | ε 55.7%, α 44.3% |

| Nuclide Z El A | Jπ | Δ (MeV) | T½, Γ, or Abundance | Decay Mode |
|---|---|---|---|---|
| 72 Hf 159 | 7/2− | −42.85 | 5.6 s 4 | ε 65%, α 35% |
| 160 | 0+ | −45.94 | 13.6 s 2 | ε 99.3%, α 0.7% |
| 161 | | −46.32 | 18.2 s 5 | ε 99.87%, α<0.13% |
| 162 | 0+ | −49.173 | 39.4 s 9 | ε 99.99%, α 8.0×10⁻⁴% |
| 163 | | −49.29 | 40.0 s 6 | ε, α<1.0×10⁻⁴% |
| 164 | 0+ | −51.82 | 111 s 8 | ε |
| 165 | (5/2−) | −51.64 | 76 s 4 | ε |
| 166 | 0+ | −53.86 | 6.77 m 30 | ε |
| 167 | (5/2)− | −53.47 | 2.05 m 5 | ε |
| 168 | 0+ | −55.36 | 25.95 m 20 | ε |
| 169 | (5/2)− | −54.72 | 3.24 m 4 | ε |
| 170 | 0+ | −56.25 | 16.01 h 13 | ε |
| 171 | 7/2(+) | −55.43 | 12.1 h 4 | ε |
| 171m 1/2− | | −55.41 | 29.5 s 9 | IT≤100%, ε |
| 172 | 0+ | −56.40 | 1.87 y 3 | ε |
| 173 | 1/2− | −55.41 | 23.6 h 1 | ε |
| 174 | | −55.847 | 2.0×10¹⁵ y 4 | α |
| | | | 0.16% 1 | |
| 175 | 5/2(−) | −54.484 | 70 d 2 | ε |
| 176 | 0+ | −54.577 | 5.26% 7 | |
| 177 | 7/2− | −52.890 | 18.60% 9 | |
| 177m | 23/2+ | −51.574 | 1.09 s 5 | IT |
| 177m | 37/2− | −50.150 | 51.4 m 5 | IT |
| 178 | 0+ | −52.444 | 27.28% 7 | |
| 178m | 8− | −51.297 | 4.0 s 2 | IT |
| 178m | 16+ | −49.998 | 31 y 1 | IT |
| 179 | 9/2+ | −50.472 | 13.62% 2 | |
| 179m | 1/2− | −50.097 | 18.67 s d | IT |
| 179m | 25/2− | −49.366 | 25.05 d 25 | IT |
| 180 | 0+ | −49.788 | 35.08% 16 | |
| 180m | 8− | −48.647 | 5.47 h 4 | IT 99.7%, β− 0.3% |
| 181 | 1/2− | −47.412 | 42.39 d 6 | β− |
| 182 | 0+ | −46.059 | 8.90×10⁶ y 9 | β− |
| 182m | 8− | −44.886 | 61.5 m 15 | β−58%, IT 42% |
| 183 | (3/2−) | −43.29 | 1.067 h 17 | β− |
| 184 | 0+ | −41.50 | 4.12 h 5 | β− |
| 184m | 8− | −40.23 | 48 s 10 | β− |
| 185 | | −38.4s | 3.5 m 6 | β− |
| 186 | | −36.4s | 2.6 m 12 | β− |
| 187 | | −33.0s | 30 s sy | β−? |
| 188 | 0+ | −30.9s | 20 s sy | β−? |
| 73 Ta 155m 11/2− | | −23.7s | 12 μs +4 −3 | p |
| 156 | (2−) | −25.8s | 144 ms 24 | p=100%, ε |
| 156m | 9+ | −25.7s | 0.36 s 4 | ε 95.8%, p 4.2% |
| 157 | 1/2+ | −29.6 | 10.1 ms 4 | α 96.6%, p 3.4% |
| 157m | 11/2− | −29.6 | 4.3 ms 1 | α |
| 157m(25/2−) | | −28.0 | 1.7 ms 1 | α |
| 158 | | −31.0s | 55 ms 15 | α 91%, ε 9% |
| 158m | (9−) | −30.9s | 36.7 ms 15 | α 95%, ε 5% |
| 159 | (1/2−) | −34.45 | 0.83 s 18 | ε 68%, α 34% |
| 159m (11/2−) | | −34.38 | 515 ms 20 | α 55%, ε 45% |

## Nuclear Wallet Cards

| Nuclide Z El A | Jπ | Δ (MeV) | T½, Γ, or Abundance | Decay Mode |
|---|---|---|---|---|
| 73 Ta 160 | | −35.88 | 1.55 s 4 | ε 66%, α 34% |
| 160m | | −35.88 | 1.7 s 2 | α? |
| 161 | | −38.73s | 2.89 s 12 | ε 99%, α? |
| 162 | | −39.78 | 3.57 s 12 | ε 99.93%, α 0.07% |
| 163 | | −42.54 | 10.6 s 18 | ε ≈99.8%, α≈0.2% |
| 164 | (3+) | −43.28 | 14.2 s 3 | ε |
| 165 | | −45.86 | 31.0 s 15 | ε |
| 166 | (2+) | −46.10 | 34.4 s 5 | ε |
| 167 | (3/2+) | −48.35 | 80 s 4 | ε |
| 168 | (2−,3+) | −48.39 | 2.0 m 1 | ε |
| 169 | (5/2+) | −50.29 | 4.9 m 4 | ε |
| 170 | (3+) | −50.14 | 6.76 m 6 | ε |
| 171 | (5/2−) | −51.72 | 23.3 m 3 | ε |
| 172 | (3+) | −51.33 | 36.8 m 3 | ε |
| 173 | 5/2− | −52.40 | 3.14 h 13 | ε |
| 174 | 3+ | −51.74 | 1.14 h 8 | ε |
| 175 | 7/2+ | −52.41 | 10.5 h 2 | ε |
| 176 | (1−) | −51.37 | 8.09 h 5 | ε |
| 177 | 7/2+ | −51.724 | 56.56 h 6 | ε |
| 178 | 1+ | −50.51 | 9.31 m 3 | ε |
| 178 | (7−) | −50.51 | 2.36 h 8 | ε |
| 179 | 7/2+ | −50.366 | 1.82 y 3 | ε |
| 180 | 1+ | −48.936 | 8.154 h 6 | ε 86%, β− 14% |
| 180m | 9− | −48.859 | >1.2×10¹⁵ y | 2ε? |
| 181 | 7/2+ | −48.442 | **99.988% 2** | |
| 182 | 3− | −46.433 | 114.43 d 3 | β− |
| 182m | 5+ | −46.417 | 283 ms 3 | IT |
| 182m | 10− | −45.913 | 15.84 m 10 | IT |
| 183 | 7/2+ | −45.296 | 5.1 d 1 | β− |
| 184 | (5−) | −42.84 | 8.7 h 1 | β− |
| 185 | (7/2+) | −41.40 | 49.4 m 15 | β− |
| 186 | (2−,3−) | −38.61 | 10.5 m 3 | β− |
| 186m | | −38.61 | 1.54 m 5 | β− |
| 187 | | −36.8s | ≈2 m | β−? |
| 188 | | −33.8s | ≈20 s | β−? |
| 189 | (7/2+) | −31.8s | 3 s s y | β−? |
| 190 | | −28.7s | 0.3 s sy | β−? |
| 74 W 158 | 0+ | −23.7s | 1.25 ms 21 | α |
| 159 | | −25.2s | 7.3 ms 27 | α ≈99.9%, ε ≈0.1% |
| 160 | 0+ | −29.4 | 91 ms 5 | α 87% |
| 161 | | −30.4s | 409 ms 18 | α 73% |
| 162 | 0+ | −34.00 | 1.36 s 7 | ε 54.8%, α 45.2% |
| 163 | | −34.91 | 2.8 s 2 | ε 87%, α 13% |
| 164 | 0+ | −38.23 | 6.3 s 2 | ε 96.2%, α 3.8% |
| 165 | (5/2−) | −38.86 | 5.1 s 5 | ε, α<0.2% |
| 166 | 0+ | −41.89 | 19.2 s 6 | ε 99.97%, α 0.04% |
| 167 | (+) | −42.09 | 19.9 s 5 | ε 99.96%, α 0.04% |
| 168 | 0+ | −44.89 | 53 s 2 | ε =100%, α 3.2×10⁻⁵% |
| 169 | (5/2−) | −44.92 | 74 s 6 | ε |
| 170 | 0+ | −47.29 | 2.42 m 4 | ε |
| 171 | (5/2−) | −47.09 | 2.38 m 4 | ε |

| Nuclide Z El A | Jπ | Δ (MeV) | T½, Γ, or Abundance | Decay Mode |
|---|---|---|---|---|
| 74 W 172 | 0+ | −49.10 | 6.6 m 9 | ε |
| 173 | 5/2− | −48.73 | 7.6 m 2 | ε |
| 174 | 0+ | −50.23 | 33.2 m 21 | ε |
| 175 | (1/2−) | −49.63 | 35.2 m 6 | ε |
| 176 | 0+ | −50.64 | 2.5 h 1 | ε |
| 177 | 1/2− | −49.70 | 132 m 2 | ε |
| 178 | 0+ | −50.42 | 21.6 d 3 | ε |
| 179 | (7/2)− | −49.30 | 37.05 m 16 | ε |
| 179m | (1/2)− | −49.08 | 6.40 m 7 | IT 99.72%, ε 0.28% |
| 180 | 0+ | −49.645 | 1.8×10¹⁸ y 2 | α |
| | | | **0.12% 1** | |
| 181 | 9/2+ | −48.254 | 121.2 d 2 | ε |
| 182 | 0+ | −48.248 | >8.3×10¹⁸ y | α |
| | | | **26.50% 16** | |
| 183 | 1/2− | −46.367 | >1.3×10¹⁹ y | α |
| | | | **14.31% 4** | |
| 183m | 11/2+ | −46.057 | 5.2 s 3 | IT |
| 184 | 0+ | −45.707 | >2.9×10¹⁹ y | α |
| | | | **30.64% 2** | |
| 185 | 3/2− | −43.390 | 75.1 d 3 | β− |
| 185m | 11/2+ | −43.192 | 1.67 m 3 | IT |
| 186 | 0+ | −42.509 | >2.7×10¹⁹ y | α |
| | | | **28.43% 19** | |
| 187 | 3/2− | −39.905 | 23.72 h 6 | β− |
| 188 | 0+ | −38.667 | 69.78 d 5 | β− |
| 189 | (3/2−) | −35.5 | 10.7 m 5 | β− |
| 190 | 0+ | −34.3 | 30.0 m 15 | β− |
| 191 | | −31.1s | >300 ns | β− |
| 192 | 0+ | −29.6s | >300 ns | β−? |
| 75 Re 160 | (2−) | −16.7s | 0.82 ms +15−9 | p 91%, α 9% |
| 161 | 1/2− | −20.9 | 0.37 ms 4 | p |
| 161m | 11/2− | −20.8 | 15.6 ms 9 | α 95.2%, p 4.8% |
| 162 | (2−) | −22.4s | 107 ms 13 | α 94%, ε 6% |
| 162m | (9+) | −22.2s | 77 ms 9 | α 91%, ε 9% |
| 163 | (1/2+) | −26.01 | 390 ms 72 | ε 68%, α 32% |
| 163 (11/2−) | | −25.89 | 214 ms 5 | α 66%, ε 34% |
| 164 | | −27.6s | 0.53 s 23 | α ≈58%, ε ≈42% |
| 165 | (1/2+) | −30.6f | ≈1 s | ε |
| 165 (11/2−) | | −30.61 | 2.1 s 3 | ε 87%, α 13% |
| 166 | | −31.85s | 2.8 s 3 | α ≥8% |
| 167 | (9/2−) | −33.86 | 5.9 s 3 | ε ≈99%, α ≈1% |
| 167m | | −34.84s | 3.4 s 4 | α ≈100% |
| 168 | (5+,6+,7+) | −35.79 | 4.4 s 1 | ε ≈100%, α ≈5.0×10⁻³% |
| 169 | (9/2−) | −38.39 | 8.1 s 5 | ε, α<0.01% |
| 169m | | −38.39 | 15.1 s 15 | α ≈0.2% |
| 170 | (5+) | −38.92 | 9.2 s 2 | ε |
| 171 | (9/2−) | −41.25 | 15.2 s 4 | ε |
| 172 | (5) | −41.52 | 15 s 3 | ε |
| 172m | (2) | −41.52 | 55 s 5 | ε |
| 173 | (5/2−) | −43.55 | 1.90 m DC | ε |
| 174 | | −43.67 | 2.40 m 4 | ε |

## Nuclear Wallet Cards

| Nuclide Z El A | Jπ | Δ (MeV) | T½, Γ, or Abundance | Decay Mode |
|---|---|---|---|---|
| 75 Re 175 | (5/2−) | −45.29 | 5.89 m 5 | ε |
| 176 | 3+ | −45.06 | 5.3 m 3 | ε |
| 177 | 5/2− | −46.27 | 14 m 1 | ε |
| 178 | (3+) | −45.65 | 13.2 m 2 | ε |
| 179 | (5/2)+ | −46.59 | 19.5 m 1 | ε |
| 180 | 1+ | −45.84 | 2.44 m 6 | ε |
| 181 | 5/2+ | −46.51 | 19.9 h 7 | ε |
| 182 | 7+ | −45.4 | 64.0 h 5 | ε |
| 182m | 2+ | −45.4 | 12.7 h 2 | ε |
| 183 | 5/2+ | −45.811 | 70.0 d 14 | ε |
| 184 | 3(−) | −44.227 | 38.0 d 5 | ε |
| 184m | 8(+) | −44.039 | 169 d 8 | IT 75.4%, ε 24.6% |
| 185 | 5/2+ | −43.822 | **37.40% 2** | |
| 186 | 1− | −41.930 | 3.7186 d 5 | β− 92.53%, ε 7.47% |
| 186m | (8+) | −41.781 | 2.0×10⁵ y | IT |
| 187 | 5/2+ | −41.216 | 4.12×10¹⁰ y 11 | β− |
| | | | **62.60% 2** | α<1.0×10⁻⁴% |
| 188 | 1− | −39.016 | 17.003 h 3 | β− |
| 188m | (6)− | −38.844 | 18.59 m 4 | IT |
| 189 | 5/2+ | −37.978 | 24.3 h 4 | β− |
| 190 | (2)− | −35.6 | 3.1 m 3 | β− |
| 190m | (6−) | −35.4 | 3.2 h 2 | β− 54.4%, IT 45.6% |
| 191 | (3/2+,1/2+) | −34.35 | 9.8 m 5 | β− |
| 192 | | −31.7s | 16 s 1 | β− |
| 193 | | −30.3s | 30 s sy | β− |
| 194 | | −27.6s | >300 ns | β− |
| 76 Os 162 | 0+ | −14.5s | 1.9 ms 2 | α |
| 163 | | −16.1s | 5.5 ms 6 | α ≈100%, ε |
| 164 | 0+ | −20.5 | 21 ms 1 | α 98%, ε 2% |
| 165 | (7/2−) | −22.6s | 71 ms 3 | α>60%, ε<40% |
| 166 | 0+ | −25.44 | 181 ms 38 | α 72%, ε 18% |
| 167 | | −26.50 | 0.81 s 6 | α 57%, ε 43% |
| 168 | 0+ | −29.99 | 2.1 s 1 | α 40%, ε |
| 169 | | −30.72 | 3.40 s 9 | ε 88.8%, α 11.2% |
| 170 | 0+ | −33.93 | 7.46 s 23 | ε 91.4%, α 8.6% |
| 171 | (5/2−) | −34.29 | 8.3 s 2 | ε 98.2%, α 1.8% |
| 172 | 0+ | −37.24 | 19.2 s 5 | ε 1.1%, ε |
| 173 | (5/2−) | −37.44 | 22.4 s 9 | α 0.4%, ε |
| 174 | 0+ | −40.00 | 44 s 4 | ε 99.98%, α 0.02% |
| 175 | (5/2−) | −40.10 | 1.4 m 1 | ε |
| 176 | 0+ | −42.10 | 3.6 m 5 | ε |
| 177 | 1/2− | −41.95 | 3.0 m 2 | ε |
| 178 | 0+ | −43.55 | 5.0 m 4 | ε |
| 179 | (1/2−) | −43.02 | 6.5 m 3 | ε |
| 180 | 0+ | −44.36 | 21.5 m 4 | ε |
| 181 | 1/2− | −43.55 | 105 m 3 | ε |
| 181m | 7/2− | −43.50 | 2.7 m 1 | ε =100%, IT≤3% |
| 182 | 0+ | −44.61 | 22.10 h 25 | ε |
| 183 | 9/2+ | −43.66 | 13.0 h 5 | ε |
| 183m | 1/2− | −43.49 | 9.9 h 3 | ε 85%, IT 15% |
| 184 | 0+ | −44.256 | >5.6×10¹³ y | α |
| | | | **0.02% 1** | |

## Nuclear Wallet Cards

| Nuclide Z El A | Jπ | Δ (MeV) | T½, Γ, or Abundance | Decay Mode |
|---|---|---|---|---|
| 76 Os 185 | 1/2− | −42.809 | 93.6 d 5 | ε |
| 186 | 0+ | −43.000 | 2.0×10¹⁵ y 11 | α |
| | | | **1.59% 3** | |
| 187 | 1/2− | −41.218 | **1.6% 3** | |
| 188 | 0+ | −41.136 | **13.29% 8** | |
| 189 | 3/2− | −38.985 | **16.21% 5** | |
| 189m | 9/2− | −38.955 | 5.81 h 6 | IT |
| 190 | 0+ | −38.706 | **26.36% 2** | |
| 190m | (10−) | −37.001 | 9.9 m 1 | IT |
| 191 | 9/2− | −36.394 | 15.4 d 1 | β− |
| 191m | 3/2− | −36.320 | 13.10 h 5 | IT |
| 192 | 0+ | −35.881 | **40.93% 19** | |
| 192m | (10−) | −33.865 | 5.9 s 1 | IT>87%, β−<13% |
| 193 | 3/2− | −33.393 | 30.11 h 1 | β− |
| 194 | 0+ | −32.433 | 6.0 y 2 | β− |
| 195 | | −29.7 | ≈9 m | β−? |
| 196 | 0+ | −28.28 | 34.9 m 2 | β− |
| 197 | | −27.6s | 2.8 m 6 | β− |
| 77 Ir 164 | (9+) | −7.3s | 0.11 ms +6−3 | p?, α? |
| 165 | (1/2+) | −11.6s | <1 μs | p? |
| 166 | (2−) | −13.2s | 10.5 ms 22 | α 93.1%, p 6.9% |
| 166m | (9+) | −13.2s | 15.1 ms 9 | α 98.2%, p 1.8% |
| 167 | 1/2+ | −17.08 | 35.2 ms 20 | α 48%, p 32%, ε 20% |
| 167m | 11/2− | −16.90 | 25.7 ms 8 | α 80%, ε 20%, p 0.4% |
| 168 | | −18.7s | 0.161 ms 21 | α 82% |
| 169 | (1/2+) | −22.08 | 0.64 s +46−24 | ε 50%, ε, p |
| 169m | (11/2−) | −21.93 | 0.308 s 22 | α 81% |
| 170 | | −23.3s | 0.87 s +18−12 | ε 94.8%, α 5.2% |
| 170m | | −23.3s | 0.44 s 6 | ε≤64%, IT<64%, α 36% |
| 171 | (1/2+) | −26.43 | 3.2 s +13−7 | ε 99%, p |
| 171m | (11/2−) | −26.43 | 1.40 s 10 | α 58%, ε≤42%, p≤42% |
| 172 | (3+) | −27.5s | 4.4 s 3 | ε 98%, α ≈2% |
| 172m | (7+) | −27.4s | 2.0 s 1 | ε 77%, α 23% |
| 173 | (3/2+,5/2+) | −30.27 | 9.0 s 8 | α 7%, ε |
| 173m | (11/2−) | −30.27 | 9.0 s 8 | ε>93%, α<7% |
| 174 | (3+) | −30.87 | 7.9 s 6 | ε 99.5%, α 0.5% |
| 174 | (7+) | −30.68 | 4.9 s 3 | ε 97.5%, α 2.5% |
| 175 | (5/2−) | −33.43 | 9 s 2 | ε 99.15%, α 0.85% |
| 176 | | −33.86 | 8.3 s 6 | ε 96.9%, α 3.1% |
| 177 | 5/2− | −36.05 | 30 s 2 | ε 99.94%, α 0.06% |
| 178 | | −36.25 | 12 s 2 | ε |
| 179 | (5/2−) | −38.08 | 79 s 1 | ε |
| 180 | (4,5) | −37.98 | 1.5 m 1 | ε |
| 181 | 5/2− | −39.47 | 4.90 m 15 | ε |
| 182 | (5+) | −39.05 | 15 m 1 | ε |
| 183 | 5/2− | −39.61 | 57 m 4 | ε |
| 184 | 5− | −39.61 | 3.09 h 3 | ε |
| 185 | 5/2− | −40.34 | 14.4 h 1 | ε |
| 186 | 5+ | −39.17 | 16.64 h 3 | ε |
| 186m | 2− | −39.17 | 1.90 h 5 | ε =75%, IT=25% |

## Nuclear Wallet Cards

| Nuclide Z El A | Jπ | Δ (MeV) | T½, Γ, or Abundance | Decay Mode |
|---|---|---|---|---|
| 77 Ir 187 | 3/2+ | −39.716 | 10.5 h 3 | ε |
| 188 | 1− | −38.328 | 41.5 h 5 | ε |
| 189 | 3/2+ | −38.45 | 13.2 d 1 | ε |
| 190 | 4− | −36.751 | 11.78 d 10 | ε, ε < 2.0×10⁻³% |
| 190m | (1−) | −36.725 | 1.120 h 3 | IT |
| 190m | (11−) | −36.375 | 3.087 h 12 | ε 91.4%, IT 8.6% |
| 191 | 3/2+ | −36.706 | 37.3% 2 | |
| 191m | 11/2− | −36.535 | 4.94 s 3 | IT |
| 191m | | −34.659 | 5.5 s 7 | IT |
| 192 | 4+ | −34.833 | 73.827 d 13 | β− 95.13%, ε 4.87% |
| 192m | 1− | −34.777 | 1.45 m 5 | IT 99.98%, β− 0.02% |
| 192m | (11−) | −34.665 | 241 y 9 | IT |
| 193 | 3/2+ | −34.534 | 62.7% 2 | |
| 193m | 11/2− | −34.454 | 10.53 d 4 | IT |
| 194 | 1− | −32.529 | 19.28 h 13 | β− |
| 194m | (10,11) | −32.339 | 171 d 11 | β− |
| 195 | 3/2+ | −31.690 | 2.5 h 2 | β− |
| 195m | 11/2− | −31.590 | 3.8 h 2 | β− 95%, IT 5% |
| 196 | (0−) | −29.44 | 52 s 1 | β− |
| 196m | (10,11−) | −29.03 | 1.40 h 2 | β− 100%, IT<0.3% |
| 197 | 3/2+ | −28.27 | 5.8 m 5 | β− |
| 197m | 11/2− | −28.15 | 8.9 m 3 | β− 99.75%, IT 0.25% |
| 198 | | −25.8s | 8 s 1 | β− |
| 199 | | −24.40 | 20 s sy | β− |
| 78 Pt 166 | 0+ | −4.8s | 300 μs 100 | α |
| 167 | | −6.5s | 0.9 ms 3 | α |
| 168 | 0+ | −11.0 | 2.1 ms 2 | α ≤100% |
| 169 | | −12.4s | 7.0 ms 2 | α |
| 170 | 0+ | −16.31 | 14.0 ms 2 | α |
| 171 | | −17.47 | 51 ms 2 | α=99%, ε 2% |
| 172 | 0+ | −21.10 | 104 ms 1 | α=94%, ε 6% |
| 173 | | −21.94 | 370 ms 13 | α 83%, ε |
| 174 | 0+ | −25.32 | 0.889 s 17 | α 76%, ε 24% |
| 175 | (7/2−) | −25.69 | 2.53 s 6 | α 64%, ε 36% |
| 176 | 0+ | −28.93 | 6.33 s 15 | ε 62%, α 38% |
| 177 | 5/2− | −29.37 | 10.6 s 4 | ε 94.3%, α 5.7% |
| 178 | 0+ | −32.00 | 21.1 s 6 | ε 92.3%, α 7.7% |
| 179 | 1/2− | −32.264 | 21.2 s 4 | ε 99.76%, α 0.24% |
| 180 | 0+ | −34.44 | 56 s 2 | ε, α=0.3% |
| 181 | 1/2− | −34.37 | 52.0 s 22 | ε, α=0.08% |
| 182 | 0+ | −36.17 | 3.0 m 2 | ε 99.96%, α 0.04% |
| 183 | 1/2− | −35.77 | 6.5 m 10 | ε, α=1.3×10⁻³% |
| 183m | (7/2−) | −35.74 | 43 s 5 | ε=100%, ε<4.0×10⁻⁴%, IT |
| 184 | 0+ | −37.33 | 17.3 m 2 | ε, α=0.001% |
| 185 | 9/2+ | −36.68 | 70.9 m 24 | ε |
| 185m | 1/2− | −36.58 | 33.0 m 8 | ε 99%, IT<2% |
| 186 | 0+ | −37.86 | 2.08 h 5 | ε, α=1.4×10⁻⁴% |
| 187 | 3/2− | −36.71 | 2.35 h 3 | ε |
| 188 | 0+ | −37.823 | 10.2 d 3 | ε, α 2.6×10⁻⁵% |
| 189 | 3/2− | −36.48 | 10.87 h 12 | ε |

| Nuclide Z El A | Jπ | Δ (MeV) | T½, Γ, or Abundance | Decay Mode |
|---|---|---|---|---|
| 78 Pt 190 | 0+ | −37.323 | 6.5×10¹¹ y 3  0.014% 1 | α |
| 191 | 3/2− | −35.698 | 2.862 d 7 | ε |
| 192 | 0+ | −36.293 | 0.782% 7 | |
| 193 | 1/2− | −34.477 | 50 y 6 | ε |
| 193m | 13/2+ | −34.327 | 4.33 d 3 | IT |
| 194 | 0+ | −34.763 | 32.967% 99 | |
| 195 | 1/2− | −32.797 | 33.832% 10 | |
| 195m | 13/2+ | −32.537 | 4.010 d 5 | IT |
| 196 | 0+ | −32.647 | 25.242% 41 | |
| 197 | 1/2 | −30.422 | 19.8915 h 19 | β− |
| 197m | 13/2+ | −30.023 | 95.41 m 18 | IT 96.7%, β− 3.3% |
| 198 | 0+ | −29.908 | 7.163% 55 | |
| 199 | 5/2− | −27.392 | 30.80 m 21 | β− |
| 199m | (13/2)+ | −26.968 | 13.6 s 4 | IT |
| 200 | 0+ | −26.60 | 12.5 h 3 | β− |
| 201 | (5/2−) | −23.74 | 2.5 m 1 | β− |
| 202 | 0+ | −22.6s | 44 h 15 | β− |
| 79 Au 169 | | −1.8s | 150 μs sy | α?, p? |
| 170 | (2−) | −3.6s | 286 μs +50−40 | p 89%, α 11% |
| 171 | (1/2+) | −7.56 | 22 μs +3−2 | p=100% |
| 171m | (11/2−) | −7.31 | 1.09 ms 3 | α 66%, p 38% |
| 172 | | −9.3s | 6.3 ms 15 | α ≤100%, p<2% |
| 173 | (1/2+) | −12.82 | 25 ms 1 | α 94%, ε, p |
| 173m | (11/2−) | −12.61 | 14.0 ms 9 | α 92%, ε, p |
| 174 | | −14.2s | 139 ms 3 | α > 0% |
| 175 | (1/2+) | −17.44 | 0.1 s sy | α?, ε? |
| 175m | (11/2−) | −17.44 | 156 ms 5 | α 94%, ε 6% |
| 176 | | −18.5s | 0.84 s +17−14 | α, ε |
| 177 | (1/2+,3/2+) | −21.55 | 1462 ms 32 | α ≤100%, ε |
| 177m | 11/2− | −21.39 | 1180 ms 12 | α ≤100%, ε |
| 178 | | −22.33 | 2.6 s 5 | α?, ε ≥40% |
| 179 | | −24.95 | 3.3 s 13 | ε 78%, α 22% |
| 180 | | −25.60 | 8.1 s 3 | ε ≤98.2%, α ≥1.8% |
| 181 | 1/2+ | −27.87 | 13.7 s 14 | ε 97.3%, α 2.7% |
| 182 | (3/2−) | −28.30 | 15.6 s 4 | ε 99.87%, α 0.13% |
| 183 | (5/2)− | −30.19 | 42.8 s 10 | ε 99.45%, α 0.55% |
| 184 | 5+ | −30.32 | 20.6 s 9 | α ≤0.02%, ε |
| 184m | 2+ | −30.25 | 47.6 s 14 | ε 70%, IT 30%, α ≤0.02% |
| 185 | 5/2− | −31.87 | 4.25 m 6 | ε 99.74%, α 0.26% |
| 185m | | −31.87 | 6.8 m 3 | ε<100%, IT |
| 186 | 3− | −31.71 | 10.7 m 5 | ε, α 8.0×10⁻⁴% |
| 187 | 1/2+ | −33.01 | 8.4 m 3 | ε, α 3.0×10⁻³% |
| 187m | 9/2− | −32.88 | 2.3 s 1 | IT |
| 188 | 1(−) | −32.30 | 8.84 m 6 | ε |
| 189 | 1/2+ | −33.58 | 28.7 m 3 | ε, α<3.0×10⁻⁵% |
| 189m | 11/2− | −33.33 | 4.59 m 11 | ε |
| 190 | 1− | −32.88 | 42.8 m 10 | ε, α<1.0×10⁻⁶% |
| 190m | (11−) | −32.88 | 125 ms 20 | IT=100% |
| 191 | 3/2+ | −33.81 | 3.18 h 8 | ε |
| 191m(11/2−) | | −33.54 | 0.92 s 11 | IT |

## Nuclear Wallet Cards

| Nuclide Z El A | Jπ | Δ (MeV) | T½, Γ, or Abundance | Decay Mode |
|---|---|---|---|---|
| 79 Au 192 | 1− | −32.78 | 4.94 h 9 | ε |
| 192m | (11−) | −32.34 | 160 ms 20 | IT |
| 193 | 3/2+ | −33.39 | 17.65 h 15 | ε |
| 193m | 11/2− | −33.10 | 3.9 s 3 | IT 99.97%, ε = 0.03% |
| 194 | 1− | −32.26 | 38.02 h 10 | ε |
| 194m | (5+) | −32.15 | 600 ms 8 | IT |
| 194m | (11−) | −31.79 | 420 ms 10 | IT |
| 195 | 3/2+ | −32.570 | 186.098 d 47 | ε |
| 195m | 11/2− | −32.251 | 30.5 s 2 | IT |
| 196 | 2− | −31.140 | 6.1669 d 6 | ε 92.8%, β− 7.2% |
| 196m | 5+ | −31.055 | 8.1 s 2 | IT |
| 196m | 12− | −30.544 | 9.6 h 1 | IT |
| 197 | 3/2+ | −31.141 | 100% | |
| 197m | 11/2− | −30.732 | 7.73 s 6 | IT |
| 198 | 2− | −29.582 | 2.6956 d 3 | β− |
| 198m | (12−) | −28.770 | 2.27 d 2 | IT |
| 199 | 3/2+ | −29.095 | 3.139 d 7 | β− |
| 200 | 1(−) | −27.27 | 48.4 m 3 | β− |
| 200m | 12− | −26.31 | 18.7 h 5 | β− 82%, IT 18% |
| 201 | 3/2+ | −26.401 | 26 m 1 | β− |
| 202 | (1−) | −24.4 | 28.8 s 19 | β− |
| 203 | (3/2+) | −23.143 | 60 s 6 | β− |
| 204 | (2−) | −20.8s | 39.8 s 9 | β− |
| 205 | (3/2+) | −18.8s | 31 s 2 | β− |
| 80 Hg 171 | | 3.5s | 59 μs +36−16 | α=100% |
| 172 | 0+ | −1.1 | 0.25 ms +35−9 | α |
| 173 | | −2.6s | 0.6 ms +5−2 | α=100% |
| 174 | 0+ | −6.65 | 2.1 ms +18−7 | α=99.6% |
| 175 | (7/2−,9/2−) | −8.0 | 10.8 ms 4 | α |
| 176 | 0+ | −11.78 | 20 ms 2 | α=100% |
| 177 | (13/2+) | −12.78 | 127.3 ms 18 | α=85%, ε 15% |
| 178 | 0+ | −16.32 | 0.269 s 3 | α=70%, ε=30% |
| 179 | | −16.92 | 1.08 s 9 | α=53%, ε=47%, εp=0.15% |
| 180 | 0+ | −20.24 | 2.58 s 1 | ε 62%, α 48% |
| 181 | 1/2− | −20.66 | 3.6 s 1 | ε 73%, α 27%, εp 0.01%, α 9.0×10⁻⁵% |
| 182 | 0+ | −23.576 | 10.83 s 6 | ε 84.8%, α 15.2% |
| 183 | 1/2− | −23.800 | 9.4 s 7 | ε 88.3%, α 11.7%, εp 2.6×10⁻⁴% |
| 184 | 0+ | −26.35 | 30.9 s 3 | ε 98.89%, α 1.11% |
| 185 | 1/2− | −26.18 | 49.1 s 10 | ε 94%, α 6% |
| 185m | 13/2+ | −26.08 | 21.6 s 15 | IT 54%, ε 46%, α=0.03% |
| 186 | 0+ | −28.54 | 1.38 m 6 | ε 99.98%, α 0.02% |
| 187 | 13/2+ | −28.12 | 2.4 m 3 | ε, α > 1.2×10⁻⁴% |
| 187m | 3/2− | −28.12 | 1.9 m 3 | ε, α > 2.5×10⁻⁴% |
| 188 | 0+ | −30.20 | 3.25 m 15 | ε, α 3.7×10⁻⁵% |
| 189 | 3/2− | −29.63 | 7.6 m 1 | ε, α < 3.0×10⁻⁵% |
| 189m | 13/2+ | −29.63 | 8.6 m 1 | ε, α < 3.0×10⁻⁵% |
| 190 | 0+ | −31.37 | 20.0 m 5 | ε, α < 3.4×10⁻⁵% |

| Nuclide Z El A | Jπ | Δ (MeV) | T½, Γ, or Abundance | Decay Mode |
|---|---|---|---|---|
| 80 Hg 191 | (3/2−) | −30.59 | 49 m 10 | ε |
| 191m | 13/2+ | −30.59 | 50.8 m 15 | ε |
| 192 | 0+ | −32.01 | 4.85 h 20 | ε |
| 193 | 3/2− | −31.05 | 3.80 h 15 | ε |
| 193m | 13/2+ | −30.91 | 11.8 h 2 | ε 92.8%, IT 7.2% |
| 194 | 0+ | −32.19 | 444 y 77 | ε |
| 195 | 1/2− | −31.00 | 10.53 h 3 | ε |
| 195m | 13/2+ | −30.82 | 41.6 h 8 | IT 54.2%, ε 45.8% |
| 196 | 0+ | −31.827 | 0.15% 1 | |
| 197 | 1/2− | −30.541 | 64.14 h 5 | ε |
| 197m | 13/2+ | −30.242 | 23.8 h 1 | IT 91.4%, ε 8.6% |
| 198 | 0+ | −30.954 | 9.97% 20 | |
| 199 | 1/2− | −29.547 | 16.87% 22 | |
| 199m | 13/2+ | −29.015 | 42.67 m 9 | IT |
| 200 | 0+ | −29.504 | 23.10% 19 | |
| 201 | 3/2− | −27.663 | 13.18% 9 | |
| 202 | 0+ | −27.346 | 29.86% 26 | |
| 203 | 5/2− | −25.269 | 46.595 d 6 | β− |
| 204 | 0+ | −24.690 | 6.87% 15 | |
| 205 | 1/2− | −22.288 | 5.14 m 9 | β− |
| 206 | 0+ | −20.95 | 8.15 m 10 | β− |
| 207 | (9/2+) | −16.2 | 2.9 m 2 | β− |
| 208 | 0+ | −13.1s | 41 m +5−4 | β− |
| 209 | | −8.3s | 37 s 8 | β− |
| 210 | 0+ | −5.1s | >300 ns | β−? |
| 81 Tl 176 | (3−,4−,5−) | 1.4s | 5.2 ms +30−14 | p=100% |
| 177 | (1/2+) | −3.33 | 18 ms 5 | α 73%, p 27% |
| 178 | | −4.8s | ≈60 ms | α?, ε? |
| 179 | (1/2+) | −8.30 | 0.42 s 6 | α<100%, ε |
| 179m | (11/2−) | −8.30 | 1.7 ms 2 | α<100%, IT, ε |
| 180 | | −9.4s | 1.5 s 2 | α 7%, εSF=1.0×10⁻⁶%, ε |
| 181 | 1/2+ | −12.801 | 1.4 ms 5 | ε, α≤10% |
| 181m | 9/2− | −11.951 | 3.2 s 3 | α |
| 182 | (7+) | −13.35 | 3.1 s 10 | ε 96%, α<4% |
| 183 | (1/2+) | −16.587 | 6.9 s 7 | ε > 0%, α |
| 183m | (9/2−) | −15.957 | 53.3 ms 3 | α 2%, ε, IT |
| 184 | (2+) | −16.89 | 11 s 1 | ε 97.9%, α 2.1% |
| 185 | (1/2+) | −19.76 | 19.5 s 5 | ε |
| 185m | (9/2−) | −19.30 | 1.93 s 8 | IT, α |
| 186 | (7+) | −20.2 | 27.5 s 10 | ε, α 6.0×10⁻³% |
| 186m | (10−) | −19.8 | 2.9 s 2 | IT |
| 187 | (1/2+) | −22.444 | ≈51 s | ε<100%, α>0% |
| 187m | (9/2−) | −22.108 | 15.60 s 12 | ε<99.9%, IT<99.9%, α 0.15% |
| 188 | (2−) | −22.35 | 71 s 2 | ε |
| 188m | (7+) | −22.35 | 71 s 1 | ε |
| 189 | (1/2+) | −24.60 | 2.3 m 2 | ε |
| 189m | (9/2−) | −24.34 | 1.4 m 1 | ε<100%, IT<4% |
| 190 | 2(−) | −24.33 | 2.6 m 3 | ε |
| 190m | 7(+) | −24.33 | 3.7 m 3 | ε |
| 191 | (1/2+) | −26.281 | ? | ε? |

# APPENDIX B

## Nuclear Wallet Cards

| Nuclide Z El A | Jπ | Δ (MeV) | T½, Γ, or Abundance | Decay Mode |
|---|---|---|---|---|
| 81 Tl 191m | 9/2(−) | −25.982 | 5.22 m 16 | ε |
| 192 | (2−) | −25.87 | 9.6 m 4 | ε |
| 192m | (7+) | −25.72 | 10.8 m 2 | ε |
| 193 | 1/2+ | −27.3 | 21.6 m 8 | ε |
| 193m | 9/2− | −27.0 | 2.11 m 15 | IT≤75%, ε≤25% |
| 194 | 2− | −26.8 | 33.0 m 5 | ε, α <1.0×10⁻⁵% |
| 194m | (7+) | −26.8 | 32.8 m 2 | ε |
| 195 | 1/2+ | −28.16 | 1.16 h 5 | ε |
| 195m | 9/2− | −27.67 | 3.6 s 4 | IT |
| 196 | 2− | −27.50 | 1.84 h 3 | ε |
| 196m | (7+) | −27.10 | 1.41 h 2 | ε 95.5%, IT 4.5% |
| 197 | 1/2+ | −28.34 | 2.84 h 4 | ε |
| 197m | 9/2− | −27.73 | 0.54 s 1 | IT |
| 198 | 2− | −27.49 | 5.3 h 5 | ε |
| 198m | 7+ | −26.95 | 1.87 h 3 | ε 54%, IT 46% |
| 199 | 1/2+ | −28.06 | 7.42 h 8 | ε |
| 200 | 2− | −27.048 | 26.1 h 1 | ε |
| 201 | 1/2+ | −27.18 | 72.912 h 17 | ε |
| 202 | 2− | −25.98 | 12.23 d 2 | ε |
| 203 | 1/2+ | −25.761 | **29.524% 14** | |
| 204 | 2− | −24.346 | 3.78 y 2 | β− 97.1%, ε 2.9% |
| 205 | 1/2+ | −23.821 | **70.476% 14** | |
| 206 | 0− | −22.253 | 4.200 m 17 | β− |
| 206m | (12−) | −19.610 | 3.74 m 3 | IT |
| 207 | 1/2− | −21.034 | 4.77 m 2 | β− |
| 207m | 11/2− | −19.686 | 1.33 s 11 | IT |
| 208 | 5(+) | −16.750 | 3.053 m 4 | β− |
| 209 | (1/2−) | −13.638 | 2.161 m 7 | β− |
| 210 | (5+) | −9.25 | 1.30 m 3 | β−, β−n 7.0×10⁻³% |
| 211 | | −6.1s | >300 ns | β− |
| 212 | | −1.7s | >300 ns | β−? |
| 82 Pb 178 | 0+ | 3.57 | 0.23 ms 15 | α, ε? |
| 179 | | 2.0s | 3 ms sy | α? |
| 180 | 0+ | −1.94 | 4.5 ms 11 | α≤100% |
| 181m(13/2+) | | −3.14 | 45 ms 20 | α<100% |
| 182 | 0+ | −6.83 | 55 ms +40 −35 | α≤100% |
| 183 | (3/2−) | −7.57 | 535 ms 30 | α~90% |
| 183m(13/2+) | | −7.47 | 415 ms 20 | α~100% |
| 184 | 0+ | −11.05 | 490 ms 25 | ε 77%, α 23% |
| 185m 13/2+ | | −11.54 | 4.24 s 17 | α~50%, ε ? |
| 185 | 3/2− | −11.54 | 6.3 s 4 | α~50%, ε ? |
| 186 | 0+ | −14.68 | 4.82 s 3 | ε 60%, α 40% |
| 187 | (3/2−) | −14.980 | 15.2 s 3 | ε 93%, α 7% |
| 187m(13/2+) | | −14.899 | 18.3 s 3 | ε 88%, α 12% |
| 188 | 0+ | −17.82 | 25.1 s 1 | ε 90.7%, α 9.3% |
| 189 | (3/2−) | −17.88 | 51 s 3 | ε >99%, α <0.4% |
| 190 | 0+ | −20.42 | 71 s 1 | ε 99.6%, α 0.4% |
| 191 | (3/2−) | −20.25 | 1.33 m 8 | ε 99.99%, α 0.01% |
| 191m(13/2+) | | −20.11 | 2.18 m 8 | ε, α=0.02% |
| 192 | 0+ | −22.56 | 3.5 m 1 | ε 99.99%, α 5.9×10⁻³% |
| 193 | (3/2−) | −22.19 | 5 m sy | ε |

| Nuclide Z El A | Jπ | Δ (MeV) | T½, Γ, or Abundance | Decay Mode |
|---|---|---|---|---|
| 82 Pb 193m(13/2+) | | −22.19 | 5.8 m 2 | ε |
| 194 | 0+ | −24.21 | 10.7 m 6 | ε, α 7.3×10⁻⁶% |
| 195 | 3/2− | −23.71 | ~15 m | ε |
| 195m 13/2+ | | −23.51 | 15.0 m 12 | ε |
| 196 | 0+ | −25.36 | 37 m 3 | ε=100%, α≤3.3×10⁻⁵% |
| 197 | 3/2− | −24.749 | 8.1 m 17 | ε |
| 197m 13/2+ | | −24.429 | 42.9 m 9 | ε 81%, IT 19% |
| 198 | 0+ | −26.05 | 2.4 h 1 | ε |
| 199 | 3/2− | −25.23 | 90 m 10 | ε |
| 199m(13/2+) | | −24.80 | 12.2 m 3 | IT<100%, ε>0% |
| 200 | 0+ | −26.24 | 21.5 h 4 | ε |
| 201 | 5/2− | −25.26 | 9.33 h 3 | ε |
| 201m 13/2+ | | −24.63 | 61 s 2 | IT>99%, ε<1% |
| 202 | 0+ | −25.934 | 52.5×10³ y 28 | ε, α<1% |
| 202m | 9− | −23.764 | 3.53 h 1 | IT 90.5%, ε 9.5% |
| 203 | 5/2− | −24.787 | 51.92 h 3 | ε |
| 203m 13/2+ | | −23.961 | 6.21 s 8 | IT |
| 203m 29/2− | | −21.837 | 480 ms 7 | IT |
| 204 | 0+ | −25.110 | ≥1.4×10¹⁷ y **1.4% 1** | α? |
| 204m | 9− | −22.924 | 1.14 h 4 | IT |
| 205 | 5/2− | −23.770 | 1.73×10⁷ y 7 | ε |
| 206 | 0+ | −23.785 | **24.1% 1** | |
| 207 | 1/2− | −22.452 | **22.1% 1** | |
| 207m 13/2+ | | −20.819 | 0.806 s 6 | IT |
| 208 | 0+ | −21.749 | **52.4% 1** | |
| 209 | 9/2+ | −17.614 | 3.253 h 14 | β− |
| 210 | 0+ | −14.728 | 22.20 y 22 | β−, α 1.9×10⁻⁶% |
| 211 | 9/2+ | −10.491 | 36.1 m 2 | β− |
| 212 | 0+ | −7.547 | 10.64 h 1 | β− |
| 213 | (9/2+) | −3.184 | 10.2 m 3 | β− |
| 214 | 0+ | −0.181 | 26.8 m 9 | β− |
| 215 | | 4.5s | 36 s 1 | β− |
| 83 Bi 184m | | 1.0s | 6.6 ms 15 | α=100% |
| 184m | | 1.0s | 13 ms 2 | α=100% |
| 185 | 1/2− | −2.21 | 63 μs 3 | p 90%, α 10% |
| 186 | (3+) | −3.17 | 15.0 ms 17 | α=100% |
| 186m | (10−) | −3.17 | 9.8 ms 13 | α=100% |
| 187 | (9/2−) | −6.37 | 32 ms 3 | α |
| 188 | (3+) | −7.20 | 60 ms 3 | α, ε? |
| 188m | (10−) | −7.20 | 265 ms 15 | α, ε? |
| 189 | (9/2−) | −10.06 | 674 ms 11 | α≥50%, ε≤50% |
| 189m | (1/2+) | −9.88 | 5.0 ms 1 | α≤50%, ε≤50% |
| 190m | (10−) | −10.9 | 6.2 s 1 | α 70%, ε 30% |
| 190 | (3+) | −10.9 | 6.3 s 1 | α 90%, ε 10% |
| 191 | (9/2−) | −13.240 | 12.4 s 4 | α 51%, ε 49% |
| 191m | (1/2+) | −13.000 | 121 ms 8 | α 68%, ε 32% |
| 192 | (3+) | −13.55 | 34.6 s 9 | α 88%, ε 12% |
| 192m | (10−) | −13.55 | 39.6 s 4 | α 10% |
| 193 | (9/2−) | −15.873 | 63 s 3 | ε 96.2%, α 3.8% |
| 193m | (1/2+) | −15.566 | 3.2 s 6 | α 84%, ε 16% |

| Nuclide Z El A | Jπ | Δ (MeV) | T½, Γ, or Abundance | Decay Mode |
|---|---|---|---|---|
| 83 Bi 194 | (3+) | −15.99 | 95 s 3 | ε 99.54%, α 0.46% |
| 194m | (10−) | −15.99 | 115 s 4 | ε 99.8%, α 0.2% |
| 194m(6+,7+) | | −15.99 | 125 s 2 | ε |
| 195 | (9/2−) | −18.024 | 183 s 4 | ε 99.97%, α 0.03% |
| 195m | (1/2+) | −17.623 | 87 s 1 | ε 67%, α 33% |
| 196 | (3+) | −18.01 | 308 s 12 | ε=100%, α 1.2×10⁻⁵% |
| 196m | (7+) | −17.84 | 0.6 s 5 | IT, ε |
| 196m | (10−) | −17.74 | 240 s 3 | ε 74.2%, IT 25.8%, α 3.8×10⁻⁴% |
| 197 | (9/2−) | −19.688 | 9.33 h 50 | ε, α 1.0×10⁻⁴% |
| 197m | (1/2+) | −19.188 | 5.04 m 16 | α 55%, ε 45%, IT<0.3% |
| 198 | (2+,3+) | −19.37 | 10.3 m 3 | ε |
| 198m | (7+) | −19.37 | 11.6 m 3 | ε |
| 198m | 10− | −19.12 | 7.7 s 5 | IT |
| 199 | 9/2− | −20.80 | 27 m 1 | ε |
| 199m | (1/2+) | −20.12 | 24.70 m 15 | ε ≥98%, IT≤2%, α=0.01% |
| 200 | 7+ | −20.37 | 36.4 m 5 | ε |
| 200m | (2+) | −20.37 | 31 m 2 | ε>90%, IT<10% |
| 200m | (10−) | −19.94 | 0.40 s 5 | IT |
| 201 | 9/2− | −21.42 | 108 m 3 | ε, α<1.0×10⁻⁴% |
| 201m | 1/2+ | −20.57 | 59.1 m 6 | ε>93%, IT≤6.8%, α=0.3% |
| 202 | 5+ | −20.73 | 1.72 h 5 | ε, α<1.0×10⁻⁵% |
| 203 | 9/2− | −21.54 | 11.76 h 5 | ε |
| 203m | 1/2− | −20.44 | 305 ms 5 | IT |
| 204 | 6+ | −20.67 | 11.22 h 10 | ε |
| 205 | 9/2− | −21.062 | 15.31 d 4 | ε |
| 206 | 6(+) | −20.028 | 6.243 d 3 | ε |
| 207 | 9/2− | −20.054 | 32.9 y 14 | ε |
| 208 | (5)+ | −18.870 | 3.68×10⁵ y 4 | ε |
| 209 | 9/2− | −18.258 | 1.9×10¹⁹ y 2 **100%** | α |
| 210 | 1− | −14.792 | 5.012 d 5 | β−, α 1.3×10⁻⁴% |
| 210m | 9− | −14.521 | 3.04×10⁶ y 6 | α |
| 211 | 9/2− | −11.858 | 2.14 m 2 | α 99.72%, β− 0.28% |
| 212 | 1(−) | −8.117 | 60.55 m 6 | β− 64.06%, α 35.94% |
| 212m | (8−,9−) | −7.867 | 25.0 m 2 | α 67%, β− 33%, β−α 30% |
| 212m | ≥16 | −6.207 | 7.0 m 3 | β− ~100% |
| 213 | 9/2− | −5.231 | 45.59 m 6 | β− 97.91%, α 2.09% |
| 214 | 1− | −1.20 | 19.9 m 4 | β− 99.98%, α 0.02% |
| 215 | (9/2−) | 1.65 | 7.6 m 2 | β− |
| 215m(25/2−) | | 3.00 | 36.4 s | IT, β− |
| 216 | (1−) | 5.87 | 2.17 m 5 | β−≤100% |
| 217 | | 8.8s | 98.5 s 8 | β− |
| 218 | | 13.3s | 33 s 1 | β− |
| 84 Po 188 | 0+ | −0.54 | 0.40 ms +20 −15 | ε<100%, α>0% |
| 189 | | −1.42 | 5 ms 1 | α |
| 190 | 0+ | −4.56 | 2.46 ms 5 | α |

| Nuclide Z El A | Jπ | Δ (MeV) | T½, Γ, or Abundance | Decay Mode |
|---|---|---|---|---|
| 84 Po 191 | (3/2−) | −5.05 | 22 ms 1 | α |
| 191m(13/2+) | | −4.92 | 93 ms 3 | α |
| 192 | 0+ | −8.07 | 33.2 ms 14 | α=99.5%, ε=0.5% |
| 193m(13/2+) | | −8.36 | 243 ms +11 −10 | α≤100% |
| 193m (3/2−) | | −8.36 | 370 ms +46 −40 | α≤100% |
| 194 | 0+ | −11.01 | 0.392 s 4 | α=100%, ε |
| 195 | (3/2−) | −11.07 | 4.64 s 9 | ε 75%, 25% |
| 195m(13/2+) | | −10.84 | 1.92 s 2 | α~90%, ε~10%, IT<0.01% |
| 196 | 0+ | −13.47 | 5.8 s 2 | α=98%, ε=2% |
| 197 | (3/2−) | −13.36 | 84 s 16 | ε 56%, α 44% |
| 197m(13/2+) | | −13.15 | 32 s 2 | α 84%, ε 16%, IT 0.01% |
| 198 | 0+ | −15.47 | 1.77 m 3 | α 57%, ε 43% |
| 199 | (3/2−) | −15.22 | 4.58 ms 52 | ε 92.5%, α 7.5% |
| 199m(13/2+) | | −14.90 | 4.13 m 43 | ε 73.5%, α 24%, IT 2.5% |
| 200 | 0+ | −16.54 | 10.9 m 11 | ε 88.8%, α 11.1% |
| 201 | 3/2− | −16.525 | 15.3 m 2 | ε 98.4%, α 1.6% |
| 201m 13/2+ | | −16.101 | 8.9 m 2 | IT 56%, ε 41%, α=2.9% |
| 202 | 0+ | −17.98 | 44.7 m 5 | ε 98.08%, α 1.92% |
| 203 | 5/2− | −17.31 | 36.7 m 5 | ε 99.89%, α 0.11% |
| 203m 13/2+ | | −16.67 | 45 s 2 | IT |
| 204 | 0+ | −18.33 | 3.53 h 2 | ε 99.34%, α 0.66% |
| 205 | 5/2− | −17.51 | 1.74 h 8 | ε 99.96%, α 0.04% |
| 206 | 0+ | −18.182 | 8.8 d 1 | ε 94.55%, α 5.45% |
| 207 | 5/2− | −17.146 | 5.80 h 2 | ε 99.98%, α 0.02% |
| 207m 19/2− | | −15.763 | 2.79 s 8 | IT |
| 208 | 0+ | −17.469 | 2.898 y 2 | α |
| 209 | 1/2− | −16.366 | 102 y 5 | α 99.52%, ε 0.48% |
| 210 | 0+ | −15.953 | 138.376 d 2 | α |
| 211 | 9/2+ | −12.432 | 0.516 s 3 | α |
| 211m(25/2+) | | −10.970 | 25.2 s 6 | α 99.98%, IT 0.02% |
| 212 (18+) | | −7.447 | 45.1 s 6 | α 99.93% |
| 213 | 9/2+ | −6.653 | 3.65 μs 4 | α |
| 214 | 0+ | −4.470 | 164.3 μs 20 | α |
| 215 | 9/2+ | −0.540 | 1.781 ms 4 | α, β− 2.3×10⁻⁴% |
| 216 | 0+ | 1.784 | 0.145 s 2 | α |
| 217 | (9/2+) | 5.901 | 1.53 s 5 | α |
| 218 | 0+ | 8.358 | 3.10 m 2 | α 99.98%, β− 0.02% |
| 219 | | 12.8s | =2 m | α?, β−? |
| 220 | 0+ | 15.5s | >300 ns | β−? |
| 85 At 191 | (1/2+) | | 1.7 ms +11 −5 | α |
| 191m | (7/2−) | | 2.1 ms +4 −3 | α |
| 193 | (1/2+) | −0.15 | 28 ms +5 −4 | α=100% |
| 193m | (7/2−) | −0.14 | 21 ms 5 | α=100% |
| 194m(13/2+) | | 0.17 | 27 ms +4 −5 | α 24% |
| 194 | (3−) | −1.2 | =40 ms | α, ε |
| 194m | | −1.2 | =250 ms | α, ε, IT |
| 195 | (1/2+) | −3.476 | 328 ms +20 −18 | α |
| 195m | (7/2−) | −3.439 | 147 ms +5 −4 | α |

## Nuclear Wallet Cards

| Nuclide Z El A | Jπ | Δ (MeV) | T½, Γ, or Abundance | Decay Mode |
|---|---|---|---|---|
| 85 At 196 | | −3.92 | 0.39 s 5 | α 94%, ε |
| 197 | (9/2−) | −6.34 | 0.390 s 16 | α 96.1%, ε 3.9% |
| 197m | (1/2+) | −6.29 | 2.0 s 2 | α≤100%, IT≤4.0×10⁻³%, ε |
| 198 | (3+) | −6.67 | 4.2 s 3 | α 90%, ε 10% |
| 198m | (10−) | −6.57 | 1.0 s 2 | α 84%, ε 16% |
| 199 | (9/2−) | −8.82 | 6.92 s 13 | α 90%, ε 10% |
| 200 | (3+) | −8.99 | 43 s 1 | α 57%, ε 43% |
| 200m | (7+) | −8.88 | 47 s 1 | ε 57%, α 43% |
| 200m | (10−) | −8.65 | 3.5 s 2 | IT∼84%, α∼10.5%, ε∼4.5% |
| 201 | (9/2−) | −10.790 | 89 s 3 | α 71%, ε 29% |
| 202 | (2,3)+ | −10.59 | 184 s 1 | β 82%, α 18% |
| 202m | (7+) | −10.59 | 182 s 2 | ε 91.3%, α 8.7% |
| 202m | (10−) | −10.20 | 0.46 s 5 | IT 99.7%, ε 0.25%, α 0.1% |
| 203 | 9/2− | −12.16 | 7.37 m 13 | ε 69%, α 31% |
| 204 | 7+ | −11.88 | 9.2 m 2 | ε 96.2%, α 3.8% |
| 204m | (10−) | −11.29 | 108 ms 10 | IT |
| 205 | 9/2− | −12.97 | 26.9 m 8 | ε 90%, α 10% |
| 206 | (5)+ | −12.42 | 30.6 m 13 | ε 99.11%, α 0.89% |
| 207 | 9/2− | −13.24 | 1.80 h 4 | ε 91.4%, α 8.6% |
| 208 | 6+ | −12.49 | 1.63 h 3 | ε 99.45%, α 0.55% |
| 209 | 9/2− | −12.880 | 5.41 h 5 | ε 95.9%, α 4.1% |
| 210 | (5)+ | −11.972 | 8.1 h 4 | ε 99.82%, α 0.18% |
| 211 | 9/2− | −11.647 | 7.214 h 7 | ε 58.2%, α 41.8% |
| 212 | (1−) | −8.621 | 0.314 s 2 | α, ε<0.03%, β−<2.0×10⁻⁶% |
| 212m | (9−) | −8.398 | 0.119 s 3 | α>99%, IT<1% |
| 213 | 9/2− | −6.580 | 125 ns 6 | α |
| 214 | 1− | −3.380 | 558 ns 10 | α |
| 215 | 9/2− | −1.255 | 0.10 ms 2 | α |
| 216 | 1− | 2.257 | 0.30 ms 3 | α, β−<6.0×10⁻³%, ε<3.0×10⁻⁷% |
| 217 | 9/2− | 4.396 | 32.3 ms 4 | α 99.99%, β− 7.0×10⁻³% |
| 218 | | 8.10 | 1.5 s 3 | α 99.9%, β− 0.1% |
| 219 | | 10.397 | 56 s 3 | α 97%, β− 3% |
| 220 | 3 | 14.35 | 3.71 m 4 | β− 92%, α 8% |
| 221 | | 16.8s | 2.3 m 2 | β− |
| 222 | | 20.8s | 54 s 10 | β− |
| 223 | | 23.5s | 50 s 7 | β− |
| 86 Rn 195 | | 5.07 | 6 ms +3−2 | α |
| 195m | | 5.12 | 5 ms +3−2 | α |
| 196 | 0+ | 1.97 | 4.4 ms +13−9 | α |
| 197 | (3/2−) | 1.48 | 65 ms +25−14 | α∼100% |
| 197m | (13/2+) | 1.48 | 19 ms +8−4 | α∼100% |
| 198 | 0+ | −1.23 | 65 ms 3 | α, ε |
| 199 | (3/2−) | −1.52 | 0.62 s 3 | α 94%, ε 6% |
| 199m | (13/2+) | −1.52 | 0.32 s 2 | α 97%, ε 3% |
| 200 | 0+ | −4.01 | 0.96 s 3 | α 98%, ε 2% |
| 201 | (3/2−) | −4.07 | 7.1 s 8 | α∼80%, ε∼20% |

## Nuclear Wallet Cards

| Nuclide Z El A | Jπ | Δ (MeV) | T½, Γ, or Abundance | Decay Mode |
|---|---|---|---|---|
| 86 Rn 201m | (13/2+) | −3.79 | 3.8 s 1 | α∼90%, ε∼10%, IT∼0% |
| 202 | 0+ | −6.28 | 10.0 s 3 | α 86%, ε 14% |
| 203 | (3/2−) | −6.16 | 44.2 s 16 | α 66%, ε 34% |
| 203m | (13/2+) | −5.80 | 26.9 s 5 | α 75%, ε 25% |
| 204 | 0+ | −7.98 | 1.17 m 18 | α 73%, ε 27% |
| 205 | 5/2− | −7.71 | 170 s 4 | ε 75.4%, α 24.6% |
| 206 | 0+ | −9.12 | 5.67 m 17 | α 62%, ε 38% |
| 207 | 5/2− | −8.63 | 9.25 m 17 | ε 79%, α 21% |
| 208 | 0+ | −9.65 | 24.35 m 14 | α 62%, ε 38% |
| 209 | 5/2− | −8.93 | 28.5 m 10 | ε 83%, α 17% |
| 210 | 0+ | −9.598 | 2.4 h 1 | α 96%, ε 4% |
| 211 | 1/2− | −8.756 | 14.6 h 2 | ε 72.6%, α 27.4% |
| 212 | 0+ | −8.660 | 23.9 m 12 | α |
| 213 | (9/2+) | −5.698 | 19.4 ms 1 | α |
| 214 | 0+ | −4.320 | 0.27 μs 2 | α |
| 215 | 9/2+ | −1.169 | 2.30 μs 10 | α |
| 216 | 0+ | 0.256 | 45 μs 5 | α |
| 217 | 9/2+ | 3.659 | 0.54 ms 5 | α |
| 218 | 0+ | 5.218 | 35 ms 5 | α |
| 219 | 5/2+ | 8.831 | 3.96 s 1 | α |
| 220 | 0+ | 10.613 | 55.6 s 1 | α |
| 221 | 7/2(+) | 14.472 | 25.7 m 5 | β− 78%, α 22% |
| 222 | 0+ | 16.374 | 3.8235 d 4 | α |
| 223 | 7/2 | 20.3s | 24.3 m 4 | β− |
| 224 | 0+ | 22.4s | 107 m 3 | β− |
| 225 | 7/2− | 26.5s | 4.66 m 4 | β− |
| 226 | 0+ | 28.8s | 7.4 m 1 | β− |
| 227 | | 33.0s | 20.8 s 7 | β− |
| 228 | 0+ | 35.4s | 65 s 2 | β− |
| 87 Fr 199 | | 6.76 | 12 ms +10−4 | α>0%, ε |
| 200 | (3+) | 6.12 | 49 ms 4 | α |
| 200m | (10−) | 6.32 | 0.57 s +27−14 | α, ε<1% |
| 201 | (9/2−) | 3.60 | 67 ms 3 | α |
| 201m | | 3.60 | 19 ms +19−6 | α |
| 202 | (3+) | 3.14 | 0.23 s +8−4 | α∼97%, ε∼3% |
| 202m | (10−) | 3.24 | 0.23 s +14−5 | α∼97%, ε∼3% |
| 203 | (9/2−) | 0.86 | 0.55 s 2 | α∼100% |
| 204 | (3+) | 0.61 | 1.7 s 3 | α∼80%, ε∼20% |
| 204m | (7+) | 0.65 | 2.6 s 3 | α≤100% |
| 204m | (10−) | 0.92 | ≈1 s | α≤100%, IT |
| 205 | (9/2−) | −1.310 | 3.80 s 3 | α∼100% |
| 206 | (2+,3+) | −1.24 | ≈16 s | α∼84%, ε∼16% |
| 206m | (7+) | −1.24 | 15.9 s 4 | α 84%, ε 16% |
| 206m | (10−) | −0.71 | 0.7 s 1 | α∼12%, IT |
| 207 | 9/2− | −2.84 | 14.8 s 1 | α 95%, ε 5% |
| 208 | 7+ | −2.67 | 59.1 s 3 | α 90%, ε 10% |
| 209 | 9/2− | −3.77 | 50.0 s 3 | α 89%, ε 11% |
| 210 | 6+ | −3.35 | 3.18 m 6 | α 60%, ε 40% |
| 211 | 9/2− | −4.16 | 3.10 m 2 | α>80%, ε<20% |
| 212 | 5+ | −3.54 | 20.0 m 6 | ε 57%, α 43% |
| 213 | 9/2− | −3.550 | 34.6 s 3 | α 99.45%, ε 0.55% |

## Nuclear Wallet Cards

| Nuclide Z El A | Jπ | Δ (MeV) | T½, Γ, or Abundance | Decay Mode |
|---|---|---|---|---|
| 87 Fr 214 | (1−) | −0.958 | 5.0 ms 2 | α |
| 214m | (8−) | −0.836 | 3.35 ms 5 | α |
| 215 | 9/2− | 0.318 | 86 ns 5 | α |
| 216 | (1−) | 2.98 | 0.70 μs 2 | α, ε<2.0×10⁻⁷% |
| 217 | 9/2− | 4.315 | 19 μs 3 | α |
| 218 | 1− | 7.059 | 1.0 ms 6 | α |
| 218m | | 7.145 | 22.0 ms 5 | α≤100%, IT |
| 219 | 9/2− | 8.618 | 20 ms 2 | α |
| 220 | 1+ | 11.483 | 27.4 s 3 | α 99.65%, β− 0.35% |
| 221 | 5/2− | 13.278 | 4.9 m 2 | α, β−<0.1%, ¹⁴C 9.0×10⁻¹³% |
| 222 | 2− | 16.35 | 14.2 m 3 | β− |
| 223 | 3/2(−) | 18.384 | 22.00 m 7 | β− 99.99%, α 6.0×10⁻³% |
| 224 | 1− | 21.66 | 3.33 m 10 | β− |
| 225 | 3/2− | 23.81 | 4.0 m 2 | β− |
| 226 | 1− | 27.4 | 49 s 1 | β− |
| 227 | 1/2+ | 29.7 | 2.47 m 3 | β− |
| 228 | 2− | 33.3s | 38 s 1 | β−≤100% |
| 229 | (1/2+) | 35.82 | 50.2 s 4 | β− |
| 230 | | 39.6s | 19.1 s 5 | β− |
| 231 | (1/2+) | 42.3s | 17.6 s 6 | β− |
| 232 | | 46.4s | 5 s 1 | β− |
| 88 Ra 202 | 0+ | 9.21 | 0.7 ms +33−3 | α |
| 203 | (3/2−) | 8.64 | 1.0 ms +50−5 | α∼100% |
| 203m | (13/2+) | 8.64 | 33 ms +22−10 | α∼100% |
| 204 | 0+ | 6.05 | 59 ms +12−9 | α |
| 205 | (3/2) | 5.84 | 210 ms +60−40 | α≤100%, ε |
| 205m | (13/2+) | 5.84 | 170 ms +60−40 | α≤100%, ε |
| 206 | 0+ | 3.57 | 0.24 s 2 | α |
| 207 | (5/2−,3/2−) | 3.54 | 1.3 s 2 | α∼90%, ε∼10% |
| 207m | (13/2+) | 4.01 | 55 ms 10 | IT 85%, α 15%, ε 0.35% |
| 208 | 0+ | 1.71 | 1.3 s 2 | α 96%, ε 5% |
| 209 | 5/2− | 1.86 | 4.6 s 2 | α 90%, ε 10% |
| 210 | 0+ | 0.46 | 3.7 s 2 | α∼96%, ε∼4% |
| 211 | 5/2(−) | 0.84 | 13 s 2 | α>93%, ε<7% |
| 212 | 0+ | −0.19 | 13.0 s 2 | α∼85%, ε 15% |
| 213 | 1/2− | 0.36 | 2.74 m 6 | α 80%, ε 20% |
| 213m | | 2.13 | 2.1 ms 1 | IT∼99%, α∼1% |
| 214 | 0+ | 0.101 | 2.46 s 3 | α 99.94%, ε 0.06% |
| 215 | (9/2+) | 2.533 | 1.55 ms 7 | α |
| 216 | 0+ | 3.291 | 182 ns 10 | α, ε<1.0×10⁻⁶% |
| 217 | (9/2+) | 5.887 | 1.6 μs 2 | α∼100% |
| 218 | 0+ | 6.65 | 25.2 μs 3 | α |
| 219 | (7/2)+ | 9.394 | 10 ms 3 | α |
| 220 | 0+ | 10.273 | 18 ms 2 | α |
| 221 | 5/2+ | 12.964 | 28 s 2 | α, ¹⁴C 1×10⁻¹²% |
| 222 | 0+ | 14.321 | 38.0 s 5 | α, ¹⁴C 3.0×10⁻⁸% |
| 223 | 3/2+ | 17.235 | 11.43 d 5 | α, ¹⁴C 8.9×10⁻¹⁰% |
| 224 | 0+ | 18.827 | 3.6319 d 23 | α, ¹⁴C 4.0×10⁻⁹% |
| 225 | 1/2+ | 21.994 | 14.9 d 2 | β− |

## Nuclear Wallet Cards

| Nuclide Z El A | Jπ | Δ (MeV) | T½, Γ, or Abundance | Decay Mode |
|---|---|---|---|---|
| 88 Ra 226 | 0+ | 23.669 | 1600 y 7 | α, ¹⁴C 3.2×10⁻⁹% |
| 227 | 3/2+ | 27.179 | 42.2 m 5 | β− |
| 228 | 0+ | 28.942 | 5.75 y 3 | β− |
| 229 | 5/2(+) | 32.56 | 4.0 m 2 | β− |
| 230 | 0+ | 34.52 | 93 m 2 | β− |
| 231 | (5/2+) | 38.4s | 103 s 3 | β− |
| 232 | 0+ | 40.6s | 250 s 50 | β− |
| 233 | | 44.8s | 30 s 5 | β− |
| 234 | 0+ | 47.2s | 30 s 10 | β− |
| 89 Ac 206m | | 13.51 | 11 ms +9−3 | α |
| 206m | (3+) | 13.51 | 22 ms +9−5 | α |
| 206m | (10−) | 13.51 | 33 ms +22−9 | α |
| 207 | 9/2− | 11.13 | 27 ms +11−6 | α |
| 208 | (3+) | 10.76 | 95 ms +24−16 | ε 1%, α |
| 208m | (10−) | 11.27 | 25 ms +9−5 | IT<10%, ε 1%, α |
| 209 | (9/2−) | 8.84 | 0.10 s 5 | α∼99%, ε∼1% |
| 210 | | 8.79 | 0.35 s 5 | α 91%, ε∼9% |
| 211 | | 7.20 | 0.21 s 3 | α∼100% |
| 212 | | 7.28 | 0.93 s 5 | ε∼57%, α∼43% |
| 213 | | 6.16 | 0.731 s 17 | α≤100% |
| 214 | | 6.43 | 8.2 s 2 | α∼89%, ε≤11% |
| 215 | 9/2− | 6.01 | 0.17 s 1 | α 99.91%, ε 0.09% |
| 216 | (1−) | 8.12 | 0.440 ms 16 | α |
| 217 | 9/2− | 8.71 | 69 ns 4 | α∼100%, ε≤2% |
| 218 | (1−) | 10.84 | 1.08 μs 9 | α |
| 219 | 9/2− | 11.57 | 11.8 μs 15 | α |
| 220 | (3−) | 13.75 | 26.4 ms 2 | α, ε 5.0×10⁻⁴% |
| 221 | | 14.52 | 52 ms 2 | α |
| 222 | 1− | 16.621 | 5.0 s 5 | α 99%, ε 1% |
| 222m | | 16.621 | 63 ms 3 | α≥88%, IT≤10%, ε≥0.7% |
| 223 | (5/2−) | 17.826 | 2.10 m 5 | α 99%, ε 1% |
| 224 | 0− | 20.235 | 2.78 h 17 | ε 90.9%, α 9.1%, β−<1.6% |
| 225 | (3/2−) | 21.638 | 10.0 d 1 | α, ¹⁴C 5×10⁻¹⁰% |
| 226 | (1) | 24.310 | 29.37 h 12 | β− 83%, ε 17%, α 6.0×10⁻³% |
| 227 | 3/2− | 25.851 | 21.772 y 3 | β− 98.62%, α 1.38% |
| 228 | 3+ | 28.896 | 6.15 h 2 | β− |
| 229 | (3/2+) | 30.75 | 62.7 m 5 | β− |
| 230 | (1+) | 33.8 | 122 s 3 | β− |
| 231 | (1/2+) | 35.9 | 7.5 m 1 | β− |
| 232 | (1+) | 39.1 | 119 s 5 | β− |
| 233 | (1/2+) | 41.5s | 145 s 10 | β− |
| 234 | | 45.1s | 44 s 7 | β− |
| 235 | | 47.7s | ≈40 s | β−? |
| 236 | | 51.5s | ≈2 m | β−? |
| 90 Th 209 | (5/2−) | 16.50 | 3.8 ms +69−15 | α |
| 210 | 0+ | 14.04 | 9 ms +17−4 | α 99%, ε 1% |
| 211 | | 13.91 | 0.04 s +3−1 | α |
| 212 | 0+ | 12.09 | 30 ms +20−10 | α, ε∼0.3% |
| 213 | | 12.12 | 140 ms 25 | α≤100% |

## Nuclear Wallet Cards

| Nuclide Z El A | Jπ | Δ (MeV) | T½, Γ, or Abundance | Decay Mode |
|---|---|---|---|---|
| 90 Th 214 | 0+ | 10.71 | 100 ms 25 | α |
| 215 | (1/2−) | 10.93 | 1.2 s 2 | α |
| 216 | 0+ | 10.30 | 0.028 s 2 | α, ε=0.01% |
| 217 | (9/2+) | 12.22 | 0.241 ms 5 | α |
| 218 | 0+ | 12.37 | 109 ns 13 | α |
| 219 | | 14.47 | 1.05 µs 3 | α |
| 220 | 0+ | 14.67 | 9.7 µs 6 | α, ε 2.0×10⁻⁷% |
| 221 | (7/2+) | 16.938 | 1.73 ms 3 | α |
| 222 | 0+ | 17.20 | 2.237 ms 13 | α |
| 223 | (5/2)+ | 19.386 | 0.60 s 2 | α |
| 224 | 0+ | 20.00 | 0.81 s 10 | α |
| 225 | (3/2)+ | 22.310 | 8.72 m 4 | α≈90%, ε≈10% |
| 226 | 0+ | 23.197 | 30.57 m 10 | α |
| 227 | 1/2+ | 25.806 | 18.68 d 9 | α |
| 228 | 0+ | 26.772 | 1.9116 y 16 | α, ²⁰O 1×10⁻¹¹% |
| 229 | 5/2+ | 29.587 | 7340 y 160 | α |
| 229m | | 29.590 | 13.9 h 30 | α |
| 230 | 0+ | 30.864 | 7.538×10⁴ y 30 | α, SF<4.×10⁻¹²% |
| 231 | 5/2+ | 33.817 | 25.52 h 1 | β−, α=4×10⁻¹¹% |
| 232 | 0+ | 35.448 | 1.405×10¹⁰ y 6 | α, 100% SF 1.2×10⁻⁹%, Ne |
| 233 | 1/2+ | 38.733 | 21.83 m 4 | β− |
| 234 | 0+ | 40.614 | 24.10 d 3 | β− |
| 235 | (1/2+) | 44.26 | 7.2 m 1 | β− |
| 236 | 0+ | 46.5s | 37.5 m 2 | β− |
| 237 | (5/2+) | 50.2s | 4.7 m 6 | β− |
| 238 | 0+ | 52.6s | 9.4 m 20 | β− |
| 91 Pa 212 | | 21.61 | 5.1 ms +61−19 | α 100% |
| 213 | (9/2−) | 19.66 | 5.3 ms +40−16 | α |
| 214 | | 19.49 | 17 ms 3 | α |
| 215 | | 17.87 | 14 ms 2 | α |
| 216 | | 17.80 | 105 ms 12 | α=98%, ε=2% |
| 217 | | 17.07 | 3.6 ms 8 | α |
| 217m | | 18.92 | 1.2 ms 2 | α 73%, IT 27% |
| 218 | | 18.67 | 0.113 ms 1 | α |
| 219m | 9/2− | 18.52 | 53 ns 10 | α |
| 220? | | 20.38 | 0.78 µs 16 | α, ε 3.0×10⁻⁷% |
| 221 | 9/2− | 20.38 | 4.9 µs 8 | α |
| 222 | | 22.12s | 3.3 ms 3 | α |
| 223 | | 22.32 | 5.1 ms 6 | α |
| 224 | | 23.87 | 0.85 s 2 | α |
| 225 | | 24.34 | 1.7 s 2 | α |
| 226 | | 26.03 | 1.8 m 2 | α 74%, ε 26% |
| 227 | (5/2−) | 26.832 | 38.3 m 3 | α 89%, ε 15% |
| 228 | 3+ | 28.924 | 22 h 1 | ε 98%, α 2% |
| 229 | (5/2+) | 29.898 | 1.50 d 5 | ε 99.52%, α 0.48% |
| 230 | (2−) | 32.174 | 17.4 d 5 | ε 91.6%, β− 8.4%, α 3.2×10⁻³% |
| 231 | 3/2− | 33.426 | 3.276×10⁴ y 11 | α, SF≤3×10⁻¹⁰% |
| 232 | (2−) | 35.948 | 1.31 d 2 | β−, ε 3.0×10⁻³% |
| 233 | 3/2− | 37.490 | 26.975 d 13 | β− |
| 234 | 4+ | 40.341 | 6.70 h 5 | β− |

| Nuclide Z El A | Jπ | Δ (MeV) | T½, Γ, or Abundance | Decay Mode |
|---|---|---|---|---|
| 91 Pa 234m | (0−) | 40.415 | 1.17 m 3 | β− 99.84%, IT 0.16% |
| 235 | (3/2−) | 42.33 | 24.44 m 11 | β− |
| 236 | 1(−) | 45.3 | 9.1 m 1 | β− |
| 237 | (1/2+) | 47.6 | 8.7 m 2 | β− |
| 238 | (3−) | 50.77 | 2.27 m 9 | β−, SF<2.6×10⁻⁶% |
| 239 | (3/2) | 53.3s | 1.8 h 5 | β− |
| 240 | | 56.8s | −2 m | β−? |
| 92 U 217 | | 22.70 | 16 ms +21−6 | α 100% |
| 218 | 0+ | 21.92 | 1.5 ms +73−7 | α |
| 219 | | 23.21 | 42 µs +34−13 | α |
| 220 | 0+ | 23.0s | ~60 ns | α?, ε? |
| 221 | | 24.6s | ~0.7 µs | α?, ε? |
| 222 | 0+ | 24.3s | 1.0 µs +10−4 | α |
| 223 | | 25.84 | 18 µs +10−5 | α, ε 0.2% |
| 224 | 0+ | 25.71 | 0.9 ms 3 | α |
| 225 | | 27.38 | 84 ms 4 | α |
| 226 | 0+ | 27.33 | 0.26 s 2 | α |
| 227 | (3/2+) | 29.02 | 1.1 m 1 | α |
| 228 | 0+ | 29.22 | 9.1 m 2 | α 95%, ε≤5% |
| 229 | (3/2+) | 31.211 | 58 m 3 | ε≈80%, α≈20% |
| 230 | 0+ | 31.615 | 20.8 d | α, SF<1×10⁻¹⁰% |
| 231 | (5/2−) | 33.807 | 4.2 d 1 | ε, α=4.0×10⁻³% |
| 232 | 0+ | 34.611 | 68.9 y 4 | α, ²⁴Ne 5×10⁻¹²%, Mg<5×10⁻¹³%, SF 3×10⁻¹²% |
| 233 | 5/2+ | 36.920 | 1.592×10⁵ y 2 | α, SF<6.0×10⁻⁹% |
| 234 | 0+ | 38.147 | 2.455×10⁵ y 6 | α, 0.0054% 5 SF 1.6×10⁻⁹%, Mg 1×10⁻¹¹%, Ne 9×10⁻¹²% |
| 235 | 7/2− | 40.921 | 7.04×10⁸ y 1 | α, 0.7204% 6 SF 7.0×10⁻⁹%, Ne=8.×10⁻¹⁰%, 28.8.×10⁻¹⁰% |
| 235m | 1/2+ | 40.921 | ≈26 m | IT |
| 236 | 0+ | 42.446 | 2.342×10⁷ y 3 | α, SF 9.4×10⁻⁸%, ³⁰Mg |
| 237 | 1/2+ | 45.392 | 6.75 d 1 | β− |
| 238 | 0+ | 47.309 | 4.468×10⁹ y 3 | α, 99.2742% 10 SF 5.5×10⁻⁵% |
| 239 | 5/2+ | 50.574 | 23.45 m 2 | β− |
| 240 | 0+ | 52.715 | 14.1 h 1 | β− |
| 241 | | 56.2s | ≈5 m | β−? |
| 242 | 0+ | 58.6s | 16.8 m 5 | β− |
| 93 Np 225 | (9/2−) | 31.59 | >2 µs | α |
| 226 | | 32.74s | 35 ms 10 | α |
| 227 | | 32.56 | 0.51 s 6 | α |
| 228 | | 33.7s | 61.4 s 14 | α 60%, ε 40% |
| 229 | | 33.78 | 4.0 m 2 | α 68%, ε 32% |
| 230 | | 35.24 | 4.6 m 3 | ε 97%, α 3% |
| 231 | (5/2) | 35.63 | 48.8 m 2 | ε 98%, α 2% |
| 232 | (4+) | 37.4s | 14.7 m 3 | ε |

| Nuclide Z El A | Jπ | Δ (MeV) | T½, Γ, or Abundance | Decay Mode |
|---|---|---|---|---|
| 93 Np 233 | (5/2+) | 37.95 | 36.2 m 1 | ε, α≤1.0×10⁻³% |
| 234 | (0+) | 39.957 | 4.4 d 1 | ε |
| 235 | 5/2+ | 41.045 | 396.1 d 12 | ε, α 2.6×10⁻³% |
| 236 | (6−) | 43.38 | 154×10³ y 6 | ε 87.3%, β− 12.5%, α 0.16% |
| 236m | 1 | 43.44 | 22.5 h 4 | ε 52%, β− 48% |
| 237 | 5/2+ | 44.873 | 2.144×10⁶ y 7 | α, SF≤2×10⁻¹⁰% |
| 238 | 2+ | 47.456 | 2.117 d 2 | β− |
| 239 | 5/2+ | 49.312 | 2.356 d 3 | β− |
| 240 | (5+) | 52.31 | 61.9 m 2 | β− |
| 240m | (1+) | 52.31 | 7.22 m 2 | β− 99.89%, IT 0.12% |
| 241 | (5/2+) | 54.26 | 13.9 m 2 | β− |
| 242 | (1+) | 57.4 | 2.2 m 2 | β− |
| 242m | (6+) | 57.4 | 5.5 m 1 | β− |
| 243 | (5/2) | 59.88s | 1.85 m 15 | β− |
| 244 | (7−) | 63.2s | 2.29 m 16 | β− |
| 94 Pu 228 | 0+ | 36.09 | 1.1 s +20−5 | α |
| 229 | (3/2+) | 37.40 | >2 µs | α |
| 230 | 0+ | 36.93 | 1.70 m 17 | α 84%, ε 16% |
| 231 | (3/2+) | 38.29 | 8.6 m 5 | ε 90%, α>0.2% |
| 232 | 0+ | 38.37 | 33.1 m 8 | ε 80%, α 20% |
| 233 | | 40.05 | 20.9 m 4 | ε 99.88%, α 0.12% |
| 234 | 0+ | 40.350 | 8.8 h 1 | ε≈94%, α=6% |
| 235 | (5/2+) | 42.18 | 25.3 m 5 | ε, α 2.8×10⁻⁵% |
| 236 | 0+ | 42.903 | 2.858 y 8 | α, SF 1.9×10⁻⁷% |
| 237 | 7/2− | 45.093 | 45.2 d 1 | ε, α 4.2×10⁻³% |
| 237m | 1/2+ | 45.239 | 0.18 s 2 | IT |
| 238 | 0+ | 46.165 | 87.7 y 1 | α, SF 1.9×10⁻⁷% |
| 239 | 1/2+ | 48.590 | 24110 y 30 | α, SF 3.×10⁻¹⁰% |
| 240 | 0+ | 50.127 | 6561 y 7 | α, SF 5.7×10⁻⁶% |
| 241 | 5/2+ | 52.957 | 14.290 y 6 | β−, α 2.5×10⁻³%, SF 2.×10⁻¹⁴% |
| 242 | 0+ | 54.718 | 3.75×10⁵ y 2 | α, SF 5.5×10⁻⁴% |
| 243 | 7/2+ | 57.756 | 4.956 h 3 | β− |
| 244 | 0+ | 59.806 | 8.00×10⁷ y 9 | α 99.88%, SF 0.12% |
| 245 | (9/2−) | 63.18 | 10.5 h 1 | β− |
| 246 | 0+ | 65.40 | 10.84 d 2 | β− |
| 247 | | 69.0s | 2.27 d 23 | β− |
| 95 Am 231 | | 42.4s | ≈10 s | ε?, α? |
| 232 | | 43.4s | 79 s 2 | ε≈98%, α≈2% |
| 233 | | 43.2s | 3.2 m 8 | α>3%, ε |
| 234 | | 44.5s | 2.32 m 8 | ε 99.96%, α<0.04% |
| 235m | | 44.7s | 10.3 m 6 | ε 99.6%, α 0.4% |
| 236? | | 46.2s | 3.6 m 1 | α 0.004%, ε |
| 236m | | 46.2s | 0.6 y 2 | ε |
| 237 | 5/2(−) | 46.57s | 73.0 m 10 | ε 99.96%, α 0.03% |
| 238 | 1+ | 48.42 | 98 m 2 | ε, α 1.0×10⁻⁴% |
| 238m | | 48.42 | 3.8 y | α |
| 239 | (5/2−) | 49.392 | 11.9 h 1 | ε 99.99%, α 0.01% |
| 240 | (3−) | 51.51 | 50.8 h 3 | ε, α 1.9×10⁻⁴% |
| 241 | 5/2− | 52.936 | 432.2 y 7 | α, SF 4×10⁻¹⁰% |
| 242 | 1− | 55.470 | 16.02 h 2 | β− 82.7%, ε 17.3% |

| Nuclide Z El A | Jπ | Δ (MeV) | T½, Γ, or Abundance | Decay Mode |
|---|---|---|---|---|
| 95 Am 242m | 5− | 55.518 | 141 y 2 | IT 99.55%, α 0.45%, α<4.7×10⁻⁹% |
| 242m(2+,3−) | | 57.670 | 14.0 ms 10 | SF=100%, α<5.×10⁻³%, IT |
| 243 | 5/2− | 57.176 | 7370 y 40 | α, SF 3.7×10⁻⁹% |
| 244 | (−) | 59.881 | 10.1 h 1 | β− |
| 244m | | 59.881 | 0.90 ms 15 | SF≤100% |
| 244m 1+ | | 59.967 | 26 m 1 | β− 99.96%, ε 0.04% |
| 245 | (5/2)+ | 61.900 | 2.05 h 1 | β− |
| 246 | (7−) | 64.99 | 39 m 3 | β− |
| 246m | 2(−) | 64.99 | 25.0 m 2 | β−, IT<0.02% |
| 247 | (5/2) | 67.2s | 23.0 m 13 | β− |
| 248 | | 70.6s | ≈10 m | β− |
| 249 | | 73.1s | ≈2 m | β−? |
| 96 Cm 232 | 0+ | 47.29 | 1 m ? | SF<30.3% |
| 233 | | 47.29 | | α, ε |
| 234 | 0+ | 46.72 | ≈2 m | ε?, α? |
| 235 | | 47.9s | 5 m sy | ε?, α? |
| 236 | 0+ | 47.9s | ≈10 m | ε, α |
| 237 | | 49.3s | ≈20 m | ε?, α? |
| 238 | 0+ | 49.40 | 2.4 h 1 | ε≥90%, α≤10% |
| 239 | (7/2−) | 51.2s | ≈2.9 h | ε, α<0.1% |
| 240 | + | 51.725 | 27 d 1 | α, Ne 99.5%, ε<0.5%, SF 3.9×10⁻⁶% |
| 241 | 1/2+ | 53.703 | 32.8 d 2 | ε 99%, α 1% |
| 242 | 0+ | 54.805 | 162.8 d 2 | α, SF 6.2×10⁻⁶%, 34.1.×10⁻¹⁴% |
| 243 | 5/2+ | 57.184 | 29.1 y 1 | α 99.71%, ε 0.29%, SF 5.3×10⁻⁹% |
| 244 | 0+ | 58.454 | 18.1 y 1 | α, SF 1.37×10⁻⁴% |
| 245 | 7/2+ | 61.005 | 8500 y 100 | α, SF 6.1×10⁻⁷% |
| 246 | 0+ | 62.618 | 4760 y 40 | α 99.97%, SF 0.03% |
| 247 | 9/2− | 65.534 | 1.56×10⁷ y 5 | α |
| 248 | 0+ | 67.392 | 3.48×10⁵ y 6 | α 91.61%, SF 8.39% |
| 249 | 1/2(+) | 70.750 | 64.15 m 3 | β− |
| 250 | 0+ | 72.99 | ≈8.3×10³ y | SF=74%, α=18%, β−≈8% |
| 251 | (1/2+) | 76.65 | 16.8 m 2 | β− |
| 252 | 0+ | 79.1s | <2 d | β− |
| 97 Bk 235 | | 52.7s | ≈20 s | ε?, α? |
| 236 | | 53.4s | ≈42 s | α?, ε? |
| 236m | | 46.2s | ≥30 d | ε |
| 237 | | 53.1s | ≈1 m | ε?, α? |
| 238 | | 54.3s | 144 s 5 | ε, αSF 0.048% |
| 240 | | 55.7s | 4.8 m 8 | εSF 2.0×10⁻³%, ε |
| 241 | (7/2+) | 56.1s | ≈3 m | α?, ε? |
| 242 | | 57.7s | 7.0 m 13 | ε≤100% |
| 243 | (3/2−) | 58.691 | 4.5 h 2 | ε=99.85%, α=0.15% |
| 244 | (4−) | 60.72 | 4.35 h 15 | ε 99.99%, α 6.0×10⁻³% |
| 245 | 3/2− | 61.815 | 4.94 d 3 | ε 99.88%, α 0.12% |
| 246m | 2(−) | 63.97 | 1.80 d 2 | ε, α<0.2% |

## Nuclear Wallet Cards

| Nuclide Z El A | Jπ | Δ (MeV) | T½, Γ, or Abundance | Decay Mode |
|---|---|---|---|---|
| 97 Bk 247 | (3/2−) | 65.491 | 1380 y 250 | α ≤ 100% |
| 248 | | 68.08s | >9 y | α |
| 248m | 1(−) | 68.08s | 23.7 h 2 | β− 70%, ε 30% |
| 249 | 7/2− | 69.850 | 330 d 4 | β−, α 1.4×10⁻³%, SF 4.7×10⁻⁸% |
| 250 | 2− | 72.951 | 3.212 h 5 | β− |
| 251 | (3/2−) | 75.23 | 55.6 m 11 | β− |
| 252 | | 78.5s | ≈2 m | β−?, α? |
| 253 | | 80.9s | ~10 m | β−? |
| 254 | | 84.4s | ≈2 m | β−? |
| 98 Cf 237 | | 57.8s | 2.1 s 3 | SF≈10%, α? |
| 238 | 0+ | 57.2s | 21 ms 2 | SF≈100% |
| 239 | | 58.1s | 39 s +37−12 | ε, α |
| 240 | 0+ | 58.0s | 0.96 m 15 | α≈98%, SF≈2%, ε |
| 241 | | 59.4s | 3.78 m 70 | ε≈75%, α≈25% |
| 242 | 0+ | 59.34 | 3.7 m 5 | α 80%, ε 20%, SF ≤ 0.01% |
| 243 | (1/2+) | 60.9s | 10.7 m 5 | ε≈86%, α≈14% |
| 244 | 0+ | 61.479 | 19.4 m 6 | α≈100% |
| 245 | (1/2+,5/2+) | 63.387 | 45.0 m 15 | ε 64%, α 36% |
| 246 | 0+ | 64.092 | 35.7 h 5 | α, ε < 4.0×10⁻³%, SF 2.5×10⁻⁴% |
| 247 | (7/2+) | 66.137 | 3.11 h 3 | ε 99.97%, α 0.04% |
| 248 | 0+ | 67.240 | 333.5 d 28 | α, SF 2.9×10⁻³% |
| 249 | 9/2− | 69.726 | 351 y 2 | α, SF 5.0×10⁻⁷% |
| 250 | 0+ | 71.172 | 13.08 y 9 | α 99.92%, SF 0.08% |
| 251 | 1/2+ | 74.135 | 898 y 44 | α, SF |
| 252 | 0+ | 76.034 | 2.645 y 8 | α 96.91%, SF 3.09% |
| 253 | (7/2+) | 79.301 | 17.81 d 8 | β− 99.69%, α 0.31% |
| 254 | 0+ | 81.34 | 60.5 d 2 | SF 99.69%, α 0.31% |
| 255 | (7/2+) | 84.8s | 85 m 18 | β− |
| 256 | 0+ | 87.0s | 12.3 m 12 | SF, α < 1%, α ≈ 1.0×10⁻⁵% |
| 99 Es 240 | | 64.2s | 1 s sy | α?, ε? |
| 241 | (3/2−) | 63.8s | 8 s +6−5 | α |
| 242 | | 65.0s | 13.5 s 25 | α > 0%, ε > 0% |
| 243 | | 64.8s | 21 s 2 | ε≈70%, α≈30% |
| 244 | | 66.0s | 37 s 4 | ε≈90%, α≈4% |
| 245 | (3/2−) | 66.4s | 1.1 m 1 | ε≈60%, α≈40% |
| 246m | | 67.9s | 7.7 m 5 | ε 90.1%, α 9.9%, ε 3.0×10⁻³% |
| 247 | (7/2+) | 68.61s | 4.55 m 26 | ε≈93%, α≈7% |
| 247m | | 68.61s | 625 d 84 | α |
| 248 | (2−,0+) | 70.30s | 27 m 5 | ε 99.7%, α≈0.25% |
| 249 | 7/2+ | 71.18s | 102.2 m 6 | ε 99.43%, α 0.57% |
| 250 | (6+) | 73.2s | 8.6 h 1 | ε≈97%, α<3% |
| 250m | 1(−) | 73.2s | 2.22 h 5 | ε ≤ 100% |
| 251 | (3/2−) | 74.512 | 33 h 1 | ε 99.5%, α 0.5% |
| 252 | (5−) | 77.29 | 471.7 d 19 | α 78%, ε 22%, β− 0.01% |
| 253 | 7/2+ | 79.014 | 20.47 d 3 | α, SF 8.7×10⁻⁶% |
| 254 | (7+) | 81.992 | 275.7 d 5 | α ≈ 100%, ε 1.7×10⁻⁴%, SF < 3.0×10⁻⁶%, ε |

| Nuclide Z El A | Jπ | Δ (MeV) | T½, Γ, or Abundance | Decay Mode |
|---|---|---|---|---|
| 99 Es 254m | 2+ | 82.072 | 39.3 h 2 | β− 98%, IT < 3%, α 0.32%, ε 0.08%, SF < 0.05% |
| 255 | (7/2+) | 84.09 | 39.8 d 12 | β− 92%, α 8%, SF 4.1×10⁻³% |
| 256 | (1+,0−) | 87.2s | 25.4 m 24 | β− |
| 256m | (8+) | 87.2s | 7.6 h | β− |
| 257 | | 89.4s | 7.7 d 2 | β−, SF |
| 258 | | 92.7s | 3 m sy | ε?, α? |
| 100 Fm 242 | 0+ | 68.4s | 0.8 ms 2 | SF ≤ 100% |
| 243 | (7/2−) | 69.3s | 0.18 s +8−4 | α ≤ 100% |
| 244 | 0+ | 69.0s | 3.3 ms 5 | SF ≤ 100% |
| 245 | | 70.2s | 4.2 s 13 | α ≤ 100%, SF ≤ 0.1% |
| 246 | 0+ | 70.14 | 1.1 s 2 | α 92%, SF 8%, ε ≤ 1% |
| 247 | (7/2+) | 71.6s | 29 s 1 | α ≥ 50%, ε ≤ 50% |
| 247m | (1/2+) | 71.6s | 4.3 s 4 | α ≤ 100% |
| 248 | 0+ | 71.91 | 36 s 2 | α 93%, ε 7%, SF 0.1% |
| 249 | (7/2+) | 73.6s | 2.6 m 7 | ε 67%, α 33% |
| 250 | 0+ | 74.07 | 30 m 3 | α ≈ 90%, ε < 10%, SF 6.9×10⁻³% |
| 250m | 0+ | 74.07 | 1.8 s 1 | IT ≥ 80%, α<20%, SF ≤ 8.2×10⁻⁵%, ε |
| 251 | (9/2−) | 75.987 | 5.30 h 8 | ε 98.2%, α 1.8% |
| 252 | 0+ | 76.817 | 25.39 h 4 | α, SF 0.0023% |
| 253 | (1/2+ | 79.350 | 3.00 d 12 | ε 88%, α 12% |
| 254 | 0+ | 80.904 | 3.240 h 2 | α 99.94%, SF 0.06% |
| 255 | 7/2+ | 83.799 | 20.07 h 7 | α, SF 2.4×10⁻⁵% |
| 256 | 0+ | 85.486 | 157.6 m 13 | SF 91.9%, α 8.1% |
| 257 | (9/2+) | 88.590 | 100.5 d 2 | α 99.79%, SF 0.21% |
| 258 | 0+ | 90.4s | 370 μs 43 | SF ≤ 100% |
| 259 | | 93.7s | 1.5 s 3 | SF |
| 260 | 0+ | 95.6s | ~4 ms | SF |
| 101 Md 245 | (1/2−) | 75.3s | 0.90 ms 25 | α, SF |
| 245m | | 75.6s | 0.35 s +23−16 | α, ε |
| 246m | | 76.3s | 1.0 s 4 | SF, α > 0%, ε > 0% |
| 247 | | 76.0s | 1.12 s 22 | α ≤ 100% |
| 248 | | 77.1s | 7 s 3 | ε 80%, α 20%, SF ≤ 0.05% |
| 249 | | 77.3s | 24 s 4 | α ≈ 60%, ε ≤ 40% |
| 250 | | 78.6s | 52 s 6 | ε 93%, α 7% |
| 251 | | 79.0s | 4.0 m 5 | ε ≈ 90%, α ≤ 10% |
| 252 | | 80.6s | 2.3 m 8 | ε ≤ 100% |
| 253 | (1/2−) | 81.3s | 6 m +12−3 | ε ≤ 100%, α |
| 254 | | 83.5s | 10 m 3 | ε ≤ 100% |
| 254m | | 83.5s | 28 m 8 | ε ≤ 100% |
| 255 | (7/2−) | 84.843 | 27 m 2 | ε 92%, α 8%, SF < 0.15% |
| 256 | (1−) | 87.62 | 77 m 2 | ε 90.8%, α 9.2%, SF < 3% |
| 257 | (7/2−) | 88.996 | 5.52 h 5 | ε 85%, α 15%, SF < 1% |
| 258 | | 91.4s | 51.5 d 3 | α, SF |
| 258m | | 91.688 | 57.0 m 9 | ε ≥ 70%, SF |

| Nuclide Z El A | Jπ | Δ (MeV) | T½, Γ, or Abundance | Decay Mode |
|---|---|---|---|---|
| 101 Md 259 | | 93.6s | 96 m 3 | SF ≈ 100%, α < 1.3% |
| 260 | | 96.6s | 31.8 d 5 | SF ≥ 42%, α ≤ 25%, ε ≤ 23%, β− ≤ 10% |
| 261 | | 98.5s | 40 m sy | α? |
| 262 | | 101.4s | 3 m sy | SF?, α? |
| 102 No 248 | 0+ | 80.7s | <2 μs | SF? |
| 249 | | 81.8s | 54 μs +15−10 | SF |
| 250 | 0+ | 81.5s | 6 μs 1 | SF ≤ 100%, α 0.1%, ε 1.0×10⁻⁵% |
| 251 | (7/2+) | 82.9s | 0.78 s 2 | α ≤ 100%, SF ≤ 8%, ε |
| 251m | (1/2+) | 83.0s | 0.93 s 6 | α ≤ 100% |
| 252 | 0+ | 82.88 | 2.27 s 14 | α 58%, ε 23%, SF 19% |
| 252m | | 82.88 | 26 d 7 | α |
| 253 | (9/2−) | 84.5s | 1.62 ms 15 | α ≤ 100%, ε |
| 254 | 0+ | 84.72 | 51 s 10 | α 90%, ε 10%, SF 0.17% |
| 254m | | 84.72 | 0.28 s 4 | IT ≈ 80% |
| 255 | (1/2+) | 86.85 | 3.1 m 2 | α 61%, ε 39% |
| 256 | 0+ | 87.824 | 2.91 s 5 | α 99.47%, SF 0.53% |
| 257 | (7/2+) | 90.24 | 25 s 3 | α ≤ 100%, SF ≤ 1.5% |
| 258 | 0+ | 91.5s | 1.2 ms 2 | SF ≤ 100% |
| 259 | | 94.1s | 58 m 5 | α 75%, ε 25%, SF < 10% |
| 260 | 0+ | 95.6s | 106 ms 8 | SF |
| 261 | | 98.5s | | β−, α |
| 262 | 0+ | 100.0s | ~5 ms | SF |
| 263 | | 103.0s | 20 m sy | α?, SF? |
| 264 | 0+ | 104.6s | 1 m sy | α? |
| 103 Lr 251 | | 87.9s | | ε?, α? |
| 252 | | 88.8s | 0.36 s +11−7 | α ≈ 90%, ε ≈ 10%, SF < 1% |
| 253 | | 88.7s | 0.57 s +7−6 | α, SF < 2% |
| 253m | | 88.7s | 1.5 s +3−2 | α 90%, SF < 2% |
| 254 | | 89.8s | 13 s 3 | α 76%, ε 24% |
| 255 | | 90.1s | 22 s 4 | α 85%, ε < 30%, SF ≤ 0.1% |
| 256 | | 91.9s | 27 s 3 | α 85%, ε 15%, SF < 0.03% |
| 257 | | 92.7s | 0.646 s 25 | α ≤ 100%, SF ≤ 0.03% |
| 258 | | 94.8s | 4.1 s 3 | α > 95%, SF < 5% |
| 259 | | 95.8s | 6.2 s 3 | α 78%, SF 22% |
| 260 | | 98.3s | 180 s 30 | α 80%, ε < 40%, SF < 10% |
| 261 | | 99.6s | 39 m 12 | SF |
| 262 | | 102.1s | ~4 h | SF ≈ 10%, ε, α |
| 263 | | 103.7s | 5 h sy | α? |
| 264 | | 106.2s | 10 h sy | α?, SF? |
| 265 | | 107.9s | 10 h sy | α?, SF? |
| 266 | | 111.1s | 1 h sy | α?, SF? |
| 104 Rf 253m | | 93.8s | 48 μs +17−10 | SF ≤ 100%, α |
| 253m | | 93.8s | ≈1.8 s | SF ≈ 50%, α ≈ 50% |

## Nuclear Wallet Cards

| Nuclide Z El A | Jπ | Δ (MeV) | T½, Γ, or Abundance | Decay Mode |
|---|---|---|---|---|
| 104 Rf 254 | 0+ | 93.3s | 23 μs 3 | SF ≤ 100% |
| 255 | (9/2−) | 94.4s | 1.64 s 11 | SF 52%, α 48% |
| 255m | | 94.4s | 0.8 s +5−2 | α ≤ 100% |
| 256 | 0+ | 94.24 | 6.4 ms 2 | SF 99.68%, α 0.32% |
| 257 | (1/2+) | 95.9s | 4.7 s 3 | α ≤ 100%, SF ≤ 1.4%, ε > 0% |
| 257m | | 95.9s | 3.9 s 4 | α ≤ 100%, SF ≤ 1.4%, ε > 0% |
| 258 | 0+ | 96.4s | 12 ms 2 | SF 87%, α 13% |
| 259 | | 98.40s | 3.2 s 6 | α 92%, SF 8% |
| 260 | 0+ | 99.1s | 21 ms 1 | SF ≤ 100%, α? |
| 261 | | 101.32 | 65 s 10 | α ≥ 80%, ε < 15%, SF < 10% |
| 262 | 0+ | 102.4s | 2.3 s 4 | SF ≤ 100%, α < 3% |
| 263 | | 104.8s | 10 m 2 | SF ≈ 100%, α |
| 264 | 0+ | 106.2s | 1 h sy | α? |
| 265 | | 108.7s | ~13 h | α? |
| 266 | 0+ | 109.9s | 10 h sy | α?, SF? |
| 267? | | 113.2s | 2.3 h +980−17 | α?, SF? |
| 268 | 0+ | 115.2s | 6 h sy | α?, SF? |
| 105 Db 255 | | 100.0s | 1.6 s +6−4 | α ≈ 80%, SF ≈ 20% |
| 256 | | 100.7s | 1.6 s +5−3 | α ≈ 64%, ε ≈ 36%, SF < 0.02% |
| 257 | | 100.3s | 1.50 s +19−15 | α > 94%, SF < 6% |
| 257m | | 100.3s | 0.76 s +15−11 | α ≥ 87%, SF ≤ 13% |
| 258 | | 101.7s | 4.0 s 10 | α 67%, ε 33%, SF < 1% |
| 258m | | 101.7s | 20 s 10 | ε ≈ 100% |
| 259 | | 102.1s | 0.51 s 16 | α |
| 260 | | 103.7s | 1.52 s 13 | α ≈ 90.4%, SF ≤ 9.6%, ε < 2.5% |
| 261 | | 104.4s | 1.8 s 4 | α ≥ 82%, SF ≤ 18% |
| 262 | | 106.3s | 35 s 5 | α ≈ 67%, SF |
| 263 | | 107.1s | 27 s +10−7 | SF 59%, α 41%, ε 3% |
| 264 | | 109.4s | 3 m sy | α? |
| 265 | | 110.5s | 15 m sy | α? |
| 266 | | 112.7s | 20 m sy | α?, SF? |
| 267? | | 114.0s | 73 m +350−33 | SF > 0% |
| 268? | | 116.9s | 32 h +11−7 | SF |
| 269 | | 118.7s | 3 h sy | α?, SF? |
| 106 Sg 258 | 0+ | 105.4s | 2.9 ms +13−7 | SF ≤ 100%, α? |
| 259 | (1/2+) | 106.7s | 0.48 s +28−13 | α 90%, SF < 20% |
| 260 | 0+ | 106.58 | 3.6 ms 9 | SF ≤ 100%, SF 50% |
| 261 | | 108.2s | 0.23 s 6 | α ≤ 100%, SF < 1% |
| 262 | 0+ | 108.4s | 6.9 ms +38−18 | SF ≥ 78%, α ≤ 22% |
| 263 | | 110.2s | 1.0 s 2 | α ≥ 70%, SF < 30% |
| 263m | 0+ | 110.2s | 0.12 s | α, IT |
| 264 | 0+ | 110.8s | 0.4 s sy | α? |
| 265 | (9/2+) | 112.82 | 8 s 3 | SF ≤ 57%, α ≥ 43% |
| 266 | 0+ | 113.7s | 21 s +20−12 | SF ≥ 50%, α ≤ 50% |
| 268 | 0+ | 117.0s | 30 s sy | α?, SF? |
| 270 | 0+ | 121.4s | 10 m sy | α?, SF? |
| 271? | | 124.3s | 2.4 m +43−10 | α 50%, SF 50% |

## Nuclear Wallet Cards

| Nuclide Z El A | | Jπ | Δ (MeV) | T½, Γ, or Abundance | Decay Mode |
|---|---|---|---|---|---|
| 106 Sg | 272 | 0+ | 125.9s | 1 h sy | α?, SF? |
| | 273 | | 128.8s | 1 m sy | SF? |
| 107 Bh | 260 | | 113.6s | 0.3 ms sy | α ≤ 100% |
| | 261 | | 113.3s | 12 ms +5−3 | α 96%, SF < 10% |
| | 262m | | 114.5s | 8.0 ms 21 | α ≤ 100% |
| | 262m | | 114.5s | 102 ms 26 | α ≤ 100% |
| | 263 | | 114.6s | 0.2 ms sy | α? |
| | 264 | | 116.1s | 0.44 s +60−16 | α ≤ 100% |
| | 265 | | 116.6s | 0.9 s +7−3 | α |
| | 266? | | 118.2s | 1.7 ms +82−8 | α |
| | 267? | | 118.9s | 17 s +14−6 | α |
| | 271? | | 125.9s | 40 s sy | α? |
| | 272? | | 128.6s | 10 s +12−4 | α |
| | 273 | | 130.1s | 90 m sy | α?, SF? |
| | 274 | | 132.7s | 90 m sy | α?, SF? |
| | 275 | | 134.4s | 40 m sy | SF? |
| 108 Hs | 263 | | 119.8s | | α ≤ 100% |
| | 264 | 0+ | 119.60 | ~0.8 ms | α ≈ 50%, SF ≈ 50% |
| | 265 | | 121.2s | 2.0 ms +3−2 | α ≈ 100%, SF ≤ 1% |
| | 266 | 0+ | 121.2s | 2.3 ms +13−6 | α ≈ 100%, SF < 1.4% |
| | 267 | | 122.8s | 52 ms +13−8 | α ≈ 80%, SF < 20% |
| | 267m | | 122.8s | 0.80 s +380−37 | α > 0% |
| | 269? | | 124.9s | 9.7 s +97−33 | α |
| | 270? | 0+ | 125.4s | 3.6 s +8−14 | α |
| | 271 | | 128.2s | 40 s sy | α?, SF? |
| | 272 | 0+ | 129.5s | 40 s sy | α?, SF? |
| | 273 | (3/2+) | 132.3s | 50 s sy | α? |
| | 274 | 0+ | 133.3s | 1 m sy | α?, SF? |
| | 275? | | 136.0s | 0.15 s +27−6 | α |
| | 276 | 0+ | 137.1s | 1 h sy | α?, SF? |
| 109 Mt | 265 | | 126.8s | 2 m sy | α? |
| | 266? | | 127.9s | 1.7 ms +18−16 | α ≤ 100% |
| | 267 | | 127.9s | 10 ms sy | α? |
| | 268? | | 129.2s | 21 ms +8−5 | α |
| | 270? | | 131.0s | 4.96 ms | α |
| | 271 | | 131.5s | 5 s sy | α? |
| | 272 | | 133.9s | 10 s sy | α?, SF? |
| | 273 | | 135.0s | 20 s sy | α?, SF? |
| | 274 | | 137.4s | 20 s sy | α?, SF? |
| | 275? | | 138.5s | 9.7 ms +460−44 | α |
| | 276? | | 140.8s | 0.72 s +87−25 | α |
| | 279 | | 145.5s | 6 m sy | α?, SF? |
| 110 Ds | 267? | | 134.5s | 3 μs +6−2 | α |
| | 268 | 0+ | 133.9s | 100 μs sy | α |
| | 269? | | 135.2s | 179 μs +245−66 | α |
| | 270 | 0+ | 134.8s | 0.10 ms +14−4 | α, SF < 0.2% |
| | 270m | | 135.9s | 6.0 ms +82−22 | α > 70%, IT ≤ 30% |
| | 271 | | 136.1s | 1.63 ms +44−29 | α |
| | 271m | | 136.1s | 69 ms +56−21 | α > 0%, IT? |
| | 272 | 0+ | 136.3s | 1 s sy | SF |
| | 273 | | 138.7s | 0.17 ms +17−6 | α |

| Nuclide Z El A | | Jπ | Δ (MeV) | T½, Γ, or Abundance | Decay Mode |
|---|---|---|---|---|---|
| 110 Ds | 274 | | 139.3s | 2 s sy | α?, SF? |
| | 275 | | 141.8s | 2 s sy | α? |
| | 276 | 0+ | 142.6s | 5 s sy | α?, SF? |
| | 277 | | 145.0s | 5 s sy | α? |
| | 278 | 0+ | 145.8s | 10 s sy | α?, SF? |
| | 279? | | 148.0s | 0.18 s +5−3 | SF 90%, α 10% |
| | 281? | | 151.0s | 11.1 s +50−27 | SF |
| 111 Rg | 272? | | 143.1s | 3.8 ms +14−8 | α |
| | 273 | | 143.2s | 5 ms sy | α? |
| | 274? | | 145.1s | 6.4 ms +307−29 | α |
| | 275 | | 145.4s | 10 ms sy | α? |
| | 276 | | 147.6s | 100 ms sy | α?, SF? |
| | 277 | | 148.6s | 1 s sy | α? |
| | 278 | | 150.5s | 1 s sy | α?, SF? |
| | 279? | | 151.3s | 0.17 s +81−8 | α |
| | 280? | | 153.2s | 3.6 s +43−13 | α |
| | 281 | | 154.0s | 1 m sy | α?, SF? |
| | 282 | | 156.0s | 4 m sy | α?, SF? |
| | 283 | | 156.9s | 10 m sy | α?, SF? |
| 112 | 277 | | 152.7s | 0.69 ms +69−24 | α |
| | 278 | 0+ | 153.1s | 10 ms sy | α?, SF? |
| | 279 | | 155.1s | 0.1 s sy | α?, SF? |
| | 280 | 0+ | 155.6s | 1 s sy | α?, SF? |
| | 282? | | 158.1s | 0.50 ms +33−14 | SF |
| | 283? | | 160.0s | 4.0 s +13−7 | α ≥ 99%, SF ≤ 1% |
| | 284? | | 160.6s | 97 ms +31−19 | SF |
| | 285? | | 162.2s | 29 s +13−7 | α |
| 113 | 278? | | 160.2s | 0.24 ms +114−11 | α |
| | 283? | | 164.4s | 100 ms +490−45 | α |
| | 284? | | 165.9s | 0.48 s +58−17 | α |
| | 285 | | 166.5s | 2 m sy | α?, SF? |
| | 286 | | 168.1s | 5 m sy | α?, SF? |
| | 287 | | 168.6s | 20 m sy | α?, SF? |
| 114 | 286? | | 171.3s | 0.16 ms +7−3 | SF 60%, α 40% |
| | 287? | | 172.9s | 0.51 s +18−10 | α |
| | 288? | | 173.0s | 0.8 s +27−16 | α |
| | 289? | | 174.4s | 2.6 s +12−7 | α |
| 115 | 287? | | 178.1s | 32 ms +155−14 | α |
| | 288? | | 179.3s | 87 ms +105−30 | α |
| | 289 | | 180s | 10 s sy | α?, SF? |
| | 290 | | 180.8s | 10 s sy | α?, SF? |
| | 291 | | 181.1s | 1 m sy | α?, SF? |
| 116 | 290 | 0+ | 185.0s | 15 ms +26−6 | α |
| | 291 | | 186.3s | 6.3 ms +116−25 | α |
| | 292? | 0+ | 186.1s | 18.0 ms +16−6 | α |
| | 293? | | | 61 ms +57−20 | α |
| 117 | 291 | | 192.4s | 10 ms sy | α?, SF? |
| | 292 | | 193.3s | 50 ms sy | α?, SF? |
| 118 | 294? | 0+ | | 1.8 ms +84−8 | α > 0%, SF? |

# APPENDIX C

# PERIODIC TABLE OF ELEMENTS

*Modern Nuclear Chemistry*, by W.D. Loveland, D.J. Morrissey, and G.T. Seaborg
Copyright © 2006 John Wiley & Sons, Inc.

| IA | IIA | IIIB | IVB | VB | VIB | VIIB | --- | VIII | --- | IB | IIB | IIIA | IVA | VA | VIA | VIIA | VIIIA |
|---|---|---|---|---|---|---|---|---|---|---|---|---|---|---|---|---|---|
| H 1 | | | | | | | | | | | | | | | | | He 2 |
| Li 3 | Be 4 | | | | | | | | | | | B 5 | C 6 | N 7 | O 8 | F 9 | Ne 10 |
| Na 11 | Mg 12 | | | | | | | | | | | Al 13 | Si 14 | P 15 | S 16 | Cl 17 | Ar 18 |
| K 19 | Ca 20 | Sc 21 | Ti 22 | V 23 | Cr 24 | Mn 25 | Fe 26 | Co 27 | Ni 28 | Cu 29 | Zn 30 | Ga 31 | Ge 32 | As 33 | Se 34 | Br 35 | Kr 36 |
| Rb 37 | Sr 38 | Y 39 | Zr 40 | Nb 41 | Mo 42 | Tc 43 | Ru 44 | Rh 45 | Pd 46 | Ag 47 | Cd 48 | In 49 | Sn 50 | Sb 51 | Te 52 | I 53 | Xe 54 |
| Cs 55 | Ba 56 | * 57– | Hf 72 | Ta 73 | W 74 | Re 75 | Os 76 | Ir 77 | Pt 78 | Au 79 | Hg 80 | Tl 81 | Pb 82 | Bi 83 | Po 84 | At 85 | Rn 86 |
| Fr 87 | Ra 88 | ** 89– | Rf 104 | Db 105 | Sg 106 | Bh 107 | Hs 108 | Mt 109 | Ds 110 | Rg 111 | 112 | 113 | 114 | 115 | 116 | 117 | 118 |
| * | | La 57 | Ce 58 | Pr 59 | Nd 60 | Pm 61 | Sm 62 | Eu 63 | Gd 64 | Tb 65 | Dy 66 | Ho 67 | Er 68 | Tm 69 | Yb 70 | Lu 71 | Lanthanides |
| ** | | Ac 89 | Th 90 | Pa 91 | U 92 | Np 93 | Pu 94 | Am 95 | Cm 96 | Bk 97 | Cf 98 | Es 99 | Fm 100 | Md 101 | No 102 | Lr 103 | Actinides |

# APPENDIX D

# LIST OF ELEMENTS

| Name | Sym | Z | Name | Sym | Z |
| --- | --- | --- | --- | --- | --- |
| Actinium | Ac | 89 | Chlorine | Cl | 17 |
| Aluminum | Al | 13 | Chromium | Cr | 24 |
| Americium | Am | 95 | Cobalt | Co | 27 |
| Antimony | Sb | 51 | Copper | Cu | 29 |
| Argon | Ar | 18 | Curium | Cm | 96 |
| Arsenic | As | 33 | Darmstadtium | Ds | 110 |
| Astatine | At | 85 | Dubnium | Db | 105 |
| Barium | Ba | 56 | Dysprosium | Dy | 66 |
| Berkelium | Bk | 97 | Einsteinium | Es | 99 |
| Beryllium | Be | 4 | Erbium | Er | 68 |
| Bismuth | Bi | 83 | Europium | Eu | 63 |
| Bohrium | Bh | 107 | Fermium | Fm | 100 |
| Boron | B | 5 | Fluorine | F | 9 |
| Bromine | Br | 35 | Francium | Fr | 87 |
| Cadmium | Cd | 48 | Gadolinium | Gd | 64 |
| Cesium | Cs | 55 | Gallium | Ga | 31 |
| Calcium | Ca | 20 | Germanium | Ge | 32 |
| Californium | Cf | 98 | Gold | Au | 79 |
| Carbon | C | 6 | Hafnium | Hf | 72 |
| Cerium | Ce | 58 | Hassium | Hs | 108 |

*Modern Nuclear Chemistry*, by W.D. Loveland, D.J. Morrissey, and G.T. Seaborg
Copyright © 2006 John Wiley & Sons, Inc.

| Name | Sym | Z | Name | Sym | Z |
|---|---|---|---|---|---|
| Helium | He | 2 | Protactinium | Pa | 91 |
| Holmium | Ho | 67 | Radium | Ra | 88 |
| Hydrogen | H | 1 | Radon | Rn | 86 |
| Indium | In | 49 | Rhenium | Re | 75 |
| Iodine | I | 53 | Rhodium | Rh | 45 |
| Iridium | Ir | 77 | Roentgenium | Rg | 111 |
| Iron | Fe | 26 | Rubidium | Rb | 37 |
| Krypton | Kr | 36 | Ruthenium | Ru | 44 |
| Lanthanum | La | 57 | Rutherfordium | Rf | 104 |
| Lawrencium | Lr | 103 | Samarium | Sm | 62 |
| Lead | Pb | 82 | Scandium | Sc | 21 |
| Lithium | Li | 3 | Seaborgium | Sg | 106 |
| Lutetium | Lu | 71 | Selenium | Se | 34 |
| Magnesium | Mg | 12 | Silicon | Si | 14 |
| Manganese | Mn | 25 | Silver | Ag | 47 |
| Meitnerium | Mt | 109 | Sodium | Na | 11 |
| Mendelevium | Md | 101 | Strontium | Sr | 38 |
| Mercury | Hg | 80 | Sulfur | S | 16 |
| Molybdenum | Mo | 42 | Tantalum | Ta | 73 |
| Neodymium | Nd | 60 | Technetium | Tc | 43 |
| Neon | Ne | 10 | Tellurium | Te | 52 |
| Neptunium | Np | 93 | Terbium | Tb | 65 |
| Nickel | Ni | 28 | Thallium | Tl | 81 |
| Niobium | Nb | 41 | Thorium | Th | 90 |
| Nitrogen | N | 7 | Thulium | Tm | 69 |
| Nobelium | No | 102 | Tin | Sn | 50 |
| Osmium | Os | 76 | Titanium | Ti | 22 |
| Oxygen | O | 8 | Tungsten | W | 74 |
| Palladium | Pd | 46 | Uranium | U | 92 |
| Phosphorus | P | 15 | Vanadium | V | 23 |
| Platinum | Pt | 78 | Xenon | Xe | 54 |
| Plutonium | Pu | 94 | Ytterbium | Yb | 70 |
| Polonium | Po | 84 | Yttrium | Y | 39 |
| Potassium | K | 19 | Zinc | Zn | 30 |
| Praseodymium | Pr | 59 | Zirconium | Zr | 40 |
| Promethium | Pm | 61 | | | |

# APPENDIX E

# ELEMENTS OF QUANTUM MECHANICS

Quantum mechanics provides a correct description of phenomena on the atomic or subatomic scale, where the ideas of classical mechanics are not generally applicable. As we describe nuclear phenomena, we will use many results and concepts from quantum mechanics. Although it is our goal not to have the reader, in general, perform detailed quantum mechanical calculation, it is important that the reader understand the basis for many of the descriptive statements made in the text. Therefore, we present, in this Appendix, a brief summary of the essential features of quantum mechanics that we shall use. For more detailed discussion of these features, we refer the reader to the references at the end of this Appendix.

## E.1 WAVE FUNCTIONS

All the knowable information about a physical system (i.e., energy, angular momentum, etc.) is contained in the wave function of the system. We shall restrict our discussion to one-body systems for the present. (We could easily generalize to many body systems.) The wave function can be expressed in terms of space coordinates and time or momenta and time. In the former notation we write,

$$\psi(x, y, z, t) \quad \text{or just } \psi \tag{E.1}$$

---

*Modern Nuclear Chemistry*, by W.D. Loveland, D.J. Morrissey, and G.T. Seaborg
Copyright © 2006 John Wiley & Sons, Inc.

These wave functions, must be "well-behaved," that is, they (and their derivatives with respect to the space coordinates) must be continuous, finite, and single valued. The functions $\psi$ are solutions to a second-order differential equation called the Schrödinger equation (see below).

The probability of finding a particle within a volume element $dx\,dy\,dz$, $W\,dx\,dy\,dz$, is given by

$$W\,dx\,dy\,dz = \psi^*\psi\,dx\,dy\,dz \qquad (E.2)$$

where $\psi^*$ is the complex conjugate of $\psi$. [To form the complex conjugate of any complex number, replace all occurrences of $i$ (where $i = \sqrt{-1}$) with $-i$. Real numbers are their own complex conjugates. $6 - 5i$ is the complex conjugate of $6 + 5i$. So $(a + ib)^*(a + ib) = (a - ib)(a + ib) = a^2 + b^2$.] The probability per unit volume (the probability density) is $W = \psi^*\psi$. If we look everywhere in the system, we must find the particle so that

$$\int \psi^*\psi\,d\tau = 1 \qquad (E.3)$$

where $d\tau$ is a volume element $dx\,dy\,dz$. Wave functions possessing this numerical property are said to be normalized. If the value of some physical quantity $P$ is a function of the position coordinates, the average or expectation value of $P$ is given by

$$\langle P \rangle = \int \psi^* P \psi\,d\tau \qquad (E.4)$$

This expectation value represents the average outcome of a large number of measurements.

## E.2 OPERATORS

Often we must compute values of quantities that are not simple functions of the space coordinates, such as the $y$ component of the momentum, $p_y$, where Equation (E.4) is not applicable. To get around this, we say that corresponding to every classical variable, there is a quantum mechanical operator. An operator is a symbol that directs us to do some mathematical operation. For example, the momentum operators are

$$\hat{p}_x = -i\hbar\frac{\partial}{\partial x}$$
$$\hat{p}_y = -i\hbar\frac{\partial}{\partial y} \qquad (E.5)$$
$$\hat{p}_z = -i\hbar\frac{\partial}{\partial z}$$

while the total energy operator $\hat{E}$ is given as

$$\hat{E} = i\hbar \frac{\partial}{\partial t} \tag{E.6}$$

Thus, to calculate the expectation value of the $x$ component of the momentum, $p_x$, we write

$$\langle p_x \rangle = \int \psi^* \left( -i\hbar \frac{\partial}{\partial x} \right) \psi \, d\tau$$

$$= -i\hbar \int \psi^* \frac{\partial \psi}{\partial x} \, d\tau \tag{E.7}$$

Similarly, the classical expression for the kinetic energy is

$$T = p^2/2m \tag{E.8}$$

which, translated to quantum mechanics terms, means the kinetic energy operator, $\hat{T}$, is, in Cartesian coordinates,

$$\hat{T} = \frac{-\hbar^2}{2m} \left( \frac{\partial^2}{\partial x^2} + \frac{\partial^2}{\partial y^2} + \frac{\partial^2}{\partial z^2} \right) \tag{E.9}$$

or, using the Laplacian operator, $\nabla^2$

$$\hat{T} = \frac{-\hbar^2}{2m} \nabla^2 \tag{E.10}$$

where

$$\nabla^2 = \frac{\partial^2}{\partial x^2} + \frac{\partial^2}{\partial y^2} + \frac{\partial^2}{\partial z^2} \tag{E.11}$$

## E.3 SCHRÖDINGER EQUATION

In 1926, Schrödinger found that behavior on the atomic or subatomic scale was correctly described by a differential equation of the form

$$\frac{-\hbar^2}{2m} \nabla^2 \psi + V\psi = i\hbar \frac{\partial \psi}{\partial t} \tag{E.12}$$

where $V$ represents the potential energy and $\psi$ the wave function of the system. Substituting from Equation (E.6), we can write

$$\frac{-\hbar^2}{2m} \nabla^2 \psi + V\psi = E\psi \tag{E.13a}$$

This equation is an example of a general class of equations called eigenvalue equations of the form $\Omega\psi = \omega\psi$ where $\Omega$ is an operator and $\omega$ is the value of an observable corresponding to that operator. (The mathematical expression $\psi$ is referred to as an eigenfunction of the operator $\Omega$).

To use the Schrödinger equation to gain information about a physical system, we must perform a set of steps that are as follows:

1. Specify the potential energy function of the system, that is, specify the forces acting (Section 1.6.1).
2. Find a mathematical function, $\psi$, which is a solution to the differential equation, the Schrödinger equation.
3. Of the many functions that satisfy the equation, reject those that do not conform to certain physical constraints on the system, known as boundary conditions.

Before illustrating this procedure for several cases of interest to nuclear chemists, we can point out another important property of the Schrödinger equation. If the potential energy $V$ is independent of time, we can separate the space and time variables in the Schrödinger equation by setting

$$\Psi(x, y, z, t) = \psi(x, y, z)\tau(t) \tag{E.13b}$$

Substituting this expression into Equation (E.13), and simplifying, we have

$$\frac{1}{\Psi}\frac{-\hbar^2}{2m}\nabla^2\Psi + V = \frac{i\hbar}{\tau}\frac{\partial\tau}{\partial t} \tag{E.14}$$

The only way this equation can be true is for both sides to equal a constant. If we call this "separation constant" $E$, we can write

$$\frac{-\hbar^2}{2m}\nabla^2\psi + V\psi = E\psi \tag{E.15}$$

and

$$i\hbar\frac{\delta\tau}{\delta t} = E\tau \tag{E.16}$$

Equation (E.15) is the time-independent Schrödinger equation. The solution to Equation (E.16) is

$$\tau(t) = e^{-i(E/\hbar)t} \tag{E.17}$$

Using the Euler relation ($e^{i\theta} = \cos\theta + i\sin\theta$), we can write

$$\tau(t) = \cos\omega t - i\sin\omega t \tag{E.18}$$

where $\tau(t)$ is a periodic function with angular frequency $\omega = E/h$. The separation constant $E$ can be shown to be the total energy, that is, the sum of the kinetic and potential energies, $T + V$.

## E.4 THE FREE PARTICLE

To illustrate how the Schrödinger equation might be applied to a familiar situation, consider the case of a "free" particle, that is, a particle moving along at a constant velocity with no force acting on the particle ($V = 0$) (Fig. E.1). For simplicity, let us consider motion in one dimension, the $x$ direction. For the time-independent Schrödinger equation, we have

$$\frac{-\hbar^2}{2m}\frac{d^2\psi}{dx^2} = E\psi \qquad (E.19)$$

or

$$\frac{d^2\psi}{dx^2} = k^2\psi \qquad (E.20)$$

where the constant $k$ is given by

$$k^2 = \frac{2mE}{\hbar^2} \qquad (E.21)$$

The allowed values of the energy $E$ are [Equation (E.21)]

$$E = \frac{\hbar^2 k^2}{2m} \qquad (E.22)$$

where $k$ can assume any value ($E$ is not quantized). Since $V = 0$, $E$ is the kinetic energy of a particle with momentum $p = \hbar k$. From de Broglie, we know that

$$\lambda = \frac{h}{p} \qquad (E.23)$$

so that we can make the association that

$$k = \frac{1}{\lambda}$$

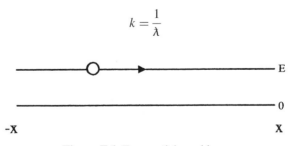

**Figure E.1** Free-particle problem.

## APPENDIX E

The solution for the Schrödinger equation, including the time-dependent part is

$$\Psi(x, t) = A \exp[ikx - i\omega t] + B \exp[-ikx - i\omega t] \tag{E.24}$$

where $k$ and $\omega$ are given (E.21) as

$$k = \frac{\sqrt{2mE}}{\hbar} \tag{E.25}$$

$$\omega = \frac{E}{\hbar} \tag{E.26}$$

This solution is the equation for a wave traveling to the right (+$x$ direction, the first term) and to the left (−$x$ direction, second term). We can impose a boundary condition, namely, we can specify the particle is traveling in the +$x$ direction. Then we have

$$\Psi(x, t) = A \exp[ikx] \exp[-i\omega t] \tag{E.27}$$

We can now calculate the values of any observable. For example, to calculate the value of the momentum $p$, we write [see Equation (E.7)]

$$\langle p \rangle = \int_{-\infty}^{\infty} \psi^* \left( -i\hbar \frac{\partial}{\partial x} \right) \psi \, dx = \sqrt{2mE} \tag{E.28}$$

which agrees, of course, with the classical result.

### E.5 PARTICLE IN A BOX (ONE DIMENSION)

Continuing our survey of some simple applications of wave mechanics to problems of interest to the nuclear chemist, let us consider the problem of a particle confined to a one-dimensional box (Fig. E.2). This potential is flat across the bottom of the box and then rises at the walls. This can be expressed as:

$$\begin{array}{ll} V(x) = 0 & 0 \leq x \leq L \\ V(x) = \infty & x < 0, x > L \end{array} \tag{E.29}$$

The particle moves freely between 0 and $L$ but is excluded from $x < 0$ and $x > L$. Inside the box, the Schrödinger equation has the form of Equation (E.19) (the free particle). The time-independent solution can be written

$$\psi(x) = A \sin kx + B \cos kx \tag{E.30}$$

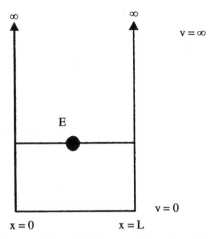

**Figure E.2** Schematic diagram of a particle in a one-dimensional box. The particle is free to move between $x = 0$ and $x = L$, but not allowed to have $x < 0$ or $x > L$.

But we know that $\psi(x) = 0$ at $x = 0$ and $x = L$. Thus $B$ must be 0 and

$$A \sin kL = 0 \tag{E.31}$$

To have $\sin kL = 0$, we must have

$$kL = n\pi \qquad n = 1, 2, 3 \tag{E.32}$$

and, using the result (E.22), we have

$$E_n = \frac{\hbar^2 k^2}{2m} = \frac{\hbar^2 \pi^2}{2mL^2} n^2 \tag{E.33}$$

In this case, the energy is quantized. Only certain values of the energy are allowed. One can show the normalization condition is satisfied if

$$\psi_n(x) = \sqrt{\frac{2}{L}} \sin \frac{n\pi x}{L} \tag{E.34}$$

The allowed energy levels, the probability densities, and the wave functions are shown for the first few levels of this potential in Figure E.3.

**Sample Problem** Suppose a neutron is confined to a one-dimensional box that is the size of a nucleus, $10^{-14}$ m. (a) What is the energy of the first excited state? (b) What is the probability of finding the neutron within a region corresponding to

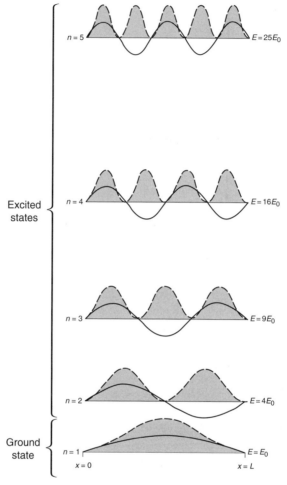

**Figure E.3** Allowed energy levels of a particle in a one-dimensional box. The wave function is shown as a solid line for each level while the shaded area gives the probability density.

20% of the width of the box, that is, between $0.4 \times 10^{-14}$ m and $0.6 \times 10^{-14}$ m in the fourth excited state?

### Solution

1. $E_0$ (the energy of the ground state) $= \dfrac{\hbar^2 \pi^2 (n)^2}{2mL^2} = \dfrac{(1.05 \times 10^{-34} \text{ Js})^2 (3.14)^2 (1)^2}{2(1.66 \times 10^{-27} \text{ kg})(10^{14} \text{ m}^2)}$

$= 3.3 \times 10^{-13} \text{ J} = 2.0 \text{ MeV}$

The energy of the first excited state, $n = 2$, will be $4E_0$, and the energy spacing between the first excited state and the ground state will be $3E_0 = 6$ MeV.

2. Probability $= \int_{x1}^{x2} \psi^2(x)\,dx = \frac{2}{L}\int_{x1}^{x2} \frac{\sin^2 5\pi x}{L}\,dx = \left(\frac{x}{L} - \frac{1}{5\pi}\sin\frac{5\pi x}{L}\right)_{x1}^{x_2} = 0.20$

which is the result obtained by inspection of the $\psi^2$ curve in Figure E.3.

## E.6  LINEAR HARMONIC OSCILLATOR (ONE DIMENSION)

One of the classic problems of quantum mechanics that is very important for our study of nuclei is the harmonic oscillator. For a simple harmonic oscillator, the restoring force is proportional to the distance from the center, that is, $F = -kx$, so that $V(x) = kx^2/2$. The Schrödinger equation is

$$\frac{-\hbar^2}{2m}\frac{d^2\psi}{dx^2} + \frac{1}{2}kx^2\psi = E\psi \tag{E.35}$$

The solution of this equation is mathematically complicated and leads to wave functions of the form

$$\psi_n(x) = N_n e^{\beta^2/2} H_n(\beta) \tag{E.36}$$

where

$$\alpha \equiv \frac{2\pi m}{\hbar}\nu_0$$

$$\beta \equiv \sqrt{\alpha} x \tag{E.37}$$

$$\nu_0 \equiv \frac{1}{2\pi}\sqrt{\frac{k}{m}} \quad \text{(the oscillator frequency)}$$

with a normalization constant of

$$N_n = \left(\sqrt{\frac{\alpha}{\pi}}\frac{1}{2^n n!}\right)^{1/2} \tag{E.38}$$

The expression $H_n(\beta)$ is the $n$th Hermite polynomial (which can be found in handbooks of mathematical functions). The energy eigenvalues can be shown to be

$$E_n = \left(n + \tfrac{1}{2}\right)h\nu_0 \tag{E.39}$$

where $m = 0, 1, 2, 3 \ldots$.

Thus, the energy levels are equally spaced starting with the zero-point energy $h\nu_0$ (Fig. E.4). Note the solutions have the property that there is some probability of finding the particle in classically forbidden regions, that is, the particle penetrates into the walls.

**654** APPENDIX E

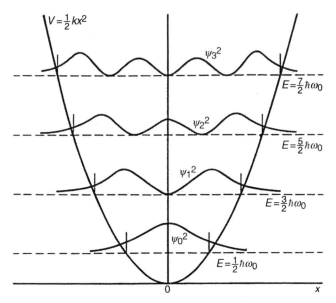

**Figure E.4** Low-lying levels and associated probability densities for the harmonic oscillator.

## E.7 BARRIER PENETRATION (ONE DIMENSION)

Another important quantum mechanical problem of interest to nuclear chemists is the penetration of a one-dimensional potential barrier by a beam of particles. The results of solving this problem (and more complicated variations of the problem) will be used in our study of nuclear α decay and nuclear reactions. The situation is shown in Figure E.5. A beam of particles originating at $-\infty$ is incident on a barrier of thickness $L$ and height $V_0$ that extends from $x = 0$ to $x = L$. Each particle has a total energy $E$. (Classically, we would expect if $E < V_0$, the particles would bounce off the barrier, whereas if $E > V_0$, the particles would pass by the barrier

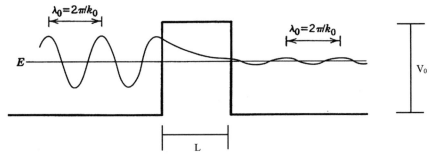

**Figure E.5** Schematic diagram of a particle of energy $E$ incident on a barrier of height $V_0$ and thickness $L$. The wave function $\psi$ is also shown.

with no change in their properties. Both conclusions are altered significantly in quantum mechanics.)

It is conventional to divide the space into three regions I, II, and III, shown in Figure E.5. In regions I and III, we have the "free-particle" problem treated in Section E.4. In region I, we have particles moving to the left (the incident particles) and particles moving to the right (reflected particles). So we expect a wave function of the form of Equation (E.24), whose time-independent part can be written

$$\psi_I = a_I e^{ik_1 x} + b_I e^{-ik_1 x} = \psi_{I \rightarrow} + \psi_{I \leftarrow} \tag{E.40}$$

where $k_1 = \sqrt{2mE}/\hbar$. In region III, we have no particles incident from $+\infty$, so, at best, we can only have particles moving in the $+x$ direction ($b = 0$). Thus

$$\psi_{III} = a_{III} e^{-k_1 x} = \psi_{III \rightarrow} \tag{E.41}$$

In region II, the time-independent Schrödinger equation is

$$\frac{d^2 \psi_{II}}{dx^2} = k_2 \psi_{II} \tag{E.42}$$

where $k_2 = [2m(V_0 - E)]^{1/2}/\hbar$, assuming $V_0 > E$. The solution is

$$\psi_{II} = a_{II} e^{k_2 x} + b_{II} e^{-k_2 x} \tag{E.43}$$

Notice that the wavelength $\lambda$ is the same in regions I and III, but the amplitude of the wave beyond the barrier is much less than in front of the barrier. It can be shown that the probability of transmitting particles through the barrier is

$$T = \frac{|\psi_{III \rightarrow}|^2 V}{|\psi_{I \rightarrow}|^2 V} = \frac{|a_{III}|^2}{|a_I|^2} \tag{E.44}$$

where $V$ is the particle speed. To determine the value of $a_{III}/a_I$, we eliminate the other constants $b_I$, $a_{II}$, $b_{II}$ by applying the conditions that $\psi$ and $d\psi/dx$ must be continuous through all space. After much algebra (see, e.g., the textbook by Evans), we have

$$T = \left[ 1 + \frac{V_0^2}{4E(V_0 - E)} \sinh^2 K_2 L \right]^{-1} \tag{E.45}$$

For nuclear applications, the barriers are quite thick ($k_2 L \gg 1$), in which case, $\sinh^2 k_2 L \approx \frac{1}{4} e^{2k_2 L}$, thus

$$T \approx 16 \frac{E}{V_0} \left( 1 - \frac{E}{V_0} \right) e^{-2k_2 L} \tag{E.46}$$

# APPENDIX E

The dominant term in this expression is the exponential. For a 6-MeV $\alpha$ particle, $V_0 = 20\,\text{MeV}$, $L = 10^{-14}$ m, we have

$$k_2 \approx \frac{[2 \times 4 \times 1.6 \times 10^{-27} \times (20-6) \times 1.6 \times 10^{-12}]^{1/2}}{1.05 \times 10^{-34}}$$

$$\approx 5.1 \times 10^{15}\,\text{m}^{-1}$$

Thus

$$e^{-2k_2 L} = e^{-102} = 5.1 \times 10^{-45}$$

and

$$T = 16 \times \tfrac{8}{20} \times \left(1 - \tfrac{8}{20}\right)(5.1 \times 10^{-45}) = 1.9 \times 10^{-44}$$

So we ignore the preexponential term, and write

$$T \approx e^{-2G} \tag{E.47}$$

where $2G = 2k_2 L = 2[2\,m(V_0 - E)]^{1/2}/\hbar$. For an arbitrarily shaped potential that would be more pertinent to nuclear $\alpha$ decay, one can show

$$2G = \frac{2}{\hbar} \int_{x_1}^{x_2} [2\,m(V(x) - E)]^{1/2}\,dx \tag{E.48}$$

where $x_1$ and $x_2$ are the points where $E = V(x)$.

What about the case where $E > V_0$. In regions I and III, the situation is the same. In region II, the wave functions will be given as

$$\psi_{\text{II}} = a_{\text{II}} e^{ik_2 x} + b_{\text{II}} e^{-ik_2 x} \tag{E.49}$$

where

$$k_2 = \frac{[2m(E - V_0)]^{1/2}}{\hbar} \tag{E.50}$$

Since the wavelength $\lambda_2$ is $1/k_2$, we can note by comparing equations that $\lambda_2 > \lambda_1$, and the momentum $[p(=(2mk_2)^{1/2})]$ becomes less. In other words, the particle is scattered.

## E.8 SCHRÖDINGER EQUATION IN SPHERICAL COORDINATES

Many problems in nuclear physics and chemistry involve potentials, such as the Coulomb potential, that are spherically symmetric. In these cases, it is advantageous to express the time-independent Schrödinger equation in spherical coordinates (Fig. E.6). The familiar transformations from a Cartesian coordinate system $(x, y, z)$ to spherical coordinates $(r, \theta, \varphi)$ are (Fig. E.6)

$$x = r \sin \theta \cos \phi \qquad y = r \sin \theta \sin \phi \qquad z = r \cos \theta \qquad (E.51)$$

The time-independent Schrödinger equation becomes

$$\frac{-\hbar^2}{2m}\left[\frac{1}{r^2}\frac{\partial}{\partial r}\left(r^2 \frac{\partial \psi}{\partial r}\right) + \frac{1}{r^2 \sin^2 \theta}\frac{\partial}{\partial \theta}\left(\sin \theta \frac{\partial \psi}{\partial \theta}\right) + \frac{1}{r^2 \sin^2 \theta}\frac{\partial^2 \psi}{\partial \phi^2}\right] + v\psi = E\psi \quad (E.52)$$

When the potential is spherically symmetric, $v = v(r)$, then the wave function can be written as

$$\psi(r, \theta, \phi) = R(r)Y_{lm}(\theta, \phi)$$

where $Y_{lm}$ are the spherical harmonic functions.

If we substitute this wave function in Equation (E.52) and collect terms, we find that all functions of $r$ can be separated from the functions of $\theta$ and $\varphi$:

$$\frac{1}{R}\frac{d}{dr}\left(r^2 \frac{dR}{dr}\right) + \frac{2mr^2}{\hbar^2}(E - V) = \frac{1}{Y_{lm}}\left[\frac{1}{\sin \theta}\frac{\partial}{\partial \theta}\left(\sin \theta \frac{\partial Y}{\partial \theta}\right) + \frac{1}{\sin^2 \theta}\frac{\partial^2 Y}{\partial \phi^2}\right] \quad (E.53)$$

Setting both sides of the equation equal to a separation constant, $l(l+1)$, where $l = 0, 1, 2\ldots$, we have

$$\frac{1}{R}\frac{d}{dr}\left(r^2 \frac{dR}{dr}\right) + \frac{2mr^2}{\hbar^2}(E - V) = l(l+1) \quad (E.54)$$

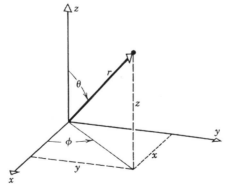

**Figure E.6** Spherical polar coordinates.

and

$$\frac{1}{\sin\theta}\frac{\partial}{\partial\theta}\left(\sin\theta\frac{\partial Y}{\partial\theta}\right) + \frac{1}{\sin^2\theta}\frac{\partial^2 Y}{\partial\phi^2} = l(l+1)Y \qquad (E.55)$$

Working on Equation (E.54), it is convenient to change variables

$$R(r) = \frac{u(r)}{r} \qquad (E.56)$$

$$\frac{d^2u}{dr^2} + \frac{2m}{\hbar^2}\left[\left(E - V(r) - \frac{l(l+1)\hbar^2}{2mr^2}\right)\right]u = 0 \qquad (E.57)$$

This is called the radial wave equation. Apart from the term involving $l$, it is the same as the one-dimensional time-independent Schrödinger equation, a fact that will be useful in its solution. The last term is referred to as the centrifugal potential, that is, a potential whose first derivative with respect to $r$ gives the centrifugal force.

It is important to note that Equation (E.55) does not contain the potential energy term, and thus once we have solved it, the solutions will supply to all cases where $V$ does not depend on $\Theta$ and $\varphi$, that is, all so-called central potentials. The wave functions $Y_{lm}(\theta, \varphi)$ are known as the spherical harmonic functions and are tabulated in many references. The indices $l$ and $m$ are related to the orbital angular momentum, $L$, of the particle relative to the origin. The magnitude of $L$ is $[l(l+1)]^{1/2}$ h and its $2l+1$ possible projections on the $z$ axis are equal to $m\hbar$ ($m = 0$, $\pm 1$, $\pm 2 \ldots \pm l$).[1] The term $l$ is called the orbital angular momentum quantum number while $m$ is the magnetic quantum number, in reference to the different energies of the $m$ states in a magnetic field (the Zeeman effect). It follows, therefore, that the specification of a particular spherical harmonic function (as a solution to the angular equation) uniquely specifies the particle's orbital angular momentum and its $z$ component.

## E.9 INFINITE SPHERICAL WELL

As an application of the Schrödinger equation, expressed in spherical coordinates, to a problem of interest in nuclear chemistry, let us consider the problem of a particle in an infinite spherical well (Fig. E.7). This potential can be defined as

$$\begin{aligned} V(r) &= 0 & r < a \\ V(r) &= \infty & r > a \end{aligned} \qquad (E.58)$$

Following our discussion in Section E.8, we expect the solution of the Schrödinger equation to be

$$\psi = R_l(r)Y_{lm}(\theta, \phi) \qquad (E.59)$$

---

[1] In more formal language, $\langle l^2 \rangle = \hbar^2 l(l+1)$ and $\langle l_z \rangle = m\hbar$.

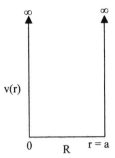

**Figure E.7** Schematic diagram of the infinite square well potential.

where the radial wave function $R_l(r)$ is a solution to the equation

$$\frac{1}{r^2}\frac{d}{dr}\left(r^2\frac{dR_l}{dr}\right) + \frac{2m}{\hbar^2}\left[E - \frac{l(l+1)\hbar^2}{2mr^2}\right]R_l = 0 \quad (E.60)$$

inside the well. The solutions of this equation are the spherical Bessel functions

$$R_l(r) = j_l(kr)\left(-\frac{r}{k}\right)^l\left(\frac{1}{r}\frac{d}{dr}\right)^l\left(\frac{\sin kr}{kr}\right) \quad (E.61)$$

where $k = \sqrt{2mE}/\hbar$. The boundary conditions require $\psi = 0$ at $r = 0$, and $r = a$, which will occur for values of $ka$ that make the Bessel functions have a value of 0 (the "zeros" of these functions). (Each $l$ value will have its own set of zeros.) These resulting values of $k$ can be used to calculate the allowed energy levels (Fig. E.8). Each level is labeled with a number $(1, 2, 3 \ldots)$ and a letter (s, p, d, e, etc.). The letter follows the usual spectroscopic notation of $l$ ($l = 0$, s; $l = 1$, p, etc.) while the number designates how many times that letter has occurred (the first d level is 1d; the second 2d, etc.).

## E.10 ANGULAR MOMENTUM

Classically, the angular momentum of a particle can be written as $\mathbf{l} = \mathbf{r} \times \mathbf{p}$. (Section 1.6.2). From this classical expression, we can write down the classical components of the vector $l$:

$$\begin{aligned} l_x &= yp_z - zp_x \\ l_y &= zp_x - xp_z \\ l_z &= xp_y - yp_x \end{aligned} \quad (E.62)$$

**660** APPENDIX E

These classical expressions can be converted to the operator language of quantum mechanics by substitutions [such as $x \to x$, $p_x \to i\hbar\,(\partial/\partial x)$, etc.]

$$l_x = -i\hbar\left(y\frac{\partial}{\partial z} - z\frac{\partial}{\partial y}\right) = -i\hbar\left(-\sin\phi\frac{\partial}{\partial\theta} - \cot\theta\cos\phi\frac{\partial}{\partial\phi}\right)$$

$$l_y = -i\hbar\left(z\frac{\partial}{\partial x} - x\frac{\partial}{\partial z}\right) = -i\hbar\left(\cos\phi\frac{\partial}{\partial\theta} - \cot\theta\sin\phi\frac{\partial}{\partial\phi}\right) \quad \text{(E.63)}$$

$$l_z = -i\hbar\left(x\frac{\partial}{\partial y} - y\frac{\partial}{\partial x}\right) = -i\hbar\frac{\partial}{\partial\phi}$$

As remarked earlier (Section E.9), the expectation values of $\langle l_z \rangle$ and $\langle l^2 \rangle$ for a central potential are

$$\langle l_z \rangle = m_l \hbar \qquad m_l = 0, \pm 1, \pm 2, \ldots, \pm l \quad \text{(E.64)}$$

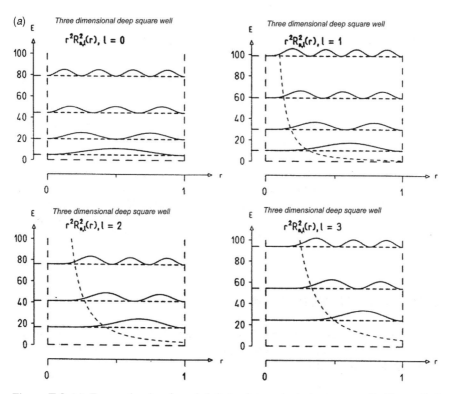

**Figure E.8** (*a*) Energy levels of an infinitely deep spherical square well. The radical probability density functions $r^2 R_{n,l}^2(r)$ are shown for different values of $l$. (*b*) The three dimensional probability densities, $\int_{n,l,m}(r,\theta)$ for an infinitely deep three-dimensional square well.

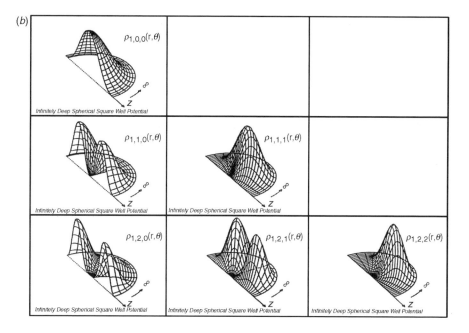

**Figure E.8** *Continued.*

and

$$\langle l^2 \rangle = l(l+1)\hbar^2 \tag{E.65}$$

We can give these results a pictorial interpretation that is worth noting. Consider a state of definite orbital angular momentum $l$. Then

$$l = \sqrt{l(l+1)}\hbar$$

The $z$ component of $l$ may have any value up to $\pm l\hbar$. The possible values of $l_z$ can be represented as the projection of a vector of length $l$ on the $z$ axis (Fig. E.9). This situation is referred to as spatial quantization. Only certain values of $l_z$ are allowed. Due to the uncertainty principle, the values of $l_x$ and $l_y$ are completely uncertain. In the language of Figure E.9, the vector representing $l$ is rotating about the $z$ axis, so that $l$ and $l_z$ are fixed, but $l_x$ and $l_y$ are continuously changing.

In chemistry, we found that to describe the complete quantum state of an electron in an atom, we had to introduce another quantum number, the intrinsic angular momentum or spin. This quantum number is designated as $s$. By analogy to the orbital angular momentum quantum number $l$, we have

$$\langle s^2 \rangle = s(s+1)\hbar^2$$
$$\langle s_z \rangle = m_s \hbar \qquad m_s = \pm \tfrac{1}{2} \tag{E.66}$$

Nucleons also have values of the spin quantum number of $s = \tfrac{1}{2}$, like electrons. The total angular momentum of a nucleon $j$ can be written as

$$\mathbf{j} = \mathbf{l} + \mathbf{s} \tag{E.67}$$

The usual quantum mechanical rules apply to $j$, that is,

$$\langle j \rangle = j(j+1)\hbar^2$$
$$\langle j_z \rangle = m_j \hbar = \langle l_z + s_z \rangle$$

where $m_j = -j, -j+1, \ldots, j-1, j$. Thus we have

$$m_j = m_l + m_s = m_l \pm \tfrac{1}{2} \tag{E.68}$$

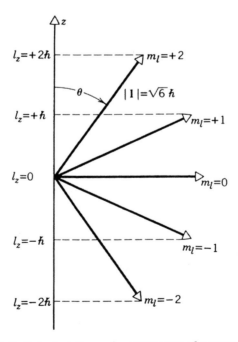

**Figure E.9** Spatial orientation and $z$ components of a vector with $l = 2$.

Since $m_l$ is always an integer, then $m_j$ must have a half integer and $j$ must be a half integer, either $j = l - \frac{1}{2}$ or $j = l + \frac{1}{2}$. Alternatively, for a given $l$ value, we have two possible values of $j$, $j = l - \frac{1}{2}$ or $j = l + \frac{1}{2}$. For example, for $l = 1$ ($p$ state), we have $j = l - \frac{1}{2} = \frac{1}{2}$ or $j = l + \frac{1}{2} = \frac{3}{2}$. We designate these states as $p_{1/2}$ and $p_{3/2}$, respectively.

## E.11 PARITY

A wave function has positive (or even) parity if it does not change sign by reflection through the origin.

$$\psi(-x, -y, z) = \psi(x, y, z) \text{ positive parity, } \pi = + \quad (E.69)$$

Alternatively, if reflection through the origin produces a change of sign, the parity of the wave function is negative ($-$).

$$\psi(-x, -y, z) = -\psi(x, y, z) \text{ negative parity, } \pi = - \quad (E.70)$$

When $\psi$ is expressed in spherical coordinates as $\psi(r, \theta, \varphi)$, then "reflection through the origin" is accomplished by replacing $\theta$ and $\varphi$ by $(\pi - \theta)$ and $(\pi + \varphi)$, respectively. ($r$ cannot change sign as it is just a distance.) In other words, the parity of the wave function is determined only by its angular part. For spherically symmetric potentials, the value of $l$ uniquely determines the parity as

$$\pi = (-1)^l \quad (E.71)$$

A corollary of this is that for a system of particles the parity is even if the sum of the individual orbital angular momentum quantum numbers $\Sigma l_i$ is even; the parity is odd if $\Sigma l_i$ is odd. Thus, the parity of each level depends on its wave function. An excited state of a nucleus need not have the same parity as the ground state.

Parity will be valuable to us in our discussion of nuclei because it is not conserved in $\beta$ decay, which will tell us that a different force, the weak interaction, is acting in $\beta$ decay compared to nuclear reactions. Also the rates of the $\gamma$-ray transitions between nuclear excited states depend on the changes in parity and can be used to determine the parity of nuclear states.

## E.12 QUANTUM STATISTICS

The parity of a system is related to the symmetry properties of the spatial portion of the wave function. Another important quantum mechanical property of a system of two or more identical particles is the effect on the wave function of exchanging the coordinates of two particles. If no change in the wave function occurs when the spatial and spin coordinates are exchanged, we say the wave function is symmetric

**664** APPENDIX E

and the particles obey Bose–Einstein statistics. If upon exchange of the spatial and spin coordinates of the two particles the wave function changes sign, the wave function is said to be antisymmetric and the particles obey Fermi–Dirac statistics. The "statistics" these particles follow profoundly affects the property of an assembly of such particles. Particles with half-integer spins, such as neutrons, protons, and electrons, are fermions, obey Fermi–Dirac statistics, have antisymmetric wave functions, and as a consequence, obey the Pauli principle. (No two particles can have identical values of the quantum numbers, $m$, $l$, $m_l$, $s$, and $m_s$.) Photons, or other particles with integer spins, such as the $\pi$ meson, are bosons, obey Bose–Einstein statistics, have symmetric wave functions, and do not obey the Pauli principle.

This difference between fermions and bosons is reflected in how they occupy a set of states, especially as a function of temperature. Consider the system shown in Figure E.10. At zero temperature ($T = 0$), the bosons will try to occupy the lowest energy state (a Bose–Einstein condensate) while for the fermions the occupancy will be one per quantum state. At high temperatures the distributions are similar and approach the Maxwell Boltzman distribution.

The Fermi–Dirac distribution can be described by the equation

$$f_{\text{FD}}(E) = \frac{1}{e^{-(E-E_{\text{F}})kT} + 1} \tag{E.72}$$

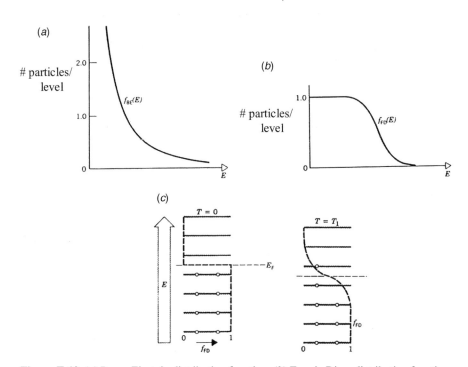

**Figure E.10** (*a*) Bose–Einstein distribution function, (*b*) Fermi–Dirac distribution function, and (*c*) filling of levels by fermions at $T = 0$ and $T = T_1 > 0$. The dashed line indicates the Fermi energies $E_{\text{F}}$.

where $f_{FD}$ is the number of particles per quantum state, $k$ is Boltzman's constant, and $E_F$ is the Fermi energy. At $T = 0$, all energy levels up to $E_F$ are occupied ($f_{FD} = 1$) and all energy levels above $E_F$ are empty ($f_{FD} = 0$). As $T$ increases, some levels above $E_F$ become occupied at the expense of levels below $E_F$.

## BIBLIOGRAPHY

Cohen-Tannoudji, C., B. Diu, and F. Laloe. *Quantum Mechanics*, Wiley, New York, 1977. An encyclopedic treatment.

Eisberg, R. M. *Fundamentals of Modern Physics*, Wiley, New York, 1961. A comprehensive treatment of modern physics.

Krane, K. S. *Modern Physics*, Wiley, New York, 1983. A well-written introductory treatment of quantum physics.

Merzbacher, E. *Quantum Mechanics*, Wiley, New York, 1961. Another treatment with several nuclear physics applications.

Scharff, M. *Elementary Quantum Mechanics*, Wiley, London, 1969. A very lucid, elementary treatment of quantum mechanics, emphasizing physical insight rather than formal theory.

Schiff, L. I. *Quantum Mechanics*, McGraw-Hill, New York, 1955. An old classic treatment that contains several applications of interest.

# INDEX

Abnormal data, 574
Absorbed dose, 531
Accelerator mass spectrometry (AMS), 84
Achromatic separation, 419
Actinide contraction, 452
Activation analysis, 366
Activity, 10
Allowed beta decay, 212
Alpha decay
   equation, 178
   recoil, 180
   coulomb energy, 182
   Gamow factor, 186
   hindrance factors, 192
   mass parabola, 183
   preformation factor, 186, 188
   Q-value, 178
   tunneling, 184
Alvarez accelerator, 405
Angular momentum
   beta decay, 200
   definition, 12
   intrinsic, 46
Anionic extractant, 597
Antineutrino, 202

Areal density, 498
Astrophysical S-factor, 343
Atomic mass unit, 19
Attenuation length, 526
Autoradiography, 107

B(E2), 231, 227
Background radiation, 606
Barium fluoride, 562
Barn (unit), 52
Baryons, 23
Bateman Equations, 74
Becquerel (unit), 63
Beer–Lambert Law, 256, 518
Beta decay
   angular momentum, 200
   equation, 202
   Fermi function, 207
   Fermi integral, 209
   parity, 214
   Q-value, 203
   selection rules, 211
Beta stability, 40
Bethe–Bloch stopping
   power, 502

*Modern Nuclear Chemistry*, by W.D. Loveland, D.J. Morrissey, and G.T. Seaborg
Copyright © 2006 John Wiley & Sons, Inc.

**668** INDEX

Binding energy
  Electron Thomas–Fermi, 30, 178
  nuclear, 32
Binomial distribution, 567
Bohr independence hypothesis, 272
Bohr magneton, 48
Bohr stopping power, 502
Bohr velocity, 505
Boiling water reactor (BWR), 391, 465
Boson, 20
Bragg counter, 542
Bragg peak, 506
Bragg's rule, 506
Branching ratio, 76
Breit–Wigner function, 274, 344
Bremsstrahlung, 514, 517
Brennan–Bernstein rules, 146

Cadmium ratio, 370
CAMAC, 565
Carrier, 94
Carrier free, 64, 583
Cationic extractant, 598
Center of mass, 528
Centrifugal potential, 190, 263
Cerenkov radiation, 518, 539
Charge distribution, 45
Charged particle activation analysis (CPAA), 370
Chemical blank, 604
Chernobyl reactor, 394, 457
CNO cycle, 347
Cockcroft–Walton accelerator, 400
Cold fusion, 435
Compound nucleus, 254, 272
Compton edge, 523
Compton scattering, 522
Conversion factors, 613
Coulomb barrier, 261, 397
Coulomb excitation, 280
Critical mass, 423
Critical state, 388
Cross section, 254
Cumulative yield, 321
Curie (unit), 64
Cyclotron, 406

Data rejection, 574
Daughter activity, 68
De Broglie wavelength, 17, 386, 641
Dead time, 545, 548
Decay constant, 58
Decay law, 59
Deep inelastic reaction, 280, 286

Delayed neutrons, 390
Detection limit, 608
Deuterium, characteristics, 131, 138
Differential cross section, 256
Diffractive scattering, 268
Direct reaction, 254
DNA analysis, 108
Double beta decay, 41, 217

Effective length, 415
Electric moments, 50
Electrodeposition, 587
Electromagnetic radiation
  angular correlation, 236
  lifetime, 242
  multipolarity, 224
  selection rules, 225
  stretched transition, 224
  transition rate, 226
Electron cyclotron resonance (ECR), 398
Electronic stopping, 500, 514
Element synthesis, 434
Energy straggling, 508
Energy width, 63, 76
Entropy, 167
Epithermal neutrons, 370
Euler relation, 640
Evaporation residue, 434, 591
Exchange capacity, 581
Excitation function, 277

Faraday cup, 589
Fermi
  distribution, 43
  energy, 165
  function, 207
  gas, 163
  golden rule, 204
  integral, 209
Fermi decay, 201, 204
Fermion, 20
Fission
  barrier, 170, 304
  charge distribution, 318
  isomer, 308
  mass distribution, 316
  TKE, 316
  width, 311
Fissionability parameter, 302
Forensic activation analysis, 372
  ft value, 209
Fuel cycle, 467
Fundamental constants, 613

Gamow factor, 186, 343
Gamow peak, 344
Gamow–Teller decay, 201
Gas amplification, 543
Gas flow detector, 546
Gas quenching, 548
Gaseous diffusion, 476
Gaussian distribution, 570
Geiger–Müller detector, 544, 547
Geiger–Nuttall Law, 185
Geochronometer, 83
Germanium detector, 557
Giant dipole resonance, 278
Gray (unit), 532
G-value, 582
Gyromagnetic ratio, 49

Hahn and Strassmann, 299
Half life, 10, 60
Halo nucleus, 44
Heavy cluster emission, 193
Heisenberg Uncertainty
    Relationship, 19, 63, 129
Hertzsprung–Russell diagram, 340
Hill–Wheeler formula, 306
Hindrance factors, 192
Hydrogen burning, 345

Impact parameter, 12, 259
Independent yield, 320
In-flight separation, 419
Interaction barrier, 264, 282
Interval distribution, 571
Intranuclear cascade, 290
Ion chamber, 540
Ion exchange, 599
Ion pair, 540
Ionization potential, 502
ISOL, 287, 417
Isomeric states, shell model, 149
Isomers, 221
Isospin, 133
Isotope dilution analysis, 122
Isotope effect, 92, 104
Isotopic abundances, 332

Kerma, 532
Kinetic isotope effect, 106
Kurie plot, 207

Lambert–Beers Law, 256, 518
Laplacian, 639
Leptons, 20
Level density, 168, 275

Level width, 273
Linear absorption coefficient, 518
Linear energy transfer (LET), 531
Liquid scintillation, 560
Lorentz force, 407
Lorentz transformations, 12
Low-level counting, 605
Low-level waste, 489

Magnetic moment, 48, 147
Magnetic rigidity, 420
Magnetic sector, 413
Mass equation
    Myers–Swiatecki, 39
    Semi-empirical, 37
    Weizsacher, 36
Mass excess, 32
Mass parabola, 183
Mass stopping power, 504
Mean free path, 289, 387
Mean life, 62
Mesons, 23
Mirror nuclei (table), 133
Mirror nuclei, coulomb, 150
Moment of inertia, 156
Monitor reaction, 589
Monopole moment, 51
Mosely, 5
Mossbauer effect, 241

Natural decay chain, 73, 77
Natural radioactivity, 77
Natural reactor, 395
Neptunium, 439
Neutral extractant, 597
Neutrino, 200, 215
    detector, 357
    oscillation, 359
    solar, 355
Neutron activation analysis
    (NAA), 370
Neutron moderator, 388
Neutron scattering, 528
Nilsson model, 160
Nuclear decay
    characteristics, 9
    equilibria, 70
    general, 8
    rate, 10
Nuclear density, 7, 43
Nuclear fallout, 80
Nuclear force, 11
Nuclear instrumentation module
    (NIM), 565

# 670 INDEX

Nuclear magneton, 48
Nuclear mass, 30
Nuclear potential, 139
Nuclear radii, 42
Nuclear reaction Q-value, 251
Nuclear shapes, 155
Nuclear skin, 43
Nuclear surface, 154
Nuclear temperature, 167
Nuclear waste, 484

Oklo natural reactor, 395
Optical model, 269

Packing fraction, 32
Pair production, 524
Parity, 47, 214, 655
Penning ion guage (PIG), 398
pep reactions, 345
Photoelectric absorption, 520
Photomultiplier, 562
Pickup reaction, 270
PIN diode, 555
PIXE, 373
Plutonium, 439
Poisson distribution, 569
Positron emission tomography (PET), 117
Potential Energy, Coulomb, 12
pp1 reaction, 346
pp2 reaction, 346
p-process, 353
Preformation factor, 186, 188
Pressurized water reactor (PWR), 391, 465
Primordial nucleosynthesis, 337
Proton activation analysis (PAA), 370
Proton emission, 195

Quadrupole moment, 50
Quarks, 23
Q-value
  alpha decay, 178
  beta decay, 203
  equation, 252
  nuclear reaction, 31, 251

Rad, 531
Radiation exposure, 531
Radiative stopping, 514
Radioactive decay law, 59
Radiocarbon, 79
Radioimmunoassay, 108
Radiolysis, 101, 493

Radiopharmaceuticals (table), 116
Radiotracer label, 99
Radiotracers (table), 98, 111
Range, 513
Range straggling, 509
Reactor poisons, 390
Relativistic mass, 14
REM (unit), 533
Residual interaction, 153
Resonance, 274
Roentgen (unit), 531
Rotational energy, 155
r-process, 352
rp process, 353
Rutherford backscattering, 376
Rutherford scattering, 265

Saddle point, 300
Saturation activity, 258
Scavengers, 583
Schmidt limit, 148
Scintillation, 559
Scission point, 300
Secular equilibrium, 72
Segre diagram, 36
Selection rules, 225
Semiconductor detector, 549
Semi-empirical mass, 37
Separation energy, 32
Shell model, 141
Sievert (unit), 533
Silicon surface detector, 554
Sodium iodide, 561
Solar
  abundances, 335
  neutrinos, 355
  reactions, 346
Spallation, 288
Specific activity, 64
Spent fuel, 479, 484, 488
Spherical coordinates, 649
Spontaneous fission, 306
s-process, 351
Stable nuclides, 36
Standard model, 20
Statistical equilibrium, 169
Stellar populations, 340
Stopping power, 500
Straggling, 508
Stripping reaction, 270
Strong force, 130
Strutinsky method, 305
Subcritical state, 390
Superallowed beta decay, 212

Supercritical mass, 423
Superdeformed nuclei, 157, 161
Superheavy nuclei, 447
Synchrocyclotron, 410
Synchrotron, 410

Tandem accelerator, 402
Thermonuclear reactions, 342, 424
Three-Mile Island reactor, 394, 457
Threshold energy, 253
Time projection chamber (TPC), 542
Townsend avalanche, 543
Track detectors, 499, 564
Transient equilibrium, 70
Transmutation, 491
Triple alpha reaction, 348
Tunneling, 184, 646

Uncertainties, 573
Uranium enrichment, 475
Uranium metal, 470

Van de Graff accelerator, 400
Vibrational motion, 158
Virtual photon, 130

Water dilution volume, 485
Watt energy spectrum, 324, 387
Wave function, 637
Weisskopf transition rate, 227, 229
Weizsacher mass equation, 36
Wheeler angular distribution, 326
Wideroe accelerator, 405
Wilzbach reaction, 94, 101
Woods–Saxon potential, 132, 153, 263

X-ray, 4

Yellowcake, 474
Yrast states, 286
Yucca mountain, 487
Yukawa potential, 132

Zeeman splitting, 236